RUSSIAN – ENGLISH

MILITARY DICTIONARY

Russian – English

Military Dictionary

LONDON
HER MAJESTY'S STATIONERY OFFICE

FOREWORD

This dictionary will prove a valuable and reliable aid of long service to students of the Russian language. Nothing like it was available when some members of the British Armed and Civil Services, who have spent years contributing to this dictionary, first began studying the language under my direction at the University of Cambridge in 1945.

The dictionary contains vocabulary encountered in a wide variety of modern Russian texts covering military, technical and general subjects.

I therefore warmly commend this work to translators, interpreters and other users of Russian and assure them that they could have no better guide.

Dame Elizabeth Hill, M.A., Ph.D., D.Litt.
Emeritus Professor of Slavonic Studies,
University of Cambridge.
Fellow of University College, London, and
Girton College, Cambridge.

First published 1983

ISBN 0 11 230019 7

PREFACE

This dictionary was produced by the Joint Technical Language Service for use by military interpreters and translators, and other members of the armed forces. Because of its success it is now for the first time made available to the public through Her Majesty's Stationery Office and booksellers.

With some 35,000 entries, the dictionary expands the Russian-English section of the single-volume *Russian-English and English-Russian Military Dictionary*. The latter is likewise produced by the Joint Technical Language Service and is now available to the public (HMSO, 1983).

Naval and air terminology is incorporated only where its use is sufficiently general as to be naturally encountered in military writings, or where the meaning of a word differs in a military context from that in naval or air contexts.

As far as possible all equivalents have been thought out afresh and conform with current usage on both sides of the curtain.

For compactness and ease of reference, adjective and noun combinations are given only in the order: noun, adjective. Many short phrases, such as commands, are listed under the prepositions which govern them; see, for example, phrases under the preposition '**На**.'

LIST OF ABBREVIATIONS

The main abbreviations used in the dictionary to provide a general explanation of words or phrases, are set out below.

AA	—anti-aircraft
AAFCE	—anti-aircraft fire control equipment
AAM	—air-to-air missile
ac	—aircraft
AD	—air defence
admin	—administr-/ation/ative
aeron	—aeronautical
AF	—air force
AFV	—armoured fighting vehicle
ammo	—ammunition
amph	—amphibious
anat	—anatom-/y/ical
A pers	—anti-personnel
artic	—articulated
arty	—artillery
ARV	—armoured recovery vehicle
astron	—astronom-/y/ical
ASW	—anti-submarine warfare
A tk	—anti-tank
auth	—authority
ball	—ballistics
BL	—breech loading
bldg	—building
br	—bridg-/e/ing
Brit	—British
BW	—biological warfare
cam	—camouflage and concealment
carb	—carburettor
carto	—cartographic
CD	—Civil Defence
cf	—compare
chem	—chemi-/cal/stry
civ	—civilian
coll	—colloquial
comms	—communications
constr	—construction
CRT	—cathode ray tube
crypt	—cryptography
CW	—chemical warfare
C.W.	—continuous wave; carrier wave
det	—detachment
DF	—direction finding; defensive fire
diff	—difference
div/Div	—division/-al
DME	—distance measuring equipment
dml	—demolitions
doc	—documents
ECCM	—electronic counter-countermeasures
ECM	—electronic countermeasures
eg	—for example
elec	—electric/-al
electron	—electronic
emf	—electro-motive force
eng	—engineering
eqpt	—equipment
ESV	—earth satellite vehicle
EW	—electronic warfare

ex —exercise
expl —explosives
FC —fire control
FCE —fire control equipment
fd —field
fig —figure
flt —flight
fmn —formation
GCI —ground controlled interception
geog —geograph-/y/ical
geol —geology
geom —geometry
GHQ —General Headquarters
gr —group
gren —grenades
gunn —gunnery
GW —guided weapons
HE —high explosive
hosp —hospital
how —howitzer
HQ —headquarters
hydr —hydraulic/s
i/c —in command
ICE —internal combustion engine
IFF —identification, friend or foe
ILS —instrument landing system
imp —imperfective
incl —including
inf —infantry
instr —instrument/s
int —intelligence
intrans —intransitive
IR —infra-red
log —logistics
lit —literally
math —mathematics
m/c —machine
mech —mechanical
med —medical
met —meteorolog-/y/ical
metall —metallurgy
MG —machine gun
mil —military
min —mines/mining
mor —mortar
msg —message
msl —missile
MT —mechanical transport
MV —muzzle velocity
naut —nautical
nav —naval
navig —navigation/-al
NBC —nuclear, biological and chemical
no —number
nucl —nuclear
NW —nuclear warfare
obs —obsolete
obsn —observation
offr —officers
OP —observation post
opp —opposite
ops —operations
opt —optics
ord —ordnance
org —organisation
para —parachut-/e/ing
per —perfective
pers —personnel
photo —photograph-/y/ic
phys —physic-/s/al
pl —platoon

POL —petrol, oil and lubricants
posn —position
PPI —plan position indicator
prec —precedence
PT —physical training
pt —point
QF —quick firing
qty —quantity
radiac —radioactivity detection, indication and computation
RCM —radar countermeasures
rd —road
regt —regiment/-al
revs —revolutions
rly —railway
rpm —revolutions per minute
RX —receiver
SA —small arms
SAM —surface-to-air missile
servo —servomechanism
SHF —super-high frequency, band 10
sigs —signals
SL —self-loading
sl —slang
smb —somebody
SMG —sub-machine gun
smth —something
Sov —Soviet
SP —self-propelled
sp —support
Srg —sound ranging
SSM —surface-to-surface missile
Stn —station
STOL —short take-off and landing
sup —supplies
surv —survey
svc —service
tac —tactic-/s/al
tech —technical
teleg —telegraphy
tgt —target
tk —tanks
topo —topography
t/p —teleprinter
tps —troops
tpt —transport
trans —transitive
trg —training
TV —television
TW —travelling wave
TX —transmitter
UHF —ultra-high frequency, band 9
u/i —unidentified
u/s —unserviceable
US —United States
veh —vehicle
VF —voice frequency
VHF —very high frequency, band 8
VT —variable time
VTOL —vertical take-off and landing
wks —works
wksp —workshop
wpn —weapon/s

А

аберра́ция aberration
~, хромати́ческая chromatic aberration
абля́ция ablation (ESV)
абоне́нт subscriber, party, user
~ телефо́нной се́ти telephone subscriber, telephone party
абрази́в abrasive, abradant
абрази́вный abrasive, abradant
абра́зия abrasion
а́брис 1. description, sketch 2. contour, outline
абсо́рбент absorbent
абсо́рбер absorber
абсорби́ровать to absorb, to pick up
абсорбцио́метр absorptiometer
абсо́рбция absorption
абсци́сса abscissa, x-coordinate
аванбе́к launching nose (br)
аванга́рд advance guard
~, боково́й flank guard
авансодержа́тель imprest holder (accounts)
авари́йка breakdown vehicle
авари́йно-восстанови́тельный emergency repair
авари́йно-спаса́тельный salvage and rescue
авари́йность accident (breakdown, damage, fault, failure) rate
авари́йный critical, breakdown, emergency, rescue, stand-by
ава́рия 1. breakdown, serious breakdown 2. accident
~, доро́жная road accident
~ с челове́ческими же́ртвами fatal accident
"а́виа" "air met" (message designation)
авиаба́за air base
~, морска́я naval air base
авиабензи́н aviation gasoline, AVGAS
авиабо́мба aerial bomb
~, аэронавигацио́нная aerial navigation flare
~, бетонобо́йная concrete-piercing bomb
~, бронебо́йная armour-piercing aerial bomb
~, зажига́тельная incendiary bomb
~, куря́щаяся toxic smoke bomb
авиабо́мба, ночна́я ориенти́рно-сигна́льная illuminating flare - night-position marker
~, оско́лочная HE fragmentation bomb
~, оско́лочно-фуга́сная fragmentation-demolition bomb
~, плани́рующая glide bomb
~ пра́ктики practice bomb
~, противоло́дочная anti-submarine bomb
~, ротати́вно-рассе́ивающая rotating cluster bomb (contents of bomblets, of various types, released by centrifugal force)
~, светя́щаяся illuminating flare
~, управля́емая guided bomb
~, фуга́сная 1. demolition bomb 2. HE bomb
~, хими́ческая chemical bomb
авиагоризо́нт artificial horizon, gyro horizon
авиагру́ппа airforce tactical formation
авиадеса́нт airborne assault force
авиадеса́нтник paratrooper, member of airborne assault force
авиадеса́нтный airborne assault
авиадиви́зия, истреби́тельная fighter division (AF)
авиазвено́ flight (section)
авиаинжене́р air engineer, aeronautical engineer
авиако́рпус, истреби́тельный fighter corps (AF)
авиамая́к aircraft beacon, air beacon, aerial beacon
авиами́на parachute mine
авианаблюде́ние aerial observation
авианаво́дчик air support controller
~, передово́й forward air controller, FAC
авиано́сец aircraft carrier
авиапредстави́тель air liaison officer
авиапулемёт aircraft machine gun
авиапу́шка aircraft cannon
авиаразве́дка 1. air observation 2. air reconnaissance
~, морска́я maritime air reconnaissance
авиаресу́рс aircraft life
авиато́пливо aviation fuel
авиатра́нспорт air transport
авиатранспорта́бельный air-transportable
авиацио́нный 1. air, aviation 2. aircraft

авиа́ция 1. aviation
2. aircraft
3. air force
~, арме́йская army aviation, army air force
~, арме́йская разве́дывательная army reconnaissance aviation
~, артиллери́йская artillery observation aircraft
~, беспило́тная 1. pilotless aircraft
2. drones
~, боева́я 1. operational aircraft
2. operational air force
~, бомбардиро́вочная 1. bomber aircraft
2. Bomber Command (Brit)
3. bomber air force
~, вое́нная military aviation
~, вое́нно-тра́нспортная military transport aviation
~, войскова́я air force element (forming organic part of army formation for direct support)
~, вспомога́тельная auxiliary aircraft
~, грузова́я airline freight service
~ да́льнего де́йствия 1. long range aircraft
2. long range air force
~, да́льняя 1. long range aircraft
2. long range air force
~, да́льняя бомбардиро́вочная long-range bomber aviation
~, истреби́тельная 1. fighter aircraft
2. Fighter Command (Brit)
3. fighter aviation
~, истреби́тельно-бомбардиро́вочная 1. fighter-bomber aircraft
2. fighter-bomber aviation
~, корректиро́вочная air observation post aircraft, spotter reconnaissance aircraft (arty)
~, корректиро́вочно-разве́дывательная air observation post aircraft, spotter reconnaissance aircraft (arty)
~, ми́нно-торпе́дная mine laying and torpedo aircraft
~, морска́я 1. naval air force
2. naval aircraft
~ подде́ржки 1. support aviation
2. support aircraft
~, разве́дывательная 1. reconnaissance aviation
2. reconnaissance aircraft
3. Air Reconnaissance Command
авиа́ция, санита́рная aero-medical evacuation units, aero-medical units, aero-medical aircraft
~ свя́зи intercommunication aircraft
~, стратеги́ческая 1. Strategic Air Force
2. strategic aviation
~, стратеги́ческая бомбардиро́вочная strategic bombers
~, такти́ческая 1. tactical aviation
2. Tactical Air Force
~, тра́нспортная military transport aviation (Sov)
~, тра́нспортно-деса́нтная Air Force of the Airborne Forces (Sov)
~, фоторазве́дывательная photographic reconnaissance aircraft
~, фронтова́я 1. Tactical Air Force
2. tactical aviation
~, ча́ртерная airline charter service
~, штурмова́я 1. ground attack aircraft
2. ground attack airforce
авиача́сть air force unit
авиаэндоско́п aviation endoscope
авиаэскадри́лья air squadron
автоба́за MT depot
автобатальо́н MT battalion
автобензи́н motor vehicle fuel
автобензозапра́вщик 1. refuelling truck
2. petrol bowser
автобензоцисте́рна petrol tanker, fuel tanker truck
автоблокиро́вка 1. automatic block system (ry)
2. automatic traffic control system (air)
автобронемаши́на armoured car
автоброneта́нковый armoured (units)
авто́бус bus
автобуфе́т mobile canteen
автоводоцисте́рна water tanker, water truck
автовы́шка tower vehicle
автогенера́тор self-excited oscillator (electron)
~, диапазо́нный broad band self-excited oscillator (electron)
~ с горю́чим во́здухом hot air generator (used in NBC) decontamination)
автогре́йдер motor grader, motorized grader
автогудрона́тор motorized tar-spreader
автодегаза́тор CW decontamination vehicle

автоди́н autodyne, autoheterodyne
автодро́м vehicle testing ground
автозаво́д selfwinding movement
автозапра́вщик refuelling truck, bowser
~ **раке́тным то́пливом** propellant truck (GW)
автоиму́щество 1. MT equipment
2. MT stores
3. MT vehicles
автока́мера inner tube (veh)
автокла́в 1. pressure cooker
2. autoclave (tech)
автоколеба́ние, неусто́йчивое unstable oscillation
автоколо́нна convoy, mechanized column
~ **подво́за горю́чего** petrol-carrying echelon
автокра́н vehicle-mounted crane
автоку́хня mobile kitchen
авто́л automobile lubricant
автолаборато́рия field laboratory, mobile laboratory
автола́вка mobile shop
автоларёк mobile canteen
автолебёдка motor cable winch
автолесово́з straddle carrier (veh)
автолиха́чество reckless driving
автомагистра́ль motor road, motorway
~ **для скоростно́го движе́ния** motorway, high-speed freeway
автомаслозапра́вщик oil bowser
автомастерска́я 1. MT workshop
2. machinery truck, truck-mounted workshop
автома́т 1. automatic weapon
2. sub-machine gun, machine carbine
3. automatic loader (ord)
4. robot
~, **боево́й** service sub-machine gun
~ **за́пуска** automatic starting control (e.g. for gas turbine engine)
~ **защи́ты генера́торов** automatic generator cut-out
~ **защи́ты се́ти** automatic circuit breaker (elec)
~ **подры́ва** self-destruction unit
~ **раскры́тия парашю́та** automatic parachute device
~ **стабилиза́ции** autopilot (GW)
~ **стрельбы́, зени́тный** anti-aircraft firing data calculator
~ **стрельбы́, центра́льный** gunfire control computer (arty)
автома́т управле́ния да́льностью automatic range control unit (GW)
автоматиза́ция automation
~ **управле́ния огнём** fire control automation
автоматизи́ровать to automate, to automatize
автома́тика 1. automatics
2. automatic mechanism (SA)
~ **с дли́нным хо́дом ствола́** long-recoil mechanism
~ **с полусвобо́дным затво́ром** delayed blow-back action
~ **самозаря́дного ору́жия** self-loading mechanism
~ **со свобо́дным затво́ром** inertia blow-back action
автомати́ческий automatic, self-acting
автома́тчик sub-machine gunner
автомаши́на motor vehicle, car (see also avtomobil')
~, **водохо́дная** amphibious vehicle, (DUKW)
~ **высо́кой проходи́мости** cross country vehicle, high mobility vehicle
~, **снегоубо́рочная** snow clearing vehicle, snowplough
автомаши́на-амфи́бия 1. amphibious vehicle, wheeled amphibian
2. wheeled landing vehicle
автомеха́ник vehicle mechanic
автомоби́ль motor vehicle, car (see also avtomashina)
~, **авари́йный** breakdown vehicle
~, **авари́йный грузово́й** breakdown truck
~, **аппара́тный** 1. radio vehicle
2. special equipment vehicle
~, **боево́й** "fighting vehicle" (Sov veh classification includes all wheeled vehs mounting weapons, wheeled prime movers and some special vehs such as signals vehs)
~, **большегру́зный** heavy hauler
~, **большо́й пла́вающий** large wheeled amphibian
~, **бортово́й** lorry with sides
~, **бортово́й грузово́й** lorry with sides
~, **брониро́ванный** armoured car, armoured vehicle
~, **букси́рный** towing vehicle
~, **вое́нный** military vehicle
~, **вое́нный грузово́й** military truck

автомоби́ль высо́кой проходи́мости high mobility vehicle, cross-country vehicle
~, головно́й leading vehicle of convoy
~, грузово́й lorry, truck
~, гу́сеничный tracked vehicle
~, гу́сеничный пла́вающий tracked amphibian
~, деса́нтный пла́вающий amphibious assault vehicle
~ для перево́зки боеприпа́сов, грузово́й ammunition truck
~ для перево́зки ли́чного соста́ва personnel carrier
~ для перево́зки ору́дий gun truck
~ для перево́зки раке́т missile transporter
~ для разво́зки труб по тра́ссе трубопрово́да pipe-laying truck
~ для тяжёлых усло́вий эксплуата́ции, грузово́й heavy-duty truck
~, земново́дный amphibious truck
~, колёсный wheeled car, wheeled truck
~, команди́рский command vehicle, commander's vehicle
~, кры́тый closed vehicle
~, легково́й motor car, passenger car
~, ма́лый пла́вающий small wheeled amphibious vehicle
~, многоколёсный multi-wheeled vehicle
~, многоо́сный multi-axle vehicle
~ на возду́шной поду́шке air cushion vehicle, hovercraft
~ на подво́дных кры́льях, пла́вающий hydrofoil vehicle
~, низкора́мный грузово́й low loader
~ о́бщего назначе́ния general service truck, all purpose vehicle
~ пе́нного туше́ния, пожа́рный foam tender
~, пла́вающий wheeled amphibian
~, пла́вающий грузово́й wheeled amphibious load carrier, wheeled truck
~, пла́вающий гу́сеничный tracked amphibian
~ повы́шенной проходи́мости 1. cross country vehicle
2. half-track vehicle (Sov)
~, пожа́рный fire-engine
~, полугрузово́й cargo and personnel vehicle
~, полугу́сеничный half-track vehicle
автомоби́ль, полузакры́тый semi-enclosed vehicle
~, разведыва́тельный reconnaissance vehicle, scout car
~ с безбортово́й платфо́рмой platform truck, flat-bed truck
~ с откидны́ми борта́ми drop-side truck
~ с пусково́й раке́тной устано́вкой missile launching vehicle, launcher truck
~ с раке́тной устано́вкой guided missile truck
~, санита́рный ambulance
~, снегоубо́рочный snowplough truck
~ со все́ми веду́щими колёсами all-wheel drive vehicle
~, строево́й first line vehicle
~ техни́ческой по́мощи emergency service truck, recovery (rescue, repair) truck, vehicle
~, трёхо́сный грузово́й three-axle truck, three-axle load carrier, six-wheeled truck, six-wheeled cargo carrier
~, тяжёлый heavy vehicle, heavy car, heavy truck
~, универса́льный грузово́й general purpose cargo truck
~, уче́бный training vehicle
~, шестиколёсный six-wheeled vehicle
~, штабно́й staff vehicle, command vehicle
автомоби́ль-амфи́бия amphibious vehicle, amphibian
автомоби́ль-бензозапра́вщик petrol refuelling truck, bowser
автомоби́ль-вездехо́д 1. cross country vehicle
2. amphibian
3. half-track vehicle
автомоби́ль-водомаслозапра́вщик oil and water servicing truck
автомоби́ль-мастерска́я machinery truck, mobile workshop
автомоби́льный vehicular, motor
автомоби́ль-самосва́л dump truck, tipper
автомоби́ль-снегоочисти́тель snow-clearing vehicle
автомоби́ль-тяга́ч tractor truck, prime mover, towing truck
~ с полуприце́пом, грузово́й tractor truck semitrailer

автомоби́ль-цисте́рна tanker
автомодули́ровать to self-modulate, to automodulate
автомодуля́ция self-modulation, automodulation
автонастро́йка autotuning
автоно́мность autonomy (length of time for which a vessel can stay underway at sea & carry out operational tasks without receiving supplies)
автоно́мный autonomous, self-contained
автопа́рк 1. vehicle park, MT park 2. vehicle pool
~, грузово́й 1. fleet of trucks 2. MT park
автоперево́зка motor transportation
автопереключа́тель automatic switch
автопило́т automatic pilot
автопистоле́т 1. automatic pistol 2. self-loading pistol
автопогру́зчик fork-lift truck, automatic loader, auto-loader, loader truck
~, ви́лочный fork-lift truck
автоподразделе́ние 1. MT detachment 2. MT sub-unit
автоподстро́йка, фа́зовая automatic phase synchronization (electron)
~ частоты́ automatic frequency control
автопо́езд tractor-trailer unit
автопокры́шка tyre casing, tyre cover
автопорожня́к "empties" (MT)
автоприце́п vehicle trailer
автопроисше́ствие traffic accident, vehicle accident
автопрокла́дка automatic position tracing, plotting
автопрокла́дчик automatic position plotter, automatic position marker
авторези́на vehicle tyre complete with inner tube
авторефрижера́тор refrigerator truck
авториза́ция authorization
авторите́т authority
авторо́та 1. MT company 2. Soviet civilian transport unit
авторота́ция autorotation (helicopters)
автосамосва́л tipper truck, dump truck, dumper
автоса́ни motor sled
автосанро́та motorized medical company
автоси́н synchro data transmission system
автоскрепле́ние process of autofrettage, cold working
автоскреплённый autofretted (piece-ord)
автосопровожде́ние auto-tracking (radar)
~ по да́льности automatic range tracking
~ по угловы́м координа́там automatic angle tracking
автоста́ртер self-starter
автостра́да motorway
автосце́пка automatic coupling (rly)
автотеле́жка internal combustion engine truck
автоте́хника MT equipment
автотопливозапра́вщик fuel bowser
автотопливоцисте́рна fuel tanker truck
автотра́кторный MT, motor transport, mechanical transport
автотра́нспорт MT, motor transport, mechanical transport
автотра́нспортный MT, motor transport, mechanical transport
автотрансформа́тор autotransformer (electron)
автотяга́ч tractor-truck
автоустано́вка, дезинфекцио́нная vehicle-mounted disinfection unit
автофретта́ж auto-frettage, cold forging of gun barrels (gunn)
автохлебопека́рня, полева́я mobile field bakery
автоцисте́рна tanker
~ высо́кой проходи́мости cross-country tanker truck
~ для горю́чего fuel tanker truck
автошко́ла driver training school (MT)
автоэшело́н serial (road mov)
аге́нт agent, secret agent informer
~, антипе́нный defoaming agent, anti-foamer
~ для удале́ния сма́зки degreasing agent
~, консервацио́нный corrosion preventive, conserving agent
~, охлажда́ющий cooler agent, coolant, refrigerant
~, очища́ющий cleaning agent
~ разве́дки intelligence agent
аге́нт-вербо́вщик recruiting agent
агенту́ра secret agents, secret service, net of agents
агитпро́п agitation and propaganda

агитпу́нкт propaganda centre
агрега́т assembly, plant unit, apparatus, unit, set, outfit
~, аэродро́мный пусково́й ground mobile starter unit (airfield)
~, бензоэлектри́ческий portable generator set
~, газогенера́торный 1. gas producer unit
2. gas generator unit (GW)
~, генера́торный generating set, generating unit
~, ди́зельный generator set
~, запа́сный stand-by set, stand-by unit
~, заря́дный battery charging plant, charging set
~ пита́ния power supply unit
~ пита́ния ору́дия gun power unit (arty)
~ поворо́тного механи́зма traversing assembly (turret)
~ прибо́ра управле́ния артиллери́йским зени́тным огнём AA FC director (FCE)
~, ручно́й запра́вочный hand operated refuelling unit
~, сва́рочный welding equipment, welding set
~, силово́й power unit
~, ста́ртовый launcher unit
~ та́нка tank component
~, тестомеси́льный kneader (field bakery)
~, турбонасо́сный turbo-pump unit
~, фильтровентиляцио́нный filter-ventilator unit
~, электрогенера́торный generating set
агрега́тчик computer and range finder operator (AD)
агресси́вность, коррозио́нная corrosion activity
агресси́вный 1. aggressive
2. corrosive
агре́ссия aggression
агре́ссор aggressor
адамси́т adamsite (CW)
адапта́ция adaptation
~, тёмная adaptation of eyes to darkness
ада́птер adapter
адге́зия adhesion
адиаба́та adiabat, adiabatic curve
адиабати́ческий adiabatic
адиаба́тный adiabatic
администрати́вно-промы́шленный administrative-industrial
администрати́вно-территориа́льный administrative-territorial
администрати́вно-хозяйственный administrative, quartermaster general's, quartermaster general's and adjutant general's
администрати́вно-экономи́ческий administrative-economical
администрати́вный administrative
администра́ция administration
~, вое́нная 1. military government
2. administration (mil)
администри́рование bureaucracy, red tape
адмира́л admiral
адмира́льский admiral's
а́дрес address (computer tech)
адреса́т addressee
адсо́рбция adsorption
адъю́нкт service student (officer studying at university, military college etc)
адъюнкту́ра graduate work
~, зао́чная correspondence course at higher military training establishment, higher military correspondence course
~, о́чная full time study
адъюта́нт aide-de-camp
~, ста́рший adjutant
адъюта́нтский adjutant general (branch)
а́збука alphabet
~ мо́рзе morse code
ази́д рту́ти mercury azide
~ свинца́ lead azide (expl)
а́зимут azimuth, bearing
~, астрономи́ческий astronomical azimuth
~, геодези́ческий azimuth
~, и́стинный true bearing, true azimuth
~, ко́мпасный magnetic bearing, azimuth, compass bearing, azimuth
~ ку́рса це́ли target course bearing
~, магни́тный magnetic bearing, azimuth, compass bearing, azimuth
~ направле́ния наступле́ния compass course of the advance
~ направле́ния стрельбы́ firing azimuth, firing bearing
~ обра́тного направле́ния back-angle azimuth, back-angle bearing
~, обра́тный back bearing, back azimuth
~ про́йденного направле́ния back bearing, back azimuth
~ свети́ла star bearing, star azimuth

áзимут упреждённой тóчки collision heading, collision point bearing
~, упреждённый future bearing
~ цéли target bearing, azimuth of target
азимутáльный azimuthal
áзимут-квадрáнт azimuth quadrant
азимýтчик navigator (soldier detailed to guide by compass through trackless terrain)
азóт nitrogen
азотúрованный nitrated, nitrided
азóтистый nitrous
акадéмия, воéнно-юридúческая academy of military law
акваплáн aquaplane
акваланúрование aquaplaning
акватóрия water area (port, river etc)
акведýк aquaduct
акклиматизáция acclimatization
акклиматизúровать to acclimatize
аккумулúрование storing of energy
аккумулúровать to accumulate, to store
аккумулятóр accumulator, battery, secondary cell, storage cell
~ аварúйного питáния emergency battery
~ давлéния, жúдкостный liquid propellant gas generator (GW)
~ давлéния, пороховóй solid propellant gas generator (GW)
~, желéзо-кáдмиево-нúкелевый nickel-iron-cadmium accumulator
~, пороховóй solid fuel gas generator
~, свинцóвый lead battery
~, свинцóвый кислóтный lead acid accumulator
~, серéбряно-цúнковый silver-zinc battery
~, стáртерный starting battery
~, сухóй dry battery
~, щелочнóй alkaline accumulator
аккумуля́торный accumulator, battery
акселерáтор accelerator
акселерóграф accelerograph
акселерóметр accelerometer
~, линéйный linear accelerometer
~, угловóй angular accelerometer
аксельбáнт aiguillette, aglet
аксиáльный axial
аксиóметр steering wheel indicator
акт 1. record, document, attested statement, certificate 2. legal act
акт, дефéктный list of defects, defects certificate
~, диверсиóнный act of sabotage
~, законодáтельный legislative measure
~ на списáние condemnation certificate (of article)
~ на списáние агрегáта в результáте износа certificate of fair wear and tear
~ о капитуля́ции surrender document
~ о непригóдности к эксплуатáции certificate of unserviceability
~ приёмки acceptance certificate
~, приёмный acceptance certificate
~, приёмо-сдáточный transfer certificate, acceptance certificate, handover/takeover certificate, handover certificate
~, прихóдный certificate receipt voucher
~ ревúзий proceedings of audit board
~, рекламациóнный unsatisfactory equipment report
~ техничéского испытáния test report
~ техничéского осмóтра inspection report
~ техничéской провéрки inspection report
~ технологúческого состоя́ния mechanical condition report
~ экспертúзы examination report
актúв, партúйный party militants
активáтор catalyst
~ воспламенéния ignition promoter
активáция 1. activation 2. catalysation
~ нейтрóнами neutron activation
активизúровать to activate
активúровать to activate
активúст activist
актúвность 1. activeness, activity, energy 2. aggressiveness
~, боевáя combat activity
~, коррозиóнная corrosive action
~ оборóны positive activities by the defence
~, химúческая chemical activity
~, я́дерная nuclear activity
актúвный 1. active, energetic 2. aggressive
актинúзм actinism
актиногрáмма actinograph record (met)
актинóграф actinograph
актинóметр actinometer
актúровать to put on record

актохрани́лище records department
акт-реклама́ция written claim under guarantee (MT admin)
аку́стика acoustics
акусти́ческий acoustic
алгори́тм decimal notation
алида́да alidade (surv)
~, проста́я open sight alidade (surv)
алка́ли alkali
алкализи́ровать to alkalize
алкало́ид alkaloid
алкого́ль ethyl alcohol
~, эти́ловый ethyl alcohol
аллю́вий alluvium
альбе́до albedo (astron, nucl)
альбо́м album
альми́на training aid showing where shot would have landed
альмука́нтарат almucanter (surv)
альпеншто́к alpenstock
альпини́зм mountain climbing, mountaineering
альпини́ст mountaineer
альтерна́тор alternator, AC generator
альтигра́ф altigraph
альтиме́тр altimeter
~, радиолокацио́нный radar altimeter
а́льфа-излуче́ние alpha radiation
а́льфа-лучи́ alpha rays
альфаме́тр 1. fuel mixture indicator 2. exhaust analyser
а́льфа-распа́д alpha decay
а́льфа-части́ца alpha particle
алюми́ниевый aluminium
алюми́ний aluminium
алюминотерми́тный aluminothermic
амальга́ма amalgam
амато́л amatol
амбразу́ра port(hole), loophole, embrasure, firing port, firing slit
~ ба́шни turret port
~, смотрова́я observation port
амбулато́рия 1. dispensary 2. walking wounded collecting post
амбулато́рный outpatient, walking wounded cases
амианти́н amiantin (ammo)
аммиа́к ammonia
аммокси́л ammoxyl
аммона́л ammonal
аммо́ний, азотноки́слый ammonium nitrate
аммо́ний, пикриновоки́слый ammonium picrate, explosive D
аммо́ний-гелигни́т ammon-gelignite
аммони́т ammonite
~, патрони́рованный cartridge-packed ammonite
аммота́л ammotal
амортиза́тор 1. shock absorber 2. buffer, recoil spring
~, арми́рованный strengthened shock absorber
~, гидравли́ческий hydraulic shock absorber
~, гидропневмати́ческий oleo-pneumatic (air-oil) shock absorber
~, двусторо́нний two-way (double-acting) shock absorber
~, жи́дкостный hydraulic shock absorber
~ колеба́ний vibration damper, vibration absorber
~, ма́сляный oil buffer, oil shock absorber
~ нака́та counter recoil buffer, recuperator buffer (gunn)
~, пружи́нный (coil) spring shock absorber
~ прямо́го де́йствия direct-acting shock absorber
амортиза́ция shock absorption, checking
амортизи́рованный cushioned, shock proof
ампе́р ampere
ампера́ж amperage
ампервитки́ ampereturns (elec)
ампериме́тр ammeter
ампе́р-ча́с ampere hour
амплиди́н amplidyne (elec)
амплитро́н amplitron (electron)
амплиту́да amplitude
~ измене́ния частоты́ frequency swing, maximum instantaneous frequency excursion, maximum instantaneous frequency deviation
~ прили́ва tidal range (difference between successive high and low waters)
амплиту́дно-модули́рованный amplitude-modulated
а́мпула 1. ampoule (med) 2. ampoule (incendiary projectile)
~, воспламени́тельная glass ampoule primer (gren)

áмпула ýровня levelling glass, spirit level bulb
áмпула-запáл glass ampoule primer (gren)
ампуломёт ampoule launcher (obs Sov for Molotov cocktail type projectiles)
амуни́ция 1. ammunition, munitions 2. individual equipment, accoutrements (excluding clothing & wpns, obs) 3. harness (mil) (obs)
амфиби́йный amphibious
амфи́бия 1. amphibian, amphibious vehicle 2. amphibious aircraft
~, трáнспортная amphibious load carrier
~, трáнспортная брони́рованная armoured amphibious load carrier
амфотéрный amphoteric
анáлиз 1. appreciation 2. analysis
~, весовóй gravimetric analysis, analysis by weight
~, визуáльный visual analysis, routine analysis
~ гáза gas analysis
~, кáчественный qualitative analysis
~, коли́чественный quantitative analysis
~ крóви blood test
~ мочи́ urine test
~ обстанóвки appreciation of the situation
~, хими́ческий chemical analysis
~ цéлей target analysis
анализáтор 1. analyser 2. component generator (computers)
~ выхлопны́х гáзов exhaust gas analyser
анализи́ровать/проанализи́ровать to analyse, to appreciate, to assess
анали́н analine
аналити́ческий analytical, analytic
аналоги́чно similar(ly) to
аналóгия, электротепловáя electrothermal analogy
анатокси́н antitoxin
анаэрóбный anaerobic
ангáр hangar
ангидри́д, сéрный sulphur trioxide
анги́на Винцéнта trench mouth
áнглдозер angle-dozer
Ангстрем Angstrom
áндшпуг crowbar, handspike
анеми́я anaemia
анемогрáмма anemogram
анемóграф anemograph
анемóметр anemometer
~, крыльчáтый vane anemometer
~, ручнóй hand anemometer
~, чáшечный cup anemometer
анемотахóметр wind speed indicator, anemometer
анерóид aneroid barometer
анестези́я anaesthetic
анизотрóпный anisotropic (elec)
áнкер stay, tie rod
анкерóвка anchoring, fixing
анкéта questionnaire, poll
аннули́ровать to annul, to cancel, to rescind, to abolish, to nullify; **аннули́ровать акт** to annul a statement; **аннули́ровать прикáз** to revoke an order
анóд anode
~, вторóй second anode
~, пéрвый first anode
аноди́рование anodic oxidation
аноди́ровать to anodize
аномáлия, магни́тная magnetic anomaly (local distortion of earth's magnetic field)
антáбка 1. lanyard ring (pistol) 2. sling swivel (SA)
~, вéрхняя upper sling swivel
~ на наконéчнике лóжи stock ferrule swivel
~, ни́жняя lower sling swivel, butt swivel
~, перéдняя передвижнáя adjustable upper sling swivel, adjustable front sling swivel
~, шарни́рная swivel
антáпка 1. lanyard ring (pistol) 2. sling swivel (SA)
антéнна antenna, radiator, array (electron)
~, авари́йная emergency antenna
~, азимутáльная azimuth antenna
~, аксиáльно-симметри́ческая axisymmetrical antenna
~, акти́вная active antenna
~, апериоди́ческая aperiodic antenna, untuned antenna
~ Беверéджа Beverage antenna, wave antenna

антéнна бегýщей волны́ travelling wave radiator, non-resonant radiator, TW antenna, travelling wave antenna
~, **биконúческая** biconical antenna
~, **бортовáя** onboard antenna
~ **брóдсайд** broadside array
~, **вéерная** fan antenna, harp antenna
~, **вертикáльная** vertical antenna
~, **вибрáторная** dipole antenna
~, **висячая** trailing antenna
~, **возбуждáемая в цéнтре** centre-driven antenna
~, **волновáя** Beverage antenna, wave antenna
~, **волновóдная** wave-guide antenna
~ **"волновóй канáл"** Yagi antenna
~, **воронкообрáзная** funnel antenna
~, **вращáющаяся** rotatable antenna, steerable antenna
~, **всенаправленная** omnidirectional antenna
~, **встрóенная** built-in antenna
~, **встрóенная в диэлектрúческий обтекáтель** radome antenna
~, **выпускнáя** trailing antenna
~, **высóтная** height-finding antenna
~, **гиперболóидная лúнзовая** hyperboloid lens antenna
~, **Г-обрáзная** inverted-L antenna, gamma antenna
~ **голóвки самонаведéния, приёмная** homing-head receiver dish
~, **дальномéрная** range measurement antenna
~, **двойнáя ромбúческая** double rhombic (horizontal) antenna
~, **двузóнтичная** double umbrella antenna
~, **двунапрáвленная** bidirectional antenna, bilateral antenna
~, **двупитáемая** double feed antenna
~, **двускáтная** inverted-V antenna
~, **двухкóнусная** biconical antenna, double-cone antenna
~, **диапазóнная** wide band antenna
~, **дипóльная** doublet antenna
~, **дирéкторная** Yagi antenna
~, **дисконúческая** discone antenna
~, **дифракциóнная** slot antenna
~, **диэлектрúческая** dielectric antenna
~, **длинопроводнáя** long wire antenna
~, **ёлочная** fishbone antenna
~, **жёсткая** fixed antenna

антéнна зенúтного излучéния sky-wave antenna
~, **зéт-обрáзная** Z-shaped antenna
~, **зóнтичная** umbrella antenna
~ **из прóволочной сéтки, лúнзовая** wire-grid lens antenna
~, **излучáющая** radiating antenna
~, **изотрóпная** isotropic antenna
~, **Кассегрéйна** Cassegrain(ian) antenna
~, **квадрáтная рýпорная** pyramidal horn radiator
~, **компенсúрованная** balanced antenna
~, **контрóльно-измерúтельная** control and measuring equipment
~, **крýглая рáмочная** circular loop antenna
~, **крýглая рýпорная** conical horn radiator, conical horn antenna
~, **линéйная** uniform antenna
~, **Л-обрáзная** L-type antenna (usually, written with latin L)
~, **логопериодúческая** log periodic antenna
~, **лучевáя** Beverage antenna, wave antenna
~, **магнúтная** ferrite rod antenna
~, **мискообрáзная** dish(-shaped) antenna
~, **многовибрáторная** multi-dipole antenna
~, **многолучевáя** multi-element beam, multi-unit antenna, multi-beam antenna
~, **многополóсная** multi-band antenna
~, **многорядная** mattress antenna
~, **многоштыревáя** multiple rod antenna, polyrod antenna
~, **многоэлемéнтная** multiunit, multiple-unit, multielement antenna
~, **надувнáя** inflatable antenna
~, **назéмная** ground antenna
~, **наклóнная** slant antenna
~, **напрáвленная** directional antenna
~ **напрáвленного дéйствия** directional antenna
~, **настрóенная** tuned antenna
~, **ненапрáвленная** non directional antenna (omnidirectional)
~, **нерезонáнсная** non-resonant antenna
~, **ножевúдная** fin type antenna, blade antenna
~, **однонапрáвленная** unilateral antenna, unidirectional antenna
~, **остролучевáя** pencil-beam antenna, (high-directional)

антéнна, остронапрáвленная sharply-beamed antenna, highly- directional antenna
~, откры́тая external antenna
~, отражённая image antenna
~, параболи́ческая parabolic reflector antenna
~, параболóидная paraboloidal reflector antenna
~, пара́боло-цилиндри́ческая cylindrical parabola antenna
~, пасси́вная parasitic antenna, non-emitting antenna
~, пеленгáторная direction-finding antenna, DF antenna
~, передаю́щая transmitting antenna
~ Пи́столькорса V-antenna
~ повéрхностной волны́ surface wave antenna
~, поворóтная rotatable antenna, (steerable)
~, полистирóловая polyrod antenna, polystyrene antenna
~, полуволновáя half-wave antenna
~, приёмная receiving antenna
~, приёмопередаю́щая transceiving antenna
~, продóльно-напрáвленная endfire antenna
~, прямоугóльная ру́порная rectangular horn radiator
~ радиовзрывáтеля radio fuze antenna
~ радиолокáтора обнаружéния warning radar antenna
~, радиолокациóнная radar antenna
~ радиолокациóнного прицéла radar sight antenna
~ радиолокациóнной голóвки самонаведéния homing dish
~, радиопеленгáторная directional antenna
~, рáмочная loop antenna, frame antenna
~, резонáнсная resonant antenna
~, резонáнсная продóльно-напрáвленная resonant endfire antenna
~, рефлéкторная reflector antenna
~, ромби́ческая rhombic antenna
~, ру́порная horn antenna
~, ру́порно-зеркáльная horn reflector antenna

антéнна с боковы́м излучéнием, логопериоди́ческая broadside log-periodic antenna
~ с волновы́м канáлом Yagi antenna
~ с концевы́м питáнием end-fed antenna
~ с ни́жним питáнием base-fed antenna
~ с обрáтным излучéнием, укорóченная short backfire antenna
~ с параболи́ческим отражáтелем parabolic reflector antenna
~ с питáнием в срéдней тóчке centre-fed antenna
~ с подáвленными боковы́ми лепесткáми диагрáммы напрáвленности side-lobe suppressed antenna
~ с синтези́рованным раскры́вом synthetic antenna, synthetic array
~ с электрóнным пóиском electronic scanning antenna
~ с электрóнным скани́рованием electronic scanning antenna
~, сегмéнтно-параболи́ческая cheese antenna
~ систéмы наведéния guidance antenna
~ систéмы самонаведéния homing antenna
~, склáдывающаяся collapsible antenna
~, стержневáя ферри́товая ferrite-rod antenna
~, "сырови́дная" cheese antenna
~ ти́па "волновóй канáл" Yagi antenna
~ ти́па "дипóль" dipole antenna
~, Т-обрáзная T-antenna
~, три́плексная triplex antenna
~, тру́бчатая sleeve antenna
~, турникéтная turnstile antenna
~, убирáющаяся folding antenna, collapsible antenna
~ углá возвышéния elevation antenna
~ углá мéста elevation antenna
~ Удá-Я́ги Yagi antenna
~, ферри́товая ferrite rod antenna
~, ферромагни́тная рáмочная ferromagnetic loop antenna
~, фи́дерная TL antenna
~, цилиндри́ческая cage antenna
~, чéтверть-волновáя quarter-wave antenna
~, штыревáя whip antenna
~, щелевáя slotted antenna, slot antenna

антéнна, щелевáя волновóдная slotted waveguide antenna, slotted waveguide array
антéнна-решётка, напрáвленная directive antenna array
~, прямоугóльная rectangular antenna array
~ с осевы́м излучéнием end-fire array
антéнный 1. antenna
2. array
антибактериáльный anti-bacterial
антибиóтики antibiotics
антивещество́ anti-matter
антивибрáтор snubber, damper (GW)
антидетонáтор 1. anti-knock compound, anti-detonating compound
2. anti-knock component
антидетонациóнный anti-knock
антидетонáция anti-knock
антидóт antidote
антикоагулáнт anticoagulant
антикоррозиóнный anti-corrosive, corrosion resisting, anti-rust, rust-preventive
антимóний antimony
антиобледени́тель de-icer
антиокисли́тель antioxidant
антипóд antipode
антиракéта anti-missile missile
антисéптика 1. antisepsis, science of antiseptics
2. antiseptic
антисепти́ческий antiseptic
антисы́воротка, иммýнная antiserum
антитéло antibody
антитокси́н antitoxin
антифебри́н antifebrin (med)
антифри́з antifreeze
~, этиленгликóлевый ethylene glycol antifreeze
антициклóн anticyclone, high pressure area
апертýра aperture
~ антéнны antenna aperture
~ лучá beam aperture
~, приёмная collecting aperture
апланáт aplanat (photo)
апогéй apogee
апохромáт apochromatic lens
аппарáт 1. instrument, device, apparatus
2. machine, machinery
3. camera
4. organization
аппарáт автомати́ческой телефóнной стáнции dial telephone set
~, буквопечáтающий телегрáфный printing telegraph apparatus
~, водолáзный diving apparatus, diving gear
~, дистанциóнно-пилоти́руемый remotely piloted vehicle
~, кинетеодоли́тный cine-theodolite apparatus
~, летáтельный flying apparatus, flying machine
~ на воздýшной подýшке air-cushion vehicle, hovercraft
~, парти́йно-полити́ческий (communist) party political machinery, (communist) party political control organs
~, перегóнный distilling equipment
~, пескострýйный sandblaster
~, посрéднический team of umpires, directing staff (on exercise)
~ развéдки intelligence organisation
~, рентгéновский X-ray generating and control equipment
~, рулóнный телегрáфный page-printer telegraph
~, спускáемый re-entry vehicle
~, стартстóпный start-stop teleprinter
~, телегрáфный telegraph instrument, teleprinter
~, телефóнный 1. telephone set
2. subscriber's set
~ управлéния administrative machinery
~, фони́ческий buzzer telephone
~, фототелегрáфный facsimile equipment (phototelegraphic)
аппарáтная 1. instrument room
2. operating room (sigs)
аппарáтная-крóсс set and frame room (teleg)
аппарáтный apparatus, instrument
аппаратýра apparatus, gear, equipment
~ автомати́ческого управлéния automatic control equipment
~, акти́вная инфракрáсная active infra-red equipment
~, бортовáя onboard equipment
~ встрóенного контрóля built-in check-out equipment
~, высоковóльтная high-voltage equipment
~, высокочастóтная high-frequency equipment

аппарату́ра, гидроакусти́ческая asdic equipment, sonar equipment, echo-sounding equipment
~, дальноме́рная 1. distance measuring equipment, DME (radar) 2. range finding equipment
~, дистанциометри́ческая distance measuring equipment, DME
~ для по́иска мин mine hunting equipment
~, дымова́я smoke generating apparatus
~ засекре́чивания scrambler, secrecy system
~, звукометри́ческая sound ranging equipment
~, каналообразу́ющая channelling equipment (sigs)
~, ла́зерная laser equipment
~ наведе́ния 1. laying devices (arty) 2. guidance equipment (GW)
~ наведе́ния, бортова́я onboard guidance equipment (GW)
~, назе́мная ground equipment, ground support equipment
~ обнаруже́ния acquisition set, acquisition radar
~ обнаруже́ния ми́нного по́ля minefield marking equipment
~ обнаруже́ния ми́нного по́ля, инфракра́сная infra-red minefield marking equipment
~, оконе́чная terminal equipment
~, пасси́вная инфракра́сная passive infra-red equipment
~, пеленгацио́нная DF equipment
~, поме́ховая jamming equipment (EW)
~ радиолокацио́нного опознава́ния це́лей identification friend or foe equipment, IFF equipment
~ радиопоме́х electronic countermeasures equipment, ECM equipment
~ радиопротиводе́йствия electronic countermeasures equipment, ECM equipment
~, разве́дывательная reconnaissance equipment, surveillance equipment
~ самонаведе́ния homing equipment
~ свя́зи, быстроде́йствующая high-speed signalling apparatus
~ свя́зи, засекре́ченная communications scrambler equipment
~ слеже́ния tracking equipment
аппарату́ра сопровожде́ния tracking equipment
~, счётно-реша́ющая computer
~, та́бельная authorized equipment, authorized apparatus, equipment laid down in establishment table
~, та́нковая навигацио́нная AFV navigation eqpt
~, термодымова́я thermal smoke generating equipment
~, топливопода́ющая fuel feed equipment
~, топливопода́ющая "Lock-on!" (radar)
~ уплотне́ния multiplexing equipment
~ управле́ния control equipment
~ управле́ния огнём fire control equipment
~, шта́тная organic equipment
~ электро́нной разве́дки electronic surveillance equipment
аппара́тчик 1. operator 2. communist party functionary
аппаре́ль ramp
~, вы́садочная landing ramp
~, погру́зочная loading ramp
~, разгру́зочная unloading ramp
~, спускова́я launching ramp
апте́ка dispensary
~, похо́дная first aid kit, medicine chest
апте́чка 1. medical chest, first aid kit 2. repair kit
~, индивидуа́льная похо́дная first aid kit
арго́н argon
арео́метр areometer, hydrometer
аре́ст 1. arrest 2. detention
~, дома́шний confinement to quarters, house arrest
~, каза́рменный open arrest
~, просто́й detention
~ с содержа́нием на гауптва́хте confinement in guardroom
~, стро́гий detention with punishment diet (sov), close arrest
арестова́ние arrest, arresting
аресто́ванный man under arrest
арестова́ть (см аресто́вывать)
аресто́вывать/арестова́ть to arrest
арифмо́метр calculating machine
а́рка 1. arc 2. upper cord of bowstring truss (eng)
~, доща́тая planked arch, laminated arch

армáда 1. armada (nav)
2. large formation of aircraft
3. large formation of tanks
арматýра 1. fittings, accessories, equipment
2. reinforcement (of concrete)
~ **в вúде прóволочной сéтки** wire-mesh reinforcement
~, **предварúтельно напряжённая** pre-stressed reinforcement
армéйский army
армúрование reinforcing, reinforcement of materials, armour protection
армúрованный 1. armoured
2. reinforced (concrete)
áрмия 1. army
2. armed forces
~, **воздýшная** Air Army (Sov fmn)
~ **вторжéния** invasion army
~, **дéйствующая** field army, army on active service
~, **механизúрованная** mechanised army
~, **милициóнная** militia
~, **общевойсковáя** all arms army, combined arms army
~, **оккупациóнная** occupying army, army of occupation
~, **отдéльная** independent army
~ **прикрытия** covering army
~, **регулярная** regular army
~, **Совéтская** 1. Soviet Army
2. Soviet Armed Forces
~, **тáнковая** tank army
~, **удáрная** shock army
~, **экспедициóнная** expeditionary force, expeditionary army
арретúр checking device, stop, detent, catch
caging (of a gyroscope)
арретúрование caging (of a gyroscope)
арретúровать to lock, to rest, to cage; **арретúровать гироскóп** to cage a gyroscope
арсенáл arsenal
арсенúд arsenide (chem)
~ **алюмúния** aluminium arsenide
~ **гáллия** gallium arsenide
~ **úндия** indium arsenide
арсúн arsine (chem)
арт- (артиллерúйский) artillery (as a prefix)
артбатарéя artillery battery
артвооружéние ordnance and ammunition
артвыстрел artillery round
артвьюк artillery pack load
артдивизиóн artillery battalion
артериáльный arterial (med)
артериосклерóз arteriosclerosis
артикуляция articulation (comms and gen)
артиллерúйский artillery
артиллерúйско-миномётный artillery-mortar
артиллерúйско-техничéский ordnance
артиллерúст gunner, artilleryman
артиллерúст-самохóдчик SP gun crewman
артиллéрия artillery
~, **армéйская** army artillery
~, **áтомная** nuclear artillery
~, **батальóнная** battalion artillery (arty organic to non-arty bn)
~, **береговáя** coastal artillery
~ **береговóй оборóны** coastal artillery
~, **бесствóльная** rocket artillery
~ **ближнего бóя, зенúтная** 1. close range anti-aircraft artillery
2. close range anti-aircraft armament (nav)
~ **большóй мóщности** heavy artillery
~, **буксúруемая** towed artillery
~, **воздýшно-десáнтная** airborne assault artillery
~, **возúмая** towed artillery
~, **войсковáя** organic artillery
~, **вьючная** pack artillery
~ **вьючной тяги** pack artillery
~, **гáубичная** howitzer artillery, howitzers
~, **гладкоствóльная** smoothbore artillery (Sov term includes mortars)
~, **гóрная** mountain artillery
~, **гóрно-вьючная** pack mountain artillery, pack artillery
~ **дáльнего бóя, зенúтная** 1. long range anti-aircraft artillery
2. long range anti-aircraft armament (nav)
~ **дáльнего дéйствия** long-range artillery
~, **дальнобóйная** long-range artillery
~ **дивúзии** divisional artillery
~, **дивизиóнная** divisional artillery
~, **железнодорóжная** railway artillery

артилле́рия, зени́тная 1. anti-aircraft artillery
2. anti-aircraft armament (nav)
3. air defence artillery
~, **истреби́тельно-противота́нковая** anti-tank artillery
~, **казема́тная** garrison artillery, casemate artillery
~, **класси́ческая** conventional artillery
~, **корпусна́я** 1. corps artillery
2. corps troops Royal Artillery (Brit)
~, **крепостна́я** fortress artillery
~, **кру́пная** heavy-calibre armament (nav, tk), heavy artillery
~ **кру́пного кали́бра** 1. heavy artillery
2. heavy-calibre armament (nav)
~, **крупнокали́берная** 1. heavy artillery
2. heavy-calibre armament (nav, tk)
~, **лёгкая** 1. light artillery
2. light armament (shipboard)
~ **ма́лого кали́бра** 1. light artillery
2. small-calibre armament (nav)
~, **малокали́берная** 1. light artillery
2. small-calibre armament (nav)
~, **малокали́берная зени́тная** light anti-aircraft artillery
~, **морска́я** naval gunnery, naval guns
~ **на автомоби́льной тя́ге** truck-drawn artillery (wheels)
~ **на гу́сеничной тя́ге** tractor-drawn artillery (tracks)
~ **на железнодоро́жных устано́вках** railway artillery
~ **на ко́нной тя́ге** horse-drawn artillery
~ **на механи́ческой тя́ге** tractor-drawn artillery
~ **на тра́кторной тя́ге** tractor-drawn artillery
~, **назе́мная** surface-to-surface artillery
~, **нарезна́я** rifled artillery (as opposed to smoothbore)
~, **нарезна́я зени́тная** rifled anti-aircraft artillery, rifled air defence artillery
~ **непосре́дственной подде́ржки** close support artillery, direct support artillery
~ **о́бщей подде́ржки** general support artillery
~, **обы́чная** conventional artillery
~, **оса́дная** siege artillery
~ **осо́бого назначе́ния** special purpose artillery
~ **ПВО** AA defence artillery
артилле́рия, подви́жная берегова́я mobile coast artillery
~, **подде́рживающая** supporting artillery
~ **подде́ржки** artillery in support
~, **подчинённая** artillery under command
~, **полева́я** field artillery
~ **полка́** artillery organic to a rifle regiment
~, **полкова́я** regimental artillery, artillery organic to the regiment
~, **при́данная** artillery under command
~, **противота́нковая** anti-tank artillery
~, **пу́шечная** cannon artillery
~ **разруше́ния** artillery carrying out initial massed strike
~, **раке́тная** rocket artillery
~, **реакти́вная** rocket artillery
~ **резе́рва верхо́вного гла́вного кома́ндования** army artillery group
~ **резе́рва гла́вного кома́ндования** general headquarters reserve artillery
~, **самохо́дная** 1. self-propelled artillery
2. SU artillery (Sov)
~, **самохо́дная зени́тная** self-propelled AA artillery
~, **скоростре́льная** quick-firing artillery, rapid-firing artillery
~ **сопровожде́ния** close support artillery
~ **сре́днего кали́бра** 1. medium artillery
2. medium-calibre armament (nav)
~, **среднекали́берная** 1. medium artillery
2. medium-calibre armament (nav)
~, **сре́дняя** 1. medium artillery
2. medium-calibre armament (nav)
~, **стациона́рная** artillery position
~, **ство́льная** tube artillery
~, **тяжёлая** heavy artillery
~, **универса́льная** dual-purpose armament (nav)
~ **усиле́ния** artillery in support
арткома́нда artillery fire order
артма́стер gun fitter
артмастерска́я artillery workshop
~, **полкова́я** regimental artillery workshop
артминвы́стрел artillery-mortar round
артминого́нь artillery-mortar fire
артналёт artillery strike
артнаступле́ние support fire plan for the assault
артнача́льник artillery commander

артобстре́л artillery quick concentration
артогнево́й artillery-fire
артого́нь artillery fire
артпа́рк artillery ordnance park
артподгото́вка artillery preparation, softening up
артпо́езд truck and towed artillery piece
артполиго́н artillery range
артрасчёт gun detachment
артсисте́ма gun and mounting
~, ба́шенная turret mounting (tk gun)
~, спа́ренная twin gun mounting (arty)
артснаря́д artillery projectile
артстрельба́ artillery fire
артустано́вка gun and mounting
артхимобстре́л 1. chemical shoot
2. gas shelling
артча́сть artillery unit
артшту́рм artillery strike
архи́в, вое́нно-учётный military records office
ары́к irrigation ditch
арьерга́рд rearguard
асбе́ст asbestos
асбе́стовый asbestos
асидо́л acidol (metal polish used by servicemen)
асимметри́ческий asymmetric
асимметри́чный asymmetric
асинхро́нный asynchronous
аспира́нт post-graduate student
ассениза́ция 1. sanitation
2. sewage disposal
ассисте́нт 1. colour officer (mil)
2. colour guard (mil)
3. subordinate teaching post or functionary (university)
ассисте́нтская working area, packing room, working room
ассортиме́нт range of commodities, range of items
астродина́мика astrodynamics
астроко́мпас astrocompass
астронаведе́ние stellar guidance, celestial guidance
астронавига́ция astronavigation, celestial navigation, star navigation
астрона́втика astronautics
астроориента́тор star tracker (instrument)
асфа́льт asphalt
асфальтобето́н bituminous concrete
асфальтобетоноукла́дчик paver, road layer
ата́ка 1. attack
2. assault, charge
~ в лоб frontal attack
~ в штыки́ bayonet charge
~, внеза́пная surprise attack
~, газобало́нная cloud gas attack, cylinder attack (CW)
~, га́зовая gas attack
~, газоволнова́я cloud gas attack (CW)
~, дру́жная co-ordinated attack
~, кинжа́льная hand-to-hand attack, assault
~, концентри́ческая converging attack
~ кру́пными си́лами assault in strength, attack in force
~, лобова́я frontal attack
~, ло́жная feint attack, false attack
~, ми́нная underground mining (mil eng)
~ на бре́ющем полёте hedge-hopping attack, low level attack
~, неме́дленная immediate attack
~, непреры́вная continuous attack, uninterrupted attack
~, обхо́дная flank attack
~, обы́чная conventional attack
~, плани́рованная deliberate attack
~, предвари́тельная preparatory attack (air)
~, психи́ческая psychological attack
~ с ты́ла attack from the rear
~ с хо́да attack (from the line of march)
~, сквозна́я penetration attack, deep penetration attack
~, та́нковая tank attack
~, та́нково-пехо́тная infantry-tank attack
~, фланго́вая flank attack
~, фронта́льная frontal attack
~, фронтова́я army group attack
~, хими́ческая gas attack
~, ядовитодымова́я toxic smoke attack
ата́ки, неоднокра́тные repeated attacks
~, повто́рные repeated attacks
атакова́ть to attack, to assault, to charge
атаку́ющий attacking
атмосфе́ра 1. atmosphere (air)
2. atmosphere (unit of pressure)
~, ве́рхняя upper atmosphere
~, защи́тная protective atmosphere
~, междунаро́дная станда́ртная standard international atmosphere

атмосфéра, ни́жняя lower atmosphere
~, **окружáющая** ambient atmosphere
~, **разрежённая** rarefied atmosphere, tenuous atmosphere
~, **слои́стая** stratified atmosphere
атмосфéрный atmospheric
áтом atom
"ÁТОМ" radioactive contamination (warning signal)
áтом, возбуждённый excited atom
~, **дéлящийся** fissile atom, fissioned atom
~, **ионизи́рованный** ionized atom
~, **лёгкий** light atom
~, **нейтрáльный** neutral atom
~, **распáвшийся** decayed atom
~, **тяжёлый** heavy atom
атомизáция atomization
атоми́стика atomic theory and atomics
áтомный atomic, nuclear
атомоплáн nuclear-powered spacecraft
атропи́н atropine (med)
атташé attache
~, **воéнный** military attache
аттенюáтор attenuator
аттестáт 1. certificate 2. certificate of service, testimonial
~, **вещевóй** individual clothing and equipment record
~, **дéнежный** pay certificate
~, **продовóльственный** ration certificate
аттестовáть to recommend
áутборт outboard motor
áутригер outrigger
ахромáт achromatic lens
ахромати́ческий achromatic (photo)
ацетилéн acetylene
~, **четырёххлóристый** acetylene tetrachloride
ацетóн acetone
аэрáция aeration
аэробаллистика aeroballistics
аэровокзáл air terminal
аэродинáмика aerodynamics
аэродинами́ческий aerodynamic
аэродрóм airfield
~ **бази́рования** base airfield (airborne ops)
~, **воéнный** military airfield (used exclusively by Armed Services)
~, **войсковóй** operational airfield
~, **вспомогáтельный** auxiliary airfield
~ **вы́грузки** airhead
аэродрóм вы́лета departure airfield
~, **дéйствующий** operational airfield
~, **запаснóй** alternative airfield, diversion airfield
~ **исхóдного райóна** departure airfield (airborne ops)
~, **лóжный** dummy airfield
~ **манёвра** diversion airfield (for GCI fighters)
~, **основнóй** major airfield
~, **передовóй** forward airfield
~ **погрýзки** air embarkation centre
~, **полевóй** temporary airfield
~, **постоя́нный** permanent airfield
~ **специáльного назначéния** special purpose airfield
~, **трассовóй** intermediate airfield
~, **тыловóй** rear airfield
~, **шкóльный** training airfield
аэродрóмный airfield
аэрозóль aerosol
аэрозóльный aerosol
аэромая́к airfield beacon
аэромоби́льный air-mobile
аэронавигациóнный aeronautical, air navigation
аэронавигáция air navigation
аэронáвтика aeronautics
аэропáуза aeropause
аэроплёнка aerial camera film
аэропóрт airport
аэросáни propeller-driven sledge
аэросни́мок aerial photograph
~, **крупномасштáбный** large-scale aerial photograph
~, **мелкомасштáбный** small-scale aerial photograph
~, **нетрансформи́рованный** unrectified air photograph
~, **перспекти́вный** oblique aerial photograph
~, **плáново-перспекти́вный** vertical-oblique aerial photograph
~, **плáновый** vertical aerial photograph
~ **с координáтной сéткой, плáновый** gridded vertical aerial photograph
~, **стереоскопи́ческий** stereoscopic aerial photograph
~, **трансформи́рованный** rectified aerial photograph
аэростáт balloon, air balloon
~ **заграждéния** barrage balloon
~, **наблюдáтельный** observation balloon

аэроста́т, привязно́й captive balloon
аэроста́тика aerostatics
аэросфе́ра aerosphere
аэрото́рий aerodrome traffic zone
аэротранспорта́бельность 1. air-portability (eqpt)
2. air-transportability (units)
аэротранспорта́бельный 1. air portable (eqpt)
2. air transportable (units)
аэроу́зел airport complex
аэроупру́гость aeroelasticity (GW)
аэрофотоаппара́т aerial camera, aircraft camera
~, **короткофо́кусный** short focal-length aircraft camera
~, **ночно́й** aerial camera for night photography
~, **среднефо́кусный** medium focal-length aircraft camera (F=15Ømm-3ØØmm)
~, **универса́льный** general-purpose aerial camera, medium focal-length aerial camera
~, **щелево́й** continuous-strip aerial camera
аэрофотограмме́трия aerial photogrammetry
аэрофотографи́рование aerial photography
аэрофотогра́фия aerial photography
аэрофотомонта́ж aerial mosaic
аэрофоторазве́дка aerial photo-reconnaissance
аэрофотосни́мок aerial photograph (for types see aehrosnimok)
аэрофотосъёмка aerial photography
~, **маршру́тная** strip aerial photography
~, **перспекти́вная** oblique aerial photography
~, **пла́новая** vertical aerial photography
~, **пла́ново-перспекти́вная** low oblique aerial photography
~, **стереоскопи́ческая** stereoscopic aerial photography
аэрофототопогра́фия aerial phototopography

Б

ба́ба 1. monkey (piling rig)
2. drop weight, ram
3. peasant woman
~, **ручна́я** hand pile-driver, mallet
баббит babbitt, bearing alloy
баби́на 1. bobbin, dispenser (GW)
2. ignition coil (ICE) (see also bobina)
ба́бка 1. monkey, drop weight
2. chuck, headstock
3. sheaf (of corn)
4. pastern (of horse)
~, **за́дняя** tailstock (lathe)
~, **пере́дняя** headstock (lathe)
~, **шпи́ндельная** headstock (lathe)
бага́ж luggage, baggage
бага́жник luggage carrier (boot, rack, hold), trunk
баго́р boathook, hook, gaff
ба́за base, basis, datum level, foundation
~, **авиацио́нная** air base
~, **авторемо́нтная** MT repair base
~, **акусти́ческая** sound ranging base
~, **вое́нная** 1. military base
2. main base (staff term)
~, **вое́нно-возду́шная** air force base
~, **вое́нно-морска́я** naval base
~, **вре́менная** 1. temporary base
2. temporary dump (airborne troops)
~, **вспомога́тельная** sub base (topo)
~, **геометри́ческая** geometric base (sound ranging)
~, **гла́вная передова́я** main advanced base
~ **глаз** interpupillary distance (obsn)
~, **горнолы́жная** base for mountain ski troops
~ **горю́чего** fuel base, fuel terminal
~ **для веде́ния опти́ческой разве́дки, коро́ткая** abbreviated flash base (arty)
~, **заря́дно-техни́ческая** battery charging shop
~ **засе́чки** base of intersection
~, **звукова́я** sound-ranging base (arty)
~, **звукометри́ческая** sound-ranging base (arty)
~ **звукоула́вливателя** sound-ranging base
~, **коро́ткая** short base (arty)
~ **маши́ны** wheel base

ба́за, морска́я naval base
~ **наблюде́ния** true observation base (arty)
~, **норма́льная** normal base (arty)
~ **окуля́ров** interocular distance, width between eyepieces
~, **операцио́нная** base of operations, operational base
~ **пеленга́ции** DF base-line, DF base
~, **перева́лочная** trans-shipment point, trans-shipment base, shipping terminal
~ **перелива́ния кро́ви** base blood transfusion unit
~ **погру́зки морско́го деса́нта** mounting base (amph ops)
~, **подви́жная авторемо́нтная** mobile MT repair workshop
~, **подви́жная ракéтно-техни́ческая** mobile rocket-technical base (SSMs)
~, **полева́я та́нко-ремо́нтная** field tank repair workshop
~, **ракéтная** missile base, rocket base
~ **ремо́нта автобронета́нковой тéхники** AFV repair depot
~, **ремо́нтная** base workshop, repair base
~ **снабжéния** supply base
~ **снабжéния, армéйская** army supply base
~ **снабжéния горю́чим** fuel supply base
~ **снабжéния фро́нта, головна́я** forward army group base
~ **со ста́ртовыми пози́циями ша́хтного ти́па, ракéтная** silo base
~ **сопряжённого наблюдéния** cross observation base (arty)
~, **ста́ртовая** launching base
~, **та́нковая** tank park
~ **техни́ческого обслу́живания и теку́щего ремо́нта** maintenance depot
~, **учéбно-материа́льная** training aids supply, supply of training equipment, training equipment and stores
~ **фро́нта, госпита́льная** front base hospital
~ **ша́сси** wheelbase
~, **я́вочная** local reporting office for recruits
бази́рование basing
бази́роваться to base (oneself on), to be based on
ба́зис basis, base (line)
~ **триангуля́ции** triangulation base
ба́зовый basic, base
базу́ка bazooka
ба́йковый baize
байпа́с by-pass (line)
бак tank, canister
~, **авари́йный то́пливный** emergency fuel tank
~, **бензи́новый** petrol tank, fuel tank
~, **брониро́ванный то́пливный** bullet-proof fuel tank, armoured steel fuel tank
~, **быстросъёмный** quick-detachable tank
~, **вне́шний** external tank, outboard tank
~, **вспомога́тельный** auxiliary tank
~, **гла́вный то́пливный** main fuel tank
~ **для гидросмéси** hydraulic tank
~ **для горю́чего** fuel tank
~, **дополни́тельный** additional tank, supplementary tank, auxiliary tank, extra tank, long-range tank
~, **запасно́й** reserve tank, emergency tank
~, **ма́сляный** oil tank
~, **мя́гкий** flexible tank, bag-type tank, cell-type tank
~, **напо́рный** header tank, supply tank
~, **нару́жный** external tank
~, **основно́й** main tank
~, **пита́ющий** feed tank
~, **протекти́рованный то́пливный** bullet-proof fuel tank
~, **расхо́дный** feed tank, supply tank, sump tank, service tank, collector tank
~, **резéрвный** reserve tank
~, **рези́новый** rubber (bag) tank
~, **самозатя́гивающийся** self-sealing tank
~, **сбра́сываемый** drop tank (ac)
~, **то́пливный** fuel tank
ба́кан beacon, buoy
бакели́товый bakelite
ба́кен beacon, buoy
бак-коллéктор accumulator tank
бактериа́льный bacterial
бактéрии bacteria
бактери́йный bacterial
бактерио́лог bacteriologist
бактериологи́ческий bacteriological, BW
бактериофа́г bacteriophage
бактéрия bacterium
бак-храни́лище storage tank
бала́нда slops, kitchen waste
бала́нс, энергети́ческий energy balance (nucl)

баланси́р 1. lever, arm
2. axle arm, suspension arm, rocker
~ ма́лых катко́в bogie bracket (tk)
~ опо́рного катка́ track wheel (support) arm, road wheel (support) arm (tk)
~, попере́чный lateral axle
~, продо́льный longitudinal beam
~ теле́жки 1. bogie arm, fork, axle
2. suspension lever (tk)
баланси́ровать/сбаланси́ровать to balance
балансиро́вка balancing, trim
~ ши́ны balancing of tyre
ба́лка 1. small ravine
2. beam, girder
3. chess (pontoon eqpt)
~, двутавро́вая "I" beam, "H" beam (eng)
~, железобето́нная reinforced concrete beam, reinforced concrete girder
~ жёсткости stiffening girder
~, консо́льная cantilever beam
~, коро́бчатая box girder
~ коро́бчатого сече́ния box section girder, box girder
~, кра́новая (crane) jib
~, мостова́я bridge beam
~, опо́рная bearing beam, load beam, support beam
~, попере́чная 1. cross member (chassis)
2. cross beam, cross girder
~ похо́дного крепле́ния cradle clamp (ord)
~, продо́льная 1. side member (chassis)
2. stringer (br)
~, пускова́я launching beam
~ с предвари́тельно напряжённой армату́рой, железобето́нная pre-stressed reinforced concrete beam
~, связна́я cradle clamp (ord)
~, тавро́вая "T" beam
~, тротуа́рная footwalk beam (br)
~, хвостова́я tail boom (helicopter)
~, шве́ллерная channel section beam
~, шворнева́я draw bar
~, широкопо́лочная двутавро́вая H-beam
~, шкворнева́я draw bar
балла́ст ballast
балластиро́вка ballasting
балли́стика ballistics
~, вне́шняя external ballistics
~, вну́тренняя internal ballistics
балли́стика коне́чных скоросте́й terminal ballistics
~, промежу́точная intermediate ballistics
~ проникнове́ния penetration ballistics
~ раке́т rocket ballistics
~ сверхвысо́ких скоросте́й hyperballistics
баллисти́т ballistite (gunn)
баллисти́ческий ballistic
балло́н 1. air bottle, cylinder, tank
2. accumulator (hydraulics etc)
3. bulb (electric lamp)
4. envelope (thermionic valve)
5. pneumatic tyre
~, азо́тный nitrogen bottle
~, кислоро́дный oxygen cylinder
~ с углекислото́й carbon dioxide bottle
~ сжа́того во́здуха compressed air bottle, high-pressure air bottle
~, стекля́нный glass envelope
балло́н-зонд weather balloon, test balloon
ба́мпер bumper
банда́ж 1. truss, taping up (med)
2. tyre (rly)
3. band, binding
~ колеса́ wheel tyre, wheel tread
ба́нить/вы́банить to sponge out, to wash out; **ба́нить/вы́банить ору́дие** to sponge out a gun (arty)
ба́нка 1. jar, can
2. thwart
~, аккумуля́торная accumulator jar (elec)
~, бомба́жная консе́рвная blown can (food supply)
~, дегазацио́нная CW decontaminant container
~, консе́рвная tin, can, tin can
~, ми́нная cluster of mines, mine cluster
ба́нка-сиде́ние thwart
банке́т fire step (fd wks)
ба́нник bore cleaner (with brush attached)
ба́нник-протира́льник piasaba brush (ord)
ба́нник-разря́дник cleaner and ejector rod
ба́нно-дезинфекцио́нный bath and decontamination
ба́нно-пра́чечный laundry and bath
бар 1. bar (met)
2. bar, sandbar (topo)

бараба́н 1. drum
2. cylinder (pistol)
~ **буссо́ли** screw drum on the director (gunn)
~, **возвра́тный** retracting drum (arty)
~ **дistáнции** range scale drum
~, **дистанцио́нный** range drum (gun sight)
~ **кату́шки** reel drum
~ **лебёдки** winch drum
~ **магази́на** winding device (drum magazine)
~ **монокуля́ра** elevation micrometer drum (director)
~, **откидно́й** swing-out cylinder (revolver)
~ **отража́теля** micrometer cowl (dial sight)
~ **прице́ла** sight drum
~ **револьве́ра** cylinder (revolver)
~ **скоросте́й** target speed drum (AA ord)
~, **тормозно́й** brake drum
~ **угломе́ра** deflection drum, azimuth drum, deflection micrometer (gunn)
~, **угломе́рный** clinometer head, micrometer head (ord)
бараба́нчик knob, drum
~ **отража́теля** elevation micrometer (gunn)
~ **то́чной наво́дки** micrometer (arty)
~ **то́чной устано́вки прице́ла** elevation micrometer
~ **угломе́ра** deflection micrometer
~, **устано́вочный** range drum (back sight)
~ **червяка́ боково́го у́ровня** elevation micrometer
бараба́нщик drummer
бара́к barrack hut
~, **сбо́рно-разбо́рный каза́рменный** standard hutting, sectional hutting
бара́ки hutments
бара́нка 1. steering wheel (slang)
2. ring shaped roll
барато́л baratol (mixture of trotyl and nitrate of barium) (expl)
бара́шек wing nut
~, **зажи́мный** butterfly nut, wing nut
бара́шки fleecy clouds
барбе́т barbette
баре́ттер barretter (radio)
ба́ржа barge
ба́ржа, артиллери́йская ammunition lighter (vessel)
~, **быстрохо́дная деса́нтная** high-speed landing craft
~ **вторже́ния** assault landing craft
~, **вы́грузочная** landing barge (eqpt)
~, **вы́садочная** landing barge, assault landing craft (personnel)
~, **деса́нтная** assault landing craft
~ **для та́нков и мототе́хники, уда́рная** mechanized landing craft, LCM
~, **наливна́я** tank barge, tanker barge
~, **несамохо́дная** dumb barge
~, **нефтеналивна́я** fuel oil barge, petrol barge
~, **речна́я** river barge
~, **самохо́дная деса́нтная** landing craft - mechanised, LCM
~, **та́нко-деса́нтная** landing craft - utility, LCU
~, **штурмова́я пехо́тно-деса́нтная** assault landing craft (infantry)
ба́рий barium
ба́рка flat bottomed barge
барка́с launch, long boat
барн barn (nucl)
баро́граф barograph
бароќа́мера altitude chamber, pressure chamber
баро́метр barometer
~, **рту́тный** mercury barometer
баро́метр-анеро́ид aneroid barometer
барометри́ческий barometric, barometrical
барра́ж 1. barrage
2. boom defence
баррaжи́рование standing patrol (air)
баррика́да barricade
баррикади́ровать/забаррикади́ровать to barricade, to block
барха́н sand hill, inland sand dune
барье́р barrier
~, **авари́йный** crash barrier
~, **глухо́й** barricade block (fd eng)
~, **звуково́й** sound barrier
~, **отко́сный** anti-tank slope
~, **противоба́шенный** anti-turret barrier (anti-tk obstacle)
~, **сне́жно-ледяно́й** snow and ice barrier
~, **теплово́й** heat barrier, thermal barrier
~, **энергети́ческий** energy barrier
бассе́йн tank, basin, reservoir
бастио́н bastion, bulwark

батальо́н battalion
~, **авторемо́нтный** MT repair battalion
~, **автотра́нспортный** MT battalion
~, **возду́шно-деса́нтный сапёрный** airborne engineer battalion
~, **дисциплина́рный** penal battalion
~, **доро́жно-строи́тельный** road construction battalion
~, **доро́жно-эксплуатацио́нный** road construction battalion
~, **доро́жный-коменда́нтский** provost traffic control battalion
~, **инжене́рно-строи́тельный** engineer construction battalion
~, **ме́дико-санита́рный** medical battalion
~, **миномётный** mortar battalion (non-arty unit)
~ **морско́й пехо́ты** marine infantry battalion
~, **мотострелко́вый** motor rifle battalion
~, **мотоцикле́тный** motorcycle battalion
~, **отде́льный** independent battalion
~, **отде́льный авторемо́нтный** independent MT repair battalion
~, **отде́льный автотра́нспортный** independent MT battalion
~, **отде́льный разве́дывательный** independent reconnaissance battalion
~, **парашю́тно-деса́нтный** parachute battalion
~, **понто́нно-мостово́й** pontoon bridge battalion, Sov
~, **ремо́нтно-восстанови́тельный** repair and refitting battalion
~, **санита́рно-тра́нспортный** ambulance battalion
~, **санита́рный** medical battalion
~, **сапёрно-штурмово́й** assault sapper battalion, assault engineer battalion
~, **сапёрный** sapper battalion
~, **сво́дный** composite battalion
~ **свя́зи** signal battalion
~, **стрелко́вый** rifle battalion
~, **та́нковый** tank battalion
~ **техни́ческого обслу́живания, аэродро́мный (АВАТОБ)** airfield technical servicing battalion
~ **тра́нспортно-деса́нтных вертолётов** helicopter assault battalion
~, **уче́бный** training battalion
~, **штрафно́й** penal battalion
батальо́н электро́нного подавле́ния electronic jamming battalion
батальо́нный battalion
батаре́йка small electric battery
батаре́йный battery (arty, elec)
батаре́я 1. battery (arty)
2. battery (elec)
3. bank (of items)
~, **аккумуля́торная** accumulator battery
~, **ано́дная** anode battery
~, **артиллери́йская** artillery battery
~ **артиллери́йской инструмента́льной разве́дки** locating battery (arty)
~, **ба́шенная** turret battery (coastal arty)
~, **берегова́я** coastal battery (arty)
~, **бу́ферная** buffer battery (elec)
~, **га́убичная** howitzer battery
~, **го́рно-вью́чная** mountain battery (arty)
~ **звуково́й разве́дки** sound-ranging battery (arty)
~, **звукометри́ческой разве́дки** sound ranging battery (arty)
~, **зени́тная** air defence battery, anti-aircraft battery (arty)
~, **зени́тная артиллери́йская** anti-aircraft artillery battery
~, **кисло́тно-свинцо́вая аккумуля́торная** lead-acid storage battery
~, **кочу́ющая** roving battery (arty)
~ **ма́лого кали́бра, зени́тная** light anti-aircraft battery, LAA battery
~, **миномётная** mortar battery
~ **на огнево́й пози́ции** battery in fire position
~, **ни́келево-ка́дмиевая аккумуля́торная** nickel-cadmium storage battery (elec)
~ **опти́ческой разве́дки** flash-spotting battery (arty)
~, **подручна́я** stand-by battery
~, **противота́нковая** battery in fire position
~ **противота́нковых управля́емых реакти́вных снаря́дов** anti-tank guided missile battery
~, **пу́шечная** gun battery
~, **раке́тная** missile battery, rocket battery
~ **светово́й разве́дки** flash ranging battery
~, **свинцо́во-кисло́тная аккумуля́торная** lead-acid storage battery (elec)

батаре́я, сере́бряно-ци́нковая аккумуля́торная silver-zinc storage battery
~, со́лнечная solar battery
~, ста́ртовая launching battery, firing battery (GW)
~, суха́я dry battery (elec)
~, термоэлектри́ческая thermopile (elec)
~ то́пливных элеме́нтов fuel-cell battery
~ топографи́ческой слу́жбы survey battery
~, щелочна́я аккумуля́торная alkaline accumulator (elec)
батожо́к rifle support (in form of a cross)
бату́д safety net (airborne trg)
ба́фтинг buffeting (GW)
бахро́мчатость fringing (electron)
бачко́вый galley hand (naval sl)
бачо́к 1. small tank (container)
2. mess kettle
~, расшири́тельный expansion tank, overflow tank
ба́шенка cupola
~, бокова́я side turret (MG)
~, бронева́я armoured cupola
~, враща́ющаяся revolving cupola
~, команди́рская commander's cupola
ба́шенный turret
башлы́к pointed cloth hood
башма́к 1. boot
2. socket
3. chock
~ гу́сеницы grouser (shoe), caterpillar block, tread block
~, металли́ческий steel shoe (timber pile)
~ оруди́йного квадра́нта gunner's quadrant shoe
~, по́люсный pole piece, pole shoe (elec)
~, ре́жущий detachable bit (drill)
~ со́шки bipod shoe (SA)
~, тормозно́й 1. brake shoe
2. brake scotch, chock
~ электромагни́тного то́рмоза electro-magnetic brake shoe
ба́шня 1. turret
2. tower
~, артиллери́йская gun turret
~, бронева́я armoured turret
~, водонапо́рная water tower
ба́шня, враща́ющаяся revolving turret, rotating turret, traversing turret
~, кача́ющаяся oscillating turret
~, лита́я cast turret (tk)
~, оруди́йная gun turret
~ с круговы́м враще́нием full-revolving turret, 360-degree traversing turret
~ с круговы́м обстре́лом full-revolving turret, 360-degree traversing turret
~ с механи́ческим приво́дом power traversed turret
~, цельноли́тая бронева́я one-piece cast armour steel turret
ББ, обы́чное conventional explosive
бди́тельность alertness, vigilance
бди́тельный alert, vigilant, security minded
бег run, double time, quick time
~, военизи́рованный military obstacle course race
~ рывка́ми alternate running and marching
бегле́ц fugitive, escapee, deserter (from post)
"Его́м!" "At the Double!"
"Его́м Марш!" "Double March!"
бе́гство hasty retreat, flight
бегуно́к runner, roller
бе́дствие distress, misfortune, disaster, calamity
~, стихи́йное natural calamity, act of god
бежа́вший escapee
бе́з вести пропа́вший missing (in action)
безавари́йный accident-free
безба́шенный turretless
безве́тренный calm (met)
безве́трие still, calm (met)
безво́дный waterless
безвозвра́тно irrevocably
безвозвра́тный irrecoverable
безгироско́пный non-gyroscopic
безде́йствие inaction
безде́йствовать to be inactive, to do nothing (imp only)
бездоро́жный roadless
бездоро́жье 1. absence or insufficiency of roads
2. season of bad roads
3. bad or impassable roads (due to weather or surface conditions)
безкурко́вый hammerless (SA)
безла́мельный sintered-plate

безло́шадник pilot without an aircraft
безме́н, пружи́нный spring balance
безопа́сность safety, security
~ на полиго́не range safety
~, противопожа́рная fire prevention
безопа́сный safe, secure
~ при неуме́лом обраще́нии fool-proof
безостано́вочно without halting
безотка́зность reliability
безотка́зный trouble-free, reliable
безотка́тность absence of recoil
безотка́тный recoilless
безуда́рный shockless, smooth, steady
безуте́чный leakproof
безъёмкостный non-capacitive (elec)
безъя́дерный non-nuclear
бели́ла white lead
~, ци́нковые zinc white, zinc oxide
бельё, нате́льное underwear
~, посте́льное bed linen
~ ти́па "се́тка" string underwear
бе́нзель seizing (means of fastening two ropes together to prevent them moving in relation to each other)
бензи́л, бро́мистый benzyl bromide
бензи́н petrol, benzine, gasoline
~, авиацио́нный aviation gasoline, AVGAS
~, автомоби́льный MT gasoline, motor gasoline, ground gasoline
~, высокоокта́новый high-octane gasoline
~, зи́мний winter-grade gasoline
~, ле́тний summer-grade gasoline
~, неэтили́рованный non-leaded gasoline, unleaded gasoline
~, синтети́ческий synthetic gasoline
бензиново́д petrol pipe
бензиново́з petrol carrier
бензиноме́р petrol gauge, gasoline gauge
бензинопрово́д petrol feed pipe
бензоагрега́т petrol unit, petrol assembly, petrol power unit
бензоба́к petrol tank, fuel tank
бензово́з petrol tanker (veh)
бензозапра́вщик petrol bowser
бензоколо́нка petrol pump
бензо́л benzol(e), benzine
бензоме́р fuel level gauge
бензомотопо́мпа petrol-driven power pump
бензонасо́с petrol pump
бензообогрева́тель gasoline pre-heater
бензоочисти́тель petrol filter
бензопрово́д petrol pipe, fuel line
бензорука́в petrol-dispensing hose
бензоскла́д petrol depot, petrol dump
~, полево́й petrol dump, field petrol depot
бензосто́йкий petrol-proof, petrol-resistant
бензотрубопрово́д petrol pipe-line
бензохрани́лище petrol storage tank
бензоцисте́рна petrol tanker (veh)
~, железнодоро́жная bulk petrol waggon (rly)
бергштри́х hachure (topo)
бе́рег shore, coast, bank, beach
~, исхо́дный near bank
~, кома́ндующий commanding bank, higher bank (topo)
~, круто́й steep bank, steep beach, steep coast
~, обрезно́й steep beach
~, обры́вистый cliff shore, steep bank
~, отло́гий sloping shore, sloping bank, shallow sloped bank
~, пло́ский flat beach, flat shore
~, противополо́жный far bank, opposite shore
берегово́й bank, beach, coast, coastal
берёза birch (tree and timber)
бери́ллий beryllium
бе́рма berm, elbow rest (in a trench)
бесе́дка boatswain's chair
бескла́панный valveless
бесперебо́йно without interruption, regularly
бесперебо́йный continuous, ceaseless, unbroken
беспило́тный pilotless
беспла́менный flashless
беспоко́ить to worry, to harass, to keep on the alert (imp only)
беспоря́дки unrest, riots, disturbances
беспрепя́тственный unhindered, unimpeded
беспреры́вный continuous, ceaseless, unbroken
беста́рный in bulk, loose
"бесхво́стка" tailless configuration (GW)
бесшо́вный weldless
бе́та-излуче́ние beta radiation
бе́та-лучи́ beta-rays

бе́та-распа́д beta decay (Russian term also includes emission of beta plus particles)
бетатро́н betatron
бе́та-части́ца beta particle
бето́н concrete
~, арми́рованный reinforced concrete
~, быстросхва́тывающийся fast-setting concrete
~, ли́той cast concrete, poured concrete
~, напряжённо арми́рованный pre-stressed reinforced concrete
~, пласти́чный plastic concrete
бетони́рование concreting
бето́нный concrete
бетонобо́йный concrete piercing
бетономеша́лка concrete mixture
бетонье́рка concrete mixer
бечева́ twine, cord, rope
бечёвка cord
бече́вник tow path
бе́шенство rabies, hydrophobia
библиоте́ка-передви́жка mobile library
бива́к bivouac
бива́чный bivouac
бидо́н can, container (liquids)
бие́ние 1. wobble (wheel) 2. beat (elec, acoustic)
~ колеса́ wheel wobble
биле́т, вое́нный record of service
~, во́инский record of service, certificate of service
~, комсомо́льский Komsomol membership card
~, отпускно́й leave pass
~, парти́йный (communist) party membership card
биметалли́ческий bimetal, bimetallic
бимото́рный twin-engined
бино́кль binoculars, field glasses
~ для вожде́ния, инфракра́сный driving infra-red binoculars
~, дневно́й day glasses, prism binoculars
~, ночно́й night glasses, night binoculars
~, полево́й field glasses
~, призмати́ческий prism binoculars, prismatic binoculars
~, при́зменный prism binoculars, prismatic binoculars
~ с се́ткой graticule glasses
~, специа́льный special purpose binoculars (infra-red)
бино́м binome (math)
бинт bandage
биологи́ческий biological
би́рка label, disc, identity tag
~, шта́бельная stack label
би́смут bismuth
биссектри́са bisector, bisectrix
би́тва major battle, decisive battle
биту́м bitumen
~, доро́жный asphalt, paving bitumen
битумино́зный bituminous
биту́мный bituminous, bitumen
битумово́з bitumen distributor
бить/поби́ть to beat, to hit, to fight, to kill, to gush out, to shoot; **бить по частя́м** to destroy piecemeal
бич bang (naval sl)
бичёвка twine
благовре́менный punctual, timely, in good time
благоприя́тный favourable, advantageous, propitious
бланк form, pro-forma blank, blank form
~ донесе́ния message form (mil)
~ за́писи координа́т це́лей, оруди́йный target record form
~ попра́вок data sheet (arty)
~, телегра́фный message form
~ тре́бований requisition form
~ формуля́ра ми́нного по́ля minefield record form
бла́нковка ghost outline map
бла́нковка-увеличи́вка photographic enlargement of a ghost outline map
блат pull, influence
бленд, противосо́лнечный anti-flash compound, anti-glare compound
бле́нда shade, blind
бле́нкер flag indicator, dolls-eye indicator (on switchboard)
блеск flash, spurt of flame
бле́скость brilliance (radar)
ближа́йший nearest, immediate
бли́жний near, close
близлежа́щий near-by, neighbouring, adjacent
близору́кость пусто́го по́ля empty visual field effect
блинда́ж 1. dug out, shelter, covered trench 2. butts (on firing range)
~, подбру́стверный trench shelter
~ подзе́много ти́па underground dug-out

бли́нкер flag indicator, dolls-eye indicator (on switchboard)
блок 1. bloc (political)
2. pulley block (mechanical)
3. unit (eqpt)
4. bay, section (fd wks)
~ **автомати́ческого сопровожде́ния** automatic tracking unit (radar)
~, **автоно́мный** self-contained unit
~, **бето́нный** concrete block
~, **боково́й** booster engine (of space-veh)
~, **вычисли́тельный** computer unit, calculator
~, **гетероди́нный** 1. oscillator unit (mine detector)
2. heterodyne unit
~ **гиростабилиза́ции** gyrostabilization unit (tk gunn)
~ **да́льности** range unit (radar)
~ **да́тчиков** sensor unit
~ **деле́ния** divider (computer)
~ **дистанцио́нного управле́ния** remote-control unit
~, **замедля́ющий** moderating block (nucl)
~ **индика́ции** display unit
~, **ка́бельный** multiple duct (sigs)
~, **ка́менный** stone block
~ **коммута́ции** switching unit (electron)
~, **ледяно́й** ice block
~, **мо́дульный** modular unit
~ **настро́йки анте́нны** antenna-tuning unit
~, **неподви́жный** standing block (tackle)
~, **одношки́вный** single block (tackle)
~ **ответвле́ния** channel drop-off unit (RR)
~ **отрабо́тки да́нных** data processing unit
~ **переключа́телей** switch unit
~ **пита́ния** power pack, power supply unit (electron)
~, **подвесно́й** snatch block (tackle)
~, **подвижно́й** moving block (tackle)
~, **подъёмный** tackle
~, **понто́нный** pontoon section
~, **поршнево́й** anvil block (on pile driver)
~, **прице́льно-вычисли́тельный** gunsight computer
~ **радиоуправле́ния** radio control unit
~, **разго́нный** final stage of space-vehicle propulsion
~, **раке́тный** final stage of space-vehicle propulsion
блок, резе́рвный standby unit, spare unit
~ **с га́ком** hook block (tackle)
~ **самоликвида́ции** self-destruction unit
~, **селе́кторный** strobe unit (radio)
~ **сервоприво́да** servo unit
~ **слеже́ния** tracking unit
~, **сме́нный** plug-in unit, interchangeable unit (electron)
~ **смеси́теля** mixing unit (radio)
~, **телегра́фный** telegraph adapter (radio)
~, **трансформа́торно-выпрями́тельный** transformer-rectifier unit
~ **умноже́ния** multiplier (computer)
~ **уплотне́ния** channelling unit, multiplexing unit (radio)
~ **управле́ния** control unit
~ **усили́телей прице́ла** sight amplifier unit
~, **центра́льный** main engine section (of space-veh)
~ **цили́ндров** cylinder block, cylinder bank
~ **шестерён** gear block, gears
~, **шифра́торно-дешифра́торный** coder-decoder unit
~, **шта́бельный** block stack
блока́да blockade
~, **возду́шная** aerial blockade, aerial siege
блокга́ус blockhouse
бло́кинг-генера́тор blocking oscillator (electron)
~, **жду́щий** triggered blocking oscillator (electron)
~ **с самовозбужде́нием** self-excited blocking oscillator (electron)
блоки́рование sealing off (tac)
~ **аэродро́ма** denying enemy use of aerodrome
блоки́ровать/сблоки́ровать to blockade, to block, to interlock (elec)
блокиро́вка 1. sealing off, siege (tac)
2. interlocking, interlock (elec eng)
3. block system (rly)
блоки́рующий blockading, blocking
блокно́т sketch pad
блок-схе́ма 1. block diagram, routing diagram
2. flow chart
бля́ха 1. metal badge
2. belt buckle

боби́на 1. bobbin, dispenser GW
2. ignition coil (ICE) (see also babina)
бобы́шка boss, lug, piston boss
"Бог войны́" "God of War" (name given to artillery in Sov army)
бод baud (unit of telegraphic speed)
боди́ст Baudot telegraph operator
бодо́ Baudot
бо́дрость cheerfulness, vigour, courage
боеви́к primer (expl)
боево́й battle, combat, fighting, martial, military, operational, warlike
боеголо́вка warhead
боеготóвность 1. battle readiness, combat readiness, stand to, readiness for action, alert
2. fitness for war (troops)
3. operational fitness (eqpt)
боеготóвый combat-ready
боезапа́с battle reserve
боёк 1. firing pin
2. striker stud (mor)
~ уда́рника firing pin
боекомпле́кт, та́бельный basic ammunition load
~, та́бельный вози́мый organic load of ammunition
боепита́ние 1. supply in the field
2. ammunition supply (arty)
боеприпа́с, а́томный 1. atomic warhead
2. nuclear weapon
~, я́дерный atomic warhead
боеприпа́сы ammunition
~, агитацио́нные propaganda leaflet ammunition
~, артиллери́йские artillery ammunition
~, а́томные atomic ammunition
~ безотка́тных ору́дий recoilless ammunition
~, бронебо́йные armour-piercing ammunition
~, винто́вочные rifle ammunition
~ для стрельбы́ из вкладны́х ство́ликов sub-calibre ammunition
~, дымовы́е smoke ammunition
~, зажига́тельные incendiary ammunition
~ карту́зного заряжа́ния breech loading ammunition
~, нали́чные ammunition in hand, available ammunition
~, обы́чные conventional ammunition
боеприпа́сы патро́нного заряжа́ния fixed ammunition
~ подкали́берные sub-calibre ammunition
~ полуунита́рного заряжа́ния semi-fixed ammunition
~, пристре́лочные observing ammunition
~ разде́льного ги́льзового заряжа́ния separate-case ammunition, separate loading ammunition
~ разде́льного заряжа́ния separate (-loading) ammunition, unfixed ammunition
~ с ине́ртным снаряже́нием, учёбные inert ammunition
~ с переме́нными заря́дами, унита́рные semi-fixed ammunition
~ с уменьшённым заря́дом reduced charge ammunition
~, снаряжённые фо́сфором phosphorus ammunition
~ стрелко́вого ору́жия small-arms ammunition
~, трасси́рующие tracer ammunition
~ унита́рного заряжа́ния fixed ammunition
~, учёбные dummy ammunition, drill ammunition, simulated ammunition, training ammunition
~, хими́ческие chemical ammunition, chemical warfare ammunition
~, холосты́е blank ammunition
~ шпи́лечного воспламене́ния pin-fire ammunition (arty)
~, я́дерные 1. nuclear warheads
2. nuclear weapons
3. nuclear shells
боеспосо́бность fighting efficiency, operational efficiency, combat effectiveness
боеспосо́бный fit for war, battleworthy
боеукла́дка ammunition stowage
~, снаря́дная shell rack
бое́ц 1. soldier
2. private soldier
бое́ц-истреби́тель infantryman armed with anti-tank weapon
бое́ц-лы́жник ski-trooper
бое́ц-носи́льщик stretcher bearer
бое́ц-парашюти́ст paratrooper
бое́ц-пехоти́нец infantryman

бой 1. action, battle, engagement, encounter (tac)
2. action (weapons)
~, арьерга́рдный rearguard fighting, rearguard action
~, бараба́нный beat of drums, roll of drums
~ без примене́ния я́дерного ору́жия nuclear-free battle
~, бли́жний close combat, close fighting
~, быстроте́чный fast moving battle, action
~ в глубине́ оборони́тельной полосы́ dog fight (the second of three stages in a deliberate attack)
~ в населённом пу́нкте fighting in a built-up area
~ в усло́виях примене́ния я́дерного ору́жия atomic battle, atomic combat, nuclear battle, nuclear combat
~ возду́шно-деса́нтных во́йск airborne forces battle
~, встре́чный encounter battle, meeting engagement
~, глубо́кий battle in depth
~, городско́й street fighting
~, да́льний long range battle, long range combat
~, дли́тельный prolonged fighting
~ за вы́садку морско́го деса́нта amphibious assault
~, затяжно́й drawn-out battle
~, манёвренный mobile battle, fluid battle
~ на окруже́ние battle of encirclement
~ накоротке́ close battle of short duration
~, наступа́тельный attack, offensive battle, offensive, attacking battle
~, норма́льный zero, harmonisation, the zero (wpn, SA)
~, ночно́й night fighting
~, оборони́тельный defence, defensive battle, defensive action
~, общевойсково́й integrated battle of all arms, integrated all-arms battle
~, огнево́й fire fight
~, предстоя́щий forthcoming battle
~, рукопа́шный hand-to-hand fighting
~ с переме́нным успе́хом fluctuating battle
~, сде́рживающий delaying action, containment battle
бой, та́нковый tank battle, tank combat, tank duel, tank versus tank action, engagement
~, у́личный street fighting
~, центра́льный centrefire (ammo)
бо́йлер heater boiler
бойни́ца loop-hole
~, откры́тая embrasure
~, пистоле́тная 1. pistol port (fd wks etc)
2. revolver port (tk)
~, покры́тая covered embrasure
~ растру́бом в о́бе сто́роны loophole (wide at back and front, narrow at centre)
~ растру́бом к проти́внику loophole (wide at front)
~ растру́бом к себе́ loophole (wide at back)
бо́йцы деса́нта assault landing troops (air or water crossing)
бок side, flank
бокови́к crosswind (on runway)
боково́й side, flank, lateral
бокс place set apart (for something)
~, испыта́тельный test bay (missile, rocket)
~, огнево́й firing bay, firing pit (missile)
~ стоя́нка parking bay
болва́нка hard steel plug (gunn), solid anti-tank shot
болва́нки dummies, drill rounds, drill mines
боле́знь disease, illness, sickness, ailment
~, венери́ческая venerial disease
~, возду́шная air sickness
~, го́рная mountain sickness
~, зара́зная infectious disease, contagious disease
~, инфекцио́нная infectious disease
~, лучева́я radiation sickness
~, морска́я sea sickness
болеутоля́ющий pain relieving, analgesic
боло́метр bolometer (radiant-heat measuring device) (GW)
боло́тистый marshy, swampy, boggy
боло́то marsh, swamp, bog
~, верхово́е upland moor, high moor
~, ле́сное forest bog
~, мохово́е moss bog
~, непроходи́мое impassable marsh
~, низи́нное low-land bog
~, проходи́мое passable marsh

болóто, сапропéлевое sapropelic bog (compound of aquatic ooze rich in non-shelly remains of aquatic organisms)
~, **сплáвинное** peat bog
~, **торфянóе** peat bog
~, **травянúстое** grassy bog, swamp
~, **труднопроходúмое** barely passable marsh
болóто-зыбýн spongy marsh
болт bolt
~, **áнкерный** foundation bolt, anchor bolt
~, **барáшковый** butterfly bolt
~, **взрывнóй** explosive bolt
~ **дрóссельной заслóнки, ограничúтельный** throttle stop screw
~, **зажúмный** jamming bolt, clamp bolt
~, **закрепля́ющий** mounting pin
~, **предохранúтельный** safety bolt
~, **разрывнóй** explosive bolt
~ **рáмы пружúны** spring frame bolt
~, **распóрный** distance bolt
~ **регулирóвки тормознýх колóдок** brake adjustment nut
~, **регулирóвочный** adjusting bolt, adjusting screw
~ **с квадрáтной голóвкой** square-head bolt
~ **с шестигрáнной голóвкой** hexagon(-head) bolt
~ **свя́зи** connecting bolt
~, **серьгообрáзный** U-bolt
~, **скрепля́ющий** retaining bolt, fastening bolt
~, **соединúтельный** coupling pin
~, **стрúнгерный** stringer bolt
~, **стяжнóй** tie bolt
болт-барáшек butterfly bolt
больнóй 1. sick person, sick case, patient 2. sick, ill, diseased
~, **инфекциóнный** infectious case
большáк main road (coll)
большóй big, large, great, high
бóмба bomb, mortar bomb
~, **авиациóнная** aerial bomb (for types see aviabomba)
~, **агитациóнная** propaganda bomb, leaflet bomb
~, **áтомная** atomic bomb
~, **аэрозóльная дезинфекциóнная** aerosol disinfectant sprayer
~, **бактериологúческая** bacteriological bomb
~, **бронебóйная** armour-piercing bomb
бóмба, водорóдная hydrogen bomb
~, **водорóдно-урáновая** hydrogen-uranium bomb
~, **глубúнная** depth charge
~, **дейтéриевая** deuterium bomb
~, **дейтéриево-трúтиевая** deuterium-tritium bomb
~ **дистанциóнного дéйствия** air-burst bomb
~, **дымовáя** smoke bomb
~, **дымовáя авиациóнная** smoke bomb (aerial)
~, **зажигáтельная** incendiary bomb
~, **кóбальтовая** cobalt bomb
~, **мáгниевая светя́щая** magnesium flare bomb
~, **манометрúческая** closed explosive bomb (device for measuring the pressure of gases formed on detonation) (expl)
~, **мегатóнная я́дерная** megaton bomb
~, **напáлмовая** napalm bomb
~, **нейтрóнная** neutron bomb
~, **неразорвáвшаяся** unexploded bomb
~, **номинáльная áтомная** nominal atomic bomb (twenty kiloton)
~, **оскóлочная** anti-personnel bomb, fragmentation bomb
~, **оскóлочно-химúческая** combined fragmentation and chemical bomb
~, **планúрующая** glide bomb
~, **плутóниевая** fission bomb
~, **противосамолётная** anti-aircraft bomb
~, **противотáнковая** anti-tank bomb
~, **радиоактúвная** radioactive bomb
~, **реактúвная** rocket-propelled bomb
~ **с тротúловым эквивалéнтом мéнее одногó миллиóна тóнны, водорóдная** sub-megaton hydrogen bomb
~, **светя́щая** aircraft flare, illuminating flare (ac)
~, **светя́щая(ся) авиациóнная** illuminating flare (ac), aircraft flare
~ **свобóдного падéния, я́дерная** free fall nuclear bomb
~, **термúтная** thermite bomb
~, **термоя́дерная** thermonuclear bomb
~, **тонкостéнная авиациóнная** light-case bomb
~, **фотоавиациóнная** photographic flash bomb

бо́мба, фуга́сная high-explosive bomb, blast bomb, demolition bomb
~, хими́ческая chemical bomb, gas bomb
~, хими́ческая авиацио́нная chemical bomb
~, ша́риковая anti-personnel (ball-bearing) bomb (obs)
~, я́дерная nuclear weapon, nuclear bomb
бомбарди́р 1. bombardier
2. bomb aimer controller (GW)
бомбардирова́ние bombing, bombardment (especially air bombardment)
бомбардирова́ть to bomb, to bombard (imp only)
бомбардиро́вка bombing, bombardment
~, возду́шная aerial bombing
~, ночна́я night bombing, night bombardment
~, стратеги́ческая strategic bombing
~, такти́ческая tactical bombing
бомбардиро́вочный bomber, bombing, bombardment
бомбардиро́вщик bomber
~, бли́жний light bomber
~, высо́тный high-level bomber
~, да́льний long-range bomber
~, двухмото́рный twin-engined bomber
~, лёгкий light bomber
~, ночно́й night bomber
~, одномото́рный single-engined bomber
~, пики́рующий dive-bomber
~, реакти́вный jet bomber
~, сре́дний medium bomber
~, тяжёлый heavy bomber
~, четырёхмото́рный four-engined bomber
бомби́ть to bomb
бомбобоя́знь panic fear of bombing
бо́мбовый bomb, bombing
бомбодержа́тель bomb rack
бомбоиска́тель bomb detector
бомболю́ки bomb doors
бомбомета́ние bombing
~ вперёд forward toss bombing
~, горизонта́льное horizontal bombing
~, группово́е pattern bombing
~ за́лпами cluster bombing
~ наза́д backward toss bombing, over the shoulder bombing
~ по пло́щади pattern bombing
бомбомета́ние с бре́ющего полёта low-level skip bombing
~ с кабри́рования toss bombing
~ с пики́рования dive bombing
~ с плани́рования glide bombing
~ с хо́да direct approach bombing
~, сери́йное stick bombing
~, усло́вное simulated bombing (e.g. with radar bomb-sight recorder)
~ че́рез плечо́ backward toss bombing, over the shoulder bombing
бомбомёт-миномёт, дальнобо́йный rocket launcher
бомбоотсе́к bomb-bay
бомбопри́це́л bombsight
бомбосто́йкий bomb-resistant, bomb-proof
бомбоубе́жище bomb-proof shelter
бомбоусто́йчивый bomb-resistant, bomb-proof
бомбоштурмово́й bombing and strafing
бон boom
бо́ны boom, boom defence (nav)
бор 1. forest (particularly a pine forest)
2. boron (metal)
бора́ bora (cold, dry N.E. wind - upper Adriatic) (met)
боргидри́д алюми́ния aluminium borohydride
борн battery terminal (elec)
бороводоро́д boron hydride
борода́ 1. breech lug, gun lug (ord)
2. beard
боро́дка уда́рника cocking piece
бородо́к punch (tool)
борозда́ groove, furrow
~, закры́тая covered furrow (caused underground by a ricochet shell)
борозды́ crazy cracking (in gun bore)
боро́ться to fight, to wrestle, to struggle, to combat (imp only)
борт 1. edge, rim
2. side (of vessel or vehicle)
~, боково́й side board (motor vehicle)
~, грузоподъёмный combined tailboard and hoist
~, за́дний tailboard (veh)
~, ле́вый port (side) (nav)
~, надво́дный freeboard
~, откидно́й за́дний folding tailboard
~, пра́вый starboard (side) (nav)
~, свобо́дный freeboard
бортинжене́р flight engineer

бортовóй 1. side
2. borne, onboard (ship, ac, vehicle, missile)
бортради́ст radio operator (on board ship, vehicle, ac etc)
бортреду́ктор final drive gear (tracked vehicle)
борштáнга boring bar (eng)
борьбá struggle, fight
~, контрбатарéйная counter-bombardment, counter-battery bombardment
~, контрминомётная counter-mortar fire
~, ми́нная mine warfare, underground mining (mil eng)
~, непосрéдственная direct contact
~, подзéмно-ми́нная underground mine warfare, sapping activity
~, радиоэлектрóнная electronic warfare (Sov)
~ с радиоэлектрóнными срéдствами проти́вника electronic counter-counter measures, ECCM
бот small boat, small ship
~, водолáзный diving boat
~, вы́садочно-вы́грузочный landing boat
~, городскóй street fighting
~ поддéржки, брони́рóванный support landing craft
~, удáрный десáнтный assault landing craft
боти́нки, похóдные ammunition boots, cold-wet-weather boots, boots CWW
ботули́зм botulism, sausage poisoning (med)
бóты overshoes
бóча boatswain (nav sl)
бочёнок small barrel (container)
бóчка barrel (container)
бочóнок small barrel (container)
брак 1. marriage
2. scrap, defective part, defect, flaw
бракoвáть/забраковáть to condemn (eqpt, stores etc), to scrap, to reject, to discard, to refuse to accept
бракóвочный unserviceable
брандвáхта понтонёров bridge protection party (a detachment to keep a crossing-place free from floating mines and other obstacles)
брандмáуэр bund wall, fireproof wall
брандспóйт 1. nozzle
2. two-man manual pump
3. projector barrel (flame thrower)
брандтру́бка nipple, tube (arty)
брать/взять to take, to undertake, to adopt; **брать/взять в ви́лку** to bracket (arty); **брать/взять в клéщи** to carry out a pincer movement; **брать/взять в плен** to take prisoner, to capture; **брать/взять в положéние приклáдом вверх** to reverse arms; **брать/взять на букси́р** to take in tow; **брать/взять на изготóвку** to adopt the ready position; **брать/взять на му́шку** to take aim; **брать/взять на передки́** to limber, to limber up; **брать/взять на плечó** to shoulder; **брать/взять на себя́ ответственность** to assume responsibility; **брать/взять направлéние** to take a bearing, to take a course; **брать/взять óтпуск** to take leave; **брать/взять пéленг** to take a bearing; **брать/взять попрáвку на** to correct for, to make allowance for; **брать/взять с бóя** to take by assault; **брать/взять шту́рмом** to take by assault, to take by storm, to carry by assault
бревéнчатый log
брёвна, въездны́е timber approach (br)
бревнó log, tree trunk
~, áнкерное anchorage log (br)
~, вéрхнее продóльное capsill (br)
~, ни́жнее продóльное groundsill, groundsel (br)
~, опóрное support log
~, помóчное lower member of stiffening transom (br)
~, продóльное опóрное waling (revetment) (eng)
~, упóрное stop beam (br)
брегери́т broegerite
брезéнт tarpaulin
~, укры́вочный canvas cover
брешь gap, breach
бригáда 1. brigade
2. gang, team of workers
3. crew (manning one machine or engine)
4. squadron of ships
~, артиллери́йская artillery brigade
~, гáубичная howitzer brigade
~, инженéрно-сапёрная sapper brigade
~, поезднáя train crew

бригáда, понтóнно-мостовáя pontoon bridge brigade (Sov)
~, рабóчая work gang
~ разбóрщиков team of dismantlers
~ разграждéния obstacle clearance brigade (special eng formation trained & equipped for clearing minefields & other obstacles)
~, ремóнтная maintenance crew
~, тáнковая tank brigade
~, тяжёлая пýшечная артиллерúйская heavy gun artillery brigade
~, тяжёло-гáубичная артиллерúйская heavy howitzer artillery brigade
~, штурмовáя инженéрно-сапёрная assault engineer brigade
бригадúр 1. ganger, foreman
2. Brigadier (Brit)
~ путú foreman ganger
бригáдный brigade
брúджи riding breeches
бризáнтный high explosive
брикéт slab, briquette
брóвка 1. edge, shoulder
2. brow, lip (of a dell, hollow)
~, внýтренняя inner edge (roadworks)
~, нарýжная outer edge (roadworks)
брод ford
бром bromine
бромбензилцианúд brombenzyl cyanide
бромúд bromide
брóмо-серéбряный silver bromide
бронеавтомобúль armoured car
~, развéдывательный armoured reconnaissance car, armoured scout car
~, штабнóй armoured command vehicle, armoured control vehicle
бронеамфúбия armoured amphibian
бронебóйка anti-tank (coll)
бронебóйно-зажигáтельный armour-piercing incendiary
бронебóйность performance against armour
бронебóйный armour-piercing
бронебóйщик anti-tank gunner, anti-tank rifleman (coll)
броневúк 1. armoured car (coll)
2. member of armoured forces (coll)
броневúст member of an AFV crew
броневóй armoured, armour
бронегрузовúк armoured truck
бронедрезúна armoured railway trolly
бронеединúца unit of armour (one tank or one scout car etc)
бронежилéт bullet-proof vest
бронекавалерúйский armoured
бронекáтер armoured launch, motor gunboat
бронеколпáк 1. armoured cupola
2. bullet-proof shield (on AFV optical instruments)
бронекóрпус armoured hull (AFV)
бронекýпол armoured cupola
бронелéнта armoured tape (cable protection)
бронемашúна 1. armoured vehicle
2. armoured car (more usual meaning)
броненóсный armoured
бронеобъéкт armoured vehicle
бронеодéжда armoured garment
бронеотря́д armoured detachment
бронепаровóз armoured locomotive
бронеплитá armour plate
~, цементúрованная case-hardened armour plate, carburised armour plate
бронеплощáдка armoured railway truck
бронепóезд armoured train
бронепробивáемость armour defeating performance, armour penetration
бронепробивáющий armour-piercing
бронепрóволока armoured wire (cable protection)
бронестеклó bullet-proof glass, armoured glass, bullet-resistant glass
бронетáнковый armoured (pertaining to troops and not to armour)
бронетранспортёр armoured personnel carrier, APC
~, грузовóй armoured cargo carrier
~, гýсеничный tracked armoured personnel carrier
~, колёсный wheeled armoured personnel carrier
~, колёсный двухóсный/ трёхóсный/ четырёхóсный two/six/eight-wheeled armoured personnel carrier
~, плáвающий amphibious armoured personnel carrier
~, полугýсеничный half-track armoured personnel carrier
бронетягáч 1. armoured recovery vehicle, ARV
2. armoured artillery prime mover
бронецéль armoured target
бронечáсть armoured unit

бро́нза bronze
~, пу́шечная gun metal
~, свинцо́вистая copper lead alloy
~, фо́сфористая phosphor bronze
бро́нзовый bronze
брони́к armoured personnel carrier, APC (coll)
брони́рование reservation of something in short supply
бронирова́ние 1. armoured protection, armour plating
2. inhibiting (rocket propellant)
~, противопу́льное protection against small arms fire
брониро́ванный 1. armoured, armour plated
2. inhibited (rocket propellant)
брони́ровать/бронирова́ть to reserve, earmark
брониро́вка armouring, armour plating
~ порохово́го заря́да inhibitor (rocket propellant)
~ пу́шки gun mount shield
броня́ 1. armour
2. umbrella (in the sense of "put up an umbrella")
~, ба́шенная turret armour
~, бортова́я side armour
~, вертика́льная vertical armour
~, гетероге́нная heterogeneous armour plate
~, гомоге́нная homogeneous armour plate
~, ка́таная rolled armour
~, клинови́дная wedge-shaped armour
~ ко́рпуса hull armour
~, листова́я plate armour
~, лита́я armour casting, cast armour
~, лобова́я frontal armour, breast plate
~, мо́щная thick armour
~, накло́нная inclined armour, glacis
~, одноро́дная homogeneous armour plate
~, печна́я light armour
~, противопу́льная bullet-proof armour, light-weight armour
~, ребри́стая ribbed armour
~, тяжёлая heavy armour
~, уси́ленная reinforced armour
~, цементи́рованная case-hardened armour, carburised armour
~, экрани́рованная spaced armour
бросо́к 1. wave (of assault)
2. bound, rush
3. throw (gren)
~ в ата́ку assault, charge
~ в штыки́ bayonet charge
~ морско́го деса́нта amphibious assault wave
бро́шенный thrown, dropped
брошю́ра pamphlet
брус beam, baulk
~, бу́ферный 1. bumper bar (motor vehs)
2. headstock (rly wagons)
~, двутавро́вый I-beam, H-beam
~, колесоотбо́йный riband, wheelguide, curbing (br)
~, ограничи́тельный stop, chock
~, подфе́рменный ledger (br)
~, помо́чный transom (br)
~, прицепно́й draw-bar (arty)
~, продо́льный longitudinal member, longeron (aeron constr)
~, сцепно́й draw-bar
~, тя́говый draw-bar
брусо́к 1. bar
2. three inch square timber
~, нажи́мный pressure block (min)
~, направля́ющий side chock
~, точи́льный whetstone, abrasive bar
~, упо́рный scotch, chock
бру́ствер breastwork, parapet
~, кругово́й all-round parapet
~, повы́шенный raised parapet
~, ты́льный parados, rear parapet (fd wks)
брусча́тка stone blocks, paving blocks
бру́сья parallel bars
бру́тто gross weight
бры́зги свинца́ bullet splash
брызгови́к mudguard
брю́ки trousers
брюшко́ ло́пасти blade face
бряца́ние rattle, clash, clink
бу́гель 1. hoop, band, yoke
2. cut-off gear actuating collar (ord)
~ подши́пника bearing cap
буго́р knoll
бугоро́к hillock
бу́дка hut, cabin, box
~, железнодоро́жная signal box (rly)
~, карау́льная sentry box
~ машини́ста (driver's) cab
~, постова́я sentry box
~, проходна́я entrance gate check point

бу́дка, съёмная утеплённая portable insulated hut
~, телефо́нная telephone box
буй buoy
~, огражда́ющий beacon buoy
~, про́блесковый flashing beacon
~, спаса́тельный lifebuoy
бук beech (timber and tree)
бу́ква, исхо́дная index letter (codes and ciphers)
бу́квенно-цифрово́й alpha-numeric
буквопеча́тание printing telegraphy
~ по ра́дио radio printing telegraphy
бу́кса axle-box
букси́р 1. tow
2. tow rope
3. tug
~, жёсткий tow-bar, towing bar
~, мя́гкий flexible tow-bar, rope
букси́рна скоба́ towing shackle
букси́рный tow, towing
буксирова́ние tow, towing
букси́ровать to tow (imp only)
буксиро́вка tow, towing
буксиро́вщик 1. towing vehicle
2. towing aircraft,target towing aircraft
букси́руемый towed, drawn
буксова́ние 1. wheel spin
2. skidding
буксова́ть to spin (wheels of vehicle), to skid (imp only)
булы́жный cobbled
бульдо́зер 1. bulldozer attachment
2. bulldozer
~ с неповоро́тным отва́лом straight blade dozer
~, та́нковый универса́льный dozer tank (tk fitted with bulldozer attachment)
бульдозери́ст bulldozer operator
бульдо́зер-снегоочисти́тель snow dozer
бум horizontal beam (physical trg)
бума́га paper
~, асбе́стовая asbestos paper
~, биту́мная pitch paper, bitumen paper
~, диагра́мная instrument chart paper
~, защи́тная protective paper
~, индика́торная indicator paper
~ на ОВ ко́жно-нарывно́го де́йствия, индика́торная blister gas paper detector
~, нажда́чная emery paper
~, обёрточная brown paper, wrapping paper
бума́га, парафини́рованная waxed paper, paraffined paper
~, прозра́чная tracing paper
~, прома́сленная oiled paper
~, разграфлённая squared paper, graph paper
~, телефо́нная telephone cable paper
~, хлопча́тая cotton, cotton fabric, cotton fibre
бума́жный paper; cotton
бу́нкер 1. bunker (fd wks)
2. coal hopper
бунт 1. riot, mutiny, rebellion
2. bale, bundle, packet
бунтова́ть to riot, to mutiny, to rebel (imp only)
бур auger, borer, drill
~, враща́тельный rotating drill, rotary drill
~, земляно́й earth auger
~, ручно́й hand drill
~, уда́рный hammer drill, punch drill
бура́в auger, gimlet, borer
бура́вчик gimlet
бура́н snowstorm
бурдю́к, ра́нцевый knapsack water container
бурело́м windfelled trees
буре́ние boring, drilling
бури́льный drilling, boring
бури́льщик borer
бури́ть/пробури́ть to bore, to drill
бур-столбоста́в post-hole auger
~, тра́кторный mobile post-hole auger
бурт collar (eng)
бу́ртик crimp, rib
~ вту́лки bushing shoulder
~ ги́льзы rim (cartridge case)
~, кольцево́й rear part of driving band (ammo)
~, устано́вочный centralising collar (eng)
~, центри́рующий centralising collar (eng)
буру́н surf, breakers
бу́ря storm, gale
~, магни́тная magnetic storm
~, о́гненная fire storm
~, песча́ная sand-storm
буссо́ль 1. artillery director
2. aiming circle (US)
3. surveying instrument

буссо́ль, артиллери́йская artillery director, aiming circle (US) (gunn, arty, surv)
~, артиллери́йская перископи́ческая artillery director, aiming circle (US) (gunn, surv)
~, и́стинная true bearing, true azimuth
~, карма́нная pocket compass
~, перископи́ческая артиллери́йская artillery director, aiming circle (US) (gunn, surv)
~ с призмо́й prismatic compass
~ теодоли́та compass box (theodolite)
~, топографи́ческая surveyor's compass
~ це́ли firing azimuth, target bearing
бу́стер booster
бути́лмеркаптан butyl mercaptan
буты́лка bottle
~, зажига́тельная Molotov cocktail
~ с горю́чей жи́дкостью Molotov cocktail
~ с горю́чей сме́сью Molotov cocktail
бу́фер buffer, bumper, shock absorber, dash-pot
~, за́дний rear bumper
~ нака́тника recoil buffer (arty)
~, пере́дний front bumper
~ што́ка по́ршня piston buffer (gun recoil)
бу́ферный bumper, buffer
бу́хта 1. bay
2. coil
бучи́льник boiler (used in NBC decontamination)
бушла́т duffle coat
~, капко́вый kapok life-jacket
бык buttress, pier
~ моста́ bridge pier (br)
быстрогоря́щий quick-burning
быстроде́йствующий quick-acting, quick-operating, quick-response
быстросъёмный quick-detachable
быстрота́ заря́дки charging rate
~ отби́ва quickness of parry (bayonet drill)
~ переда́чи transmitting speed, sending speed
быстрохо́дность agility, rapidity
быт everyday life, routine
быть начеку́ to be on one's guard (imp only)
бычо́к combat unit commander (nav sl)
бьеф reach, pond
бьеф, ве́рхний upper water, head race, head water (of dam sluice gate)
~, ни́жний lower water, tail race, tail water (of dam sluice gate)
бюллете́нь bulletin
~, метеого́рный meteor telegram for mountain artillery
~ метеозву́к meteor telegram for sound ranging
~, метеозени́тный anti-aircraft meteor telegram
~, метеоогнево́й meteor telegram for ground artillery
~, метеорологи́ческий 1. weather report
2. meteor telegram (arty)
~, метеосре́дний meteor telegram for medium artillery
бюро́, констру́кторское design bureau
~, парти́йное (communist) party bureau

В

в алфави́тном поря́дке alphabetically, in alphabetical order
"В Ата́ку!" "Charge!"
в боево́й гото́вности under arms, ready for action
в боево́й обстано́вке 1. under battle conditions, in the field
2. on active service (mil law)
в бою́ in action
"В Две Шере́нги Стро́йся!" "Form Two Ranks!"
в движе́нии on the move
в де́йствии in action
в заты́лок in line, one behind the other
в испра́вном состоя́нии 1. in working order
2. roadworthy (veh)
"В Коло́нну Марш!" "Form Column!" (order on radio to AFVs etc)
"В Коло́нну По Два Станови́сь!" "In File - Fall In!"
"В Коло́нну По Одному́ Станови́сь!" "In Single File - Fall In!"
в ли́нию abreast
в меридиона́льном направле́нии longitudinally

в момéнт вы́стрела at the instant of firing
в неисправном состоя́нии 1. out of order 2. unroadworthy (veh)
в нóгу in step
"В однý Шерéнгу Стрóйся!" "Form Single Rank!"
в пéшем поря́дке on foot
в пéшем строю́ on foot
в пóлной боевóй готóвности ready for instant action
в пóлный рост upright
в полуразóбранном ви́де semi-dismantled
в поря́дке слýжбы in the course of duty, officially
в предéлах within the limits of, within the tolerances
в предéлах досягáемости within range
в разгáре бóя in the heat of battle
в разгáре лéта in the height of summer
в разóбранном ви́де dismantled
в рост full height
"В Ружьё!" "Take Up Arms!" (from ground arms or pile arms position)
"В Ряды́ Стрóйся!" "Form Two Deep!"
в сбóре group, assembly, assembled
в спéшенном поря́дке dismounted
в ствóре in line
в строю́ in the ranks
в строя́х closed up (troops)
"В Укры́тие!" "Take Cover!"
в упóр point blank (lit touching)
в ширóтном направлéнии transversally
вáга crow-bar, lever
вагóн 1. coach wagon (rly) 2. a wagon load 3. a cubic measure (38 cubic metres)
~, багáжный luggage van, parcel wagon (rly)
~, баллáстный ballast wagon, gravel wagon (rly)
~, грузовóй wagon, freight car (rly)
~, изотерми́ческий temperature controlled wagon (rly)
~, клáссный passenger coach
~, кры́тый covered wagon, van, box car
~, пассажи́рский passenger coach (rly)
~, прицепнóй trailer car
~, рефрижерáторный refrigerator wagon
~, санитáрный ambulance coach (rly)
вагóн, спáлный sleeping car, sleeping carriage (rly)
~, тележечный bogie car (rly)
~, товáрный goods wagon (Brit), freight car (US)
~, тормознóй brake van, brake car (rly)
вагóн-лéдник refrigerator wagon
вагóн-платфóрма flat-car
вагóн-холоди́льник refrigerator wagon
вагóн-хóппер hopper wagon (rly)
вагóн-цистéрна tank wagon
вагóны, товáрные freight stock, goods stock, wagons
вазели́н vaseline, petroleum jelly
~, техни́ческий petroleum jelly, low-quality petroleum jelly
ваксини́ровать to vaccinate to vaccinate
вáкуум vacuum
~, абсолю́тный absolute vacuum, perfect vacuum
~, непóлный partial vacuum
вáкуум-аппарáт vacuum tank (veh)
вакуумбачóк autovac
вакуумме́тр vacuum gauge
вáкуум-насóс vacuum pump
вáкуум-цили́ндр, тормознóй brake vacuum cylinder
вакци́на vaccine
вакцинáция vaccination, immunization
вал shaft, axle, spindle, embankment, wall (eng), barrage (arty)
~, береговóй shore bank (topo)
~, ведóмый 1. driven shaft 2. layshaft (gearbox)
~, ведýщий 1. driving shaft 2. primary shaft (gear box)
~, вéрхний распредели́тельный overhead camshaft (ICE)
~ ви́лки включéния clutch fork shaft
~ выжимнóй ви́лки clutch fork shaft
~ выжимнóй отвóдки clutch fork shaft
~, ги́бкий flexible shaft
~ глáвной передáчи final drive shaft
~, глáвный main shaft
~, гребнóй propeller shaft (water craft)
~, двойнóй double barrage, barrage in depth, double moving barrage (arty)
~, двойнóй огневóй 1. barrage in depth, double barrage, parallel barrage (arty) 2. fire obstacle (burning material)
~, дéрево-землянóй earth-timber breastwork

вал дифференциа́ла, веду́щий differential drive shaft
~, **земляно́й** earth bank, earth rampart
~, **карда́нный** propeller shaft, Cardan shaft, universal shaft
~, **коле́нчатый** crankshaft
~, **коренно́й** elevating pinion shaft (arty)
~, **кривоши́пный** crankshaft
~ **кулака́ то́рмоза** brake camshaft, expander shaft
~, **кулачко́вый** camshaft
~, **межмостово́й карда́нный** interaxle propeller shaft
~, **мише́нный** rifle target, butt (on rifle range)
~, **насыпно́й** rampart
~, **огнево́й** 1. moving barrage (arty) 2. fire obstacle, fire barrier (burning material)
~ **отбо́ра мо́щности** power take-off shaft
~ **отво́дки фрикцио́на** clutch operating shaft, withdrawal shaft
~, **отключа́емый** declutch shaft
~, **первы́чный** primary shaft, clutch shaft (gearbox)
~, **по́лый** quill, (hollow) shaft
~, **приводно́й** actuating shaft, drive shaft
~, **промежу́точный** intermediate shaft, connecting shaft, layshaft (gearbox)
~, **просто́й** moving HE barrage, HE barrage
~ **пружи́ны** spring shaft
~, **распредели́тельный** camshaft
~, **распредели́тельный ве́рхний** overhead camshaft
~, **резцо́вый** boring bar
~, **рулево́й** steering shaft, steering-wheel shaft (veh)
~, **сквозно́й** through shaft
~, **снегово́й** snow bank, obstacle made up of packed snow
~, **сне́жный** snow bank, packed snow obstacle
~ **со́шки руля́** steering arm, steering shaft, steering lever
~ **стабилиза́ции ку́рса, телескопи́ческий** telescopic shaft for stabilising course (course and speed sight)
~ **стре́льбища, земляно́й** butt (on range)
~, **стре́льбищный** stop butt
~, **торсио́нный** torsion bar
вал, трансмиссио́нный transmission shaft, transmission line
~, **фасо́нный** splined shaft, spline shaft
~ **фрикцио́на** clutch shaft
~, **шлицево́й** splined shaft
~, **эксце́нтриковый** eccentric shaft
~ **я́коря** armature shaft
валёк 1. roller 2. swingletree
ва́ленки felt snow boots
вале́нтность valency
ва́лик 1. shaft, small shaft (mech) 2. bank, slope
~, **вертика́льный** vertical shaft
~, **кулачко́вый** camshaft
~, **осево́й** rocker shaft (valve gear)
~, **па́зовый** splined shaft
~, **поводко́вый** selector rod (tk gear box)
~, **приводно́й** actuating shaft, drive shaft
~ **с ве́рхним расположе́нием, кулачко́вый** overhead camshaft
~ **с шестерне́й** pinion shaft
~, **спусково́й** actuating spindle
~, **сто́порный** locking pin (engine draught connection)
~, **торсио́нный** torsion bar
ва́лик-крестови́на governor, governor camshaft
ва́лик-сте́нка low wall, low banking
ва́лка tree chopping, tree clearing
~ **взры́вом** tree blow-down
валу́н boulder
валю́та currency
вана́дий vanadium
ва́нты shrouds (naut)
вар pitch, tar
вара́ктор voltage-variable capacitor, variable-voltage capacitor
ва́режки mittens
ва́режки-вкла́дыши inner mitts, inner mittens
вариа́нт variation
вариоку́плер variocoupler (elec)
варио́метр variometer (elec)
ва́та cotton wool
~, **стекля́нная** glass fibre, fibre glass
ватерли́ния waterline
~, **грузова́я** load line (of amphibian or ship)
ватерпа́с level (instrument)
ватерпасо́вка levelling (action of using instrument)

ва́тман Whatman paper, high quality drawing paper
ва́тник quilted jacket
ватт watt
ватт-час watt-hour
ва́хта watch, tour of duty
вбива́ть/вбить to drive in; **вбива́ть/вбить клин** to drive in a wedge (mech), to drive in a salient
вбить (см вбива́ть)
вброд by ford, fording
ВВ, обы́чное conventional explosive
введе́ние introduction
~ **в заблужде́ние** deception
~ **в ми́ну взрыва́теля** mine arming
~ **инги́битора** inhibiting (propellant, expl)
вверну́ть (см ввёртывать)
ввёртывать/вверну́ть to screw in
ввести́ (см вводи́ть)
ввинти́ть (см вви́нчивать)
вви́нчивать/ввинти́ть to screw in
ввод 1. input (radio)
2. gathering (GW)
3. commitment (tac)
4. lead (elec)
~, **анте́нный** antenna lead-in
~ **в заблужде́ние** deception
~ **в строй** 1. taking on charge (eqpt)
2. introduction (of new equipment)
~ **да́нных** data input
~ **ка́беля** cable terminal
вводи́ть/ввести́ to bring in, to introduce, to commit, to gather (GW); **вводи́ть/ввести́ в бой** to bring into action, to commit to action; **вводи́ть/ввести́ в де́йствие** to put into effect, to make effective, to bring into action; **вводи́ть/ввести́ в заблужде́ние** to mislead, to deceive; **вводи́ть/ввести́ в курс** to brief, to bring up to date; **вводи́ть/ввести́ в ми́ну взрыва́тель** to fuse a mine, to prime a mine; **вводи́ть/ввести́ в соста́в** to introduce into, to allot; **вводи́ть/ввести́ в строй** to take on charge (eqpt), to introduce (of new equipment); **вводи́ть/ввести́ в эксплуата́цию** to put into service; **вводи́ть/ввести́ да́нные** to introduce data, to insert data
вво́дная 1. exercise setting, narrative
2. hypothetical situation introduced into an exercise
вво́дная, такти́ческая tactical orientation, narrative
ввяза́ться (см ввя́зываться)
ввя́зываться/ввяза́ться to get involved in; **ввя́зываться/ввяза́ться** to become engaged
вглубь in depth
вда́вленность фро́нта enemy salient
вда́вливание cave-in
вда́влина 1. dent, impression (physical)
2. salient
вдыха́ние inhalation
веде́ние бо́я conduct of battle
~ **ка́рты** maintaining a map
~ **огня́** delivery of fire, firing
~ **пристре́лки** ranging
~ **радиообме́на** handling of radio traffic
ве́домость list, register
~, **бала́нсовая** list of balances (accounting)
~, **дефе́ктная** repair list
~, **доро́жная** freight bill (rly)
~, **инвентаризацио́нная** inventory (document)
~, **инвента́рная** inventory list, equipment record sheet
~, **комплекто́вочная** list of standard equipment, delivery list
~, **контро́льная** check list, inspection sheet
~ **на ГСМ, расхо́дная** record of POL consumption
~ **на ЗИП** vehicle kit list
~, **отпра́вочная** packing note
~, **отчётно-тре́бовательная** combined stock return and requisition
~, **переда́точная** transfer freight bill
~, **платёжная** acquittance roll, pay roll, pay-sheet
~, **постова́я** guard report
~, **приёмо-сда́точная** 1. receipt and issue voucher
2. handover/takeover certificate
~, **прихо́дная** receipt, receipt voucher
~, **прихо́до-расхо́дная** issue and receipt register
~ **рабо́т** schedule of work, list of jobs
~ **рабо́т** schedule of work, list of jobs
~, **разда́точная** 1. distribution list
2. issue slip
~ **расхо́да боеприпа́сов** record of ammunition expenditure

ве́домость, расхо́дная 1. list of issues
2. record of expenditure
~, су́точная 1. unit daily report book
2. daily strength return
ве́домство the authorities
~, вое́нное War Department
ведо́мый 1. driven (mech)
2. slave (servo, comms)
3. wingman (AF)
ведро́ bucket, pail
~, брезе́нтовое canvas bucket
~, пожа́рное fire bucket
веду́щий 1. leading
2. predominant
3. driving (mech)
4. leader (AF)
ве́ер 1. sheaf (US) (arty)
2. lines of fire (arty)
~ действи́тельного пораже́ния fire for effect (arty)
~, паралле́льный guns parallel (Brit), parallel sheaf (US) (arty)
~ разры́вов sheaf of bursts (US), simple concentration (Brit)
~, расходя́щийся distributed fire
~, сосредото́ченный concentrated fire
~, сходя́щийся concentrated fire, guns concentrated (Brit), converging sheaf (US) (arty)
веерообра́зный fan-shaped
вездехо́д cross-country vehicle
~, гу́сеничный tracked cross-country vehicle
~, колёсный wheeled cross-country vehicle
~, снегово́й snow-going vehicle, oversnow vehicle
~, снегохо́дный snow-going vehicle, oversnow vehicle
~, тра́нспортный cross-country load carrier
вездехо́дность cross-country capability
ве́ктор vector
~, едини́чный unit vector (GW)
~ тя́ги thrust line (GW)
величина́ size, value, magnitude, quantity
~, абсолю́тная absolute value
~ взрыва́теля, дро́бная fractional fuse range
~ вы́носа то́чки прице́ливания aim-off allowance (gunn)
~, зави́симая от частоты́ frequency response (electron)
величина́, за́данная given value
~, заме́ренная measured value
~ измене́ния высоты́ height rate (AA gunn)
~ измене́ния направле́ния bearing rate (rate of change in the bearing of the target from the gun)
~ измене́ния пе́ленга bearing rate (AA gunn)
~ измене́ния расстоя́ния range rate (AA gunn)
~ измене́ния расстоя́ния в результа́те движе́ния та́нка rapid correction (tank) (applied to allow for movement of tank)
~ измене́ния расстоя́ния це́ли rate of change of range to target (tank gunn)
~ измене́ния сближе́ния closing rate (AA gunn)
~, оптима́льная optimum value
~ разнобо́я ору́дий calibration error (arty)
~, расчётная computed value, calculated value
~ то́ка заря́да charging rate (elec)
~, углова́я angular rate, angular unit (AA gunn)
~ устано́вки взрыва́теля fuse length
~ устано́вки тру́бки fuse length
велоапте́чка cycle repair outfit
велосиме́тр velocimeter
велосипе́д bicycle
велосипеди́ст cyclist
велоситро́н velocitron
ве́на vein
вене́рик venerial disease case
вене́ц ring, hub, rim, crown
~, зубча́тый toothed ring
~ махови́ка́, зубча́тый starter ring, starter gear ring
вено́зный venous (med)
вентили́ровать to ventilate
ве́нтиль valve, rectifier, isolator, gate
~, впускно́й intake valve, inlet valve, admission valve
~, дро́ссельный butterfly valve, throttle valve, choke (mech)
~, запо́рный stop cock, stop valve
~, обводно́й by-pass valve, release valve
~, откидно́й flap valve
~, проходно́й straight-through valve, in-line valve, through-way valve

ве́нтиль, регули́рующий control valve, regulating valve
~, спускно́й discharge valve, escape valve, outlet valve, blow valve, drainage valve, release valve
~, таре́льчатый disc valve, poppet valve, baffle valve, plate valve
~, уравни́тельный pressure equalising valve, relief valve
~, шарни́рный flap valve
~, шарово́й ball valve
~, язычко́вый reed valve
вентиля́тор 1. ventilator
2. fan (engine)
~ дви́гателя engine fan, motor fan
~, осево́й axial-flow fan
~, охлажда́ющий cooling fan
~ радиа́тора radiator fan
~, ручно́й hand-operated ventilator
~, центробе́жный centrifugal fan
вентиляцио́нный ventilating, ventilation
вентиля́ция ventilation
~, есте́ственная natural ventilation
~, иску́сственная artificial ventilation, mechanical ventilation, positive draught
~, принуди́тельная forced air supply, induced ventilation
~, сквозна́я through-flow ventilation
венту́ри труба́ venturi tube
ве́нчик rim, halo, crown, collar
~ боево́й личи́нки rim of bolt head (SA)
~ сте́ржня return spring tube cup (MG)
~ уда́рника strike collar
верблю́д 1. camel
2. type of towing vehicle
3. double peaked pulse (radar sl)
вербова́ть to recruit
вербо́вка recruiting
вербо́вщик recruiter
верёвка cord, rope
~, альпи́йская climbing rope
~, вытяжна́я static line (para)
~ раскры́тия парашю́та, вытяжна́я parachute rip cord
верёвочный rope
веретено́ spindle
~ нака́тника control plunger (recoil system)
~ то́рмоза отка́та control rod (on recoil buffer)
ве́рность воспроизведе́ния fidelity of reproduction
вероя́тность probability
вероя́тность безотка́зной рабо́ты probability of no failure
~ попада́ния hit probability
~ пораже́ния kill probability
~ уничтоже́ния probability of destruction
вероя́тный probable
верпова́ние warping
верста́ verst (roughly equiv to 3500 feet)
верста́к work bench
versто́вка map to scale 1 : 48,000
верте́ться (см враща́ться)
вертикализа́ция vertical adjustment (msl)
вертлю́г 1. swivel
2. saddle (ord)
3. crosshead (on machine gun)
4. yoke (of mortar)
5. central pivot (of gun mounting)
вертодро́м heliport
вертолёт helicopter
~, боево́й combat helicopter, gunship helicopter
~ больщо́й грузоподъёмности heavy lift helicopter
~, вое́нно-тра́нспортный troop-carrying helicopter
~, вооружённый armed helicopter
~, грузово́й freight helicopter, cargo helicopter
~, двухвинтово́й twin rotor helicopter
~ наведе́ния pathfinder helicopter
~ огнево́й подде́ржки helicopter gunship
~, одновинтово́й single rotor helicopter
~, па́лубный ship-borne helicopter
~, поисково-спаса́тельный search and rescue helicopter
~, противоло́дочный anti-submarine helicopter (ASW)
~, санита́рный ambulance helicopter
~ свя́зи liaison helicopter
вертолёт-корректиро́вщик artillery spotting helicopter
вертолётный helicopter
вертолётоно́сец, деса́нтный commando ship, helicopter carrier, commando helicopter carrier
вертолёт-целеуказа́тель target indicator helicopter (pathfinder)
вертолётчик helicopter pilot

верту́шка 1. vane
2. helicopter (slang)
3. revolving radar antenna (slang)
4. current speed meter (liquids)
~, гидрометри́ческая hydrometric vane (eng, br)
верх top, head
~ ба́шни turret top (tk)
~ ку́зова, брезе́нтовый canvas top, tarpaulin top
верхня́к cap (in mining)
верхо́вный supreme
верхово́й 1. mounted, riding
2. upper, upstream
3. mounted man, rider
верхо́вье 1. headwater (of a river)
2. head (of a valley)
верши́на crest, peak
~ му́шки tip of foresight
~ снаря́да tip of shell
~ траекто́рии vertex of trajectory
верьне́р vernier
вес weight
~, боево́й combat weight, weight combat loaded
~ га́зов, уде́льный gas density
~ заря́да charge weight
~ заря́да, относи́тельный load ratio, capacity, charge weight ratio, CWR (ratio of weight of charge to weight of projectile)
~, коне́чный final weight (GW)
~, мёртвый net weight, dead weight
~, о́бщий total weight, all-up weight
~, объёмный weight per unit volume
~ поле́зной нагру́зки payload weight
~, по́лный боево́й all-up weight
~, поро́жный tare weight
~, расчётный design weight
~ снаря́да shot weight
~, со́бственный 1. unladen weight (fuelled)
2. kerb weight (of vehicle)
~, ста́ртовый launch weight, take-off weight, all-up weight
~, уде́льный 1. specific gravity
2. density (of GW fuel)
~, устано́вочный installation weight
~, факти́ческий actual weight
~, чи́стый net weight
~, эксплуатацио́нный weight in running order
весло́ oar, paddle
весово́й weight, gravimetric
вести́ маши́ну (цм води́ть маши́ну)
вести́/провести́ to conduct, to carry on, to carry out, to be engaged in; **вести́ бой** to conduct a battle; **води́ть/вести́ маши́ну** to drive; **вести́ ого́нь** to fire, to execute fire; **вести́ ого́нь в промежу́тки** to fire through the gaps (of own troops); **вести́ ого́нь на пораже́ние** to fire for effect; **вести́ ого́нь на уничтоже́ние** to conduct destruction fire; **вести́ ого́нь прямо́й наво́дкой** to fire over open sights; **вести́ ого́нь че́рез го́лову** to fire over the heads of; **вести́ одино́чный ого́нь** to fire at will; **вести́ опро́с** to interrogate, to conduct an interrogation; **вести́ отве́тный ого́нь** to return fire; **вести́ перегово́ры** to conduct negotiations, to carry on a conversation (comms); **вести́ переда́чу** to transmit
вестово́й messenger, orderly
ве́сы scales, balance, weighing machine
~, автомоби́льные weighbridge
~, рыча́жно-опти́ческие optical beam balance (instrument measuring jet engine thrust)
"весьма́ сро́чная" "Most immediate" (message precedence designation)
ветвь branch, arm
~, ве́рхняя top run (of tank track)
~, восходя́щая ascending branch (of a trajectory)
~, нисходя́щая descending branch (of a trajectory)
~, опо́рная bottom run (of tank track)
ве́тер wind
~, баллисти́ческий ballistic wind, equivalent constant wind
~, боково́й crosswind
~ в призе́мном сло́е во́здуха surface wind conditions (airborne ops)
~, ве́кторный vector wind
~, встре́чный head wind
~, госпо́дствующий prevailing wind
~, действи́тельный effective wind
~, обли́ческий oblique wind
~, относи́тельный relative wind
~, переме́нный variable wind
~, попу́тный following wind, rear wind
~, постоя́нный constant wind
~, призе́мный surface wind

ве́тер, продо́льный wind along line of sight, range wind (gunn)
~, проти́вный head wind
~, си́льный fresh breeze (force 5 Beaufort scale)
~, сла́бый light breeze (force 2 Beaufort scale)
~, уме́ренный gentle breeze (force 3 Beaufort scale)
~, урага́нный hurricane (force 12 Beaufort scale)
~, циклострофи́ческий cyclostrophic wind
ветера́н veteran
ветерина́рный veterinary
ве́тка, железнодоро́жная branch line (rly)
ветлазаре́т veterinary hospital
ве́тошь cotton waste, rags, swab
~, хлопчатобума́жная cotton waste
ветперсона́л veterinary personnel
ветрово́й wind
ветроме́р wind gauge
ветроуказа́тель wind direction indicator
ветсре́дства veterinary equipment
ве́ха 1. stake, marker, pole
2. aiming post (arty)
~, бли́жняя near aiming post (arty)
~, да́льняя far aiming post
~, прице́льная aiming post
~, ство́рная ranging pole, survey pole, stadia rod
~, топографи́ческая surveyor's pole
вещево́й clothing (pertaining to clothing and equipment)
вещества́, лету́чие volatile matter
~, фосфорорга́ни́ческие отравля́ющие organophosphorus war gases
вещество́ substance, agent
~, активи́рующее activating agent
~, а́льфа-акти́вное alpha particle emitter
~, аммони́йно-сели́тренное взры́вчатое ammonium nitrate explosive
~, антигризу́тное взры́вчатое permissible explosive (min)
~, антикоррозио́нное anti-corrosive agent, anti-rust agent, corrosion inhibitor
~, бе́та-акти́вное beta particle emitter
~, боево́е отравля́ющее war gas
~, боево́е радиоакти́вное radiological warfare agent, radio-active war material
вещество́, боево́е хими́ческое war gas, chemical warfare agent
~, взры́вчатое explosive, blasting agent
~, взры́вчатое бриза́нтное high explosive
~, воспламеня́щееся inflammable substance
~, высокосто́йкое отравля́ющее highly persistent war gas
~, газообра́зное отравля́ющее vapour type gas, vapour gas (CW)
~, горю́чее combustile substance
~, дегазацио́нное decontaminant, decontamination agent
~, дегази́рующее decontaminant, decontamination agent
~, дезактиви́рующее radiological decontaminant
~, деля́щееся fissile material
~, детони́рующее взры́вчатое detonating explosive, detonating powder, exploder
~ для промежу́точного детона́тора, взры́вчатое booster charge
~, дробя́щее взры́вчатое high explosive
~, дымообра́зное отравля́ющее dust-type gas (irritant smoke or nose gas)
~, дымообразу́ющее smoke agent, smoke composition, smoke-producing substance, smoke generator
~, жи́дкое боево́е отравля́ющее liquid war gas
~, зажига́тельное incendiary agent
~, излуча́ющее га́мма-лучи́ gamma-active material
~, изоли́рующее insulating material, insulation material
~, иниции́рующее взры́вчатое initiating (high) explosive, initiator, primer, priming powder, detonating powder
~, ка́пельно-жи́дкое отравля́ющее droplet gas (CW)
~, кле́йкое adhesive material
~ ко́жно-нарывно́го де́йствия, боево́е отравля́ющее blister gas
~, ко́жно-нарывно́е отравля́ющее blister gas, vesicant gas
~ крапи́вного де́йствия, отравля́ющее irritant agent (CW)
~, легковоспламеня́ющееся highly inflammable matter

вещество́, лету́чее volatile matter, volatile substance
~, лету́чее взры́вчатое volatile explosive
~, малочувстви́тельное взры́вчатое inert explosive
~, мета́тельное взры́вчатое low-order explosive, burning explosive, non-shattering explosive, propellant, propellant agent
~, мышьякови́стое отравля́ющее arsenical gas (CW)
~, нарывно́е отравля́ющее blister gas
~ не́рвно-паралити́ческого де́йствия, отравля́ющее nerve agent (CW)
~, не́рвно-паралити́ческое отравля́ющее nerve gas
~, несто́йкое отравля́ющее non-persistent war gas, non-persistent chemical agent
~, нитри́рованное взры́вчатое nitroexplosive
~, нитроглицери́новое взры́вчатое nitroglycerine explosive
~, общеядови́тое отравля́ющее nerve and paralysant gases (Sov term includes both categories)
~, обы́чное взры́вчатое conventional explosive
~, огнесто́йкое fire-proof substance, fire-proof material
~, отбра́сываемое ejected matter
~, отравля́ющее 1. war gas, toxic agent, chemical warfare agent 2. poisonous material
~, отравля́ющее в аэрозольном состоя́нии aerosol agent
~, отравля́ющее газообра́зное toxic agent
~, отравля́ющее для заражéния мéстности ground contaminating agent
~, отравля́ющее жи́дкое liquid toxic agent, toxic liquid
~, отравля́ющее имитацио́нное simulated agent
~, отравля́ющее ко́жно-нарывно́го де́йствия blister agent, vesicant agent
~, отравля́ющее несмертоно́сное non-lethal agent
~, отравля́ющее нестóйкое non-persistent agent

вещество́, отравля́ющее приводя́щее к потéре боеспосо́бности incapacitating agent
~, отравля́ющее раздража́ющего де́йствия irritant agent
~, отравля́ющее слезоточи́вого де́йствия tear agent
~, отравля́ющее слезоточи́вое tear agent
~, отравля́ющее смерте́льного де́йствия, хими́ческое lethal chemical agent
~, отравля́ющее смерте́льное lethal agent
~, отравля́ющее сто́йкое persistent agent
~, отравля́ющее удуша́ющего де́йствия choking agent, pulmonary irritant, asphyxiant
~, отравля́ющее учéбное training chemical agent
~, отравля́ющее ядови́тое toxic chemical agent
~, охлажда́ющее cooling agent
~ паралити́ческого де́йствия, отравля́ющее paralysing agent (CW)
~, парообра́зное отравля́ющее vapour gas (CW)
~, пенообразу́ющее foaming agent
~, пласти́ческое взры́вчатое plastic explosive
~, пласти́чное взры́вчатое plastic explosive
~, плеснезащи́тное mould preventive agent
~, повéрхностно-акти́вное surface active agent (CW)
~, предохрани́тельное взры́вчатое permitted explosive (Brit), permissible explosive (US), safety explosive
~, психохими́ческое psychochemical gas
~, психохими́ческое отравля́ющее psychochemical war gas
~, радиоакти́вное radioactive substance, radioactive matter, radioactive material
~, радиоакти́вное отравля́ющее radioactive matter, radioactive material
~, раздража́ющее отравля́ющее sternutator, nose gas, tear gas (CW) (Sov term includes all these irritants)
~, расщепля́ющееся fissionable material

веществó рвóтного дéйствия, отравля́ющее vomiting agent
~, связу́ющее 1. bonding agent, cement, agglutinant
2. binder (in propellant)
~, свя́зывающее binding agent, binder
~, слезоточи́вое отравля́ющее tear gas, lachrymator
~, смáзочное lubricant
~, стóйкое отравля́ющее persistent war gas, persistent chemical agent
~, твёрдое взры́вчатое solid explosive
~ типа В, отравля́ающее V-agent, written with latin V
~ токси́ческого дéйствия, боевóе toxic warfare agent
~, туманообрáзное droplet gas
~ удушáющего дéйствия choking war gas
~, удушáющее отравля́ющее choking gas, suffocating gas
~, универсáльное взры́вчатое all-purpose explosive
~, учéбно-боевóе стóйкое отравля́ющее persistent gas training mixture
~, учéбное отравля́ющее gas training mixture
~, фосфоресци́рующее phosphorescent material, radiolite
~, фосфоросодержáщее war gas containing phosphorus (Sov term includes some of the nerve gases and also some of the compounds such as DFP)
~, хими́ческое chemical matter, chemical substance
~, чихáтельное отравля́ющее nose gas, sternutator
~, я́дерное взры́вчатое nuclear explosive
~, ядови́тое toxic agent, toxic substance
вéщи, бессрóчные non-expendable items
вещиму́щество kit
взаи́мный mutual, reciprocal
взаимодéйствие 1. interaction
2. mutual support, cooperation
~, боевóе combat team-work
~, огневóе mutual fire support
~ огня́ и движéния combination of fire and movement
~, такти́ческое tactical cooperation
взаимодéйствовать to cooperate
взаимозави́симость interdependence
взаимозави́симость дáльности и углá возвышéния range-elevation relationship
взаимозаменя́емость interchangeability
взаимозаменя́емый interchangeable
взаимоиндýкция mutual induction
взаимоотношéние relationship, interrelationship
взаимопóмощь mutual aid
взаимосвя́зь interconnection
взведéние 1. winding
2. cocking (wpn)
3. arming (detonator, fuse)
~, замéдленное delayed arming (ammo)
~ удáрника arming
взведённый 1. cocked (wpn)
2. primed, armed (detonators, grenades etc)
взвéсить (см взвéшивать)
взвести́ (см взводи́ть)
взвéшивание weighing
взвéшивать/взвéсить to weigh, to weigh up, to evaluate
взвод 1. platoon
2. cocking (wpn)
~, авторемóнтный MT repair platoon
~, автотрáнспортный MT platoon
~, администрати́вно-хозя́йственный administrative platoon
~, артиллери́йский artillery troop (Brit), artillery platoon
~ боевóго питáния ammunition supply platoon
~, боевóй sear notch
~ звуковóй развéдки sound ranging platoon
~, звукометри́ческий sound ranging platoon
~, зени́тно-пулемётный anti-aircraft machine-gun platoon
~, знамённый colour platoon (ceremonial)
~, инженéрно-дорóжный road engineer platoon
~, инженéрно-сапёрный sapper platoon
~ инструментáльной развéдки observation platoon
~, комендáнтский headquarters defence platoon
~, мéдико-санитáрный medical platoon
~, мотопехóтный motorized infantry platoon

взвод, направля́ющий guiding platoon, leading platoon
~, огнево́й 1. gun platoon, fire platoon (arty)
2. launcher platoon (GW)
~ опти́ческой разве́дки 1. visual observation platoon (consists of three OPs & a computing centre)
2. flash spotting platoon
~, парашю́тно-деса́нтный parachute platoon
~, па́рковый MT park platoon
~ подвижны́х средств свя́зи mobile liaison platoon
~, понто́нный pontoon platoon
~, предохрани́тельный half-bent, automatic safety (on pistol hammer)
~, пулемётный machine-gun platoon
~ разве́дки reconnaissance platoon
~ раке́тно-техни́ческого обеспе́чения missile technical maintenance platoon
~, санита́рный medical platoon
~, сапёрный sapper platoon, pioneer platoon, engineer platoon
~ свя́зи signals platoon
~ снабже́ния supply platoon
~ соба́к-миноиска́телей dog platoon (mine clearance)
~, стрелко́во-парашю́тный parachute rifle platoon
~, та́нковый tank platoon
~, телефо́нно-ка́бельный telephone-cable platoon (mil)
~ техни́ческого и хозя́йственного обеспе́чения technical and supply maintenance platoon
~ техни́ческого обеспе́чения technical maintenance platoon
~ уда́рника inner cocking lever (of breech mechanism)
~ управле́ния headquarters platoon
~, фотограмметри́ческий photogrammetric platoon (at arty Group HQ - obtains coordinates of targets from air photographs)
~ фотозатво́ра shutter trigger (photo)
~ хими́ческой защи́ты chemical defence platoon
~ хими́ческой и радиацио́нной разве́дки chemical and radiation recce platoon
~ хими́ческой разве́дки и наблюде́ния chemical reconnaissance and observation platoon
взвод, хозя́йственный administrative platoon
~ хране́ния горю́чего petroleum storage platoon
~, штабно́й staff platoon, HQ platoon
взводи́ть/взвести́ to cock (wpn), to arm, to prime (fuse, detonator)
взво́дный 1. platoon commander (sl)
2. platoon
взгляд вперёд foresight (surv)
~ наза́д backsight (surv)
вздва́ивание рядо́в forming fours (drill)
взлёт take-off
~, вертика́льный vertical take off
~, группово́й formation take-off
~ для висе́ния take-off and hover (helicopters)
~ и поса́дка, коро́ткий short take-off and landing STOL (ac)
~ с по́мощью ста́ртового реакти́вного дви́гателя jet-assisted take-off
взлётный take-off
взлом break-in (tac)
взмо́рье coastal waters
взму́чивание stirring up, making turbid, making cloudy, making muddy
взо́рванный blown-up, blasted
взорва́ть (см взрыва́ть)
взрыв 1. explosion, burst, blast
2. armed (fuse setting)
~ (а́томного) ядра́ nuclear explosion, fragmentation of nucleus
~, а́томный atomic explosion, nuclear explosion
~, возду́шный airburst
~, возду́шный а́томный nuclear airburst, atomic airburst
~, детонацио́нный sympathetic detonation
~, заме́дленный delayed explosion, delayed contact fire, hang fire
~, запозда́лый delayed explosion, hang fire
~, затяжно́й delayed explosion, hang fire
~, конта́ктный назе́мный ground level burst, ground level explosion
~ на пове́рхности surface burst, surface explosion
~, назе́мный ground burst
~, назе́мный я́дерный nuclear ground burst
~, непо́лный low-order detonation

взрыв, подво́дный 1. underwater explosion
2. underwater burst (nucl)
~, подво́дный а́томный underwater atomic explosion
~, подзе́мный 1. underground explosion
2. underground burst (nucl)
~, радиоуправля́емый radio detonation
~, самопроизво́льный spontaneous explosion
~ такти́ческого я́дерного боеприпа́са tactical nuclear burst, tactical nuclear explosion
~ термоя́дерной бо́мбы thermo-nuclear (bomb) burst, thermo-nuclear (bomb) explosion
~, управля́емый controlled explosion
~, я́дерный nuclear burst, nuclear explosion
взрыва́ние explosion, bursting, blasting, demolishing, blowing up, blasting operation, firing, shooting
~, огнево́е fuse firing
~, одновре́менное instantaneous firing
~, подво́дное underwater blasting
взрыва́тель fuse, blasting fuse, detonating fuse, firing device, explosive device, firing assembly, explosive assembly, firing gear, explosive gear, fuse mechanism, exploder
~, акти́вный неконта́ктный active proximity fuse
~, акусти́ческий acoustic proximity fuse
~, акусти́ческий неконта́ктный acoustic proximity fuse
~, барометри́ческий barometric fuse
~ без замедли́тельного механи́зма non-delay fuse
~ без замедли́тельного устро́йства non-delay fuse
~, боково́й body fuse
~, взведённый armed fuse
~ возду́шно-уда́рного де́йствия concussion fuse, air-burst fuse, air-pressure fuse
~ всесторо́ннего де́йствия all-ways fuse
~, высокочасто́тный индукцио́нный radio proximity fuse
~, высокочувстви́тельный уда́рный sensitive percussion fuse
~, гальваноуда́рный electrical and percussion fuse
взрыва́тель, гидродинами́ческий liquid fuse
~, гидростати́ческий hydrostatic fuse
~, головно́й nose fuse
~ двойно́го де́йствия double-action fuse
~ дистанцио́нно-уда́рного де́йствия time and percussion fuse
~, дистанцио́нный time fuse
~, до́нный base fuse
~ заме́дленного де́йствия delayed action fuse
~ заме́дленного де́йствия, хими́ческий chemical delayed action fuse
~, заме́дленный delayed action fuse
~, зара́нее устано́вленный pre-set fuse
~, звукочасто́тный audio-frequency (type) fuse
~, зени́тный anti-aircraft fuse
~, индукцио́нный frequency-induction fuse, frequency-induction igniter
~, ине́ртный inertia fuse (ammo)
~ инерцио́нного де́йствия, уда́рный graze fuse
~, инерцио́нный graze fuse
~, инфракра́сный infra-red fuse, infra-red sensitive fuse
~ кольцево́го ти́па ring-type fuse
~ комбини́рованного де́йствия combined action fuse, combined action igniter
~, конта́ктный percussion fuse
~, ла́зерный laser fuse
~, магни́тный magnetic fuse, magnetic proximity fuse
~, магнитоиндукцио́нный magnetic induction (type) fuse
~, магнитостати́ческий dip-needle mechanism
~ мгнове́нного де́йствия instantaneous fuse, direct action fuse
~ мгнове́нного де́йствия, уда́рный direct-action impact fuse, instantaneous percussion fuse
~, мгнове́нный 1. instantaneous fuse, direct action fuse
2. super-quick fuse (US)
~, мембра́нный diaphragm control fuse, diaphragm control igniter
~, механи́ческий mechanical fuse, clockwork fuse
~, механи́ческий дистанцио́нный time mechanical fuse, clockwork time fuse

взрыва́тель механи́ческого де́йствия mechanical-action fuse
~, механи́ческо-хими́ческий mechanical-chemical fuse
~, ми́нный fuse, mine igniter, mine switch
~ наведённого де́йствия target-influence fuse
~ нажи́много де́йствия pressure operated fuse, pressure operated igniter
~ натя́жного де́йствия pull-firing device, trip-wire actuating device, pull-action fuse
~ неизвлека́емости anti-lift fuse, anti-lift igniter
~, неконта́ктный 1. proximity fuse, VT fuse (ammo)
2. influence igniter, influence detonator
~, необезвре́живаемый anti-handling fuse
~ непредохрани́тельного ти́па non-boresafe fuse
~, нета́бельный improvised fuse
~, опти́ческий photoelectric proximity fuse
~ оско́лочного де́йствия disruptive detonator fuse
~, основно́й primary fuse
~, пасси́вный passive fuse
~, пасси́вный неконта́ктный passive proximity fuse
~, поездно́й train fuse
~ полевы́х фуга́сов plate withdrawal igniter, VPF igniter, plate withdrawal fuse, VPF fuse (Sov)
~, полуакти́вный semi-active fuse
~, полуакти́вный неконта́ктный semi-active proximity fuse
~, полупасси́вный semi-passive fuse
~, полупасси́вный неконта́ктный semi-passive proximity fuse
~, порохово́й дистанцио́нный powder-burning time fuse, combustion time fuse
~, постоя́нный standard HE fuse
~ предохрани́тельного ти́па bore-safe fuse
~ противота́нковой грана́ты, до́нный anti-tank projectile base fuse
~, пружи́нный spring fuse
~, пулево́й bullet-sensitive fuse
~, пьезоэлектри́ческий piezo-electric fuse
взрыва́тель, радиолокацио́нный radar fuse, radar proximity fuse, radar pulse fuse
~, радиоуправля́емый frequency induction fuse
~, разгру́зочный release switch, release operated fuse, release operated igniter
~, рыча́жный tilt rod fuse, tilt rod igniter
~ с самоликвида́тором auto-destructive fuse, self-destruction type fuse
~, сверхчувстви́тельный super-sensitive fuse
~, снаряжённый fully-equipped fuse, detonator and fuse assembly
~, телеуправля́емый remote-controlled fuse
~, теплово́й infra-red fuse
~, тёрочный friction igniter, friction fuse
~ уда́рного де́йствия impact fuse, percussion fuse
~, уда́рный impact fuse, percussion fuse
~, универса́льный all-purpose fuse
~, упрощённый simplified fuse
~, уче́бный dummy fuse
~, фотоэлектри́ческий photoelectric fuse
~, хими́ческий chemical igniter, chemical fuse
~, часово́й ми́нный mine mechanical time fuse, clockwork mine fuse
~, штырево́й rod-type fuse
~, электри́ческий electrically-operated fuse
~, электростати́ческий electrostatic proximity fuse
~, электростати́ческий неконта́ктный electrostatic proximity fuse
~, электрохими́ческий electro-chemical fuse
взрыва́тель-лову́шка anti-disturbance mine
взрыва́ть/взорва́ть to explode, to burst, to blast, to blow up, to execute demolitions, to carry out demolitions, to detonate, to fire; **взрыва́ть/взорва́ть боеприпа́сы** to set off ammunition
взрывно́й explosive
взрывобезопа́сный explosion-proof
взрывоопа́сность explosion hazard, explosion risk
взрывоопа́сный liable to explode, having explosion risk

взрывоусто́йчивый explosion-proof, blast-proof
взрывочувстви́тельный blast-sensitive
взрывпаке́т thunder flash, smoke-puff charge
взрывча́тка explosives
взры́вчатый explosive
взыска́ние punishment, penalty
~, **де́нежное** stoppage of pay, forfeiture of pay, fine
~, **дисциплина́рное** summary punishment, disciplinary award
взыска́ть (см взы́скивать)
взы́скивать/взыска́ть to punish, to impose a penalty
"Взять в Но́гу!" 1. "Keep Step!" 2. "Pick Up the Step!"
взять (см брать)
виаду́к viaduct
вибра́тор 1. vibrator (elec) 2. vibration contact, vibration igniter 3. resonant antenna 4. vibrator (for concrete)
~, **акти́вный** active element (Brit), driven element (US) (of antenna)
~ **анте́нны, акти́вный** active element of antenna
~, **кулачко́вый** cam and spring vibrator
~, **петлеобра́зный** folded-dipole antenna
~, **полуволново́й** half-wave radiator (of antenna)
~, **симметри́чный** dipole antenna
вибра́ция vibration, oscillation, buffeting, jitter
~ **при круче́нии** torsional vibration
~ **та́нка при движе́нии** pitching motion of tank
вибри́ровать to vibrate
вибро́метр vibrometer
вибропреобразова́тель vibrator converter (elec)
вибропро́чность vibration resistance
~, **высо́кая** good vibration resistance
виброусто́йчивость vibration resistance
вид 1. view 2. kind, form, sort, type, shape 3. Service
~ **авиа́ции** element of air force (e.g. tactical, long range, naval)
~ **бо́я** type of battle
~ **в пла́не** plan view
~ **в разре́зе** sectional view, cut-away view
вид, вне́шний outward appearance
~, **вну́тренний** inside view
~ **вооружённых сил** Service (Army, Navy, Airforce)
~ **колеба́ния магнетро́на** magnetron mode
~ **ми́нного загражде́ния** pattern of minefield
~, **о́бщий** general view
~ **огня́** type of fire
~ **с части́чным вы́резом** part-sectional view
~ **сбо́ку** side elevation, side view, lateral view
~ **све́рху** top view
~ **сза́ди** back elevation, back view, rear view
~ **сни́зу** bottom view
~ **спе́реди** front view, front elevation
~, **увели́ченный** enlarged view
виде́ние, ночно́е night vision
видеодете́ктор video detector
видеои́мпульс video pulse
видеопереда́тчик video transmitter
видеосигна́л video signal
видеоусили́тель video amplifier (electron)
видеочастота́ video frequency
видико́н vidicon, photo-conductive camera tube
ви́димость visibility
~, **больша́я** good visibility
~, **взаи́мная** intervisibility
~, **дневна́я** day visibility
~, **исключи́тельная** exceptional visibility
~, **никуды́шная** appalling visibility
~, **ночна́я** night visibility
~, **нулева́я** zero visibility
~, **ограни́ченная** limited visibility
~, **плоха́я** bad visibility
~, **пони́женная** reduced visibility
~, **пряма́я** line of sight
~, **хоро́шая** clear visibility, fair visibility
ви́димый visible, visual
видоизмене́ние modification
видоизмени́ть (см видоизменя́ть)
видоизменя́ть/видоизмени́ть to modify
видоиска́тель view finder
ви́ды огня́ types of fire (gunn)
визг тормозо́в brake squeal
визи́р viewer, view-finder, sight, peep-sight, backsight

визи́р а́зимута, за́дний lateral backsight (AA ord)
~ а́зимута, пере́дний lateral foresight (AA ord)
~, азимута́льный eyepiece of the azimuth sight, bearing sight (arty)
~ гру́бой наво́дки open sight (AA)
~ для наво́дки че́рез кана́л ствола́ boresight
~, опти́ческий optical sight
~, откры́тый open sight, auxiliary sight
~ угла́ возвыше́ния, за́дний vertical backsight, elevation backsight (AA ord)
~ угла́ возвыше́ния, пере́дний vertical foresight, elevation foresight (AA ord)
~ угло́в ме́ста elevation sight
~ целеуказа́ния target indication sight
~ центра́льной наво́дки director sight (nav)
визи́р-иска́тель open sight
визи́рование 1. sighting, alignment
2. approval, passing (by higher authority)
3. tracking (radar)
~ анте́нны antenna boresight
~, обра́тное backsighting (surv)
~, прямо́е foresighting (surv)
визи́ровать to align (arty), to sight (surv)
визуа́льный visual
ви́лка 1. fork
2. plug
3. bracket (gunn)
4. spider (flexible fabric covering)
5. yoke (universal joint)
~ выключе́ния сцепле́ния clutch withdrawal fork
~, двухделённая short bracket (arty)
~ для досыла́теля bracket for rammer (ord)
~, доро́жная road fork
~ за́днего хо́да gun stay (on AA gun carriage)
~, нагру́зочная battery tester
~, необеспе́ченная unverified bracket (gunn)
~, нулева́я straddle, virtual straddle (gunn)
~, обеспе́ченная established bracket (gunn)
~, пе́рвая long bracket (nav gunn)
~, после́дняя short bracket (gunn)
~ похо́дного крепле́ния cradle clamp, barrel clamping gear, gun stay
ви́лка, предохрани́тельная safety fork, safety clip (of anti-tk mine)
~, пристре́лочная bracket (gunn)
~ соедини́тельного механи́зма fork end
~ спусково́го рычага́ striker retaining prongs
~ тормозно́й тя́ги brake clevis, brake-rod yoke
~ туре́ли pivot yoke
~, у́зкая short bracket (gunn)
~ фрикцио́на, выжимна́я clutch shaft fork, clutch withdrawal fork
~, четырёхделённая long bracket (gunn)
~,широ́кая long bracket (gunn)
~, ште́псельная two-pin plug (elec)
ви́ллис "Willys", jeep
виля́ние wander, wobble, rocking motion, side motion, hunting
~ колёс wheel wobble
~ прице́па jack-knifing (of articulated vehicle)
вини́л vinyl
вино́вность guilt
винт screw, propeller
~ алида́ды, микроме́тренный upper tangent screw (on theodolite)
~, бара́шковый butterfly screw, wing screw
~ визи́ра, зажи́мный peep-sight clamping screw
~, возду́шный airscrew, aircraft propeller
~, вы́верочный adjusting screw
~, горизонта́льный traversing screw (mor)
~, гребно́й screw propeller
~ жиклёра, регулиро́вочный jet adjuster
~, за́дний ложево́й butt screw (SA)
~ зажи́ма ли́мба body clamping screw
~, зажи́мный clamping screw
~, нажи́мный adjusting screw
~, несу́щий main rotor (of helicopter)
~, подъёмный levelling screw
~, потайно́й grub screw
~, предохрани́тельный safety screw
~, регулиро́вочный adjusting screw
~, рулево́й tail rotor (of helicopter)
~ с квадра́тной голо́вкой square head screw
~ с ле́вой резьбо́й left-hand screw
~ с несплошно́й наре́зкой slotted screw
~ с пло́ской голо́вкой flat-head screw

винт с полукру́глой голо́вкой button-head screw, cup-head screw, round-head screw
~ **с потайно́й голо́вкой** countersunk head screw
~ **с пра́вой резьбо́й** right-hand screw
~ **с шестигра́нной голо́вкой** hexagonal head screw
~, **соедини́тельный** body locking pin (SA)
~, **сто́порный** set screw, anchoring screw
~ **то́рмоза, упо́рно-регулиро́вочный** brake adjusting screw
~, **трёхло́пастный** three-blade propeller
~, **трёхло́пастный гребно́й** three-bladed propeller (naut)
~ **упо́ра** front trigger-guard screw (SA)
~, **устано́вочный** set screw
~, **уто́пленный** recessed screw
~, **ходово́й** lead screw, leader, guide screw
~, **элевацио́нный** levelling screw (surv)
винто́вка rifle
~, **автомати́ческая** 1. automatic rifle 2. self-loading rifle
~, **боева́я** service rifle
~ **для рукопа́шного бо́я, уче́бная** dummy rifle
~ **"к бо́ю"** rifle at alert position
~, **крупнокали́берная** large bore rifle, large calibre rifle
~, **малокали́берная** small bore rifle, small calibre rifle
~, **пневмати́ческая** air rifle
~ **под уси́ленный патро́н** high-powered rifle
~, **полуавтомати́ческая** self-loading rifle, semi-automatic rifle
~, **пристре́лочная** spotting rifle
~ **с затво́ром рыча́жного ти́па** lever-action rifle
~ **с опти́ческим прице́лом** rifle with telescopic sight
~, **самозаря́дная** self-loading rifle, semi-automatic rifle
~, **самостре́льная** automatic rifle
~, **сва́ленная** canted rifle (failure to keep foresight blade vertical)
~, **сна́йперская** sniper rifle
~ **со скользя́щим затво́ром** bolt-action rifle
~, **спорти́вная** match rifle, sporting rifle
винто́вка, укоро́ченная short rifle
~, **фехтова́льная** dummy rifle, fencing rifle
~, **целева́я** precision rifle, target rifle
~, **штурмова́я** assault rifle
винтово́й screw, screwed, spiral, helical
винто́вочный rifle
винтокры́л Rotodyne, convertiplane, compound helicopter
винтпатро́н rifle round, rifle cartridge
винтполиго́н rifle range
вира́ж 1. banking (of road curve) 2. banking turn (ac)
ви́рус virus
висе́ние hovering (of helicopter)
вискозиме́тр viscosity meter, viscometer
витами́н vitamin
витки́ breech threads (BL breech block)
вито́к coil, loop, turn
~ **свя́зи** coupling loop (electron)
вихрево́й turbulent
вихреобразова́ние turbulence
вихрь vortex, eddy, whirl
~, **га́зовый** gas eddy
вища́ twisted brushwood used as rope
вка́пываться/вкопа́ться to dig oneself in
вклад, де́нежный money deposit
вкла́дыш bushing, bush (mech), liner
~ **затво́ра** breech-block bush (arty)
~ **колпака́** insert of cap, insert of dome, shim of cap, shim of dome (ammo)
~ **поддо́на** insert of sabot, shim of sabot (ammo)
~ **подши́пника** bearing liner, bearing bush, bearing shell, bearing insert
вклине́ние penetration, break-in (tac)
~, **части́чное** partial penetration, partial break-in
вкли́ниваться/вклини́ться to penetrate, to break in (tac), to wedge between; **вкли́ниваться/вклини́ться в оборо́ну** to break into the position
вклини́ться (см вкли́ниваться)
включа́тель switch (elec)
~, **курко́вый** cocking switch
включа́ть/включи́ть to connect, to plug in, to switch on, to switch in, to include; **включа́ть/включи́ть в прика́з** to include in an order, to incorporate in an order; **включа́ть/включи́ть дви́гатель** to start the engine, to switch on the engine;

включа́ть/включи́ть предохрани́тель to apply the safety catch; **включа́ть/включи́ть сцепле́ние** to engage clutch
включа́ться/включи́ться to join in; **включа́ться/включи́ться в коло́нну** to join in the column; **включа́ться/включи́ться в ли́нию** to tap a wire (comms)
включе́ние 1. connection, switching on, plugging in
2. inclusion, incorporation
3. cut-in
~, паралле́льное shunt connection (elec)
~ переда́ч механи́зма поворо́та ба́шни engagement of turret traverse mechanism
~, пла́вное smooth engagement (eng)
~, ре́зкое fierce engagement (eng)
~, сери́йное series connection (elec)
~ то́рмоза, э́кстренное emergency application of the brake
~ фрикцио́на clutch engagement
включи́ть (см включа́ть)
включи́ться (см включа́ться)
вкопа́тьця (см вка́пываться)
вла́га moisture, humidity, dampness
влагоизоля́ция waterproofing
влагоме́р humidity meter, hydrometer
влагонепроница́емый moisture-proof, moisture-resistant, damp-proof
влагоотдели́тель moisture separator
влагоотта́лкивающий moisture-repellent
влагопоглоти́тель drying cell
влагосодержа́ние moisture content, humidity
влагосто́йкость moisture resistance, wet strength
вла́жность humidity
~, относи́тельная relative humidity
~, уде́льная specific humidity
власть authority, power
~, дисциплина́рная power of command, disciplinary powers
"Вле́во На ...!" "Bear Left Towards ...!"
влива́ние pouring in, injection, infusion
~, внутриве́нное intravenous injection
влия́ние influence, effect
~ ве́тра wind effect
~, экрани́рующее screening effect (electron)
вмени́ть (см вменя́ть)
вменя́ть/вмени́ть to hold responsible; **вменя́ть/вмени́ть в вину́** to hold responsible (for an offence)
вмести́мость capacity, holding capacity, storage capacity, seating capacity
~, грузова́я loading capacity, load capacity, freight capacity
вмеша́тельство, вооружённое armed intervention
~, ли́чное personal intervention
вмонти́рованный built-in
вмя́тина dent, dig, pit, scar
"вне вся́ких катего́рий" "above all categories" (message precedence designation)
"вне вся́кой о́череди" "above all precedence" (message precedence designation)
вне доро́г(и) off-road, off the road
~ досяга́емости out of range, out of reach
"вне очерёдности" "extra precedence" (message category - includes all top priority designations)
вне преде́лов действи́тельного огня́ out of effective range
~ слу́жбы off duty
~ укры́тия in the open
~ ча́сти away from unit
внедре́ние introduction (of something)
внеза́пность surprise
~, такти́ческая tactical surprise
внеза́пный surprise
внемасшта́бный not to scale
"внеочередна́я" "above precedence" (message precedence designation)
внесро́чный special
вне́шта́тный non-organic, surplus-to-establishment, non-established (admin)
внима́ние attention
"Внима́ние Де́лай что Я!" "Follow Me!"
"Внима́ние Стой!" "Attention Halt!"
внуше́ние admonition
во взаимоде́йствии in cooperation (with)
во вре́мя бо́я before the enemy (mil law)
во́гнутый concave
вода́ water
~, артезиа́нская artesian well water
~, бытова́я water for personal use
~, ве́рхняя headwater
~, верхова́я upstream water
~, водопрово́дная town water, tap water, mains water
~, грунтова́я ground water

водá, дистиллирóванная distilled water
~ для мытья́ bath water, washing water
~, дождевáя rain water
~, жёсткая hard water
~, загрязнённая polluted water
~, заражённая infected water, contaminated water, polluted water
~, кипячёная boiled water
~, колóдезная well water
~, леднико́вая glacier water
~, мáлая low water, low tide
~, морскáя sea water
~, мýтная turbid water
~, мя́гкая soft water
~, низовáя downstream water
~, облучённая activated water
~, опрессóванная prepressurized water
~, охлаждáющая cooling water
~, питьевáя drinking water
~, повéрхностная surface water
~, подзéмная subterranean water, underground water, subsurface water
~, пóлная high tide
~, прéсная fresh water, sweet water
~, протóчная running water
~, роднико́вая spring water
~, свéжая fresh water
~, снеговáя snow water
~, солёная salt water, brine
~, спóрная race, swift current
"ВОДÁ СПУЩЕНÁ" "NO WATER", "RADIATOR EMPTIED" (written on empty radiators)
водá, срéдняя mean sea level
~, стóчная sewage
~, стоя́чая stagnant water
~, тáлая thaw water, water from thawing ice
~, чи́стая open water (in ice)
води́ло 1. planet carrier (epicyclic gear train)
2. tow-bar
води́льце operating rod, operating lever, connecting rod
води́тель driver (MT or AFV)
"Води́тель Слезáй!" "Driver Dismount!"
води́тель-дизели́ст driver of diesel-engined vehicle
води́тель-мехáник driver mechanic
води́ть маши́ну to drive
вóдный water
водобоя́знь hydrophobia, rabies
водогрéйка hot water heater, boiler
водоём waterhole, water reservoir, source of water supply
~, искýсственный artificial pond, man-made pond
водозапрáвщик water bowser
водоизмещéние displacement, water displacement
водоистóчник source of water
водокáчка water pumping station
водолáз diver
водолáз-подры́вник frogman
водомаслогрéйка water and oil heater
водомаслозапрáвщик water-oil bowser, oil-and-water servicing lorry, oil-and-water servicing truck
водомéр water meter, flow gauge
водомёт hydro-jet
водонепроницáемость impermeability to water, watertightness, waterproofness
водонепроницáемый watertight
водонóс water carrier (veh)
водонóсный water-bearing, aquiferous
водоопресни́тель distilling plant
водоотвóд drainage channel, drain
водоотвóдный overflow, drainage
водоотдели́тель water trap
водоотт́алкивающий water-repellant
водоохлади́тель water-cooling unit
водоочисти́тель water purifier
водопáд waterfall
водоподъёмник, ячéйсто-лéнточный belt and bucket water elevator, bucket lift
водопóй watering place, water supply
водоприёмник 1. intake (hydr eng)
2. water inlet (of veh drainage system)
водопровóд water conduit, water supply
водопроницáемость water permeability
водопротóк water course
водораздéл watershed
водорóд hydrogen
~, газообрáзный gaseous hydrogen
~, жи́дкий liquid hydrogen
~, сверхтяжёлый tritium
~, свя́занный combined hydrogen
~, тяжёлый heavy hydrogen
~, хлóристый hydrogen chloride
водосбóрник water catchment basin
водосбрóс spillway, terrace outlet
водосли́в spillway, weir
водоснабжéние water supply
водосодержáние water content

водосодержа́щий containing water, hydrous, aqueous
водосто́йкий water-resistant
водосто́йкость water resistance
водосто́к drain, sump, water course
водосто́чный drainage
водоумягче́ние water softening
водоумягчи́тель water softener
водоупо́рность water resistance
водоупо́рный water-resistant
водохрани́лище reservoir, water storage tank
водоцисте́рна water tanker (veh)
воды́, подмерзло́тные subpermafrost water
воева́ть to make war, to wage war, to be at war (imp only)
военача́льник military commander, military leader
военизи́рованный militarized, armed, para-military (civ org)
военкома́т military commissariat (soviet organ of local military control within the territory of a civ admin area)
военко́р war correspondent
вое́нно-морско́й naval, navy
военнообя́занный 1. person subject to military law
2. reservist
военнопле́нный prisoner of war
вое́нно-полево́й field
вое́нно-промы́шленный military-industrial
военнослу́жащий serviceman, member of the armed forces
вое́нный 1. serviceman, member of the armed forces
2. military, war
военпре́д military representative
военто́рг Military Trading Organization (Soviet equivalent to NAAFI)
военфе́льдшер medical assistant (trained junior medical officer, but not fully-qualified as doctor)
вожа́тый driver, dog handler, handler of pack animals
вожде́ние 1. driving (veh)
2. flying, pilotage
~ по пересечённой ме́стности cross-country driving
возбуди́тель 1. causative agent (med)
2. stimulant (med)
3. exciter (elec)
4. master oscillator (rad)
~ боле́зни pathogenic agent
~ боле́зни, бактериа́льный bacterial agent, bacteriological warfare agent, biological agent
~ боле́зни расте́ний, бактериа́льный anti-crop biological agent
возбуди́ть (см возбужда́ть)
возбужда́ть/возбуди́ть to excite, to actuate (relay, magnet)
возбужде́ние 1. excitation (elec)
2. initiation (expl)
возведе́ние erection, construction, building
возвести́ (см возводи́ть)
возводи́ть/возвести́ to construct, to erect, to build, to create
возвра́т 1. return
2. fly back, retrace (electron)
~ в подчине́ние reversion to command
возврати́ть (см возвраща́ть)
возвра́тно-поступа́тельный reciprocating (mech)
возвраща́ть/возврати́ть to regain; **возвраща́ть/возврати́ть инициати́ву** to regain the initiative; **возвраща́ть/возврати́ть на ро́дину** to repatriate
возвраще́ние в подчине́ние reversion to command
возвыше́ние elevation, rise
возвы́шенность high ground, elevation
возвы́шенный raised
возгла́вить (см возглавля́ть)
возглавля́ть/возгла́вить to head, to lead; **возглавля́ть/возгла́вить ата́ку** to lead the assault, to spearhead the assault
возде́йствие action, effect, influence, pressure
~, артиллери́йское effect of artillery fire
~, биологи́ческое biological effect
~, дли́тельное long term effect
~, кратковре́менное short term effect, immediate effect, temporary effect
~, непосре́дственное direct influence, immediate action, direct action
~, огнево́е fire, fire power, effect of fire
~ отравля́ющих веще́ств effect of CW weapons, effect of toxic agents
~, теплово́е thermal effect (nucl)

возде́йствие уда́рной волны́, ко́свенное indirect action of blast
возде́йствовать to influence, to exert pressure on
во́здух air
"во́здух" 1. "AIR" (signal of approach of en ac)
2. "air" (msg precedence designation)
во́здух, аркти́ческий arctic air
~, атмосфе́рный atmospheric air
~, вла́жный humid air, moist air
~, заражённый contaminated air
~, засо́санный induced air
~, ионизи́рованный ionized air
~, окружа́ющий ambient air, surrounding air
~, охлажда́ющий cooling air
~, проду́вочный scavenge air
~, рабо́чий actuating air
~, разрежённый rarefied air
~, сжа́тый compressed air
во́здух-во́здух air-to-air (wpn)
во́здух-земля́ air-to-surface (wpn)
воздухово́д air vent, air duct
воздуходу́вка air blower
воздухозабо́р air inlet, air intake (in shelter, bldg)
воздухозабо́рник air duct, air inlet, air intake, air scoop, intake
воздухоохлади́тель air cooler
воздухоочисти́тель air cleaner
~ с ма́сляной ва́нной oil-bath air cleaner
~ ти́па мельтицикло́н multi-cyclone-type air cleaners, bank of centrifugal pre-cleaners
воздухопито́к air intake
воздухопрово́д air duct
~, гла́вный тормозно́й main brake air pipe
воздухопрово́ды, подводя́щие air intake pipes
воздухопу́ск compressed air starting system
воздухохрани́тель air cylinder, air reservoir
возду́шно-гидравли́ческий hydro-pneumatic
возду́шно-деса́нтный airborne
возду́шный air, pneumatic, atmospheric
возлага́ть/возложи́ть to place, to entrust; **возлага́ть/возложи́ть зада́чу** to set a mission, to give a task
возложи́ть (см возлага́ть)
возме́здие retaliation
~, масси́рованное massive retaliation
возмеще́ние поте́рь loss replacement
возмо́жности, боевы́е fighting capabilities, combat and technical capabilities
~, материа́льные resources, material resources
~ нанесе́ния стратеги́ческих я́дерных уда́ров strategic nuclear capabilities
~, огневы́е fire capabilities, firepower
~ та́нка, наступа́тельные offensive potential of tank
~ та́нка, оборони́тельные defensive potential of tank
возмо́жность possibility, capability
~ нанесе́ния такти́ческих я́дерных уда́ров tactical nuclear capability
~, поража́ющая offensive capability, destructive capability, destructive effect (wpn)
возмуще́ние 1. indignation, resentment
2. perturbation, wash (GW)
~ ионосфе́ры ionospheric disturbance
вознагражде́ние, единовре́менное one-time grant
"Возобнови́те Переда́чу!" "Resume Transmitting!" (radio)
возобнови́ть (см возобновля́ть
возобновля́ть/возобнови́ть to renew, to replace; **возобновля́ть/возобнови́ть наступле́ние** to renew an offensive, to renew an attack
во́зраст, преде́льный age limit
~, призывно́й call-up age
во́ин serviceman, soldier (lit warrior)
~, уве́чный war cripple
во́инский soldierly, military, martial, pertaining to military personnel
во́инственный warlike, martial
во́йлок 1. felt
2. bratticing canvas
~, изолиро́вочный insulating felt
~, стекля́нный подкла́дочный surfacing mat, overlay mat
во́йлочный felt
война́ war, warfare
~, а́томная nuclear war
~, бактериологи́ческая bacteriological warfare
~, биологи́ческая biological warfare

Война́, Вели́кая Оте́чественная Great Fatherland War (World War 11)
война́, всео́бщая я́дерная general nuclear war
~, **глоба́льная** global war
~, **гражда́нская** civil war
~, **зи́мняя** winter warfare
~, **лока́льная** local war, regional war
~, **манёвренная** mobile warfare
~, **молниено́сная** lightning war, "Blitzkrieg"
~ **на истоще́ние** war of attrition
~ **на уничтоже́ние** war of annihilation
~, **назе́мная ми́нная** mine warfare
~, **неограни́ченная** total war
~, **неограни́ченная я́дерная** unrestricted nuclear warfare
~, **ограни́ченная** limited war
~, **око́пная** trench warfare
~, **оса́дная** siege warfare
~, **пе́рвая империалисти́ческая** First World War
~, **пе́рвая мирова́я** First World War
~, **подвижна́я** fluid warfare
~, **позицио́нная** position warfare, static warfare
~, **превенти́вная** preventive war
~, **психологи́ческая** psychological warfare
~, **радиолокацио́нная** electronic warfare
~, **радиоэлектро́нная** electronic countermeasures (ECM)
~, **раке́тно-я́дерная** nuclear missile war
~ **с примене́нием я́дерного ору́жия** nuclear warfare
~, **та́нковая** armoured warfare, tank warfare
~, **тота́льная** total war
~, **тота́льная я́дерная** total nuclear war
~, **транше́йная** trench warfare
~, **хими́ческая** chemical warfare, CW
~, **экономи́ческая** economic warfare
~, **электро́нная** electronic warfare
~, **я́дерная** nuclear warfare
войска́ troops, forces
~, **авиадеса́нтные** airborne forces, airborne assault forces
~, **автомоби́льные** 1. MT troops 2. MT units
~, **аэромоби́льные** air cavalry (US)
~, **бронета́нковые** armoured troops, armour
~, **вну́тренные** internal security troops
войска́, вое́нно-деса́нтные airborne troops
~, **возду́шно-деса́нтные** airborne forces, airborne troops
~, **вооружённые** armed troops
~, **де́йствующие** fighting troops (including services)
~, **доро́жные** road repair and traffic control troops
~, **железнодоро́жные** railway troops
~, **запа́сные** replacement troops
~, **зени́тные раке́тные** surface-to-air missile forces
~, **инжене́рные** engineers, sappers, engineer troops
~, **истреби́тельно-противота́нковые** anti-tank troops, tank destroyer troops
~, **механизи́рованные** mechanised troops
~ **министе́рства вну́тренних дел и госуда́рственной безопа́сности** Internal Security Troops (Sov) (subordinate to Ministry of Interior)
~, **мотомеханизи́рованные** mechanized troops
~, **мотопехо́тные** motorized infantry
~, **мотострелко́вые** motorised rifle troops
~, **назе́мные** ground forces
~, **небоевы́е** 1. non-combatant troops 2. rear-services personnel
~, **необстре́лянные** raw troops
~, **обеспе́чиваемые** supported troops
~, **оккупацио́нные** occupation troops
~ **операти́вно-такти́ческого назначе́ния** tactical forces
~, **парашю́тно-деса́нтные** parachute troops
~ **ПВО страны́** air defence forces (in USSR a separate arm comprising radar early warning system, observer corps, searchlights, AA units and fighter aircraft)
~, **перебра́сываемые по во́здуху** air-transported troops
~, **пограни́чные** border troops, frontier troops
~, **подви́жные** mobile troops
~, **подде́рживающие** supporting troops
~, **полевы́е** combat troops
~, **поса́дочно-деса́нтные** air-landing assault troops
~, **поса́дочные** air-landed forces

войска́, потребля́ющие consumer troops (opposite of supplying troops)
~ **прикры́тия плацда́рма** bridgehead covering force
~, **прожéкторные** searchlight troops
~, **радиотехни́ческие** radar troops, radar equipment operators
~, **ракéтные** rocket troops, missile troops, missile forces
~, **ракéтные стратеги́ческие** Strategic Rocket Forces (Sov)
~, **регуля́рные** regular troops
~ **свя́зи** signal troops
~, **специа́льные** special troops (supporting arms etc)
~ **стратеги́ческого назначéния, Ракéтные** Strategic Rocket Forces (Sov), strategic rocket troops
~, **сухопу́тные** land forces
~, **та́нковые** tank troops, armoured troops
~, **топографи́ческие** survey troops
~, **уда́рные** assault troops
~, **хими́ческие** chemical warfare troops, CW troops
~, **экспедицио́нные** expeditionary force
во́йско forces, troops (see more usual vojska)
войсково́й army, military
войти́ (см входи́ть)
волево́й determined, resolute
волна́ 1. wave
2. wavelength, frequency (radio)
3. assault wave (tac)
~ **а́томного взры́ва, уда́рная** atomic blast
~, **ба́зисная** base surge (NBC)
~, **баллисти́ческая** ballistic wave, shell wave, compression wave (short range arty)
~, **бегу́щая** travelling wave (radar, radio, electron)
~, **взрывна́я** blast wave, shock wave
~, **возду́шная уда́рная** air shock wave
~, **гектометри́ческая** hectometric wave (100 - 1,000 metres)
~, **головна́я** 1. mach wave (nucl)
2. shell wave (gunn)
~, **головна́я уда́рная** mach front (nucl)
~ **давлéния** pressure wave
~, **декаметри́ческая** decametric wave (10 - 100 metres, radio)
~ **детона́ции** detonation wave
волна́, детонацио́нная detonation wave
~, **дли́нная** long wave (1,000 - 10,000 metres, radio)
~, **ду́льная** 1. gun wave
2. muzzle wave (US)
~, **запа́сная** 1. reserve wavelength
2. alternative frequency
~, **засо́пловая уда́рная** tail blast
~, **звукова́я** sound wave
~, **зéмная** ground wave (radio)
~, **коро́ткая** short wave, decametric wave (10 - 100 metres, radio)
~, **метро́вая** metric wave
~, **миллиметро́вая** millimetric wave
~, **незатуха́ющая** undamped wave, unattenuated wave
~, **несу́щая** carrier wave
~, **обра́тная взрывна́я** back blast
~, **основна́я** 1. main frequency
2. basic wavelength
3. primary frequency (US)
~, **отражённая** 1. reflected wave (radar)
2. reflected shock wave (nucl)
~, **отражённая от земли́** ground reflected wave
~, **па́дающая** primary shock wave (nucl)
~, **па́дающая уда́рная** primary shock wave (nucl)
~ **падéния снаря́да** impact wave (wave set up by impact of projectile)
~, **повéрхностная** ground wave (radio)
~, **подзéмная уда́рная** subsurface shock wave
~, **поперéчная магни́тная** transverse magnetic wave, TM wave
~, **поперéчная электри́ческая** transverse electric wave, TE wave
~, **поперéчная электромагни́тная** transverse electromagnetic wave, TEM wave
~, **преоблада́ющая** dominant wave (GW)
~, **простра́нственная** ionospheric wave, indirect wave, sky wave (radio)
~, **пряма́я** direct wave (radio)
~, **рабо́чая** working frequency, operating frequency, operating wavelength
~ **разры́ва снаря́да** burst wave (acoustic wave produced by bursting shell)
~, **сантиметро́вая** centimetric wave
~, **сжима́ющая** compression wave
~, **соплова́я** back blast
~, **срéдняя** medium wave, hectometric wave (100 - 1,000 metres, radio)

волна́, стоя́чая standing wave (elec)
~, тропосфе́рная space wave
~, уда́рная blast wave, shock wave
~, ультразвукова́я ultrasonic wave
~, ультракоро́ткая VHF/microwave, "ultra short wave" (0.1 centimetre - 10 metres, radio)
~, этало́нная standard wave (antenna polarization)
~ я́дерного взры́ва, уда́рная atomic blast
~, ядовитодымова́я toxic smoke wave
волни́стость corrugation
волново́д waveguide (radar)
~, акусти́ческий acoustic waveguide
~, возду́шный atmospheric duct (radio)
~, ги́бкий flexible waveguide, interlocked type waveguide, corrugated waveguide
~, диафрагми́рованный septate guide (electron)
~, диэлектри́ческий dielectric waveguide
~, измери́тельный slotted waveguide
~, коаксиа́льный coaxial waveguide
~, кру́глый circular waveguide
~, ли́нзовый lens waveguide
~, многомо́довый oversized waveguide
~ основно́го ти́па колеба́ний dominant mode waveguide
~ основно́й мо́ды dominant mode waveguide
~ откры́того ти́па free-space waveguide
~, по́лый hollow-type waveguide, pipe-type waveguide
~, прямоуго́льный rectangular waveguide/grid
~, регуля́рный air-filled waveguide
~ с бегу́щими во́лнами travelling waveguide
~ со мно́гими мо́дами oversized waveguide
~ со мно́гими ти́пами колеба́ний oversized waveguide
~ тра́кта нака́чки pump waveguide
~, широкополо́сный broad-band waveguide
~, щелево́й slotted waveguide
волногаси́тель blast damper
~, грави́йный gravel blast damper
волноло́м breakwater
волноме́р wavemeter (radio)
волноме́р, гетероди́нный heterodyne wavemeter (radio)
~, резона́нсный resonance wavemeter (radio)
волнообра́зный undulatory
волнообразова́тель, контро́льный calibrator (radio)
волноотво́д breakwater
волноотража́тель breakwater (on motor launches etc)
волноре́з breakwater
во́лны, декаметри́ческие decametric waves (10 - 100 metres)
~, дециметро́вые decimetric waves (0.1 - 1 metre, radio)
~, дли́нные long waves, kilometric waves (1,000 - 10,000 metres, radio)
~, коро́ткие short waves, decametric waves (10 - 100 metres, radio)
~, метро́вые metric waves (1 - 10 metres, radio)
~, миллиметро́вые millimetric waves (0. 1 - 1 centimetre, radio)
~, сантиметро́вые centimetric waves (1 - 10 centimetres, radio)
~, сверхдли́нные myriametric waves (10,000 metres and above, radio)
~, сре́дние medium waves, hectometric waves (100 - 1,000 metres, radio)
волны́, электромагни́тные electromagnetic wave
волокни́стый fibrous
волокно́ 1. grain (timber) 2. fibre
~, асбе́стовое asbestos fibre
~, древе́сное wood fibre
~, синтети́ческое synthetic fibre
во́локом by dragging
волоку́ша akja, travois (type of sledge using two crossed poles)
~, специа́льная special sled, special travois
"Во́льно!" 1. "Stand at Ease!" 2. order given to men to come from ceremonial step to ordinary step (Sov)
вольнонаёмный civilian employee (at mil establishment)
вольноопределя́ющийся volunteer
вольт volt
вольт-ампе́р volt-ampere
вольтме́тр voltmeter
~, компенсацио́нный slide back voltmeter, self-balancing voltmeter

вольтме́тр, многошка́льный multi-range voltmeter
~ переме́нного то́ка alternating current voltmeter, A.C. voltmeter
вольфра́м tungsten, wolfram
вольфра́мовый tungsten, wolfram
воодушеви́ть (см воодушевля́ть)
воодушевля́ть/воодушеви́ть to inspire, to inspirit, to raise the spirits
вооружа́ть(ся)/вооружи́ть(ся) to arm
вооруже́ние arms, weapons, armament (Sov term may include auxiliary eqpt such as radar, rangefinders, ammo etc)
~, авиацио́нное aircraft armament
~, артиллери́йское artillery equipment, gun armament
~, бронета́нковое armament of AFVs (Sov term includes sights and ancillary eqpt)
~, гла́вное main armament
~, оборони́тельное defensive armament
~, основно́е primary weapon, major weapon
~, реакти́вное rocket weapons, missiles (Sov term includes rocket- and jet-propelled missiles)
~ та́нка, артиллери́йское gun armament, main armament (tk)
~ та́нка, основно́е main armament of tank, tank's main armament
~, та́нковое tank armament
~, шта́тное standard weapon, organic weapon
вооружённый armed
вооружи́ть(ся) (см вооружа́ть(ся))
вопреки́ прика́зу contrary to orders
вопро́с, уче́бный lesson
ворва́ться (см врыва́ться)
вороне́ние browning, oxide coating
воро́нка 1. cone
2. funnel (for filling)
3. shell hole, crater
4. belt guide (MG)
~ для наполне́ния filling funnel
~, кумуляти́вная cone of a hollow-charge shell
~, мёртвая 1. dead space, dead zone (AA/AD gunn)
2. dead space (radio/radar transmitter)
во́рот 1. winch, windlass
2. collar
3. hoist
~, подъёмный capstan
воро́та, габари́тные loading gauge (rly)
~, задвижны́е sliding caisson (dock, lock)
~, одностворчатые single-leaf gate (dock, lock)
~, створчатые leaf-type gate (dock, lock)
воротни́к collar (dress, mech)
~, ко́жаный 1. U-leather packing (stuffing box)
2. leather collar (eng)
~ распредели́тельного ва́ла camshaft collar
воск wax
~, парафи́новый paraffin wax
~, уплотня́ющий sealing wax
ВОСО (слу́жба Вое́нных Сообще́ний) Military Transport Organisation (rly) (Sov)
воспале́ние inflammation
~ лёгких pneumonia
воспита́ние 1. education
2. upbringing
3. indoctrination
воспламене́ние ignition
~, заме́дленное delayed ignition
~, запозда́лое retarded ignition
~ заря́да, затяжно́е hangfire (gunn)
~, мгнове́нное instantaneous ignition
~, многото́чечное multipoint ignition
~ от сжа́тия compression ignition
~ от уда́ра impact ignition, shock ignition, percussion ignition
~, после́довательное multistage ignition (GW)
~, уда́рное percussion ignition
~, цепно́е chain ignition
~, электри́ческое electric firing
~, электромагни́тное electro-magnetic firing
воспламени́тель igniter, igniting fuse, ignition cartridge, igniter assembly, blasting cap, primer cap, priming element
~, вспомога́тельный sub-igniter
~, дополни́тельный booster charge (ammo)
~, механи́ческий igniter mechanism, safety fuse, percussion igniter
~ на осно́ве чёрного по́роха black-powder base igniter
~ нажи́много де́йствия, тёрочный pressure-friction fuse
~, основно́й main igniter

воспламенитель, пиротехнический pinwheel igniter
~, пороховой powder-charge igniter
~, тёрочный friction fuse, friction igniter, friction primer
~, ударный percussion igniter
~, фрикционный friction igniter, friction fuse
воспламениться (см воспламеняться)
воспламеняемость ignition quality, flammability, inflammability
воспламеняемый combustible, ignitable, inflammable
воспламеняться/воспламениться to ignite
восполнение reinforcement, replenishment, replacement of casualties
воспрепятствовать (несож.) to hinder, to impede
воспретить (см воспрещать)
воспрещать/воспретить to forbid, to prohibit, to interdict
воспрещение 1. prohibition 2. interdiction (tac)
восприимчивость, магнитная susceptibility (magnetism, elec)
восприятие, вкусовое sense of taste, taste
~, зрительное sense of sight, sight
~, обонятельное sense of smell, olfactory perception
~, осязательное sense of touch, touch
~, слуховое sense of hearing, hearing, aural perception
воспроизведение reproduction
~ данных, автоматическое automatic data presentation
воспроизвести (см воспроизводить)
воспроизводить/воспроизвести to breed (nucl)
восстанавливать/восстановить to restore, to recondition, to re-establish, to rehabilitate
восстание rebellion, revolt, insurrection
~, вооружённое armed insurrection
восстановитель импульсов regenerator
восстановить (см восстанавливать)
восстановление restoration, reconstruction, renovation, rehabilitation, re-establishment, repair and recovery, re-adjustment
восстановление, полное complete overhaul, complete rebuild
~ постоянной составляющей DC restoration (elec)
~ протектора re-treading
восток east
восточный eastern
восход rising
"восьмёрка" "figure of eight" (cleaning rag folded & twisted in the middle, for small arms bore)
восьмёрка, габаритная figure of eight track used for MT driver training, serpentine
воюющий belligerent
впадина dent, hollow, depression (geog)
~ зуба "tooth" space (space between adjacent gear teeth)
впай lead wire (electron tube)
впайка, стеклянная glass seal (electron)
"Вперёд!" "Advance!"
впередилежащий forward, in front, forward of
вперекрест in a crossed fashion, lying across one another
вписать в прицел to line up in the sights (air gunnery)
~ (см вписывать)
вписывать/вписать to enter, to insert, to inscribe
впитывание absorption
вполдерева half lap joint (carpentry)
впотай flush, countersunk
"Вправо На ---!" "Bear Right Towards ---!"
впрыск injection (fuel)
~ бензина petrol injection
~ воды water injection
~ горючего fuel injection
~, непосредственный direct injection (diesel), direct fuel injection (petrol engine)
~, поздний late injection (fuel)
~, раний early injection (fuel)
~ топлива, непосредственный direct fuel injection
впрыскивание injection (gen)
впрыскивать/впрыснуть to inject, to spray in
впрыснуть (см впрыскивать)
впуск intake, inlet
враг enemy
враждебный hostile

вра́жеский enemy
врасплóх by surprise
враспóр in a braced (or wedged) position
врач medical officer, doctor
~, **ветерина́рный** veterinary surgeon, veterinary officer
~, **воéнно-ветерина́рный** military veterinary surgeon
~, **воéнный** army doctor, medical officer, MO
~, **гарнизóнный** garrison medical officer
~, **дивизиóнный** 1. division medical officer
2. Assistant Director Medical Services at Div HQ (Brit)
~, **корпуснóй** 1. corps medical officer
2. Deputy Director Medical Services (Brit)
врачéбный medical
враща́ться/вертéться to rotate, to revolve, to spin; **враща́ться/вертéться вхолостýю** to run idle, to idle (engine); **враща́ться/вертéться по хóду часовóй стрéлки** to turn clockwise; **враща́ться/вертéться прóтив хóда часовóй стрéлки** to turn anti-clockwise
враща́ющийся revolving, rotary, rotating, rotatable, turning
вращéние revolution, rotation, spin, turning
~ **по часовóй стрéлке** clockwise rotation
~, **прáвое** clockwise rotation
~, **прецессиóнное** precession spin, precession (GW)
~ **прóтив часовóй стрéлки** counter-clockwise rotation
~ **снаря́да** spin of projectile
вреди́тель saboteur, wrecker
вреди́тельство sabotage
врезáние снаря́да в нарéзы spinning of projectile by rifling
временнóй time
врéменный temporary
врéмя безоткáзной рабóты, срéднее mean time between failures
~, **внеслужéбное** free time, off duty hours, spare time
~, **внеурóчное** free time, off duty hours, out-of-lesson time, spare time
~, **воéнное** war-time, time of war
~ **вхóда ПУА́ЗО в режи́м** AAFCE settling time
врéмя горéния delay time (igniter set, fuse)
~ **горéния зажигáтельной трýбки** fuse time
~ **горéния трасси́рующего состáва** tracer burning time
~ **деионизáции** de-ionization time
~, **декрéтное** daylight-saving time
~ **закры́тия** change time (servo mech)
~ **запáздывания** time lag
~ **запáздывания вы́стрела** drill time (arty) (period between moment when calculation of fuse and deflection begins, and moment when gun fires)
~ **запáздывания регули́рования** control system lag
~, **звёздное** sidereal time
~, **и́стинное мéстное** apparent local time
~, **исхóдное** starting time (tac)
~ **мéжду откáзами, срéднее** mean time between failures
~, **мéстное** local time
~, **наблюдáтельное** observing interval, observation interval
~ **начáла наступлéния** time of attack, H-hour
~ **облучéния** exposure time (nucl)
~ **обнаружéния** detection time, acquisition time
~ **оборóта** turn-around time
~ **огневóго налёта** time on target
~ **отправлéния** departure time
~ **падéния бóмбы, характеристи́ческое** time of fall of a particular bomb
~ **передáчи** time of dispatch
~ **подлёта** arrival time (ac)
~, **подлётное** approach time (ac, AA gunn)
~ **подписáния** date-time group, DTG (sigs)
~ **полёта** time of flight
~ **полёта, пóлное** range table time of flight (gunn)
~ **полёта электрóна** electron transit time
~, **полётное** time of flight (GW, arty)
~, **пояснóе** zone time
~, **пояснóе декрéтное** standard time (zone time adjusted to geographical boundaries)
~ **прибы́тия** arrival time, arriving time
~ **прибы́тия, предполагáемое** expected time of arrival

вре́мя прибы́тия, расчётное estimated time of arrival, ETA
~ **прибы́тия, факти́ческое** actual time of arrival
~ **приёма** 1. filing time , time handed in (sigs)
2. resting time (radar)
~ **пробе́га, группово́е** group velocity (elec)
~ **пробе́га по автомати́ческой характери́стике** free-running time
~, **пролётное** transit time (electron)
~, **рабо́тное** prediction time (AD)
~ **сближе́ния** time to impact (missile)
~ **сбра́сывания парашюти́стов** jump time
~, **свёртывания** closing down time, time for closing down
~, **све́тлое** daylight hours
~, **служе́бное** on-duty time
~ **снаря́да, полётное** time of flight of projectile
~, **со́лнечное** solar time
~ **сраба́тывания** response time (servomech), time of action (relay)
~, **сре́днее гринви́ческое** Greenwich Mean Time, GMT
~, **упреди́тельное** 1. prediction interval, time of flight of projectile (AD gunn)
2. travel time of target from position at observation to future posn (tk)
~, **штрафно́е** time penalty
время́нка 1. temporary structure
2. portable heater, makeshift heater
вруб 1. notch (carpentry joint)
2. cut, cut-hole (mining)
~, **боково́й** side cut
~, **клиново́й** wedge cut (mining)
~, **пирамида́льный** pyramid cut (mining)
вру́бка cutting, notching, channelling
вруча́ть/вручи́ть to hand over, to deliver, to entrust
вруче́ние 1. presentation (award etc)
2. handing over, delivery
вручну́ю by hand, manually
врыва́ться/ворва́ться to penetrate, to break into (tac); **врыва́ться/ворва́ться на пози́ции** to break into the position (tac)
всад helve socket (of a pickaxe)
вса́сывание suction
~ **во́здуха** air induction
~ **то́плива** fuel induction
вса́сывающий suction
"все налицо́" "all present and correct"
"всё по́нял" "Roger", "Message received and understood" (sigs)
всеарме́йский all-army
всепого́дность all-weather capability
всепого́дный all-weather
всесезо́нный all-weather, all-season
всесторо́нний all-round
всея́дный 1. omnivorous
2. multi-fuel (sl)
вскрыва́ть/вскрыть to open, to reveal, to disclose
вскры́тие реки́ break-up of river ice
вскрыть (см вскрыва́ть)
вспе́нивание foaming, frothing
всплеск 1. splash
2. blip (on radar screen)
~ **разве́ртки** a burst of scanning (on a previously empty display screen)
вспомога́тельный auxiliary
вспры́скивание injection
~, **интравено́зное** intravenous injection
вспы́хивать/вспы́хнуть to break out (war), to catch fire, to burst into flames
вспы́хнуть (см вспы́хивать)
вспы́шка 1. flash, flare, blaze
2. ignition, spark, burst
~ **вы́стрела** muzzle flash
~, **обра́тная** back-fire
~, **огнева́я** burst of fire
вста́вить (см вставля́ть)
вста́вка insert
~ **направля́ющей теле́жки** bogie guide block
~, **настро́ечная** tuning slug (elec)
~, **соедини́тельная** plate shackle, fish-plate
вставкодержа́тель fuse carrier, fuse holder
вставле́ние ка́псюля priming
вставля́ть/вста́вить to insert, to introduce, to fix in, to put in, to install, to bed; **вставля́ть/вста́вить взрыва́тель** to set a fuse, to prime a fuse; **вставля́ть/вста́вить запа́л** to prime; **вставля́ть/вста́вить ка́псюль** to prime (fuse)
вставно́й insertable, plug-in
"Встать!" "Get Up!"
встра́ивать/встро́ить to build in, to incorporate
встре́тить (см встреча́ть)
встре́ча contact, link up (tac)

встре́ча, пе́рвая first contact, initial contact
встреча́ть/встрети́ть to meet, to encounter
встро́енный built-in
встро́ить (см встра́ивать)
вступа́ть/вступи́ть to enter; **вступа́ть/вступи́ть в бой** to engage, to get in to action, to move into action, to join battle
вступи́ть (см вступа́ть)
вступле́ние entering, joining
вта́скивание dragging (into, onto)
вторже́ние invasion (of), incursion into
втори́чный secondary
вту́лка bush, bung, slug, sleeve
~, боева́я firing hole bush (ord)
~, ввинтна́я screw bushing
~, вставна́я cylinder liner
~ га́йки рукава́ tail (hose coupling)
~, головна́я nose fuse head (ammo)
~ дистанцио́нного жа́ла time needle plug
~, до́нная base plug (ammo)
~, ду́льная nose plug (ammo)
~ замедли́теля detonator sleeve (gren)
~ звена́ гу́сеницы track link bush
~, ка́псюльная primer cup (ammo)
~ кла́пана, направля́ющая valve guide
~, кони́ческая cone bush
~, лату́нная brass bushing
~ мо́стика bridge plug (ord)
~ наду́льника muzzle plug
~, нажи́мная gland (stuffing box)
~, направля́ющая 1. valve guide
2. barrel bushing (pistol)
~, отража́тельная spill deflector (diesel injector)
~, перехо́дная tapered clamping sleeve, Morse taper sleeve
~, пла́вающая floating bush (tech)
~, поворо́тная control sleeve (diesel fuel pump)
~ пружи́ны затво́ра, опо́рная bolt-spring stop
~, разрезна́я split bearing
~, распо́рная 1. distance bush, distance piece
2. gear retainer (diesel injector)
~, регули́рующая 1. throttling bush (recoil cylinder)
2. rotating valve (ord buffer)
~, рези́новая rubber grommet
вту́лка с уда́рным механи́змом firing mechanism block
~ са́льника packing bush
~, соедини́тельная connecting bush
~ ствола́ barrel collar
~ ствольно́й коро́бки standing breech (SL rifle)
~ толка́теля, направля́ющая follower guide (diesel injector)
~ тра́ка track bushing, link bushing
~ уда́рника striker case, firing-pin housing
~ уда́рного уда́рника screwed plug of percussion pellet (time/percussion fuse)
~, уплотня́ющая gland bush
~, холоста́я 1. fuse-hole plug (ammo, arty)
2. bush, bushing (roller chain)
3. liner (eng)
~, цили́ндровая cylinder liner
~ ша́риковая ball race
втя́гивать/втяну́ть to pull (in), to retract, to draw (into)
втяну́ть (см втя́гивать)
вулканиза́тор, кольцево́й annular vulcanizer
вулканиза́ция vulcanization, curing
~ в автокла́ве autoclave vulcanization
~ в пре́ссе press curing
~ горя́чим во́здухом hot-air curing, dry heat vulcanization
~, ка́мерная hot-air chamber vulcanization
~, холо́дная vulcanization at normal temperature, cold cure
вход 1. entry, entrance, access
2. input (elec)
~ в атмосфе́ру re-entry (missile)
~, горизонта́льный drift, adit (mil eng)
"ВХОД ЗАПРЕЩЁН" "OUT OF BOUNDS", "NO ENTRY"
вход, коле́нчатый dog-leg entrance (mining)
~, парога́за steam inlet
~ ПУА́ЗО в режи́м AAFCE settling
~, пу́льный leed, barrel throat (SA)
~ сопла́ nozzle entrance
~, углово́й angle input
входи́ть/войти́ to enter, to go in, to be included in; **входи́ть/войти́ в боево́й взвод** to engage the sear notch; **входи́ть/войти́ в луч** to intercept the beam;

входи́ть/войти́ в связь to establish contact, to make contact, to open up communications; входи́ть/войти́ в соприкоснове́ние to establish contact, to make contact

входно́й 1. entrance
2. input

входя́щий incoming (signals, mail etc)

вхожде́ние в связь netting (sigs)

въезд, аппаре́льный 1. landing bay (br)
2. half-floating bay (br)
3. ramp (approach) (eng)

"ВЪЕЗД ВОСПРЕЩЁН" "NO ENTRY", "DO NOT ENTER"

выба́нить (см ба́нить)

вы́бег rundown (ICE), stopping distance, braking distance

выбива́ть/вы́бить to dislodge, to knock out; **выбива́ть/вы́бить из стро́я** to put out of action

выбира́ть/вы́брать to choose, to select, to elect, to take out (patent), to take everything out, to haul, to haul in, to pull; **выбира́ть/вы́брать зазо́р** to take up the backlash, to take up the clearance

вы́бить (см выбива́ть)

вы́бойна dent, dint

вы́бор 1. selection, choice, option
2. sample, sampling

~ це́лей target selection

вы́борка sample

вы́борочный selective, sampling

выбрако́вка condemnation, rejection (of sub-standard or unserviceable items)

выбра́сывание ejection, extraction

~ стре́ляных гильз, автомати́ческое self-ejecting

выбра́сыватель extractor (SA), extractor lever (ord)

~ гильз shell extractor

~ обо́ймы clip ejector

~, приводи́мый в де́йствие затво́ром bolt-actuated ejector

~, приводи́мый в де́йствие си́лой отка́та recoil-actuated ejector

~, приводи́мый в де́йствие ство́льной коро́бкой receiver-actuated ejector

~, стержнево́й rod ejector (revolver)

выбра́сывать/вы́бросить to throw out, to eject

вы́брать (см выбира́ть)

вы́брос pip, blip (radar)

вы́бросить (см выбра́сывать)

вы́броска возду́шного деса́нта parachute drop, paradrop

~ грýзов с самолёта supply drop

~, ма́ссовая mass airdrop

~ парашю́тного деса́нта parachute drop, paratroop dropping, paradrop

~ с во́здуха airdrop

выведе́ние из стро́я putting out of action, becoming a casualty

вы́веренный zeroed (arty)

вы́верить (см выверя́ть)

вы́верка adjustment, zeroing-in, calibration, lining-up

~ дальноме́ра по да́льности coincidence adjustment

~ нулево́й ли́нии прице́ливания collimation

~ по да́льности range calibration

выверя́ть/вы́верить to adjust, to align, to regulate

вы́вести (см выводи́ть)

выве́тривание weathering, erosion

выве́шивание ва́гой levering

вы́винтить (см выви́нчивать)

выви́нчивать/вы́винтить to unscrew, to screw off

вы́вих dislocation (med)

вы́вод 1. conclusion, deduction, inference
2. withdrawal
3. outlet

выводи́ть/вы́вести to lead out, to withdraw, to remove, to conclude, to destroy, to hatch, to portray, to eliminate; **выводи́ть/вы́вести из бо́я** to bring out of action, to take out of action; **выводи́ть/вы́вести из стро́я** to put out of action, to take out of action, to knock out of action, to force out of action, to bring out of action; **выводи́ть/вы́вести карау́л** to turn out the guard

выводно́й escorting sentry, escort (prisoner's)

вы́воз evacuation

вы́говор, стро́гий severe reprimand

вы́годный advantageous, favourable

выгора́ние burning out (ICE valve etc), fading (colour), pitting (elec)

выгружа́ть/вы́грузить to unload, to offload, to discharge cargo

вы́грузить (см выгружа́ть)

вы́грузка 1. unloading, offloading
2. detraining, debussing

вы́грузочный unloading, debussing, detraining
выдава́ть/вы́дать to issue, to distribute, to hand out, to betray, to give away, to extradite; **выдава́ть/вы́дать да́нные** to furnish data, to deliver data, to turn out data
вы́дать (см выдава́ть)
вы́дача issue
~ взаме́н 1. issue in lieu of
2. issue in exchange for
~ да́нных presentation of information
~, непла́новая extraordinary issue
~, повто́рная reissue
выдвига́ться/вы́двинуться to advance, to move (forward), to rise, to rise from the ranks
выдвиже́ние move forward, movement
~ в глубь побере́жья break out of the beach-head
выдвижно́й telescopic, sliding
вы́двинуться (см выдвига́ться)
выделе́ние allocation
вы́деленный allotted, detailed, detached, specified
вы́делить (см выделя́ть)
вы́делиться (см выделя́ться)
выделя́ть/вы́делить to detail, to allot, to make available; **выделя́ть/вы́делить в распоряже́ние** to place at disposal of; **выделя́ть/вы́делить тра́нспорт** to allocate transport
выделя́ться/вы́делиться to show up, to stand out
вы́держать (выде́рживать)
выде́рживать/вы́держать to withstand, to endure, to persist, to maintain, to mature, to store (nucl); **выде́рживать/вы́держать направле́ние** to keep direction; **выде́рживать/вы́держать темп наступле́ния** to maintain the momentum of the attack
вы́держка 1. stamina, tenacity, staying power
2. excerpt
3. exposure (photo)
~ вре́мени time lag
вы́дохнуться (см выдыха́ться)
выдыха́ние exhalation
выдыха́ться/вы́дохнуться to have no fight left
вы́езд, полево́й field exercise
вы́емка 1. cutting (rd, rly)
2. recess (SA)
~ гру́нта excavation, dredging
~, кольцева́я cannelure, groove (ammo)
~, кумуляти́вная focusing hollow (shell, ammo)
выжига́ние scorching
вы́звать (см вызыва́ть)
выздора́вливающий convalescent
вы́зов 1. call, request
2. call (radio, telep)
3. calling method (telep)
~, авари́йный distress call
~, избира́тельный selective ringing, selective call signal
~, инду́кторный magneto calling, inductive calling, inductive ringing (comms)
~, непосре́дственный direct request (air sp etc)
~ огня́ call for fire, request for fire (arty)
~ по ра́дио radio call
~ радиоста́нции call, call up (radio)
~, телефо́нный telephone call
~, тона́льный audio-frequency ringing, voice-frequency ringing
~, фони́ческий buzzer call, buzzer calling
~, циркуля́рный collective call (comms)
~, э́кстренный emergency call
вызыва́ть/вы́звать to call, to call for; **вызыва́ть/вы́звать карау́л в ружьё** to turn out the guard; **вызыва́ть/вы́звать ого́нь** to call for fire support, to call down fire; **вызыва́ть/вы́звать ого́нь проти́вника** to draw the enemy's fire
вы́играть (см вы́игрывать)
вы́игрывать/вы́играть to win; **вы́игрывать/вы́играть бой** to win the battle
вы́игрыш анте́нны antenna gain
~ мо́щности power gain
~ трансду́ктора transducer gain
вы́йти в отста́вку to retire
~ (см выходи́ть)
выка́чивание pumping (out), evacuation, exhaustion, deflation
вы́кладка, по́лная field service marching order, FSMO
~, по́лная боева́я field service marching order, FSMO
~, по́лная похо́дная field service marching order, (FSMO)

выключа́тель switch, cut-out, releasing device
~, авари́йный emergency switch, safety switch
~, автомати́ческий automatic switch
~ бо́йка firing mechanism cut-out (mor)
~, быстроде́йствующий high-speed switch, high-speed circuit breaker
~ высо́кого напряже́ния high-voltage switch
~ зажига́ния ignition switch
~ магази́на magazine safety (pistol)
~ ма́ссы battery switch, master switch
~ механи́зма взаи́мной за́мкнутости eccentric bolt thumb lever (AA ord)
~, перекидно́й tumbler switch, toggle switch
~ пита́ния main supply switch
~, поворо́тный rotary switch
~, предохрани́тельный safety switch, emergency governor
выключа́ть/вы́ключить to take out (gear), to switch off, to disconnect, to turn off, to disengage; **выключа́ть/вы́ключить дви́гатель** to stop the engine, to switch off the engine; **выключа́ть/вы́ключить предохрани́тель** to release the safety catch; **выключа́ть/вы́ключить ток** to switch off current, to cut off current, to deenergise (relay etc); **выключа́ть/вы́ключить сцепле́ние** to disengage clutch
выключе́ние switching-out, cut-out, cut-off, releas(ing), disengagement, uncoupling, declutching, shut-off, shut-down
~ дви́гателя, самопроизво́льное engine stall
~ сцепле́ния declutching
~ сцепле́ния, двойно́е double-declutching
вы́ключить (см выключа́ть)
вы́колотка drift (tool)
~ зве́ньев гу́сениц track pin removing drift
~ па́льцев тра́ков track pin removing drift
вы́копировка copying, duplicating
выкра́шивание брони́ armour scabbing
~ подши́пника bearing disintegration (mech)
вы́лазка raid, sortie
выле́живание ageing (substances, materials)
вы́лет 1. flight
2. overhang
3. departure (projectile
4. jib radius, reach (crane)
~, боево́й operational flight
~, небоево́й non-operational flight
~, чи́стый clear outreach (crane)
выма́чивание soaking
вы́мпел pennant, streamer, nun's kit
вы́нести (см выноси́ть)
вы́нос 1. fall of shot to side of target (gunn)
2. evacuation (wounded) (by stretcher bearers)
~ зна́мени, торже́ственный trooping the colour
~ то́чки прице́ливания aim off, aiming off
~ то́чки прице́ливания вы́ше це́ли aiming-up
~ то́чки прице́ливания ни́же це́ли aiming-down
выноси́ть/вы́нести to carry away, to take off, to endure, to bear, to announce, to bring out; **выноси́ть/вы́нести то́чку прице́ливания** to aim off
выно́сливость endurance
вы́пад lunge, thrust
выпаде́ние fall
~ радиоакти́вного дождя́ rainout (nucl)
~ радиоакти́вных веще́ств fallout (nucl)
~ радиоакти́вных оса́дков radioactive fall-out
выпа́ривание evaporation
выпа́ривать/вы́парить to evaporate
вы́парить (см выпа́ривать)
вы́печка batch (qty of bread, loaves)
выпира́ние jack-knifing (artic veh)
вы́писать (см выпи́сывать)
вы́писка 1. excerpt, extract (doc)
2. discharge (from hospital)
выпи́сывать/вы́писать to discharge, to copy out, to extract
вы́плавка smelting
вы́плата payment
вы́платить (см выпла́чивать)
выпла́чивать/вы́платить to issue (pay, allowances, pension), to pay
выполне́ние execution, carrying out, fulfilment, achievment, accomplishment
вы́полнить (см выполня́ть)

выполня́ть/вы́полнить to carry out, to fulfil, to accomplish; **выполня́ть/вы́полнить зада́чу** to accomplish a mission, to carry out a task; **выполня́ть/вы́полнить прика́з** to carry out an order, to execute an order
вы́править (см выправля́ть)
вы́правка bearing, carriage
~, вое́нная military bearing
~, строева́я military bearing
выправля́ть/вы́править to correct, to set right, to smooth, to straighten
выпрями́тель rectifier (elec)
~, ано́дный plate rectifier, anode rectifier
~, ве́нтильный valve rectifier
~, герма́ниевый germanium rectifier
~, двухполупериóдный full-wave rectifier
~, заря́дный battery charger, charging rectifier
~, кенотро́нный kenotron
~, кре́мниевый silicon rectifier
~, купро́ксный copper-oxide rectifier
~, ла́мповый valve rectifier
~, однополуперио́дный half-wave rectifier
~, полупроводнико́вый semi-conductor rectifier, solid-state rectifier
~, рту́тный mercury rectifier
~, селе́новый selenium rectifier (electron)
~, сухо́й dry rectifier
~, электро́нный electronic rectifier
вы́прямить (см выпрямля́ть)
выпрямле́ние rectification (elec)
~, двухполуперио́дное full-wave rectification
вы́прямленный rectified (elec)
выпрямля́ть/вы́прямить to straighten, to rectify (elec)
вы́пуклость convexity
~ ба́шни turret bulge
вы́пуклый convex
вы́пуск 1. output, discharge, exhaust 2. model (veh etc) 3. issue, number (journal, book etc)
~ во́здуха из тормозно́й систе́мы brake bleeding
выпуска́ющий stick commander (parachuting)
выпускни́к graduating student
выпускно́й 1. exhaust, discharge, outlet 2. final (exam)
выраба́тывать/вы́работать to work out, to draw up, to generate (missile commands)
вы́работать (см выраба́тывать)
вы́работка manufacture, manufacturing, production, development, generation (elec)
~ реше́ния working out a solution (Sov commander's reconnaissance and planning)
выра́внивание alignment, dressing
~ да́нных data smoothing (electron)
выра́вниватель 1. levelling mechanism 2. levelling screw
выра́внивать(ся)/вы́ровнять(ся) to straighten (ranks), to align, to level off, to equalize; **выра́вниваться/вы́ровняться в заты́лок** to cover off (drill)
вы́рез cut, groove, notch, opening, indentation
~ взво́да уда́рника cock notch
~ встре́чным ко́нусом two-sided chamfer cavity (tyre repair)
~ для вы́ступов ла́пок экстра́ктора groove for cartridge extractor pin (SA)
~ для досыла́теля clearance for feed piece (SA)
~ для курка́, боево́й cocking-piece groove, cocking-piece notch
~ для постано́вки курка́ на предохрани́тельный взвод half-cock notch
~ нару́жным ко́нусом outside chamfer cavity (tyre repair)
~ стрелко́вого ору́жия slot (SA)
~ што́ка camming recess
вы́резка вну́тренним ко́нусом inside chamfer cutting (tyre repair)
~ встре́чным ко́нусом two-sided chamfer cutting (tyre repair)
~ нару́жным ко́нусом outside chamfer cutting (tyre repair)
вы́рисоваться (см вырисо́вываться)
вырисо́вываться/вы́рисоваться to silhouette oneself, to stand out (against background), to be silhouetted
вы́ровнять(ся) (см выра́внивать(ся))
вы́рубка clearing (in a wood)
~ кружко́в blanking (ammo manufacture)
~, лесна́я felling area in wood, clearing in wood

выруча́ть/вы́ручить to relieve
вы́ручить (см выруча́ть)
вы́ручка relief, rescue, assistance
вырыва́ть/вы́рыть to dig, to dig out, to dig up, to exhume, to excavate
вы́рыть (см вырыва́ть)
вы́садить(ся) (см выса́живать(ся))
вы́садка 1. landing, disembarkation (from ships)
2. landing (tps from ac)
3. detraining
~ возду́шного деса́нта 1. airborne assault landing, airborne landing
2. airborne assault
~ возду́шного деса́нта с вертолётов helicopter-borne landing
~ деса́нта assault landing
~ морско́го деса́нта 1. seaborne assault landing, amphibious landing
2. amphibious assault
~ парашю́тного деса́нта parachute landing
~ поса́дочного деса́нта 1. airborne landing (of air-landing forces)
2. air-landed operation
вы́садочный landing
выса́живать(ся)/вы́садить(ся) to land, to disembark, to detrain, to detruck; **выса́живать/вы́садить морско́й деса́нт** to carry out an amphibious landing
высве́рливать/вы́сверлить to drill out (mech)
вы́сверлить (см высве́рливать)
высвобожде́ние liberation, assistance in escaping
вы́следить (см высле́живать)
высле́живать/вы́следить to track, to stalk
вы́слуга лет length of service
~ лет, о́бщая total length of service
высо́кий high, tall
высоково́льтный high voltage
высокодобро́тный high-Q (elec)
высокока́чественный high-quality, high-grade
высокоманёвренный highly manoeuvrable
высокоокта́новый high-octane
высокоо́мный high resistance
высокоподви́жный highly mobile
высокорадиоакти́вный highly radioactive, "hot"
высокоскоростно́й high-speed
высокоцета́новый high-cetane
высокочасто́тный high-frequency
высота́ height, altitude
~ апоге́я apogee altitude
~, безопа́сная safety height
~ боево́го примене́ния operational altitude
~ взведе́ния arming altitude
~ вы́броски парашюти́стов jumping altitude
~, вы́численная computed height, computed altitude
~, габари́тная overhead clearance, clearance height
~, госпо́дствующая commanding height, dominating height
~, де́йствующая effective height
~, и́стинная true height (height above the ground as distinct from height above sea level)
~, кома́ндная commanding height, dominating height
~, кома́ндующая commanding height, dominating height
~ ли́нии це́ли height of line of fire
~, максима́льная peak altitude
~, метацентри́ческая metacentric height
~, минима́льная minimum altitude
~ мо́ста, строи́тельная height of deck above clearance (br)
~ му́шки "height of the foresight" (SA) (distance from centre of bore to top of foresight)
~ над у́ровнем мо́ря height above sea level, altitude
~ ни́жней грани́цы о́блака cloudbase height, cloud ceiling
~ о́блака cloud height
~ о́блачности cloud height
~ орби́ты orbital altitude
~, относи́тельная relative height
~, поле́зная clear height (stores)
~ прице́ла "height of the rear sight" (SA) (difference between the "height of the foresight" and the distance from centre of bore to the notch in the rear sight)
~ разры́ва height of burst
~ сече́ния contour interval, vertical interval
~ сече́ния релье́фа vertical interval

высота́, скома́ндованная ordered height, determined height, laid down height
~ сраба́тывания взрыва́теля fuse arming altitude
~ съёмки height of exposure (air photo)
~, топографи́ческая spot height
~ то́чки spot height
~ траекто́рии greatest height of trajectory (gunn)
~, углова́я angular altitude, elevation angle (met obsn)
~ це́ли target height
~ шта́беля, безопа́сная safe stacking height
высо́тность height capability, height range (AD)
высо́тный high-altitude, high-level
высотобо́йность effective ceiling (AD wpn)
высотоме́р 1. altimeter
2. height finder (optical instr)
3. height gauge (eng instr)
~, зени́тный air defence height finder
~, короткоба́зовый short base height finder
~, ла́зерный laser altimeter
~, опти́ческий optical height finder
~, радиолокацио́нный 1. height finding radar
2. radar altimeter, radio altimeter
высотоме́рщик height finder operator
вы́ставить (см выставля́ть)
вы́ставка вертика́ли vertical indication (inertial navigation)
~ по звёздам star tracking (inertial navigation)
выставле́ние posting (of sentries, guards)
выставля́ть/вы́ставить to post (sentries, guards); **выставля́ть/вы́ставить а́зимут** to set in the azimuth
вы́стилка paving, flooring, corduroy road (log road laid across swampland)
~, фаши́нная fascine corduroy (rd constr)
~, хворостяна́я brushwood corduroy
выстра́иваться/вы́строиться to fall in, to form up, to draw up
вы́стрел 1. shot, round
2. report of a shot
3. backfire (ICE)
~, артиллери́йский artillery round, round of artillery ammunition
~, беспла́менный flashless discharge
~, боево́й service round (SA)
вы́стрел, винто́вочный rifle shot
~ для прове́рки ли́нии прице́ливания zeroing shot
~, затяжно́й hangfire (gunn)
"вы́стрел и забы́л" "fire and forget" (A tk GW)
вы́стрел карту́зного заряжа́ния round of separate ammunition, BL or bag-charged ammunition
~, контро́льный sighting shot, "sighter"
~, миномётный mortar bomb, mortar round
~, навесно́й high shot, curved-trajectory shot
~ навски́дку snap shot
~, неприце́льный random shot, pot shot
~, одино́чный single round, single shot
~ о́пытного назначе́ния experimental round
~ патро́нного заряжа́ния round of QF fixed ammunition
~, по́лностью скомплекто́ванный complete round
~ полуунита́рного заряжа́ния semi-fixed round
~, практи́ческий practice round
~, предупреди́тельный warning shot
~, преждевре́менный "premature" (gunn)
~, пристре́лочный sighting shot, ranging round, "sighter"
~, прице́льный aimed shot
~, про́бный proof shot
~, прогрева́ющий conditioning shot, warming shot
~, прогре́вный "warmer"
~, прямо́й direct fire shot, one round of direct fire
~, раздельноги́льзовый QF separate loading round
~ разде́льного гильзово́го заряжа́ния QF separate loading round
~, рикоше́тный ricochet
~ с по́лным заря́дом full-charge round
~, случа́йный accidental discharge, accidental shot
~, трениро́вочный practice shot
~ унита́рного заряжа́ния QF fixed round
~, уче́бный drill round
~, холосто́й blank round
~, шта́тный standard round
вы́стрелить (см стреля́ть)

вы́стрелы, боевы́е service ammunition
вы́строиться (см выстра́иваться)
вы́ступ 1. projection, protuberance
2. bulge, salient (tac)
3. lug, cam
~, боево́й bolt camming lug (SA)
~ вкла́дыша затво́ра breech-lock cam
~ для дна ги́льзы, направля́ющий cartridge stop (MG)
~ для пу́ли, направля́ющий bullet stop (MG)
~, запира́ющий locking lug (pistol)
~ затво́ра bolt arm
~ затво́ра, боево́й locking lug
~ магази́на, опо́рный retaining lug of magazine
~, направля́ющий guide rib (MG)
~, ограничи́тельный 1. pip (on clip of MG metal belt)
2. stop lug
~, опо́рный stop lug
~ по́ршня затво́ра breechblock lug
~, предохрани́тельный safety lug
~ самовзво́да hammer cocking lug (SA)
~ уда́рника, боево́й firing-pin plug
~, упо́рный retaining lug
~ шепта́ла, боево́й sear nose (SA)
вы́ступы, нарезны́е се́кторные barrel-locking thread
высыха́ние drying, evaporation
выта́скивание extrication
вытесне́ние displacement, dislodgement, substitution
вытесни́тель displacer
выто́птанность trampled down area
вы́точка recess, groove, hollow chamfer
~ боево́й личи́нки затво́ра bolt recess
~ для спусково́й защёлки sear recess
~ запа́ла, кольцева́я neck of striker (gren)
вы́точки, кольцевы́е cradle clamping grooves (mor barrel)
вытя́гивание коло́нны time a column (tps etc) takes to pass a given point
вытя́гивать/вы́тянуть to stretch, to elongate, to extricate (veh recovery)
вы́тяжка загото́вки ги́льзы case drawing
~, маневро́вая shunting line (rly)
вы́тянуть (см вытя́гивать)
вы́учка training (condition, standard)
~, полева́я field training
~, такти́ческая tactical training
вы́хлоп exhaust (ICE)
~, ды́мный smoky exhaust
выхлопно́й exhaust
вы́ход 1. breakout (tac)
2. discharge, exhaust
3. yield (nucl)
4. outcome
~, авари́йный emergency exit
~ бо́йка уда́рника protrusion of firing pin
~ в по́ле field exercise
~ в по́ле, кома́ндно-штабно́й command staff field exercise without troops
~, запа́сный emergency exit
~ из бо́я disengagement
~ из окруже́ния break out of encirclement
~ из стро́я break down
~ на цель in-flight
~, пожа́рный fire escape
~, полево́й field exercise
~ проду́ктов деле́ния fission yield
~ раке́ты в зо́ну излуче́ния радиолокацио́нной ста́нции missile gathering (radar controlled GW)
выходи́ть/вы́йти to go out, to emerge, to leave, to be published, to be issued, to run out, to turn out, to overlook; **выходи́ть/вы́йти из бо́я** to disengage, to break off action; **выходи́ть/вы́йти из соприкоснове́ния** to break contact; **выходи́ть/вы́йти из стро́я** to break ranks, to go out of action, to break down; **выходи́ть/вы́йти с бе́рега** to exit from beach (amph ops); **выходи́ть/вы́йти с бе́рега плацда́рма** to break out of a bridgehead, to break out of a beachhead
вы́чертить (см выче́рчивать)
выче́рчивать/вы́чертить to draw, to trace, to plot; **выче́рчивать/вы́чертить конве́рт** to draw an envelope (S)ov tank gunn exercise without firing)
вычисле́ние calculation, computation
вычисли́тель 1. computer, computing device
2. wind graph operator (sound ranging)
3. computer operator
~, анало́говый analogue computer
~, анало́го-цифрово́й hybrid computer (analog-digital)
~ для определе́ния вре́мени взры́ва fuse-time computer

вычисли́тель модули́рующего ти́па analogue computer
~ **отклоне́ний** deviation computer, deflection computer
~ **раке́ты, бортово́й** missile computer (on board)
~ **систе́мы наведе́ния** guidance computer
~ **систе́мы слеже́ния** tracking computer
~ **систе́мы сопровожде́ния** tracking computer
~ **систе́мы управле́ния огнём** fire control computer
~, **цифрово́й** digital computer
вы́числить (см вычисля́ть)
вычисля́ть/вы́числить to calculate, to compute; **вычисля́ть/вы́числить а́зимут** to derive (calculate) an azimuth; **вычисля́ть/вы́числить да́нные** to compute data, to calculate data
выше́дший из стро́я out of action, unserviceable
вышестоя́щий superior, higher
вышибно́й propelling (gunn)
вы́шка tower
~, **автомоби́льная** tower wagon
~, **артиллери́йская** artillery observation tower
~, **водонапо́рная** elevated water storage tank, water tower
~, **деревя́нная** observation tower of wooden scaffolds (surv)
~, **кома́ндная** control tower
~, **наблюда́тельная** watch tower, observation tower
~, **парашю́тная** parachute training jump tower
~, **пускова́я** missile launching tower
~, **ста́ртовая** missile launching tower
~, **сторожева́я** watch tower
~ **управле́ния** control tower (range)
вы́явить (см выявля́ть)
выявле́ние discovery (of target), indication (of target)
выявля́ть/вы́явить to make clear, to expose, to show up; **выявля́ть/вы́явить цель** to discover a target, to indicate a target
вьюга snow storm
вьюк pack
~, **ко́нский** pack (animal)
~, **перено́сный** carrying harness (mor)
~, **спинно́й** man-pack
вьюк, хозя́йственный domestic load (pack)
вьюковожа́тый animal transport driver, horse transport driver (pack)
вью́чка loading (pack animals)
вью́чный pack
вью́шка reel
вяз elm
вя́зка binding, joining, tying
вя́зкий viscous
вя́зкость 1. viscosity (fluids) 2. toughness (solids)
~, **уда́рная** impact resistance (metallurgy)
вязкотеку́чий viscous-flow, plastic

Г

габари́т clearance, external dimension, limiting dimension, overall dimension, loading gauge (rly)
~ **нагру́зки** clearance limit, loading guage
габари́тный overall, clearance
габари́ты dimensions
га́вань harbour
газ gas
~, **балло́нный** bottled gas
~, **боло́тный** marsh gas
~, **веселя́щий** laughing gas
~, **выхлопно́й** exhaust gas
~, **генера́торный** producer gas
~, **горчи́чный** mustard gas
~, **горю́чий** fuel gas, combustion gas, gas from combustion, gaseous fuel
~, **идеа́льный** perfect gas
~, **ине́ртный** inert gas
~, **кова́рный** insidious gas
~, **наполня́ющий** filling gas
~, **отрабо́танный** waste gas, used gas
~, **отравля́ющий** war gas, poisonous gas, toxic gas
~, **по́лный** full throttle, full power
~, **порохово́й** powder gas, powder fumes
~, **рво́тный** vomiting gas
~, **реа́льный** imperfect gas
~, **рудни́чный** firedamp
~, **сгоре́вший** burnt gas, exhaust gas
~, **сжи́женный** liquefied gas

газ, смертоно́сный lethal gas
~, уга́рный carbon monoxide
~, углеки́слый carbon-dioxide
~, уду́шливый choking gas, suffocating gas
~, чиха́тельный nose gas, sternutator (CW)
~, ядови́тый toxic gas, poisonous gas
газго́льдер gas holder, gas container, gas tank
газе́та, светова́я newsreel
га́зик "Gazik" (Sov equivalent of jeep)
газифика́тор gasificator, gasifier
газифика́ция gasification, carburation, production of gas
газоанализа́тор gas analyser, gas tester gas analysis apparatus
газобалло́н gas cylinder, gas bottle
га́зовый gas
газогенера́тор 1. gas producer 2. gas generator (GW)
газо́йль gas oil, diesel fuel
газомёт gas projector
газонепроница́емый gas-proof, gas-proofed
газообра́зный gaseous
газообразова́ние gas generation, gas production
газооку́ривание gas chamber training
газоопредели́тель gas detector, gas detector kit
газоотво́д gas escape pipe
газоотво́дный gas escape, gas outlet
газоотража́тель jet deflector, flame deflector, blast deflector
газопрово́д gas pipe, gas main
газопроница́емость gas pepmeability, gas penetrability
газопроница́емый gas permeable
газопу́ск release of gas
газоразря́дный gas discharge (elec)
газораспределе́ние valve gear (ICE)
~, золотнико́вое sleeve valve gear (ICE)
~, кла́панное valve gear (ICE)
газосва́рка gas welding, oxyacetyline welding
газосто́йкий gas-proof
газотро́н hot-cathode gas-filled diode (elec)
газоубе́жище gas-proof shelter
газохрани́лище gas holder, gas storage tank

"Га́зы!" "Gas!" (verbal alarm for gas and/ or radioactivity)
га́зы, порохові́е propellant gases, powder gases
~, сгоре́вшие burnt gases, exhaust gases
~, сжа́тые compressed gases
га́йка nut
~, бара́шковая butterfly nut, finger nut, wing nut
~, быстросмыка́ющаяся instantaneous coupling (on hose)
~, винтова́я screw coupling (on hose)
~, вы́верочная adjusting nut, regulating nut
~, герматизи́рующая packing nut
~, дли́нная sleeve nut
~, ду́льная sealing collar (gun barrel)
~, зажи́мная lock-nut, locking nut
~, замко́вая lock-nut, locking nut
~, кле́ммовая split nut
~, колпачко́вая dome nut
~, кольцева́я circular nut
~, контря́щая lock-nut, locking nut
~, коро́нчатая castellated nut
~, крыла́тая wing nut, butterfly nut
~, крыльча́тая wing nut, butterfly nut, finger nut
~, му́фтовая sleeve nut
~, нажи́мная 1. cap (fuse) 2. lock nut
~ нака́тника, суха́рная recuperator collar (ord)
~, опо́рная lock nut
~, передвижна́я sliding nut
~, получи́стая semi-bright nut, semi-finished nut
~, потайна́я countersunk nut
~, регулиро́вочная adjusting nut
~, рифлёная knurled nut, milled nut
~ рукава́ hose coupling
~ с нака́ткой knurled nut, milled nut
~ с отве́рстиями capstan nut
~ с у́шком wing nut
~, са́льника gland nut
~, самоконтря́щаяся self-locking nut
~, соедини́тельная 1. union nut 2. coupling (hose)
~, сто́порная lock-nut, locking nut
~, стяжна́я 1. tightening nut, turnbuckle 2. spray valve nut (diesel ICE)
~, упо́рная back nut, stop nut
~, устано́вочная adjusting nut
~, цементи́рованная case-hardened nut

га́йка, шестигра́нная hexagon nut
~, шестигра́нная прорезна́я slotted hexagon nut
~, юстиро́вочная adjusting nut
га́йка-бара́шек butterfly nut, wing nut, finger nut
гайковёрт (pneumatic) nut wrench
~, ручно́й portable wrench
гак hook
~, букси́рный tow attachment
галере́я 1. adit, shelter adit (min) 2. culvert (dry dock)
~, ми́нная 1. gallery (min) 2. underground mining tunnel 3. sap
~, соедини́тельная connecting gallery (min)
гале́ты plain biscuits, hard tack
галифе́ riding breeches
галло́н, америка́нский US gallon (3.78 litres)
~, англи́йский imperial gallon (4.546 litres)
гало́ halo (met)
галопи́рование pitching
гальвано́метр galvanometer
~, баллисти́ческий ballistic galvanometer
гальваностеги́ческий electroplating
га́лька pebble, shingle
га́мма-акти́вность gamma-activity
га́мма-излуче́ние gamma radiation
~, жёсткое hard gamma radiation
~, захва́тное induced gamma radiation
~, мгнове́нное prompt gamma radiation, gamma flash, instantaneous gamma radiation
~, мя́гкое soft gamma radiation
~, оско́лочное delayed gamma radiation
га́мма-квант gamma quantum
га́мма-лучи́ gamma rays
га́мма-фон background gamma radiation
га́ндшпуг handspike
га́ншпуг handspike
гара́ж garage
гара́нтия warranty, guarantee
гармо́ника harmonic
гарнизо́н garrison
~, отдалённый outstation, outpost, outlying garrison
гарнизо́нный garrison
гарниту́р, телефо́нный operator's telephone head set
гаси́тель 1. extinguisher 2. damper
~ колеба́ний shock absorber, damper
гаси́ть/погаси́ть to put out, to extinguish; **гаси́ть/погаси́ть парашю́т** to expel air from parachute on landing
гастроентери́т gastroenteritis
гать corduroy road (log road laid across swampland)
~, хворостяна́я brushwood road
га́убица howitzer
~, букси́руемая towed howitzer
~, го́рная mountain howitzer
~, го́рно-вью́чная pack howitzer, pack mountain howitzer
~ на гусени́чном ходу́ tracked howitzer
~, оса́дная siege howitzer
~, полева́я field howitzer
~, самодви́жущаяся howitzer with auxiliary propulsion
~, самохо́дная self-propelled howitzer
га́убица-пу́шка gun-howitzer
га́убичный howitzer
гауптва́хта guardroom (as place of detention)
~, войскова́я unit guardroom (place of detention)
~, гарнизо́нная garrison guard room, town guardroom
Га́усса распределе́ние Gaussian distribution
га́фний hafnium
гаше́ние 1. blanking 2. slaking (lime)
гаше́тка trigger, trigger button, push button trigger, key trigger
гварде́ец 1. member of Guards unit (Sov) 2. guardsman
гварде́йский Guards
гва́рдия guards
гвоздодёр claw, nail extractor
гексами́н hexamine
гексахлорме́ламин hexachloromelamine
гексахлорэта́н hexachloroethane
гексоге́н hexogen (expl)
ге́лий helium
гелио́граф heliograph
генера́л general
~ а́рмии general (army commander)
~, бriга́дный brigadier general (Brit)
генера́л-адъюта́нт aide-de-camp general

Генерали́ссимус Сове́тского Сою́за Generalissimo of the Soviet Union (hitherto held only by Stalin)
генералите́т general officers, "brass hats"
генера́л-лейтена́нт lieutenant-general
генера́л-майо́р major-general
генера́л-полко́вник colonel-general
генера́льский general's
генера́тор 1. generator 2. oscillator
~, **авари́йный** emergency generator
~, **автомодули́руемый** self-pulsed oscillator
~, **автотрансформа́торный** tapped-coil oscillator, Hartley oscillator (electron)
~, **ацетиле́новый** acetylene generator
~ **бие́ний** beat frequency oscillator
~ **временно́й развёртки** time base generator
~, **вспомога́тельный** auxiliary generator
~ **высо́кой частоты́** high frequency oscillator
~, **гетероди́нный** beat frequency oscillator
~, **двухко́нтурный** tuned-anode tuned-grid oscillator
~, **двухта́ктный** push-pull oscillator
~, **ёмкостный** tapped capacitor oscillator, split-capacitance oscillator, Colpitts oscillator
~, **задаю́щий** master oscillator
~, **звуково́й** audio-frequency oscillator
~ **и́мпульсных поме́х** discontinuous jammer, pulse jammer
~ **и́мпульсов** pulse generator
~, **интерполяцио́нный** interpolation oscillator (elec)
~ **инфракра́сного диапазо́на, ква́нтовый** infra-red maser, IR maser
~, **ква́нтовый опти́ческий** laser
~, **ква́рцевый** quartz oscillator, crystal-controlled oscillator
~, **клистро́нный** klystron oscillator
~ **колеба́ний** oscillator
~ **колеба́ний ни́зкой частоты́** audio frequency oscillator
~, **компа́ундный** compound wound generator
~, **ла́зерный** laser generator
~, **ла́мповый** valve oscillator
~, **магнетро́нный** magnetron oscillator
~, **микромо́дульный** micromodule oscillator
генера́тор, многомо́довый multimode generator
~ **на боково́й полосе́** side-band generator
~ **на полупроводнико́вых прибо́рах** transistor oscillator
~, **настра́иваемый** tuneable oscillator
~, **незатуха́ющих колеба́ний** continuous-wave oscillator
~ **непреры́вных колеба́ний** continuous wave oscillator, CW oscillator
~ **огнетуши́тельной пе́ны** foam generator
~, **опти́ческий ква́нтовый** laser
~ **переме́нного то́ка** alternator, alternating-current generator
~ **переме́нной частоты́** cycle generator
~ **пилообра́зных сигна́лов** sawtooth generator
~ **постоя́нного то́ка** direct-current generator, continuous-current generator
~, **преры́вно-де́йствующий** chopping oscillator, chopper oscillator
~ **развёртки** sweep generator (CRT)
~ **развёртки да́льности** range-sweep generator
~ **разрывны́х колеба́ний** relaxation oscillator
~, **резе́рвный** reserve generator
~, **релаксацио́нный** relaxation oscillator
~ **с отсе́чкой** squegging oscillator, squegger oscillator (electron)
~ **с самовозбужде́нием** self-excited generator
~ **с электро́нной пересто́йкой** voltage tuning oscillator
~, **самовозбужда́емый** autodyne
~, **самопрерыва́ющийся** squegging oscillator, squegger oscillator (electron)
~, **се́риесный** series-wound generator
~ **сигна́лов** signal generator
~ **синусоида́льных колеба́ний** sine-wave generator
~, **спа́ренный** coupled oscillator
~, **стабилизи́рованный ква́рцем** crystal-stabilized oscillator
~, **транзи́сторный** transistor generator
~, **трёхто́чечный** Hartley, Colpitts or Meissner oscillator
~ **фу́нкции** function generator
~ **шу́ма** noise generator
~, **шунтово́й** shunt generator, shunt wound generator

генера́тор, электро́нный valve oscillator
генера́ция generation
генери́рование generating, generation
ге́нри henry (elec eng, unit of measurement of inductance)
геншта́б general staff
генштаби́ст General Staff Academy graduate (Sov)
геодези́ческий geodetic (surv)
геоде́зия geodesy, geodetic surveying
геомагни́тный geomagnetic
геоме́трия geometry
гепати́т hepatitis
гепта́н heptane
гепто́д heptode, pentagrid (elec)
гербици́д herbicide, weed killer
герма́ний germanium
герметиза́тор sealer
герметиза́ция 1. sealing
2. waterproofing
3. gas proofing
герметизи́рованный pressurized, sealed
гермети́ческий hermetic, hermetical, air-tight, hermetically-sealed
гермети́чный air-tight
Геро́й Сове́тского Сою́за Hero of the Soviet Union (highest award for bravery in the USSR)
герц hertz, cycles per second
гетероге́нный heterogeneous
гетероди́н local oscillator, heterodyne
~, калибрацио́нный calibrating oscillator
~, ме́стный local oscillator, heterodyne unit
~ телегра́фного приёма beat-frequency oscillator
гетеросфе́ра heterosphere
ге́тры gaiters
ги́бель destruction, loss
ги́бка bending
ги́бкий flexible, elastic, pliable, supple, yielding
ги́бкость flexibility, pliability
~ огня́ flexibility of fire
гигие́на hygiene
~, ли́чная personal hygiene
гигиени́ческий hygienic, sanitary
гигро́граф hygrograph
гигро́метр hygrometer
гигроскопи́ческий hygroscopic, absorbent
гидра́влика hydraulics
гидравли́ческий hydraulic
гидрази́н hydrazine
гидразингидра́т hydrazine hydrate
гидрази́новый hydrazine, diamine
гидразо́н hydrazone
гидра́нт hydrant
гидра́т hydrate
~ о́киси hydroxide
гидри́д ли́тия lithium hydride
гидроавиа́ция seaplanes
гидроаку́стика echo sounding, sonar
гидроаэродро́м seaplane base
гидробрю́ки waders
гидро́граф 1. hydrograph
2. hydrographer
гидрогра́фия hydrography
гидродина́мика hydrodynamics
гидроизоляцио́нный waterproof
гидроизоля́ция waterproofing, hydraulic sealing
гидрокомбинезо́н diving dress, diving suit (shallow water)
гидрокостю́м diving suit
гидро́лиз hydrolysis
гидроло́гия hydrology
гидролока́тор asdic set, sonar set
гидромеха́ника hydromechanics
гидромехани́ческий hydromechanical
гидромото́р hydraulic motor
гидрому́фта 1. hydraulic coupling
2. fluid flywheel
3. fluid clutch
гидронасо́с hydraulic pump
гидроо́кись hydroxide
гидропереда́ча hydraulic transmission, hydraulic drive
гидроприво́д hydraulic drive
гидропу́льт stirrup pump
гидропу́льт-ведро́ fixed-bucket type stirrup pump
гидропу́льт-косты́ль stirrup pump
гидроразъём hydraulic coupling (GW)
~, автомати́ческий self-sealing hydraulic coupling (GW)
гидросамолёт seaplane
гидросамолёт-разве́дчик reconnaissance seaplane
гидроспидо́метр water flow speed indicator, log, current meter (nav)
гидроста́тика hydrostatics
гидрото́рмоз 1. hydraulic brake
2. water brake, water dynamometer
гидротрансформа́тор fluid converter, hydraulic torque converter

гидротренажёр under-water trainer: used to train crews in tank-snorkel operations
гидроусилúтель hydraulic actuator, servo control unit, power control jack, booster
гидрофóбия hydrophobia
гидрофóн hydrophone
гидроэлектрúческий hydroelectric
гúльберт gilbert (elec eng, unit of measurement)
гúльза 1. cartridge case, shell case 2. liner, barrel sleeve
~ **без закрáины** rimless cartridge case
~, **бесшля́почная** rimless cartridge case
~, **бутылочная** bottle-necked cartridge case
~, **вставнáя** cylinder liner, insertion liner
~, **заклинúвшаяся** jammed case (ammo)
~ **запáла** detonator tube
~ **запáльной трýбки** primer tube
~, **заря́дная** cartridge case (ammo)
~, **картóнная** cardboard cartridge case
~, **латýнная** brass cartridge case (ammo)
~ **лю́льки** cradle liner (ord)
~, **мóкрая** wet liner (cylinder)
~ **насóса-форсýнки** plunger bushing (diesel ICE)
~, **обтюрúрующая** self-sealing cartridge case
~, **отожённая** annealed case
~, **патрóнная** cartridge case
~ **патрóнного заряжáния** QF fixed cartridge case
~, **перфорúрованная** perforated cartridge case
~, **пластмáссовая** plastic case
~ **раздéльного заряжáния** QF separate loading cartridge case
~, **разорвáвшаяся** ruptured cartridge case
~ **с закрáиной** rimmed cartridge case
~ **с тугúм обжáтием дýльца** tightly crimped cartridge case
~ **с умéншенной закрáиной** semi-rimless case
~, **сгорáющая** combustible cartridge case
~ **снаря́да** shell case
~, **снаря́дная** shell case
~ **со шля́пкой** rimmed cartridge case
~, **стальнáя** steel cartridge case
гúльза, стрéляная empty case, fired case (ammo)
~, **сухáя** dry liner (cylinder)
~ **тóпливного насóса** barrel of fuel pump (diesel ICE)
~ **тормознóго барабáна** brake drum liner
~, **учéбная** drill cartridge case
~ **цилúндра** cylinder liner, liner sleeve (eng)
~ **цилúндра, мóкрая** wet cylinder liner (eng)
~, **частúчно сгорáющая** semi-combustible cartridge case
гильзоотвóд cartridge case deflector
гильзоотражáтель cartridge case ejector
гильзоулáвливатель spent cartridge-case bag
гильзоуловúтель spent cartridge-case bag
гимнастёрка "gimnasterka" (Soviet service pattern blouse)
гипербаллúстика hyperballistics
гипéрбола hyperbole
гиперболúческий hyperbolic
гиперзвуковóй hypersonic
гипосульфúт hyposulphate
гипохлорúт кáльция calcium hypochlorite
гипсомéтрия hypsometry (carto)
гировертикáнт vertical gyroscope (aeron)
гирогоризóнт 1. horizontal gyroscope (aeron) 2. artificial horizon
гироинтегрáтор gyroscopic integrator, integrated gyroscope
гирокóмпас gyro compass
~, **артиллерúйский** artillery gyrocompass
гирокомпасúрование gyrocompassing (inertial navigation)
гирокурсоуказáтель gyroscopic course indicator
гиромотóр gyro motor, gyro torque motor
гироплáн gyroplane
гироплатфóрма gyroscope-stabilized platform, inertially stabilized platform
гирополукóмпас azimuth gyro, directional gyroscope
гироприбóр gyro, gyroscopic device, gyroscope
гироскóп gyroscope
~ **крéна** roll gyroscope

гироско́п, курсово́й directional gyroscope
~, позицио́нный displacement gyroscope
~ рыска́ния yaw gyroscope
~, свобо́дный attitude gyroscope
~, скоростно́й rate gyroscope
~ танга́жа pitch gyroscope
гироскопи́ческий gyroscopic
гиростабилизи́рованный gyro-stabilized
гироу́зел gyro unit
гистерези́с hysteresis (elec)
гита́ра gear train (tracked vehicle)
главнокома́ндование, верхо́вное (ВГК) supreme high command (Sov)
главнокома́ндующий commander-in-chief
~, верхо́вный supreme commander
~ сухопу́тными войска́ми вооружённых сил Commander-in-Chief Land Forces
главпуро́вец Member of Chief Political Directorate (Sov)
гла́дкий smooth, flush
гладкостbо́льный smoothbore
глаз, невооружённый naked eye
глазо́к периско́па periscope head
глазоме́р judgement by eye (range etc)
глазоме́рный by eye
гла́сис 1. glacis
2. forward slope of parapet
гле́тчер glacier
глико́ль glycol
гли́на clay
~, сла́нцевая slaty clay
гли́нистый clay, clayey
глисса́да 1. glide path (ac)
2. landing beam (radar)
глицери́н glycerine, glycerol
гло́хнуть/загло́хнуть to stall (engine), to go out (fire)
гло́хнуть/огло́хнуть to become deaf
глубина́ depth
~ боево́го поря́дка depth of deployment
~ боевы́х зада́ч depth of objectives
~ бро́да fording depth
~ бро́да, допуска́емая fordable depth
~ досыла́ния снаря́да seating distance (in breech) (ord)
~ коло́нны depth of column
~ модуля́ции modulation depth
~ наре́за depth of rifling
~ оборо́ны depth of defence
~ оборо́ны, операти́вная operational depth of defence (Sov)
глубина́ подво́дного вожде́ния deep wading depth
~, преде́льная limiting depth, maximum depth
~ проника́ния depth of penetration
~ проникнове́ния depth of penetration
~ реза́ния cutting depth
~ стро́я depth of body of troops
глубиноме́р depth gauge
глубоме́р, гидрофони́ческий sonic depth finder
глубь оборо́ны depth of defence
глуше́ние damping
"Глуши́ Мото́р!" "Stop Engine!" "Switch Off!" (MT)
глуши́тель 1. silencer, baffle, muffler
2. jammer, interference generator
~ вы́хлопа exhaust silencer
~ радиолока́тора radar jammer
~ радиолока́тора, модули́рованный шу́мами noise-modulated radar jammer
~ свя́зи communication jammer
~ шу́ма noise suppressor, silencer, muffler
глы́ба, ледяна́я ice block
гнездо́ 1. nest, socket, recess
2. jack (teleph)
3. housing, casing
~ бараба́на револьве́ра revolver chamber
~, бетони́рованное concrete emplacement
~, боково́е side fuse-well, side igniter-well (ammo, mine)
~ в казённике breech ring bush
~ в прикла́де butt trap
~ голо́вки трено́ги tripod socket
~ для детона́тора detonator socket, detonator housing
~ для запа́ла igniter set recess
~ для затво́ра breech bush
~ для затво́рного кли́на breech recess
~ для ка́псюля cap chamber
~ для ка́псюля-детона́тора detonator socket
~ для уда́рного механи́зма firing mechanism housing
~ для элеме́нта неизвлека́емости anti-lift device well (min)
~, до́нное base fuse well, base igniter well (ammo)
~, замо́чное breech bush

гнездо́, запа́льное igniter set recess, primer hole
~ **казённика, поршнево́е** breech-block housing, breech opening, breech recess
~ **кла́пана** valve seat, valve seating
~, **кла́панное** valve seat, valve seating
~, **клиново́е** breech mortice (ord)
~, **ключево́е** tommy hole (of fuse)
~ **крепле́ния по-похо́дному, ца́пфенное** gun crutch (tk, arty)
~, **поршнево́е** breech opening
~ **прикла́да** butt trap
~, **пулемётное** machine-gun nest, machine-gun pit
~ **ра́неных** casualty collecting point
~, **ца́пфенное** trunnion bed, trunnion seat, trunnion housing
~ **ца́пфы** trunnion housing
~, **штати́вное** tripod socket (eg on base of camera)
гнето́к plunger
~ **выбра́сывателя** extractor plunger
~ **пружи́ны** spring compressor, spring plunger
гной pus (med)
гну́щийся flexible
год, бюдже́тный fiscal year
~ **вы́пуска** 1. year of manufacture 2. year of completion (school etc)
~, **отчётный** fiscal year
~ **снаряже́ния** year of filling (ammo)
~, **фина́нсовый** financial year
го́дность fitness
го́дный fit
~ **к вое́нной слу́жбе** fit for military service
голова́ head (anat)
голо́вка 1. head 2. spur (pistol)
~, **акти́вная радиолокацио́нная** active radar homing head (GW)
~ **бло́ка цили́ндров** cylinder head
~, **боева́я** warhead
~ **бо́лта** bolt head, screw head
~, **бронебо́йная** armour-piercing cap
~, **ве́рхняя** small end (ICE)
~ **взрыва́теля** fuse head, nose cap
~ **ви́нта** screw head
~, **воспроизводя́щая** playback head (tape recorder)
~ **впота́й** countersunk head
~ **дви́гателя** 1. cylinder head 2. combustion chamber head (GW)
голо́вка, дете́кторная detector head
~, **дымова́я** smoke head (ammo)
~, **закладна́я** head (rivet)
~, **запи́сывающая** recording head (tape recorder)
~ **звукоснима́теля** sound head, pick-up
~ **извлека́теля** clearing plug head (SA)
~ **инфракра́сного прице́ла** infra-red sight head, IR sight head
~ **ка́меры** combustion chamber head
~ **ка́меры сгора́ния дви́гателя** top of engine combustion chamber (ICE)
~, **кассе́тная боева́я** warhead containing bomblets, cluster bomb warhead
~ **кла́пана** valve head
~ **ключа́** knob of morse key
~, **кривоши́пная** big end (ICE)
~, **ла́зерная скани́рующая** laser scanning head
~, **ла́зерная следя́щая** laser tracking head
~, **магни́тная** magnetic head
~, **нажи́мная** pressure plunger (min)
~, **ни́жняя** big end (ICE)
~, **обы́чная боева́я** conventional warhead
~, **окуля́рная** eyepiece end (periscope)
~, **опти́ческая** optical mounting
~, **пиротехни́ческая** pyrotechnic head
~, **поворо́тная** 1. dial-sight sighting head (arty) 2. swivel head (GW), rotating head
~, **по́исковая** search head (mine detector)
~, **поршнева́я** connecting-rod small end, piston head
~, **потайна́я** countersunk head
~, **приви́нтная** separate head (ammo)
~, **распыли́тельная** 1. burner cup (GW) 2. injector (fuel system)
~, **расто́чная** boring head
~, **резцо́вая** cutter head
~ **ре́льса** rail head
~ **с обы́чным соударе́нием струй** conventional impinging head (GW)
~ **самонаведе́ния** homing head, target seeker
~ **самонаведе́ния, акти́вная** active homing head (GW)
~ **самонаведе́ния, инфракра́сная** infra-red homing head, infra-red seeker (GW)

голо́вка самонаведе́ния, пасси́вная passive homing head, passive seeker
~ самонаведе́ния, полуакти́вная semi-active homing head, semi-active seeker
~ самонаведе́ния, радиолокацио́нно-опти́ческая electro-optical homing head (GW)
~ самонаведе́ния, теплова́я infra-red homing head, thermal radiation homing head (GW)
~, самонаводя́щаяся seeker head (GW)
~, сва́рочная welding blowpipe
~ слеже́ния tracking head
~ снаря́да missile nose
~, счи́тывающая detecting head
~, телевизио́нная television "eye"
~ уда́рника striker head
~, форсу́ночная burner head, injector head (liquid propellant rocket)
~, фуга́сная high explosive warhead
~, хими́ческая chemical warfare warhead
~ цили́ндра cylinder head
~ шатуна́ connecting-rod end
~ шатуна́, ве́рхняя connecting-rod small end
~ шатуна́, ни́жняя connecting-rod big end
~, я́дерная боева́я nuclear warhead
головно́й leading, head
"Головно́й Убо́р Наде́ть!" "Replace Head-dress!"
"Головно́й Убо́р Снять!" "Remove Head-dress!"
головокруже́ние dizziness, vertigo
гололёд ice (on road)
гололе́д ice deposits
гололе́дица ice (on road)
го́лубь, почто́вый carrier pigeon
голубя́тня, вью́чная pack pigeon-cote
гомоге́нный homogeneous
гондо́ла 1. nacelle
2. open high-side wagon (rly)
гонио́метр goniometer
~, ми́нный mining goniometer
го́нка вооруже́ний arms race
гонорре́я gonorrhoea
гопкали́т hopcalite
гора́, ни́зкая hill up to 800 metres
горе́лка burner
~, однопла́менная single-flame orifice, single-flame burner
горе́лка, сва́рочная welding torch, welding blowpipe
горе́ние combustion, burning
~, дегресси́вное degressive burning (ammo)
горизо́нт horizon, ground line
~, ви́димый visible horizon, apparent horizon, terrestrial horizon
~ высо́ких вод high-water line
~ меже́нных вод low-water level (river)
~ ни́зких вод low-water line
~ ору́жия horizontal, horizontal plane (arty)
~ па́водка high-water level (river)
~ радиолокацио́нной ста́нции radar horizon
~, радиолокацио́нный radar horizon
горизонта́ль contour
~, вспомога́тельная form line (carto), half contour line
~, дополни́тельная form line, half contour line (carto)
~, основна́я principal contour (carto)
~, полови́нная form line, half contour line (carto)
горизонта́льный horizontal
горизонти́рование levelling
~, попере́чное cross levelling
горизонти́ровать to level
го́рка 1. small hill, hillock
2. zoom (aeron)
горлови́на 1. manhole, crater, orifice, mouth, vent, throat, neck
2. breech opening
3. input coupling (gearbox)
~, гильзоотво́дная cartridge-case deflector chute, shell deflector chute (ord)
~ магази́на magazine housing
~, маслозаливна́я oil filler
~ пита́ния feed mouth
~ приёмника belt guide, feed block platform (SA)
~, топливозапра́вочная fuel filler
горн charged chamber (mining)
~, кузне́чный forge, forging furnace
горни́ло сраже́ния heat of battle
горни́ст bugler
го́рный mountain(ous)
го́род, большо́й town with more than 100,000 inhabitants
~, ма́лый town with less than 50,000 inhabitants

го́род, сре́дний town with between 50,000 and 100,000 inhabitants
городо́к, бара́чный hutted camp
~, вое́нный cantonment, military station, permanent military camp
~, инжене́рно-хими́ческий NBC warfare engineer training wing
~, инжене́рный engineer training establishment
~, каза́рменный barrack area, barracks
~, карау́льный guard duties training area
~, ко́мплексный training establishment for several types of training
~, огнево́й artillery training school
~, парашю́тно-деса́нтный parachute ground-training wing
~, парашю́тный parachute ground-training wing
~, уче́бный training wing, training ground
~, штурмово́й assault course, obstacle course
го́ры, средневысо́тные mountains of height 800 metres - 2,000 metres (Sov)
горю́чее fuel
~, авиацио́нное aviation fuel
~, автомоби́льное MT fuel, ground fuel
~, антидетонацио́нное anti-knock fuel
~, взрывоопа́сное explosive fuel
~, водоро́дное hydrogen fuel
~, высококалори́йное exotic fuel, high-energy fuel (GW)
~, высокосо́ртное high-grade fuel
~, высокоэнергети́ческое high-energy fuel
~, газообра́зное gaseous fuel
~, жи́дкое liquid fuel
~ и сма́зочные материа́лы petrol, oil and lubricants, POL
~, кероси́новое kerosene fuel
~, неуглеводоро́дное non-hydrocarbon fuel
~, нефильтро́ванное unfiltered fuel
~, пусково́е starting fuel
~, разливно́е fuel in bulk
~, раке́тное rocket fuel
~, самовоспламеня́ющееся autoigniting fuel, hypergolic fuel, hypergol, spontaneously ignitable propellant
~, синтети́ческое synthetic fuel
~, спиртово́е alcohol fuel
горю́чее, хими́ческое chemical fuel
~, этили́рованное ethylized fuel
~, я́дерное nuclear fuel
горю́честь combustibility, inflammability
горю́чий combustible, inflammable
горя́чий 1. hot
2. hot-tempered
госграни́ца national boundary
госиспыта́ние state trials (of new eqpt)
го́спиталь hospital
~, арме́йский авиацио́нный army air hospital
~, инфекцио́нный полево́й подви́жный mobile field isolation hospital
~, полево́й 1. field hospital
2. field ambulance (Brit med unit)
~, полево́й подви́жный mobile field hospital (roughly equates to main dressing station (Brit))
~, райо́нный резе́рвный regional reserve hospital
~, транзи́тный эвакуацио́нный transit hospital
~, эвакуацио́нный hospital for evacuated personnel
госпо́дство domination
~ в во́здухе air supremacy, mastery in the air
госпо́дствовать to dominate, to command
госпо́дствующий dominating, commanding
госуда́рственный state, national, governmental
"гото́в к труду́ и оборо́не" "fit for labour and defence" (PT standard)
готова́льня set of drawing instruments
гото́вность, боева́я 1. battle readiness, combat readiness, stand-to, readiness for action, alert
2. readiness for war (troops)
3. operational fitness (eqpt)
~ к отраже́нию хими́ческого нападе́ния gas alert
~, мобилизацио́нная mobilization preparedness
~, техни́ческая 1. fitness for work
2. roadworthiness (veh)
"Гото́во!" "Ready!"
гофр corrugation
гофри́рованный corrugated
гра́бли rake
гра́вий gravel

гравита́ция gravitation
градие́нт gradient
~, бари́ческий barometric gradient
~, барометри́ческий barometric gradient
~ давле́ния pressure gradient
~, температу́рный temperature gradient
гради́рня tower-cooler, cooling tower
градуи́рование calibration, graduation
градуи́рованный graduated
градуи́ровать to calibrate, to graduate
градуиро́вка calibration
гра́дус degree
гра́дусник thermometer
"гражда́нка" "Civvy Street", civilian life (sl)
грамм-а́том gram-atom
грамм-кало́рия gramme-calorie
грамм-моле́кула gram-molecule
гра́мота, охра́нная safe conduct
~, почётная certificate of merit
гра́мотный competent
грана́та 1. grenade 2. shell
~, briза́нтная high explosive shell
~, бронебо́йная armour-piercing rocket, armour-piercing shell (arty)
~, винто́вочная rifle grenade
~, винто́вочная дымова́я rifle smoke grenade
~, винто́вочная кумуляти́вная hollow charge anti-tank rifle grenade
~, винто́вочная противота́нковая rifle anti-tank grenade
~, винто́вочная хими́ческая rifle gas grenade
~, дистанцио́нная time-fused HE shell
~ дистанцио́нного де́йствия, оско́лочная time-fused fragmentation shell
~, дымова́я ручна́я smoke hand grenade
~, зажига́тельная incendiary grenade
~, кумуляти́вная 1. hollow-charge shell 2. hollow-charge grenade
~, наступа́тельная offensive grenade
~, неоконча́тельно снаряжённая unfused shell
~, оборони́тельная defensive grenade
~, оконча́тельно снаряжённая fused shell
~, освети́тельная illuminating shell
~, оско́лочная 1. fragmentation grenade 2. fragmentation shell
грана́та, оско́лочная дистанцио́нная fragmentation time-fused shell
~, оско́лочно-фуга́сная fragmentation HE shell
~, оско́лочно-шрапне́льная high-explosive shrapnel shell
~, практи́ческая ручна́я practice hand grenade
~, прилипа́ющая противота́нковая anti-tank "sticky" grenade
~, противопехо́тная anti-personnel grenade
~, противота́нковая 1. anti-tank grenade 2. anti-tank rocket (inf)
~, противота́нковая винто́вочная anti-tank rifle grenade
~, противота́нковая реакти́вная HE anti-tank rocket (of A-tk rocket launcher)
~, противота́нковая ручна́я зажига́тельная incendiary anti-tank hand grenade, "Molotov Cocktail"
~, реакти́вная HEAT rocket (of A-tk rocket launcher)
~, руже́йная rifle grenade
~, ручна́я hand grenade
~, ручна́я дымова́я smoke hand grenade
~, ручна́я оско́лочная fragmentation hand grenade
~, ручна́я противота́нковая anti-tank hand grenade
~ с радиовзрыва́телем shell with radio-proximity fuse
~ с ру́чкой, ручна́я stick grenade
~, тонкосте́нная зажига́тельная incendiary frangible grenade, "Molotov Cocktail"
~ уда́рного де́йствия percussion-fused grenade
~ уда́рного де́йствия, фуга́сная percussion-fused HE grenade
~, уче́бная practice grenade, drill grenade
~, фо́сфорная phosphorous grenade
~, фуга́сная ручна́я fragmentation hand grenade
~, хими́ческая chemical grenade, gas grenade
~, ядовитоды́мная toxic smoke grenade
~, яйцеви́дная ручна́я spherical hand grenade egg-shaped hand grenade
грана́та-болва́нка drill grenade
"Грана́тами Ого́нь!" "Throw!" (gren)

гранатомёт 1. grenade projector, grenade launcher
2. rocket launcher (anti-tk)
~, винтóвочный rifle grenade projector
~, полуавтомати́ческий реакти́вный semi-automatic rocket launcher
~, противотáнковый 1. anti-tank rocket launcher
2. anti-tank grenade launcher
~, противотáнковый реакти́вный anti-tank rocket launcher
~, ружéйный rifle grenade projector, grenade discharger cup
~, ручнóй hand-held anti-tank rocket launcher
~, ручнóй противотáнковый hand-held infantry anti-tank rocket launcher, hand-held anti-tank grenade launcher
~, станкóвый противотáнковый infantry anti-tank rocket launcher (on mounting)
гранатометáние grenade throwing
гранатомётчик 1. No 1 of rocket launcher team
2. grenade thrower (soldier)
грани́ца boundary, bound, limit, frontier, border, perimeter
~ безопáсной зóны safety limit (range firing)
~ зóны действи́тельного огня́ maximum effective range
~ зóны поражéния, ближáйшая inner limit of impact zone (SAM)
~ зóны поражéния, дáльняя outer limit of impact zone (SAM)
~ сéктора обстрéла limit of arc of fire
~, тыловáя rear boundary
~ усилéния gain margin (GW)
грань, боевáя driving edge (rifling)
~ боевóго устýпа locking recess face
~ нарéза edge of land (rifling)
~ пóля нарéзов edge (rifling)
~, холостáя loading edge (rifling of gun barrel)
графá поправóк дáльности range correction column
~ прихóда receipt column (ledger)
грáфик 1. chart
2. time schedule, radio schedule, programme
~, врéменной time-table
~ движéния schedule of operations, traffic timetable
грáфик движéния подвижны́х средств связи signal dispatch service timetable, SDS timetable
~ дежýрства duty roster
~, диспéтчерская progress chart
~ контрóля control chart
~ линéйного масштáба scale line
~ отправлéния документов check list of traffic due to be dispatched (part of schedule of timed operational traffic)
~ очерёдности priority schedule
~ переправы river crossing movement table, detailed plan for assault crossing
~ погрýзки embarkation schedule, loading schedule
~ подáчи заявок schedule of indents
~ подхóдного движéния movement planning chart
~ поправок correction chart
~ поступлéния докумéнтов check list of traffic due to be received (part of schedule of timed operational traffic)
~ прицéлов elevation chart, elevation graph
~, стоя́чий round-the-clock work schedule
~ траектóрий graphical firing chart, trajectory chart, trajectory diagram, ranging chart, ranging diagram
~ хóда инвентаризáции inventory chart
графи́т graphite
графи́ческий graphic(al)
гребёнка 1. rack (rack and pinion)
2. screw-thread chaser
грéбень 1. crest
2. cam, toe, rib
~, артиллери́йский false crest (arty)
~, боевóй false crest (arty)
~ колесá wheel flange
~, направля́ющий guide horn (tk track)
~, огневóй false crest (arty)
~, такти́ческий false crest (arty)
~, топографи́ческий crest line, ridge line (carto)
~ трáка, ведýщий driving lug
~ трáка, направля́ющий track guide
~, укрывáющий intervening crest
гребéц oarsman, paddle number
грéбля rowing
грéйдер 1. grader, road grader
2. earth track, dirt road
~, самохóдный motor (road) grader
грéйдер-элевáтор elevating grader

гре́йфер scoop, grab, grapple
гре́лка heater
~, хими́ческая chemical heater (chem)
гриб, постово́й mushroom-shaped sentry shelter
грибо́к 1. mushroom head (valve)
2. fungus
гри́венник 1. 1Ø kopek piece
2. indicates number 1Ø
гри́вка backsight blade
~ с про́резью battle aperture sight
гри́длик grid leak (elec)
гриф rubber stamp
~ секре́тности degree of classification
гроза́ storm (thunder and lightning)
грозоразря́дник lightning arrester, lightning conductor
громи́ть/разгроми́ть to raid, to sack, to loot, to smash up, to pound (from the air)
громкоговори́тель loudspeaker
~, выносно́й extension loudspeaker
~, динами́ческий moving coil loudspeaker
~, пьезоэлектри́ческий piezoelectric loudspeaker, crystal loudspeaker
~, электродинами́ческий moving-coil loudspeaker
громоотво́д lightning protector, lightning conductor
~, безвозду́шный gas-discharge lightning conductor
гро́хот 1. roar
2. sieve
~, ручно́й hand sieve
грохоти́ть/прогрохоти́ть to screen, to sieve, to sift
грохоче́ние screening, sieving, sifting
груз load, cargo, consignment, freight, shipment
~, абсолю́тно необходи́мый essential cargo
~, возду́шный air delivered stores
~, во́инский military stores (carried by train, ship, veh, ac), military load, military consignment
~, габари́тный load within loading gauge
~, громо́здкий bulk freight, bulk shipment
~, деса́нтный air-delivered stores
~, ёмкий bulky load
~ на вне́шней подве́ске slung load (below a helicopter)
груз нава́лом bulk goods, bulk freight
~ нали́вом bulk liquid load
~, насыпно́й bulk load
~, на́сыпью free flowing load, free flowing cargo
~, негабари́тный oversize load
~, обра́тный return load, return shipment
~, па́лубный deck load, deck cargo
~, поле́зный 1. payload (GW)
2. useful load
~ регуля́тора governor flyweight
~, сме́шанный mixed load
~, составно́й composite load
~, сре́дний average load
~, сро́чный urgent shipment, high priority shipment, priority freight
~, та́бельный laid-down load, standard load
~, типово́й standard load
грузи́к распредели́теля distributor weight
~ регуля́тора governor flyweight
грузи́ться/погрузи́ться to embark, to entrain, to embus
грузови́к lorry, truck
~ для перево́зки та́нков tank transporter
~, заря́дный ammunition truck
грузови́к-платфо́рма flat bed truck
грузови́к-самосва́л tipper truck, dump truck
грузовмести́мость cargo capacity
грузово́й goods, cargo, freight, loading
грузонапряжённость traffic density
грузооборо́т freight turn-over, freight traffic
грузоотправи́тель consignor
грузоподъёмник hoist
грузоподъёмность 1. lifting capacity (crane)
2. cargo capacity, load-carrying capacity, payload
~ кра́на, номина́льная crane rating
~ ледяно́го покро́ва ice-bearing capacity
~ мо́ста bridge capacity
~ паро́ма ferry capacity
~, по́лная deadweight capacity
грузополуча́тель consignee
грузопото́к (freight) traffic flow, (freight) traffic volume

грузопото́ки, эвакуацио́нные flow-back (eqpt)
гру́зы, тяжелове́сные, доставля́емые по во́здуху беспоса́дочным спо́собом heavy airdrop supplies
грунт 1. ground, soil
2. primer (paint)
~, аллювиа́льный alluvial soil
~, боло́тистый boggy ground
~, водонепроница́емый impervious bed, impervious stratum, impervious layer
~, водоно́сный water-bearing ground
~, вы́нутый spoil, excavated soil
~, гли́нистый clayey soil
~ дна nature of bed (river etc)
~, и́листый slimy ground, muddy ground
~, камени́стый stony soil
~, лёссовый loess soil
~, мя́гкий soft soil, soft ground
~, нано́сный alluvial soil
~, неусто́йчивый unstable ground
~, песча́ный sandy soil
~, пло́тный compact soil
~, пылева́тый silt soil, powdery soil
~, свя́зный cohesive soil
~, сла́бый soft ground
~, слежа́вшийся compacted soil
~, сугли́нистый loamy soil
~, супесча́ный sandy-loam soil
~, сыпу́чий cohesionless soil
~, та́лый thawed ground
~, торфяни́стый peaty soil
~, утрамбо́ванный packed earth, packed soil
грунто́вка primer, priming coat (paint)
грунтово́й ground, soil
грунтозаце́п grouser (eng)
гру́ппа group. party. team. force
~, авари́йная damage control party
~, авиацио́нная 1. air force task force
2. ad hoc air force tactical formation
~, арме́йская артиллери́йская army artillery group
~, арме́йская зени́тная артиллери́йская army anti-aircraft artillery group
~ а́рмий army group
~ артилле́рии бли́жнего бо́я close support artillery group
~, артиллери́йская artillery group
~, артиллери́йская разве́дывательная artillery reconnaissance party
~, берегова́я beach group
гру́ппа, боева́я battle group
~, брига́дная brigade group
~, возду́шно-деса́нтная диверсио́нная special forces airborne group
~, возду́шно-деса́нтная такти́ческая airborne battle group, combat team
~, временна́я date-time group
~ да́льнего де́йствия long range group (sub-division of div arty gp)
~, диверсио́нная sabotage group
~, дивизио́нная артиллери́йская divisional artillery group
~ загражде́ния obstacle construction group
~ загражде́ния, подви́жная mobile obstacle group
~ захва́та "snatch" group (of fighting patrol)
~, зени́тная артиллери́йская anti-aircraft artillery group
~, инжене́рная разве́дывательная engineer reconnaissance party
~, инжене́рно-разве́дывательная engineer reconnaissance party
~, инспекцио́нная inspection team, inspection party
~, испыта́тельная test team
~ истреби́телей та́нков tank hunting team, tank hunting party
~, корпусна́я артиллери́йская corps artillery group
~ кро́ви blood group
~ мин mine cluster
~ мини́рования mine-laying party, mine-laying team
~, мото́рная power unit, main engine (tk etc)
~, накрыва́ющая straddle, observer-target bracket, OT bracket, bracketing salvo (gunn)
~, нейрохирурги́ческая neurosurgical team
~ обеспе́чения 1. covering party
2. protective party (eng tac)
~, обеспе́ченная накрыва́ющая confirmed target round (arty)
~ обслу́живания уча́стка вы́садки beach group
~ огнево́й подде́ржки ВМФ Gunfire Support Group (Naval Forces)
~, операти́вная 1. tactical HQ
2. operations group

гру́ппа парашю́тов cluster parachute arrangement (supply dropping)
~ подде́ржки пехо́ты, артиллери́йская infantry support artillery group
~ подрыва́ния заря́да а́томного ВВ atomic demolition mine firing party, ADM firing party
~, по́исковая "snatch" group, raiding party
~, по́исково-сортиро́вочная casualty search and classification group (CD)
~, полкова́я артиллери́йская regimental artillery group
~ проде́лывания прохо́дов в ми́нном по́ле minefield breaching party, minefield breaching team
~ проры́ва assault group
~ противота́нкового наступле́ния anti-tank force
~ прямо́й наво́дки direct fire group
~ радиацио́нной разве́дки radiation detection team
~ разве́дки и удале́ния мин mine-detecting and clearing party
~, разве́дывательная reconnaissance party
~, разве́дывательно-диверсио́нная reconnaissance and sabotage group
~ разгражде́ния breaching team
~ размини́рования mine-clearing party
~, ре́йдерская raiding party
~, рекогносциро́вочная reconnaissance party
~, ремо́нтная repair team
~, ремо́нтно-эвакуацио́нная repair and recovery party, repair and recovery section
~ самопривя́зки artillery survey group
~, ско́вывающая holding force (tac)
~ сопровожде́ния ВМС escort group (Naval Forces)
~ та́нков подде́ржки пехо́ты tank group in direct support of infantry
~ техни́ческой по́мощи и эвакуа́ции technical aid and evacuation group
~, това́рно-техни́ческая marshalling yards section (rly stn)
~, уда́рная assault group
~ управле́ния command group
~ управле́ния батаре́и battery command group, battery HQ
~ усиле́ния reinforcement group
гру́ппа устано́вки и маскиро́вки мин mine-laying and burying party
~ устано́вщиков мин mine-laying party
~ форсу́нок injector set
~, штурмова́я assault group (arty)
~, эвакуацио́нная recovery detachment
группирова́ние bunching (electron)
группирова́тель buncher cavity (electron)
группирова́ться/сгруппирова́ться to bunch (electron)
группиро́вка grouping
~, уда́рная battle group
группово́й group
гру́ша прице́ла sight adjusting yoke screw
гряда́ 1. ledge (river bed) 2. ridge
грязеви́к sump, sludge pan, grease trap
гу́бка 1. cellular rubber, sponge 2. jaw (vice)
~, накладна́я vice clamp
~, подвижна́я sliding jaw (vice)
гудро́н asphaltic bitumen
гудрона́тор asphalt spreader, bitumen spreader
гужево́й animal-drawn
гужо́н screw rivet, tap
гул roar (engine)
~ мото́ра engine roar
гу́сеница track (veh)
~, осла́бленная loose track
~, подрессо́ренная live track
~, резинометалли́ческая rubber block track
~, уши́ренная extra-wide track
~, цельнометалли́ческая all-metal track
гу́сеничный tracked
гусмати́к self-sealing tyre
густота́ движе́ния traffic density
гусько́м in file, in single file

Д

давáть/дать to give; **давáть/дать газ** to open throttle; **давáть/дать зáдний ход** to reverse; **давáть/дать залп** to fire a salvo; **давáть/дать отбóй** to sound "all clear", to ring off (teleph)
давлéние pressure
~, атмосфéрное atmospheric pressure
~, барометри́ческое barometric pressure
~ в заря́дной кáморе, высóкое high breech pressure
~ в канáле стволá bore pressure, gun-barrel pressure
~, возду́шное air pressure
~ гáзов в кáмере сгорáния combustion pressure (gunn)
~, гидравли́ческое hydraulic pressure
~, гидродинами́ческое hydrodynamic pressure
~, дóнное base pressure
~, ду́льное muzzle pressure
~, избы́точное 1. pressure above atmospheric pressure 2. over-pressure (gunn)
~ колесá на грунт wheel pressure, ground pressure (wheels)
~ на грунт ground pressure
~ на опóру bearing pressure
~ на ось axle loading
~ надду́ва boost pressure, supercharge pressure
~, осевóе axial thrust
~, отрицáтельное negative pressure
~, повы́шенное boost pressure
~, рабóчее working pressure, operating pressure, operational pressure, effective pressure
~, равновéсное equilibrium pressure
~, срéднее индикáторное indicated mean effective pressure (eng)
~, срéднее эффекти́вное (brake) mean effective pressure (eng)
~, стати́ческое static pressure
~, удéльное 1. specific pressure 2. ground pressure (vehs)
~ форси́рования shot start pressure (ball)
~, форси́рующее shot start pressure (ball)
далёкий 1. distant, far 2. long-range
дáльнего дéйствия long-range
дáльний 1. distant, far 2. long-range
дальнобóйность maximum range (wpn)
дальнобóйный long-range
дальномéр range-finder
~, акусти́ческий acoustic range finder, sound range finder
~, вертикальнобáзный vertical base rangefinder
~, внутрибáзовый fixed base rangefinder
~, зени́тный anti-aircraft range-finder
~, лáзерный laser range-finder
~ на руби́новом лáзере ruby laser range-finder
~ на совмещéние coincidence range-finder
~, опти́ческий optical range-finder
~, пасси́вный инфракрáсный passive infra-red range-finder
~, радиолокациóнный radar range-finding equipment
~ с окуля́рной сéткой stadiametric range-finder
~, сапёрный 1. engineer range-finder 2. gap-measuring instrument
~, стереоскопи́ческий stereoscopic range-finder
дальномéр-высотомéр combined height and range-finder
дальномéр-прицéл combined rangefinder and sight
дальномéрщик range-finder operator
дáльность range, distance
~ автосопровождéния range at which equipment is "locked on" (gunn)
~, баллисти́ческая ballistic range
~, безопáсная safety range
~ боевóго применéния combat range
~ ви́димости visibility, range of visibility
~, горизонтáльная 1. ground range (arty) 2. plan range (nav gunn) 3. horizontal range (radar, arty) 4. range to point of graze (gunn)
~, действи́тельная range to point of impact
~ действи́тельного огня́ effective range
~ дéйствия operating range, operational range, range of action

да́льность де́йствия, максима́льная maximum radius of action (arty)
~ **де́йствия, преде́льная** maximum operating range
~ **до це́ли, исчи́сленная** standard predicted horizontal range
~ **засе́чки** distance from base of intersection to target
~, **испра́вленная** corrected range
~, **и́стинная** true range
~, **исхо́дная** opening range (arty)
~, **исчи́сленная** computed range, calculated range
~ **команди́р-цель** observer-target distance, OT distance
~, **максима́льная горизонта́льная** maximum horizontal range (ord)
~, **наблюдённая** observed range (arty)
~, **назе́мная** ground range
~ **наибо́лее действи́тельного огня́** maximum effective range
~, **накло́нная** slant range
~, **настоя́щая** present range (nav gunn)
~ **обнаруже́ния** 1. range of first sighting (AA)
2. detection range (radar)
~, **ограни́ченная** limited range
~ **ору́дие-цель** battery-target distance, BT distance
~ **оруди́йного огня́** gun range
~ **паде́ния** range to point of graze
~ **переда́чи** transmission range, transmission distance (radio)
~ **перехва́та, накло́нная** range of engagement
~ **по ка́рте** map range
~ **по ре́перу, пристре́лянная** record range, range by shooting
~ **полёта** range of flight, range ability
~ **полёта, максима́льная** maximum range (missile etc)
~ **полёта раке́ты** missile range
~ **полёта снаря́да** range of fall of shot (nav gunn)
~, **по́лная горизонта́льная** base of trajectory, range to point of graze (ball)
~, **пристре́лянная** reduced range, adjusted range
~, **прице́льная** 1. gun range (SA)
2. map range, range to target (arty)
3. range up to which a weapon is sighted
~ **прямо́го вы́стрела** flat-trajectory range, point blank range
да́льность прямо́й ви́димости line-of-sight range, optical range, visibility range
~ **радиолокацио́нного обнаруже́ния** radar acquisition range, detection range
~ **разры́ва** range achieved, range to point of burst
~, **растя́нутая** expanded range (radar)
~, **расчётная** estimated range
~ **стрельбы́** range (of fire)
~ **стрельбы́, максима́льная** maximum range of engagement
~ **стрельбы́, минима́льная** minimum range of engagement
~, **теку́щая горизонта́льная** present horizontal range (gunn)
~, **теку́щая накло́нная** present slant range (gunn)
~, **топографи́ческая** map range
~, **убо́йная** killing range
~, **упреждённая** future range
~, **упреждённая горизонта́льная** future horizontal range, future ground range
~, **упреждённая накло́нная** future slant range
дальтони́зм colour blindness, Daltonism
да́мба causeway, dam, dike, dyke, embankment, sea wall
~, **берегова́я** embankment
~, **струенаправля́ющая** guide-bank (br)
да́нные data, information
~, **аэрологи́ческие** weather data
~, **баллисти́ческие** ballistic data, gun data
~, **входны́е** data-in, input data, incoming data
~, **выдава́емые** data-out, output data, out-going data
~, **выходны́е** data-out, output data, out-going data
~ **для наведе́ния** guidance intelligence, guidance data
~ **для наво́дки ору́дия** gun laying data
~, **запомина́емые** stored data
~, **исхо́дные** initial data, basic data, opening data
~, **исчи́сленные** predicted data
~, **ито́говые** summarizing data, summarized data
~, **картографи́ческие** map data
~, **курсовы́е** course data
~, **метеорологи́ческие** meteorological data (arty)

да́нные, необрабо́танные raw data
~, непреры́вно-упреждённые continuously-predicted data
~, отправны́е starting data, basic data
~, подтвержда́ющие confirming evidence, supporting evidence, corroborating data
~ пристре́лки fired gun data (arty)
~, пристре́лочные ranging data
~, пристре́лянные fired gun data (arty)
~, прове́ренные verified information, verified data
~, прое́ктные design data
~ разве́дки intelligence, reconnaissance data
~ разве́дки ме́стности terrain data, terrain intelligence
~, разве́дывательные intelligence, reconnaissance data, information
~ сопровожде́ния tracking data (target)
~, та́ктико-техни́ческие technical specifications (eqpt, mech)
~, такти́ческие operational performance data (eqpt)
~, телеметри́ческие telemetry data
~, техни́ческие engineering data, specifications
~, топографи́ческие map data
~, упреждённые predicted data, future data
~, усто́йчивые steady-state data
~, цифровы́е digital data, numerical data, figures
Да́рси Darcy (nucl)
да́та date
~ изготовле́ния date of manufacture
да́тчик 1. pick-up
2. transducer, sensor, transmitter
~ бензоме́ра fuel gauge transmitter
~ вертика́льной наво́дки elevation transmitter (arty)
~, гру́бый coarse transmitter (servo mech)
~ давле́ния pressure transducer, pressure pick-up (GW)
~ давле́ния ма́сла oil pressure gauge sender unit
~, инфракра́сный infra-red sensor, IR sensor
~, клавиату́рный morse keyboard transmitter
~ ко́да мо́рзе morse keyboard transmitter
да́тчик кома́нд command transmitter
~, опти́ческий optical sensor
~, оруди́йный gun data transmitter
~, потенциометри́ческий potentiometer pick-off, potentiometric transducer
~, телевизио́нный television sensor, TV sensor
~, температу́рный temperature sensing device
~ температу́ры охлажда́ющей жи́дкости coolant temperature sender unit
~, теплово́й heat sensor
~, фотоэлектри́ческий photo-electric sensor
~, электромагни́тный electro-magnetic transducer
~, электро́нный electronic sensor, electronic pick-up
дать (см дава́ть)
да́ча 1. ration, food
2. cottage
3. (piece of) woodland
~, су́точная day's ration
~, фура́жная animal ration, forage ration
две́рца exit, jump door (para)
~ деса́нтного лю́ка bottom escape hatch (tk)
~ запа́ла detonator sleeve cover
~ лю́ка access door
~, опускна́я trap door
~, откидна́я trap (revolver)
дверь, брониро́ванная armoured door
~, взрывозащи́тная blast door
~, водонепроница́емая water-tight door
~, воздухонепроница́емая air-tight door
~, газонепроница́емая gas-tight door
~, гермети́ческая hermetically-sealed door, air-tight door, gas-proof door
~, запа́сная emergency door, exit door
~, защи́тная sealed door, blast door (atomic shelter)
~, клинке́тная sliding door
~, опускна́я trap door
~, противопожа́рная fire-proof door, fire-resisting door
~, раздвижна́я sliding door
~, руло́нная зака́тывающаяся roll-up door
~, сдвижна́я sliding door
дви́гатели, спа́ренные coupled engines, twin engines
дви́гатель engine

двигатель, бензиновый petrol engine
~, **бесклапанный** valveless engine, sleeve-valve engine
~, **бесшумный** silent-running engine
~, **боковой стартовый** wrap around boost, wrapped round booster, wrap around booster (GW)
~, **быстроходный** high-speed engine
~ **в сборе** engine assembly, motor assembly
~, **верньерный** vernier engine (GW)
~ **внутреннего сгорания** internal combustion engine
~, **B-образный** "V" engine usually written with latin V
~, **водомётный** hydro-jet propulsion unit
~ **водяного охлаждения** water-cooled engine
~ **воздушного охлаждения** air-cooled engine
~, **воздушно-реактивный** jet engine, air-breathing jet engine
~, **воздушный пусковой** air starting motor (of diesel)
~, **восьмицилиндровый** eight-cylinder engine
~, **всеядный** multifuel engine
~, **вспомогательный** auxiliary engine
~, **вспомогательный стартовый** booster
~, **вспомогательный танковый** tank auxiliary engine
~, **газотурбинный** gas turbine engine
~, **гидрореактивный** hydrojet engine
~ **двойного действия** double-acting engine
~ **двойного действия, ротативный** double-acting rotary engine
~, **двухконтурный** turbofan-type jet engine
~, **двухконусный турбореактивный** bypass turbojet engine, ducted fan jet engine
~, **двухрядный** double-row engine
~, **двухтактный** two-stroke engine
~, **дефорсированный** derated engine
~ **дизеля** diesel, compression ignition engine
~, **длинноходный** long-stroke engine
~ **жидкостного охлаждения** liquid-cooled engine
~, **жидкостный ракетный** liquid-propellant rocket engine

двигатель, жидкостный реактивный liquid-propellant rocket engine
~, **забортный** outboard motor, outboard engine
~, **звездообразный** radial engine
~, **исполнительный** actuating motor, drive motor (servo), servo mechanism
~, **коловратный** rotary engine
~, **линейный асинхронный** linear induction motor
~, **маршевый** sustainer engine
~, **многоблочный** multibank engine
~, **многокамерный ракетный** multi-chamber rocket engine
~, **многотопливный** multi-fuel engine
~ **на смешанном топливе, ракетный** hybrid rocket engine
~, **навесной** outboard motor
~, **неотделяемый стартовый** integral booster, integral rocket unit
~ **H-образный** "H" engine
~, **однорядный** in-line engine
~, **оппозитный** opposed-cylinder engine
~, **основной танковый** tank main engine
~, **отремонтированный** reconditioned engine
~ **повышенной мощности** uprated engine
~, **подвесной** outboard motor
~, **пороховой ракетный** solid-propellant rocket motor
~, **пороховой реактивный** solid-propellant rocket motor
~, **поршневой** piston engine
~, **прямоточный** ramjet engine, resonance duct engine
~, **пульсирующий воздушно-реактивный** pulsejet engine, intermittent-impulse duct engine
~ **работает с перебоями** the engine is misfiring
~, **работающий на различных видах топлива** multi-fuel engine
~, **ракетный** rocket engine, rocket motor
~, **реактивный** 1. reaction motor 2. rocket motor, rocket engine 3. jet engine
~ **с боковым расположением клапанов** side valve engine
~ **с верхним расположением клапанов** overhead valve engine

дви́гатель с водяны́м охлажде́нием water-cooled engine
~ **с возду́шным охлажде́нием** air-cooled engine
~ **с воспламене́нием от сжа́тия** diesel engine, compression ignition engine
~ **с впры́ском то́плива** fuel-injection engine
~ **с лине́йным расположе́нием цили́ндров** in-line engine
~ **с надду́вом** supercharged engine
~ **с непосре́дственным распы́ливанием, ди́зельный** diesel engine with direct injection
~ **с переме́нной сте́пенью сжа́тия** variable-compression engine
~ **с противополо́жно дви́гающимися по́ршнями** opposed-piston engine
~ **с турбонадду́вом** turbo-charged engine
~, **сери́йный** series-production engine
~, **ста́ртовый** boost motor, booster (GW)
~ **стеклоочисти́теля** windscreen wiper motor
~ **та́нка** tank power pack
~, **та́нковый** tank engine
~, **та́нковый газотурби́нный** tank gas turbine
~ **твёрдого то́плива** solid propellant motor
~ **твёрдого то́плива, раке́тный** solid-propellant rocket motor
~, **теплово́й** heat engine
~, **турбовинтово́й** turboprop engine
~, **турбовентиля́торный** ducted-fan type engine, turbo-fan engine
~, **турбореакти́вный** gas-turbine jet engine, turbojet
~, **тя́говый** traction engine
~, **четырёхта́ктный** four-stroke engine, four-cycle engine
~, **шестицили́ндровый** six-cylinder engine
~, **электри́ческий раке́тный** electric rocket engine
~, **я́дерный раке́тный** nuclear rocket engine
дви́гатель-генера́тор motor generator
дви́гаться/дви́нуться to move; **дви́гаться/дви́нуться нака́том** to coast, to freewheel; **дви́гаться/дви́нуться по ине́рции** to coast; **дви́гаться/дви́нуться по цели́не** to move across country; **дви́гаться/дви́нуться скачка́ми** to move by bounds; **дви́гаться/дви́нуться цельно́ю** to move across country
движе́ние 1. movement, motion 2. traffic
~ **безре́льсового тра́нспорта** vehicular traffic
~, **беспрепя́тственное** unhindered movement, unimpeded movement
~ **в две сто́роны** two-way traffic
~ **вне доро́г** off-road movement
~, **возвра́тно-поступа́тельное** shuttle, reciprocating motion
~, **встре́чное** two-way traffic
~, **высокоскоро́стное** high-speed running
~, **грузово́е** freight traffic
~, **двусторо́ннее** two-way traffic
~, **двухпу́тное** two-way traffic
~, **двухсторо́ннее** two-way traffic
~, **железнодоро́жное** railway traffic, rail movement
~ **за́дним хо́дом** reversing
~, **интенси́вное** heavy traffic
~, **и́стинное** true motion (radar)
~, **колеба́тельное** oscillation, oscillatory motion
~, **косо́е** slant movement, movement across a front
~ **нака́том** free wheeling, coasting
~, **невозмущённое** undisturbed motion
~, **неинтенси́вное** light traffic
~, **неусто́йчивое** unstable motion
~, **обра́тное** return-flow, return movement, reverse motion
~, **обхо́дное** outflanking movement, bypassing movement
~, **однопу́тное** single line traffic
~, **односторо́ннее** one-way traffic
~, **охва́тывающее** enveloping movement
~, **пассажи́рское** passenger traffic
~ **перека́том** leap-frogging (tac)
~, **перекрёстное** intersecting traffic
~ **по а́зимуту** march on compass bearing
~ **по гра́фику** running to schedule
~ **по доро́ге** road running
~ **по ине́рции (с вы́ключенными дви́гателями)** coasting, free wheeling
~ **по лучу́** beam riding (GW)
~ **по пересе́ченной ме́стности** cross-country movement

движéние, поступáтельное translation movement, reciprocating movement (GW)
~, похóдное movement in column
~, прецессиóнное precession (of a gyroscope)
~, прúгородное commuter traffic
~, реактúвное reaction propulsion, jet propulsion, rocket propulsion
~, сúльное heavy traffic
~ скачкáми movement by bounds (tac)
~ снарядá в канáле стволá shot travel
~ трáнспорта traffic flow
~, турбулéнтное turbulence, eddying, vortex motion
~, ýличное street traffic
~, флангóвое lateral movement, movement across the front
~, фронтáльное move by a front (of armies)
движúтель propeller, propulsive device, propelling device
~, гýсеничный track drive
движóк 1. slide, feed block actuating slide (SA)
2. shutter (fuse)
3. small portable engine
~ крышки магазúна magazine-cover slide (SA)
двúнуться (см двигáться)
двор, грузовóй freight yard (rly)
~, товáрный freight yard (rly)
"двóрник" windscreen wipers (coll)
двояковóгнутый concave
двояковыпуклый convex
двукóлка, санитáрная ambulance cart (two-wheeled)
двунóга bipod
двунóга-лафéт bipod mounting (mor)
двуóкись углерóда carbon dioxide
двуствóльный double-barelled
двутáвр I-beam, H-beam
двухдвúгательный twin-engined
двухкилометрóвка map with scale 1/200,000
двухколёсный two-wheeled
двухмéрный two-dimensional
двухмéстный two-seater
двухмотóрный twin-engined
двухóсный two-axled
двухполупериóдный full-wave
двухпоршневóй double-piston
двухствóльный double-barrelled, twin-barrelled
двухступéнчатый two-stage
двухтáктный 1. two-stroke
2. push-pull (electron)
двухцилúндровый two-cylinder
двухъя́русный two-tier
дебаркáдер landing stage, platform jetty
дéбет charge, debit
деблокúрование relief of beleaguered troops
деблокúровать 1. to clear, to relieve, to unblock
2. to raise a blockade
девиáция, кóмпасная compass deviation
~, магнúтная magnetic deviation
дегазáтор CW decontamination agent
~, жúдкий CW decontamination liquid
дегазациóнный CW decontamination
дегазáция CW decontamination
~, лúчная personal cleansing (CW)
~ мéстности area decontamination
~, пóлная detailed CW decontamination
~, частúчная partial decontamination
дегазúровать to decontaminate CW
дегерметизáция unsealing (hermetic)
дежýрный duty officer, duty NCO, duty man, duty
~, заступáющий next for duty
~, оператúвный duty officer (ops)
~, ответственный duty officer (ops)
~ по караýлам 1. field officer of the day
2. officer in charge of guards (in multi-unit garrison)
~ по кýхне NCO i/c cookhouse
~ по лáгерному сбóру camp duty officer (in multi-unit camp)
~ по пáрку vehicle-park orderly
~ по полкý regimental duty officer
~ по рóте company orderly sergeant
~ по связи duty signals officer
~ по стрельбé range officer
~ по чáсти 1. unit duty officer
2. unit duty NCO
~ по штáбу duty staff officer
~ по эшелóну troop-train duty officer
дежýрство 1. duty, being on duty
2. tour of duty
~, боевóе operational readiness (missiles)
~, непрерывное continuous manning, continuous duty

дезактива́тор radioactive decontamination apparatus
дезактива́ция radioactive decontamination
~ **ме́стности** area decontamination
~ **пове́рхности** surface decontamination
~, **по́лная** detailed radioactive decontamination
~, **части́чная** partial radioactive decontamination
дезакти́ровать to decontaminate (radioactivity)
дезерти́р deserter
дезерти́ровать to desert
дезерти́рство desertion
дезинсе́кция disinfestation, delousing, fumigation
дезинфе́ктор 1. disinfection orderly 2. sterilizer, disinfector
дезинфекцио́нный disinfestation, delousing, fumigation
дезинфе́кция fumigation, disinfection, disinfestation, sterilization
~, **по́лная** detailed disinfestation, detailed disinfection
~, **части́чная** partial biological decontamination, tactical biological decontamination
дезинфици́ровать to disinfect
дезинформа́ция 1. deception 2. dissemination of false information
дезка́мера disinfecting chamber
дезока́мера disinfecting chamber
дезорганиза́ция disorganisation, disruption
дезорганизова́ть to disorganize, to disrupt
дезориента́ция 1. loss of bearings 2. disorientation, confusion, demoralization
дезориенти́рование deception (effort)
дезориенти́ровать to confuse, to mislead
дезустано́вка decontaminating equipment (CW)
де́йствие 1. action 2. operation 3. effect
~, **безотка́зное** trouble-free operation
~, **боево́е** action, combat operation
~, **бронебо́йное** armour-piercing effect
~ **взры́ва** blast action, blast effect
~ **взры́ва, дробя́щее** shattering effect
де́йствие взры́ва, ко́свенное secondary blast effect
~, **демонстрати́вное** deception measure, feint
~, **дли́тельное** prolonged action
~, **заме́дленное** 1. delay (fuse setting) 2. delayed action
~, **замедля́ющее** retarding effect
~, **иниции́рующее** initiating action
~, **карте́чное** canister effect (ammo)
~, **ко́жно-нарывно́е** blistering effect, blistering action, vesication
~, **мгнове́нное** instantaneous action, direct action (fuse)
~, **нарывно́е** vesicant action, blistering action
~, **оско́лочное** fragmentation effect, fragmentation action
~, **отве́тное** counteraction
~, **отвлека́ющее** feint, diversionary attack, diversionary operation
~ **отка́та** recoiling action
~ **перека́тами** leap-frogging (tac)
~ **по це́ли** target effect (ammo)
~, **поража́ющее** casualty effect, lethality
~, **предстоя́щее** future operation
~, **пробивно́е** penetration (SA)
~, **противозако́нное** offence (legal), illegal action, crime
~, **пружиня́щее** cushion effect
~, **раздража́ющее** irritant action, irritant effect
~, **сде́рживающее** containment, containment battle, delaying action
~, **слезоточи́вое** lacrimatory effect
~, **смерте́льное** lethal effect
~ **тормозо́в** braking action
~, **тормозя́щее** braking effect
~, **убо́йное** lethal effect
~, **уда́рное** percussion action (ammo, fuze)
~ **уда́рной волны́ взры́ва** blast effect
~ **уда́рной волны́, поража́ющее** blast damage
~, **удуша́ющее** asphyxiation effect
~, **фуга́сное** HE effect, blast effect
~, **шрапне́льное** shrapnel effect
действи́тельность огня́ fire effect
~ **огня́, наибо́льшая** maximum fire effect

действи́тельный 1. real, actual, factual, valid
2. active
3. operating
4. in force, effective
де́йствия, беспоко́ящие harassing action, nuisance operation
~, **боевы́е** action, fighting, combat operations, warfare
~ **бронета́нковых войск, боевы́е** armoured warfare
~ **в гора́х, боевы́е** mountain warfare
~ **в пусты́нях, боевы́е** desert warfare
~ **возду́шно-деса́нтных войск, боевы́е** airborne operations
~, **диверсио́нные** subversive operations
~ **зимо́й, боевы́е** winter warfare
~ **инжене́рных войск** engineer activity, engineer operations
~, **нереши́тельные** indecisive actions
~, **обма́нные** subversive activities
~, **оборони́тельные** defensive operations
~, **отве́тные** counteraction, operations in retaliation
~, **партиза́нские** guerilla activities, partisan activities
~, **подрывны́е** subversive activities
~ **с примене́нием раке́т, боевы́е** missile operations
~ **с примене́нием я́дерного ору́жия, о́бщие наступа́тельные** general thermonuclear offensive
~, **сде́рживающие** delay(ing) actions, delay(ing) operations
~, **совме́стные** joint operations, combined operations
~, **согласо́ванные** co-ordinated efforts, co-ordinated operations
~, **такти́ческие оборони́тельные** tactical defence
де́йствовать/поде́йствовать to act, to operate, to function, to have an effect (upon); **де́йствовать/поде́йствовать в пе́рвом эшело́не** to spearhead the assault
де́йствующий 1. acting
2. operating
~ **си́лой отда́чи** blowback-operated, recoil-operated
дейте́рий deuterium
дейтеро́н deuteron (nucl)
деклино́метр declinometer
декодирова́ть to decode
декомпре́ссия decompression
декомпре́ссор decompressor
декора́ция garnish (cam)
декреме́нт затуха́ния damping ratio
декре́т decree
де́лать/сде́лать to make, to do; **де́лать/сде́лать внуше́ние** to admonish, award an admonition; **де́лать/сде́лать замеча́ние** to admonish; **де́лать/сде́лать за́пись** to record, to post an entry; **де́лать/сде́лать зару́бки** to blaze (a trail); **де́лать/сде́лать перекли́чку** to call the roll
делега́т liaison officer
~ **кома́ндования** liaison officer from higher formation
~ **свя́зи** liaison officer in the field
деле́ние 1. graduation (instr)
2. fission (nucl)
3. division (maths)
~ **взрыва́теля, скома́ндованное** ordered fuse setting
~ **изображе́ния на две ча́сти** halving (range-finding)
~ **прице́ла** backsight line
~, **скома́ндованное** ordered setting
~ **угломе́ра** 1. graduation of dial sight
2. mil (unit of angular measure used by Soviet arty)
~ **угломе́ра, большо́е** major division on scale of dial sight, one hundred mils (Sov)
~ **частоты́** frequency division (electron)
~ **шкалы́** graduation, (scale) division, notch, index, mark, reading
~ **ядра́** nuclear fission
дели́тель мо́щности, невзаи́мный non-reciprocal power divider (radar)
~ **напряже́ния** voltage divider
~ **частоты́** frequency divider
де́ло 1. affair, deed, act, activity
2. file (documents)
3. engagement (mil)
~, **артиллери́йское** gunnery
~, **вое́нное** military affairs, military matters
~, **вое́нно-инжене́рное** military engineering
~, **вое́нно-хими́ческое** chemical warfare
~, **доро́жно-мостово́е** road and bridge engineering
~, **инжене́рное** engineering
~, **ли́чное** personal record, personal file

де́ло, ми́нно-подрывно́е mine warfare
~, подзе́мно-ми́нное tunnelling (mil eng)
~, подрывно́е practice of demolition, technique of demolition
~, полево́е вое́нно-инжене́рное field engineering
~, сапёрное combat engineering, field engineering, combat engineer technique
~, складско́е warehousing, storehousing
~, стрелко́вое musketry
делопроизводи́тель clerk
делопроизво́дство documentation, clerical duties
де́льта delta
де́льта-модуля́ция delta modulation
дельтови́дный delta-shaped, triangular
демаски́рование 1. breach of camouflage discipline
2. detection of camouflage
демаски́ровать to give away, to disclose (position)
демобилиза́ция demobilisation
демодуля́тор demodulator
~, амплиту́дный amplitude demodulator
~, часто́тный frequency demodulator
демодуля́ция demodulation
демонстра́ция 1. feint
2. demonstration (tac)
демонта́ж dismantling
демонти́ровать to dismantle
деморализова́ть to demoralise
де́мпфер damper, buffer (mech and elec)
~, жи́дкостный liquid damper
демпфи́рование damping
демультиплика́тор reduction gear, underdrive, auxiliary gear box
денсито́метр densitometer
денщи́к batman
день, па́рковый servicing day, vehicle maintenance day
~, паркохозя́йственный servicing day, vehicle maintenance day
де́ньги, кварти́рные lodging allowance
~, командиро́вочные travel allowance, detached duty allowance
~, кормовы́е ration allowance
~, пайко́вые ration allowance, ration cash allowance
~, подъёмные 1. disturbance allowance
2. appointment allowance (Sov)
~, поши́вочные clothing allowance
де́ньги, прого́нные mileage allowance, travelling allowance
~, продово́льственные ration allowance, subsistence allowance, messing cash allowance
~, проездны́е travel allowance
~, путевы́е travel allowance
~, столо́вые ration allowance, subsistence allowance, messing cash allowance, messing expenses
~, су́точные daily allowance, daily subsistence allowance
~, экипиро́вочные outfit allowance
депе́ша dispatch, message
~, многоа́дресная multi-address message, circular message
депо́ 1. depot
2. repair shop (rly)
~, локомоти́вное locomotive depot
~, парово́зное engine shed, locomotive depot, locomotive servicing depot
~, пожа́рное fire station
деполяриза́тор depolarizor
дератиза́ция rat extermination
де́рево, ли́ственное deciduous tree
~, хво́йное coniferous tree
деревообраба́тывающий woodworking
деревообрабо́тка woodworking
деревоплита́ timber slab
держа́ва, сухопу́тная land power
держа́вка carrier, holder
~ рукоя́тки затво́ра breech mechanism lever retaining catch
держа́к handle
держа́тель 1. holder, carrier
2. support
~ запа́ла detonator holder (gren)
держа́ть to hold, to keep; **держа́ть курс на (пози́цию)** to head for (a position); **держа́ть направле́ние по а́зимуту** to keep direction by compass bearing; **держа́ть связь** to maintain communication(s), to keep in touch
де́рзость, боева́я audacity (in battle)
growth
дерива́ция drift (gunn)
дермати́н leatherette
дёрн turf
деса́нт 1. assault landing
2. assault landing force
~, вертолётный heliborne assault, heliborne attack
~, возду́шно-поса́дочный air-mobile assault

деса́нт, возду́шный 1. airborne assault force
2. airborne assault landing
~, демонстрати́вный diversionary assault landing
~, демонстрати́вный возду́шный diversionary airborne landing
~, корабе́льный assault landing party (nav)
~, ло́жный simulated assault landing, feint assault landing
~, морско́й 1. amphibious assault landing
2. amphibious assault landing force
~, операти́вный возду́шный 1. major independent airborne assault landing
2. major independent airborne assault landing force
~, операти́вный морско́й 1. major seaborne assault landing
2. major seaborne assault landing force
~, парашю́тно-поса́дочный combined parachute and air-landed assault force
~, парашю́тный 1. parachute assault landing
2. parachute assault landing force
~, поса́дочный 1. airlanded assault
2. airlanded assault force
~, разве́дывательно-диверсио́нный commando-type raid
~ с бо́ем opposed landing
~, стратеги́ческий морско́й independent seaborne assault of strategic importance
~, такти́ческий возду́шный 1. tactical airborne assault landing
2. tactical airborne assault landing force (usually below divisional level)
~, та́нковый tank-borne infantry
~, фланго́вый flank assault (amph ops)
десанти́рование airborne assault
десанти́ровать to make an assault landing
деса́нтник 1. paratrooper
2. member of assault landing force
~, обу́ченный затяжны́м прыжка́м sky-diver
деса́нтник-парашюти́ст paratrooper
деса́нтник-сапёр assault engineer, assault pioneer
деса́нтно-досту́пность уча́стка побере́жья suitability of beach for seaborne landing
деса́нтно-опа́сный susceptible to an assault landing
деса́нтно-перепра́вочный assault river-crossing (eqpt, etc)
деса́нтный 1. assault landing
2. assault
"деса́нту пригото́виться!" "assault crews to stations!" (naval command on landing ship)
десятикилометро́вка 1/1 million scale map
детализа́ция detailed treatment
дета́ль component part, part
~, второстепе́нная minor part
~ дви́гателя engine part
~, запасна́я spare part
~, изно́шенная worn-out part
~, крепёжная attachment, fastening, securing part
~, несу́щая bearer (eng, mech)
~, отве́тственная vital part, essential part
детекти́рование detection (elec)
~, ано́дное anode rectification, anode bend detection
~, дио́дное diode detection
~, се́точное grid detection (elec)
детекти́ровать to detect (radio)
дете́ктор detector
~ дифференциа́льного ти́па discriminator (radio)
~, дро́бный ratio detector (radio)
~ излуче́ния radiation detector
~, и́мпульсный pulse detector
~, инфракра́сный infra-red detector
~, кре́мниевый silicon detector (radio)
~ мостово́го ти́па ratio detector (radio)
~ опознава́ния identity detector (distance measuring eqpt)
~ отноше́ний ratio detector
~ рассогласова́ния error detector
~, се́точный grid detector, grid circuit detector (elec)
~ сигна́ла оши́бки error detector
~, часто́тный frequency discriminator (radio)
детона́тор 1. detonator
2. magazine (fuse)
3. gaine (priming tube; fuse)
~, втори́чный secondary detonating charge
~, динами́тный dynamite primer

детона́тор, дополни́тельный 1. exploder (ammo)
2. booster charge (igniter set)
~ **заме́дленного де́йствия** delayed action detonator
~ **мгнове́нного де́йствия** instantaneous detonator
~, **перви́чный** primary detonating charge
~, **промежу́точный** 1. relay detonator
2. booster charge
~, **уда́рный** air-burst fuse, air-pressure fuse, concussion detonator, impact detonator
детона́ция 1. detonation
2. knock, knocking (engine)
~ **горю́чей сме́си** knocking, pinking
~, **индукцио́нная** sympathetic detonation
~, **индуци́рованная** induced detonation
~ **на расстоя́нии** .sympathetic detonation, sympathetic concussion
~, **непо́лная** low order detonation (arty)
~ **че́рез влия́ние** sympathetic detonation
детони́ровать to detonate
дефе́кт defect, deficiency, fault
~ **ма́ссы** mass defect (nucl)
~, **опа́сный** critical defect
~, **скры́тый** hidden weakness
дефекта́ция fault finding, detection of defects
дефекто́вка fault finding, detection of defects
дефектоско́п flaw detector, defectoscope
~, **магни́тный** magnetic flaw detector, defectoscope, ultrasonic flaw detector, ultrasonic defectoscope, magnetic crack detector
дефекту́ра deficiencies
дефиле́ defile
~, **межозерно́е** isthmus between lakes
дефици́тный in short supply
дефле́ктор deflector, baffle
деформа́ция deformation, strain, distortion
~ **круче́ния** torsion strain
~, **пласти́ческая** plastic deformation
~ **растяже́ния** tensile strain
~ **сжа́тия** compressive strain
~ **ствола́, кольцева́я** hoop strain
децибе́л decibel
дешифра́тор 1. decoder
2. cine-gun photograph accessor

дешифра́тор, стру́йный running record plotter (sound ranging)
~, **электро́нный** electronic decoder
дешифри́рование deciphering, decoding, interpretation
~ **аэросни́мков** air photo interpretations
~, **стереоскопи́ческое** interpretation of stereopairs
~, **фотографи́ческое** air photo interpretation
дешифри́ровать to decipher, to interpret an aerial photograph
дешифрова́льщик photographic interpreter
дешифро́вка 1. deciphering, decoding, interpretation
2. photographic interpretation
дешифро́вщик 1. decipherer
2. film reader (arty)
де́ятельность, боева́я combat activity, action
~, **вое́нно-хозя́йственная** administration, logistic functions
~ **войск, боева́я** combat activity of troops
~, **диверсио́нная** guerilla activities
~, **огнева́я** fire activity
~, **рационализа́торская** improvement by modification, innovation
джип jeep
диагра́мма diagram, chart, scheme, graph
~, **индика́торная** indicator diagram
~ **напра́вленности** polar diagram (radio, radar)
~ **напра́вленности излуче́ния** polar diagram (radio, radar)
~ **обнаруже́ния** coverage diagram (radar)
диамагни́тный diamagnetic
диа́метр, вну́тренний 1. bore diameter, bore (ord)
2. internal diameter
~ **основа́ния ба́шни** turret ring diameter
~ **поворо́та, наиме́ньший** minimum turning circle
~ **пого́на ба́шни** turret ring diameter
~ **по́ршня ме́жду кана́вками** land diameter (gunn)
~ **цили́ндра** cylinder bore
дианизиди́н dianisidine (CW)
диапазо́н band, frequency range (radio)
~ **а́зимута** azimuth range
~ **амплиту́д** amplitude range

диапазóн, ближний инфракрáсный near infra-red band, near IR band
~, ближний ультрафиолéтовый near ultraviolet band, near UV band
~ волн waveband (radio)
~ высóт altitude range
~ гектометрóвых волн medium-frequency band (from 300 to 3,000 kiloHertz)
~ декаметрóвых волн high-frequency band (from 3 to 30 megaHertz)
~ дециметрóвых волн ultrahigh-frequency band (from 300 to 3,000 megaHertz)
~, динами́ческий dynamic range (radio, radar)
~ звуковы́х частóт audio-frequency range, tonal range
~, инфракрáсный infra-red band, IR band
~ километрóвых волн low-frequency band (from 30 to 300 kiloHertz)
~ метрóвых волн very high-frequency band (from 30 to 300 megaHertz)
~ миллиметрóвых волн extremely high-frequency band (from 30 to 300 gigaHertz)
~ мириаметрóвых волн very low-frequency band (from 3 to 300 kiloHertz)
~ настрóйки tuning range
~ обзóра search band
~, опти́ческий optical band
~ пóиска search band
~, растя́нутый spread band (radio)
~ сантиметрóвых волн superhigh-frequency band (from 3 to 30 gigaHertz)
~ скоростéй velocity range
~, срéдний инфракрáсный middle infra-red band, middle IR band
~, срéдний ультрафиолéтовый middle ultraviolet band, middle UV band
~, ультрафиолéтовый ultraviolet band, UV band
~ частóт frequency band
~ частóт, рабóчий working frequencies, operating frequencies (radio)
диапозити́в diapositive, positive
диапроéктор film-strip projector, slide projector
диафаноскóп visibility meter
диафи́льм diapositive film, slide
диафрáгма 1. diaphragm, membrane 2. iris (waveguide)
диафрáгма, апертýрная aperture stop (photo)
~, дéйствующая aperture stop (photo)
~, дóнная base cover (ammo)
~, звукопередаю́щая speech valve (respirator)
~, кольцевáя circular aperture
~, опти́ческая optical diaphragm
~ пóля зрéния field stop (optics)
~, предохрани́тельная разрывнáя safety burst diaphragm (arty)
~ ракéты grain holder (GW)
~ тóрмоза откáтки recoil buffer compensator valve
дивергéнция divergence
диверсáнт saboteur
диверсиóнный diversionary, sabotage
дивéрсия 1. sabotage 2. diversion
дивизиóн battalion (artillery)
~, артиллери́йский artillery battalion
~, гáубичный howitzer battalion
~, гáубичный артиллери́йский howitzer battalion
~ десáнтных кораблéй amphibious group (Naval Forces)
~, зени́тно-артиллери́йский anti-aircraft battalion
~, истреби́тельно-противотáнковый tank-destroyer battalion
~, миномётный mortar battalion (arty unit)
~ óбщей поддéржки general support battalion
~, отдéльный independent battalion
~, ракéтный missile battalion
~, реакти́вный rocket launcher battalion
~ самохóдной артиллéрии self-propelled artillery battalion
дивизиóнный 1. divisional, division 2. battalion (arty)
диви́зия division
~, авиациóнная 1. Sov air division 2. Brit AF group
~, артиллери́йская artillery division
~, бронетáнковая armoured division
~, воздýшно-десáнтная airborne division
~, горнострелкóвая mountain rifle division
~, запáсная мотострелкóвая reserve motor rifle division
~, зени́тная anti-aircraft division

диви́зия, кавалери́йская cavalry division
~, мотопехо́тная motorized infantry division
~, моторизо́ванная motorized division
~, мотострелко́вая motor rifle division
~, пехо́тная infantry division (non-Sov armies)
~ проры́ва, артиллери́йская breakthrough artillery division
~, стрелко́вая rifle division
~, та́нковая tank division
дигидрофенарсазинхлори́д dihydrophenarsazine chloride, adamsit, DM gas (CW)
диетвра́ч dietician
ди́зель diesel engine
~, бескомпре́ссорный solid injection diesel engine
~, компре́ссорный air-injection diesel engine
~ с непосре́дственным впры́ском diesel engine with direct injection, diesel engine with mechanical injection
ди́зель-мо́лот diesel power hammer
~, сва́йный diesel piling hammer
ди́зель-мото́р diesel engine
ди́зельный diesel
дизенте́рия dysentry
дизто́пливо diesel oil, diesel fuel
диизопропилфторфосфа́т di-isopropyl phosphoro-fluoridate
димеркаптопропано́л dimercaptopropanol, British Anti-Lewisite, BAL (CW)
диметиланили́н dymethylaniline
дина́мик moving-coil loudspeaker
дина́мика dynamics
~ бо́я fluidity of the battle
~ жи́дкости fluid dynamics
~ проры́ва momentum of the breakthrough
~, тя́говая dynamics of traction
динами́т dynamite
~, высокопроце́нтный high-grade dynamite
динами́ческий fast-moving (tac)
динами́чность dynamism
дина́мо dynamo
дина́мо-маши́на dynamo
~ постоя́нного сме́шанного возбужде́ния compound-wound direct current generator
дина́мо-маши́на постоя́нного то́ка dynamo, direct current generator
динамо́метр dynamometer dict75
~, тормозно́й brake dynamometer
динамо́н dynammon (expl)
динами́ческий dynamic
динатро́н dynatron
динафтали́т dinaphthalite (expl)
динитронафтали́н dinitronaphthaline (expl)
дино́д dynode (electron)
дио́д diode
~, вы́ращенный grown junction diode
~, герма́ниевый germanium diode
~, демпфи́рующий damping diode
~, диффузио́нный diffused junction diode
~, кристалли́ческий crystal diode
~, плоскостно́й junction diode
~, сплавно́й alloy-junction diode, fused-junction diode
дио́птр 1. diopter (optics) 2. aperture (sight)
дипо́ль dipole
~, аксиа́льный axial dipole
~, ве́ерный fan dipole
~ Ге́рца Hertz antenna
~, двойно́й double dipole
~, иго́льчатый needle dipole
~, магни́тный magnetic dipole
~, направля́ющий пасси́вный director (antenna)
~, отража́тельный chaff dipole
~, полуволново́й half-wave dipole
~, приёмный pick-up dipole
~, то́нкий thin dipole
~, широкополо́сный wide-band dipole
~, электри́ческий electric dipole
директи́ва directive
дире́ктор 1. director (antenna) 2. director (teleph) 3. transmitter (data transmission)
директри́са 1. directrix (gunn) 2. moving target gunnery range (for direct fire practice)
~ акусти́ческой ба́зы sub-base centre line
~ огня́, основна́я centre of arc (gunn)
~ прямо́й наво́дкой moving target gunnery range (for direct fire practice)
~ стрельбы́ 1. directrix (gunn) 2. gun centre of arc

директри́са, та́нковая moving-target tank gunnery range (for direct fire practice)
дирижа́бль dirigible
дирижёр, вое́нный bandmaster
дисбала́нс out-of-balance, unbalance
диск 1. disc, plate
2. drum magazine
3. plate (multiplex telegraph distributor)
~, ведо́мый driven plate (of clutch)
~, веду́щий driving disc plate
~, дистанцио́нный range disc
~ колёса wheel disc (veh)
~, нажи́мный pressure plate (clutch)
~, спусково́й actuating disc, trigger actuator
~ сцепле́ния clutch plate disc
~ сцепле́ния, ведо́мый clutch driven plate disc
~ сцепле́ния, нажимно́й clutch pressure plate disc
~ то́рмоза, опо́рный brake anchor plate
~, тормозно́й brake disc
~, траля́щий mine clearing roller disc
~ тре́ния friction plate
~, упру́гий flexible fabric disc (for coupling)
~ фрикцио́на clutch disc
~ фрикцио́на, ведо́мый clutch driven plate, clutch pressure plate
~ фрикцио́на, веду́щий clutch driving plate
~ фрикцио́на му́фты clutch plate
~ фрикцио́на, нажимно́й clutch pressure plate
~ фрикцио́на, опо́рный clutch cover plate
~ фрикцио́на, отжи́мный clutch release plate
~, фрикцио́нный friction disc, clutch plate
~ храпово́го колеса́ ratchet plate
ди́ски, дешифра́торные combiner wheels (Baudot receiver)
дискримина́тор discriminator (electron)
~ и́мпульсов pulse discriminator
~, ко́довый code discriminator
~, фа́зовый phase discriminator
~ часто́тной модуля́ции frequency-modulation discriminator
~, часто́тный frequency discriminator
дискримина́тор электро́нного дешифра́тора телегра́фного ко́да electronic code translator discriminator
дислока́ция 1. deployment garrisoning
2. disposition
дислоци́ровать to deploy
дисперги́рование dispersion, scattering
диспе́рсия dispersion
диспе́тчер, поездно́й train movements controller
диспетчериза́ция 1. progress system
2. dispatching (US)
диспе́тчерская dispatcher's office, control room, traffic controller's room
диспози́ция disposition
диссе́ктор dissector (tube)
~ с полупрозра́чным фотокато́дом image dissector
дистанцио́нный 1. remote (control)
2. delay (fuse)
диста́нция 1. range (weapon)
2. distance (drill)
~, безопа́сная safe range, safe distance
~ взведе́ния взрыва́теля arming distance
~, взлётная take-off distance
~ горе́ния трасси́рующего соста́ва range of trace (before burn out)
~ действи́тельного огня́ maximum effective range
~ до це́ли, и́стинная true distance to target
~, убо́йная killing range, lethal range
~, упреждённая future range
дистилли́рование distillation
дистилля́ция distillation
дисципли́на 1. discipline
2. subject (branch of learning)
~ в эфи́ре wireless discipline, radio discipline
~, во́инская military discipline
~ движе́ния march discipline
~ движе́ния по доро́гам road discipline
~ ма́рша march discipline
~, маскиро́вочная camouflage discipline
~ перегово́ров voice procedure
~ свя́зи communications discipline, radio discipline
~, специа́льная specialized subject
дисциплини́рованность degree of discipline
ди́тчер drag shovel (eng)
дифосге́н diphosgene

дифтери́т diphtheria
дифферéнт trim (ship or amphibian)
~ на кóрму down by the stern
~ нá нос down by the head (nav)
дифференциáл differential
~ зáднего мóста rear-axle differential
~, компенси́рующий comparison differential
~, промежýточный interaxle differential
~ с кони́ческими шестерня́ми bevel(-gear) differential
~ с шестерня́ми gear differential
~, самоблоки́рующий self-blocking differential
~, червя́чный worm gear differential
дифференциáтор differentiator (radio)
диффýзор 1. diffusor
2. choke tube, choke, venturi (ICE)
3. air duct
4. cone (loud-speaker)
~, первúчный primary venturi
дихлорами́н dichloramine
дихлордифенилтрихлорэтáн DDT
дихлордиэтилсульфи́д dichlorodiethyl sulphide, mustard gas
дихлорэтáн dichloroethane
диэлéктрик dielectric, non-conductor
длинá волны́ wavelength
~ волны́, крити́ческая cut-off wavelength, critical wavelength
~ волны́, предéльная cut-off wavelength, critical wavelength
~, габари́тная overall length, clearance length
~ гýсеницы, опóрная length of track on the ground
~ заря́дной кáморы seating distance
~ канáла стволá length of bore (gunn)
~, наибóльшая length overall
~ откáта length of recoil
~, относи́тельная length in calibres, relative length
~ с пýшкой вперёд, óбщая overall length with gun forward (tk)
~ хóда lead (rifling, thread etc)
длинноствóльный long-barrelled
длиннофóкусный long-focus
дли́тельность вспы́шки flash duration
~ и́мпульса pulse length, pulse width, pulse duration
~ развёртки sweep interval
~ спáда и́мпульса pulse decay time
дли́тельность счи́тывания reading duration, persistence (radar)
~ фрóнта и́мпульса pulse rise time
~ хранéния storage period
для служéбного пóльзования 1. "for official use only"
2. restricted (security classification)
дневáльный orderly
~ по рóте company orderly (responsible for arms and eqpt - assists orderly sergeant)
днéвка day's halt
дневнóй day, daylight, daytime
дни́ще 1. bottom (river, trench etc)
2. crown (piston)
3. hull floor (tk)
~ пóршня piston head
~ рва траншéй trench bottom
дно боевóй чáсти warhead bottom (arty)
~, ввинтнóе screwed base (ammo)
~, вкладнóе pinned base (ammo)
~ ги́льзы base (cartridge case)
~ кóроба bottom plate (MG)
~ нарéза bed of groove
~ пóршня piston crown, piston head
~ снаря́да base of shell
~ снаря́да, ввинтнóе base plug
до нóрмы up to scale
добáвка addition
~, стабилизи́рующая stabilizing agent (propellants etc)
добровóлец volunteer
добровóльный voluntary
добрóтность quality factor, Q factor (elec)
добы́ча 1. yield, output
2. extraction
3. loot, booty, plunder
довéсок заря́да incremental charge
довести́ (см доводи́ть)
доводи́ть/довести́ to lead; **доводи́ть/довести́ до штáтной чи́сленности** to bring up to establishment; **доводи́ть/довести́ задáчу** to set a task
довóльствие rations, pay and allowances
~, вещевóе clothing allowance
~, дéнежное pay and allowances, pay
~, добáвочное дéнежное extra cash allowance, additional pay
~, квартúрное 1. barrack services
2. lodging allowance
~, котловóе collective messing, communal messing
~, натурáльное rations in kind

дово́льствие, су́точное one day's ration
~, фина́нсовое cash allowance
дово́льствовать to administer, to feed, to sustain, to subsist (imp only)
довора́чивать/доворoти́ть to "screw home" (coll - of a nut etc), to complete a turn
доворо́т switch (gunn)
~, исчи́сленный predicted switch (gunn)
~ от репе́ра на цель, исчи́сленный corrected map shift (gunn)
~, топографи́ческий map line (gunn)
доворoти́ть (см довора́чивать)
догово́р treaty, agreement, contract
~, ми́рный peace treaty
~ о ненападе́нии non-agression treaty
дождеме́р rain gauge
дожига́ние afterburning
до́за, безопа́сная safe dose
~, допусти́мая permissible dose
~ излуче́ния radiation dose
~ излуче́ния, поглощённая absorbed radiation dose
~, интегра́льная accumulated dose (nucl)
~, лета́льная lethal dose, fatal dose
~ облуче́ния radiation dose
~ облуче́ния, допусти́мая биологи́ческая biological radiation tolerance
~ облуче́ния, сумма́рная total radiation dose
~, поглощённая absorbed dose
~, преде́льно-допусти́мая maximum permissible dose (eg of radiation)
~ радиа́ции radiation dose
~, смерте́льная fatal dose, lethal dose
~, сумма́рная total dose (nucl)
~, токси́ческая toxic dose
~, чрезме́рная overdose
~, эквивале́нтная equivalent dose
дозапра́вить (см дозаправля́ть)
дозапра́вка "topping up" (coll), refuelling
дозаправля́ть/дозапра́вить to top up, to refuel
дозаряди́ть (см дозаряжа́ть)
дозаря́дка charging, charging up
дозаряжа́ть/дозаряди́ть to (re)load with full magazine, to (re)load to capacity
доза́тор 1. weight batcher 2. automatic cartridge filling machine
дозвуково́й subsonic
дозиме́тр dosimeter
~ а́льфа-излуче́ния alpha-survey meter
дозиме́тр для измере́ния га́мма-радиа́ции gamma-radiation monitoring instrument
~, индивидуа́льный individual dosimeter
~ каранда́шного ти́па pen-type dosimeter
~, карма́нный pocket dosimeter
~, кольцево́й плёночный film ring dosimeter (radiac)
~, перено́сный portable dosimeter
~, плёночный film dosimeter, radiation exposure strip
~ с переда́тчиком transmitting dosimeter (radiac)
дозиметри́ст radiological monitor
дозиметри́ческий monitoring, dosimetric, radiac
дозиме́трия radiac dosimetry, radiation dosimetry, monitoring
дозиро́вка 1. proportioning, metering 2. dosage
дознава́тель investigating officer
дозна́ние legal inquiry
дозо́р patrol
~, боево́й разве́дывательный battle reconnaissance element, fighting and reconnaissance patrol
~, боково́й flank guard, flank patrol
~, головно́й point of vanguard, forward patrol of advance guard
~, головно́й разве́дывательный leading reconnaissance patrol
~, инжене́рно-разве́дывательный engineer reconnaissance patrol
~, маскиро́вочный camouflage patrol (checks own camouflage)
~, отде́льный разве́дывательный 1. independent reconnaissance patrol 2. independent reconnaissance group
~, па́рный two-man patrol
~, постоя́нный standing patrol
~ похо́дного охране́ния mobile reconnaissance troops of protective force
~ радиацио́нной и хими́ческой разве́дки NBC reconnaissance patrol
~, радиолокацио́нный radar picket
~, разве́дывательный reconnaissance patrol (up to pl strength)
~, ты́льный rear party of rear guard
~, хими́ческий разве́дывательный CW reconnaissance patrol

дозо́рный 1. member of patrol
2. patrol
дозо́рный-раке́тчик patrol rocket launcher number
док, плаву́чий floating dock
докла́д report
докла́дывать/доложи́ть to report
доктри́на doctrine
~ бронета́нковых войск, такти́ческая armoured tactical doctrine
докуме́нт document
~, засвиде́тельствованный оправда́тельный certificate receipt voucher, C.R.V.
~, прихо́дный receipt, voucher
~, прихо́до-расхо́дно- оправда́тельный issue and receipt supporting voucher
~, проездно́й travel document
документа́ция documents, documentation, papers
~, медици́нская medical records
~ ми́нных поле́й recording of minefields
~, отчётная technical papers, record(s)
~, эксплуатацио́нная operating instructions, operating documents
докуме́нты, ли́чные personal documents
~, учётно-отчётные accounting records, accounting instruments
долг, во́инский military duty
~ слу́жбы call of duty
~, служе́бный call of duty
долгове́чность length of life, durability
долговре́менный permanent, lasting
долгота́ longitude
~, географи́ческая longitude
до́лжность appointment post
~, исполня́ющая acting status
~, штабна́я staff duty, staff appointment
долива́ние topping up
доли́вка topping up
доли́на valley, vale
доложи́ть обстано́вку to report on the situation
~ (см докла́дывать)
долото́ 1. chisel
2. drill bit
дом офице́ров officer's club
до́мик счётной устано́вки castle (geiger counter housing)
домкра́т jack
~, буты́лочный screw jack
~, винтово́й screw jack
~, гидравли́ческий hydraulic jack
домкра́т, ре́ечный rack-and-lever jack
~, ручно́й hand jack
донесе́ние report
~, боево́е action report
~ о чрезвыча́йном происше́ствии
1. accident report
2. report on unusual occurrence
~ об обстано́вке situation report
~, пи́сьменное written report
~ по ли́чному соста́ву, су́точное personnel report
~ по свя́зи, боево́е signals completion report (Sov - rendered to signals officers giving time of establishment of signal comms and details)
~, разве́дывательное 1. intelligence report
2. reconnaissance report
~, спе́шное urgent message
~, сро́чное priority message
донести́ (см доноси́ть)
до́нный base, bottom
доноси́ть/донести́ to report, to inform
доно́счик informer
до́нышко ги́льзы base of cartridge case
дообеспе́чение completion of deficiencies, making up of deficiencies
дообеспе́чивать/дообеспе́чить to make up deficiencies, to bring up to scale
дообеспе́чить (см дообеспе́чивать)
дооборудова́ть to complete fitting out, to complete equipping
До́плер Doppler
дополне́ние supplement, appendix
допра́шивать/допроси́ть to interrogate
допризы́вник man undergoing pre-conscription military training (between registering & call-up)
допро́с interrogation
допроси́ть (см допра́шивать)
до́пуск 1. admittance
2. tolerance (mech)
~, жёсткий разме́рный close dimensional tolerance
~ к соверше́нно секре́тной рабо́те top secret clearance
~ по ве́сам weight allowance
допуска́емый permissible, allowable
до́пуски, угловы́е angular tolerances
дорабо́тка modification, alternations
доразве́дка "final" reconnaissance
доразве́довать to make final reconnaissance

дорн mandrel
доро́га road, roadway
~, **автогужева́я** motor road
~, **автолежнева́я** corduroy motor road, logging road
~, **автомоби́льная** (motor) highway, (motor) road, autoroad, motor way, road traffic route
~, **асфальти́рованная** asphalt road, bituminous road
~, **бреве́нчатая** corduroy road (log road laid over swampy ground)
~ **в две полосы́ движе́ния** dual carriage-way
~, **вое́нная** 1. military road 2. line of communication, L of C
~, **вое́нно-автомоби́льная** maintenance road
~, **войскова́я** tactical road
~, **вре́менная** temporary road
~, **вре́менная желе́зная** field railway, light railway
~, **всепого́дная** all-weather road
~, **вью́чная** bridle path
~, **гатева́я** corduroy road
~, **грави́йная** gravel road
~, **гре́йдерная** earth track, dirt road
~, **грунтова́я** dirt road, earth road, unsurfaced road, cross-country track, natural soil road, unimproved road
~, **грунтова́я улу́чшенная** earth road, mechanically-stabilized earth road
~, **двухпо́лосная автомоби́льная** dual carriage-way
~, **двухпу́тная** two-lane road, two-track road, two-way road
~, **желе́зная** railway, railroad
~, **жердева́я** corduroy road (log road laid over swampy ground)
~, **за́городная** country road
~, **кана́тная подвесна́я** aerial cableway, cable railway, aerial ropeway
~, **кана́тно-подвесна́я** aerial cableway, cable railway, aerial ropeway
~, **капита́льная** first-class all-weather road with 100 ton capacity bridges and at least 14 metre carriageway
~, **коле́йная** treadway road
~, **коле́йно-лежнева́я** treadway road
~, **лежнева́я** corduroy road (log road laid across swamp land)
~, **магистра́льная** main highway
доро́га, магистра́льная желе́зная main-line railway, trunk-line railway
~ **ме́стного значе́ния** country road, country route, district road, district route, local road, local route, regional road, regional route
~, **многопу́тная** multi-lane road
~, **многоря́дная** multi-lane road
~, **мощёная** paved highway
~ **на косого́ре** hillside road
~, **надзе́мная** elevated road, elevated roadway
~, **насыпна́я** causeway
~, **непрое́зжая** impassable road
~, **объездна́я** bypass
~, **оде́тая** surfaced road, paved road, made-up road
~, **однопу́тная** one-track road, one-way road
~, **подвесна́я кана́тная** aerial cableway, cable railway, aerial ropeway
~, **подъездна́я** access road, approach road
~, **полева́я** cart track
~, **приго́дная для испо́льзования то́лько в суху́ю пого́ду** fair-weather road
~, **приго́дная для испо́льзования в любу́ю пого́ду** all-weather road
~, **просёлочная** dirt road, country road, side road
~, **профили́рованная** earth road
~, **радиа́льная** radial road, front to rear road
"ДОРО́ГА РЕМОНТИ́РУЕТСЯ" "ROAD UNDER REPAIR"
доро́га, рока́дная lateral road, lateral route, lateral line of communication (see also rokada)
~, **рулёжная** taxiway (airfield)
~ **с вы́пуклым про́филем** cambered road
~ **с металли́ческим покры́тием** metalled road
~ **с ограни́ченным число́м въе́здов** limited access road
~ **с одно́й полосо́й движе́ния** single-track road
~ **с односторо́нним движе́нием** one-way road
~ **с поса́дкой** road with surface vegetation on verges
~ **с твёрдым покры́тием** hard road, hard-surfaced road

доро́га, сквозна́я through road
~, стратеги́ческая strategic road
~, то́рная beaten track
~, трёхпу́тная three-lane road
~, тупико́вая "dead-end" road, cul-de-sac
~, узкоколе́йная полева́я желе́зная light military railway, narrow gauge railway
~, фронта́льная axial road
~, цеме́нтно-бето́нная concrete road
~, чёрная tarred surface road, black-top road
~, шоссе́йная 1. highway
2. bituminous road
~, шосси́рованная tarmac road
~, щебёночная ballast road, broken-stone road
доро́га-зи́мник road used only in winter
доро́жка 1. path, track
2. runway
~, бегова́я 1. wheel path (tk)
2. raceway, ball raceway
~, жердева́я corduroy track
~ каче́ния raceway
~, поса́дочная landing strip
~, рулёжная taxiway (airfield)
~, ста́ртовая runway
доро́жка-насти́л beach roadway
доро́жно-коменда́нтский provost traffic-control
доска́ board, plate
~, винтова́льная screw plate
~, инструмента́льная instrument panel
~, кро́вельная roofing board
~ насти́ла floor board
~, насти́лочная chess (br)
~, ни́жняя опо́рная groundsill, groundsel
~, обши́вочная spile, spiling board (mining), sheeting, lagging
~, полова́я flooring, floor board
~ почёта "board of honour" (exhibited list of pers of unit or sub-unit who have distinguished themselves)
~, поясна́я chord of plank web girder (br)
~, прибо́рная instrument panel, dashboard
~, прижи́мная pressure plate
~, распредели́тельная switchboard panel, distribution board
~ свя́зи connecting panel
доска́, чёрная 1. blackboard
2. "black list"
досла́ть (см досыла́ть)
досло́вный verbatim
доста́вить (см доставля́ть)
доста́вка delivery, forwarding
~ гру́зов беспоса́дочным спо́собом airborne supply drop
~ к це́ли стратеги́ческого я́дерного ору́жия strategic nuclear delivery
~ материа́льних средств delivery of stores and equipment
~ по во́здуху air delivery
~ средств материа́льно-техни́ческого обеспе́чения поса́дочным спо́собом air-landed logistic support
~, сро́чная emergency delivery, express shipment
доставля́ть/доста́вить to deliver, to supply
достига́ть/дости́чь to reach, to attain; **достига́ть/дости́чь превосхо́дства над проти́вником** to gain advantage over the enemy
достиже́ние attainment, achievement, progress
дости́чь (см достига́ть)
достове́рность све́дений reliability of information, credibility of information
до́ступ approach, access
досту́пность accessibility
досыла́ние вручну́ю hand ramming
досыла́тель 1. feed horn (SMG)
2. rammer
~ затво́ра feed piece
~, механи́ческий power rammer (arty)
~, пневмати́ческий pneumatic rammer
досыла́ть/досла́ть to ram (gunn)
досыла́ющий "rammer number" (arty)
досы́лка ramming (gunn)
досы́льник "rammer number" (arty)
досяга́емость range, reach, extent
~, наибо́льшая maximum effective range (arty)
~ огня́ range of fire
~ по высоте́ 1. ceiling (arty)
2. maximum height (GW)
~ по высоте́, преде́льная maximum vertical range, ceiling (AA gunn)
доукомплектова́ние making up of deficiencies, bringing up to scale (eqpt)

доукомплектова́ть to make up deficiencies,, to bring up to scale (eqpt), to bring up to strength (pers)
дра́га dredge, drag
драги́рование dredging
драги́ровать to dredge, to drag
драгле́йн dragline (excavator)
древеси́на wood, timber
~, делова́я (commercial) timber, workable wood
~, мелкосло́йная fine-grained wood
~ мя́гких поро́д soft wood
~, сухосто́йная dead wood
~, сыра́я unseasoned wood
~ твёрдых поро́д hard wood
древеси́на-ба́льза balsa wood
древе́сный wood
дре́вко flag staff, pike (for banners)
дрези́на line inspection trolley (rly)
дрейф leeway (navigation), drift
дрель bit brace, brace
дрена́ж 1. drainage
2. fuel tank venting system
дрени́рование drainage, draining
дрени́ровать to drain
дроби́лка crusher
дроби́ть/раздроби́ть to split up
дробле́ние 1. splitting up
2. breaking up (of shell)
~ частоты́ frequency splitting
дробови́к shotgun
дробс drop indicator (teleph)
дробь 1. small shot, canister (ammo)
2. rattle (MG fire)
3. fraction (maths)
~, бараба́нная ruffle (low beat of drum)
дробя́щий blasting
дро́ссель 1. throttle, choke (carb)
2. choke, choke coil (elec)
~, блокиро́вочный bypass choke (elec, electron)
~, по́лностью откры́тый fully open throttle
дружи́на 1. bodyguard (obs)
2. volunteer squad
~, доброво́льная пожа́рная voluntary fire fighting group
~, наро́дная volunteer public order patrol (civil defence)
~, санита́рная voluntary aid detachment
дружи́нник противовозду́шной оборо́ны air raid warden
дуб oak
дубли́рование duplication
~ свя́зи duplication of communication facilities
дубли́ровать to parallel, to duplicate, to double (imp only)
дуга́ arc (electron, welding)
~ траекто́рии arc of trajectory
~, электри́ческая arc (elec)
дугово́й arc
ду́ло muzzle
ду́льце mouth of cartridge case (ammo)
ду́мпкар dump car, tipping car
ду́плекс, дифференциа́льный differential duplex (teleg)
~ по схе́ме мо́стика bridge duplex (comms)
дуст fumigant
дух, боево́й fighting spirit
~, мора́льный morale
душева́я shower room
дуэ́ль, артиллери́йская artillery duel
дым smoke
~, маскиру́ющий obscuring smoke, screening smoke
~, нейтра́льный non-toxic smoke
~, цветно́й coloured smoke
~, ядови́тый toxic smoke
дымзаве́са smoke screen
ды́мка dry haze (met)
~, пульси́рующая mirage
дымле́ние smoking, fuming
ды́мность вы́хлопа exhaust smoke
дымово́й smoke
дымогенера́тор smoke generator
дымомаскиро́вка smoke screening, laying of smoke screen
дымообразова́ние smoke formation
дымообразова́тель smoke generator
дымопу́ск smoke production, smoke generation
дымохо́д flue
дымснаря́д smoke shell
дыха́ние, иску́сственное artificial respiration
дышло́ 1. pole beam
2. tow bar
дюйм inch
дю́на dune, sand drift
дюра́ль duralumin
дюралюми́ниевый duralumin
дюралюми́ний duralumin

Е

единица 1. unit, element
2. unity
3. figure one
~, британская тепловая British thermal unit, BTU
~ измерения unit of measure
~ информации, двоичная (information) bit, binary digit
~ количества теплоты heat unit, thermal unit, unit of heat quantity
~ массы 1. unit of mass
2. mass unit (nucl)
~ площади unit area (math)
~, расчётно-снабженческая unit of supply (standard measure or unit)
~, тактическая tactical unit
единицы, тактические tactical units
единоборство single combat
единовременный 1. one-time
2. simultaneous
единогласный unanimous
единоначалие sole responsibility (of commander for both military and political matters)
единоначальник "sole commander" (responsible for both military and political matters)
единственный sole, only
единство unity
единый united, common
едкий 1. caustic, corrosive
2. pungent, acrid
ёж hedgehog (obstacle)
~, противотанковый anti-tank hedgehog (obstacle, eng)
ежегодник almanac, annual publication, year-book
ёлочкой herring-boning (skiing)
ель fir, spruce, white deal
ёмкости containers, tanks
~, заправочные refueling tanks
ёмкость 1. capacity
2. capacitance (elec)
~ аккумулятора storage battery capacity
~ бака tank capacity
~ в ампер-часах amp-hour capacity
~ в ватт-часах watt-hour capacity
~, входная input capacitance
~, выходная output capacitance
ёмкость госпиталя, коечная hospital bed capacity
~, динамическая capacitance reactance (elec)
~ запоминающего устройства memory capacity
~, заправочная tank capacity (POL)
~ ленты belt capacity
~, междувитковая interwinding capacitance (elec)
~, междуэлектродная inter-electrode capacitance (elec)
~ памяти storage capacity
~, паразитная stray capacitance (elec)
~, подземная underground storage tank
~, проходная grid-anode capacitance (elec)
~, эластичная dracone
ёршик cleaning brush (SA)
~, проволочный wire brush
"Есть!" "Yes Sir!"
ефрейтор lance-corporal

Ж

жалин jaline, picric powder
жало striker needle (ammo)
~, дистанционное time needle (fuse)
~, ударное percussion needle (fuse)
жалоба complaint
жалюзи 1. shutters
2. engine louvres
~, вентиляторные air outlet louvres (tk)
~, впускные air intake louvres, air inlet louvres
~, входные air inlet louvres, air intake louvres
~, выпускные air outlet louvres
~ прожектора searchlight shutters
~, радиаторные radiator air inlet louvres (tk)
жаропрочный heat-proof, refractory
жаростойкий heat-proof
жаростойкость heat resistance
жароупорный heat-proof, heat-resistant
жгут 1. twisted material, braid
2. tourniquet
3. packing material (stuffing box)
ждущий triggered (electron)

желати́н, гремýчий blasting gelatine
желати́н-динами́т gelatine-dynamite
железá gland (anat)
железнодорóжный railway
желéзный iron
желéзо iron
~, **арматýрное** concrete reinforcement bars, concrete reinforcement wire
~, **волни́стое** corrugated iron
~, **корытное** channel-section steel
~, **котéльное** boiler plate
~, **крóвельное** sheet iron, sheet roofing iron
~, **листовóе** sheet iron
~, **литóе** ingot iron, cast iron
~, **оцинкóванное** galvanized iron
~, **полосовóе** strip iron
~, **свáрочное** wrought iron
~, **сортовóе** section iron, structural iron
~, **техни́ческое** mild steel
~, **угловóе** angle iron
~, **швéллерное** channel iron
~, **Ш-обрáзное** E-beam
железобетóн ferro-concrete, reinforced concrete
~, **предвари́тельно-напряжённый** prestressed reinforced concrete
~, **сбóрный** precast concrete
железобетóнный reinforced concrete
жёлоб 1. gutter, trough
2. barrel groove (ord)
3. rolling bearing track
4. narrow steep gully (topo)
~, **направля́ющий** groove (on running surface of ski)
~ **цевья́ лóжи** barrel channel, barrel bed
~, **шаровóй** roller path
желобóк groove (rifling)
~, **прицéльный** sighting groove (gunn)
"жёлудь" acorn valve (electron)
жéнщина-денщи́к batwoman (servant)
жердь pole, picket, rod
~, **прижи́мная** revetment picket
~, **продóльная** revetment lacing board
~, **расщеплённая** longitudinally-split pole
жёсткость hardness, stiffness, rigidity
~ **на кручéние** torsional rigidity
~ **подвéски** suspension rate
жесть tin
живóтное, вью́чное pack animal
живýчесть 1. "working life" (eqpt)
2. ability to survive (pers)
живýчесть ствола́ barrel life
~ **цéли** target vulnerability
жи́дкость, агресси́вная corrosive fluid
~, **вя́зкая** viscous fluid
~, **гидротормозна́я** hydraulic brake fluid
~, **горю́чая** inflammable liquid
~, **дегазациóнная** liquid decontaminating agent
~, **дегази́рующая** personal CW decontamination kit
~, **дезактивациóнная** decontamination liquid
~ **для гидросистéм** hydraulic fluid
~, **незамерзáющая** antifreeze liquid
~, **низкозамерзáющая** antifreeze liquid
~, **низкозамерзáющая охлаждáющая** antifreeze liquid
~, **охлаждáющая** 1. cutting lubricant, cutting fluid (machine tools)
2. coolant, engine cooling liquid
~, **пенообразýющая** foam-forming liquid
~, **промы́вочная** cleansing fluid, flushing fluid
~, **самовоспламеня́ющаяся** self-igniting incendiary fuel
~, **смáзочно-охлаждáющая** cutting oil
~ **стéол** "steol" hydraulic fluid (solution of glycerine, ethyl alcohol and anti-corrosion additive)
~, **тормозна́я** brake fluid
жизнь, боевáя life in the field
жиклёр jet, carburettor jet
~ **бы́стрых оборóтов** accelerator jet (carb)
~, **глáвный** main jet (carb)
~, **дози́рующий** metering jet (carb)
~, **дополни́тельный** auxiliary jet (carb)
~, **компенсациóнный** compensating jet (carb)
~ **мáлых оборóтов** idle jet, idling jet (carb)
~, **пусковóй** starting jet, starter jet (carb)
~, **тóпливный** fuel(-injection) nozzle, fuel(-injection) jet, fuel-spray nozzle, fuel-spray jet
~ **ускори́тельного насóса** accelerating jet (carb)
~ **холостóго хóда** idling jet, idle running jet (carb)
~ **экономáйзера** economizer jet (carb)

жи́ла 1. core
2. strand (wire)
3. vein
~, изоли́рованная core (of elec cable)
~, ме́дная copper core (cable)
~ про́вода cable core
~, токопроводя́щая conductor (cable)
~, уплотнённая segmental-strand conductor
жиле́т, бронево́й body armour
~, компенси́рующий pressure breathing waistcoat
~, плава́тельный swimming jacket
~, спаса́тельный life-jacket
жи́лка 1. strand (wire)
2. core (cable)
жи́мки bulldog grip (br)
жир fat, grease
жи́тель, ме́стный local inhabitant
жнивьё 1. stubble
2. field containing stubble
жо́лоб 1. gutter, trough
2. barrel groove (ord)
3. rolling bearing track
4. narrow steep-sided gully (topo) (see also "zhelob")
жужжа́лка buzz bomb
журна́л log book, register
~, аппара́тный operator's log
~ боевы́х де́йствий war diary
~, входя́щий signal log "in"
~ вы́хода и возвраще́ния маши́н vehicle in-and-out register
~ испыта́ний trials log
~, исходя́щий "out" register (sigs)
~ наблюде́ния observer's log
~ облуче́ния radiation exposure record (nucl)
~, рабо́чий work sheet, work log
~ разве́дки observation post log book
~ состоя́ния свя́зи radio station log, signal centre log, communication centre log
~ учёта register
~ учёта батаре́й проти́вника counter-battery log
~ учёта огнево́й де́ятельности артилле́рии и миномётов проти́вника log of hostile fire
~ це́лей target log

З

"За Мной!" "Follow Me!"
"За Мной в Коло́нну по Одному́ Марш!" "In Single File behind me Move!"
"За Мной к Бо́ю!" "Extended Line on me!"
"За Спину́!" "Sling Arms!" (in Sov army wpn is slung diagonally across back)
забаррикади́ровать (см баррикади́ровать)
забива́ние jamming, blocking (comms, tac)
забива́ть/заби́ть to drive in, to hammer in, to stop up, to fill up, to jam (radio); **забива́ть/заби́ть сва́ю** to drive a pile
заби́вка tamping (expl)
~ свай pile driving
заби́ть (см забива́ть)
заблаговре́менность the interval between the issue of a meteorological forecast and the start of the period to which it applies
заблуди́ться to lose one's way (perf only)
заблужда́ться to err, to be mistaken (imp only)
забо́й face (mining)
~ скота́ butchering, slaughtering, cattle slaughtering
забо́йна dent
забо́йщик face worker (mining)
заболева́емость sick rate
заболева́ние, инфекцио́нное infectious disease
~, лучево́е radiation sickness
заболо́ченность swamp, condition of marshy ground
забо́р 1. fence
2. embankment (A tk obstacle)
~ на высо́ких ко́льях, про́волочный high wire fence
~, про́волочный wire fence, wire-fence obstacle
~, уси́ленный про́волочный double apron fence
~, электризу́емый про́волочный electrified fence
забо́рник filler neck
~ во́здуха air intake
забракова́ть (см бракова́ть)
зава́л barrier, block

зава́л, лесно́й tree barrier, tree blowdown
~ **на доро́ге** road block
~, **прибре́жный** tree barrier at water's edge
заведе́ние institution, establishment
~, **вое́нно-уче́бное** military educational establishment
~, **вы́сшее уче́бное (ВУЗ)** higher educational establishment
~, **уче́бное** training establishment
заве́дующий скла́дом officer (or NCO) i/c store
заве́ренный attested, certified correct
заве́рить (см заверя́ть)
заверша́ть/заверши́ть to complete; **заверша́ть/заверши́ть ата́ку** to press home the assault
заверше́ние completion, conclusion, pressing home (of the assault)
заверши́ть (см заверша́ть)
заверя́ть/заве́рить to certify
заве́са, артиллери́йская curtain of fire
~, **дымова́я** smoke screen
~, **маски́рующая дымова́я** screening cloud (smoke)
~, **назе́мная дымова́я** ground smoke screen
~, **огнева́я** curtain of fire
~, **ослепля́ющая дымова́я** blanketing cloud (smoke), smoke screen
~, **подви́жная дымова́я** rolling smoke screen
~, **прикрыва́ющая** protective screen
~, **противоракетная заградительная** aerospace defence shield (US)
~, **противота́нковая огнева́я** anti-tank barrage
завести́ (см заводи́ть)
завизи́ровать to approve (requisition etc) (perf only)
зависа́ющий hovering
зави́симость, механи́ческая mechanical interlocking
~, **перекрёстная** cross coupling (GW)
завихре́ние backwash, wake turbulence
завихри́тель atomizer (fuel burner)
заво́д, артиллери́йский ordnance factory
~ **боеприпа́сов** munition factory
~, **вое́нный** munition factory
~ **вооруже́ния** armament factory
~, **локомоти́вный** locomotive repair works
~, **оруди́йный** ordnance plant
заво́д, патро́нный cartridge factory, ammunition factory
~ **по снаряже́нию боеприпа́сов** ammunition filling plant
~, **снаря́дный** shell factory, ammunition factory
~, **снарядоснаряжа́тельный** shell-filling factory
~, **снаряжа́тельный** loading plant, filling plant
~, **та́нковый** tank(-production) plant, tank(-production) factory
~, **ши́нный** tyre factory
"Заводи́!" "Start Up!"
заводи́ть/завести́ to start up (engine, veh), to wind up (clockwork)
заво́дка starting up (engine)
за́водь creek, backwater
завоева́ние 1. conquest 2. achievements, gains
завоева́ть (см завоёвывать)
завоёвывать/завоева́ть to conquer, to win, to gain
завуали́рованный fogged (photo), veiled
завяза́ть (см завя́зывать)
завяза́ть/завя́знуть to bog down
завя́зка бо́я initial stages of encounter battle, engagement of the enemy
завя́знуть (см завяза́ть)
завя́зывать/завяза́ть to tie, to tie up, to knot, to join; **завя́зывать/завяза́ть бой** to join battle
загаса́ть/загасну́ть to go out, to die out (intrans)
загаси́ть (см загаша́ть)
загасну́ть (см загаса́ть)
загаша́ть/загаси́ть to extinguish, to put out, to turn off (trans)
загерметизи́ровать to waterproof, to seal
заги́б bend, fold, crease, bead flange
~ **ги́льзы** case rim (ammo)
~, **носово́й** turn-up (tip of ski)
~, **подаю́щий** feed lip (ord)
заги́бы lips (magazine)
загло́хнуть (см гло́хнуть)
заглубле́ние digging in, sinking
заглублённый dug in, sunken
заглуша́ть/заглуши́ть to jam, to blackout, to bar, to damp, to attenuate (radio, radar), to muffle, to deaden (sound)
заглуше́ние muffling, silencing, suppression, quenching, jamming
заглуши́ть (см заглуша́ть)

заглýшка blank, core plug, plug, stopper, end cap
заголóвок preamble (comms)
загорáние lighting up (lamps), catching fire
загорáющийся inflammable, igniting
заготáвливать/заготовля́ть to procure, to stock up, to prepare
заготóвка 1. acquisition, procurement, provision
2. blank (metal for turning)
заготовлéние, интендáнтское ordnance and stores supply
заготовля́ть (см заготáвливать)
заградúтель rejector (electron)
~, мúнный mine layer (nav)
заградúть (см заграждáть)
заграждáть/заградúть to barricade, to obstruct, to stop, to block
заграждéние artificial barrier, artificial obstacle, engineer obstacle
~, аэростáтное balloon barrage
~, бетóнное concrete obstacle
~, бревéнчатое log obstacle
~, воздýшное balloon barrage
~, дорóжное road obstruction, road obstacle, road-block
~, дорóжное охранúтельное security road-block
~, защúтное прóволочное protective wire entanglement
~ из колю́чей прóволоки barbed wire obstacle
~ из пирамидáльных железобетóнных нáдолб, противотáнковое dragon's teeth, conical tank obstacle
~, инженéрное artificial obstacle
~, инженéрно-химúческое chemical obstacle, NBC obstacle
~, искýсственное artificial obstacle
~, малозамéтное low wire entanglement
~, манёвренное tactical minefield
~, мúнно-взрывнóе mixed minefield
~, мúнное minefield
~ на дорóге road-block
~, наводнóе boom, protective obstacle on surface of water, temporary boom obstacle
~, невзрывнóе противотáнковое unmined anti-tank obstacle
~, огневóе barrier of fire, curtain of fire
~, перенóсное portable obstacle, movable obstacle
заграждéние под напряжéнием, электризýемое прóволочное electrified wire (entanglement)
~, подвóдное underwater obstacle
~, постоя́нное fixed obstacle
~, прóволочное barbed wire entanglement
~ прóтив воздýшного десáнта anti-airborne assault obstacle
~, противодесáнтное 1. beach obstacle
2. anti-assault-landing obstacle
~, противотáнковое anti-tank obstacle
~, противотáнковое огневóе anti-tank indirect defensive fire
~, сетевóе net defence
~, стационáрное fixed obstacle
~, тактúческое tactical obstacle
~, фортификациóнное fortified construction, fortified obstacle
~, химúческое chemical obstacle, gas obstacle, gas barrier
~, электризýемое electrified obstacle
заграждéния, берегoвые beach obstacles
~, взрывные demolition obstacle belts
~, инженéрные artificial obstacles (constructed by engineers)
~, инженéрные подвóдные и берегoвые artificial underwater and beach obstacles
~, электризóванные electrified wire obstacles
~, электризóванные прóволочные electrified wire obstacles (fences etc)
заграничный foreign
загружáть/загрузúть to load, to fill, to charge
загрузúть (см загружáть)
загрýзка loading, load (elec etc)
загрýзчик (conveyor) loader, belt loader, roller loader, chain loader
загрязнéние fouling, pollution
~,повéрхностное surface contamination
загýбник mouthpiece (for frogmen)
загýстевание thickening, solidification, condensation
загущённость thickness
загущённый thick, dense, thickened
задáние task
~, плáновое target, objective
~ по подвóзу conveyance task
~, сýточное daily task, daily target
задáча task, aim, mission, intention, problem

задáча, ближáйшая immediate objective, first objective
~, боевáя mission, task, combat task (Sov term includes peacetime guard duties)
~, внеплáновая opportunity task
~ встрéчи "the Ack-Ack problem" (coll), prediction problem, anti-aircraft problem
~, неплáновая огневáя impromptu fire mission
~, огневáя
~, основнáя main task
~, основнáя боевáя fundamental operational task
~, очереднáя immediate task
~, послéдующая subsequent task
~, постáвленная set task, allotted task
~, самостоя́тельная independent task
~, тактúческая tactical mission
~, чáстная боевáя particular operational task
~ этáпа, боевáя operational phase objective
задáча-летýчка short exercise
задвúжка 1. slide (MG), bolt, slide valve 2. locking strip (mine)
задевáть/задéть to catch, to graze, to foul
задéлка, контáктная contact piece (ord)
задéлывание sealing
задержáть (см задéрживать)
задéрживать/задержáть to stop, to halt, to contain, to stem, to delay, to retard, to hinder, to hold up, to detain, to arrest, to apprehend
задéржка 1. stoppage, jamming 2. delay, hold-up
~ воспламенéния ignition delay time
~, запирáющая breech lock (SA)
~, затвóрная 1. hold-open catch (SA) 2. slide stop (pistol) 3. bolt stop (rifle)
~ úмпульса в приёмнике receiver delay
~ по врéмени time lag
~, трéбующая значúтельного врéмени для её устранéния prolonged stoppage
~, фиксúрованная set delay
задéржник retainer
задéть (см задевáть)
задúр scoring (in gun barrel)
"Зáдний Ход!" "Reverse!" (veh)
задóлженность liability
задымúть (см задымля́ть)
задымлéние smoke screening
задымля́ть/задымúть to lay a smoke screen
заедáние jamming, catching, blocking, seizure
~ подшúпника bearing seizure
~ тóрмоза brake binding
заéзд run
~, зачётный classification run (driver trg)
~, контрóльный test run (driver trg)
~, тренирóвочный practice run (veh)
зажжéние initiation (ignition)
зажигáние ignition
~, аккумуля́торное battery ignition
~, батарéйное battery ignition
~, батарéйно-катýшечное battery-and-coil ignition
~, внéшнее external pilot-flame ignition (msl)
~, внýтреннее internal pilot-flame ignition (msl)
~, затяжнóе hangfire (rockets)
~, искровóе spark(-plug) ignition
~ на отры́в coil ignition, make-and-break ignition
~, неправúльное imperfect ignition, imperfect sparking
~, обрáтное backfiring (arc rectifier, electron)
~, одноврéменное instantaneous ignition
~ от магнéто magneto ignition
~, пиротехнúческое pyrotechnic ignition, cartridge ignition
~, пóзднее late ignition, delayed ignition, retarded ignition
~, предкáмерное precombustion chamber ignition
~, рáннее advanced ignition
~, фáкельное hot-bulb ignition (diesel)
~, форкáмерное pilot-flame ignition, precombustion chamber ignition (GW)
~, электрúческое electrical ignition
~ электрозапáлом glow-plug ignition
~, электрóнное electronic ignition
зажигáтель 1. starter electrode 2. igniter
зажигáтельный incendiary
зажúм 1. clamping screw 2. terminal (elec)
~ аккумуля́торной батарéи pole terminal

зажи́м вертлю́га cradle pintle clamp (ord)
~, винтово́й screw terminal (elec)
~, выходно́й output terminal (elec)
~ кру́га body clamping screw
~, носово́й nose clip
~, пружи́нный snap terminal
~, хвостово́й tail grab
зажима́ние gripping, clamping, clipping, cramping
зажо́р ice jam
зазвуково́й supersonic
заземле́ние earth, earth connection (elec)
заземлённый earthed
заземли́тель earthing electrode (elec)
заземли́ть (см заземля́ть)
заземля́ть/заземли́ть to earth (elec)
зазо́р gap, clearance (eng, elec)
~, боково́й side play
~, искрово́й spark gap, sparking plug gap
~ кла́пана valve clearance
~, кла́панный valve clearance
~ ме́жду пере́дним срé́зом затво́ра и дном ги́льзы cartridge head space
~, осево́й end play
~, теплово́й expansion clearance
зазо́рник shim
заи́ндевение formation of hoar frost
зайти́ (см заходи́ть)
за́йчик 1. night firing training lamp (attached to a rifle)
2. blip (on radar screen)
зака́з order
заказа́ть (см зака́зывать)
зака́зывать/заказа́ть to order (goods)
закали́ть (см закаля́ть)
зака́лка 1. chilling, quench hardening, tempering (metall)
2. toughening (pers)
~, физи́ческая 1. physical training
2. toughening
зака́лывать/заколо́ть to stab, to pin (up); **зака́лывать/заколо́ть штыко́м** to bayonet
закаля́ть/закали́ть to temper (metall), to harden, to toughen, to season (pers)
зака́пывать/закопа́ть to dig in (trans)
за́кись азо́та nitrous oxide
закла́дка 1. laying
2. backing, packing
закладна́я mortgage
заклёпка rivet
заклине́ние jamming, wedging, sticking
закли́нивание jamming, wedging, ramming (projectile)
заключа́ть/заключи́ть to conclude, to infer, to close, to wind up; **заключа́ть/заключи́ть догово́р** to conclude a contract
заключе́ние, одино́чное solitary confinement
заключи́ть (см заключа́ть)
закоди́ровать to encode; **закоди́ровать ме́стность** to name reference points (target indication)
заколо́ть (см зака́лывать)
зако́н law, code
~, вое́нно-уголо́вный military penal code
~ всеми́рного тяготе́ния law of gravitation
~ обра́тных квадра́тов inverse square law
~ подо́бия 1. scaling law (Brit)
2. law of similarity (US)
~ рассе́ивания law of dispersion
законода́тельство legislation
законтри́ровать to lock (with lock nut)
зако́панный dug-in
закопа́ть (см зака́пывать)
закра́ина flange, rim, neck, collar, bead
~ ги́льзы cartridge rim, rimmed base
~ покры́шки bead of tyre
закрепи́ть (см закрепля́ть)
закрепи́ться (см закрепля́ться)
закрепле́ние 1. fastening, fixing, attaching
2. consolidation (tac)
~, клиново́е cotton fastening
~ успе́ха consolidation (tac)
закрепля́ть/закрепи́ть to secure, to fix, to consolidate (tac), : **закрепля́ть/закрепи́ть за** to allot, to assign
закрепля́ться/закрепи́ться to consolidate; **закрепля́ться/закрепи́ться на плацда́рме** to consolidate a bridgehead, to secure the air-head, to consolidate the air-head
закругле́ние 1. curve, curvature
2. rounding off
закру́тень windlassing stick
закру́тка tourniquet, artery clamp
~ из про́волоки wire tie
~ торсио́на torsion bar spring tension

закру́чивание 1. windlassing
2. twisting
закрыва́ть/закры́ть to close, to shut, to cover; **закрыва́ть/закры́ть счёт** to close an account, to settle an account
закры́лок flap (ac, GW)
закры́тие cover
~, есте́ственное natural cover
закры́ть (см закрыва́ть)
заку́поривать/заку́порить to plug up, to block, to obturate
заку́порить (см заку́поривать)
заку́порка seal, plug, clogging, obstruction
зал, обе́денный dining hall
~, огнево́й indoor range
залега́ть/зале́чь to go to ground
зале́чь (см залега́ть)
зали́вка, бабби́товая babbitt lining
~ боеприпа́сов pouring method of filling ammunition
~ горю́чим priming (fuel)
заложе́ние horizontal equivalent, H.E. (topo)
зало́жник hostage
залп salvo, volley
~, накрыва́ющий straddle, observer-target bracket, O-T bracket, bracketing salvo
~, недолётный salvo short (gunn)
~, ориентиро́вочный ranging salvo
~, перелётный salvo over
~, пристре́лочный ranging salvo
~, про́бный ranging salvo
за́лповый salvo
"За́лпом Ого́нь!" "Salvo Fire!"
за́лпы, дру́жные co-ordinated salvoes
зама́зка cement, putty, adhesive compound, paste, plaster, lute
зама́нивать/замани́ть to entice, to lure, to attract, to decoy; **зама́нивать/замани́ть в заса́ду** to draw into ambush
замани́ть (см зама́нивать)
замаскиро́ванный camouflaged
замаскирова́ть (см маскирова́ть)
зама́х hook (tac)
замедле́ние retardation, moderating
~ полёта снаря́да deceleration of projectile, retardation of projectile
замедли́тель 1. detonator delay charge, delay element (fuse)
2. delay mechanism (SA)
3. moderator (nucl reactor)
замедли́тель, автомати́ческий automatic retarder
~ взрыва́теля fuse delay
~, порохово́й 1. powder delay assembly (fuse)
2. safety fuse (gren)
замедли́ть (см замедля́ть)
замедли́ться (см замедля́ться)
замедля́ть/замедли́ть to slow down, to hinder
замедля́ться/замедли́ться to decelerate, to slow down
заме́на exchange, replacement, relief
~ колеса́ wheel changing
~, многоальфави́тная polyalphabetic substitution (crypt)
~, сезо́нная seasonal changing (oil etc)
~ ствола́ rebarrelling
замени́тель 1. substitute item
2. sub-calibre tube (ord)
~ а́томной бо́мбы atomic bomb simulator
замени́ть (см заменя́ть)
заменя́емость 1. replaceability
2. interchangeability
заменя́емый interchangeable
заменя́ть/замени́ть to substitute, to relieve, to replace, to supersede
заме́р measurement, gauging
замере́ть (см замира́ть)
замерза́емость liability to freezing
замерза́ние freezing, congealing, solidifaction, chilling
замести́тель deputy
~ команди́ра second in command
~ команди́ра по .... officer in charge of .... (internal command)
~ команди́ра по техни́ческой ча́сти technical supply officer
замеча́ние admonition
замини́рованный mined
замини́ровать (см мини́ровать) to mine, to close by mining (perf only)
замира́ние fading (radio)
~, избира́тельное selective fading
~ при со́лнечных возмуще́ниях Dellinger fade out
замира́ть/замере́ть to fade (radio), to fade, to die away (sound)
замкну́ть (см замыка́ть)
замко́вый No.2 (Brit), breech number (arty)

замо́к 1. locking device, lock, fastener 2. breech mechanism
~ блокиро́вки рулево́й коло́нки steering column lock
~ зажига́ния ignition switch (key operated)
~ кла́пана valve lock ring
~, клиново́й wedge-type breech mechanism (ord)
~ коро́бки переда́ч selector lock (gear box)
~ откидно́го бараба́на crane (revolver)
~ парашю́та, двухко́нусный twin-cone quick-release mechanism (for releasing parachute canopy cover)
~, самозапира́ющийся automatic lock
~, стыково́й jointing tie
~, суха́рный latch (RL)
~ таре́лки кла́пана valve-spring retaining cap
~, тричéтверти-автомати́ческий three-quarter-automatic breech mechanism
~, чéтверть-автомати́ческий quarter-automatic breech mechanism
~, шарни́рный articulated connector
замора́живание freezing, refrigerating, chilling
замора́живать/заморо́зить to refrigerate, to freeze, to get frostbitten
заморо́женный frozen, chilled
заморо́зить (см замора́живать)
замо́чный breech worker (nav gunn)
зампоте́х technical officer, technical adjutant
замыка́ние 1. closing (of circuit) 2. locking 3. (technical maintenance) echelon
~, коро́ткое short circuit, shorting (elec)
~, механи́ческое mechanical interlocking, latching
~ на ма́ссу earthing
~, техни́ческое technical maintenance echelon (brings up rear of MT column)
замыка́тель contact breaker, contactor
~ взрыва́теля firing switch
~ вибрацио́нного де́йствия vibration igniter, vibration circuit- closure fuse
~, вибрацио́нный vibration igniter, vibration circuit closure fuse
~ заме́дленного де́йствия delay initiator
~, магни́тный magnetic firing unit
замыка́тель мгнове́нного де́йствия instantaneous initiator
~ ствола́ barrel lock (MG)
~, часово́й clockwork circuit-closer fuse
~, часово́й вибрацио́нный clockwork vibration circuit-closer fuse
~, электри́ческий electric switch
~, эле́ктро-механи́ческий electro-mechanical circuit-closer fuse
~, эле́ктро-хими́ческий electro-chemical circuit-closer fuse
замыка́ть/замкну́ть to make contact (elec); **замыка́ть/замкну́ть нако́ротко** to short circuit; **замыка́ть/замкну́ть цепь** to make a circuit
замыка́ющий 1. file closer (US, drill) 2. last person, vehicle or section in column 3. "tail-end charlie" (AF)
за́мысел aim, intention, project, plan, design, scheme
замы́чка lock, locking piece
за́навес, газонепроница́емый gas curtain, gas-proof curtain
~, гермети́ческий air-tight curtain, gas-proof curtain
занести́ (см заноси́ть)
занима́ть/заня́ть to occupy
занима́ться/заня́ться to be occupied, to be engaged; **занима́ться/заня́ться на ку́рсе** to attend a course
зано́с skid, drift
~, сне́жный snowdrift
заноси́ть/занести́ to drop in, to note down; **заноси́ть/занести́ в рее́стр** to register; **заноси́ть/занести́ счёт** to credit
занумерова́ть (см занумеро́вывать)
занумеро́вывать/занумерова́ть to number
заня́тие 1. occupation, seizure, capture 2. exercise, training period
~ в по́ле exercise in the field
~, группово́е training exercise (for officers)
~, инстру́кторско-методи́ческое instructional methods exercise
~, кла́ссное indoor lesson, indoor lecture
~, ко́мплексное multi-theme exercise
~, показно́е demonstration (instruction)
~, семина́рское group study
~, строево́е drill instruction

заня́тие та́ктико-специа́льное specialist arms tactical exercise, special-to-arm tactical exercise
~, **та́ктико-строево́е** battle drill exercise
заня́тия, артиллери́йские строевы́е artillery drill
~, **команди́рские** exercises for officers
~, **полити́ческие** political instruction of NCOs and men
~ **с офице́рами, команди́рские** officers' training days
~, **такти́ческие** tactical training
~, **учéбно-трениро́вочные** training exercises
~, **учéбные** training exercise, training lessons
заня́ть (см занима́ть)
заня́ться (см занима́ться)
заострённый pointed, sharp, tapered
за́пад west
за́падный west, western
западня́ trap, pitfall
запажи́ливать to fit wheelguides (ribands) to a bridge
запа́здывание delay, retardation, time lag
~ **вы́стрела** firing interval, dead time (gunn)
~ **зажига́ния** ignition lag, ignition delay
~ **ору́жия** mechanical interval (gunn)
~ **откры́тия кла́пана** valve lag, valve delay
~ **стрелка́** personal interval (gunn)
запа́л fuse, primer, igniter, detonator, exploder, igniter set (gren)
~, **боево́й** service primer, live primer
~, **гальвани́ческий** electric detonator, electric igniter
~ **заме́дленного де́йствия** delay primer
~, **искрово́й** induction exploder, spark fuse
~ **к ручны́м грана́там, унифици́рованный** standard igniter set for hand grenades
~ **Кове́шникова** Koveshnikov igniter set (gren)
~, **механи́ческий** percussion detonator (mine)
~ **нака́ливания** flash igniter
~, **электри́ческий** electric detonator
запа́льник igniter
~, **пиротехни́ческий** pyrotechnic igniter
~, **порохово́й** powder charge igniter
запа́с supply, reserve, store, stock
~, **авари́йний** emergency reserve
~ **а́рмии и фло́та** army and naval reserve
~ **артснабже́ния, неприкоснове́нный** first line ammunition holdings (arty)
~, **боево́й** battle reserve
~ **боеприпа́сов** ammunition holding, ammunition reserve
~ **боеприпа́сов, нали́чный** ammunition on hand
~, **вози́мый** reserve on wheels
~ **горю́чего** reserve of fuel, fuel reserve, fuel stock
~ **моточасо́в** overhaul life of a vehicle (overhaul interval - in Sov army assessed in hours)
~, **неприкоснове́нный** 1. emergency reserve
2. emergency stores, emergency rations
~, **неприкоснове́нный продово́льственный** emergency rations
~, **неснижа́емый** minimum permitted holding
~, **носи́мый** individual reserve, first line scale (ammo)
~, **первонача́льный** initial reserve, initial stock
~, **первонача́льный теку́щий** initial operating stock
~ **плаву́чести** reserve buoyancy
~, **подвижно́й** 1. mobile stores holding
2. mobile reserve
~ **продово́льствия, авари́йный** emergency rations
~ **продово́льствия, расхо́дный** day-to-day supply of rations
~ **про́чности** safety factor
~, **ремо́нтный** housewife (sewing kit - cotton, needles etc)
~, **стратеги́ческий** strategic reserve
~, **теку́щий** working stock, operating reserve
~ **то́плива** 1. fuel reserve
2. fuel stock
~, **устано́вленный** authorized level (stores)
~ **хо́да** 1. range (running time on one fuelling)
2. radius of action
~, **храни́мый** reserve stock
запасно́й reserve, spare, alternate

запа́сный spare, reserve (see zapasnoj)
запа́сы вое́нного вре́мени, неприкоснове́нные mobilization equipment
~ вооруже́ния arsenal
~ материа́льных средств, вози́мые first line reserve of ammo, POL and foodstuffs
~ материа́льных средств, передовы́е forward holdings
~, нали́чные stocks on hand, available stocks
запеленгова́ть (см запеленго́вывать)
запеленго́вывать/запеленгова́ть to take a bearing; **запеленго́вывать/запеленгова́ть радиоста́нцию** to take a DF bearing
запира́ние locking, closing
~ затво́ра locking (breech)
~ кривоши́пно-шату́нного ти́па toggle lock
~ переко́сом затво́ра tilting block lock (breech)
~ поворо́том затво́ра rotating bolt locking mechanism (breech)
запира́ть/запере́ть to clamp, to lock
запира́ющий 1. blocking (electron) 2. locking
запере́ть (см запира́ть)
записа́ть (см запи́сывать)
"Записа́ть Устано́вки!" "Record as Target!" (arty)
"Записа́ть Устано́вку!" "Record as Target!" (arty); **записа́ть цель** to record as target (arty)
записа́ться в а́рмию to enlist
~ (см запи́сываться)
за́писи стрельбы́ gunnery records
запи́ска, объясни́тельная explanatory report
~, служе́бная memorandum
~, сопроводи́тельная cover sheet, cover note
~, строева́я personnel status record
~, уволни́тельная leave pass
запи́сывать/записа́ть to write down, to note, to jot down; **запи́сывать/записа́ть в расхо́д** to debit
запи́сываться/записа́ться to register, enter one's name, to join, to make an appointment (with), to subscribe to
за́пись 1. entry (conduct sheet etc), record, log 2. break (sound ranging)
~, дебето́вая debit entry
~, звукова́я sound recording, transcription
~ о прихо́де credit entry
~ о расхо́де debit entry
заплани́рованный planned, pre-planned
запле́чик shoulder (mech)
заподлицо́ flush (with)
заподо́зрить (см подозрева́ть)
заполне́ние filling (in, up), charging, priming
заполни́тель filler
заполя́рье the polar regions
запо́р catch, seal
~, гидравли́ческий oil seal
запотева́ние misting-up, fogging (eyepieces, optical instruments etc)
запотева́ть/запоте́ть to mist up, to fog up (eye pieces, optical instruments etc)
запоте́ть (см запотева́ть)
запра́вить (см заправля́ть)
"Запра́виться!" "Stand Easy!"
запра́виться (см заправля́ться)
запра́вка 1. fill, filling, fuelling, refuelling 2. charging 3. scale of normal fuel issue 4. amount necessary to fill with fuel or charge up
~ доло́т bit sharpening
~ жи́дким кислоро́дом filling with liquid oxygen (GW)
~ посте́лей bedmaking
заправля́ть/запра́вить to set, to season; **заправля́ть/запра́вить горю́чим** to refuel, to fill up with fuel, to fuel **заправля́ть/запра́вить обмундирова́ние** to adjust clothing
заправля́ться/запра́виться to refresh oneself; **заправля́ться/запра́виться горю́чим** to refuel
запра́вщик fuel tanker, filling truck, refueller, fuel servicing truck
запра́шивать/запроси́ть to interrogate (radar), to enquire
запрессо́вка, порохова́я fuse composition
запреще́ние открыва́ть ого́нь conservation of fire
~ радиопереда́ч radio ban
запрещённый out of bounds, forbidden

заприхо́дование debiting
запроекти́рованный rated, planned
запроси́ть (см запра́шивать)
запро́счик interrogator (radar)
~ радиолока́ции radar interrogator
~, радиолокацио́нный radar interrogator
~, самолётный радиолокацио́нный aircraft interrogator (IFF)
запро́счик-отве́тчик interrogator-transponder, interrogator-responder (radar)
запру́да dam, dyke
запря́жка, соба́чья dog team
запуга́ть (см запу́гивать)
запу́гивание intimidation
запу́гивать/запуга́ть to intimidate, to frighten
за́пуск 1. launching, launch (missile) 2. start up, starting (motor)
~, автоно́мный starting engine (off ac batteries)
~, возду́шный compressed air starting
~ дви́гателя engine starting
~ дви́гателя буксиро́вкой tow starting
~, реа́льный live firing (missile)
запуска́емый из ша́хтной пусково́й устано́вки silo-launched
запуска́ть/запусти́ть to launch (missile), to start (motor)
запусти́ть (см запуска́ть)
~ (см запуска́ть)
запча́сть spare part
"запята́я" "comma" (parachuting manoeuvre)
запята́я, пла́вающая floating point (computer)
заража́ть/зарази́ть to contaminate, to infect
зараже́ние contamination
~, бактериа́льное bacteriological contamination
~, бактериологи́ческое bacteriological contamination
~ га́мма-излуче́нием, радиоакти́вное gamma-ray contamination
~ ме́стности, радиоакти́вное radioactive contamination of ground
~ (ме́стности) с во́здуха airborne contamination
~, радиоакти́вное atomic radiation contamination, radioactive contamination
зараже́ние радиоакти́вными оса́дками fallout contamination
~, хими́ческое gas contamination, chemical contamination
заражённость contamination
~ вреди́телями pest infestation
заражённый насеко́мыми insect-infested
зарази́ть (см заража́ть)
зара́нее вы́численный precomputed
~ наведённый preset
~ устано́вленный preset
за́рин sarin (gas)
зарисо́вка sketch
за́росль undergrowth
зарыва́ть/зары́ть to bury, to dig in
зары́ть (см зарыва́ть)
заря́ 1. call. bugle call (reveille etc) 2. daybreak, dawn 3. sunset. nightfall
~, вече́рняя lights out
~, торже́ственная ceremonial tattoo, beating retreat
~, у́тренняя "reveille"
заря́д 1. charge (expl, elec etc) 2. squall (met) 3. filling (ammo)
~, акти́вный initial charge
~, а́томный atomic warhead
~, баллисти́товый (вышибно́й) ballistite cartridge
~, безды́мный боево́й smokeless propellant
~, беспла́менный flashless charge, antiflash charge
~, боево́й 1. warhead 2. propellant charge 3. full charge (arty)
~, боево́й кумуляти́вный shaped-charge missile, hollow charge missile (A tank missile)
~, боево́й я́дерный nuclear warhead
~, бурово́й подрывно́й borehole charge
~, быстрогоря́щий fast-burning charge
~ в су́мке, подрывно́й satchel charge
~, воспламени́тельный primer charge
~, воспламеня́ющий igniting charge, flash charge
~, второ́й second charge
~, вышибно́й 1. bursting charge, burster 2. propelling charge (bounding mine)
~, горя́щий по ча́сти пове́рхности restricted burning charge

заря́д, горя́щий с торца́ end burning charge
~, дви́жущий propellant charge (expl)
~, детони́рующий detonating charge, priming charge
~ для уче́бной стрельбы́, уме́ньшенный practice reduced charge
~, доба́вочный additional charge, boosting charge
~, до́нный вышибно́й base burster
~, дополни́тельный augmentary charge, secondary charge (mor)
~, заглуша́ющий secondary ice-breaking charge
~, зажига́тельный incendiary charge
~, инжене́рный engineer demolition charge
~, иницийрующий 1. initiator (expl) 2. booster charge
~ ка́псюльной вту́лки, порохово́й priming charge
~ ка́псюльной тру́бки, порохово́й priming charge
~, ка́псюльный percussion charge
~ карту́зного заряжа́ния bag cartridge
~, коло́нковый column charge
~, кольцево́й дополни́тельный circular augmenting charge (mor bomb)
~, кольцево́й фигу́рный подрывно́й necklace charge (dml)
~, комбини́рованный composite charge
~, кони́ческий кумуляти́вный подрывно́й hollow charge
~, конта́ктный contact charge
~, конта́ктный нару́жный cutting charge
~, котло́вый concentrated charge
~, кумуляти́вный shaped charge, hollow charge
~, кумуляти́вный подрывно́й hollow demolition charge, shaped demolition charge
~, лито́й разрывно́й cast filling (ammo)
~, мета́тельный propellant charge (missile)
~, многоша́шечный multiple-grain charge (rocket technology)
~, мо́дульный modular charge (rocket technology)
~, мо́щный heavy charge
~ на вы́брос cratering charge
~, напа́лмовый napalm charge
~, несгоре́вший unburnt charge
заря́д, норма́льный service charge, normal charge, standard charge
~, ныря́ющий anti-submarine charge, depth charge. depth bomb
~, освети́тельный illuminating charge
~, основно́й 1. primary cartridge (mortar bomb) 2. primary charge (ammo)
~, отказа́вший misfire
~, пасси́вный inert charge
~, пе́рвый first charge
~, перебива́ющий cutting charge
~, переда́точный 1. magazine detonator (fuse) 2. booster lead
~, переме́нный variable charge
~, пиротехни́ческий pyrotechnic charge
~, подо́бранный переме́нный adjusted charge
~, подрывно́й explosive charge, demolition charge
~, подрывно́й бурово́й borehole charge
~, по́лный full charge
~, по́лный боево́й equivalent full charge
~, пони́женно-боево́й reduced charge
~, порохово́й propellant charge, powder charge
~ порохово́й про́бы proof charge
~, постоя́нный fixed charge
~, практи́ческий reduced charge
~, про́бный test charge
~, простра́нственный space charge (elec)
~ размини́рования mine-clearing charge
~, разрывно́й bursting charge, HE shell filling
~, разря́дочный clearing charge
~, ра́нцевый подрывно́й satchel charge
~, реакти́вный propellant charge (GW)
~ с прогресси́вным горе́нием progressive-burning charge (msl)
~ с регресси́вным горе́нием regressive-burning charge (msl)
~ самоликвида́тора self-destruct charge
~, сне́жный snow squall
~, сосредото́ченный concentrated charge
~, составно́й composite charge
~ специа́льного назначе́ния special purpose charge
~, сплошно́й shot

заря́д твёрдого то́плива solid-propellant charge (msl)
~, термоя́дерный thermonuclear charge
~, тру́бчатый tubular charge
~, удлинённый bangalore torpedo
~, уменьшённый reduced charge
~, уси́ленно-боево́й supercharge
~, уси́ленный supercharge
~, уче́бный practice charge, dummy charge
~, фигу́рный irregularly shaped charge
~, холосто́й blank charge, blank cartridge
~, части́чный composite charge
~ чёрного по́роха black-powder charge
~, эквивале́нтный equivalent charge
~, электри́ческий electric charge
~, я́дерный боево́й nuclear warhead
~ ядра́ nuclear charge (charge on nucleus of atom)
~, я́русный spaced charge
заряди́ть (см заряжа́ть)
заря́дка 1. charge, load
2. charging (elec)
3. loosening-up exercises
~, у́тренняя morning PT
заря́д-ло́дочка augmenting cartridge
заря́дник ammunition hoist, ammunition crane
заря́дный No. 6 (Brit), ammunition number, charge handler
заря́д-торпе́да, удлинённый bangalore torpedo
заря́ды шпу́ров borehole charges
заряжа́емый с ду́ла muzzle loading (wpn)
~ с казённой ча́сти BL, breech loading (wpn)
"Заряжа́й!" "Load!"
заряжа́ние 1. priming (gren)
2. loading (wpn)
3. charging (elec)
~, безги́льзовое breech loading (ammo, ord) (pertaining to ordnance in which obturation is by means of a pad in the breech block as opposed to a cartridge case in quick-firing equipments)
~ вручну́ю hand loading
~, ги́льзовое QF loading (ammo, ord)
~, двойно́е double loading
~, карту́зное breech loading (ammo)
~, магази́нное magazine loading
~, обо́йменное clip-loading
заряжа́ние, патро́нное QF fixed loading ammunition
~, полуавтомати́ческое semi-automatic loading
~, разде́льное ги́льзовое QF separate loading (ammo)
~, ручно́е hand loading (SA)
~ с казённой ча́сти breech loading (wpn) .
заряжа́тель, бу́ферный trickle-charger
заряжа́ть/заряди́ть to load (ammo, wpn), to prime (gren), to charge (elec, ammo);
заряжа́ющий 1. No. 2 (Brit) (mor)
2. loading number, loader, No. 4 (arty)
заряжённый 1. loaded (wpn)
2. primed (gren)
3. charged (elec, ammo)
заса́да ambush
засвети́ть (см засве́чивать)
засве́тка 1. response, blip (radar)
2. gating (elec)
за́светло in daylight, during daylight
засве́чивать/засвети́ть to gate, to strobe (elec), to illuminate, to affect with light (radar screens)
засе́ка abatis (tree obstacle)
засека́ть/засе́чь to plot, to "fix" (targets etc); **засека́ть/засе́чь визи́рованием с двух то́чек** to locate by intersection
засекрети́ть (см засекре́чивать)
засекре́чивание security classification
засекре́чиватель ре́чи speech scrambler, speech secrecy eqpt
засекре́чивать/засекрети́ть to classify (for security)
засе́сть to bog down, to take up a position (perf only)
засе́чка 1. intersection (surv)
2. fix (radar)
~ высоты́ height fix (radar)
~, звукова́я intersection by sound ranging, sound ranging location
~, ко́мпасная compass cross bearing
~, обра́тная resection (surv)
~ огневы́х средств по звуку вы́стрелов sound ranging
~, опти́ческая flash spotting, flash observation, flash ranging (arty), observed bearing
~ по бле́ску flash spotting
~ по звуку sound ranging
~, полуобра́тная resection from a known line (surv)

засéчка, прямáя graphic intersection (arty obsn)
~, световáя flash ranging
~ цéли target location by intersection
засéчь (см засекáть)
заслóн 1. screen (tac)
2. road block
~, противотáнковый anti-tank screen
заслóнка 1. damper, choke
2. slide, flap, baffle
~, броневáя armoured flap
~, дрóссельная butterfly throttle, butterfly valve
~ карбюрáтора, воздýшная choke (ICE)
~ клáпана flap valve
~ лю́ка, броневáя armoured hatch door
~, модули́рующая chopper disc (infra-red homing system)
~, светомаскирóвочная black-out blind
~, шаровáя ball mounting (MG)
заслýга merit
засóв bolt, bar
засорéние 1. fouling
2. grass, clutter (radar)
застáва outpost, security detachment, frontier picquet, sub-unit of frontier troops
~, боковáя похóдная flank party, detachment of flank guard (tac)
~, головнáя point unit
~, головнáя похóдная point of vanguard (tac)
~, неподви́жная боковáя flank picquet
~, отдéльная сторожевáя outlying sentry post
~, полевáя protective detachment, protective outpost
~, похóдная march security patrol
~, сторожевáя outpost, bivouac guard patrol
~, ты́льная похóдная rear party (in rear guard)
застáвить (см заставля́ть)
застáвка, боковáя flank guard
заставля́ть/застáвить to force, to compel, to make; **заставля́ть/застáвить залéчь проти́вника** to cause the enemy to go to ground
застёжка catch
~, пружи́нная spring catch
застигáть/засти́гнуть to catch; **застигáть/засти́гнуть врасплóх** to take by surprise
засти́гнуть (см застигáть)
застилáть/застлáть to cover; **застилáть/застлáть мáтами** to mat
застлáть (см застилáть)
застóпоривание замкá locking (breech)
застóпорить (см стóпорить)
застревáние bogging down
застревáть/застря́ть to bog down, to get stuck, to stick
зáструга small snow drift
застря́вший bogged down
застря́ть (см застревáть)
заступáть/заступи́ть to replace, to relieve, to take over
заступи́ть (см заступáть)
заступлéние take-over (sentry posts)
~ в наря́д assumption of duties
затáпливать/затопи́ть to kindle, to light a fire (in a stove), to flood, to inundate
затвердевáние hardening
затвóр 1. breech-block (SA)
2. slide (pistol)
3. breech mechanism (ord)
4. shutter (photo)
~, автомати́ческий automatic breech mechanism
~ в сбóре bolt assembly
~ вертикáльно пáдающего ти́па, клиновóй vertical sliding(-type) breech-block
~, вертикáльный клиновóй vertically operated wedge type breech mechanism (ord)
~, гильоти́нный vertical focal-plane shutter (photo)
~, горизонтáльный клиновóй horizontal sliding breech, horizontal sliding wedge type breech
~, клиновóй wedge-type breech mechanism (ord)
~, откиднóй swinging breech mechanism
~, откиднóй шарни́рный hinged breech-block mechanism
~, перекáшивающийся tip-up type breech-block
~, полуавтомати́ческий semi-automatic breech mechanism
~, поршневóй screw-type breech mechanism, slotted screw breech mechanism
~, продóльно скользя́щий sliding breech bolt

затвóр прямóго дéйствия без поворóта при запирáнии straight-pull bolt
~ с поворóтом при запирáнии turning bolt (Brit), turn bolt
~, самовзвóдный repeating trip lock (arty)
~, свобóдный 1. unlocked breech mechanism
2. blow-back breech-block
~, скользя́щий sliding breech-block
~ со стреля́ющим механи́змом breech-block with firing mechanism
~, состоя́щий из двух частéй two-piece bolt
~, центрáльный diaphragm shutter (photo)
~, шарни́рный hinged breech-block
~, ши́рмовый focal-plane shutter (photo)
~, штóрный roller-blind shutter (photo)
~, эксцентри́ческий eccentric screw breech mechanism (ord)
затвóр-удáрник sleeve-bolt and spigot, combined bolt and striker
затемнéние blackout
~, пóлное full blackout
~, части́чное partial blackout, dim-out
затемни́ть (см затемня́ть)
затемня́ть/затемни́ть to black out, to blind
затени́ть (см затеня́ть)
затеня́ть/затени́ть to shade in (map)
затёс blaze (mark on trees etc)
затёска blaze (mark on trees etc)
зати́шье, боевóе lull in battle
затóн backwater
затопи́ть (см затáпливать и затопля́ть)
~ (см затопля́ть)
затоплéние flooding, inundation
затопля́ть/затопи́ть to kindle, to light a fire (in a stove) to flood, to inundate
затóр jam, congestion
~, ледянóй ice blocking, ice jamming
затормáживать/затормози́ть to brake, to apply the brakes
затормози́ть (см затормáживать)
затрéбование requisition, request
затрýбка dent
затухáние damping, attenuation
~ боковóй полосы́ sideband attenuation
~ в атмосфéре atmospheric attenuation
~, крити́ческое critical damping
~ передáчи transmission loss
затухáтель damper
затухáющий damped
заты́лок 1. butt plate (SA), back plate
2. back of the head, occiput (anat)
заты́льник 1. butt plate (SA)
2. back plate
3. rear cross piece
~ приклáда butt, butt plate
~ приклáда, откиднóй folding butt plate
заты́чка bung, plug, stopper, cork
затя́гивание tightening
~ частоты́ frequency pulling
затя́жка 1. tightening, tightness
2. backlash
3. tie rod, tie beam
заусéнец burr, barb
зафикси́рованный fixed
зафикси́ровать (см фикси́ровать)
захвáт 1. capture, seizure
2. grip, clamp, clip, clasp
3. gun guides, guide clip, clamping piece (ord)
4. coverage (air photo)
5. lock-on (radar)
~ голóвкой самонаведéния homing-head lock-on (msl)
~ и сопровождéние, автомати́ческий auto-detection and tracking (of radar echoes)
~ лучóм beam capture (missile)
~ радиолокациóнной стáнцией radar lock-on (msl)
~ ракéты missile capture
~ с хóду scooping up (tac)
~ цéли в ви́лку bracket adjustment, bracketing (arty)
~, цепнóй securing chain
захвати́ть (см захвáтывать)
захвáтчик invader, aggressor
захвáтывание 1. trapping (nucl)
2. frequency pulling
~ тóрмоза brake drag
захвáтывать/захвати́ть to seize, to capture, to lock on (radar) ; **захвáтывать/захвати́ть в ви́лку** to straddle, to bracket (gunn); **захвáтывать/захвати́ть в плен** to take prisoner, to capture
захвáтывающий bracketing (gunn)
захвáченный trapped
захлебнýться (см захлёбываться)
захлёбываться/захлебнýться to become bogged down (veh, tk; also tac)

захо́д на поса́дку landing approach, final approach (AC)
~, про́бный trial run
заходи́ть/зайти́ to circle (ac), to set (sun), to drop in, to call for; **заходи́ть/зайти́ плечо́м** to wheel (drill); **заходи́ть/зайти́ с фла́нга** to outflank
захожде́ние плечо́м wheeling (drill)
~ фла́нгом wheeling (drill, tac)
заце́п claw
~ выбра́сывателя extractor horn, extractor claw
зацепле́ние meshing, engagement, gearing, catching, hooking
~, постоя́нное constant mesh (gears etc)
~, червя́чное worm mesh, worm gearing
заце́пы движка́ feedblock slide pawls (SA)
зачекова́ть (см чекова́ть)
зачёт test, examination
зачи́слить (см зачисля́ть)
зачисля́ть/зачи́слить to include; **зачисля́ть/зачи́слить в часть** to take on strength
зачи́стка dressing, stripping, skinning, cleaning, clearing
зашифрова́ть (см зашифро́вывать)
зашифро́вка encyphering
зашифро́вывать/зашифрова́ть to encypher
зашлифова́ть (см зашлифо́вывать)
зашлифо́вывать/зашлифова́ть to hone out
защёлка 1. catch
2. contactor latch (rocket launcher)
~ заты́льника прикла́да butt-plate catch
~ капо́та bonnet catch
~ кожуха́ barrel nut catch
~ коро́бки шестерни́ pinion pawl (SA)
~ магази́на magazine catch
~ механи́зма автомати́ческого спу́ска, спускова́я rammer catch lever (ord)
~ обо́ймы clip catch
~ о́си ла́пок экстра́ктора extractor spindle arm
~ панора́мы dial sight clamp
~ предохрани́теля safety catch
~ прикла́да butt latch
~ прице́ла sight clamp
~, пружи́нная spring catch
~ рукоя́тки затво́ра LBM actuating lever
защёлка рукоя́тки перезаряжа́ния cocking handle catch
~ ручно́го спу́ска, спускова́я single-shot rammer check lever (ord)
~ соедини́тельной му́фты barrel lock catch
~ ствола́ barrel catch
~ хому́тика slide catch
~ штыка́ bayonet catch
защи́та defence, protection
~, бронева́я armour protection
~, индивидуа́льная personal protection
~, като́дная cathodic protection
~, коллекти́вная противохими́ческая collective anti-chemical protection, collective CW protection
~ от корро́зии corrosion protection
~ от ору́жия массово́го пораже́ния protection against mass destruction weapons
~ от радиоакти́вного излуче́ния radiation protection
~ от радиоакти́вных оса́дков fallout protection
~ от ржа́вчины rust prevention
~ от теплово́го излуче́ния protection against thermal radiation
~, противоа́томная protection against nuclear attack
~, противобактериологи́ческая bacteriological defence, protection against bacteriological agents
~, противобиологи́ческая protection against biological agents, biological (warfare) defence
~, противопожа́рная fire protection
~, противорадиолокацио́нная radar countermeasures, RCM
~, противохими́ческая anti-gas protection, protection against chemical agents, CW defence
~, радиолокацио́нная radar counter-countermeasures, RCCM
~, реле́йная relay protection
защи́тник 1. defender, protector
2. defending officer (court martial)
защи́тный protective
защища́ть/защити́ть to defend, to protect
защищённость protection
защищённый protected, screened
~ от насеко́мых insect-proof
~ от парази́тов vermin proof
защити́ть (см защища́ть)

заяви́ть (см заявля́ть)
~ (см заявля́ть)
заяви́ться (см заявля́ться)
~ (см заявля́ться)
зая́вка 1. indent, requisition
2. notification
~ **на боеприпа́сы** ammunition indent, ammunition demand
~ **на подря́д** tender
~ **на снабже́ние** indent for supplies
~ **на снабже́ние горю́чим** fuel indent
~, **неудовлетворённая** outstanding demand, outstanding requisition
~, **пла́новая** routine requisition
~, **по́лная** detailed indent
~, **предвари́тельная** 1. preliminary demand
2. preliminary notification
3. tentative requisition
~, **реквизицио́нная** requisition receipt
~, **сро́чная** 1. urgent indent
2. red star indent (Brit)
~, **теку́щая** current indent
~, **у́стная** verbal demand (indent)
заявле́ние 1. declaration
2. application
заявля́ть/заяви́ть to declare, to announce
заявля́ться/заяви́ться to report (for duty etc)
зва́ние 1. rank, title
2. trade grading
~, **во́инское** military rank
~, **почётное** honorary rank
звезда́ 1. star
2. star network (elec)
звёздка star gauge
~, **контрпредохрани́тельная** delayed arming shutter segment (fuse)
~, **тормозя́щая** shakeproof washer
звездообра́зный radial, star-shaped
звёздочка 1. star (officer's badge of rank)
2. sprocket
3. chain wheel
~, **веду́щая** final drive sprocket
~ **гу́сеницы** track sprocket
~ **натяжно́го приспособле́ния** tensioner sprocket
~, **парази́тная** idler
~ **фикса́тора** feed cylinder journal (ord)
звено́ 1. link, member, unit, section
2. level
3. flight (ac sub-unit)
4. track link, track shoe (tracked veh)
звено́ гу́сеницы track link
~ **гу́сеничной ле́нты** track
~, **неразбо́рное** one-piece track shoe (tracked veh)
~, **носи́лочное** stretcher party
~, **обеспе́чивающее** administrative echelon
~, **перехо́дное** chain link, spring link
~ **подве́ски** suspension link
~, **подве́шенное** suspension link
~ **систе́мы снабже́ния** echelon of supply, supply echelon, supply level
~, **такти́ческое** tactical level
звеньеотво́д link ejection chute (SA)
звеньесбо́рник used-link container
зве́рства atrocities
звоно́к переме́нного то́ка magneto bell
~, **поляризо́ванный** magneto bell
~ **постоя́нного то́ка** trembler bell
звук вы́стрела report of gun
звукоба́за sound ranging base
звукобатаре́я sound ranging battery
звуково́й sonic, acoustic, sound
звукоглуши́тель silencer, muffler
звукоза́пись sound recording, transcription, tape recording
~, **магни́тная** magnetic sound recording
~, **механи́ческая** acoustical sound recording
~, **опти́ческая** optical sound recording
~, **фотографи́ческая** optical sound recording
звукоизоля́ция soundproofing
звуколока́ция sound fixing and ranging
звукомаскиро́вка sound damping
~ **имита́цией шу́мов** noise simulation
звукометри́ст sound ranging surveyor
звукометри́ческий sound ranging
звукоме́трия sound ranging, sound location
звуконепроница́емость sound opacity, acoustic opacity
звуконепроница́емый soundproof
звукопеленга́тор sound locator
звукопи́сец sound recorder
звукоприёмник microphone (sound ranging unit)
звукорадиобу́й radio sonobuoy
звукоснима́тель sound pickup
звукоула́вливатель sound locator
звукоулови́тель sound locator
звукоусили́тель audio-amplifier, sound amplifier

зда́ние building, structure
~, беска́ркасное wall bearing structure, wall bearing building, frameless building
~, водоёмное water tower
~, ка́ркасное framed building
землеро́йный ditching, earth-moving, excavating
землесо́с dredging pump, suction dredger
землечерпа́лка dredge, dredger
земля́, ниче́йная no man's land
~, расти́тельная top soil
земля-во́здух surface-to-air, ground-to-air
земля́-земля́ surface-to-surface, ground-to-ground
земля́нка dug-out, shelter trench
земляно́й earth
земно́й ground, terrestrial
зени́т zenith
зени́тка anti-aircraft gun
зени́тно-артиллери́йский anti-aircraft
зенитнораке́тный anti-aircraft missile, SAM
зени́тный air defence, anti-aircraft
зени́тчик anti-aircraft gunner
зени́тчик-планшети́ст aircraft plotter
зени́тчик-пулемётчик anti-aircraft machine gunner
зенкова́ть to countersink, to counterbore, to rebore (ICE)
зе́ркало 1. mirror
2. reflector (radar antenna)
3. surface (river)
4. front face (breech-block)
5. wall (cylinder)
~, вну́треннее inside mirror, interior mirror
~ для осмо́тра кана́ла ствола́ barrel reflector, barrel mirror
~ за́днего ви́да driving mirror, rear-view mirror
~, нару́жное exterior mirror, external mirror, outside mirror
зерни́стость по́роха powder grain size
зерно́ grain, granule
~, многокана́льное порохово́е multi-perforated grain (ammo)
зетме́тр impedance bridge, Z meter
зимо́вка winter quarters, wintering
злоупотребле́ние misuse, abuse
змееви́к coiled pipe
змееви́к, охлажда́ющий condenser coil, cooling coil
змей-мише́нь sleeve target
знак sign, mark
~, берегово́й landmark on coast (amph ops)
~, вое́нный номерно́й military vehicle registration number plate
~, госуда́рственный опознава́тельный national marking
~, доро́жно-сигна́льный road sign, traffic sign, route sign
~, доро́жный road sign, traffic sign
~, ли́чный identity disk
~ на прохо́де в ми́нном по́ле minefield-lane marking sign
~, нагру́дный badge worn on breast
~, нарука́вный sleeve insignia
~, номерно́й licence (number) plate
~, обстано́вочный navigational beacon, navigational marker (naut)
~, огради́тельный warning sign, boundary marker
~ огражде́ния warning sign, boundary marker
~ огражде́ния уча́стка хими́ческого зараже́ния gas contamination marker, gas contamination label
~ оконча́ния ending sign (morse)
~ опа́сности danger sign
~, опознава́тельный 1. identification sign, formation sign
2. recognition signal, recognition marking
~ отклоне́ния sense of error, sense of deviation (arty)
~ оши́бки erase sign (morse)
~, предупреди́тельный warning sign, danger sign
~, предупрежда́ющий warning sign, danger sign
~ препина́ния punctuation symbol (morse)
~, путево́й wayside marker, track signal
~, раздели́тельный long break sign (morse)
~ разли́чия insignia, badge of rank
~ разры́ва sense of burst (arty)
~, светово́й traffic control lamp, marker light
~, служе́бный procedure sign (comms)
~, топографи́ческий усло́вный map symbol

знак, триангуляцио́нный triangulation point
~, усло́вный 1. conventional sign 2. pre-arranged signal
~, усло́вный внемасшта́бный conventional sign not to scale
~, усло́вный масшта́бный conventional sign to scale
зна́ки, постоя́нные служе́бные operating signals (comms)
знако́мить/познако́мить to aquaint (with), to introduce (to); **знако́мить/познако́мить с обстано́вкой** to brief, to put in the picture
знако́мство, практи́ческое working knowledge, practical experience
знаменоно́сец colour bearer, standard bearer
знамёнщик colour bearer, standard bearer
зна́мя colours, standard
значе́ние, де́йствующее effective value, root-mean square value, RMS value (elec)
~ М Mach value
~, эффекти́вное 1. effective value 2. root-mean square (RMS) value (elec)
значки́ст holder of badge, holder of medal
значо́к badge, flash
золотни́к valve
зо́на zone, area, belt
~ бактериа́льного зараже́ния zone of bacteriological contamination
~, безопа́сная safety zone
~, бли́жняя foreground
~, вероя́тная probable zone
~ вероя́тного пораже́ния zone of probable damage
~ возду́шных подхо́дов approach funnel (aeron)
~ воспроизво́дства blanket (nucl reactor)
~ горе́ния combustion zone
~, да́льняя background
~ действи́тельного пораже́ния lethal area
~ де́йствия system coverage (SAM) (volume of sky in which impact can occur)
~, демилитаризо́ванная demilitarized zone
~ досяга́емости impact zone (AA gunn)
~ загражде́ний obstacle zone
зо́на, замини́рованная mined zone
~, запре́тная prohibited area
~, запрещённая forbidden region (semiconductor etc)
~ зараже́ния area of contamination, contaminated zone
~, заражённая contaminated zone
~ зени́тного артиллери́йского прикры́тия anti-aircraft belt
~ зени́тного огня́ gun fire area (AA gunn)
~ зени́тного огня́, пло́ская engagement zone related to height of attack (AA gunn)
~ зени́тной оборо́ны, запре́тная gun fire area class A, GFA class A (AA gunn)
~, мёртвая 1. silent zone (radio) 2. blind spot (radar)
~ молча́ния silent zone (radio)
~ наблюде́ния surveillance zone
~ наведе́ния guidance zone
~ нахожде́ния contact area
~, нейтра́льная neutral zone, no-man's land
~ обнаруже́ния coverage (radar set)
~ оборо́ны, такти́ческая tactical zone of defence (Sov)
~ обстре́ла engagement zone (gunn)
~ обстре́ла, пло́ская engagement zone related to height of attack
~, огнева́я fire zone
~ огня́ zone of fire
~ огня́ зени́тной артилле́рии gun fire area (AA gunn)
~, опа́сная danger area
~ опа́сности danger area
~, операти́вная operations zone
~ о́тпуска issues area
~ по́иска search area
~ по́лного разруше́ния area of complete destruction
~ попада́ний impact area
~, поража́емая dangerous zone (SA)
~ пораже́ния 1. impact zone (SAM) 2. lethal area, killing zone 3. area under fire
~ пу́ска missile launch zone (zone in which target is located at moment of SAM launch)
~, рабо́чая service area (GW)
~, равносигна́льная equisignal zone (DF)

зо́на радиоакти́вного зараже́ния radioactive contamination zone
~ **радиолокацио́нного обнаруже́ния це́лей** coverage (radar set)
~ **разреже́ния** area of low pressure
~ **разруше́ний** demolition belt, zone of destruction
~ **разры́вов** impact area
~ **сжа́тия** area of high pressure
~ **си́льного зараже́ния** heavy contamination area, intense contamination area
~ **си́льного разруше́ния** area of heavy damage
~, **сме́жная** adjacent zone (grid system)
~ **сопровожде́ния** zone of maximum tracking range (radar)
~, **сре́дняя** middle distance
~ **стрельб** firing area
~ **такти́ческой оборо́ны** defended area
~ **тле́ния и горе́ния в зава́лах** firestorm area (NW)
~, **укреплённая** fortified zone, fortified belt
~ **уме́ренного разруше́ния** area of moderate damage
~ **усто́йчивости** stability range
~ **Фре́неля** Fresnel zone (radio comms)
~ **хране́ния** storage area
~, **шестигра́дусная** Universal Trans-Mercator six degree zone (grid system)
зона́льный zonal, regional
зонд 1. probe unit (radiac)
2. probe, sound
3. sounding balloon
зонда́ж sounding
зонди́рование sounding
~, **возвра́тно накло́нное** inclined backscatter sounding (ionosphere)
зонди́ровать to sound, to probe
зондиро́вка sounding
зрачо́к, входно́й entrance pupil (optical instr)
~, **выходно́й** exit pupil (optical instr)
зре́ние, бинокуля́рное binocular vision
~, **цветово́е** colour vision
зуб tooth, cam, cog
~ **самовзво́да** hammer cocking bent
зубе́ц 1. cog, tooth (mech)
2. pip, blip (radar)
3. dent
зуби́ло 1. cold chisel, bit, punch
2. sett (mining)
зуби́ло, ру́чное cold chisel
~, **слеса́рное** cold chisel
зубоврачéбный dental
зуботехни́ческий dental mechanical
зубча́тка 1. gear wheel
2. rack
3. ammunition rack (MG)
зу́бчики 1. hachures (topo)
2. ripple (elec)
зу́бья шестерён gear teeth
зу́ммер buzzer (comms)
~, **ла́мповый** valve-type AF oscillator
зумпф sump, pit
зыбу́н spongy marsh, moor
зы́бчатость ripple factor (elec)
зыбь ripple (elec)

И

игла́ needle, thin rod, spindle
~, **дросселиру́ющая** needle valve (ICE)
~, **запо́рная** needle valve (diesel ICE)
~ **кла́пана** needle valve
~, **нажи́мная** nozzle valve rod (diesel ICE)
~, **профили́рованная** metering needle
~, **регули́рующая** adjusting needle valve (ord buffer)
~, **то́пливная** needle valve
~ **форсу́нки** injector valve
игнитро́н ignitron (elec)
игра́ 1. exercise
2. play (eng)
~, **бокова́я** side play (eng)
~, **вое́нная** war game (the Sov term includes one-sided exercises)
~, **двухступе́нная вое́нная** two-level tactical exercise without troops
~, **двухсторо́нняя вое́нная** two-sided tactical exercise without troops
~, **кома́ндно-штабна́я вое́нная** command post exercise, CPX
~, **многостепе́нная вое́нная** multilevel tactical exercise without troops
~, **одностепе́нная вое́нная** single-level tactical exercise without troops
~, **односторо́нняя вое́нная** one-sided tactical exercise without troops

игра́, осева́я end float, end play, axial play (eng)
~, продо́льная end play
игра́ть/сыгра́ть to play, to perform, to act; **игра́ть/сыгра́ть трево́гу** to sound action stations
"игре́к" 1. "Y"
2. "eastings" (RT voice, Sov)
идентифика́ция identification
идентифици́рование identification
идти́ в но́гу to keep step; **идти́ за́дним хо́дом** to back, to reverse (veh); **"Идти́ не в Но́гу!"** "Break Step!"; **идти́ по а́зимуту** to march on bearing
~ (см ходи́ть)
изаллоба́ра isallobar (met)
избега́ть/избежа́ть to avoid, to evade, to steer clear; **избега́ть/избежа́ть пле́на** to avoid capture
избежа́ть (см избега́ть)
избира́тель selector
избира́тельность selectivity (electron)
~, амплиту́дная amplitude selectivity
~ по направле́нию directional selectivity
~ по частоте́ frequency selectivity
~, простра́нственная directional selectivity
~, часто́тная frequency selectivity
избы́ток surplus
~ энта́льпии excess enthalpy
изве́стие news, information
извести́ть (см извеща́ть)
изве́стка mortar (building)
и́звесть lime
~, гашёная slaked lime
~, хло́рная calcium hypochlorite, chloride of lime, bleach (CW)
извеща́тель announcing system, broadcasting system
~, автомати́ческий пожа́рный automatic fire alarm
~, теплово́й пожа́рный heat detector
извеща́ть/извести́ть to inform
извеще́ние notification
~, у́стное verbal intimation
изви́листый winding, twisting, tortuous
извлека́тель 1. clearing plug (SA)
2. extractor (ord)
~ разорва́вшихся гильз ruptured cartridge extractor
извлека́ть/извле́чь to extract; **извлека́ть/извле́чь из ми́ны взрыва́тель** to defuse a mine
извлече́ние extraction, pulling, withdrawal
извле́чь (см извлека́ть)
изги́б bend, bending
и́згородь fence, partition
~, жива́я hedgerow
изгото́виться (см изготовля́ться)
изгото́вка ready position
изготовле́ние processing, manufacture, manufacturing, making, preparation, production
изготовля́ться/изгото́виться to prepare, to get ready; **изготовля́ться/изгото́виться к наступле́нию** to poise for an attack
издава́ть/изда́ть 1. to publish, print
2. to promulgate, issue
3. to utter
4. to exhale; **издава́ть/изда́ть прика́з** to issue an order, to publish an order
изда́ть (см издава́ть)
изде́лие product, article, item
~, стандартизи́рованное standardized article
изде́лия, кана́тные и верёвочные cordage
изли́шек excess
изло́жница mould, casting
излуча́тель radiating element (antenna)
~, изотро́пный isotropic radiator
~, то́чечный point source radiator (nucl)
излуча́ть/излучи́ть to radiate, to emit, to beam, to ray
излуча́ющий radiant, radiating
излуче́ние radiation, emission
~ абсолю́тно чёрного те́ла black body radiation
~ а́льфа-части́ц, радиоакти́вное alpha radiation
~ бе́та-части́ц, радиоакти́вное beta radiation
~, втори́чное secondary radiation
~, вы́нужденное stimulated radiation, emission, induced radiation, emission
~, заме́дленное residual radiation (nucl)
~, запа́здывающее delayed radiation
~, индуци́рованное stimulated radiation, emission, induced radiation, emission
~, инфракра́сное infra-red radiation, IR radiation
~, ионизи́рующее ionizing radiation
~, когере́нтное coherent emission

излуче́ние, корпускуля́рное corpuscular emission, particulate radiation
~, ла́зерное laser emission
~, мя́гкое low energy radiation, soft radiation, soft energy radiation
~, напра́вленное directional radiation, beaming, beamed radiation
~, непреры́вное continuous wave transmission, C.W. transmission
~, ночно́е nocturnal radiation, terrestrial radiation
~, оста́точное residual radiation
~, отражённое back-scattered radiation
~, побо́чное spurious radiation
~, проника́ющее hard radiation, penetrating radiation, high-energy radiation
~, радиоакти́вное radioactive radiation
~, распа́дное decay radiation (nucl)
~, рентге́новское X-radiation, X-ray radiation
~, светово́е thermal radiation, radiant energy, heat radiation
~, стимули́рованное stimulated radiation, emission, induced radiation, emission
~, сумма́рное total radiation, integrated radiation
~, теплово́е 1. heat radiation, thermal radiation
2. infra-red radiation
~, ультрафиоле́товое ultra-violet radiation, emission, UV radiation, emission
~, фоново́е background radiation
~, электромагни́тное electromagnetic radiation
~ эне́ргии energy radiation
~, я́дерное nuclear radiation
излу́чина bend, curve
излучи́ть (см излуча́ть)
изма́тывание 1. harassing
2. wearing out
изма́тывать/измота́ть to harass, to wear out
изме́на treason, treachery, betrayal
измене́ние change
~ устано́вки change of setting (instr)
измени́ть (см изменя́ть)
изме́нник traitor
изменя́ть/измени́ть to change, to alter, to modify, to betray
измере́ние measuring, measurement, metering
~ изно́са кана́ла ствола́ bore wear measurement
изме́ренный measured, metered
измери́мый measurable
измери́тель gauge, meter, indicator
~, гироскопи́ческий control rate gyroscope (GW)
~ глубины́ dip-stick
~ поме́х radio noise meter, noise-to-signal meter
~ радиоакти́вности radiac meter
~ расхо́да то́плива fuel flowmeter
~ у́ровня ма́сла oil-level gauge
измери́тельный measuring, metering, gauging
и́зморозь hoarfrost, whitefrost, rime
и́зморось drizzle, sleet
измо́танность 1. harassment
2. exhaustion
измота́ть (см изма́тывать)
изна́шивание depreciation, wear and tear
изна́шивать/износи́ть to wear out
изно́с 1. usage
2. wear and tear
3. wear out
~ гу́сениц track wear
~ дви́гателя engine wear
~, есте́ственный fair wear and tear
~, преждевре́менный premature wearing out
~ при каче́нии wear due to rolling
~ при скольже́нии wear due to sliding
~, сло́жный wear due to rolling and sliding friction
~ ствола́ barrel wear
~ цили́ндра cylinder wear
~ цили́ндра дви́гателя bore wear
~ ши́ны tyre wear
износи́ть (см изна́шивать)
износосто́йкий tear-proof, wear-resistant
износоупо́рный tear-proof
износоусто́йчивость resistance to wear, wear resisting properties
износоусто́йчивый long-wearing, wear-resisting
изнури́ть (см изнуря́ть)
изнуря́ть/изнури́ть to wear down, to harass
"изо рта́ в рот" "mouth to mouth", "kiss of life" (artificial respiration)
изоба́ра isobar

изображе́ние image, representation
~ **анте́нны, зерка́льное** antenna image
~ **в разо́бранном ви́де** exploded view
~ **масшта́ба** representation of scale
~, **перевёрнутое** inverted image
~ **релье́фа** representation of relief (carto)
изоги́псы contours (topo)
изого́на isogon, isogonic line (magnetism)
изодина́ма isodynamic line (magnetism)
изодо́за isodose
изокли́на isoclinic line
изоли́ровать to isolate, to insulate; **изоли́ровать огнём** to box in (arty); **изоли́ровать плацда́рм** to seal off a beachead
изоля́тор 1. isolation ward, isolation section
2. insulator, non-conductor
~, **металли́ческий** metallic insulator
~, **оре́шковый** shell-type insulator
~, **фарфо́ровый** porcelain insulator
изоля́ция 1. isolation, quarantine
2. insulation
~, **теплова́я** thermal insulation, heat insulation
изоме́р isomer
~, **я́дерный** nuclear isomer (physics, chem)
изопропилнитра́т isopropyl nitrate
изоте́рма isotherm
изотерми́ческий isothermic
изото́п isotope
~, **радиоакти́вный** radioactive isotope
~, **усто́йчивый** stable isotope
израсхо́дование expenditure, spending, use, consumption
израсхо́дованный spent, used
израсхо́довать боеприпа́сы to expend ammunition, to run out of ammunition
~ **(см расхо́довать)**
изуче́ние ме́стности study of ground, (Brit), estimation of terrain (US)
~ **обстано́вки** appreciation of the situation
~ **проти́вника** detailed appreciation of enemy
изъя́н defect, flaw
изъя́ть (см изыма́ть)
изыма́ть/изъя́ть to withdraw, to remove, to take away
изыска́тельский prospecting, surveying
изэнтропи́ческий isentropic
иконоско́п iconoscope (electron)
ил slit, slime, sludge, mud, ooze
и́листый slimy, sludgy, muddy, silty
име́ть(ся) to have, there is, there are; **име́ть бо́льшую досяга́емость** to outrange; **име́ть бо́льшую сро́чность** to have higher priority; **име́ть в распоряже́нии** to have at one's disposal; **име́ться в избы́тке** to be in surplus supply
име́ющийся в распоряже́нии available
~ **налицо́** available
имита́тор 1. simulator, battle simulator
2. battle simulation specialist
~ **огня́ та́нкового ору́дия** tank gun simulator
имитацио́нный simulator
имита́ция imitation, simulation
имити́ровать to simulate, to imitate
иммуниза́ция immunization
импеда́нс impedance
~, **входно́й** input impedance
~, **выходно́й** output impedance
~ **излуче́ния** radiation impedance (antenna)
~ **исто́чники** source impedance
~, **нормализо́ванный** normalized impedance (elec)
импе́ллер impeller, blade
импло́зия implosion
импрегни́рованный impregnated
импровиза́ция improvisation
импровизи́рованный improvized, makeshift
и́мпульс impulse, pulse
~ **большо́го то́ка** high current pulse (elec)
~ **высо́кого напряже́ния** high voltage pulse (elec)
~, **запуска́ющий** trigger pulse
~, **зонди́рующий** transmitted pulse (radar)
~, **калибро́вочный** calibration pulse (electron)
~, **когере́нтный** coherent pulse (electron)
~, **кома́ндный** command impulse
~, **нача́льный** transmitted pulse (radar)
~, **опознава́тельный** identification pulse (radar)
~, **опо́рный** reference pulse (radar)
~ **основа́ния** pedestal pulse (electron)
~ **отда́чи** recoil momentum

и́мпульс отмéтки marker pulse, pip pulse, notch pulse (electron, radar)
~, отпирáющий gate pulse (electron)
~, отражённый echo, echo pulse (radar)
~, пилообрáзный sawtooth pulse (electron)
~, погóнный linear pulse
~, поджигáющий trigger pulse (electron)
~ покóя spacing pulse (teleprinter)
~, пропускáющий gating impulse (electron)
~, пусковóй 1. marking impulse (teleprinter)
2. action pulse, firing pulse (electron)
3. trigger signal, trigger pulse (radar)
~, световóй thermal energy received per unit area, thermal radiation intensity (nucl)
~, синхронизи́рующий synchronizing impulse, sync pulse
~, стáртовый start signal (teleprinter)
~, тепловóй amount of radiant energy absorbed, amount of thermal radiation absorbed (nucl)
~ трапецоидáльной фóрмы trapezoidal pulse
~, эталóнный calibration pulse
и́мпульс-э́хо echo (radar)
имýщество equipment, properties, property
~, бессрóчное equipment on permanent charge
~, боевóе warlike stores, war material, equipment, material, stores
~, бронетáнковое stores and equipment (for servicing and repairing AFV's)
~, вещевóе kit
~, воéнное 1. munitions, warlike stores, warlike equipment, military stores
2. service property
~, воéнно-мостовóе military bridging equipment
~, воéнно-техни́ческое military technical stores
~, воéнно-хими́ческое chemical warfare stores, chemical warfare equipment
~, воéнно-хозя́йственное Q stores
~, госпитáльное постéльное hospital bedding
~, дегазациóнное CW decontamination equipment
~, запрáвочное fuelling equipment, refuelling equipment
имýщество, зубоврачéбное dental equipment
~ и материáлы, маскирóвочное camouflage stores and equipment
~, изли́шнее surplus, surplus stores
~, инженéрное engineer equipment and materials, engineer stores
~, интендáнтское Q stores
~, исprávное serviceable equipment, useable stock
~, канцеля́рское office equipment
~, кварти́рное barrack equipment
~, лáгерное camp equipment, camp stores
~, лимити́рованное rationed stores
~, лы́жное ski equipment
~, лы́жно-сáнное ski-sledge equipment
~, мéдико-санитáрное medical stores and equipment
~, медици́нское medical stores
~, мостовóе bridging equipment
~, нали́чное stores in hand, available stores
~, недви́жимое static installations, immoveable property
~, нерасхóдное non-expendable items
~, нетáбельное non-standard equipment
~, палáточное tentage
~, парашю́тно-десáнтное parachute and supply-dropping equipment
~, переправочное water crossing equipment, wet-gap crossing equipment
~, подотчётное accountable stores, inventory asset
~, подрывнóе demolition equipment
~, полевóе field equipment, field stores
~, понтóнно-мостовóе pontoon bridging equipment
~, поступáющее incoming stock
~, расхóдное expendable equipment, expendable stores, expendable items
~, расхóдуемое expendable equipment, expendable stores, expendable items
~ свя́зи signals equipment, signals stores
~, спи́санное written-off equipment, written-off stores
~, стирáемое washing
~, тáбельное scaled equipment, authorized stores, establishment stores
~, такелáжное block and tackle
~, тáнковое tank stores
~, техни́ческое technical stores

иму́щество, топографи́ческое survey stores
~, трофе́йное captured equipment
~, труднозатопля́емое floating equipment, unsinkable equipment, floating-river crossing equipment
~ ча́сти, вози́мое unit baggage
инвали́дность disability (med)
инвалю́та foreign currency
инвентариза́ция making an inventory, stocktaking
~, периоди́ческая routine stocktaking
~, части́чная spot inventory check
инвента́рь 1. stock, equipment
2. inventory
~, каза́рменный 1. barrack inventory
2. barrack stores
~, лы́жный ski equipment
~, противопожа́рный fire-fighting equipment
инве́ртор 1. inverter (elec)
2. phase inverter (rad)
ингибити́рование inhibition
ингиби́тор inhibitor
ингредие́нт ingredient, component
и́ндекс 1. index
2. pointer
~ вя́зкости viscosity index
~ модуля́ции modulation index (electron)
~, модуляцио́нный modulation index (electron)
~, неподви́жный index mark (elevation receiver, ord)
~ пенетра́ции penetration index
~ пласти́чности plasticity index
~ преломле́ния refractive index
индивидуа́льный individual
индика́тор 1. indicator
2. display, display unit (radar)
~ а́зимута bearing indicator
~, вертика́льный vertical display
~ взры́ва detonation indicator
~, визуа́льный visual display
~, вы́несенный remote display
~, выносно́й remote display
~ высо́ты це́ли target height indicator
~, глисса́дный glide path indicator
~, горизонта́льный horizontal display
~ гру́бой да́льности coarse range tube, coarse range indicator
~ да́льности range indicator
~ дви́жущихся це́лей moving target indicator, MTI
индика́тор запро́счика interrogation indicator
~ излуче́ния radiation indicator
~, инфракра́сный infra-red indicator, IR indicator
~ кругово́го обзо́ра plan position indicator, PPI
~ кругово́го обзо́ра, вы́несенный remote plan position indicator, adjacent plan position indicator
~ ку́рса course indicator, course detector (aeron)
~, ла́зерный laser display
~ мо́щности power-level indicator
~ настро́йки tuning indicator
~, обзо́рный plan position indicator, PPI
~ обнаруже́ния warning indicator, acquisition scope
~ отноше́ния у́ровней signal-to-noise ratio indicator (electron)
~ подвижны́х це́лей moving target indicator
~, радиолокацио́нный radar display
~ с кругово́й развёрткой, одноме́рный J-scope
~ с кругово́й шка́лой dial indicator
~ с ли́мбом dial indicator
~ с растя́нутой развёрткой expanded file
~ се́кторного обзо́ра off-centred display, off-centred plan position indicator display
~, се́кторный off centred display (radar)
~ ситуа́ции, электро́нный cathode ray tube surface situation plot
~ теплово́го излуче́ния thermal detector
~ то́чной да́льности fine range tube, fine range indicator
~, трёхкоордина́тный three-dimensional display
~ угла́ возвыше́ния elevation indicator, elevation tube (radar)
~ угла́ ме́ста elevation indicator (radar)
~ упреждённой высоты́ altitude prediction indicator
~, цифербла́тный dial indicator
~, электро́нный indicator tube, electronic display
индика́ция 1. indication
2. display, presentation
~, двухме́рная two-dimensional display, two-dimensional presentation

индика́ция на стекле́ head-up display (for aircraft pilot)
~ по звуково́му нулю́ aural-null indication
~, простра́нственная three-dimensional display
~, трёхме́рная three-dimensional display
индукти́вность inductance
~, взаи́мная mutual inductance (elec)
~ вы́вода lead inductance
индукти́ровать to induce (elec)
инду́ктор 1. indicator 2. magneto generator (teleph)
инду́кция, магни́тная magnetic induction, flux density (elec)
~, электромагни́тная electromagnetic induction
ине́ртность inertness (chem)
инерцио́нность зри́тельного восприя́тия persistence of vision (med)
~ управле́ния раке́той guidance lag (GW)
ине́рция inertia
~ зри́тельного восприя́тия persistency of vision (med)
инже́ктор injector
инжене́р engineer
~, вое́нный military engineer
~, войсково́й unit engineer
~, корпусно́й corps engineer (Sov) (roughly equates to Corps Commander Royal Engineers (Brit))
~, полково́й regimental engineer, officer commanding regimental engineer personnel
инжене́р-капита́н engineer captain
инжене́р-лока́торщик radar technician
инжене́р-майо́р engineer major
инжене́р-меха́ник mechanical engineer
инжене́рно-доро́жный engineer road construction and repair
инжене́рно-строи́тельный engineer construction
инжене́рно-хими́ческий chemical, NBC, NBC engineer
инжене́рный engineer
инжене́р-подполко́вник engineer lieutenant-colonel
инжене́р-полко́вник engineer colonel
инжене́р-связи́ст signals technician
инжене́р-техно́лог process engineer, production design engineer
инжене́р-техно́лог ГЦМ fuel technologist, POL specialist
инжене́р-хи́мик chemical engineer
инжрезе́рв reserve engineer troops
инжча́сть engineer unit
инициати́ва initiative
~, ча́стная individual initiative
инициа́тор igniter
~, комбини́рованный pyrotechnic and chemical igniter
~, пиротехни́ческий pyrotechnic igniter
~, хими́ческий chemical igniter
иниции́рование initiation
иниции́ровать to initiate, to trigger; **иниции́ровать взры́вчатое вещество́** to initiate explosive
инклина́тор dip circle (instr)
инклино́метр inclinometer
инопланетя́нин being from another planet
инсе́ктицид insecticide
инспекти́ровать to inspect
инспе́ктор inspector
инспе́кция inspection
~, автотра́кторная traffic control organisation (of large garrison)
инста́нции, кома́ндные chain of command
~ кома́ндования chain of command
~ медици́нской слу́жбы medical channels
~ служб ты́ла, кома́ндные administrative chain of command, rear services chain of command
~, служе́бные service channels, chain of command, proper channels
~, техни́ческие technical channels
инста́нция authority, echelon, echelon of command, channel
~, кома́ндная level of command
~, назнача́ющая appointing authority
~, перви́чная дово́льствующая initial source of supply
~, прове́рочная controlling authority, inspecting authority
~, проверя́ющая controlling authorities, inspecting authorities
~, техни́ческая technical channel
~, утвержда́ющая approving authority, confirming authority
инструкта́ж briefing
~ пе́ред вы́летом pre-flight briefing

инструктор артиллери́йского де́ла gunnery instructor
~ по стрельбе́ артилле́рии gunnery instructor
~, хими́ческий CW officer, CW instructor
инстру́кция manual, official instructions
~ по обслу́живанию материа́льной ча́сти, постоя́нная standing operating procedure
~ по осмо́тру memorandum of examination
~ по техни́ческому обслу́живанию maintenance instruction, servicing manual
~ по управле́нию тра́нспортной маши́ной driving instruction
~ по ухо́ду и обслу́живанию routine order
~ по эксплуата́ции operating instructions
~, постоя́нно-де́йствующая standing operating instruction
инструме́нт tool, tools, instrument
~, ме́лкий small tool
~, противопожа́рный fire-point
~, ручно́й hand tool
~, хирурги́ческий surgical instrument
~, ша́нцевый entrenching tool
инструме́нты и принадле́жности, запасны́е spare parts (tools and accessories)
~ и приспособле́ния, запасны́е spare tools and accessories
инсули́н insulin
интегра́тор integrator (computer)
~ а́зимута bearing integrator
~, моделиру́ющий analogue integrator
~ перемеще́ния yaw integrator
~ ускоре́ния integrating accelerometer
интегри́рование integration (maths)
интенда́нт quartermaster
интенси́вность движе́ния traffic density, volume of traffic
~ излуче́ния energy density, radiation intensity (nucl)
~ излуче́ния, спектра́льная spectral energy density per unit wavelength
~ изна́шивания wear rate
~ лучеиспуска́ния radiation intensity
~ отка́зов failure rate
~ радиа́ции, допусти́мая radiation safety level
интенси́вность радиоакти́вности intensity of radioactivity
~ светово́го излуче́ния intensity of radiant energy
интерва́л 1. interval, gap 2. gun interval
~, безопа́сный safe interval
~, временно́й time gap, interval
~, защи́тный guard band (radio)
~ по горизо́нту horizontal interval
~ разры́ва horizontal distance between target and point of burst
~ разры́ва рикоше́та distance of burst from point of ricochet
~ сре́дней то́чки разры́вов horizontal distance between target and mean point of burst
~, убо́йный lethal radius (of shell burst)
~ экспози́ций range of exposures (nucl)
интервало́метр intervalometer
интерве́нция, вооружённая armed intervention
интерни́рование internment
интерни́ровать to intern
интерполя́ция interpolation
интерфере́нция interference fading (radio)
интерферо́метр interferometer
интерце́птор 1. interceptor 2. spoiler (GW)
инфекцио́нный infectious, contagious, communicable
информа́ция information, data
~, дискре́тная digital information
~, ло́жная false information
~, необрабо́танная raw information, raw data
~, обрабо́танная processed information, processed data
~, предвари́тельная advance information
~, цифрова́я digital information
информи́ровать/проинформи́ровать to inform
инфракра́сный infra-red
инъе́кция injection (med)
ио́д iodine
ио́н ion
иониза́ция ionization
~, оста́точная residual ionization
~, уда́рная ionization by collision
ионизи́ровать/ионизова́ть to ionize
ионизова́ть (см ионизи́ровать)

иономе́тр ionometer
ионосфе́ра ionosphere
ипри́т yperite, mustard gas
~, **азо́тистый** nitrogen mustard (CW)
~, **пере́гнанный** purified mustard gas
ипри́ты, азо́тистые nitrogen mustards (CW)
иррадиа́ция irradiation
искажа́ть/искази́ть to distort, to garble (sigs)
искаже́ние distortion
~, **амплиту́дное** amplitude distortion
~, **гармони́ческое** harmonic distortion
~ **длин** distortion of length (carto)
~ **изображе́ния** image distortion, picture distortion
~, **нелине́йное** non-linear distortion (elec)
~ **площаде́й** distortion of coordinate squares (carto)
~ **сигна́ла** signal distortion
~, **трапецоида́льное** trapezoidal distortion (electron)
~, **фа́зовое** phase distortion
~, **часто́тное** frequency distortion
искази́ть (см искажа́ть)
иска́тель 1. selector (teleph) 2. view-finder (photo)
~, **вспомога́тельный** auxiliary selector, auxiliary switch
~, **ко́довый** code selector
~ **мин, винто́вочный** rifle mine detector
исключа́ть/исключи́ть to exclude, to rule out, to eliminate
исключе́ние 1. dismissal, expulsion 2. exception
~ **со слу́жбы** dismissal from service
исключи́ть (см исключа́ть)
и́скра spark
искре́ние sparking (elec)
~ **колле́ктора** commutator sparking
~, **си́льное** arcing
искрогаше́ние spark suppression
искрообразова́ние spark formation, sparking
искроулови́тель spark trap (stove)
иску́сственный artificial
иску́сство, вое́нное art of war (the Sov term includes strategy and tactics)
~, **операти́вное** strategy (Sov)
испаре́ние evaporation
~ **горю́чего** fuel evaporation
~, **пове́рхностное** surface evaporation
испари́тель vaporizer
испари́ть (см испаря́ть)
испаря́емость 1. rate of evaporation 2. volatility
испаря́ть/испари́ть to evaporate, to vaporize
исполне́ние 1. fulfilment (of wish), execution (of order) 2. make, version, design (tech)
~ **до́лжности, вре́менное** temporary holding of appointment
исполня́ющий обя́занности acting
~ **обя́занности комба́та** acting battalion commander
испо́льзование use, utilization, employment, consumption
~ **ме́стности** use of ground
~ **не по назначе́нию** improper use
испо́льзовать to make use of, to use, to exploit; **испо́лзовать преиму́щество** to press home an advantage; **испо́льзовать результа́ты я́дерного уда́ра** to exploit the results of a nuclear strike
"Испра́вить Наво́дку!" "Relay the Aim!"
испра́вить (см исправля́ть)
исправле́ние correction, correcting
испра́вленный corrected
исправля́ть/испра́вить to correct, to repair, to mend
испра́вность good condition, good working order, serviceability
~, **техни́ческая** good mechanical working order
испра́вный serviceable, in good working order
испуска́ть/испусти́ть to emit, to give off; **испуска́ть/испусти́ть лучи́** to radiate
испусти́ть (см испуска́ть)
испыта́ние test(ing), trial, check(ing)
~, **боево́е** battle service test, service trials
~ **боево́й те́хники** troop trial
~ **в эксплуатацио́нных усло́виях** field trial
~, **войсково́е** troop trials
~, **вы́борочное** random test, random check
~, **дли́тельное** long term test
~, **доро́жное** road test
~, **климати́ческое** climatic trial
~, **контро́льное** check test
~, **лаборато́рное** laboratory test

испыта́ние на долгове́чность life (service) test, life (service) trial
~ **на изно́с** wear test
~ **на ло́мкость** brittleness test
~ **на надёжность** reliability test, reliability trial
~ **на приго́дность к эксплуата́ции** performance test
~ **на про́чность** structural test
~ **на разруше́ние** destruction test
~ **на расхо́д горю́чего** fuel-consumption test
~ **на сопротивле́ние** resistance test
~ **на сохра́нность** safety test
~ **на сте́нде** bench test
~ **на то́пливную экономи́чность** fuel consumption test
~ **на уста́лость** fatigue test
~, **пове́рочное** check test
~, **полево́е** field test, field trial
~, **приёмное** acceptance trials
~, **приёмосда́точное** acceptance trials
~, **приёмочное** acceptance trials
~ **пробе́гом** road test
~, **про́бное** trials test
~, **сда́точное** delivery trials (eqpt)
~, **сте́ндовое** bench test
~ **стрельбо́й** test firing
~, **типово́е** routine test, routine trial
~ **тормозо́в** brake test
~, **ходово́е** running test, performance test
~, **эксплуатацио́нное** in service test, in service trial
~ **я́дерного ору́жия** nuclear weapon test
иссле́дование investigation, research, analysis
иссле́довать to investigate, to explore, to analyse
истека́ть/исте́чь to elapse, to expire; **исте́кать/исте́чь кро́вью** to bleed profusely, to bleed to death
истече́ние 1. effluence, exhaust 2. expiry
исте́чь (см исте́кать)
истира́ние wear, abrasion
исто́к source
исто́чник 1. spring (topo) 2. source
~ **излуче́ния** radiation source
~ **инфе́кции** source of infection
~ **инфракра́сного излуче́ния** infra-red radiator
исто́чник нейтро́нов 1. neutron source (nucl) 2. neutron emission device (nucl)
~ **пита́ния** source of power, power supply (elec)
~ **пита́ния, вне́шний** external power source
~ **пита́ния, встро́енный** built-in power source
~ **радиа́ции** radiation source
~ **то́ка** source of power, source of current (elec)
исто́чники, ме́стные local sources
истоща́ть/истощи́ть to drain, to exhaust, to deplete, to wear out; **истощать/истощи́ть резе́рвы** to drain reserves
истоще́ние, боево́е battle fatigue
~ **запа́са** stock depletion
истощи́ть (см истоща́ть)
истреби́тель fighter (ac)
~, **одноме́стный** single-seater fighter (ac)
~, **реакти́вный** jet fighter (ac)
~ **та́нков** tank destroyer, tank hunter
~, **учё́бно-тренир́овочный** fighter trainer
истреби́тель-бомбардиро́вщик fighter-bomber
истреби́тельно-авиацио́нный fighter-aircraft, fighter
истреби́тельно-бомбардиро́вочный fighter-bomber
истреби́тельно-противота́нковый tank-destroyer (of units)
истреби́тельный fighter
истреби́тель-перехва́тчик interceptor fighter (ac)
истребле́ние та́нков tank hunting
истре́бование indenting
истре́бовать to demand and obtain, (perf only); **истре́бовать боеприпа́сы** to indent for ammunition
исхо́д outcome, result
~ **бо́я** outcome of battle
исхо́дный initial, opening, starting, basic
исходя́щий outgoing (sigs)
исчеза́ть/исче́знуть to disappear
исчезнове́ние тра́ссы це́ли target rejection
исче́знуть (см исчеза́ть)
исчисле́ние, операцио́нное operational calculus (maths)

исчи́сленный corrected, calculated, predicted
ито́г 1. total
2. summing-up (of exercise)
~ **заня́тия** summing up of exercise
~, **о́бщий** grand total
итого́ altogether, in all, total
иша́к 1. donkey
2. mule

К

"К Бо́ю!" 1. "Form extended line!"
2. "Action!" (arty)
"К Бо́ю Гото́вься!" "On Guard!"
"К Бо́ю с Ты́ла!" "Action Rear!"
"К Бо́ю с Фро́нта!" "Action Front!"
"К Маши́нам!" "Fall In by Your Vehicle!" (embussing drill)
"К Ноге́!" "Order Arms!"
"К Ору́жию!" "Fall In on Your Weapons!"
кабелеукла́дчик cable layer
кабелеукла́дыватель cable layer
ка́бель cable
~, **брониро́ванный** armoured cable (sigs)
~ **в оплётке, многожи́льный** sheathed multi-core cable
~, **вое́нно-полево́й** field cable
~, **возду́шный** overhead cable, aerial cable
~, **двухжи́льный** twin-core cable, two-core cable
~, **коаксиа́льный** coaxial cable
~, **магистра́льный** main cable
~, **многожи́льный** multicore cable
~, **одножи́льный** single-core cable
~, **оруди́йный** main distribution box to gun cable (AAFCE)
~, **освети́тельный** lighting cable
~, **освинцо́ванный** lead-covered cable
~, **паралла́ксный** predictor-to-junction-box data transmission cable (AAFCE)
~, **пита́ющий** feeder cable
~, **подво́дный** submarine cable line, underwater cable line (comms)
~, **подзе́мный** underground cable
~, **полево́й** ground cable, field cable
~, **полево́й телегра́фный** field telegraph cable
ка́бель, полево́й телефо́нный field telephone cable
~, **прибо́рный** gunnery data cable (AAFCE)
~, **пупинизи́рованный** coil-loaded cable
~, **рези́новый шла́нговый тяжёлый** heavy rubber-sheathed cable
~, **сигна́льный** signal cable
~, **силово́й** power cable
~, **соедини́тельный** connecting cable, jumper cable
~, **телефо́нный** telephone cable
~, **трёхжи́льный** three-core cable
~, **ту́мбовый** adaptor cable joining main cable to junction box (AAFCE)
~, **удлини́тельный** extension cable
~ **управле́ния** control wire
~, **шла́нговый тяжёлый** heavy sheathed cable
~, **экрани́рованный** shielded cable
кабеста́н capstan
каби́на 1. cab (veh)
2. cockpit (ac)
~, **отки́дная** tilting cabin
~ **с кондициони́рованием во́здуха** air-conditioned cabin
~ **управле́ния** driver's cab (veh)
кабине́т, вое́нный study group (for reserve officers)
~, **зубно́й** dental centre
~, **зубовраче́бный** dental centre, dental surgery
~, **методи́ческий** training aids department
кабота́ж coastal shipping
кабри́рование pitch-up, pitching (aeron)
кавалери́йский cavalry
кавале́рия cavalry
кавалье́р spoil band (road constr)
кавита́ция cavitation, set-back (ammo)
кагане́ц simple oil lamp
ка́дмий cadmium
кадр 1. frame (radar, photo)
2. cadre
кадрови́к regular soldier
ка́дровый regular, cadre
ка́дры personnel, cadres
~, **кома́ндные** officers' cadres
ка́жущийся apparent
каза́рма barracks, barrack building
казема́т 1. chamber
2. underground shelter, shelter

казема́т, боево́й casemate, casement (field works)
казённик 1. base cap, breech piece (mor)
2. breech ring (ord)
казна́ 1. breech (ord)
2. treasury
3. the State
казначе́й paymaster, finance officer
казначе́йство, полево́е field cash office
казнозаря́дный breech-loaded
казнь, сме́ртная capital punishment
ка́йла pick, pickaxe
Как Слы́шно "Signals!" "How do you hear me?"
кале́ние incandescence
ка́ли, е́дкое caustic potash
кали́бр 1. gauge
2. calibre
3. yield (nucl)
~ а́томной бо́мбы energy release of nuclear bomb (in kilotons)
~, контро́льный master gauge
~, кру́пный 1. heavy calibre, large calibre
2. high energy release (nucl) (over 100 kt)
~, ма́лый 1. small calibre
2. low energy release (nucl) (under 10 kt)
~, микрометри́ческий micrometer screw gauge
~, преде́льный limit gauge
~, рабо́чий working gauge
~, сре́дний 1. medium calibre (arty)
2. medium energy release (nucl) (10-100 kt)
~ ствола́ bore diameter
~, цилиндри́ческий plug gauge
~ я́дерного боеприпа́са nuclear yield
калибра́тор calibrator
~, ква́рцевый calibrated crystal oscillator
калибри́рование calibration testing
~ снаря́дов measuring of the calibre of projectiles
калиброва́ть to calibrate
калибро́вка calibration
~ се́тки grid sizing operation (electron)
калибромéр calibre gauge (ord)
ка́лий potassium
~, хло́ристый potassium chloride
~, хлорноватоки́слый potassium chlorate
~, циа́нистый potassium cyanide
калори́йность calorific heat value
калори́метр calorimeter
калориме́трия calorimetry
калори́фер space heater, air heater
калори́ческий calorific
кало́рия calorie
~, больша́я large calorie (one thousand calories)
~, ма́лая calorie, small calorie (as opposed to "large calorie")
калы́шка kink (rope)
ка́лька tracing paper, tracing cloth
ка́льций calcium
~, хло́ристый calcium chloride
ка́мбуз 1. cookhouse
2. galley (nav)
ка́мень 1. stone
2. crank sliding block (ord breech)
~, бу́товый rubble
~ мотыля́ crank sliding block (ord breech)
~, рва́ный rubble
ка́мера 1. cell, chamber
2. chamber (ord, SA)
3. camera
4. inner tube (tyre)
~, автомоби́льная inner tube (tyre)
~ ва́куума-аппара́та, поплавко́вая upper vacuum chamber
~, возду́шная inner tube (of tyre)
~, га́зовая gas cylinder, gas chamber (SA, MG)
~, газоку́ривательная gas chamber (gas trg)
~, горя́чая hot cell
~, дегазацио́нная CW decontamination chamber
~, дезинфекцио́нная disinfection room, delousing chamber
~, дезинфекцио́нная парова́я steam sterilizing chamber, steam chamber
~, жарова́я heat chamber
~, заря́дная 1. charge chamber (mining)
2. charging case (mine)
3. breech chamber (ord)
4. cavity (ammo)
~, ионизацио́нная ionization chamber
~, испари́тельная vaporizing chamber
~ карбюра́тора, поплавко́вая carburettor float chamber
~, кла́панная valve chamber
~, конденсацио́нная cloud chamber (physics)
~, кривоши́пная crank case (ICE)

кáмера, мáршевая cruising thrust chamber
~, мúнная charge chamber (mining)
~, напóрная pressure chamber, plenum chamber
~, огневáя combustion chamber
~ окýривания gas chamber (gas trg)
~, парашю́тная parachute container
~, паровáя steam chamber
~, поплавкóвая float chamber (carb)
~, самозаклéивающаяся self-sealing tube, puncture-proof tube
~ сгорáния combustion chamber
~, смесúтельная mixing chamber (ICE)
~, спасáтельная escape chamber (submarine)
~, стáртовая 1. booster combustion chamber (GW)
2. pre-combustion chamber (diesel)
~, сужáющаяся necked down chamber
~, сушúльная drying chamber
~, телевизиóнная television camera
~ тумáнов cloud chamber (physics)
~, форсáжная reheat unit, afterburner (ac)
камертóн tuning fork
камнедробúлка rock-crusher
камнемёт, албáнский rock avalanche booby trap
~, фугáсный stone fougasse (mine warfare)
кáмора 1. cell, chamber
2. chamber (ord) (see kamera for types)
~, гáзовая gas chamber, gas cylinder
кампáния campaign
камуфлéт camouflet
камуфля́ж camouflage, disruptive colouration
канáва 1. ditch
2. inspection pit
~, водоотвóдная drainage ditch
~, дренáжная drainage channel, drainage ditch
~, забанкéтная banked ditch
~, нагóрная водоотвóдная hillside catch-water drain
~, осмóтровая inspection pit
~, смотровáя inspection pit
канáвка 1. ditch
2. ring groove (piston, ICE)
~, винтообрáзная spiral groove
~, гáзовая gas port (SA)
~, кольцевáя annular groove, cannelure
канáвка, маслораспределúтельная oil distributing channel
~, мáсляная oil channel
~, поршневáя piston ring groove (ICE)
~ поршневóго кольцá piston-ring groove
~, смáзочная oilway (in engine)
~, шпóночная key bed, key groove, key way
канавокопáтель ditching machine, ditch digger, trencher
~, плýжный trenching plough
канáл 1. canal, channel
2. channel (comms)
3. bore (ord)
~, актинометрúческий actinometric channel (telemetry)
~, водоотвóдный drainage channel
~, воздýшный air duct
~, входнóй inlet port
~ для отвóда гáзов gas passage
~ для смáзки grease passage, oil gallery
~, дренáжный drainage channel
~, запáльный ignition tube, fire channel (ord)
~, мáсляный oil passage way
~, обвóдный bypass, bypass channel (electron)
~, обхóдный bypass channel, alternative channel
~ передáчи transmission channel
~, перепускнóй transfer passage, by-pass, passage way
~, радиорелéйный телегрáфный radio relay telegraph channel
~, резéрвный guard channel
~ свя́зи communications channel
~, сквознóй запáльный fire channel (ord)
~, смáзочный oil duct, oil way
~, соединúтельный connecting channel
~ стволá bore (ord)
~ стволá, глáдкий smooth bore
~ стволá, конúческий tapered bore
~, уплотнённый multiplexed channel
канализáция drainage, sewerage
каналообразовáние channeling
канáт rope, cable, hawser
~, буксúрный tow rope
~, несýщий main cable (suspension br)
~, подвеснóй suspension cable
~, причáльный mooring line, mooring rope

кана́т, про́волочный wire rope
~, расцепля́ющий trip rope (pile driver)
~, я́корный anchor cable
кана́тик, анте́нный antenna cable
кани́стра can, canister
канитéль 1. thin wire
2. wire mesh element (air filter)
~, мéдная copper gauze
~, металли́ческая metallic wool element
канифа́с-бло́к snatch block
канифо́ль rosin
канона́да cannonade
кант piping (uniform)
кантова́ть to cant, to tilt, to manipulate
канцеля́рия office, documentation office
~ ро́ты company office
каньо́н canyon
капасита́нс capacitance
капита́льный 1. complete, major (of an overhaul)
2. first-class, all-weather and 100-ton capacity bridges
3. with at least 14 metres of carriageway (of a road)
капита́н captain (all services)
~ второ́го ра́нга captain second rank (equates to commander R.N.)
~ пéрвого ра́нга captain first rank (equates to captain R.N.)
~ трéтьего ра́нга captain third rank (equates to Lt-Cdr R.N.)
капита́н-лейтена́нт senior lieutenant (nav)
капитули́ровать to capitulate, to lay down arms
капитуля́ция capitulation, surrender
~, безогово́рочная unconditional surrender
~, ма́ссовая mass surrender, surrender en masse
капка́н snare, trap
ка́пля drop
капони́р 1. protected or dispersed hard standing for aircraft
2. pen
3. douple caponier (eng)
капо́т 1. cowling (ac)
2. bonnet (veh)
ка́псула capsule
ка́псюль percussion cap, primer cap, detonator
~, впрессо́ванный inset primer
~ ги́льзы cartridge primer
ка́псюль, дистанцио́нный powder train, time fuse
~, прижима́ющий primer percussion cap
~, уда́рный percussion fuse detonator, detonator (fuse)
ка́псюль-воспламени́тель igniter set cap
~, патро́нный percussion cap
~, тру́бочный detonator, igniferous detonator (fuse)
ка́псюль-детона́тор 1. blasting cap (mining)
2. igniter
3. disruptive detonator
~ нажи́много дéйствия pressure cap
~ элемéнта неизвлека́емости anti-lift igniter
каптена́рмус company quartermaster sergeant, CQMS
~ ро́ты company quartermaster sergeant, CQMS
каптёр NCO in charge of store
каптёрка storeroom, store
капюшо́н hood, cowl
~, газонепроница́емый gas-proof hood
~, огнезащи́тный antiflash hood
ка́ра penalty, punishment
караби́н 1. carbine
2. snap hook, hook, clip, snap link
~, автомати́ческий automatic carbine
~, самозаря́дный self-loading carbine
карава́н convoy
~, вью́чный pack train
карактро́н charactron
каранда́ш 1. pencil
2. anti-dim stick
~, ту́шевый chinagraph pencil
каранти́н 1. quarantine
2. basic training
~, по́лный absolute quarantine
кара́ть/покара́ть to punish, to penalise
карау́л guard (body of men)
"Карау́л в Ружьё!" "Turn Out the Guard!"
карау́л, вну́тренний regimental guard
~, врéменный temporary guard
~, гарнизо́нный garrison guard
~, ефрéйторский guard commanded by lance-corporal
~, ла́герный camp guard
~, но́вый new guard
~, отдéльный detached post
~, офицéрский guard commanded by an officer

карау́л, полево́й field picquet, outguard, security standing patrol
~, постоя́нный permanent guard
~, почётный guard of honour
~, сержа́нтский guard commanded by a sergeant
~, сменя́емый old guard (being relieved)
~, ста́рый old guard
карау́лка guardroom (sl)
карау́льный 1. member of guard
2. sentry on secondary post
3. guard
карбо́лка carbolic acid (sl)
карбюра́тор carburettor
~ встре́чного ти́па down-draught carburettor
~, двухдиффу́зорный twin choke carburettor
~, двухка́мерный carburettor with two float chambers
~, зато́пленный flooded carburettor
~, необледенева́ющий non-icing carburettor
~ опроки́нутого ти́па down-draught carburettor
~ перевёрнутого ти́па down-draught carburettor
~ с восходя́щим пото́ком up-draught carburettor
~ с горизонта́льным пото́ком side-draught carburettor
~ с нисходя́щим пото́ком down draught carburettor
~ с обра́тным пото́ком down-draught carburettor
~ с па́дающим пото́ком down-draught carburettor
карбюра́ция carburation
карда́н universal joint
~, жёсткий mechanical universal joint
~, звéнчатый flexible universal joint
~, иго́льчатый needle-bearing universal coupling, needle-bearing universal joint
~, мя́гкий flexible universal joint
~, универса́льный шарово́й universal ball joint
~, упру́гий fabric joint, flexible fabric coupling
каре́та sidecar
каре́тка 1. carriage (mech, cableway)
2. bogie
3. sidecar
4. sliding gear
5. crank assembly (bicycle)
~ подве́ски bogie
каре́тка-шестерня́ cluster gear
карка́с 1. frame, framework
2. carcass, casing (tyre)
~ ба́шни turret basket
карни́з half-tunnel, cornice
~, сне́жный snow cornice
ка́рта map, chart
~, бла́нковая ghost outline map
~, боева́я battle map
~ в штриха́х hachure map
~, вое́нная military map
~ возду́шной обстано́вки air situation map
~, географи́ческая atlas map, map with scale over 1:1,000,000
~, гидрогеологи́ческая water supply map, map of water resources
~, гипсометри́ческая hypsometric map, contour map
~, доро́жная road map
~, инжене́рная операти́вная engineer operations map
~, ко́нтурная planimetric map, outline map
~ кру́пного масшта́ба large scale map
~, ледяна́я ice slab, block of ice (cut out of frozen river to make ice-free ferry channel)
~, ло́цманская pilot's chart (nav)
~, маршру́тная route card, route map
~, мелкомашта́бная small scale map (1:500,000 to 1:1,000,000 scale)
~, мерка́торская Mercator projection map, Mercator chart
~, метеорологи́ческая weather map
~, миллио́нная map with 1:1,000,000 scale
~, морска́я chart (naut)
~, нема́я outline map
~, обзо́рная atlas map, general purpose map (has scale smaller than 1:1,000,000)
~, обзо́рно-топографи́ческая general topographical map (1:500,000 to 1:1,000,000 scale)
~ обстано́вки situation map
~, по́днятая marked map

ка́рта, рабо́чая operational map, battle map (tac)
~, **радиолокацио́нная** radar map
~, **релье́фная** relief map
~ **свя́зи, рабо́чая** line route map (sigs)
~, **синопти́ческая** synoptic chart, weather map
~, **сма́зочная** lubrication chart
~, **среднемасшта́бная** medium scale map (1:300,000 - 1:1,000,000)
~, **такти́ческая** battle map
~ **техни́ческих осмо́тров автомоби́лей** record of vehicle technical inspections, workshop repairs and overhauls
~, **топографи́ческая** topographical map (1:10,000-1:1,000,000 scale)
~, **физи́ческая** physical map
ка́рта-планше́т map board
ка́ртер 1. housing, casing
2. crankcase
3. gearbox
~ **бортово́й переда́чи** final drive housing
~ **за́днего мо́ста** rear-axle housing, rear-axle casing
~ **маховика́** flywheel housing, flyweel casing
~, **мо́крый** wet sump (ICE)
~, **ни́жний** sump
~ **рулево́го механи́зма** steering box
~, **сухо́й** dry sump (ICE)
~ **сцепле́ния** clutch housing
~ **фрикцио́на** clutch housing
карте́чь shrapnel, grape shot, canister
карти́на шу́мов noise pattern
картогра́фия cartography
карто́н cardboard, millboard
~, **асбе́стовый** asbestos millboard
картоте́ка card index
картофелечи́стка potato peeler
ка́рточка, армату́рная clothing and equipment record
~, **артиллери́йско-стрелко́вая** artillery firing card
~ **взыска́ний и поощре́ний** conduct sheet
~ **води́теля, авари́йная** driver's accident identification card
~ **для стрельбы́** range card
~ **ежедне́вных прове́рок материа́льной ча́сти** daily operational checklist
~ **за́писи стрельбы́** range card
ка́рточка, зени́тная стрелко́вая anti-aircraft range card
~ **консерва́ции** maintenance record card for engines in storage
~, **медици́нская** medical record
~ **огня́** range card
~, **отчётная** sketch map showing own and enemy positions
~, **перфорацио́нная** punched card
~ **противота́нкового огня́** anti-tank range card
~, **процеду́рная** hospital treatment card
~, **служе́бная** record of service card
~, **стрелко́вая** range card
~, **техни́ческая** servicing schedule (ac)
~ **учёта** record card, stock card, registration card
~ **учёта доз облуче́ния** radiation dosage card
~, **учётная** index card, registration card, record card
~, **учётно-послужна́я** record and pay book
ка́рточка-квита́нция QSL card (radio)
карту́з bag, cartridge (arty)
~ **воспламени́теля** igniter bag, igniter cartridge
~, **заря́дный** separate-loading bag (gunn)
карту́зник cartridge box
карту́шка, ко́мпасная compass card
карцинотро́н carcinotron (electron)
карье́р quarry, sand pit
каса́тельная tangent
ка́ска steel helmet
каска́д stage (electron)
~, **бу́ферный** buffer stage (electron)
~, **вво́дный** input stage (electron)
~, **выходно́й** output stage (elec)
~, **мо́щный** power amplification stage
~, **предконе́чный** antenna driver
~, **промежу́точный** intermediate stage (elec)
ка́сса, полева́я сберега́тельная field savings bank
кассе́та 1. magazine, cassette (photo)
2. air cleaner casing
3. cluster (bombing)
~ **для зажига́тельных бомб** incendiary bomb cluster
кассе́тный 1. magazine, cassette (photo)
2. cluster (bombing)
3. pertaining to an air cleaner casing
ката́лиз catalysis

катализáтор catalyst
каталóг запасны́х частéй spare parts list
катамарáн catamaran
катапу́льта catapult
катарáкт dashpot (damper)
катастрóфа 1. catastrophe
2. fatal accident, near-fatal accident
категори́йность маши́ны classification of vehicle by condition
категóрия 1. category, class, status, grade
2. condition
~, возрастнáя age group
~ вооружéния weapon category
~ дéнежного довóльствия pay grade
~ очерёдности обеспéчения precedence code
~, штáтная rank provided for by establishment
кáтер launch, cutter
~, букси́рно-мотóрный towing launch
~, десáнтный landing craft mechanized (personnel), LCM (personnel)
~, мотóр-торпéдный motor torpedo boat, MTB
~, пехóтно-десáнтный landing craft assault
~, развéдывательный десáнтный reconnaissance landing craft
~, спасáтельный rescue launch
~, торпéдный motor torpedo boat
катери́ст member of motor launch crew
кáтер-торпедолóв torpedo recovery vessel
кáтер-трáльщик, удáрный десáнтный assault landing minesweeper
кати́ть (см качáть)
катки́, вертикáльные vertical rollers (of gun mounting)
~, горизонтáльные rollers (gun mounting)
катóд cathode (elec)
~, активизи́рованный activated cathode
~, вольфрáмовый tungsten cathode
~, накалённый heated cathode
~, окси́дный oxide-coated cathode
~, подогрéвный indirectly heated cathode
~, тори́рованный thoriated cathode
~, холóдный cold cathode
катодолюминесцéнция cathodoluminescence
катóк 1. roller
2. road roller
3. wheel (MG or mor mounting)
4. skating rink
~, вéрхний top roller
~, двойнóй опóрный double road wheel (tracked veh)
~, дорóжный road roller
~, дорóжный кулачкóвый sheepsfoot roller
~, направля́ющий idler wheel (tk) (tracked vehicle)
~, опóрный road wheel (tk)
~, поддéрживающий top roller
~ телéжки, опóрный bogie roller
~, трáльный mine clearing roller
~, упóрный bogie wheel (tk)
кату́шка 1. reel, drum
2. coil (elec)
3. bobbin dispenser (GW control wire)
~, дрóссельная choke coil
~ зажигáния ignition coil
~ индукти́вности inductor, induction coil
~, инду́кторная 1. field coil
2. induction coil
~, индукциóнная induction coil
~, кáбельная cable drum
~ магнéто, пусковáя magneto boost coil
~ настрóйки tuning coil
~ перемéнной свя́зи varicoupler
~ прерывáтеля cut-out coil
~, реакти́вная choke coil, reactance coil
~, рукáвная hose reel
~ с корзи́ночной обмóткой basket coil
~ с отвóдами tapped coil
~, секциони́рованная sectional coil
~, следя́щая servo coil
~, терми́ческая heat coil
~, фокуси́рующая focussing coil (electron)
~ электромагни́та field coil
кату́шка-искáтель search-coil (mine detector)
"катю́ша" multi-barrelled vehicle mounted rocket launcher
кáустик caustic soda
кáфедра 1. university chair professorship
2. faculty, department of learning
качáлка "rocking horse" (Sov) (vehicle-movement simulator used for SA firing training)

качáние 1. hunting (servo)
2. rolling, rocking, swinging
3. pumping
~ **лучá** beam swinging
качáть/катúть to roll
качáть/покачáть to pump, to rock, to swing
качéние rolling, rotary motion
кáчества, боевы́е fighting efficiency
~, **комáндные** leadership qualities
~, **конструктúвно-эксплуатациóнные** construction and operating characteristics
~, **механúческие** mechanical properties
~, **морáльно-боевы́е** fighting efficiency and morale
~ **пóроха, баллистúческие** ballistic characteristics of propellant
кáчественный qualitative
кáчество, баллистúческое ballistic characteristic
~, **боевóе** combat characteristic
~ **воспроизведéния** fidelity, quality of reproduction
~ **двúгателя** impulse-weight ratio (rocket technology)
~ **комáндного состáва** quality of leadership
кáчка play (eng)
~, **боковáя** rolling (motion)
~, **бортовáя** roll (aeron, nav), rolling motion
~, **вертикáльная** heaving (naut)
~, **килевáя** pitching (aeron)
~, **носовáя** pitching (motion)
~, **поперéчная** rolling (motion)
~, **продóльная** pitching (motion)
кáшица, вóдная watery paste, slurry
каю́к kayuk (small flat-bottomed rowing boat)
кая́к kayak (canoe)
квадрáнт field clinometer, clinometer
~, **орудúйный** field clinometer
квадрáт 1. square
2. grid square, grid reference
~, **мéлкий** small square (grid with six figure reference) (topo)
~, **тренирóвочный** drill square
квадратúчное, срéднее root mean square
квалификáция 1. classification
2. qualification, skill
квалифицúрованный qualified, skilled
квант quantum
квартáл street block
~, **жилóй** block of dwellings
квартúра billet, quarters
~ **офицéрского состáва** officers' quarters
квартирмéйстер quartermaster (non-Sov)
квартúрный barrack, quartering
квартúрьер billeting officer, quartering officer
квартúрьеры billeting party
квитáнция 1. receipt, voucher
2. acknowledgement (comms)
кенотрóн kenotron (elec)
керáмика ceramics
~, **огнеупóрная** fire-proof ceramics, fire-resistant ceramics, refractory ceramics
керн core
кéрнер centre punch
кернó prick punch
керогáз kerosine pressure stove
керосúн paraffin, kerosine
~, **авиациóнный** aviation kerosene
керосинозапрáвщик kerosene fueller, kerosene bowser
кессóн 1. caisson
2. coffer dam
кетгýт catgut
кибернéтика cybernetics
килогéрц kilohertz, kHz
килóметр kilometre
километрóвка 1:100,000 scale map
килотóнна kiloton, kton
киль 1. fin (aeron)
2. keel
~, **скуловóй** bilge rail
кильвáтер wake
кúнгстон Kingston valve
кинемáтика снаря́да missile motion kinematics
кинематúческий kinematic
кинескóп picture tube, kinescope
кинжáл dagger
кинокáмера cine-camera
кинолóгия study of dogs (especially war dogs)
кинопередвúжка mobile cinema
кинотеодолúт cine-theodolite
кипéние радиáтора radiator boiling
кúпп-релé Kipp relay, one-shot multi-vibrator
кипрéгель plane-table alidade, telescopic alidade (surv)
киркá pick

киркомоты́га pickaxe, mattock
кирпи́ч 1. brick
2. bricks (collective)
кирпи́чный brick
кислорóд oxygen
~, жи́дкий liquid oxygen
~, одноáтомный monatomic oxygen
кислородозапрáвщик oxygen dispenser
кислотá acid
~, азóтная nitric acid
~, аккумуля́торная battery acid
~, карбóловая carbolic acid
~, нафтéновая naphthenic acid, hexahydrobenzoic acid
~, пикри́новая picric acid, melinite, trinitrophenol (expl)
~, сéрная sulphuric acid
~, сини́льная prussic acid
~, солянáя hydrochloric acid
~, фóсфорная phosphoric acid
~, хлóрная perchloric acid, perchlorate acid
~, хлорсульфóновая chlorsulphonic acid, CSA (CW)
кислотомéр acidometer
кислотостóйкий acid-resistant
кислотоупóрный acid-proof, acid-resisting
ки́сточка, щети́нная bristle brush
кисть small brush
~, маля́рная paint brush
кисть-катóк paint roller
ки́тель single-breasted uniform jacket
~, закры́тый jacket with high stand up collar
~, откры́тый open-necked jacket
кия́нка mallet, wooden hammer
клавиату́ра keyboard
клáвиш key (instr)
клáвиша key (instr)
клáдбище cemetery, burial ground
~, вóинское military cemetery
клáдка, арми́рованная кирпи́чная reinforced brick masonry
~, бетóнная concrete masonry
~ из тёсаного кáмня dressed masonry
~, кáменная masonry, stonework
~, кирпи́чная brickwork
~ нáсухо dry laying (building)
кладовáя store, storeroom
~, подви́жная mobile store, store truck
~, продовóльственная ration store
кладовáя, расхóдная QM store, articles-in-use store
кладовщи́к storeman, storekeeper
~, стáрший senior storeman
клáпан 1. valve
2. flap
3. pack cover flap (parachute)
4. drop indicator (switchboard)
~, вдыхáтельный inlet valve (of respirator)
~, вéрхний overhead valve (ICE)
~, взрывозащи́тный blast valve
~ впры́ска injection valve
~, впускнóй inlet valve, intake valve
~, всáсывающий inlet valve, intake valve
~, выдыхáтельный outlet valve (on respirator)
~, выпускнóй exhaust valve (ICE)
~, высоторегули́рующий levelling valve
~, выхлопнóй exhaust valve (ICE)
~ горю́чего fuel valve
~, двухходовóй two-way valve
~, декомпрессиóнный decompressor (diesel ICE)
~ добáвочного вóздуха compensating air valve
~, дренáжный vent valve
~, дрóссельный butterfly valve
~, дыхáтельный atmospheric valve, breather valve, vent valve
~, запóрный stop valve
~ заправки fuelling valve
~, игóльчатый needle valve
~ избы́точного давлéния excess pressure valve
~, измери́тельный metering valve
~ мáсляного насóса, предохрани́тельный oil-pump safety valve
~, маши́нный регуля́торный throttle valve (steam engine)
~ модерáтора run-out control valve
~, нагнетáтельный delivery valve
~ насóса pump valve
~, нормáльно закры́тый normally closed valve
~, нормáльно откры́тый normally open valve
~ обрáтного дéйствия normally closed valve

кла́пан, обра́тный 1. non-return valve, check valve
2. flat check valve (diesel ICE)
~, откидно́й flap valve
~, отсе́чный 1. cut-off valve, shut-off valve, trip valve
2. check valve (diesel ICE)
~, паровозду́шный pressure and vacuum relief valve (veh radiator)
~, парово́й pressure relief valve (veh radiator)
~ перека́чки то́плива fuel-transfer valve
~, переключа́тельный change-over valve
~, перепускно́й relief valve, by-pass valve
~, пита́тельный feed-water valve
~, пласти́нчатый flat check valve (diesel ICE)
~ пода́чи бензи́на petrol valve
~ пода́чи то́плива fuel valve
~, подвесно́й overhead valve (ICE)
~, подпружинённый spring-loaded valve
~ попла́вки, иго́льчатый float needle valve
~, поплавко́вый float valve
~, предохрани́тельно-перепускно́й safety by-pass valve
~, предохрани́тельный 1. safety valve
2. relief valve (diesel ICE)
~ промы́вочной систе́мы flush valve
~ прямо́го де́йствия normally open valve
~, пусково́й starting valve (ICE)
~, разгру́зочный relief valve
~, регули́руемый adjustable valve
~, регули́рующий control valve
~, редукцио́нный 1. reduction valve, check valve
2. pressure regulator
~ с ша́риком ball valve
~ сжа́тия compression valve
~, сливно́й drain valve
~, солено́идный solenoid valve
~, ство́рчатый leaf valve
~, сто́порный check valve
~, таре́льчатый disc valve
~, телефо́нный drop indicator (teleph)
~, топливодози́рующий 1. propellant control valve (GW)
2. fuel control valve
~ то́рмоза отка́та recoil valve (arty)
кла́пан, тормозно́й brake valve
~, трёхходово́й three-way valve
~, управля́ющий control valve
~, шарово́й ball check valve, ball valve
~, экрани́рованный masked inlet valve (diesel ICE)
кла́панный 1. valve
2. flap
3. pack cover flap (parachute)
4. drop indicator (switchboard)
классифика́ция classification, grading, sorting
~ ка́рты map classification
классифици́ровать to classify, to sort
кла́ссность classification, category
кла́ссный 1. qualified
2. classroom
3. qualified tradesman or specialist
клева́ние 1. pitching (aeron)
2. hunting (electron)
клева́нт toggle
клево́к graze, groundburst (gunn)
клей, столя́рный carpenter's glue
клейме́ние stencilling, service marking
клеймо́ 1. die, stamp, stamping die
2. stamping, service marking
~ испыта́ния на про́чность proof mark
~, фабри́чное trade-mark
кле́мма 1. terminal (elec)
2. clamp
кле́мшел clamshell excavator
клёпка riveting
~ впота́й flush riveting
клер plain language (comms)
кле́тка 1. cage
2. cell (anat)
3. square (graph)
4. crib pier (br)
~, откры́тая стелла́жная open storage bin, stock bin
~, стелла́жная bin unit, stock bin
~, упако́вочная crate, packing crate
кле́щи 1. tongs, pincers
2. pincers (tac)
3. pliers
клин 1. chock, wedge, cotter
2. spearhead, salient, wedge
3. wedge-type breech-block (ord)
4. panel (parachute)
~, бронета́нковый armoured spearhead
~, генера́торный oscillating klystron
~ затво́ра wedge-type breech-block

клин повы́шенного давлéния wedge of high pressure
~, тáнковый tank spear head
~, тормознóй brake expander
~, усили́тельный amplifying klystron
~, юстирóвочный correction wedge (on range-finder)
клин-бáба ice chisel
клинóк blade
~ стóпора locking plunger
клинóметр clinometer
кли́ренс clearance, road clearance (veh)
клирфáктор non-linear distortion factor
клистрóн klystron (radio)
~, двухкóнтурный double resonator klystron, two-cavity klystron
~, двухкóнусный double resonator klystron
~, двухрезонáторный two-cavity klystron, double-resonator klystron (elec)
~, отражáтельный reflex klystron
кли́ца cleat
клиянка mallet, maul
клуб-столóвая, офицéрская officers' mess
клык бáмпера overrider
ключ 1. key
2. spanner
3. firing handle (exploder)
~, гáечный wrench, spanner
~, даю́щий impulse data transmitter
~, дистанциóнный fuze key
~ для регулирóвки натяжéния гýсеницы track adjusting spanner
~, заводнóй winding key
~ зажигáния ignition key
~, козловóй trestle spanner
~, комбини́рованный combination tool
~ коммутáторного барабáна rotary switch (AAFCE)
~, консéрвный tin opener, can opener
~ Мóрзе sender key, Morse key
~, накиднóй 1. ring spanner
2. box wrench (US)
~, откры́тый double-end spanner, open-end spanner
~, передáточный transmitting key, morse key
~, прямóй гáечный straight spanner
~, разводнóй гáечный adjustable spanner
~, раздвижнóй adjustable spanner
ключ регуля́тора gas regulator key (SA)
~, телегрáфный morse key, telegraph key
~, торцóвый 1. box spanner
2. winder (clockwork mechanism)
~, трещóтный reversible ratchet wrench, ratcheting end wrench
~, установóчный fuze setter (ammo)
~, францýзский monkey wrench
~, швéдский adjustable wrench
~ ши́фра cipher key
ключ-генерáтора automatic morse transmitter
ключ-отвёртка combination tool
кляп gag
кля́тва oath, vow
кнехт bollard
кни́га больны́х sick book
~, домовáя house register
~ основнóго учёта ledger
~ парóлей book of passwords
~ посети́телей visitors' book
~, разнóсная "IN" register (sigs)
~, счётная account book
~ увольнéний record of leave book
~ учёта ledger, register
~ учёта инвентáрного имýщества equipment ledger
~ учёта поступлéний receipts book
~ учёта ремóнта маши́н workshop journal, progress record sheet
~, шнуровáя loose-leaf file, expanding book
кни́жка, артиллери́йско-стрелкóвая target record book (arty)
~, лими́тная proportion book, scale book (stores and eqpt)
~, полевáя field message pad
~, расчётная account book
~, служéбная soldier's record of service book
~, солдáтская soldier's service and pay book
~, трудовáя personal employment book (Sov worker's document)
кнóпка push button, button, knob
~, боевáя firing button
~, вися́чая pendant switch, pendant push
~, восстанáвливающая re-set knob, re-set button
~, вызывнáя call button
~ звуковóго сигнáла horn button

кно́пка "ключ" press-to-key button (sigs)
~, нажи́мная press button, push button
~, нажи́мно-вытяжна́я push-pull knob
~, отжи́мная pull button
~ повто́рного за́пуска relight button (GW)
~, подвесна́я звонко́вая pendant switch, pendant bell-push
~ подры́ва destruct button
~ "пуск" start button, launch button (msl)
~, пускова́я starter button
~, сигна́льная calling key
~, спускова́я firing button
~ "стоп" stop button
коагули́рование coagulation
коагуля́ция coagulation
коба́льт cobalt
кобура́ holster
~, пистоле́тная pistol holster
кобура́-прикла́д holster stock
кобы́ла 1. mare
2. long vaulting horse
кова́ть to forge
ко́вка forging
ко́вкость ductility
ко́врик 1. course (rd constr)
2. mat, rug
ко́врики brushwood road matting
ковш 1. trencher bucket, scoop (eng)
2. ladle, scoop
~, гре́йферный grab bucket
~ дноуглуби́тельного снаря́да dredger bucket
~, опрокидно́й tipping bucket
~, опроки́дывающийся загру́зочный skip loader
когти́ crampons
код 1. code
2. word (computer capacity)
3. codeword
~, алфави́тный alphabetic code
~, бина́рный binary code
~, бу́квенно-цифрово́й alpha numeric code
~, группово́й group code
~, двои́чный binary code
~, десяти́чный decimal code
~ запро́са interrogator code
~, и́мпульсный pulse code
~ Мо́рзе Morse code
~ опознава́ния identification code
~, опознава́тельный station authentication code
код, сигна́льный signal code
~, сокращённый abbreviated code
~, телегра́фный telegraph code
~, управля́ющий control code
~, усло́вный pre-arranged code
~, цифрово́й digital code
~, числово́й numerical code
~, шифрова́льный cipher code
~ шифрова́ния cipher code
ко́декс code (law)
коди́рование coding
~ ме́стности naming of reference points (gunn)
коди́ровать to encode
ко́довый code
кодогра́мма coded message
ко́жа, бараба́нная batter head (drum)
~, иску́сственная artificial leather
кожу́х 1. case, casing, housing, covering
2. jacket (barrel, ord)
3. box (nucl)
~ водяно́го охлажде́ния water jacket, cooling water jacket
~ возду́шного охлажде́ния cooling air jacket, air jacket
~ дви́гателя engine cowling
~ ка́ртера, защи́тный crankcase guard
~ магази́на magazine housing
~ му́фты clutch cover plate
~ охлажде́ния cooling jacket
~ прибо́ра instrument case
~ ста́тора stator casing (eng)
кожу́х-затво́р slide (pistol)
козёл 1. goat
2. short vaulting horse (gym)
ко́злы 1. trestle
2. saw-horse, saw-buck
козырёк 1. deflector, visor, sun visor
2. peak (head dress)
3. overhead protection above loophole
~ ветрово́го стекла́ windscreen
~ каби́ны windscreen
~, предохрани́тельный baseplate safety rim (mor)
~, противосо́лнечный защи́тный sun visor
ко́йка bunk, bed, hospital bed
~, двухъя́русная double-tier bed
кока́рда cap badge
ко́кор ammunition tray, primer case (min)
~, карту́зный cartridge box
кокс coke

кол picket, stake
~, а́нкерный anchor stake, holdfast picket, guy peg
~ заземле́ния earth pin
~, ориенти́рный marker, reference stake
~, оттяжно́й picket (line comms)
ко́лба electric light bulb
~, стекля́нная glass envelope
колдо́бина small hole, pit (topo)
колеба́ние 1. oscillation, vibration
2. fluctuation, variation
~ боково́го отно́са swaying oscillation
~ боково́й ка́чки rolling oscillation
~ в попере́чной пло́скости rolling
~ в продо́льной пло́скости pitching
~ галопи́рования pitching oscillation
~, гармони́ческое harmonic motion
~, крути́льное torsional vibration
~, несу́щее carrier signal
~, прецессио́нное gyroscopic precession, precessional yaw (ballistics)
~ снаря́да в полёте keyholing, corkscrewing (gunn)
~, фугои́дное phugoid oscillation (ac)
колеба́ния Б type B waves (radio)
~ второ́го ро́да class C amplifier oscillations
~, вы́нужденные forced oscillations
~, затуха́ющие damped oscillations
~, незатуха́ющие undamped oscillations
~ пе́рвого ро́да class A amplifier oscillations
~, прили́во-отли́вные tidal variations in level
~, свобо́дные free oscillations, natural oscillations
~, со́бственные free oscillations, natural oscillations
~ ти́па А-1 type A-2 waves (radio)
~ ти́па А-2 type A-1 waves (radio)
~ ти́па А-3 type A-3 waves (radio)
~, часто́тно-модули́рованные frequency modulated oscillations
колеба́ться/поколеба́ться to oscillate, to vacillate, to undulate, to fluctuate, to waver, to hesitate
коле́йность traffic width, number of lanes in carriageway
коле́но 1. crank (mech)
2. bend
3. leg (route)
4. knee
~, соедини́тельное connecting elbow
колео́птер coleopter, annular wing aircraft
колёса, винтовы́е helical gears
~, спа́ренные coupled wheels
колёсами on wheels
колёсико castor wheel
колёсно-гу́сеничный half-track
колёсный wheeled
колесо́ wheel
~, ведо́мое driven wheel
~, веду́щее 1. driving wheel (wheeled vehicle)
2. driving sprocket (tracked veh)
~, винтово́е зубча́тое spiral gear
~, вне́шнее off-side wheel
~, гребно́е paddle wheel
~ гу́сеницы и ступи́цы в сбо́ре, натяжно́е track adjusting wheel and hub assembly
~ гу́сеницы, направля́ющее track idler wheel
~ гу́сеницы, натяжно́е track idler wheel
~, двухска́тное double wheel, twin wheel, twin-tyred wheel
~, ди́сковое disc wheel
~, запа́сное spare wheel
~, зубча́тое toothed wheel, gear wheel
~, компенсацио́нное натяжно́е track tensioner (tk), compensating idler wheel
~, кони́ческое зубча́тое bevel pinion, bevel wheel
~, махово́е fly-wheel
~, направля́ющее 1. idler wheel (tracked veh)
2. guiding wheel
~, незави́симо подве́шенное independently sprung wheel
~, пере́днее направля́ющее 1. front idler wheel (tk)
2. steerable nose-wheel (ac)
~, подде́рживающее running wheel
~, рабо́чее impeller (centrifugal pump)
~, рулево́е steering wheel
~, спи́цевое spoked wheel
~, турби́нное turbine rotor
~, ходово́е bogie wheel, road wheel
~, храпово́е 1. ratchet wheel
2. ratchet head (revolver)
~, цепно́е sprocket
~, цилиндри́ческое зубча́тое spur gear
~, червя́чное worm wheel
~, червя́чное зубча́тое worm wheel
~, эвольве́нтное зубча́тое involute gear

колесоотбо́й riband, wheel guide, curbing (br)
коле́чко, ме́дное copper ring, copper joint washer
колея́ 1. wheel track, track
2. gauge (rly)
3. rut
~, **деревя́нная** wooden trackway
~, **железнодоро́жная** 1. railway track
2. gauge (rly)
~, **защи́тная** wearing surface wheel track, protective wheel track (br)
~, **норма́льная** standard gauge (rly)
~, **у́зкая** narrow-gauge (rly)
~, **широ́кая** standard gauge (broad gauge, USSR) (rly)
коли́чественный quantitative, numerical
коли́чество quantity, amount
~ **вы́стрелов на танк** rounds per tank
~ **движе́ния** momentum
~ **движе́ния отка́тных часте́й при вы́стреле** recoil momentum
~ **ко́ек, шта́тное** authorized bed capacity (hosp)
~, **нали́чное** amount in hand, number available, quantity in stock
~, **разрешённое** level of entitlement, authorised quantity
~, **та́бельное** 1. authorized quantity
2. authorized strength, strength laid down in establishment table
коллекти́в 1. staff
2. personnel, team
~, **во́инский** military personnel
колле́ктор 1. commutator (elec)
2. manifold
3. collector (transistor)
~, **ве́рхний** top tank, header tank (of vehicle radiator)
~, **впускно́й** inlet manifold, intake manifold
~, **вса́сывающий** inlet manifold
~, **выпускно́й** exhaust manifold, exhaust outlet
~, **выхлопно́й** exhaust manifold
~, **ни́жний** bottom tank (veh radiator)
~ **радиа́тора** radiator header tank
коллима́тор collimator
~, **телескопи́ческий шарни́рный** articulated tank telescopic sight
коллима́ция collimation
коловоро́т brace, bit brace
коло́дец well, shaft, pit
коло́дец, артезиа́нский artesian well
~, **водопоглоща́ющий** soakage pit, soakaway
~, **водосбо́рный** 1. sump
2. drainage sump without soakaway (in trenches in impervious soil)
~, **водосто́чный** drainage sump (min)
~, **водяно́й** water well
~ **для отли́вки оруди́йных ство́лов** gun pit
~, **компенсацио́нный** capacity tube, compensating well (carb)
~, **ме́лкий тру́бчатый** small pipe well, small bore well
~, **мелкотру́бчатый** small bore well
~, **ми́нный** drain well
~, **ми́нный** demolition shaft (min)
~, **нортоно́вский** Norton tube well (eng)
~ **осветлённой воды́, водосбо́рный** filtered water sump
~, **осуши́тельный** drainage sump
~, **распредели́тельный** distribution chamber
~, **сбо́рный** bilge well
~, **смотрово́й** inspection well
~, **сто́чный** drain pit
~, **тру́бчатый** tube well
~, **ша́хтный** excavated well
коло́дка 1. wheel chock, chock
2. brake shoe
3. ribbon bar holder (decorations)
4. shoe tree
5. terminal block (radio)
~, **кле́ммная** terminal block
~, **нажи́мная** pressure block (mine)
~, **о́рденская** ribbon bar holder (decorations)
~, **прице́льная** backsight bed (SA)
~, **распредели́тельная** distributor segment
~, **соедини́тельная** plug and socket connector
~, **страхо́вочная** chock
~, **тормозна́я** brake shoe
~, **у́гольная** carbon block
~, **ште́псельная** plug (elec)
коло́дочка, эбони́товая ebonite head (mine detector)
ко́локол bell
колоколобра́зный bell-shaped
ко́ломазь axle grease
коло́нка column (machine tools)

колóнка, бензи́новая petrol pump (filling station)
~, бензозапрáвочная petrol pump
~, бензораздáточная petrol pump (filling station)
~, вертикáльная vertical column (of telescopic alidade)
~, водоразбóрная public water pump (on water main), stand pipe
~, запрáвочная fuel filling pump
~, маслораздáтельная oil dispenser
~, маслораздáточная oil dispenser
~, пожáрная fire standpipe
~, рулевáя steering column (veh)
~, топливораздáточная fuel pump, fuel filling pump
колóнна column
~, автомоби́льная vehicle convoy, MT column, MT convoy
~, автотрáнспортная MT convoy, vehicle convoy, MT column
~ боевóго питáния ammunition train
~, бронетáнковая armoured column
~ бронетáнковых частéй armoured column
~, взводнáя platoon column
~, головнáя тáнковая armoured spearhead
~, общевойсковáя all arms column, all arms convoy
~, охлаждáющая cooling tower
~ по два column of twos
~ по одномý single file
~ по отделéниям column of sections
~ по три column of threes
~ по четы́ре column of fours
~, похóдная column of route
~, рóтная company column
~ санитáрного трáнспорта ambulance convoy, medical convoy
~, сгруди́вшаяся bunched-up column, closed up column
колонновожáтый file leader, convoy leader, column leader
колотýшка mallet
колóть/кольнýть to stab; **колóть/кольнýть штыкóм** to thrust with a bayonet
колпáк 1. cap, protective cap, cover
2. hub cap (veh)
3. horn (mine)
4. cupola
5. mobile pill box
колпáк, баллисти́ческий ballistic cap
~, бетóнный concrete cupola, small concrete pillbox
~, броневóй armoured cupola
~, защи́тный protective cap
~ кардáна universal joint casing
~, колёсный hub cap (veh)
~ нажи́много ди́ска, опóрный clutch cover plate
~ перискóпа, брони́рованный armoured periscope cover
~, предохрани́тельный 1. safety cap (fuze)
2. fuze cover
~ стабилизáтора conical sleeve (gren stabilizer)
колпáра wheel set (veh)
колпачóк вéнтиля valve dust cap
~ взрывáтеля 1. igniter hood (mine)
2. fuze cap (ammo)
~, защи́тный protective cap
~ кáпсюля, мéдний copper primer cap
~ кáпсюля, металли́ческий metal primer cap
~, металли́ческий metal cap
~, нажи́мный pressure plate (mine)
~, предохрани́тельный 1. safety cap (fuze)
2. fuze cover
~ с кáпсюльным состáвом detonating cap
~, установóчный safety cap (fuze)
кóлышек picket, peg, small stake
~, ориенти́рный marking stake, minefield marking stake, marker peg
~, палáточный tent peg
кольнýть (см колóть)
кóльца, разрезны́е стальны́е outer steel rings (ord obturator)
кольцевóй circular, annular, ring-shaped
кольцеобрáзный circular, annular, ring-shaped
кольцó 1. ring, coil, washer, collar
2. star with lateral working (radio)
3. basket, disc (ski stick)
~, альтиметри́ческое altitude ring (radar)
~ барабáна отражáтеля dial sight reader plate
~ барабáна угломéра dial sight micrometer scale
~, беговóе race (roller bearing)
~, бортовóе base ring (tyre)

кольцо́, буссо́льное ring on the director, director ring (AAFCE eqpt)
~, ве́рхнее дистанцио́нное upper time train ring (on fuze)
~, вытяжно́е rip-cord ring (para)
~ га́зового регуля́тора gas regulating ring (SA)
~, графи́товое уплотня́ющее carbon seal ring
~, графи́товое упо́рное carbon thrust ring (clutch)
~, диоптри́ческое фокусиро́вочное focusing ring (opt)
~, дистанцио́нное time ring (fuze)
~ для ремня́ lanyard ring (revolver)
~, ду́льное muzzle ring
~ ды́ма smoke ring
~, зажи́мное chuck ring, clamp ring
~, запо́рное locking ring
~, калибрацио́нное range marker circle
~, калибро́вочное calibration ring
~ кла́пана, устано́вочное valve distance ring
~, компрессио́нное поршнево́е piston compression ring, piston pressure ring (ICE)
~, конта́ктное slip ring, contact ring
~, концентри́чное поршнево́е concentric piston ring
~, крепёжное fastening ring
~, крепи́тельное fastening ring
~, ложево́е barrel band (SA)
~, маслоотража́тельное oil thrower ring (ICE)
~, маслосбра́сывающее oil regulating ring, oil scraper ring
~, маслособира́ющее oil regulating ring, oil scraper ring
~, маслосъёмное oil regulating ring, scraper ring (ICE)
~, ма́сляное oil-ring
~, наби́вочное packing ring, obturating ring
~, нажи́мное ring clamp
~ нача́ла отсчёта zero ring (radar)
~, неподви́жное fixed ring
~, ни́жнее дистанцио́нное lower time train ring (fuze)
~ но́жки, металли́ческое header insert (electronic valve)
~, обтюра́торное binder ring, obturating ring

кольцо́, обтюри́рующее obturating washer, obturating ring
~ освобожда́ющего што́ка, вытяжно́е release-pin ring
~, подви́жное дистанцио́нное time-setting ring (fuse)
~, подде́рживающее retaining ring
~, поршнево́е piston ring
~, поршнево́е компрессио́нное piston compression ring, piston pressure ring (ICE)
~, поршнево́е уплотни́тельное piston compression ring
~, постоя́нное fixed ring (gyro sight)
~, предохрани́тельное 1. safety ring 2. breech guard (rocket launcher)
~, прижи́мное 1. snap ring 2. detonator retaining ring (pressure friction fuze)
~, прице́льное elevating dial (MG)
~, про́волочное сто́порное wire locking ring
~, прокла́дочное рези́новое rubber washer
~, промежу́точное intermediate ring
~, проста́вочное spacing ring, spacing washer
~, пружи́нное snap ring, circlip
~, пружиня́щее snap ring, circlip
~, пружиня́щее сто́порное circlip
~, разводно́е split ring
~, разрезно́е snap ring, circlip
~, распо́рное distance ring, distance collar, distance washer
~, регулиро́вочное adjusting ring, regulating ring
~, регули́рующее 1. adjusting ring, regulating ring 2. throttling ring (ord)
~, рези́новое rubber ring
~, рулево́е thrust ring (GW)
~ с затво́рным гнездо́м, ввинтно́е breech ring
~ са́льника packing ring
~, сепара́торное roller ring (mounting, ord)
~, скользя́щее slip ring
~, сма́зочное lubricating ring
~, соплово́е nozzle ring
~, составно́е поршнево́е compound piston ring
~, сто́порное snap ring, circlip
~, увя́зочное eye bolt

кольцó угломéра 1. dial sight main scale (gunn)
2. azimuth gear dial (mor)
~, угломéрное main scale of dial sight
~ уплотнéния packing ring
~, уплотни́тельное 1. oil seal ring, ring seal, sealing ring
2. compression ring (ICE)
~, уплотня́ющее ring seal, sealing ring
~, упóрное thrust collar (thrust bearing)
~, установóчное 1. positioning collar
2. adjustable ring (time fuze)
3. adjusting collar
~, фéтровое уплотня́ющее felt sealing ring
~, центри́рующее centring ring
~, шворневóе trail eye, lunette (ord)
~, эксцéнтриковое eccentric ring
кóлья, заострённые sharpened timber pickets
~ разби́вки marker pegs (trench sites etc)
колю́чка barb (wire)
комáнда 1. order, command
2. team, detachment
~, авари́йная crash crew, emergency repair crew
~, авари́йно-спасáтельная safety party (br)
~, возду́шно-десáнтная подрывнáя airborne demolition team
~, дегазациóнная decontamination party
~, дискрéтная discrete command (GW)
~ для ведéния огня́ fire order
~, дорóжно-ремóнтная road repair gang
~, имитациóнная battle simulation team
~, исполни́тельная executive word of command
~, мишéнная butt party (detail)
~, огневáя fire order
~, плáвная continuous control command, proportional control command (GW)
~, погру́зочная loading party
~ поднóсчиков боеприпáсов ammunition party
~, пожáрная fire-fighting party, fire picquet, fire fighting team, fire brigade
~, посáдочная embarkation team (amph ops)
~, похорóнная burial party
~, предвари́тельная cautionary word of command
~, противопожáрная fire-fighting team
комáнда, рабóчая fatigue party, labour detail, working party
~ радиациóнной развéдки radiation survey team
~, разби́вочно-трассирóвочная reconnaissance and tracing party (trenching)
~ размини́рования mine clearing party
~, рáзовая on-off command (GW)
~ разрушéния demolition party
~, салю́тная firing party (ceremonial)
~, санитáрная ambulance party, ambulance team
~ сапёров-подрывникóв engineer demolition party
~, сбóрная 1. ad hoc party (pers)
2. combined team (sport)
~, спасáтельная rescue team, rescue party
~, строи́тельная construction detachment, construction team
~ укáзчиков butt party, butt detail
~ установщиков мин mine-laying party
командáрм army commander
команди́р commander (of Sov Corps or of units of lower echelon)
~ батальóна battalion commander
~ батарéй battery commander
~ бригáды brigade commander
~ взвóда platoon commander
~ взвóда управлéния HQ platoon commander
~ вы́садки landing commander (nav)
~ десáнта landing force commander
~ миномёта 1. mortar detachment commander
2. No. 1 of mortar (Brit)
~, млáдший 1. subaltern
2. non-commissioned officer, NCO
3. junior commander
~, нижестоя́щий subordinate commander
~, общевойсковóй overall commander, force commander, all arms commander
~ ору́дия 1. gun detachment commander
2. No. 1 (of wpn) (Brit)
~ отделéния тя́ги NCO in charge of wagon lines (arty)
~ подразделéния detachment commander, section commander, sub-unit commander
~, подчинённый subordinate commander
~ тáнка tank commander

командир экипажа crew captain
командир-единоначальник responsible commander (Sov) (responsible for both military and political matters)
командированный detached for duty
командировать to detach, to detach for outside duty
командировка 1. temporary detachment on duty
2. duty journey, official journey
3. documentary authority for duty journey
~, служебная duty journey, official journey
командный command, control
командование 1. command
2. command group
~, армейское army command, command of army formation
~, верховное главное supreme command
~, вышестоящее higher command
~, главное high command
~, общевойсковое overall command
~, фронтовое 1. command of front (Sov)
2. command of army group (Brit)
командующий general officer commanding (of Sov Army or of higher echelon)
~ военно-воздушными силами фронта 1. commander of air army subordinate to front (Sov)
2. commander of tactical air force supporting army group (Brit)
~ воздушно-десантными войсками commander-in-chief airborne forces
~ десантными силами joint force commander, JFC
~ противовоздушной обороной района regional anti-aircraft defence commander
~ фронтом 1. commander of front (Sov)
2. army group commander (Brit)
комбат battalion commander
комбинезон overalls
~, защитный protective combination suit, protective overalls
~, маскировочный camouflage suit
~, проолифованный oilskin overalls, anti-gas overalls
комбриг brigade commander
комдив divisional commander
комель butt end of a pole, facsine
комендант commandant
~, военный 1. commandant
2. Officer Commanding troops on ship or train
3. Rail Transport/Traffic Officer
~ гарнизона camp commandant (permanent camp)
~ лагерного сбора camp commandant (temporary camp)
~ переправы bank group commander
~ погрузки embarkation officer
~ порта port security officer
~ пункта высадки морского десанта beachmaster
~ пункта переправы crossing place commander
~, районный военный area commandant
~ участка area commandant
~ участка переправы crossing control officer
~ штаба полка regimental camp commandant (Sov) (performs duties similar to those of Brigadier or Div HQ camp commandant)
комендатура commandant's office, commandant's staff
~, военная 1. military "komendatura"
2. commandant's office, commandant's staff
комиссариат, военный "Military Commissariat" (Sov organ of local military control within civil admin area)
комиссия board, commission
~, военно-врачебная medical board
~, врачебная medical board
~, врачебно-экспертизная medical board
~ для технического осмотра technical inspection team
~, квалификационная trade testing board
~, отборочная selection board
~, партийная (communist) party commission
~, приёмная selection board
~, приёмо-техническая recruiting medical and selection board
~, проверочная board of survey
комкор corps commander
коммуникации lines of communication
~, военные lines of communication
коммуникация 1. communication
2. line of communication

коммуника́ция по гру́нту road communication
коммунуслу́ги communal services
коммута́тор 1. switchboard (teleph) 2. commutator (elec)
~, анте́нный transmit-receive switch, T-R switch
~, батаре́йный battery switch
~, бесшнурово́й cordless switchboard
~, избира́тельный selector switch
~, кла́вишный cordless switchboard
~, ламе́льный elementary cordless switchboard
~, лине́йно-батаре́йный concentrator (teleg)
~, телефо́нный switchboard
~, шнурово́й cord switchboard
~, электро́нный electronic switch
коммута́тор-ключ rotary switch
коммута́ция switching (elec)
~, безы́скровая arcless commutation
~, бесконта́ктная non-contact switching, arcless switching
коммути́ровать/скоммути́ровать to switch, to interconnect
ко́мната, ле́нинская Lenin Room (study room in Sov unit)
компа́ктный compact, dense, tight, concise, rugged
компара́тор, временно́й time comparator (electron)
~ радиоизмере́ния substitution field strength measuring assembly (radio)
компари́рование напряжённости по́ля substitution method of measuring field strength (radio)
ко́мпас compass
~, гиромагнитносбо́рный gyro flux-gate compass
~, гиромагни́тный gyro-magnetic compass
~, гироскопи́ческий gyroscopic compass
~, индукцио́нный induction compass
~, карма́нный pocket compass
~, магни́тный magnetic compass
~, призмати́ческий prismatic compass
~ с отража́тельной при́змой, жи́дкостный liquid prismatic compass
компаунди́рование compounding, blending
компа́унд-маши́на 1. compound expansion engine 2. compound-wound machine (elec)
компенса́тор 1. balance, compensator 2. balancing gear, compensator (ord) 3. recoil system replenishing tank, gravity tank (ord)
~, двухклинովо́й rotating wedges (range-finder)
компенса́ция compensation, balance, balancing, equalizing
~ астигмати́зма astigmatism control, stig control
~ радиодевиа́ции quadrantal error correction (in an aircraft's radio compass)
компенси́рованный balanced
ко́мплекс 1. system 2. complex 3. site
~, автомати́ческий противораке́тный 1. automatic anti-missile system 2. automatic anti-missile complex 3. automatic anti-missile site
~, боево́й weapons system
~, дополни́тельный раке́тный 1. complementary missile system 2. complementary missile complex 3. complementary missile site
~, зени́тный раке́тный 1. anti-aircraft missile site, surface- to-air missile site 2. anti-aircraft missile complex, surface-to-air missile complex 3. anti-aircraft missile system, surface -to-air missile system
~ наведе́ния guidance system
~, назе́мный раке́тный land-based missile system
~ обору́дования complex of equipment
~, подви́жный раке́тный mobile missile system
~, полево́й раке́тный battlefield support system
~, радиолокацио́нный radar network
~, раке́тный 1. rocket complex, missile complex 2. rocket site, missile site
~, реакти́вный missile system
~, самохо́дный self-propelled weapons system
~ СО́Н-ПУА́ЗО radar and AAFCE system

ко́мплекс стрелко́вого ору́жия small-arms weapons system
~, стрелко́вый small-arms weapons system
~ управле́ния огнём integrated fire control system
комплекси́ровать to combine
ко́мплексный 1. multiple, complex
2. integrated, comprehensive
компле́кт 1. kit, outfit, set
2. assembly
3. both terminal sets (teleg, teleph)
~, артиллери́йский дегазацио́нный artillery decontamination set (CW)
~, боево́й 1. battle scale of ammunition (rounds carried per man, per gun or per unit)
2. weapon system (missiles)
~ боеприпа́сов 1. battle scale of ammunition (Brit)
2. unit of fire (US)
3. ammunition allotment
~ вы́стрела complete round
~, дегазацио́нный CW decontamination set, CW decontamination outfit, CW decontamination equipment
~ для подрывны́х рабо́т demolition kit
~ запасны́х инструме́нтов и принадле́жностей maintenance kit of tools and accessories
~ запасны́х часте́й, ремо́нтный repair set of spare parts
~, защи́тный protective clothing set
~, защи́тный фильтру́ющий permeable protective outfit (CW)
~ ЗИП, оруди́йный gun spare parts kit
~ имита́ции артиллери́йской стрельбы́ artillery fire simulation outfit
~, индивидуа́льный дегазацио́нный personal CW decontamination set
~ индивидуа́льных средств защи́ты от я́дерного, биологи́ческого и хими́ческого ору́жия personal NBC equipment
~ инструме́нтов tool kit
~ крепле́ния маши́н vehicle lashing kit (amph ops)
~ ма́сок camouflage kit
~ медици́нского иму́щества medical chest
~ мостострои́тельных средств bridge assembly unit
~ обмундирова́ния set of uniform
компле́кт обору́дования для опресне́ния морско́й воды́ sea water de-salination kit
~, общевойсково́й защи́тный all-arms protective clothing outfit
~, оруди́йный gun spare parts kit
~ пехо́тного ша́нцевого инструме́нта infantry entrenching tools
~, полево́й отопи́тельный field heating and cooking system
~, по́лный complete set
~, пулемётно-миномётный дегазаци́онный machine gun and mortar CW decontamination set
~, ремо́нтный repair kit
~ снаряже́ния set of individual equipment
~, специа́льный special outfit (decontamination kit etc)
~ столя́ного инструме́нта joiner's kit
~ та́бельной ма́ски regulation camouflage net set
~ универса́льных приспособле́нний, еди́ный standard tool set, standard set of universal tools (tk)
~, фи́льтро-вентиляцио́нный filter system of ventilation
~, шату́нный connecting rod assembly
компле́ктность обору́дования completeness of equipment
комплектова́ние building up, bringing up to prescribed strength, to equip to scale, kitted out
комплектова́ть/укомплектова́ть to build up, to bring up to prescribed strength, to equip to scale, to kit out
комплекто́вка stowage, arrangement, parcelling
компо́зитрон compositron
компоне́нт component, ingredient, constituent
компоно́вка layout, grouping, composition, arrangement, configuration
~ механи́зма mechanical layout
~, монобло́чная tandem configuration (staging in rocket technology)
~, паке́тная parallel configuration (staging in rocket technology)
~, та́ндемная tandem configuration (staging in rocket technology)
компре́ссия compression
~, высо́кая high compression

компрессо́метр compressometer
компре́ссор 1. air compressor
2. recoil cylinder, recoil absorber (ord)
~, возду́шный air compressor
~, одноступе́нчатый single-stage compressor
~, осево́й axial-flow compressor
~ противооткатных устро́йств recoil buffer
~, тормозно́й brake compressor
комприми́рование compression
ко́мптон-эффе́кт Compton effect, Compton recoil effect
комсомо́л Komsomol (Communist youth organization)
комсомо́лец Komsomol member
комсо́рг Komsomol leader
комсоста́в officers (Sov)
комсоста́вский officers' (Sov)
конве́йер conveyor, conveyor belt
~, ле́нточный belt conveyor
~, приводно́й power conveyor
~, ро́ликовый roller conveyor
конве́кция convection
~, теплова́я thermal convection, heat convection
конве́рсия conversion
~, каталити́ческая catalytic conversion
~, хими́ческая chemical conversion
конве́рт, санита́рный stretcher bag (for evacuation of wounded in cold climates)
конвои́рование escort duty, escorting
конвои́ровать to escort, to convoy
конвои́руемый escorted, under escort, in convoy
конво́й escort, convoy
конво́йный 1. escort (one man)
2. convoy
конденса́тор 1. capacitor (elec)
2. condenser (mech)
~, блокиро́вочный bypass capacitor, blocking capacitor (elec, electron)
~, керами́ческий ceramic capacitor
~ настро́йки tuning capacitor
~ переме́нной ёмкости variable capacitor, adjustable capacitor
~, подго́ночный trimmer (elec)
~, подстро́ечный tuning capacitor, trimmer
~ постоя́нной ёмкости fixed capacitor
~, раздели́тельный decoupling capacitor
~ свя́зи coupling capacitor
~, слюдяно́й mica capacitor
конденса́тор, шунти́рующий shunt capacitor
~, электролити́ческий electrolytic capacitor
конденса́ция condensation
конде́нсор optical condenser
~ тёмного по́ля dark field optical condenser
кондициони́рование conditioning
~ во́здуха air conditioning
конере́зка rootcutting machine, rooter
коне́ц 1. end, terminal
2. end station (of base) (arty)
~ горе́ния all-burnt (arty, GW)
~, моты́жный chisel end (of pickaxe)
~, пя́точный heel (ski), heel end
~ трубы́, нарезно́й threaded end of pipe
~, шлицево́й splined end
коне́чный end, terminal, termination, final
кони́ческий tapered, bevelled, conical
ко́нкурс officers' written and practical competitive examination
ко́нница cavalry
ко́нный mounted (on horseback)
коново́д horse holder (man)
ко́новязь horse lines
коно́ид, баллисти́ческий ballistic cam (computer)
конопля́ indian hemp
коносаме́нт bill of lading
консерва́ция 1. state of preservation (mech)
2. being put into reserve, preservation, mothballing
~, дли́тельная long-term corrosion proofing, long-term conservation
~, кратковре́менная short-term corrosion-proofing, short-term conservation
~ маши́н storing and preservation of vehicles
консервооткрыва́тель tin opener, can opener
консе́рвы 1. preserves, tinned food
2. protective goggles, snow goggles
консисте́нтность consistence, consistency, body
консисте́нция consistence, consistency, body
консисто́метр consistometer
консо́л console (elec)
консо́ль cantilever (br, eng)

консо́ль, берегова́я bank cantilever (br)
~, тротуа́рная footwalk cantilever (br)
консо́льный overhanging, cantilever
конста́нта constant
конструи́рование design
констру́кция design, construction
~, безра́мная 1. frameless construction (bldg)
2. integral construction (veh)
~, безру́бочная unjointed construction (timber)
~, бло́чная block unit, block construction
~, венча́тая notched construction
~, взрывоусто́йчивая blast-resistant design
~, долгове́чная long-life construction
~, жёсткая rigid structure
~, ка́менная masonry construction
~, карка́сная skeleton construction
~, мо́дульная modular construction
~, надёжная fail-safe design
~, несу́щая self-supporting structure, load-carrying structure
~, облегчённая lightweight design
~, про́чная rugged construction
~, ра́мная strutting frame revetment (trenches)
~, ра́мно-бло́чная framed strut construction
~, ра́мно-щитова́я frame and panel construction
~, сварна́я welded construction
~, слои́стая laminated construction
~, составна́я composite construction
~, сто́йчатая framed construction
~, це́льная one-piece construction
конта́кт ме́жду генера́тором и мото́ром поворо́тного механи́зма traverse rotary base junction (AFV)
~, нажимно́й press contact
~, ненадёжный poor contact
~, отжи́мный pull contact
~, охва́тываемый male contact (elec, mech)
~, охва́тывающий female contact, (elec, mech)
~, преры́вистый intermittent contact
~, пружи́нный spring contact
~, рту́тный mercury contact (elec)
~ с механи́ческим де́йствием contact actuated mechanically
~, скользя́щий sliding contact, rubbing contact
конта́ктор contactor (elec)
конта́кты прерыва́теля-распредели́теля distributor contact breaker points
конте́йнер container
~, грузово́й stores container, air drop stores container
~, изотерми́ческий heat insulated container
~, напа́лмовый napalm container, napalm tank
~, сбра́сываемый drop tank (ac)
~, складно́й collapsible container
континге́нт contingent
~, возрастно́й age group
континге́нтный rationed
контра́кт, закры́тый terminated contract
контра́ст, акусти́ческий acoustic contrast, acoustic discreteness
контра́стность изображе́ния image contrast
~ фо́на background contrast
контрата́ка counter-attack
контратакова́ть to counter-attack (imp and perf)
контратако́вывать to counter-attack (imp)
контрбатаре́йный counter-battery
контрга́йка lock nut
контрманёвр counter-manoeuvre
контрме́ра counter measure
контрмини́рование counter-mining
контрнападе́ние counter-attack
контрнаступа́ющий on the counter-offensive
контрнаступле́ние counter-offensive
контро́вка па́льца тра́ка link pin clip (tk)
контролёр monitor, controller, supervisor, checker
~, арме́йский military representative (functions include the technical inspection and oversight of expenditure of stores and funds)
~ гру́за, авиацио́нный load control officer (AF)
~ такти́ческой обстано́вки tactical control officer, TCO (GW)
контролёр-ревизо́р auditor
контроли́ровать/проконтроли́ровать to control, to check, to supervise
контро́ллер control, controller (device)
~ поворо́тного механи́зма traverse controller (AFV)

контро́ль control, monitoring, check
~ **да́льности** checking of range
~, **дозиметри́ческий** radiac monitoring, control of radiation exposure (nucl)
~ **ка́чества** quality control
~ **по стра́нам све́та** cardinal point correction
~, **после́довательный** sequential sampling
~, **приёмочный** acceptance test, certification check
~, **промежу́точный** working inspection
~, **санита́рный** sanitary surveillance, sanitary control
~ **свя́зи** monitoring (comms)
~ **стрельбы́** check fire (gunn)
~, **телевизио́нный** television guidance, television command (GW)
~, **централизо́ванный автомати́ческий** centralised monitoring
контро́льно-техпу́нкт technical control point
контро́льный control, check
контрподгото́вка, артиллери́йская artillery counter-preparation, artillery counter-preparation fire, artillery counter-bombardment
контрпредохрани́тель уда́рника percussion pellet safety spring
контрпружи́на control spring, creep spring
контрразве́дка counter-intelligence
контрразве́дчик counter-intelligence officer, counter-intelligence agent
контрре́лсь guard rail (rly)
контрсна́ряд counter shot
контрсопло́ female nozzle
контруда́р counterstroke, counterblow, counterthrust
контруда́рный counterblow, counterthrust
контршпиона́ж counter-intelligence, counter-espionage
контршто́к recoil system control plunger (ord)
контрэска́рп "one-way" tank ditch, tank step, counterscarp
конту́женный shell-shocked, concussed
конту́зия shell-shock, concussion
ко́нтур 1. circuit (elec)
2. outline, contour
~, **анте́нный** antenna tuning circuit
~, **апериоди́ческий** aperiodic circuit
ко́нтур, бала́нсный balancing network (electron)
~, **втори́чный** secondary circuit
~, **входно́й** input circuit
~, **габари́тный** outline (for clearance)(rly, veh)
~, **дифференци́рующий** differentiating circuit (electron)
~ **заде́ржки вре́мени** time delay circuit
~, **за́мкнутый** closed circuit (elec)
~, **колеба́тельный** resonant circuit, oscillatory circuit
~, **корректи́рующий** corrector circuit
~, **настро́енный** tuned circuit
~, **неколеба́тельный** aperiodic circuit, non-resonant circuit
~, **перви́чный** primary circuit
~ **рассе́ивания** shot dispersion (arty)
~, **резона́нсный избира́тельный** resonant selective circuit, resonant selective filter circuit (electron)
~, **свя́занный** coupled circuit (elec)
~, **уда́рный** ringing circuit (electron)
~ **элеро́нов, рулево́й** roll surface drive system (ac, GW)
ко́нус 1. cone
2. sleeve target (ac towed)
3. locking cone (parachute)
~ **вы́хода порохові́х га́зов, за́дний** back-blast area
~, **дополни́тельный** back cone
~, **заду́льный** blast mark (on ground in front of gun muzzle), muzzle blast
~ **излуче́ния** radiation lobe (antenna)
~ **ка́моры, снаря́дный** seat of projectile
~, **мёртвый** cone of silence, cone of nulls
~ **молча́ния** cone of silence
~ **Мо́рзе** Morse taper
~, **нару́жный** female cone, concave cone
~, **носово́й** nose cone
~ **разлёта оско́лков** fragmentation pattern cone
~ **разлёта шрапне́ли** shrapnel cone, fragmentation pattern cone
~ **рассе́ивания** cone of dispersion, cone of fire
ко́нус-ветроуказа́тель wind sock
ко́нус-мише́нь aerial sleeve target
ко́нусный tapered
конусообра́зный bell-shaped, cone-shaped
конфере́нция, техни́ческая technical conference

конфигура́ция profile, shape, configuration
концентра́т, пищево́й concentrated food
концентра́ция concentration
~, безопа́сная safe concentration (CW)
~, боева́я lethal concentration
~, допусти́мая permissible concentration
~ отравля́ющих веще́ств war gas concentration per unit volume (CW)
~ отравля́ющих веще́ств, смерте́льная lethal gas concentration, lethal vapour concentration (CW)
~ паро́в, опа́сная dangerous concentration of fumes
~, поража́ющая harmful concentration (CW), casualty producing concentration
~, токси́ческая toxic level
~, убо́йная lethal concentration (CW)
концентри́чески concentrically
концла́герь concentration camp
конь 1. horse
2. vaulting horse
координа́та co-ordinate
координатоме́р co-ordinate scale
~ це́ли tracking head (GW)
координа́тор artillery plotter, artillery co-ordinator
координа́ты, географи́ческие geographical co-ordinates, terrestrial co-ordinates (latitude and longitude)
~, исчи́сленные calculated co-ordinates
~ кони́ческой систе́мы conical co-ordinates
~ координа́т map reference, co-ordinates
~ по ка́рте grid co-ordinates, grid reference
~, поля́рные polar co-ordinates
~, прямоуго́льные rectangular co-ordinates
~ сфери́ческой систе́мы spherical co-ordinates
~, теку́щие present position data (gunn)
~, угловы́е angular co-ordinates
~ це́ли, упреждённые target predicted co-ordinates
координа́ция огня́ fire co-ordination
копа́тель ша́хтных коло́дцев well-sinking machine
копёр 1. pile driver
2. headgear, headframe (mine shaft)
3. impact testing machine
копёр, батаре́йный multiple pile-driving rig
~ для заби́вки свай pile-driving frame, piling rig
~, разбо́рный деревя́нный collapsible wooden piling rig
~, разбо́рный металли́ческий collapsible metal piling rig
~, сваебо́йный pile-driver
копи́р cam plate
ко́пия докуме́нта, уведоми́тельная advance copy
ко́поть soot
кора́бль 1. ship, vessel
2. aircraft (coll)
~, возду́шный 1. aircraft
2. air liner (Sov term includes both military and civil ac)
~, деса́нтный amphibious warfare ship
~ деса́нтных сил, штабно́й amphibious group headquarters ship, amphibious command ship
~ огнево́й подде́ржки, деса́нтный landing force support ship
~, пехо́тно-деса́нтный landing ship infantry, LSI
~, та́нко-деса́нтный landing ship tank, LST
кора́бль-док, деса́нтно-вертолётный landing ship assault, LSA
~, деса́нтный landing ship dock, LSD
корди́т cordite
корзи́на для перено́ски боеприпа́сов ammunition basket
корзи́нка для панора́мы dial sight socket (ord)
коридо́р, возду́шный air corridor
ко́рка, ледяна́я ice crust
корма́ stern (ship, boat)
кормово́й rear, stern
корнере́зка rootcutting machine, rooter
ко́роб breech casing (MG)
~ лю́льки cradle frame
~ передка́ limber box
коро́бка ба́шенного выключа́теля turret switch box
~ возвра́тной пружи́ны return spring casing (MG)
~, дверна́я door frame
~ для взрыва́телей fuse box
~, затво́рная main casing (machine carbine)
~, ка́бельная cable junction

коро́бка, кла́панная valve holder
~, кле́ммная terminal box
~, кулачко́вая cam housing, rocker box (ICE)
~, магази́нная magazine case, magazine
~, магистра́льная mains distribution box (elec)
~ механи́зма управле́ния ба́шней turret control box
~, наби́вочная stuffing box
~ отбо́ра мо́щности power take-off point (veh)
~, патро́нная ammunition box
~, переда́точная transfer box (veh)
~ переда́ч gear box
~ переда́ч, автомати́ческая automatic transmission gear box
~ переда́ч, дополни́тельная auxiliary gear box
~ переда́ч, многоступе́нчатая multiple-speed gear box
~ переда́ч, планета́рная planetary gear box
~ переда́ч, разда́точная transfer gear box
~ переда́ч с ручны́м переключе́нием manual transmission gear box
~ переда́ч со синхронизи́рованным переключе́нием synchromesh gear box
~ переда́ч, трёхступе́нчатая three-speed gear box
~ переме́ны переда́ч gear box
~ поворо́тного механи́зма traversing mechanism housing (AFV)
~, поглоти́тельная filter box (respirator, CW)
~, подба́шенная ammunition ready rack (AFV)
~ приво́да поворо́тного механи́зма traversing gear case
~ прице́ла sighting gear case
~, противога́зная 1. respirator filter canister
2. respirator container
~, противога́зовая 1. respirator filter canister
2. respirator container
~, пулемётная MG ammo box, MG belt box (containing MG ammunition in belts)
~ радиа́тора, ве́рхняя radiator header tank
коро́бка радиопрово́дки, кле́ммная radio terminal box
~, разда́точная splitter box, transfer box
~, распредели́тельная 1. junction box, distribution box
2. rocker box
~ рулево́го механи́зма steering box
~ са́льника packing box, stuffing box
~ скоросте́й gearbox
~, сма́зочная lubricator, oil can, grease cup
~, соедини́тельная junction box
~ стани́ны gun trail side plates
~, ство́льная 1. casing (sub-machine gun)
2. breech (rifle)
3. body (SA)
~ стреля́ющего механи́зма percussion lock slide box (ord)
~, фильтру́ющая respirator container
~ шестерни́ pinion casing (MG)
коробле́ние distortion, warping, twisting, buckling
коромы́сло 1. rocker, rocker arm
2. breech actuating shaft (ord)
~ уравни́теля то́рмоза brake compensator balance lever
Коро́ткая "short halt"
коротково́лновый short wave
короткостово́льный short-barrelled
коро́ты́ш 1. bearing block 2 bolster (br)
"Коро́че Шаг!" "Step Short!"
ко́рпия lint
ко́рпус 1. corps
2. body, casing, frame
3. hull (tk)
4. block of buildings
5. airframe (missile, ac)
6. shell (bomb)
~ амортиза́тора shock-absorber body
~, арме́йский army corps
~, артиллери́йский artillery corps (provided arty support in major ops during World War 11)
~ бу́фера buffer tube
~ водяно́го насо́са water-pump body
~, возду́шно-деса́нтный airborne corps
~ вытяжно́го кольца́ rip-cord ring (parachute)
~ ги́льзы walls of cartridge case
~ грана́ты grenade body
~, кавалери́йский cavalry corps
~ кла́пана valve body

ко́рпус конденса́тора condenser shell
~, механизи́рованный mechanized corps
~ миномётной ми́ны mortar bomb case
~, оско́лочный fragmenting body, fragmentation body
~, офице́рский the Officer Corps
~ прибо́ра instrument case
~ прице́ла sight body (ord)
~ раке́ты rocket body, missile body
~ са́льника stuffing box
~ снаря́да shell body
~ та́нка tank hull
~, хими́ческий chemical corps, CW corps
~, цельноли́тый all-cast body
~ цили́ндра cylinder body
корпусно́й/ко́рпусный 1. corps
2. body, case 3 hull (tk)
4. shell (of bomb)
5. pertaining to block of bldgs
корректи́рование correction, adjustment
~ да́льности adjustment in range, range adjustment
~ направле́ния adjustment in line
~ огня́ 1. observation and correction of fire (Brit)
2. observation of fire (US)
~ огня́ с самолёта aircraft fire correction
корректи́ровать/прокорректи́ровать to correct (gunn) (Brit), to adjust (gunn) (US)
корректиро́вка spotting, artillery fire correcting
корректиро́вщик 1. observer, artillery observer
2. spotter plane
~ огня́ spotter (gunn)
корре́ктор, акусти́ческий sound corrector, acoustic corrector
~ горизонта́льного паралла́кса convergence corrector (FCE)
~ дерива́ции drift corrector (gunn)
~ непаралле́льности tilt corrector (FCE)
~ прице́ла ору́дия muzzle velocity and temperature corrector (sight)
корректу́ра spotting correction
~, бокова́я correction for line
~, вертика́льная correction for range
~, горизонта́льная correction for line
~ да́льности range correction
~ направле́ния correction for line
~ по вертика́ли correction for range
~ по горизо́нту correction for line
корректу́ра угла́ прице́ливания elevation correction
корреспонде́нт 1. distant station (station being worked) (comms)
2. correspondent
~, вое́нный war correspondent
~, фронтово́й front-line reporter
корроди́ровать to corrode, to rust
коррози́йность corrosiveness
корро́зия corrosion
~, то́чечная pits, pitting, point corrosion
ко́ртик dagger, dirk
корчева́тель rooter
ко́ршун kite, hawk (also used as nickname for hostile aircraft)
коры́то, ма́сляное oil pan, sump (ICE)
коря́га sunken trees
коса́ spit of land
~, песча́ная sand bar, sand spit
коси́ть/скоси́ть to squint, to look asquint; **коси́ть/скоси́ть огнём** to mow down with fire
косого́р slope (topo)
косо́й sloping, transverse, skew, oblique
костёр bonfire, camp fire
косты́ль 1. sprag, spike
2. crutch (walking aid)
~, пучи́нный dog spike (rly)
костю́м suit, overalls
~, водола́зный diving suit
~, защи́тный gas suit, protective suit
~, камуфли́рованный garnished camouflage suit
~, компенси́рующий anti-G suit
~, лёгкий защи́тный light anti-gas suit, light protective suit
~, легководола́зный frogman's suit
~, маскиро́вочный camouflage suit
~, маскиро́вочный зи́мний white camouflage suit
~, маскиро́вочный камуфли́рованный garnished camouflage suit
~, моча́льный маскиро́ванный camouflage suit with bast garnish
~ парашюти́ста parachuting suit
~, пла́вательный flotation suit
~, противоперегру́зочный anti-G suit
~, тяжёлый защи́тный heavy anti-gas suit, heavy protective suit
костя́к оборо́ны backbone of the defence
котёл 1. cauldron, pot
2. pocket, encirclement (tac)
3. boiler

котёл, а́томный atomic pile, atomic reactor
котело́к kettle, mess tin
котело́к-фля́га combined mess tin and water bottle
коте́льная boiler house
котлова́н foundation excavation, pit, ditch, trench
котлови́на hollow (topo)
ко́уш thimble
кофферда́м cofferdam
кочене́ть/окочене́ть to become numb (with cold)
ко́чка tussock (topo)
ко́шка grapnel, grappling iron, crab
~, двухла́пная twin-fluke grappling hook
~, трёхла́пная triple-fluke grappling hook
~, четырёхла́пная four-fluke grappling hook
ко́шки crampons
коэффицие́нт 1. coefficient, factor 2. ratio, rate
~, аэродинами́ческий aerodynamic coefficient
~, баллисти́ческий ballistic coefficient
~ бегу́щей волны́ travelling-wave ratio
~ безопа́сности safety factor (tech)
~ буты́лочности chambrage
~ вероя́тности probability factor
~ вероя́тности попада́ния vulnerability factor hit probability factor
~ воспроизво́дства reproduction factor
~ вя́зкости coefficient of viscosity
~ грузоподъёмности stowage factor
~ загру́зки load factor
~ загрязне́ния contamination factor
~ запа́са safety margin, safety factor
~ затуха́ния attenuation factor, damping factor
~ искаже́ния distortion factor
~ креня́щего моме́нта rolling moment coefficient
~ лобово́го сопротивле́ния drag coefficient
~ модуля́ции modulation factor
~ надёжности reliability factor, safety factor
~ наполне́ния capacity (ammo), volumetric efficiency (shell, missile), space factor
коэффицие́нт напра́вленного де́йствия coefficient of directivity
~ нараста́ния growth factor
~ объёмного расшире́ния coefficient of volume expansion
~ освещённости luminosity factor
~ ослабле́ния reduction factor, coefficient of atmospheric attenuation (nucl)
~ ослабле́ния проника́ющей радиа́ции radiation attenuation factor
~ отклоне́ния deviation ratio
~ отраже́ния reflection coefficient
~ перево́да conversion factor
~ перегру́зки acceleration factor, load factor, dynamic factor
~, перегру́зочный acceleration factor
~, погру́зочный stowage factor
~ поле́зного де́йствия 1. efficiency 2. radiation efficiency (antenna)
~ поле́зного де́йствия, терми́ческий ideal thermal efficiency
~ поле́зного де́йствия, механи́ческий mechanical efficiency (output/input)
~ поле́зного де́йствия, о́бщий overall efficiency
~ поле́зного де́йствия, теплово́й thermal efficiency, calorific efficiency, heat efficiency, coefficient of thermal efficiency
~ поле́зного де́йствия то́рмоза, терми́ческий эффекти́вный brake thermal efficiency
~ поле́зного де́йствия то́рмоза, экономи́ческий brake thermal efficiency
~ потребле́ния consumption factor
~ преломле́ния refraction coefficient
~ прице́ла coefficient of the range component of the OT factor (gunnery)
~ пропорциона́льности coefficient of proportionality
~ пропуска́ния permittivity (elec)
~ Пу́ассона Poisson's ratio
~ радиопоме́хи jam-to-signal ratio signal-to-"noise" ratio (jamming)
~ распростране́ния propagation factor
~ рассе́ивания dispersion coefficient
~ рассе́яния scattering cross section
~ расхо́да flow coefficient, discharge coefficient
~ расшире́ния coefficient of expansion

коэффицие́нт рефра́кции index of refraction, refraction index, refractive index
~ **сжа́тия** coefficient of compression
~ **сжима́емости** coefficient of compressibility
~ **смерте́льности** lethal index (gas)
~ **спа́да радиоакти́вности** decay constant (nucl)
~ **спрямле́ния траекто́рии** coefficient of proportionality
~ **стоя́чей волны́** standing wave ratio
~ **суже́ния** contraction coefficient
~, **температу́рный** temperature coefficient
~ **теплово́го расшире́ния** coefficient of thermal expansion
~ **трансформа́ции** transformation ratio (transformer)
~ **тре́ния** coefficient of friction
~ **удале́ния** observer-target/battery-target factor, OT/BT factor
~ **управле́ния** control ratio
~ **усиле́ния** amplification factor
~ **усиле́ния анте́нны** antenna gain factor
~ **усиле́ния ла́мпы** amplification factor of valve (electron)
~ **усиле́ния мо́щности** power gain
~ **усиле́ния сту́пени** stage amplification (electron)
~ **фо́рмы** 1. factor depending on shape of shell (Brit)
2. form factor (US)
~ **чувстви́тельности** sensitivity index
~ **ша́га дальноме́ра** coefficient of the line component of the observer-target factor, OT factor
~ **шу́ма** noise factor
~ **я́ркости** relative brightness of surface (fieldcraft)
"краб" metal cupola (sl)
кра́ги leggings
край 1. edge
2. region
3. kray (USSR admin div)
~ **оборо́ны, пере́дний** forward edge of battle area, FEBA, forward defended localities, FDL's, main line of resistance
кран 1. cock, tap, faucet
2. crane
~, **автомоби́льный** vehicle mounted crane, crane truck
~, **бензи́новый** petrol cock, petrol valve
кран, водоразбо́рный water tap
~, **возду́шный** 1. air cock, air tap
2. priming pump control tap
~, **гидравли́ческий** water column, water crane (rly)
~, **гре́йферный** grab crane
~, **грузоподъёмный** hoist crane
~, **двухходово́й** two-way cock, two-way valve
~ **для спу́ска ма́сла** oil drain valve, oil drain cock
~, **запо́рный** stop-cock, shut-off valve
~, **козло́вый** overhead crane
~, **консо́льный** travelling wall crane
~, **маслоспускно́й** oil drain cock, oil-drain valve
~, **многоходово́й** multiway tap
~, **мостово́й** overhead travelling crane
~ **на гу́сеничном хо́ду** crawler crane (tracked)
~, **насте́нный** wall-mounted crane
~, **перепускно́й** release cock
~ **пода́чи** delivery cock, delivery valve
~, **подъёмный** hoisting crane, loading hoist
~, **пожа́рный** fire hydrant, fire cock
~, **полнопроходно́й** full-flow valve
~, **понто́нный** pontoon crane
~, **порта́льный** gantry crane
~, **про́бный** test cock
~, **пробоотбо́рный** test cock
~, **разда́точный** delivery nozzle
~, **регули́рующий** retarding valve (ord)
~ **с са́льником** stuffing box cock
~, **самохо́дный** mobile crane
~, **сливно́й** drain cock, drain tap
~, **сма́зочный** lubricating cock
~, **спускно́й** drain cock, pet cock
~, **стрелово́й** derrick (crane)
~, **то́пливный** fuel cock, fuel tap
~, **то́пливный распредели́тельный** fuel distribution cock
~, **тормозно́й** brake control valve (veh)
~, **трёхходово́й** three-way tap, three-way cock, three-way valve
~, **устано́вочный** fuse setting screw (ammo)
~, **четырёхходово́й** four-way cock, four-way tap, four-way valve
кран-ба́лка single-girder overhead travelling crane

кра́нец 1. ready-use ammunition locker (arty)
2. tender (boat)
кра́ник, про́бный pet cock
~, спускно́й drain tap
крановщи́к crane operator
кран-стрела́ jib crane
кран-укоси́на 1. sheers
2. derrick (eng)
крапи́вный 1. nettle
2. irritant (CW)
кра́ска, антикоррозио́нная rust-inhibitive paint, anti-corrosive paint, anti-rusting paint
~, огнезащи́тная fire-retardant paint
~, светя́щаяся luminous paint
~, фосфоресци́рующая luminous paint, phosphorescent paint
~, эма́левая enamel (paint)
кра́ски, гипсометри́ческие layer tints (carto)
краскомёт paint spray gun
кра́тер 1. crater
2. hollow, crater (arc lamp)
кремалье́ра rack and pinion (device)
кре́мний silicon
крен 1. roll, bank (aeron)
2. heel, list (veh, ship)
крени́ться/накрени́ться to roll, to bank (aeron), to heel, to list (veh, ship)
кренóметр 1. roll indicator (aeron)
2. clinometer
крепёжный fastening, securing
крепи́льщик timber man (mining)
крепи́ть to secure, to fasten, to reinforce, to timber (mining)
крепле́ние 1. securing, fastening
2. timbering (mining)
3. chock
~, а́нкерное cable anchorage, anchorage (br)
~, букси́рное tow attachment
~, деревя́нное wooden timbering (fd eng)
~, жёсткое 1. cable binding (skis)
2. rigid fastening (mech)
~, забивно́е spiling (mining)
~, закладно́е frames and sheeting timbering (mining)
~ кру́тостей транше́и оде́ждой с распо́рками trench shoring
~, лы́жное ski binding
~, похо́дное travelling crutch (ord)
крепле́ние ремня́ безопа́сности seat-belt anchorage
~, сплошно́й close timbering (mining)
кре́пость 1. fortress
2. strength (tech)
3. yield strength (of metal)
крепь timbering (mining)
~, забивна́я spiling (mining)
~, сплошна́я close timbering (mining)
крест ните́й cross wires (instr)
крестови́на 1. cross piece, cross pin, cross plate
2. spider (universal joint)
3. frog (rly)
~ карда́на universal-joint trunion
~ лафе́та spider (universal joint)
~, нажи́мная spider cover (min)
крестообра́зный cruciform
кре́шер crush gauge, crusher gauge, pressure gauge
креще́ние, боево́е baptism of fire, blooding
крива́я, баллисти́ческая ballistic curve
~ безопа́сности curve of safety (ballistics)
~ вероя́тности пораже́ния це́ли kill probability curve
~ выжива́ния survival curve
~ давле́ния pressure curve
~ да́льностей range curve
~, изотру́бочная fuse curve
~, калибро́вочная calibration curve
~ мо́щности power curve
~ нараста́ния growth curve (nucl)
~ пого́ни attack curve (ac)
~ пораже́ния а́томным взры́вом effects curve (nucl)
~ производи́тельности output curve
~ распа́да decay curve (nucl)
~ расхо́да то́плива consumption curve (fuel)
~, резона́нсная resonance curve (electron)
~ ско́ростей velocity curve
~ совмеще́ния matching curve (GW)
~ траекто́рии trajectory curve
~, трёхто́чечная three-point curve (GW)
~, упреждённая lead curve (AD FCE)
кривизна́ curvature
~ ствола́ barrel curvature
~ траекто́рии curvature of trajectory
кривоши́п crank

кривоши́п кли́на breech block actuating shaft (ord)
~, составно́й divided crank
кривоши́пный crank, cranked
кри́зис, волново́й shock stall (missile)
криптогра́мма cryptogram
криптогра́фия cryptography
кристаллиза́ция crystallization
кристалли́ческий crystalline, crystal
крова́ть, похо́дная camp bed
кровоизлия́ние haemorrhage
кровопроли́тие bloodshed
кровотече́ние bleeding, haemorrhage
кровь, консерви́рованная banked blood
кроки́ plan sketch, military sketch, sketch map
~ маршру́та road sketch
~ ме́стности area sketch
кро́мка edge, rim
~ бе́рега water's edge
~ вхо́да intake lip
~, за́дняя trailing edge (aeron)
~, ни́жняя peripheral seam (parachute)
~, пере́дняя leading edge (aeron)
~ плу́нжера, спира́льная отсе́чная plunger helix (diesel ICE)
~, ре́жущая lip, cutting edge (machine tool)
кронци́ркуль outside caliper
кроншите́йн 1. bracket
2. stay (gunshield)
3. magazine housing
~ для крепле́ния пу́шки в похо́дном положе́нии external travelling gun crutch
~ лю́льки cradle bracket (arty)
~, опо́рный support bracket
~ опти́ческого прице́ла telescopic sight mounting bracket
~ подъёмного механи́зма elevating gear bracket (arty)
~ прице́ла sight mount
~, ца́пфенный trunnion bracket
кросс 1. main distribution frame room (teleg) 2 main distribution frame (teleg)
3. cross-country running
круг 1. circle
2. star with lateral working (radio)
~, а́зимутный circular protractor with range markings
~, алида́дный vernier plate
~, артиллери́йский artillery board protractor
круг, артиллери́йский целлуло́идный artillery protractor
~, большо́й great circle (navig)
~, вертика́льный vertical circle, vertical arc (topo)
~, горизонта́льный horizontal circle (theodolite)
~ да́льности range circle
~ да́льности, подви́жный moveable range ring (radar)
~ ката́ния колеса́ rolling circle of wheel
~, нажда́чный abrasive disc, emery disc
~, поворо́тный turntable (rly)
~, спаса́тельный lifebelt, lifebuoy
~, точи́льный abrasive disc, emery disc
~, угломе́рный 1. azimuth scale (ord)
2. horizontal scale (FCE)
~, целлуло́идный celluloid protractor
~, шлифова́льный grinding wheel
круглогу́бцы round-nosed pliers
круглосу́точно round-the-clock
кругове́рть circular motion (ESV after landing in sea
кругово́й circular, all-round
кругозо́р all-round view
"Круго́м!" "About Turn!"
круг-трансформа́тор plotter
кру́жка mug
кружо́к, перга́ментный parchment washer
~, суко́нный cloth washer (eng)
крупа́, сне́жная soft hail (met)
крупногабари́тный bulky, large, space-consuming
крупнозерни́стый coarse, coarse-grain
крупнокали́берный heavy, large-calibre
крупномасшта́бный large-scale
крупнопане́льный (precast) large-panelled
крупноразме́рный large-sized
крутизна́ 1. steepness
2. sharpness (curve)
3. transconductance, slope (elec)
~ бе́рега beach gradient
~, ко́мплексная transadmittance
~ наре́зки twist of rifling, angle of rifling, angle of twist
~ наре́зов inclination of the grooves, angle of twist, angle of rifling
~, постоя́нная uniform twist (ord)
~ преобразова́ния conversion conductance
~ ска́та angle of slope (carto)

крутизна́ спу́ска downhill gradient
~ траекто́рии steepness of trajectory
~ характери́стики 1. dynamic conductance (diode)
2. mutual conductance (triode)
круто́й steep, abrupt, sharp
крутопа́дающий steep(-grade)
кру́тость 1. slope, steepness
2. sharpness (curve)
3. earth face (fd eng)
~, оде́тая reveted slope
~, пере́дняя front slope
~, ты́льная rear slope
кру́ча precipice, abrupt slope
круше́ние derailment, wreck (collision), smash, accident
крыла́тка impeller (pump)
крыла́тый winged, aerodynamic
крыло́ 1. wing, aerofoil
2. bumper, mudguard (MT)
3. leaf (br)
4. aisle (bldg)
5. impeller, vane, blade (aeron)
6. arm (semaphore)
~, за́днее rear fender
~, крестообра́зно располо́женное cruciform wing
~, лета́ющее flying wing (tailless ac)
~, пере́днее front fender
~, подво́дное hydrofoil
~, прямоуго́льное rectangular wing
~ семафо́ра semaphore arm
~, стрелови́дное swept back wing
крыльча́тка impeller (pump)
кры́ша 1. roof
2. hull roof plate (tk)
~, раздвижна́я sliding roof
кры́шка cover, lid, cap, top plate
~, бокова́я side plate (pistol)
~, гермети́ческая hermetically-sealed cover
~ голо́вки бло́ка rocker box (ICE)
~ деса́нтного лю́ка hull escape door (tk)
~, ду́льная muzzle cover
~, защи́тная protective cap
~ ко́роба top cover (gunn)
~ коро́бки подаю́щего механи́зма feed box cover (SA)
~ ла́за manhole cover
~ лю́ка hatch cover
~ лю́льки cradle cap (ord)
~ магази́на magazine bottom plate
~, нажи́мная pressure plate (min)
кры́шка наливно́й горлови́ны filler cap
~, норма́льная obturator (ammo)
~ подши́пника bearing cap (diesel ICE)
~ приёмника feed block cover (MG)
~ распредели́теля distributor cap (ICE)
~ смотрово́го лю́ка inspection cover, inspection plate
~ спусково́й коро́бки trigger mechanism cover (SA)
~ ство́льной коро́бки top plate of receiver, magazine-opening cover (SA)
~, съёмная removable cover
~ то́пливного ба́ка fuel tank cap
~ уда́рника striker cover (ord)
~, упо́рная striker spring cover (ord)
~, уси́ленная strengthening lid (QF ammo)
~ цили́ндра, за́дняя cylinder gland bush (ord)
~, цили́ндровая cylinder head (ICE)
~ шатуна́ connecting rod cap
крюк hook, staple
~, букси́рный towing hook
~, вертлю́жный swivel hook
~, грузово́й load hook
~ для подъёма дви́гателя engine lifting hook
~, зацепно́й catch hook
~ копра́, спусково́й pile-driver trip hook
~, подъёмный tackle hook
~, прицепно́й coupling hook
~, ручно́й bag hook
~, спусково́й trip hook
~, сцепно́й towing hook, coupling hook, draw-bar hook
~, тя́говый draw-bar hook, towing hook
~, уносно́й spring hook, spring trailer hook
крючо́к hook, claw, catch
~ предупреди́тельным спу́ском double-pull trigger
~ с двумя́ то́чками опо́ры, спусково́й double-pull trigger
~ с предупреди́телем, спусково́й double-pull trigger
~, соедини́тельный claw
~, спусково́й 1. trigger (SA)
2. drain cock (radiator)
~, спусково́й чувстви́тельный hair trigger (SA)
~, сцепно́й coupling hook
крю́чья, ледовы́е ice pitons

кряж ridge (topo)
ксили́л trinitroxylene, TNX (HE)
ксили́лбромид trinitroxylene bromide (HE)
"Кто Идёт?" "Who Goes There?"
куб, перего́нный still
кува́лда sledge-hammer, double-face hammer
кувырка́ние somersaulting (para)
кузне́ц blacksmith
кузне́чный blacksmith's
ку́зов body (veh)
~, безра́мный unitary body
~, бу́нкерный hopper body
~, жёсткий rigid body
~, закры́тый enclosed body
~, комбини́рованный multi-purpose body
~, кры́тый canopy body
~, опроки́дывающийся tipper body, dump body
~, откры́тый open body
~ прице́па trailer body
~, раздвижно́й extendable body
~ с откидны́ми бо́ртами drop-side body
~, саморазгружа́ющийся dump body (of truck)
~ самосва́ла dump truck body
~, цельнометалли́ческий all-metal body
кузовно́й body (veh)
ку́зов-пика́п pick-up body
ку́зов-платфо́рма platform body
ку́зов-фурго́н box-body (veh)
куку́шка 1. tree-top sniper
2. narrow-gauge locomotive (sl)
3. local train
4. cuckoo
кула́к 1. fist
2. striking force (tac)
3. cam
4. jaw (chuck)
~, бронево́й armoured strike force
~, брониро́ванный mailed fist
~, огнево́й massive concentration of fire, concentrated fire support
~, поворо́тный stub-axle assembly, swivel-axle assembly
~, разжимно́й expander cam
~, тормозно́й brake cam
кулачо́к 1. detent (ammo)
2. cam
~ выбра́сывателя extractor cam
~, отпира́ющий unlocking cam
кулачо́к подава́теля feed pawl (SA)
~ прерыва́теля contact-breaker cam, distributor cam (ICE)
~ храповика́ ratchet pawl
кули́са 1. crank
2. rocker arm, gear shift gate
3. connecting link
4. slideway
5. cam guide, cam groove (ord)
куло́н Coulomb (elec)
куль sack
кумуляти́вный 1. hollow-charge (ammo)
2. cumulative
купе́ compartment
ку́пол 1. cupola
2. canopy (parachute)
~ парашю́та (parachute) canopy
курвиме́тр curvometer
курга́н burial mound, tumulus (topo)
куро́к 1. cocking piece, hammer (SA)
2. cocking lever, firing lever (ord)
курс 1. course, path, route
2. course of instruction
~, баллисти́ческий ballistic course
~, боево́й bombing run (ac)
~ боево́й подгото́вки, основно́й basic training
~ вожде́ния driving course
~, встре́чно-пересека́ющий collision course (GW)
~, встре́чный collision course, head-on course (GW)
~, ко́мпасный compass course
~, магни́тный magnetic heading
~ мето́дики, теорети́ческий methods of instruction course
~, паралле́льный parallel course
~ переподгото́вка upgrading course
~, пересека́ющий crossing course, intersecting course
~, повто́рный refresher course
~, попу́тный pursuit course (GW)
~, поса́дочный магни́тный magnetic approach heading (air navig)
~, постоя́нный constant heading
~ пресле́дования pursuit course (GW)
~ специа́льной авари́йно-спаса́тельной подгото́вки special training course for salvage and rescue work
~ стрельб gunnery course
~ стрельб из пистоле́та pistol firing course

курс усоверше́нствования up-grading course
~ це́ли target course
курса́нт 1. person on course, student, trainee
2. officer cadet
курса́нт-стажёр officer cadet undergoing practical training with unit
курсопрокла́дчик course plotter, course recorder
курсоуказа́тель course indicator, direction indicator
ку́ртка short jacket, work jacket
~, ва́тная short coat with quilted lining
~ танки́ста tank crewman's jacket
курье́р courier, messenger
куса́чки cutting pliers, wire cutters, nippers
куста́рник bush, shrub, brushwood, undergrowth
кусторе́з brush cutter
ку́хня kitchen, cookhouse
~, вью́чная pack kitchen
~, оча́жная похо́дная Soyer stove, Soyer boiler (cooking)
~, полева́я field kitchen
~, похо́дная mobile field kitchen
ку́чность бо́я accuracy, grouping capability (weapon)
~, высо́кая pin-point accuracy
~, ни́зкая inaccurate firing
~ пробо́ин pattern of hits
~ стрельбы́ consistency, grouping (gunn)
ку́чный grouping, grouped (gunn)
кюве́т side drain (road)
кю́ри curie (nucl)
кю́ри-эквивале́нт curie equivalent

Л

лабири́нт labyrinth (gland)
лабора́нт laboratory assistant, laboratory technician
~ из полиго́нной кома́нды range technician
лаборато́рия laboratory
~ для испыта́ний test laboratory
~, испыта́тельная test laboratory
лаборато́рия, иссле́довательская research laboratory
~, передвижна́я portable laboratory
~, перено́сная portable laboratory
~, подви́жная mobile laboratory
~, полева́я хими́ческая field laboratory (CW)
лави́на avalanche
ла́герь camp
~, бара́чный cantonment, hutted camp
~, бива́чный bivouac
~, вре́менный temporary camp
~, зи́мний winter camp
~, концентрацио́нный concentration camp
~, пала́точный tented camp
~, постоя́нный standing camp
~, похо́дный bivouac
~, уче́бный training camp
лады́га trunnion bracket (ord)
лаз 1. mousehole, inter-communication hole
лаза́ние crawling, climbing
лазаре́т hospital
~, ветерина́рный veterinary hospital
~, войсково́й sick bay
~, полково́й regimental aid post (RAP)
ла́зер laser
~, арго́новый argon laser
~, арго́новый ио́нный argon ion laser
~, га́зовый gas laser
~, ге́лий-нео́новый helium-neon laser
~, дио́дный diode laser
~, и́мпульсный pulsed laser
~ инфракра́сного диапазо́на infra-red laser, IR laser
~, кра́сный red (light) laser
~ на арсени́де га́ллия gallium arsenide laser
~ на двуо́киси углеро́да carbon dioxide laser
~ на неоди́ме neodymium laser
~ на руби́новом криста́лле ruby (crystal) laser
~ на стекле́ glass laser
~ на стекловолокне́ fibre laser
~ на углеки́слом га́зе carbon dioxide laser
~ непреры́вного излуче́ния, га́зовый continuous-wave gas laser, CW gas laser
~, перено́сный man-portable laser
~, портати́вный portable laser

ла́зер, руби́новый ruby (crystal) laser
~ с высо́ким усиле́нием high-gain laser
~ свя́зи communication laser
~, ультрафиоле́товый ultra-violet laser, UV laser
лазу́рь engineer's blue
лазу́тчик scout, spy
ла́йнер liner (ship, ac)
~, возду́шный airliner
лак, ма́сляный oil varnish
~, нитроцеллюло́зный nitro-cellulose varnish
~, целлюло́зный cellulose varnish
лакрима́тор tear gas, lachrymator
ла́мели, одноимённые identical commutator bars
ла́мель commutator bar
ла́мпа 1. valve (Brit), tube (US) (electron)
2. lamp
3. light bulb
~ бегу́щей волны́ travelling wave tube, TW tube
~, ва́куумная vacuum tube, vacuum valve
~, выпрями́тельная rectifier valve (electron)
~, газоразря́дная gas-discharge lamp
~, газосве́тная gas discharge lamp (illuminating)
~, генера́торная oscillator valve
~, жёсткая hard valve (elec)
~, запира́ющая gate valve, strobe valve
~, запуска́ющая trigger valve
~, и́мпульсная flash lamp (photo)
~, инфракра́сная infra-red lamp, IR lamp
~, комбини́рованная multiple function valve
~, контро́льная control lamp, test lamp, pilot lamp
~, кристалли́ческая crystal valve
~, ксено́новая xenon lamp
~, люминесце́нтная fluorescent tube
~, маячко́вая disc-seal valve, lighthouse valve
~, миниатю́рная па́льчиковая tom-thumb valve, bantam tube miniature valve with pressed glass base
~ нака́ливания filament lamp, electric-light bulb
~, на́триевая sodium lamp
~, нео́новая neon lamp
ла́мпа обра́тной волны́ backward wave valve
~, освети́тельная (illuminating) lamp
~ освеще́ния instrument light, dial light
~, па́льчиковая miniature valve with pressed-glass base
~, пая́льная blow lamp, brazing torch, heating torch, blow torch
~, перено́сная inspection lamp, portable lamp
~, приёмно-усили́тельная receiving-amplifier tube
~ прямо́го нака́лка directly heated cathode valve
~, пу́говичная doorknob valve
~, пятисе́точная pentagrid tube, heptode
~, пятиэлектро́дная pentode
~, реакти́вная reactance valve
~, рту́тная mercury vapour lamp
~ с ги́бкой сто́йкой lamp with flexible mounting
~ с переме́нной крутизно́й variable-mu valve
~, сигна́льная 1. indicating lamp, warning light
2. signal lamp
~, сигна́льно-пробле́сковая aldis lamp
~, смеси́тельная mixing valve (elec)
~, софи́тная horizon-type lamp
~, строби́рованная gate valve, strobe valve
~, строби́рующая gate valve, strobe valve
~, три́ггерная flip-flop tube
~, указа́тельная indicating lamp
~, усили́тельная amplifier valve
~, флуоресце́нтная fluorescent lamp
~, часто́тно-преобразова́тельная frequency-changing valve
~, четырёхэлектро́дная tetrode
~, шестиэлектро́дная hexode
~, экрани́рованная screen-grid valve
~, электро́нная thermionic valve, valve
ла́мпа-вспы́шка flash tube
~, ква́рцевая quartz flash tube
~, ксено́новая xenon flash tube
~, лине́йная linear flash tube
ла́мпа-жёлудь acorn valve
ламподержа́тель valveholder
ла́мпочка 1. small lamp
2. electric bulb
~ "пожа́р" fire warning light
~, у́гольная carbon-filament lamp

ландша́фт landscape
ла́па дви́гателя engine mounting arm
~ **крепле́ния ка́ртера** crankcase arm
~, **шворнева́я** trail plate, lunette (veh), trail eye
ла́пка lug
~ **выбра́сывателя** toe of extractor
~ **экстра́ктора** extractor arm
ла́пник branch of coniferous tree
ларингофо́н laryngophone, throat microphone
ласт swim-fin (frogman's suit)
лату́нь brass
~, **отве́тсвенного назначе́ния** standard brass
~, **патро́нная** cartridge brass
лафе́т 1. carriage (ord)
2. mounting (mor)
~, **двуххо́ботовый** split-trail carriage
~, **жёсткий** fixed carriage (ord)
~, **колёсный** wheeled carriage, wheeled mounting
~, **подви́жный** mobile carriage
~ **с коро́бчатыми стани́нами** box trail carriage
~ **с поворо́тным механи́змом вертлю́жного ти́па** central pintle carriage
~ **с раздвижны́ми стани́нами** split-trail carriage
~ **с рессо́рным сошнико́м** spring-spade carriage
~, **тумбо́вый** central pedestal mounting (ord)
~, **универса́льный** all-purpose mounting (ord)
~, **упру́гий** carriage with recoil
лафе́т-двуно́га bipod mounting
лафе́т-трено́га tripod mounting
лафе́т-ту́мба column, pintle, mount
лебёдка windlass, winch, capstan, winding engine
~, **аэроста́тная** balloon winch
~, **букси́рная** towing winch
~, **погру́зочная** loading winch
~, **подъёмная** lifting winch
лебёдка-реду́ктор winch with reduction gear
"Левée!" "Bear Left!"
левофланго́вый left-flank, left-flanking
леге́нда legend (map, diagram)
леги́рование alloying
легкоартиллери́йский light-artillery
легкоброниро́ванный lightly armoured
легководола́з frogman, underwater swimmer
легковоспламеня́ющийся highly inflammable
легкопла́вкий easily melted, fusible
легкора́неный 1. lightly-wounded man
2. lightly-wounded
легкораствори́мый readily soluble
легкосжижа́ющийся easily liquefied
лёгкость управле́ния easily controllable, drivable
лёд ice
~, **берегово́й** coastal ice
~, **до́нный** anchor ice, bottom ice
~, **дрейфу́ющий** drift ice
~, **есте́ственный** natural ice
~, **кристалли́ческий** crystal ice
~, **па́ковый** pack ice
~, **пове́рхностный** surface ice
~, **прозра́чный** clear ice
~, **сухо́й** dry ice, solid carbon dioxide
~, **фи́рновый** firn ice, neve
ледни́к glacier
ледоко́л ice-breaker (ship)
ледоме́р ice measurer
ледоре́з ice cutter, ice-breaker (on bridge, pier etc)
~, **шатро́вый** knife-edge ice-breaker (on bridge, pier etc)
ледору́б ice axe
ледоста́в freezing over (rivers etc)
ледохо́д ice-gang (movement of ice between spring and autumn), drifting of ice
лёжа lying, in lying position
ле́жень sleeper, foundation beam, wall plate, stringer, groundsill, stay block
~, **берегово́й** abutment baulk, bridge seat, bank seat, bank transom (br)
~, **опо́рный** sleeper, saddle beam
~, **понто́нный** boat transom
~, **попере́чный** sleeper
~, **я́корный** anchor log
ле́звие 1. blade
2. razor blade
лейкопла́ст adhesive tape
лейкопла́стырь soil sample plate (radiac eqpt)
ле́йнер lining tube, gun liner (in gun barrel)
~, **запа́сный** replacement liner

ле́йнер, кони́ческий tapered liner (ord, gun barrel)
~, свобо́дный loose liner, removable liner
~, скреплённый inner "A" tube (ord, gun barrel)
~, цилиндри́ческий parallel liner (ord, gun barrel)
лейне́ровать to reline (of barrels)
лейни́рование relining (ord, gun barrel)
лейтена́нт lieutenant
~, мла́дший junior lieutenant
~, ста́рший senior lieutenant
лека́ло curve, template, French curve
лексико́н, вое́нный military language, military vocabulary
ле́мех сошника́ spade plate (ord)
лени́вец track idler (wheel), idler (sprocket), belt tightening wheel
~ гу́сеницы track adjusting wheel
~, пере́дний front idler wheel
ле́нта 1. ribbon, tape
2. slip (teleprinter)
3. belt (MG)
~, ги́бкая flexible belt (MG)
~, гу́сеничная track (veh)
~, жёсткая strip (MG feed)
~, зве́нчатая патро́нная linked belt (SA)
~, звенева́я разъёмная disintegrating link belt (SA)
~, звукометри́ческая recorder film, recorder tape (sound ranging)
~, измери́тельная measuring tape
~, изоляцио́нная insulating tape
~, клейкая gummed tape, water-activated tape
~, ли́пкая adhesive tape
~, магни́тная magnetic tape
~, мате́рчатая fabric belt (MG)
~, ме́рная measuring tape, surveyor's tape
~, металли́ческая 1. metal tape, metal band
2. fragmentation liner (gren)
3. metal belt (MG)
~, металли́ческая зве́нчатая metal link belt (MG)
~, нераспада́ющаяся звенева́я патро́нная non-disintegrating ammunition belt, non-disintegrating cartridge belt
~, обмо́точная binding tape
~, огнепрово́дная tape fuse
ле́нта оредно́в и меда́лей medal ribbon
~, о́рденская ribbon of order, medal ribbon
~ откры́того ти́па open belt (MG)
~, патро́нная ammunition belt, cartridge feed belt
~, перфорацио́нная perforated tape
~, пи́льная saw blade
~, пулемётная MG ammunition belt
~, разъёмная disintegrating link belt (MG)
~, разъёмная пласти́нчатая disintegrating link belt (MG)
~ с неза́мкнутыми гнёздами, патро́нная open link belt (SA)
~, стальна́я steel strip, steel band
~, телегра́фная telegraph tape, ticker tape
~, тка́невая scrim
~ то́рмоза brake band
~, тормозна́я brake band
~, холщо́вая fabric belt (MG)
~, шарни́рная articulated strip (MG feed)
лепесто́к lobe
~, боково́й side lobe, minor lobe (electron)
~, гла́вный main lobe, major lobe
~, основно́й main lobe, major lobe
~ побо́чного излуче́ния side lobe, minor lobe
лепёшка pellet (arty)
лес wood, forest, timber, lumber
~, ка́рликовый dwarf wood, stunted forest (sub-arctic topo)
~, ли́ственный deciduous wood
~, низкоро́слый stunted wood
~, пилёный sawn timber
~, ре́дкий sparse wood
~, сме́шанный mixed wood (topo)
~, сортово́й commercial timber
~, строево́й construction timber, building timber
~, хво́йный wood of coniferous trees
леса́ 1. woods, forest
2. scaffolding
~, ли́ственные desiduous wood
"ле́сенкой" side-stepping ascent on skis
лесни́чество forest area
лесно́й 1. forest
2. lumber
лесоматериа́л timber, lumber
~, вы́держанный seasoned timber
~, нестру́ганный rough lumber

лесоматериáл, пилёный precut lumber
~, пропи́танный креозóтом creosoted timber
~, свежераспи́ленный green timber
~, строевóй construction timber, building timber
~, строи́тельный construction timber, building timber
лесопи́лка saw-mill
лесостéпь wooded steppe
лесотýндра wooded tundra
лёсс loess (fine wind-blown deposit)
лéстница, верёвочная rope ladder
~, штурмовáя scaling ladder
лёт flight, flying
летнáб air observer, reconnaissance pilot, air observation post pilot
лéтний summer
лётный pertaining to flying
летóвка summer camp
летýчесть volatility
~, высóкая high volatility
летýчка 1. short exercise 2. mobile detachment, road patrol 3. unit broadsheet
~, авари́йная 1. crash tender 2. recovery vehicle
~, железнодорóжная санитáрная temporary ambulance train
~, корóткая short intensive course of instruction, short exercise, short course
~, передовáя ремóнтная 1. forward repair team 2. LAD recovery team
~, ремóнтная emergency repair truck, maintenance vehicle, mobile repair shop, workshop lorry
~, санитáрная temporary ambulance train
~, такти́чкеская short tactical exercise for officers, (Sov), short TEWT
~ ти́па "А" mobile workshop type "A"
~, хирурги́ческая mobile surgical unit
лётчик pilot, airman
лётчик-наблюдáтель air observer, reconnaissance pilot, air observation post pilot
лётчик-смéртник suicide pilot
лётчик-снáйпер classification awarded to military pilots of exceptional skills
лечéние medical treatment
~, амбулатóрное outpatient treatment, outpatient care
лечéние, стационáрное hospital treatment, in-patient treatment
лигни́т lignite
лигрои́н naphtha, ligroin
лидди́т lyddite, picric acid, melinite
ликвидáтор self-destruction device
ликвидáция 1. wiping out, mopping up 2. liquidation, elimination
ликвиди́ровать to wipe out, to destroy, to liquidate; **ликвиди́ровать скоплéние трáнспорта** to clear traffic congestion
лимáн estuary
лимб azimuth scale, azimuth scale plate
~, вертикáльный vertical scale (theodolite)
~, горизонтáльный horizontal scale (theodolite)
~ шкалы́ кýрса course bearing scale ring (AA sight)
лими́т allocation, limit, ration
~ горю́чего, вы́деленный fuel allotment
лимóнка fragmentation hand grenade (sl)
линеаризáция linearization
линéйка rule, ruler, straight edge
~, визи́рная sight rule
~ для сня́тия отсчётов recorder cursor (sound ranging)
~ дозиметри́ста, счётная radiation slide rule
~, запóрная shift-lock bar (teleprinter)
~, комбинáторная selector bar (teleprinter)
~, копи́рная swivelling slide (lathe taper attachment)
~, лáгерная lane between tents, boundary (in tented camp)
~, логарифми́ческая slide rule
~, масломéрная dip stick
~, масштáбная rule
~, мéрная measuring rule
~, металли́ческая metallic ruler, metal ruler
~, параллéльная parallel ruler
~, повéрочная straight edge
~, си́нусная sine bar
~, спусковáя universal bar (teleprinter, stop-start teleg)
~, счётная slide rule
~, штриховáя graduated rule
линéйность linearity
линéйный marker (on parade) (pers)

линейный 1. line, linear
2. capital (of ships)
3. marker (on parade) (pers)
ли́нза lens
~, **вогнутовы́пуклая** positive meniscus lens
~, **выпукловóгнутая** negative meniscus lens
~, **глáзная** eye lens
~, **двояковóгнутая** double concave lens
~, **двояковы́пуклая** double convex lens
~, **магни́тная** magnetic lens
~, **молóчная** opal lamp cap
~, **отрицáтельная** negative lens
~, **плóско-вóгнутая** plano-concave lens
~, **плóско-вы́пуклая** plano-convex lens
~, **положи́тельная** positive lens
~, **прямоугóльная отражáющая** Porro prism
~, **рассéивающая** diverging lens
~, **собирáтельная** converging lens
~, **ступéнчатая** stepped lens (corrected for spherical aberration)
~, **электрóнная** electronic lens
ли́нии, коммуникациóнные lines of communication
ли́ния 1. line
2. line (unit of measure - 2.54 mm, Ø.1 in)
~, **абонéнтская** subscriber's line
~, **агони́ческая** line of no magnetic deviation, agonic line
~ **антéнны, питáющая** antenna feeder
~ **атáки, исхóдная** assault line
~ **бáзиса** base line (minefield)
~, **бáзисная** base line
~, **балáнсная** compensation circuit
~ **без потéрь** loss-free line (elec)
~ **безопáсности** safety line
~, **береговáя** coastline
~, **бесконéчная** infinite line (electron)
~, **бесконéчно-дли́нная** infinite line (electron)
~, **беспотéрная** loss-free line (elec)
~, **боевáя** extended line (tac)
~ **бросáния** line of departure (ball)
~, **верховáя я́корная** upstream anchor line (br)
~ **взвóдных колóнн** line of platoon columns
~ **визи́рования** line of sight, sight(ing) line
~ **водосли́ва** drainage line (topo)
ли́ния возвышéния 1. axis of the bore (Brit)(gunn)
2. line of elevation (US)
3. theoretical line of fire (naval gunn)
~, **возду́шная** overhead line, open wire line, aerial line
~, **возду́шная кáбельная** aerial cable line
~, **вывéренная прицéльная** corrected line of sight
~ **вы́стрела** 1. axis of the bore (Brit) (gunn)
2. line of elevation (US)
3. theoretical line of fire (naval gunn)
~, **геодези́ческая** geodesic, geodetic line
~ **горизóнта** skyline
~ **дáльнего обнаружéния** early warning line, long range detection line
~, **демаркациóнная** 1. demarkation line
2. international date line
~, **железнодорóжная** railway line
~ **зáданного пути́** predicted or required track (air navig)
~, **иску́сственная** artificial line (elec)
~, **кáбельная** cable line
~ **канáла стволá** axis of the bore (gunn)
~, **коаксиáльная** coaxial line
~ **коллимáции** line of collimation
~ **колóнн** column (drill fmn)
~, **магистрáльная** trunk circuit, main line
~, **магни́тная силовáя** line of force (magnetic)
~, **малоканáльная** low-capacity multi-channel line (comms)
~, **манипуляциóнная** keying line (comms)
~ **маши́н** line abreast (veh fmn)
~, **международная** international circuit (comms)
~ **мини́рования, исхóдная** minefield base line
~ **наблюдéния** 1. spotting line (naval gunn)
2. line observer target, line OT (Brit arty)
~ **наимéньшего сопротивлéния** line of least resistance
~ **настоя́щего мéста цéли** line of present position
~ **начáла отсчёта** reference line
~, **ненастрóенная** nonresonant line
~, **несимметри́чная** unbalanced line

ли́ния, низова́я я́корная downstream anchor line (pontoon br)
~, нулева́я zero line
~ нулево́го а́зимута zero-bearing line
~, оборони́тельная line of defence
~, однопроводна́я single-wire line (teleg)
~, опо́рная datum line
~ ору́дие-цель gun-target line
~, осева́я centre line
~ отва́ла line of embarkation (tac)
~, отвесна́я plumb line
~ отделе́ний column of sections
~ оцепле́ния cordon
~ паде́ния 1. line of arrival (gunn, Brit) 2. line of fall (gunn, US)
~ пе́ленга line of bearing
~ переда́чи высо́кого напряже́ния high tension transmission line
~ переда́чи да́нных data line, data link
~ переда́чи, коаксиа́льная coaxial transmission line
~ переда́чи кома́нд command link
~ переда́чи, многокана́льная multichannel transmission line
~ переми́рия truce line
~ перепра́вы, исхо́дная river crossing start line
~, подво́дная ка́бельная submarine cable line, underwater cable line
~, подзе́мная ка́бельная underground cable line
~, полева́я ка́бельная ground cable line
~, полева́я телефо́нная field telephone line
~, постоя́нная permanent line (teleph, teleg)
~, при́городная suburban line
~ пристре́ляных да́льностей line of known ranges
~ пристре́ляных угломе́ров line of known angles
~ прице́ливания line of sight
~ прице́ливания, зави́симая rocking bar system (gunsight)
~ прице́ливания, незави́симая independent line of sight system
~ прице́ливания, полузави́симая semi-independent line of sight system
~, прице́льная sight radius (line joining centre of upper edge of backsight to tip of foresight)
~ прича́ла line of disembarkation (tac)
ли́ния, пропуска́емая send line (teleg)
~, равносигна́льная equisignal line, scan axis (radar)
~ ра́вных а́зимутов curve of equal bearing (radio navig)
~ радиолокацио́нных ста́нции, передова́я advanced radar belt
~, радиореле́йная radio-relay link
~, радиотелеметри́ческая radio-telemetry line, link
~ разбе́га, нулева́я zero length (GW)
~ развёртки trace (radar)
~, разграничи́тельная sector boundary, boundary between formations or units
~ разре́за изображе́ния separating line (radar)
~ ра́ннего предупрежде́ния early-warning line
~ распростране́ния line of propagation (antenna)
~, резона́нсная resonant line (electron)
~ с больши́м затуха́нием lossy line (elec)
~ сбо́рки, пото́чная continuous assembly line
~ свя́зи 1. line of communication 2. link (GW)
~ свя́зи, гла́вная main line of communication, main axis of communication
~ се́тки, вертика́льная Y-axis, Y-line
~ се́тки, горизонта́льная X-axis, X-line
~, силова́я line of force
~, силова́я магни́тная magnetic line of force
~, симметри́чная двухпроводна́я balanced pair (elec)
~, служе́бная engineering circuit (comms), engineering channel (comms)
~ снаря́д-цель line of sight (GW)
~, соедини́тельная connecting line
~, телефо́нная telephone line
~ транше́й trench line
~ трубопрово́да, обхо́дная by-pass pipe line
~ управле́ния control line
~ факти́ческого пути́ track made good (air navig)
~, фи́дерная antenna feeder
~ фро́нта forward edge of battle area, FEBA
~, ходова́я leg (survey)

ли́ния це́ли 1. line BT, line battery-target (Brit arty)
2. gun-target line
~ це́ли, упреждённая line of future position (arty)
~, центра́льная centre line
~, шестова́я poled cable (line comms)
линь, раке́тный rocket line (for saving life etc at sea)
ли́пкость stickiness, adhesiveness, tackiness
ли́ра U-shaped supporting arms (searchlight)
лиса́ 1. fox
2. small radio buoy (sl)
лист sheet, plate
~, боково́й side plate
~, бортово́й side plate (armour)
~, бронево́й armour plate
~, ваго́нный waybill
~, ве́рхний top plate, roof plate
~, ве́рхний накло́нный бортово́й
1. superstructure side plate (AFV) (Brit)
2. hull above fender (US)
~, ве́рхний накло́нный лобово́й 1. glacis plate (AFV) (Brit)
2. front plate upper (US)
~ дни́ща hull bottom plate, floor plate (AFV)
~ жалюзи́, бронево́й armoured louvres
~, за́дний rear plate (AFV)
~ ка́рты map sheet, chart sheet
~ ка́танной брони́ sheet of rolled armour plate
~, коренно́й main leaf, master leaf, top leaf (spring)
~, кормово́й rear plate (AFV)
~ ко́рпуса, ве́рхний hull roof plate (AFV)
~ ко́рпуса, кормово́й hull rear plate (AFV)
~ ко́рпуса, лобово́й hull front plate (AFV)
~ ко́рпуса, ни́жний hull bottom plate (AFV)
~ ко́рпуса, носово́й hull front plate (AFV)
~, маршру́тный route sheet
~, накло́нный sloping plate
~ наря́дов duty roster
~, ни́жний бортово́й 1. hull side plate (AFV) (Brit)
2. below fender plate (US)
лист, ни́жний накло́нный лобово́й
1. lower nose plate (AFV) (Brit)
2. front plate lower (US)
~, опро́сный questionnaire
~, откры́тый consignee receipt list
~, полётный air freight manifest, flight manifest, air freight list
~, похва́льный written commendation
~, путево́й transport work ticket (MT)
~, распо́рный tie plate
~, рессо́рный spring leaf
~ рессо́ры spring leaf
~, углово́й corner plate, gusset plate, gusset sheet
~, упако́вочный contents list, packing sheet
ли́ственница larch
листо́вка leaflet, broadsheet
листо́вка-мо́лния printed newsflash, operational news sheet
листово́й lamellar, sheet
листо́к, боево́й news bulletin operational news sheet (issued to troops)
~, больни́чный medical certificate (of fitness or unfitness for duty)
~ нетрудоспосо́бности medical certificate (of unfitness for work)
ли́тий lithium
лито́й cast
литра́ж capacity in litres (engine)
литьё под давле́нием die casting, injection-moulding
~, стально́е steel casting
~, цветно́е non-ferrous casting
~, центробе́жное centrifugal casting
лихора́дка, водяна́я marsh fever
~, жёлтая yellow fever (med)
ли́хтер, вы́садочно-вы́грузочный landing lighter
~, та́нковый tank landing craft, LCT
лицендра́т Litz wire
лицо́, вменя́емое person responsible for his actions (mil law)
~, гражда́нское civilian
~, дово́льствуемое recipient
~, должностно́е officer or soldier holding specific appointment
~, подо́пытное "guinea pig"
личи́нка, боева́я 1. bolt head (rotating bolt) (SA)
2. locking sleeve (straight pull bolt)
3. breech face (arty, ord)
лиша́ть/лиши́ть to deprive, to withhold

лишéние звáния deprivation of rank
лишúнка, запирáющая bolt lock
лишúть (см лишáть)
лобовóй frontal, front
ловýшка trap
~, взрывнáя booby trap
~ для тáнков tank trap
~, иóнная ion trap
~, механúческая baffle
~ мúны prevention of stripping equipment, PSE (min)
~, противотáнковая tank trap
~, тáнковая tank trap, tank obstacle
лóдка boat
~, десáнтная assault boat
~, десáнтная складнáя collapsible assault boat, folding assault boat
~, канонéрская gunboat
~, летáющая flying boat
~, надувнáя inflatable boat, pneumatic boat
~, надувнáя спасáтельная inflatable dinghy
~, нырúющая submersible
~, плоскодóнная flat-bottom boat
~, подвóдная submarine
~, речнáя канонéрская river gunboat
~, складнáя folding boat
~, спасáтельная lifeboat, rescue boat
~, цельноплáстиковая all-plastic boat
~, штурмовáя assault boat
лóдка-волокýша pulka, boat sledge, akja
лодыга trunnion bracket
лодыжка 1. ankle 2. firing-pin nut, cocking lever
лóжа stock (SA)
~, ружéйная rifle stock
ложбúна, барúческая trough of low pressure
лóже, речнóе river bed
ложевúк stock repairer (joiner in armourer's workshop)
ложéвщик stock maker (SA)
"Ложúсь!" "Lie Down!"
лóжка, буровáя spoon bit (earth auger)
лóжный false, dummy, feint, spurious
лóзунг slogan
локализáтор, посáдочный localizer (ac landing aid)
локáтор (см радиолокациóнная стáнция, радиолокáтор) detector, position finder, radar
локáтор, úмпульсный лáзерный pulse lidar, pulse laser radar
~, инфракрáсный infra-red radar, IR radar
~, лáзерный lidar, laser radar
~ наведéния, лáзерный guidance lidar, guidance laser radar
~ наведéния по лучý, лáзерный guidance lidar, laser beam-rider radar
~, напрáвленный вперёд, лáзерный forward-looking lidar, forward-looking laser radar
~, нацéливаемый с пóмощью инфракрáсной систéмы, лáзерный IR-aimed lidar, IR-aimed laser radar
~ непрерывного излучéния, лáзерный continuous-wave lidar, continuous-wave laser radar
~, оптúческий laser radar, optical radar
~ подсвéтки цéли, лáзерный illuminating laser radar
~ с фазирóванной решёткой, лáзерный phased-array lidar, phased-array laser radar
~ с частóтной модулúцией, лáзерный frequency-modulated lidar, FM lidar, frequency-modulated laser radar, FM laser radar
~, сканúрующий, лáзерный scanning laser radar
~, тактúческий лáзерный tactical laser radar
~, трёхмéрный лáзерный three-dimensional laser radar
~ управлéния огнём, лáзерный fire-control lidar, fire-control laser radar
локáторщик detector
локомотúв locomotive
~, дúзель-электрúческий diesel-electric locomotive
локсодрóмия rhumb line (navig)
лом crowbar
~, лáпчатый crowbar
ломáть/сломáть to break
лóмкость brittleness
лонжерóн 1. longeron, spar 2. side member (veh chassis)
лóпасть blade
лопáта shovel
~ для окóпных рабóт entrenching shovel
~, механúческая power shovel, mechanical shovel, tractor shovel

лопа́та, обра́тная back-acter, drag shovel
~, прямá́я face shovel (excavator)
~, сапёрная sapper shovel
лопа́тка 1. shovel, trowel
2. blade, vane (turbine)
3. cleat (tracked vehicle)
4. shoulderblade
~, рабо́чая moving turbine blade
лот sounding rod
лото́к 1. chute, open top drain
2. bomb carrier (mortar)
3. loading tray, ammunition tray (ord)
~ для патро́нов, направля́ющий cartridge guide (SA)
~, заря́дный loading tray
~, затво́рный carrier tray (ord)
~ кли́на loading trough (ord)
~, откидно́й pivoting loading tray (ord)
~, погру́зочный loading tray, charging tray
~ приёмника cartridge guide (SA)
лото́к-досыла́тель carrier (ord)
ло́ция sailing directions (book)
ло́шадь, вью́чная pack-horse
лощи́на hollow, dell
лубрика́тор lubricator
луг, альпи́йский alpine meadow
лук bow (wpn)
лука́ bend (river, road)
лу́нка hole
лу́па magnifying glass
"Луч!" "Switch On Searchlight!"
луч beam, ray
~ анте́нны antenna beam
~, бегу́щий scanning beam
~, земно́й ground wave (radio)
~ зре́ния line of sight
~, инфракра́сный infra-red ray
"луч, каранда́шный" pencil ray, pencil beam
луч, косми́ческий cosmic ray
~ ла́зера laser beam
~, модули́рованный modulated beam
~ наблюде́ния line of vision, line of sight
~ наведе́ния guidance beam
~ наведе́ния раке́ты missile-guidance beam
~, наводя́щий guide beam, guidance beam (radar)
~, накло́нный slant antenna
~, напра́вленный directed beam
луч, остронапра́вленный narrow beam, sharp beam, pencil beam
~, о́стрый narrow beam, sharp beam, pencil beam
~, отклонённый deflected beam
~, отражённый reflected beam
~, пове́рхностный ground wave (radar)
~, прижа́тый low-altitude beam
~, простра́нственный skywave, indirect ray (radio)
~ радиолока́тора radar beam
~, радиолокацио́нный radar beam
~, рентге́новый X-ray
~, скани́рующий scanning beam
~, сопровожда́ющий tracking beam
~ сопровожде́ния це́ли target tracking beam
~, у́зкий tight beam, narrow beam
~, ультрафиоле́товый ultraviolet ray
~ управле́ния, широ́кий gathering beam
~, управля́ющий guide beam, guidance beam
~, широ́кий wide beam
~, электро́нный electron beam
лучево́й radial, radiation
лучеиспуска́ние radiation
лучи́, инфракра́сные infra-red rays
~, рентге́новские X-rays
лы́жа ski
~, опо́рная skid, "idiot shoe" (bulldozer blade)
лы́жи trammels (for securing heavy drop equipment)
лы́жки, специа́льные special skis
лы́жник skier
лыжня́ ski track, ski run, ski trail
лы́сина bald patch
лы́ска flat (ord)
льго́та privilege
льди́на ice floe
льдобето́н ice concrete
льдобу́р ice auger
лю́ди 1. people
2. men
3. personnel
люизи́т Lewisite
люк hatch
~, авари́йный escape hatch
~, ба́шенный turret hatch
~, бронево́й armoured hatch
~ в дни́ще та́нка, деса́нтный bottom escape hatch (AFV)
~, вентиляцио́нный ventilator hatch

люк, враща́ющийся revolving hatch
~, входно́й access hatch
~, выходно́й escape hatch
~, грузово́й loading hatch
~, деса́нтный escape hatch (tk)
~ для осмо́тра inspection hatch
~ для чи́стки access plate (AFV hull floor)
~, запа́сный выходно́й emergency exit
~ ка́ртера, вентиляцио́нный crankcase breather
~ команди́ра commander's hatch (AFV)
~ (меха́ника-) води́теля driver's hatch (AFV)
~, мото́рный engine access hatch
~, ни́жний bottom escape hatch (AFV)
~, подмото́рный engine access hatch (from underneath)
~, смотрово́й inspection hole, inspection hatch
~, туре́льный turret hatch
лю́ки для осмо́тра дви́гателя и трансми́ссии engine and transmission access hatches/plates
люк-лаз small hatch
люкс lux (photometry)
лю́лька 1. breech casing (QF ord) 2. cradle (ord)
~, коры́тная trough-type cradle (ord)
~, корытообра́зная trough-type cradle (ord)
~, обо́йменная tube-type cradle (ord)
лю́лька-обо́йма cradle brass bearing ring
лю́мен lumen (photometry)
люминесце́нция luminescence
люминофо́р luminescent compound
люне́т steady, steady rest, lunette
люфт back lash, play, back play
~, боково́й side play
~, осево́й end play
~ руля́ steering backlash
лю́чек, смотрово́й visor (tk)
лязг clank, clang
ля́мка strap, band, drag rope
~, гла́вная кругова́я main strap (of training type parachute harness)
~ для перено́ски carrying sling
~, носи́лочная stretcher sling, stretcher strap
~ парашю́та harness suspension strap (parachute)
~ парашю́та, вытяжна́я static line (para)
ля́мка парашю́та, подвесна́я parachute-harness strap
~, плечева́я shoulder strap
~, подвесна́я harness suspension strap (para)

М

"М", число́ Mach number
магази́н 1. magazine (SA)
~, бараба́нный drum magazine
~, вое́нный canteen, shop
~, двухря́дный double line magazine, double-row magazine
~, ди́сковый circular magazine drum, disc magazine
~, доба́вочный secondary magazine
~ ёмкостью десяти́ патро́нов ten-round magazine
~, запасно́й spare magazine
~, коро́бчатый box magazine
~, кру́глый round magazine
~, неснаряжённый empty magazine
~, подство́льный tube magazine
~, пятизаря́дный five-shot magazine
~, снаряжённый charged magazine (SA)
~, тру́бчатый tubular magazine
~, цилиндри́ческий drum magazine
магистра́ль 1. main line (rly) 2. main road, trunk road 3. main lane (air) 4. main (water etc)
~, во́дная main waterway
~, водопрово́дная water main
~, водяна́я fire main (nav, naut)
~, вое́нная military trunk road
~, высоково́льтная high voltage main line (sigs)
~, га́зовая gas main, gas pipeline
~, пита́ющая supply line
~ подво́да во́здуха air intake pipe
~, то́пливная fuel line
~, тормозна́я brake line
~ трубопрово́да, гла́вная main pipeline
магнети́зм magnetism
~, оста́точный residual magnetism
магне́то magneto
~, четырёхпо́люсное dual magneto
магнетро́н magnetron

магнетро́н бегу́щей волны́ travelling wave magnetron
~ без свя́зок unstrapped magnetron
~, коаксиа́льный coaxial magnetron
~, многока́мерный multicavity magnetron
~, многорезона́торный multicavity magnetron, multiresonator magnetron
магне́трон, настра́иваемый tunable magnetron
магнетро́н, пакети́рованный packaged magnetron
~, паке́тный packaged magnetron
~, разнопо́лостный rising sun magnetron (electron)
~, разнорезона́торный rising sun magnetron (electron)
~, разрезно́й split anode magnetron (electron)
~, резона́торный resonant magnetron
~ с подстро́йки tunable magnetron
~ с разрезны́м ано́дом split anode magnetron
~ с резона́торами ти́па отве́рстие с ще́лью hole-and-slot magnetron
ма́гний magnesium
магни́т magnet
~, подковообра́зный horseshoe magnet
~, постоя́нный permanent magnet
~, следя́щий follow-up magnet, servo magnet
магнитоме́р magnetometer
магнитостати́ческий magnetostatic
магнитострикцио́нный magnetostriction
магнитостри́кция magnetostriction
магнитофо́н magnetic recorder, magnetic tape, recorder, tape recorder
~, двухдоро́жечный double-track tape recorder
~, многокана́льный multi-channel recorder
магни́т-уничтожи́тель magnetic compass corrector
ма́зер, аммиа́чный ammonia maser
~, га́зовый gaseous maser, gas-phase maser
~, ге́лиево-нео́новый helium-neon maser
~ на криста́лле руби́на ruby maser
~, руби́новый ruby maser
мазо́к droplet
мазу́т black oil, petroleum residue, fuel oil
мазу́т, фло́тский furnace fuel oil, FFO
мазу́т-то́пливо, нефтяно́е fuel oil
мазь, амуни́чная dubbin, saddle soap
~, защи́тная protective ointment, anti-gas ointment
~, колёсная axle grease
~, лы́жная ski wax
ма́йна gap in ice (usually man-made)
майо́р major
мака́дам, биту́мный bituminous macadam
макаро́нник extended serviceman (sl)
маке́т dummy, mock-up, model
~ в натура́льную величину́ full-scale mock-up
~ ме́стности model of the ground
~ самолёта aircraft mock-up
~ синопти́ческих сообще́ний pattern for synoptic messages for computer print out
~ та́нка a dummy tank
~ та́нка, слу́жащий мише́нью representative tank target
максвелме́тр fluxmeter
ма́ксимум слы́шимости сигна́лов maximum signal strength
малогабари́тный small-sized
малокали́берный small-calibre
малолитра́жка small car (engine capacity Ø.8 to 2 litres)
маломо́щный low power, fractional horse power
маляри́я malaria
манёвр manoeuvre
~ авиа́ции, аэродро́мный move of air force unit to another base
~ в глубину́ manoeuvre in depth
~ и ого́нь fire and movement
~, ло́жный feint, demonstration
~, маскиро́вочный deception manoeuvre
~, обхо́дный turning manoeuvre
~, огнево́й fire flexibility
~ огневы́ми сре́дствами manoeuvring of fire power, adjustment of fire
~ огнём adjustment of fire
~, отвлека́ющий diversionary action, feint manoeuvre, evasive manoeuvre
~, охва́тывающий enveloping manoeuvre
~ по фро́нту lateral movement
~, противозени́тный evasive action, avoiding action (by ac against AA)

манёвр, противоистреби́тельный evasive action, avoiding action (by ac against fighters)
~, программи́рованный programmed manoeuvre (GW)
~, широ́кий wide manoeuvre
манёвренность mobility, manoeuvrability
~ на пересечённой ме́стности cross-country manoeuvrability
~, такти́ческая tactical movement
манёвренный fluid, mobile
маневри́рование manoeuvring
маневри́ровать to manoeuvre
маневроспосо́бность manoeuvrability
манёвры manoeuvres, field exercise
~, ре́зкие accentuated manoeuvring (aeron)
мане́рка canteen (drinking vessel)
мане́тка lever, button
манже́та 1. packing ring, packing collar
2. mechanism holder (gren)
3. cuff (shirt)
~, рези́новая rubber sealing ring, packing ring
~, са́льниковая packing gland
~, уголко́вая cup packing
~, уплотня́ющая seal, gland
~, ча́шечная cup packing
манипулирова́ть to manipulate
манипуля́тор 1. manipulator
2. keyboard (baudot)
3. key (morse)
4. sub-modulator (radar)
~, тона́льный voice frequency telegraph modulator
манипуля́ция 1. keying (comms)
2. manipulating
~, двухтона́льная two-tone keying (electron)
~, дистанцио́нная remote keying
~ сдви́гом частоты́ frequency shift keying
манифе́ст, грузово́й manifest, ship cargo manifest, load manifest
мано́метр manometer, pressure gauge
~, автоши́нный tyre pressure gauge
~, возду́шный air pressure gauge, manometer
~, ма́сляный oil pressure gauge
~ нака́тника recuperator gauge (gunn)
~, пружи́нный spring pressure gauge
~, то́пливный fuel pressure gauge
мано́метр, цифербла́тный dial pressure gauge
~, ши́нный tyre pressure gauge
марафе́т "bull" (unnecessary polishing etc)
ма́рганец manganese
Марго́лин Margolin (Sov small-bore pistol)
ма́рка 1. mark
2. make (eqpt, veh)
~, азимута́льная azimuth mark
~, визи́рная sighting mark
~, почто́вая postage stamp
~, прице́льная aiming mark (on sight graticule)
~ то́плива fuel specification
~, устано́вочная adjusting mark
ма́ркер marker (beacon)
маркиро́вка marking
~ боеприпа́сов ammunition marking
~, опознава́тельная identifying marking, identification marking
~ це́ли target marking
маркиро́вщик marker (member of minefield marking team)
ма́рля gauze
мародёр looter
мародёрство looting
марчева́нка spiling, spiling board (mining)
марш march, move
~ вдоль фро́нта lateral march
~, комбини́рованный march partly on foot and partly by MT
~ на сближе́ние approach march
~, стреми́тельный forced march
~, торже́ственный 1. ceremonial march
2. march past
~, форси́рованный forced march
ма́ршал Marshal
~ авиа́ции Marshal of the Air Force
~, гла́вный Chief Marshal of Arm (Sov)
~ ро́да во́йск Marshal of Arm, Marshal of Service (arty, tks etc) (Sov)
Ма́ршал Сове́тского Сою́за Marshal of the Soviet Union
марш-бросо́к short forced march
ма́ршевый 1. pertaining to march
2. draft, reserve (of troops)
3. sustainer (of missile engine)
4. sustained
марширова́ть to march
марш-подхо́д approach march

маршру́т route, plan of march, path, itinerary
~, вариа́нтный alternate route, alternative route
~, второстепе́нный alternate route, alternative route
~ выдвиже́ния route of advance
~ движе́ния traffic route
~, за́мкнутый 1. circular march route (A to A via B), out-and-back run 2. shuttle freight train (rly)
~, запасно́й alternative routing
~, кольцево́й 1. circular route 2. shuttle freight train
~, ло́манный interrupted march
~, обходно́й detour route
~ перево́зки timetable, itinerary
~ по гла́вному пути́ main route
~, предпочти́тельный preferred route
~, сквозно́й through route
маршрутиза́ция through-freight organisation
ма́ска 1. screen, blind 2. mantlet (AFV) 3. facepiece (head-harness-type respirator)
~, автомоби́льная vehicle camouflage net
~, артиллери́йская artillery concealment set (cam nets)
~, больша́я автомоби́льная large vehicle camouflage net
~, вертика́льная vertical camouflage net
~, горизонта́льная flat overhead camouflage net
~, группова́я collective concealment, collective screen
~, зо́нтичная umbrella cover
~, ма́лая автомоби́льная small vehicle camouflage net
~, миномётная mortar concealment set (cam nets)
~, наддоро́жная overhead camouflage above road, flat top above road
~, накло́нная draped camouflage net
~ от возду́шного наблюде́ния overhead screen, overhead camouflage screen
~, придоро́жная vertical camouflage screen along roadside
~, расти́тельная natural camouflage
~, та́нковая tank camouflage net
ма́ска-гирля́нда garland camouflage
ма́ска-перекры́тие overall camouflage net
~, пло́ская flat net covering, flat top (cam)
ма́ска-экра́н vertical screen camouflage net
маскиму́щество camouflage material
маскирова́ть/замаскирова́ть to camouflage
маскиро́вка camouflage
~, возду́шная concealment against air observation, concealment from the air
~ войск field camouflage
~, декорати́вная disguise camouflage
~, дымова́я smoke screening
~, есте́ственная natural camouflage
~, звукова́я sound masking
~, индивидуа́льная individual camouflage
~, инфракра́сная infra-red camouflage, IR camouflage
~ искаже́нием фо́рмы camouflage by disruption of form
~ крыш маскиро́вочными сетя́ми roof netting, camouflage net
~ на по́ле бо́я field camouflage
~, операти́вная surprise and deception
~, опти́ческая visual camouflage
~ от визуа́льного наблюде́ния visual deception
~ от наблюде́ния в инфракра́сных луча́х infra-red camouflage, IR camouflage, infra-red masking, IR masking
~, полева́я concealment in the field
~, постоя́нная fixed camouflage
~, противолокацио́нная anti-radar camouflage
~, противорадиолокацио́нная radar camouflage
~, пятни́стая partial camouflage
~, радиолокацио́нная radar security and deception
~, светова́я blackout, blacking out (of windows etc for camouflage)
~, структу́рная structural camouflage
~, техни́ческая artificial camouflage
маскиро́вочный camouflage
маскоде́жда camouflage clothing
масксе́ть camouflage net
маскхала́т denison smock, camouflage robe

маслёнка oil bottle, oil cup, oil hole, oil plug, oil can, grease nipple, lubricator
~, двугóрлая oil bottle with two compartments
~, игóльчатая needle lubricator
~, кáпельная drip feed lubricator
~, колпачкóвая grease cup, lubricating screw
~, подýшечная pad lubricator
~, ручнáя hand lubricator, oil can
~, фити́льная capillary oiler, wick feed lubricator
~, шприцевáя oil nipple
мáсло oil
~, авиациóнное aviation oil, aviation lubrication oil
~, автомоби́льное automobile engine oil
~, автомоби́льное смáзочное automobile engine oil
~, антикоррозиóнное rust preventive oil, corrosion inhibiting oil, slushing oil
~, веретённое spindle oil
~, водораствори́мое water soluble oil
~, всесезóнное multigrade oil
~, гидравли́ческое hydraulic fluid
~, гипóидное hypoid oil
~, ди́зельное diesel engine oil
~ для амортизáторов shock absorber oil
~ для вéрхней смáзки цили́ндров upper-cylinder lubricant
~ для корóбок передáч высóкого давлéния extreme pressure transmission oil
~ для промы́вки flushing oil
~, живóтное animal oil
~, загрязнённое dirty oil
~, загущённое смáзочное multiviscosity oil
~, защи́тное 1. slushing oil 2. preservative oil
~, индустриáльное industrial oil
~, кастóровое castor oil
~, костянóе bone oil, neat's foot oil
~, лёгкое смáзочное light lubricating oil
~, льнянóе linseed oil
~, маловя́зкое low-viscosity oil
~, минерáльное смáзочное mineral lubricating oil
~, мóющее detergent oil
~, оружéйное gun oil, rifle oil
~, осевóе axle oil
~, охлаждáющее cooling oil
мáсло, очи́щенное refined oil
~, прибóрное instrument oil
~, промы́вочное flushing oil
~, расти́тельное vegetable oil
~, регенери́рованное reclaimed oil
~, ружéйное rifle oil
~, синтети́ческое synthetic oil
~, смáзочное lubricating oil
~, соля́ровое diesel oil, solar oil
~, спермацéтовое sperm oil, whale oil
~, средневя́зкое medium viscosity oil
~, терми́чески стаби́льное thermally-stable oil
~, товáрное commercial oil
~, трансмиссиóнное gear oil, transmission oil
~, тяжёлое смáзочное heavy lubricating oil
~, универсáльное автомоби́льное multi-grade automobile engine oil
маслобáк oil tank
маслогрéйка oil preheater
маслозапрáвщик oil bowser
масломéр oil level gauge
маслонепроницáемый oil-tight, oil-proof
маслообрáзный oil-like, oily
маслоотдели́тель oil separator
маслоотражáтель oilflinger, oil thrower, lubricant retainer
маслоотсáсывание oil suction
маслоотстóйник oil sump
маслоохлади́тель oil cooler
маслоочисти́тель oil cleaner
~, центробéжный centrifugal oil cleaner
маслоподводя́щий oil-feed, oil-supply
маслоподогревáтель oil heater
маслопóмпа, нагнетáтельная oil pressure pump
маслопровóд oil pipe
маслопроницáемость oil permeability
маслорадиáтор oil cooler
маслораспределéние oil distribution
маслораствори́мость oil solubility
маслосбóрник oil collector, drain pan, drip pan
маслосистéма lubrication (oil) system
маслостóйкий oil-resisting, oil-resistant
маслостóйкость oil-resistance
маслосчётчик oil meter
маслотопливозапрáвщик combination oil and fuel servicing truck
маслоуказáтель oil-level indicator, oil gauge

маслоуловитель oil trap, oil catcher, oil dish, oil drip pan, oil separator
маслофильтр oil filter
маслощуп oil gauge, oil level gauge, oil dipstick, oil measuring rod
маслоэкстрактор oil extractor
маслянистость oiliness, lubricating property
маслянистый oily
масляно-пневматический oleo-pneumatic
масляный oil(y), greasy
масса, докритическая sub-critical mass
~, кабельная cable sealing compound
~, конечная final mass (missile)
~, критическая critical mass (nucl)
~, начальная initial mass (missile)
~ откатных частей mass of recoiling parts (gunn)
~, пломбирная sealing compound
~, подкритическая subcritical mass (nucl)
~, предельная critical mass (nucl)
~, пропиточная impregnating compound
~, сверхкритическая supercritical mass (nucl)
~, стартовая firing weight, initial total mass, all-up weight (missile)
массив large tract of territory, massif
~, горный mountainous area, hilly area
~, лесисто-болотистый terrain of forest and swamp
~, лесной forest, wooded area
массирование massing
~ сил и средств massing of forces and equipment
массированный massed, massive
массовый mass, en masse
масс-спектрометр mass-spectrometer
мастер master (qualification)
~, артиллерийский master gunner
~ вождения master driver
~, оружейный armourer, artificer, gunsmith
~, старший оружейный master armourer
~, химический chemical warfare specialist, CW specialist
мастерская workshop, small workshop
~, авторемонтная MT repair workshop
~, армейская артиллерийская ремонтная army artillery repair workshop
мастерская, армейская ремонтная army repair workshop (at army level)
~, артиллерийская artillery workshop
~, базовая артиллерийско- техническая base ordnance workshop
~, базовая ремонтно-восстановительная base workshop
~, бронетанковая ремонтная AFV repair workshop
~, вещевая tailor's and cobbler's workshop
~, войсковая автомобильная ремонтно-эксплуатационная unit MT repair truck
~, заточная tool-grinding workshop
~, кузнечная forge, forge shop
~, механическая machine shop
~, монтажная assembly workshops, fitting shop
~ обозно-вещевого снабжения clothing and equipment supply workshop
~, обувная boot repair shop, cobbler's shop
~, оружейная armourer's shop, small-arms workshop
~, подвижная truck-mounted workshop. workshop truck
~, подвижная ремонтная 1. light aid detachment, LAD
2. mobile repair workshop
~, подвижная сварочная mobile welding workshop
~, подвижная танкоремонтная mobile tank repair workshop
~, полевая ремонтная field maintenance workshop, field repair workshop
~, портняжная tailor's shop
~, походная mobile repair shop
~, радиоремонтная radio repair workshop
~ ремонта бронетанковой техники armoured vehicle repair workshop, armoured workshop
~, ремонтная repair workshop
~, сборочная assembly shop
~, стационарная ремонтная static repair workshop, static repair installation
~, танко-ремонтная tank repair workshop, armoured workshop, tank repair workshop lorry
~ технического обслуживания workshop for technical servicing

мастерска́я, упако́вочная packing area
~, фронтова́я ремо́нтная army group repair workshop
~, центра́льная ремо́нтная central repair workshop
~, шо́рно-седе́льная harness shop
мастерска́я-лету́чка mobile repair workshop
мастерство́, боево́е battlecraft, battle skills, combat skills
~, во́инское military skill
~, огнево́е skilled gun-handling
~, стрелко́вое marksmanship, skill at arms
масти́ка mastic, cement, paste
масшта́б scale (map etc)
~ в ми́лях mile scale
~ вре́мени time base (sound ranging)
~ да́льности range scale
~ ка́рты map scale
~, кру́пный large scale
~, лине́йный drawn scale, graphic scale, linear scale (map)
~, ме́лкий small scale
~, метри́ческий metric scale
~, основно́й basic scale
~, переме́нный non-uniform scale
~, попере́чный interpolating scale (surv)
~, постоя́нный uniform scale
~, пропорциона́льный proportional scale
~, рабо́чий working scale
~, чи́сленный representative fraction (map scale etc)
~ шаго́в pace scale
мат, жердево́й corduroy mat
~ из хво́роста brushwood mat
~, плетёный brushwood mat
материа́л, взры́вчатый explosive material
~, вое́нный warlike stores
~, воспроизводя́щий fertile material (nucl)
~, горю́чий combustible material, fuel
~ для доро́жной оде́жды, ка́менный road metal
~ для покры́тия доро́ги paving material
~, жаропро́чный heat-proof material
~, зажига́тельный flammable material
~, защи́тный shielding material
~ като́да, норма́льный normal cathode material
~ като́да, пасси́вный passive cathode material
материа́л, невоспламеня́ющийся non-flammable material
~, обти́рочный cotton waste, rags
~, перевя́зочный dressings, bandages
~, подру́чный locally available material, improvised material
~, проти́рочный cleaning material, bore cleaning material (wpn)
~, радиоакти́вный radioactive material
~, радиопоглоща́ющий radar absorbent material
~, расхо́дный expendable material
~, ремо́нтный repair material(s)
~, слои́стый laminated material
~, сма́зочно-горю́чий petroleum oils and lubricants, POL
~, сма́зочный lubricant, lubricating material
~, строи́тельный building material
~, теплоизоли́рующий lagging, heat insulation material
~, теплоизоляцио́нный heat-insulating material
~, то́пливный fuel, combustible material
материа́лы, авиацио́нные сма́зочные aviation lubricants and greases
~, агенту́рные information from covert sources
~, амортизи́рующие упако́вочные shock absorbing materials, cushioning materials
~, вое́нные munitions
~, горю́че-сма́зочные petrol, oil and lubricants, POL
~ и обору́дования для оборони́тельных рабо́т defensive supplies and equipment, defence stores
~, кондицио́нные standard materials (up to standard)
~, некондицио́нные substandard materials
~, перепра́вочные water crossing aids, wet-gap crossing aids
~, подру́чные makeshift materials, materials to hand, improvised materials
~, сма́зочные lubricants
мате́рия, прорези́ненная rubberized fabric
ма́тка housing (mech)
~, зажи́мная elevating gear tumbler (MMG)
матра́сник idler, loafer (Brit "blanket presser") (sl)

ма́трица matrix, pattern, jig, stencil, die
матча́сть materiel, equipment, technical equipment
мах Mach
махме́тр machmeter
махови́к 1. handwheel
2. flywheel
3. rotor (gyroscope)
~ бараба́нчика прице́ла sight deflection knob
~ буссо́ли elevation micrometer drum (ord)
~ вво́да да́льности range setting handwheel (ord)
~ вво́да ско́рости це́ли target speed setting handwheel (ord)
~ вво́да угло́в пики́рования и кабри́рования angle of dive setting handwheel (ord)
~ вертика́льного наклоне́ния range handwheel (naval gunn), elevation handwheel (ord)
~ горизонта́льного наклоне́ния deflection handwheel (naval gunn)
~ горизонта́льной и вертика́льной корректу́р deflection and elevation correction knob, line and range correction knob
~ дви́гателя engine flywheel
~ для совмеще́ния ни́тей на паралла́ксе parallax correction knob
~ поворо́тного механи́зма traversing grip, traversing handwheel (ord)
~ подъёмного механи́зма elevating handwheel (ord)
~ попра́вки взрыва́теля fuse setting lever
~ совмеще́ния rate handwheel (AAFCE)
~, устано́вочный setting knob
маховичо́к knob
~ вертика́льной наво́дки elevating knob
~ вертика́льной регулиро́вки elevating knob
~ винта́ це́лика backsight windgauge setting knob
~ для устано́вки боковы́х попра́вок deflection drum (sight)
~ для устано́вки угло́в прице́ливания range drum (sight)
~ коро́бки прице́ла cross-levelling handwheel (ord)
~ подъёмного механи́зма прице́ла range handwheel (ord)

ма́чта mast (radio, ship)
~, выдвижна́я telescopic mast
~, радиолокацио́нная radar mast
~, телескопи́ческая telescopic mast
маши́на machine, vehicle
~, авари́йная recovery vehicle
~, автодегазацио́нная bulk decontamination vehicle (CW)
~, администрати́вно-хозя́йственная administrative vehicle
~, анало́говая вычисли́тельная analogue computer
~, аппара́тная 1. radio vehicle
2. operating truck (radar)
~ артиллери́йского снабже́ния ammunition supply vehicle
~, ба́зовая basic vehicle (on to which various bodies, equipment etc may be fitted)
~, ба́шенная turreted vehicle
~, боева́я 1. fighting vehicle
2. armoured fighting vehicle, AFV (Sov term covers all armed vehs, APCs, arty prime movers, sigs and radar vehs)
~, боева́я брониро́ванная armoured fighting vehicle, (AFV)
~, боева́я разве́дывательная дозо́рная armoured reconnaissance vehicle, BRDM (Sov armed recce vehicle)
~, боева́я реакти́вная rocket launcher vehicle, vehicle mounted rocket launcher
~, боева́я хими́ческая CW contamination vehicle
~, брониро́ванная armoured vehicle
~, брониро́ванная ремо́нто-эвакуацио́нная armoured recovery vehicle, ARV
~, бури́льно-кра́новая auger
~, бури́льно-кра́новая грузоподъёмная auger
~, вездехо́дная cross-country vehicle
~, вое́нная military vehicle
~, возду́шно-деса́нтная боева́я airborne assault vehicle
~, вспомога́тельная 1. auxiliary vehicle (Sov classification - same as "transport veh" - includes staff cars, ambulances & all vehicles used for carrying pers & stores other than ammo)
2. auxiliary engine
~ высо́кой проходи́мости high mobility vehicle

маши́на, вычисли́тельная 1. computer
2. calculating machine
~, вы́шедшая из стро́я immobilized vehicle
~, головна́я leading vehicle (in fmn)
~, головна́я дозо́рная forward patrol vehicle
~, грузова́я lorry, load-carrying vehicle, cargo truck
~, грузоподъёмная hoist, hoisting machine, lifting machine
~, грунтостабилизи́рующая soil stabilization plant
~, гу́сеничная tracked vehicle
~, дегазацио́нная CW decontamination vehicle
~, дискре́тная digital computer
~ дискре́тного де́йствия, вычисли́тельная digital computer
~ дискре́тного счёта, вычисли́тельная digital computer
~ для бездоро́жья cross-country vehicle
~ для перево́зки горю́чего fuel vehicle
~ для раскла́дки мин mine-laying vehicle
~ для эвакуа́ции повреждённой те́хники recovery vehicle
~, дозо́рная 1. scout vehicle, point vehicle (tac)
2. recce vehicle
~, землеро́йная earth-moving equipment, earth-moving plant
~, землеро́йно-тра́нспортная earth-moving plant
~, инжене́рная engineer plant
~, испра́вная serviceable vehicle
~ кла́сса "А" A vehicle (AFV, Brit)
~, колёсная wheeled vehicle
~, колёсно-гу́сеничная half-track vehicle
~, коло́дезная well-sinking machine
~, команди́рская command vehicle
~, команди́рская брониро́ванная armoured command vehicle, ACV
~, комбини́рованная поли́во-мо́ечная mechanical sweeper (runway clearance)
~, конденса́торная подрывна́я dynamo condenser exploder
~, корчева́льная rooter (eng)
~, лёгкая lightweight vehicle
~ лёгкого ве́са lightweight vehicle
~, лесопи́льная mechanical saw, power saw

маши́на, математи́ческая computer
~, моделі́рующая analogue computer
~, мо́ечная washing-down vehicle (CW)
~ на гу́сеничном ходу́ tracked vehicle
~ на колёсном ходу́ wheeled vehicle
~ на твёрдых схе́мах, вычисли́тельная solid-state computer
~, небронирóванная un-armoured vehicle
~ непреры́вного де́йствия, вычисли́тельная analogue computer
~ непреры́вного де́йствия analogue computer
~, норма́льная электри́ческая general-purpose electric motor (elec)
~, обору́дованная ба́шней turreted vehicle
~ о́бщего назначе́ния, вое́нная general-purpose military vehicle
~, перепра́вочно-деса́нтная assault crossing vehicle, assault amphibian
~, перфорацио́нная punch card machine
~ пехо́ты, боева́я mechanized infantry combat vehicle
~, пи́шущая typewriter
~, пла́вающая amphibious vehicle, amphibian
~, пла́вающая гу́сеничная tracked amphibian, landing vehicle tracked, LVT
~ повы́шенной проходи́мости cross-country vehicle
~, подби́тая vehicle casualty, damaged vehicle
~, подрывна́я demolition exploder
~, пожа́рная fire engine, fire tender
~, полугу́сеничная half-track vehicle
~, потерпе́вшая ава́рию immobilized vehicle
~, разве́дывательная reconnaissance vehicle
~ реакти́вной артилле́рии, боева́я rocket launcher vehicle, vehicle-mounted rocket launcher
~, ремо́нтная breakdown vehicle
~, ремо́нтно-восстанови́тельная recovery vehicle
~ с боеприпа́сами, грузова́я ammunition truck
~ с лёгкой бронево́й защи́той lightly armoured vehicle
~ с приво́дом на все колёса all-wheel drive vehicle

маши́на, самохо́дная перепра́вочно-деса́нтная assault crossing vehicle, assault amphibian, amphibious landing vehicle
~, самохо́дная тра́нспортно-заряжа́ющая self-propelled loader (missiles)
~, санита́рная medical vehicle, ambulance
~, сваебо́йная mechanical piledriver
~ снабже́ния боеприпа́сами ammunition supply vehicle
~, снегоубо́рочная snowclearing vehicle
~, снегохо́дная тра́нспортная over-snow vehicle
~ со всеми́ веду́щими колёсами all-wheel drive vehicle
~, сочленённая articulated vehicle
~, строева́я "operational support vehicle" (Sov vehicle classification-includes ammo vehs, wpn team carriers, HQ vehs, ARVs and wksp vehs)
~, счётная calculating machine
~, танкоподо́бная tank-like vehicle
~ техни́ческого обслу́живания maintenance vehicle
~ техни́ческой по́мощи recovery vehicle
~, тра́нспортная transport vehicle (Sov term includes B vehs and ARVs)
~, тра́нспортно-заряжа́ющая transporter-loader vehicle (msl)
~, транше́йная trench-digger, trencher, entrenching machine
~, универса́льная инжене́рная самохо́дная all-purpose engineer vehicle
~, уче́бная training vehicle
~, уче́бно-боева́я "operational training vehicle" (Sov classification - vehs earmarked for tac trg use)
~, уче́бно-строева́я "training support vehicle" (Sov veh classification of "operational support vehs" used for trg)
~, хими́ческая bulk decontamination vehicle
~, цифрова́я digital computer
~, цифрова́я вычисли́тельная digital computer
~, штабна́я staff vehicle, command vehicle
~, электро́нная вычисли́тельная electronic computer
маши́на, электро́нная цифрова́я electronic digital computer
маши́на-ана́лог anologue computer
машини́ст 1. engine driver (rly) 2. person working or controlling machine
машини́стка female typist
маши́нка 1. typewriter 2. small machine
~, взрывна́я electric exploder
~ для снаряже́ния магази́на magazine filler
~ для снаряже́ния патро́нной ле́нты belt-filling machine
~ для снаряже́ния патро́нов cartridge loader, cartridge loading- machine
~, конденса́торная подрывна́я dynamo condenser exploder
~, пи́шущая typewriter
~, подрывна́я demolition exploder
~, рулева́я 1. small steering engine 2. control surface servo, control actuator, wing actuating servo- mechanism, wing actuator assembly (missile etc)
~, шифрова́льная cypher machine
маши́ны, инжене́рно-строи́тельные engineer construction plant
~, инжене́рные engineering plant
~, снегоубо́рочные snow-removal equipment
мая́к marker, beacon
~, глисса́дный glide-path transmitter (airfield radio beacon)
~, приводно́й homer, homing station, homing beacon
~, радиолокацио́нный radar beacon
~, светово́й light beacon
мая́к-отве́тчик radar beacon, racon, responder
ма́ятник pendulum, bob
мгла dry haze (met)
мегаге́рц megahertz, megacycles per second
мегато́нна megaton
мегато́нный megaton
мегафо́н megaphone
ме́ггер megger (elec)
мего́м megohm
меда́ль medal
медве́дка cradle, truck carriage (arty)
медикаме́нты medicines, drugs
ме́дико-санита́рный medical, medical and sanitary

медици́на, авиацио́нная aviation medicine
медици́нский medical
медосмо́тр medical examination
медпо́мощь medical aid
медпу́нкт medical aid post
~, батальо́нный 1. battalion medical post 2. regimental aid post (Brit)
~, полково́й regimental aid post, RAP
медсанба́т medical battalion
медсестра́ nurse, nursing sister
медь copper
~, хло́ристая cuprous chloride
межгарнизо́нный inter-garrison
междуре́чье island in river, ait, eyot
меже́нь mean water level (river etc)
мезопа́уза mesopause
мезосфе́ра mesophere
мел chalk, whitening
меле́ть/обмеле́ть to become shallow (rivers etc)
мелини́т melinite
мелково́дье shallows
мелкозерни́стый fine (powder etc), fine-grain
мелкомасшта́бный small-scale
мелкосидя́щий light-draught, shallow-draught
мель shoal, sandbank
ме́льница mill
мельхио́р cupro-nickel
мельхиориза́ция nickelling (barrel fouling)
мембра́на 1. diaphragm 2. diaphragm, needle disc (fuse)
~, у́гольная carbon diaphragm
ме́нзула plane table, map board
"Ме́ньше!" "Down!" (nav gunn)
ме́ра measure, action
~ наказа́ния, вы́сшая capital punishment, supreme penalty
мерзлота́, ве́чная permafrost
меридиа́н, географи́ческий true meridian, geographical meridian
~, гринви́ческий Greenwich Meridian
~, и́стинный true meridian, geographical meridian
~, ко́мпасный compass meridian
~, магни́тный magnetic meridian
~, нача́льный initial meridian, base meridian, basic meridian
~, нулево́й zero meridian, Greenwich meridian
меридиа́н, опо́рный standard meridian, prime meridian
~, осево́й central meridian
мери́ло gauge
ме́рка, координа́тная romer, map grid
~, Л-обра́зная координа́тная L-shaped co-ordinate card
~, прямоуго́льная координа́тная rectangular co-ordinate card
~, целлуло́идная координа́тная transparent map grid, transparent romer, transparent map grid overlay
мероприя́тия, инжене́рные engineer activity, engineer operations, engineer effort
~, маскиро́вочные camouflage measures
~ по индивидуа́льной защи́те individual protective measures (NBC)
~ по предотвраще́нию пожа́ров fire prevention measures
~, предохрани́тельные precautionary measures, safety measures
~, противорадиолокацио́нные radar counter-measures, RCM
~, противохими́ческие anti-gas measures
~, профилакти́ческие 1. prophylaxis, prophylactic measures 2. preventive maintenance
~, санита́рно-профилакти́ческие 1. sanitary measures, sanitation 2. preventive medical measures
мерца́ние scintillation, flickering
ме́ры безопа́сности safety measures, safety precautions
~ борьбы́ с радиопротиводе́йствием electronic counter-countermeasures, ECCM
~ защи́ты, коллекти́вные unit protective measures
~ обеспе́чения protective measures, means of protection
~ по предотвраще́нию обледене́ния frost precautions
~ предосторо́жности precautions, safety precautions
~ принужде́ния measures of restraint, coercive measures, force
~ противоде́йствия active counter measures

ме́ры, санита́рно-профилакти́ческие 1. preventive medical measures 2. sanitary measures
~ те́хники безопа́сности safety measures
ме́стник local feature, permanent echo (radar)
ме́стность 1. locality, area 2. landscape, ground, terrain
~, боло́тистая boggy ground
~, всхолмлённая undulating ground
~, гори́стая mountainous terrain, mountainous country
~, го́рная mountainous terrain, mountainous country
~, госпо́дствующая dominating ground
~, заболо́ченная marshy ground
~, закры́тая close country
"МЕ́СТНОСТЬ ЗАРАЖЕНА́" "CONTAMINATED AREA"
ме́стность, заражённая contaminated area
~, кома́ндующая commanding ground
~, кры́тая close country
~, леси́стая wooded terrain
~, леси́сто-боло́тная marshy woodland
~, лесна́я wooded terrain
~, мини́рованная mined locality
~, непроходи́мая impassable terrain
~, неро́вная rough ground
~, пересечённая broken country, rough country
~, пока́тая sloping ground
~, полузакры́тая undulating ground, semi-close country
~, поспе́шно обору́дованная hastily prepared ground
~, проходи́мая passable terrain
~, пусты́нно-степна́я desert and steppe country
~, равни́нная flat terrain
~, резкопересечённая rugged countryside
~, си́льно-пересечённая difficult country
~, сла́бо пересечённая moderate terrain
~, сложнопересечённая heavily broken terrain
~, танкодосту́пная ground suitable for tanks, tankable ground
~, танконедосту́пная tank-proof terrain, ground unsuitable for tanks
~, труднодосту́пная adverse terrain
ме́стность, холми́стая rolling country
ме́стный local
ме́сто 1. place, spot, point, site 2. seat (tpt) 3. piece (of luggage, cargo) 4. post, station
~ ава́рии accident site, crash scene
~ бивака́ bivouac site
~, боево́е action station
~ встре́чи rendezvous point
~ вы́броски dropping zone
~ вы́броски деса́нта dropping zone
~ вы́грузки unloading zone
~, действи́тельное ground position (ac)
~ заса́ды ambush point
~, исхо́дное starting place
~ наво́дки мо́ста bridging site
~ назначе́ния destination
~ ночле́га overnight camp
~ остано́вки halt point
~ о́тдыха rest area
~ откры́того хране́ния dump
~, отхо́жее latrine
~ перепра́вы crossing place, site
~ постро́йки мо́ста bridging site
~ проры́ва area of breakthrough
~ пу́ска launch site (msl)
~, рабо́чее operating position
~ разруше́ния demolition site
~ расположе́ния location
~ расположе́ния, бива́чное bivouac camp site
~ расположе́ния та́нков tank harbour, leaguer
~ расположе́ния та́нков, подгото́вленное к кругово́й оборо́не tank leaguer (all round defence)
~ самолёта ground position (of aircraft)
~ сбо́ра assembly area, rally area
~, сбо́рное assembly point
~ свети́ла, географи́ческое sub-stellar point (navig)
~ скопле́ния concentration area
~ слу́жбы place of duty
~ сосредото́чения ра́неных casualty collecting post
~ стоя́нки 1. hardstanding 2. parking place
~ стоя́нки ору́дий gun park
~, танконедосту́пное tank-proof locality
~, уязви́мое vulnerable point
местонахожде́ние location, position
местоположе́ние location, position

месторасположéние location, position
метáлл, бéлый white metal (bearing alloy)
~, благорóдный precious metal
~ для подши́пников bearing metal
~, лёгкий light metal
~, основнóй base metal
~, тяжёлый base metal
~, цветнóй non-ferrous metal
~, чёрный ferrous metal
металлизáция 1. metal spray coating 2. bonding (elec)
металлолóм metal scrap, scrap metal
металлýргия, цветнáя non-ferrous metallurgy
~, чёрная ferrous metallurgy
метáльщик thrower (pers)
метáн methane
метáние гранáт grenade throwing
~ из положéния лёжа throwing from a lying position
~ с колéна throwing from a kneeling position, throwing from a crouched position
метанóл methanol
метаскóп metascope
метáть/метнýть to throw; **метáть/метнýть гранáту** to throw a grenade
метацéнтр metacentre
"метёлочка" artillery fire simulator (Sov)
метéль blizzard, snowstorm
~, низовáя blowing snow
мéтео-гóрный meteor telegram for mountain artillery
метеодáнные meteor data
мéтео-звýк meteor report (meteorological report for sound ranging artillery)
мéтео-зени́тный meteor report (meteorological report for AA artillery)
метеообстанóвка, факти́ческая actual conditions meteorological report, "actual" (AF)
мéтео-огневóй meteorological report for field artillery
метеорадиолокáтор meteorological radar, weather radar
метеорóлог meteorologist
метеоролóгия meteorology
~, авиациóнная air meteorology
метеосвóдка meteorological report, weather report
метеосрéдний meteor telegram for medium artillery
метеостáнция meteorological station, weather station
метилами́н methylamine
мéтка mark, sign, score, label
~ "верх" upper proximity dot (gyro gunsight) (ac)
~ дáльности range marker (instr)
~, измери́тельная measuring symbol (range)
~ "низ" lower proximity dot (gyro gunsight) (ac)
~, поперéчная cross line (bombsight), graticule
~, регулирóвочная adjustment mark
~ ýровня пóлной заправки full level mark
~, установóчная adjustment mark
мéткий accurate
мéткость accuracy
~, большáя pin-point accuracy
~, ни́зкая low accuracy
~, техни́чески допускáемая hit probability
метнýть (см метáть)
мéтод, аналити́ческий 1. analytic method 2. computation (surv)
~, баллисти́ческий ballistic method
~ искýсственного дыхáния "изо ртá в рот" mouth-to-mouth artificial respiration
~ испытáний testing method, procedure
~ испытáния, стандартизóванный standard testing method, procedure
~, исслéдовательский research method
~, когерéнтно-и́мпульсный coherent impulse method (GW) (radar)
~, контурнокомбини́рованный method of contouring from air photographs
~ наведéния guidance method (msl)
~ наведéния в мгновéнную упреждённую тóчку встрéчи lead angle method of approach guidance (GW)
~ наведéния по пересекáющемуся кýрсу deviated pursuit course guidance, fixed lead navigation (GW)
~ накры́тия line of sight system (missile guidance)
~ "нóжниц" "scissors method" (of missile guidance)
~ определéния октáнового числá octane rating method
~ охлаждéния method of cooling

ме́тод паралле́льного сближе́ния collision course method of approach (GW)
~ **пого́ни** zero bearing method, pure pursuit course method (homing guidance) (GW)
~ **программи́рованного наведе́ния** "programmed guidance method" (missile flies on a curved trajectory which satisfies one or several requirements at the same time) (Sov)
~ **произво́дства** production method, method of manufacture
~ **пропорциона́льного наведе́ния** proportional navigation method (GW)
~ **прямо́го наведе́ния** zero bearing guidance method (GW)
~ **равносигна́льной зо́ны** 1. lobe switching method, lobing (radar) 2. equisignal method, split method (DF)
~, **равносигна́льный** lobe-switching method (electron)
~ **ра́вных отклоне́ний** equal deflection method (DF)
~ **ремо́нта, агрега́тный** field repair method, assembly replacement method of repair, "field repair", unit replacement (as opposed to in situ repair of components)
~ **самонаведе́ния** homing method (msl)
~ **снаряже́ния, футля́рный** block method of filling (ammo)
~ **совмеще́ния** line of sight method (missile guidance)
~ **спрямле́ния траекто́рии** constant bearing navigation method, angle method of approach (missile guidance)
~, **стереоскопи́ческий** stereoscopic method (air photo)
~ **трёх то́чек** line-of-sight method (missile guidance)
~ **углово́го сближе́ния** "method of angular approach" (missile flies on a curved trajectory which satisfies one or several requirements at the same time) (Sov)
~ **управле́ния, интерце́пторный** bang-bang control method (GW)
мето́дика подгото́вки training methods, training technique
~ **рабо́ты** procedure
метроно́м metronome
метршто́к dipstick, measuring rod
ме́тчик tap, screw tap
механиза́ция mechanization
механизи́рованный mechanized
механизи́ровать to mechanise
механи́зм mechanism, device, working part, gear
~, **автомати́ческий подаю́щий** automatic feed mechanism (ord)
~ **автомати́ческого заряжа́ния** self-loading mechanism
~ **автомати́ческого спу́ска** automatic firing gear
~ **баллисти́ческих бараба́нов** drum gear, drum unit (AAFCE)
~ **ба́шни, поворо́тный** turret traversing mechanism
~ **без предупрежде́ния, спусково́й** single-pull trigger action
~, **блоки́рующий** locking device
~ **в сбо́ре, уда́рный** firing-pin assembly
~, **вертика́льно па́дающий клиново́й** vertical wedge type breech (ord)
~ **вертика́льного наведе́ния** elevating gear (ord)
~ **вертика́льной и горизонта́льной корректу́р** elevation and deflection correction mechanism
~ **вертика́льной стабилиза́ции** vertical stabilisation mechanism
~ **взаи́мной за́мкнутости** catch-and-release device
~ **взведе́ния** arming mechanism
~ **взрыва́теля** fuse mechanism
~ **"взрыв-невзры́в"** "armed-safe" mechanism
~, **взрывно́й** detonating mechanism
~ **включе́ния и выключе́ния** on-off mechanism
~ **возбужде́ния взры́ва** initiating mechanism
~, **возвра́тный** return mechanism
~, **воспламеня́ющий** striker mechanism (gren), igniter mechanism
~, **враща́ющийся** revolving gear
~ **враще́ния голо́вки-отража́теля** telescope head elevating gear (gunsight)
~, **временно́й** timing mechanism
~, **выбра́сывающий** extractor mechanism (SA, ord)
~ **выключе́ния** withdrawal mechanism, release mechanism
~, **выра́внивающий** 1. counterpoise (ord) 2. cross-levelling gear (mor)

механи́зм высоты́ height gear (AAFCE)
~ **газораспределе́ния** valve gear assembly (ICE)
~, **газораспредели́тельный** camshaft and valve gear, valve gear assembly (ICE)
~ **горизонта́льного наведе́ния** 1. traversing gear (arty) 2. training gear (nav gunn)
~ **горизонта́льной да́льности** ground range gear (AAFCE)
~ **горизонта́льной стабилиза́ции** horizontal stabilisation mechanism
~ **горизонти́рования** 1. levelling mechanism, equalising mechanism, equaliser 2. cross-levelling gear (mor)
~ **гу́сеницы, натяжно́й** track adjuster
~ **диста́нции** range mechanism
~ **дистанцио́нного де́йствия** time mechanism (arty)
~ **для регулиро́вки зазо́ра кла́пана** valve (clearance) adjuster
~, **заводно́й** clockwork mechanism
~ **заде́ржки** delay mechanism, time-delay mechanism
~, **зажимно́й** clamping device
~, **закрива́ющий** 1. closing mechanism 2. breech closing mechanism (ord)
~ **замедле́ния устано́вки взрыва́теля на боево́й взвод** delay arming mechanism
~, **замедля́ющий** delay mechanism
~, **замко́во-сто́порный** hold-back (GW launcher)
~, **запира́ющий** locking mechanism
~ **заряжа́ния** magazine filler
~ **затво́ра, закрыва́ющий** breech closing mechanism
~ **затво́ра, открыва́ющий** breech opening mechanism
~ **затво́ра, уравнове́шивающий** breech balancing gear (ord)
~, **затво́рный** breech mechanism
~, **инерцио́нный** graze percussion mechanism
~, **камерто́нный** sound synchronizing mechanism (sound ranging), tuning fork mechanism
~, **конта́ктный** contactor mechanism
~ **крепле́ния пу́шки в похо́дном положе́нии** (travelling) gun crutch

механи́зм, кривоши́пно-шату́нный crankshaft and connecting rod assembly (ICE)
~, **кулачко́вый** valve gear, cam gear, valve operating mechanism (ICE)
~, **курко́вый** hammer action
~, **лентопротяжно́й** paper-drive mechanism, tape transport mechanism, tape drive
~ **магази́на, пода́ющий** magazine platform mechanism
~, **магази́нный** magazine mechanism
~, **механи́ческий поворо́тный** power traverse mechanism
~, **механи́ческий стреля́ющий** mechanical firing device
~ **наво́дки** laying mechanism
~ **нажи́много де́йствия, спусково́й** pressure switch
~, **нако́льный** firing mechanism (mor bomb)
~ **натяже́ния гу́сеницы** track tensioning mechanism
~, **натяжно́й** stretching device, track tensioner (tk)
~, **ножно́й спусково́й** foot-firing mechanism (ord)
~ **облегче́ния закрыва́ния затво́ра** counterbalance (gunn)
~ **обра́тного хо́да** reversing mechanism
~ **отбо́ра мо́щности** power take-off (mech)
~ **отбра́сывания затво́ра** retracting mechanism (gunn)
~ **отка́та** recoil mechanism (gunn)
~, **открыва́ющий** 1. breech opening mechanism (ord) 2. opening mechanism
~ **отсе́чек** feed cylinder mechanism (ord)
~ **паралла́кса и ве́тра** displacement and wind gear (AAFCE)
~ **перебро́ски** slewing mechanism
~ **перезаряжа́ния** retracting mechanism (breech)
~ **переключе́ния переда́ч** speed-control mechanism, gear controls, gear lever
~ **переме́нного отка́та** cut-off gear (ord)
~ **пики́рования и кабри́рования** diving targets gear (AAFCE)
~, **пита́ющий** feed mechanism, feeder
~ **поворо́та** steering mechanism, steering gear

механи́зм поворо́та ба́шни turret traverse mechanism
~ **поворо́та, планета́рный** epicyclic steering unit (tk)
~, **поворо́тный** 1. traversing gear, traversing mechanism (gun etc) 2. steering gear, steering mechanism (veh)
~ **поглоще́ния отда́чи** recoil mechanism, recoil system (gunn)
~ **пода́чи** feed mechanism (SA)
~, **подаю́щий** feed mechanism
~, **подъёмно-поворо́тный** elevation and traverse mechanism
~ **подрессо́ривания** suspension mechanism
~ **подры́ва** destructor mechanism
~, **подъёмный** 1. elevating gear (ord) 2. hoist
~ **полуавтома́тики** semi-automatic gear (ord)
~, **предохрани́тельно-взводя́щий** safety-and-arming mechanism
~, **предохрани́тельный** safety mechanism
~, **приводно́й** drive mechanism, actuator
~ **приёмника, подаю́щий** belt-feed mechanism (SA)
~ **прице́ла** mechanism of vertical aim-off (gun sight)
~ **пу́шки, уравнове́шивающий** cradle balancing gear (ord)
~, **размыка́ющий** trip mechanism
~, **разобща́ющий** trip mechanism
~, **разрезно́й** cutaway mechanism (instructional eqpt), cutaway demonstration equipment
~, **распредели́тельный** distribution device, distribution gear
~, **реверси́вный** reversing mechanism
~ **регулиро́вки тормозо́в** brake adjuster
~, **регулиро́вочный** adjuster
~, **рулево́й** steering gear
~ **ручно́го спу́ска** hand firing gear
~, **рыча́жный** linkage
~ **с колéнчатой рукоя́ткой, запира́ющий** breech locking mechanism, breech locking lever
~ **с предохрани́телем, затво́рный** safety lock
~ **с предупрежде́нием, спусково́й** double-pull trigger action
механи́зм с ручны́м приво́дом hand traverse gear
~ **с силовы́м приво́дом, поворо́тный** power traverse mechanism
~ **самовзведе́ния** self-cocking action
~ **самоликвида́ции** self-destruct mechanism
~ **свобо́дного хо́да** free-wheel mechanism
~, **силово́й поворо́тный** power traverse mechanism
~, **следя́щий** servo-mechanism, follow-up mechanism
~ **совмеще́ния** rate gear (AAFCE)
~ **спу́ска** release mechanism
~, **спусково́й** trigger group, trigger mechanism (SA)
~, **сто́порный** locking mechanism
~ **стрельбы́** firing mechanism
~, **стреля́ющий** firing mechanism (SA)
~ **сцепле́ния** trip mechanism
~, **счётно-реша́ющий** computing mechanism, computer
~ **то́нкой вертика́льной наво́дки** fine elevating gear (MMG)
~ **то́нкой горизонта́льной наво́дки** traversing micrometer
~ **то́рмоза, храпово́й** brake ratchet
~, **тормозно́й** brake gear
~ **угло́в прице́ливания** sight setting mechanism
~, **угломе́рный** angle-measuring mechanism, azimuth mechanism
~ **уда́рницого ти́па, уда́рный** striker-type firing mechanism
~, **уда́рно-воспламеня́ющий** percussion-ignition mechanism (gren)
~ **уда́рного взрыва́теля, инерцио́нный** graze percussion mechanism
~, **уда́рно-спусково́й** 1. firing and trigger mechanism, trigger mechanism 2. hammer and trigger assemblies (pistol)
~, **уда́рный** striker mechanism
~ **управле́ния** control mechanism, control system
~ **управле́ния пово́зкой** steering link (gun platform)
~ **упреди́тельного треуго́льника** predicting unit (AAFCE)
~, **уравнове́шивающий** equilibrator, balancing gear (ord)
~ **у́ровня** levelling gear

механи́зм устано́вки угломе́ра azimuth scale (gunsight)
~ учёта ве́тра wind resolver, wind resolving mechanism, wind deflection calculator (AAFCE)
~ учёта попра́вки на паралла́кс parallax offset mechanism
~ фокусиро́вки объекти́ва objective lens adjustment
~, фрикцио́нный friction mechanism
~, фрикцио́нный множи́тельный 1. wheel and disc multiplier, ball and disc multiplier 2. friction multiplier
~ фрикцио́нов 1. wheel and disc gear, ball and disc gear 2. friction mechanism
~, храпово́й ratchet and pawl, ratchet-wheel gear
~ целика́ mechanism of lateral aim-off (gunsight)
~, часово́й clockwork mechanism
~, шарни́рный articulating mechanism
~, эжекцио́нный gas expelling device
~, электри́ческий стреля́ющий electrical firing mechanism
механи́змы, грузоподъёмные hoisting equipment
~, элемента́рные циркуляцио́нные elementary circulation mechanisms of the atmosphere
меха́ник artificer, mechanic
~ по вооруже́нию armament mechanic, armourer
~ по прибо́рам instrument mechanic
меха́ника, небе́сная celestial mechanics
меха́ник-води́тель driver, driver-mechanic
меха́ник-морори́ст motor mechanic
меха́ник-прибори́ст instrument repairer
механи́ческий power-driven
мехосоедине́ние mechanized formation (mech div etc)
мехтя́га mechanical transport
меч sword
меша́лка mixer
~ для цеме́нтного раство́ра cement mixer
мешкови́на sacking, hessian
мешо́к, вещево́й pack, kit-bag
мешо́к, деса́нтный airborne stores container, derby sack (Brit for free dropping of supplies), free drop sack (air supply container)
~, землено́сный sand-bag
~, многосло́йный laminated bag
~, многосло́йный бума́жный laminated paper-bag
~, огнево́й 1. pocket of fire 2. killing ground (tac)
~, парашю́тно-деса́нтный supply-dropping container, airborne stores container
~, парашю́тно-деса́нтный мя́гкий non-rigid supply-dropping container, soft parachute bag
~, парово́й steam pocket
~, похо́дный kitbag
~ с песко́м sand-bag
~, спа́льный sleeping bag
~, удлинённый elongated bag, long sand-bag
мига́лка 1. electric target imitator 2. flashing direction indicator, "winker" (veh)
микани́т micanite
микроампе́р microampere
микро́б, болезнетво́рный disease-carrying microbe
микроба́р microbar
микробло́к microassembly
микрокли́мат microclimate
микролитра́жка "baby" vehicle (engine capacity up to 800 ccs)
микро́метр micrometer
~, резьбово́й screw-thread micrometer
микроминиатюриза́ция subminiaturization
микромо́дуль micromodule
микро́н micron
микрооргани́зм microorganism, germ
микропереключа́тель microswitch
микрорентге́н microroentgen
микроско́п, просто́й simple microscope
~, сло́жный compound microscope
~, шка́ловый filar microscope
~, электро́нный electron microscope
микросни́мок micrograph, microphotograph
микросхе́ма microcircuit
~, интегра́льная integrated microcircuit
~, каска́дная cascade microcircuit
~, кре́мниевая silicon-base microcircuit

микросхе́ма, логи́ческая micrologic circuit
~ на кера́мике ceramic-based microcircuit
~ на керами́ческой подло́жке ceramic-based microcircuit
~, печа́тная microprinted circuit
~, тонкоплёночная thin-film microcircuit
микрофара́да microfarad
микрофо́н microphone
~, горлово́й laryngophone, throat microphone
~, динами́ческий moving coil microphone
~, ка́псюльный button microphone
~, конденса́торный condenser microphone, electronic microphone
~, ле́нточный ribbon microphone
~, миниатю́рный midget microphone, lapel microphone
~, пьезоэлектри́ческий crystal microphone
~, углезерни́стый carbon granule microphone, carbon shot microphone
~, у́гольный carbon microphone
~, электродинами́ческий moving-coil microphone
~, электромагни́тный electromagnetic microphone
микроэлектро́ника microelectronics
~, полупроводнико́вая solid-state microelectronics
~, тонкоплёночная thin-film microelectronics
милиционе́р militiaman (Sov Police)
мили́ция militia (Sov Police)
миллиампе́р milliampere
миллиампермéтр milliammeter
миллиба́р millibar
миллио́нка map to scale 1:1,000,000
миллирентге́н milliroentgen
миллиси́менс millisiemen, mS
мимео́граф mimeograph
ми́на 1. mine (ammo)
2. bomb (mor)
3. grenade (projected)
4. rocket
~, авиацио́нная aircraft-laid mine
~, автомоби́льная anti-vehicular mine
~, автоно́мная неконта́ктная self-contained influence mine
~, акусти́ческая acoustic mine
ми́на, анте́нная antenna(-type) mine
~, антикли́ренсная anti-clearance mine, tilt-rod mine
~, берегова́я beach (defence) mine
~, бронебо́йная armour-piercing mine
~ в боево́м положе́нии armed mine
~ в пла́стиковом ко́рпусе plastic-cased mine
~ взрывно́го де́йствия blast mine
~, взрывоусто́йчивая blast-resistant mine
~, вибрацио́нная vibration-action mine
~, возду́шная air mine
~, выпры́гивающая anti-personnel bounding mine
~, выпры́гивающая оско́лочная bounding shrapnel mine
~ Гали́цкого, лету́чая flying rocket mine (designed by Galitskij)
~, гидродинами́ческая pressure mine
~, деревя́нная я́щичная wooden box mine
~, десятипёрая ten-vaned mortar bomb
~, диверсио́нная sabotage device (incl - booby traps)
~, до́нная ground mine
~, доро́жная road mine
~, дрейфу́ющая drifting mine
~, дымова́я smoke bomb
~ огражде́ния obstacle mine
~ заграждéния, противота́нковая anti-tank obstacle mine
~, зажига́тельная incendiary mortar shell, incendiary mine
~, имитацио́нная battle effects simulator
~, индукцио́нная influence mine
~, ине́ртная inert mine (trg)
~, иниции́рующая initiating mine
~, конта́ктная contact mine
~, контро́льно-самоде́йствующая electro-contact mine
~, кумуляти́вная hollow charge projectile
~, лафетопро́бная mounting test bomb (mor)
~, ло́дочная submarine-laid mine
~, магни́тная dip-needle magnetic mine (nav)
~, магни́тная доро́жная magnetic road mine, magnetic anti-transport mine
~ мгнове́нного де́йствия instantaneous action mine
~, металли́ческая metallic mine

ми́на, миномётная mortar bomb
~ наведённого де́йствия influence mine
~, надкали́берная кумуляти́вная hollow-charge stick-bomb
~ нажи́много де́йствия pressure operated mine
~, назе́мная land mine
~ напра́вленного де́йствия directional mine
~, направля́емая directional mine
~ натяжно́го де́йствия pull-action mine, trip-wire mine
~, неизвлека́емая mine with anti-lift device
~, неконта́ктная influence mine
~, неметалли́ческая non-metal mine
~, необезвре́живаемая mine with anti-left device
~, неподви́жная fixed mine (sea or river)
~, нета́бельная improvised mine
~, неуправля́емая contact mine
~, обезвре́женная neutralized mine
~, объе́ктная mined demolition charge
~, оконча́тельно снаряжённая armed mine
~, оперённая finned bomb
~, освети́тельная illuminating bomb (mor)
~, оско́лочная HE bomb (mor), fragmentation mine
~, оско́лочно-фуга́сная high-explosive (fragmentation) bomb
~, парашю́тная parachute mine
~, партиза́нская доро́жная partisan anti-transport mine
~, пла́вающая floating mine
~, плову́чая floating mine
~, подви́жная противота́нковая anti-tank mine chain
~, подво́дная конта́ктная submerged contact mine
~ полко́вника Стари́нова, автомоби́льная "Starinov" anti-vehicle mine
~, практи́ческая practice shell, practice bomb, practice mine, training shell, training bomb, training mine
~, прилипа́ющая limpet mine
~, противогу́сеничная anti-track mine, track-operated A tk mine
~, противодеса́нтная 1. beach mine 2. anti-assault landing mine
ми́на, противодни́щевая belly attack mine
~, противокли́ренсная anticlearance mine, tilt-rod mine
~, противопехо́тная anti-personnel mine
~, противопехо́тная деревя́нная wooden anti-personnel mine
~, противопехо́тная карто́нная cardboard anti-personnel mine
~, противопехо́тная металли́ческая metal anti-personnel mine
~, противопехо́тная оско́лочная anti-personnel shrapnel mine
~, противота́нковая anti-tank mine
~, противота́нковая противогу́сеничная track cutting mine
~, противотра́нспортная anti-vehicle mine
~, пры́гающая противопехо́тная anti-personnel jumping mine
~, пулева́я противопехо́тная anti-personnel bullet mine
~, радиоуправля́емая radio-controlled mine
~ разгру́зочного де́йствия pressure release action mine
~, разря́женная defused mine, disarmed mine
~, реакти́вная HEAT rocket (of A tk rocket launcher)
~, речна́я river mine
~ с деревя́нным ко́рпусом wooden-box mine
~ с заря́дом бе́лого фо́сфора, зажига́тельная white phosphorous mine
~ с металли́ческим ко́рпусом metallic mine
~ с пластма́ссовым ко́рпусом plastic-cased mine
~ с прибо́ром кра́тности delay action mine
~ с рыча́жным взрыва́телем tilt-rod mine
~ с стержневы́м взрыва́телем tilt-rod mine
~ с элеме́нтом неизвлека́емости booby-trapped mine, mine fitted with anti-lift device
~, самоде́льная improvised mine
~, светосигна́льная trip flare
~ свя́зи message shell

ми́на, сигна́льная trip flare
~, снаряжённая armed mine, complete mine
~, сплавна́я floating mine
~, сухопу́тная land mine
~, та́бельная basic mine, issue mine, standard mine
~, та́нковая anti-tank mine
~, та́нковая бума́жная paper anti-tank mine
~, та́нковая деревя́нная wooden anti-tank mine
~, та́нковая смольнобума́жная tarpaper anti-tank mine
~, таре́лочная Teller-type mine
~, уда́рная contact mine
~, управля́емая controlled mine
~, уче́бная exercise mine, practice mine
~, уче́бно-имитацио́нная трениро́вочная mine simulator, simulated mine
~, хими́ческая 1. chemical mine 2. chemical mortar bomb
~, шестипёрная six-vaned mortar bomb
~, шрапне́льная shrapnel mine
~, я́корная moored mine
~, я́щичная box mine
"ми́на-ба́бочка" "butterfly bomb"
ми́на-грана́та, кумуляти́вная hollow-charge rocket (A tk)
ми́на-лову́шка booby trap
ми́на-пу́ля "bullet mine" (shrapnel mine containing bullets)
ми́на-сюрпри́з booby trap
ми́на-фуга́с, зажига́тельная HE incendiary mine
минбатаре́я mortar battery
минба́тр mortar battery
минвзво́д mortar platoon
минёр mine specialist, specialist in mines, mine-clearing and demolition
миниатюриза́ция miniaturization
миниатю́р-полиго́н miniature range
~ с обозначе́нием разры́вов puff target range
ми́нимум слы́шимости сигна́лов signal strength minimum (DF)
мини́рование mine-laying, mining
~ внабро́с scattered mine-laying
~, воспрети́тельное interdiction mining
~, ма́ссовое mass mine-laying, mechanical mine-laying
~, механизи́рованное mechanical mine-laying
мини́рование на воспреще́ние interdiction mining
~, поспе́шное hasty mine-laying
~ раскла́дчиком mining by mine-layer
~ строевы́м расчётом hand mine-laying by team drill
мини́рованный mined
мини́ровать/замини́ровать to mine, to lay mines
Министе́рство Вну́тренних Дел Ministry of Internal Affairs, MVD, (Sov)
~, Вое́нное War Ministry
~ Вооружённых Сил Ministry of the Armed Forces (Sov)
~ Госуда́рственной Безопа́сности Ministry of State Security, MGB (Sov, obs)
~ Иностра́нных Дел Ministry of Foreign Affairs (Sov), Foreign Office
~ Оборо́нной Промы́шленности (СССР) Ministry of the Defence Industry (Sov)
~ Оборо́ны Ministry of Defence
~ Путе́й Сообще́ния СССР Ministry of Rail and Road Transport (Sov)
~ Свя́зи Ministry of Communications (Sov)
мини́стр, вое́нный 1. War Minister 2. Secretary of State for War (US)
ми́нно-взрывно́й mine
ми́нный mine
миноиска́тель mine detector
~, автомоби́льный vehicular (mounted) mine detector
~ для разве́дки неметалли́ческих мин non-metallic mine detector
~, носи́мый portable mine detector
~, уче́бный training mine detector
~, электри́ческий electrical mine detector, electrical mine-detecting set
~, электро́нный electronic mine detector
миномёт mortar
~, батальо́нный battalion mortar
~, гварде́йский "guards mortar" (Sov truck-mounted multi-barrelled rocket launcher)
~, го́рный mountain mortar
~, длинноство́льный long-barrelled mortar
~, ду́льно-заря́дный muzzle loading mortar
~, заряжа́ющийся с казённой ча́сти breech-loaded mortar

миномёт, короткоство́льный short-barrelled mortar
~, кочу́ющий roving mortar
"Миномёт на Вьюки!" "Prepare for Carry!" (mor)
миномёт, нарезно́й rifled mortar
~, основно́й ranging mortar, pivot mortar
~, полково́й regimental mortar
~, реакти́вный vehicle-mounted multi-barrelled rocket launcher
~, реакти́вный противота́нковый anti-tank rocket launcher
~, ро́тный company mortar
~, самохо́дный self-propelled mortar, mobile mortar
~, хими́ческий chemical projector
миномета́ние aerial mine-laying
миномётный mortar
миномётчик mortarman
миноогнефуга́с high explosive incendiary mine (napalm)
ми́нус 1. negative current
2. minus, short (gunn)
мио́з myosis (contraction of pupils)
мир 1. peace
2. world
~, блатно́й criminal world
мира́б small magnetic ground mine (nav)
ми́ска bowl
митка́ль calico
мише́нь target
~, бегу́щая moving figure target
~, боева́я service target
~, букси́руемая towed target
~, букси́руемая возду́шная towed aerial target, drogue
~, быстродви́жущаяся fast moving target
~, вертика́льная vertical target
~, внеза́пно появля́ющаяся surprise target
~, возду́шная aerial target
~, гла́зная eye vane
~, головна́я head figure target (silhouette)
~, гру́дная head and shoulders target
~, дви́жущаяся manoeuvring target, moving target
~ для зачётных стрельб classification target
~ для холо́дной пристре́лки boresight target
мише́нь, зени́тная телеуправля́емая anti-aircraft remote-controlled target
~, контро́льная reference target
~, ко́нтурная silhouette target
~, мига́ющая "winking" target (tgt equipped with light for night firing)
~, неподви́жная stationary target
~, объёмная three-dimensional target
~, подвесна́я target drogue
~, подви́жная moving target
~, появля́ющаяся snap target (practice)
~, поясна́я waist figure target (silhouette from waist upwards)
~, пристре́лочная sighting target
~, радиолокацио́нная radar target
~, ростова́я figure target (full height silhouette)
~ с электромига́лкой "winking" target (tgt equipped with light for night firing)
~, скрыва́ющаяся disappearing target
~, спорти́вная competition target
~, та́нковая tank target
~, телеуправля́емая возду́шная drone, aerial drone
~, управля́емая drone
~, фигу́рная figure target
~, фла́жная sleeve target, banner target (AA)
мише́нь-конве́рт "envelope" target (tk gunn tgt)
мише́нь-ко́нус, букси́руемая drogue, sleeve target
~, возду́шная sleeve target (aeron)
мише́нь-парашюти́ст parachute target
мла́дший junior
многозаря́дный multi-charge
многокра́тность с часто́тным разделе́нием frequency division multiplex
многомото́рный multi-engined
многоо́сный multi-axle (veh)
многопролётный multi-span (br)
многопу́тный multi-lane
многоство́льный multi-barrelled
многото́пливный multi-fuel
многоуго́льник polygon
многофа́зный multiphase, polyphase
многоцве́тный polychromatic
многоцелево́й multi-purpose
многоя́русный multi-layered
множи́тель ослабле́ния attenuation factor (elec)
мо mho

мобилиза́ция mobilization
мобилизова́ть to mobilise
моби́льность mobility
~, **такти́ческая** tactical mobility
моби́льный mobile
моги́ла, бра́тская mass grave, communal grave
~, **о́бщая** common grave, communal grave
модели́рование simulation
~ **це́ли** target simulation
моде́ль 1. simulator
2. model
~ **в натура́льную величину́** mock-up (full size)
~, **электро́нная** electronic simulator
модера́тор 1. moderator
2. non-return valve
~ **то́рмоза отка́та** recoil buffer (gunn)
модерниза́ция modernisation
модернизи́рованный modernised
модернизи́ровать to modernise
модифика́ция modification
~, **аллотро́пная** allotropic modification
модифици́рованный modified
модули́рованный modulated
модули́ровать to modulate
мо́дуль module, modulus, modular circuit, modular package (electron)
~, **вставно́й** plug-in module
~, **дио́дно-рези́сторно-конденса́торный** capacitor-resistor-diode module
~, **интегра́льный** integrated circuit module
~, **логи́ческий** logic module
~ **преломле́ния** refractive modulus
~ **с керами́ческой подло́жкой** ceramic-base module
~, **сме́нный** plug-in module
~, **тонкоплёночный** thin-film (circuit) module
~ **упру́гости** modulus of elasticity
~, **цифрово́й** digital module
модуля́тор modulator
~, **амплиту́дный** amplitude modulator
~ **бегу́щей волны́** travelling wave modulator
~, **группово́й** group modulator (sigs)
~, **и́мпульсный** pulse modulator
~ **клистро́на** buncher (re-entrant type cavity resonator)
~ **когере́нтного излуче́ния, фа́зовый** coherent light phase modulator
модуля́тор с ли́нией заде́ржки delay-line type pulser (radar)
~, **светово́й** light modulator
~, **фа́зовый** phase modulator
~, **часто́тный** frequency modulator
~, **электромехани́ческий часто́тно-и́мпульсный** electro-mechanical frequency impulse modulator
модуля́ция modulation
~, **автоано́дная** anode self-modulation
~, **амплиту́дная** amplitude modulation
~, **амплиту́дно-и́мпульсная** pulse amplitude modulation
~, **амплиту́дно-и́мпульсная часто́тная** pulse-amplitude-modulation frequency modulation, PAM-FM
~, **ано́дная** anode modulation
~, **ано́дно-экра́нная** anode screen modulation
~, **взаи́мная** cross modulation
~, **вне́шняя** external modulation
~, **вну́тренняя** internal modulation
~, **временна́я и́мпульсная** pulse time modulation (electron)
~, **двойна́я** double modulation
~ **за защи́тную се́тку** suppressor-grid modulation
~ **излуче́ния ла́зера** laser modulation
~, **и́мпульсная** pulse modulation (elec)
~, **и́мпульсно-временна́я** pulse-time modulation
~ **интенси́вности пучка́** intensity modulation
~, **ко́дово-и́мпульсная** pulse-code modulation, pulse-count modulation, PCM
~, **многокра́тная** multiple modulation
~, **непосре́дственная** direct modulation
~, **несинхро́нная** asynchronous modulation
~ **одни́м то́ном** single-tone modulation
~, **однополо́сная** single-side-band modulation
~, **перекрёстная** cross modulation
~ **по дли́тельности, и́мпульсная** pulse-duration modulation
~ **по пло́тности** density modulation, bunching
~ **по ско́рости** velocity modulation
~, **побо́чная** spurious modulation
~ **прерыва́телем** chopper modulation
~ **с одно́й боково́й полосо́й** single-sideband modulation

модуля́ция, се́точная grid modulation (elec)
~ **се́точным смеще́нием** control-grid modulation
~, **синхро́нная** synchronous modulation
~, **тройна́я** triple modulation
~, **углова́я** angle modulation
~, **фа́зовая** phase modulation, PM
~, **фа́зово-и́мпульсная** pulse-phase modulation, pulse position modulation, PPM
~, **фазои́мпульсная** pulse-phase modulation, pulse position modulation, PPM
~, **часто́тная** frequency modulation, FM
~, **часто́тно-и́мпульсная** pulse-frequency modulation, PFM
~, **широ́тная** width modulation
~, **широ́тно-и́мпульсная** pulse duration modulation, PDM, pulse length modulation, pulse-width modulation
~, **шумова́я** noise modulation
~ **я́ркости луча́** intensity modulation
мо́йка washing (action), washing machine
молектро́ника molecular electronics
молибде́н molybdenum
молниеотво́д lightning conductor, lightning arrester
"мо́лния" "express" (message precedence designation)
молоко́, известко́вое milk of lime
~, **сухо́е** milk powder, dried milk
мо́лот, копро́вый rammer
~, **сва́йный** piling hammer, pile-driving hammer
~, **сва́йный пневмати́ческий** pneumatic piling hammer
молото́к, боево́й sledge hammer
~, **бури́льный** jack hammer, rock drill (pneumatic drill)
~, **клепа́льный** riveting hammer
~, **отбо́йный** concrete-breaker (percussion pneumatic tool)
~, **пневмати́ческий бури́льный** pneumatic drill
~, **пневмати́ческий отбо́йный** pneumatic pick
молча́ние silence, (radio) silence
моме́нт moment
~ **бо́я, перело́мный** turning point of battle
моме́нт включе́ния, нача́льный initial engagement
~, **вне́шний демпфи́рующий** aerodynamic damping moment (GW)
~, **вну́тренний демпфи́рующий** damping moment due to variation of moment of inertia (GW)
~, **восстана́вливающий** restoring moment (GW)
~, **восстана́вливающий враща́ющий** reverse torque, restoring torque (servo mech)
~ **вхо́да в зацепле́ние** point of engagement
~ **вы́лета** moment (time) the base of projectile leaves gun muzzle
~, **демпфи́рующий** damping moment (missile)
~, **дестабилизи́рующий** disturbing moment (GW, missile)
~ **зажига́ния** firing point, ignition point
~, **изгиба́ющий** bending moment (eng)
~ **ине́рции** moment of inertia
~ **кабри́рования** positive pitching moment
~ **корре́кции** corrective moment
~ **кре́на** roll moment
~, **креня́щий** rolling moment (GW)
~, **крутя́щий** torque, turning moment (engine)
~ **наступле́ния детона́ции** detonation point
~, **опроки́дывающий** disturbing moment, unstable moment
~, **продо́льный** pitching moment
~, **путево́й** yawing moment (GW)
~ **разры́ва** breaking point
~ **рыска́ния** yawing moment
~ **си́лы** moment of force
~ **соприкоснове́ния** moment of impact (projectile)
~ **сопротивле́ния** moment of inertia, moment of resistance
~, **стабилизи́рующий** stabilizing moment, restoring moment, stable moment
~, **стати́ческий** aerodynamic moment (missile)
~ **тангажа́** pitching moment
~, **туша́щий** damping moment (GW)
~, **управля́ющий** control moment
~ **усто́йчивости** moment of stability
~, **шарни́рный** hinge moment

монитóр, двернóй doorway monitor (nucl)
моноблóк monobloc
монокуля́р eye-piece (single)
монокуля́рный monocular
монoóкись фтóра fluorine monoxide
монорéльс monorail
моностани́на box trail, one-piece trail (ord) monobed (tech)
монохлорами́н Б chloramine B (decontaminant)
монтáж assembling, assembly, fitting, mounting, rigging, replacing, wiring, packaging, erection
~, беспáечный solderless wiring
~ в стóйке rack mounting
~ на панéли panel mounting
~ на пóлках shelf mounting
~ на стати́вах rack mounting
~, навеснóй point-to-point wiring
~, печáтный printed(-circuit) wiring
монти́рование аэрофотосни́мков mosaic assembly (airphoto)
монти́ровать/смонти́ровать to mount, to fit, to assemble; **монти́ровать/смонти́ровать тéхнику на платфóрме** to lash equipment on heavy-drop platform (airborne ops)
мóре sea
~, береговóе coastal waters, territorial waters
морóз frost
морóзилка freezing chamber
морозостóйкий frost-proof, winter-proof, antifreeze, non-freezing
морозостóйкость resistance to frost
мóрось drizzle
морскóй naval, nautical, sea
~ десáнтный amphibious assault
морти́ра mortar, howitzer (obs)
~, осáдная siege mortar
морти́рка, винтóвочная rifle grenade discharger
~, ружéйная grenade launcher, rifle grenade discharger
мóрфий morphine
морфи́н morphine
морфи́новый morphine
мост 1. bridge
2. axle (veh)
~, автодорóжный highway bridge, road bridge
~, автомаши́нный vehicular bridge
мост, áрочный arch bridge
~, бáлочно-консóльный girder cantilever bridge
~, бáлочно-подкóсный frame supported girder bridge
~, бáлочный girder bridge
~, бревéнчатый log trestle bridge
~, ведýщий driving axle (veh)
~, верёвочный rope footbridge, rope bridge
~, взóрванный blown bridge
~, вися́чий suspension bridge
~, воéнный низковóдный low level bridge (built by sappers, usually one way and temporary)
~, воздýшный airlift, airbridge
~, волновóдный waveguide bridge
~, выдвижнóй sliding bridge
~, высоковóдный high bridge (high enough to allow passage of spring floods, ice & ships)
~, ги́бкий вися́чий common suspension bridge, unstiffened suspension bridge
~, двухкры́лый разводнóй double-leaf swing bridge
~, двухподкóсный framed bridge with two intersecting single locks
~, двухпролётный double-span bridge
~, двухпýтный two-way bridge
~, двухъя́русный double-level bridge, double-storey bridge, double-deck bridge
~, длиннопролётный long-span bridge
~ для автомоби́лей мáлой грузоподъёмности light-duty axle
~ для безрéльсового трáнспорта road bridge
~, железнодорóжно-автодорóжный road-and-railway bridge
~, жёсткий rigid axle
~, зáдний rear axle
~, затопля́емый submersible bridge
~, канáтный rope bridge
~, колéйно-механизи́рованный truck-mounted treadway bridge
~, колéйный track bridge, treadway bridge
~, комбини́рованный floating bridge with fixed approach spans
~, комбини́рованный вися́чий stiffened suspension bridge
~, консóльный cantilever bridge
~ лафéта lift frame (missile erector)

мост, лёгкий вися́чий light suspension bridge
~, лета́ющий штурмово́й gap spanner
~, многопролётный multiple-span bridge, multispan bridge
~ на жёстких опо́рах fixed-support bridge
~ на козло́вых опо́рах trestle bridge
~ на надувны́х понто́нах air pontoon bridge
~ на опо́рах из ло́док boat bridge
~ на плову́чих опо́рах floating bridge
~ на плота́х raft bridge
~ на постоя́нных усто́ях fixed bridge
~ на ра́мных опо́рах trestle bridge
~ на ряжа́х crib bridge
~ на сва́йных опо́рах pile (trestle) bridge
~ на та́нковом мостоукла́дчике, складно́й штурмово́й tank-launched scissors bridge
~ на штурмовы́х ло́дках assault-boat bridge
~, наводно́й floating bridge
~, нака́тный sliding bridge
~, наплавно́й floating bridge
~, неведу́щий non-driving axle
~, неподви́жный fixed bridge
~, неразводно́й fixed bridge
~ неразрезно́й систе́мы continuous (one-piece) bridge
~, низково́дный low-level floating bridge
~ "но́жницы", складно́й scissors bridge
~, одноподко́сный single-lock framed bridge
~, однопролётный single-span bridge
~, однопу́тный single-way bridge
~, отка́тный pull-back draw bridge
мост. пере́дний front axle
мост, перекидно́й short gap bridge
~, переплётный single lock frame bridge
~, перехо́дный half floating bay (br)
~, пешехо́дный footbridge
~, пешехо́дный вися́чий 1. suspension footbridge
2. "PVM" bridge (Sov)
~, плашко́утный boat bridge
~, поворо́тный swing bridge
~, подвесно́й suspension bridge
~, подви́жный bascule bridge
~, подво́дный submerged bridge
мост, подко́сный frame bridge, framed bridge
~, подъёмный lift bridge, lifting bridge, draw bridge
~, понто́нный pontoon bridge
~, разбо́рный temporary bridge, equipment bridge (made of standard equipment and easily dismantled), prefabricated bridge
~, разводно́й opening bridge, swing bridge
~, раздвижно́й movable-span bridge
~, разрезно́й за́дний articulated rear axle (veh)
~ разрезно́й систе́мы simply supported bridge
~, ра́мный frame bridge
~, раско́сный truss bridge
~, решётчатый truss bridge
~, ри́гельно-подко́сный double-lock framed bridge
~ с двумя́ переда́чами, веду́щий two-speed axle
~ с двухступе́нчатой переда́чей, веду́щий double-reduction axle, dual-range axle
~ с ездо́й по́верху deck bridge
~ с ездо́й по́низу through bridge
~ с односторо́нним движе́нием one-way bridge
~ с отка́тывающим пролётом retractable span bridge
~ с прого́нами коро́бчатого сече́ния box girder bridge
~ с раскрыва́ющимся пролётом bascule bridge
~, самонаводя́щийся self-propelled bridge
~, самохо́дный self-propelled bridge
~, самохо́дный наплавно́й amphibious bridge
~, самохо́дный штурмово́й self-propelled assault bridge
~, сбо́рно-разбо́рный sectional bridge, equipment bridge, Bailey-type bridge
~, сбо́рный prefabricated bridge
~, сва́йный timber pier bridge
~, сверхтяжёлый super-heavy bridge
~, сво́дчатый arch bridge (eng)
~, складно́й folding bridge
~, складно́й штурмово́й scissors bridge, assault bridge

мост со сплошны́м насти́лом fully decked bridge
~ **сопротивле́ния** resistance bridge, Wheatstone bridge (elec)
~, **стально́й** steel bridge
~, **та́бельный** standard bridge
~ **такти́ческого назначе́ния** tactical bridge
~, **тяжёлый ба́лочный** heavy girder bridge
~, **тяжёлый понто́нный** heavy pontoon bridge
~, **тяжёлый штурмово́й** heavy assault bridge
~, **управля́емый** front axle, steering axle (veh)
~, **цепно́й** chain bridge
~, **штурмово́й коле́йный** wheel-track assault bridge
~, **эстака́дный** scaffold bridge
мост-акведу́к aqueduct bridge
мо́стик 1. light portable bridge
2. bridge (on ship)
3. bridge (elec)
~ **взрыва́теля** bridge wire (primer)
~, **коле́йный** treadway
~, **коле́йный перекидно́й** wheel track bridge
~, **лине́йный** linear circuit test set (dml)
~, **откидно́й** folding servicing platform (GW)
~, **перекидно́й** 1. folding bridge
2. portable footbridge
~, **перено́сный** portable ramp. loading ramp
~, **перехо́дный** half floating bay (br)
~, **смотрово́й** inspection ramp
~, **транше́йный** treadway
~ **Уи́тстона** Wheatstone bridge (elec)
~, **штурмово́й** assault foot-bridge
"мост-но́жницы" scissors bridge
мостова́я carriageway, paving, pavement (road surface)
~, **булы́жная** cobbled roadway, cobbled paving (road surface)
мосто́к, въездно́й loading ramp
мостострое́ние bridge building
мостоукла́дчик bridgelayer
~, **та́нковый** tank bridgelayer, armoured bridgelayer
мост-по́дмости scaffold bridge
мотово́з shunting engine (rly)
мото́к coil, skein, bundle
мотолебёдка power winch
мотолопа́та power shovel
мотомехча́сть mechanized unit
мотопехо́та motorized infantry
мотопила́ power saw
мотопланёр powered glider
мотопо́мпа motorized pump, engine-driven pump
~, **пожа́рная** fire pump
мото́р motor, engine
~, **асинхро́нный** asynchronous motor, induction motor
~ **водяно́го охлажде́ния** water-cooled engine
~ **возду́шного охлажде́ния** air-cooled engine
~, **индукцио́нный** induction motor
~, **исполни́тельный** actuating motor, driving motor
~, **навесно́й** outboard motor
~ **поворо́та ба́шни** turret traverse motor
~, **подвесно́й** outboard motor
~, **подвесно́й ло́дочный** outboard motor
~ **постоя́нной ско́рости** constant speed motor
~, **ря́дный** in-line engine
мото́р-весло́ motor oar, outboard motor
моторесу́рсы MT resources, reserves
моториза́ция motorization (of inf unit)
моторизо́ванный motorized
мотори́ст насо́са pump operator
мото́р-компре́ссор motor-compressor
мото́рно-трансмиссио́нный engine and transmission
мото́рный motor, motorized, self-propelled
моторо́ллер motor scooter
мото́рчик small motor
мотострелко́вый motor rifle
мотострело́к member of motor rifle unit, "motor rifleman"
мотохо́д motor scooter
мотоци́кл motorcycle
~ **с коля́ской** motorcycle with sidecar
мотоцикли́ст motorcyclist
~ **свя́зи** despatch-rider, DR
~, **связно́й** despatch-rider, DR
моточа́с 1. running time (engine)
2. driving hour (unit of driving experience)
моты́га mattock, hoe
моты́ль 1. crank, handle
2. toggle lever (Luger pistol)

мо́щность 1. power, horsepower, output, capacity
2. yield (nucl wpn)
~ **в лошади́ных си́лах** horsepower output
~, **входна́я** input power
~, **действи́тельная** true horsepower, actual horsepower
~, **дли́тельная** continuous rating
~ **до́зы** dose rate
~, **запасна́я** reserve capacity, reserve power
~, **индика́торная** indicated brake horse power
~, **ка́жущаяся** apparent power (elec)
~, **кратковре́менная** short-time rating
~, **литро́вая** brake horse power per litre
~, **максима́льная** maximum output
~, **номина́льная** rated power, rated output, rating (engine)
~, **огнева́я** fire power
~ **огня́** fire power
~, **отдава́емая** power output
~, **подводи́мая** input power
~, **по́лная** total output, total power, gross output, gross power
~, **преде́льная** ultimate output, ultimate capacity
~, **прое́ктная** rated power
~ **радиоста́нции** transmitter power (radio)
~, **располага́емая** available power
~, **расчётная** power rating
~, **резе́рвная** emergency capacity, reserve power
~, **сре́дняя** mean (horse)power
~, **тормозна́я** brake horsepower, bhp
~, **тре́буемая** required (horse)power
~, **уде́льная** power-to-weight ratio, bhp per ton (engines)
~, **эксплуатацио́нная** operating power
~, **эффекти́вная** engine output
~, **эффекти́вная тормозна́я** brake horsepower
~ **я́дерного боеприпа́са** atomic yield
мо́щный heavy-duty
мощь, боева́я fighting strength, deadliness
~, **огнева́я** fire power
~, **пробивна́я** penetrative power
му́жество courage, fortitude
музыка́нт орке́стра bandsman
мул mule
мультивибра́тор multivibrator (elec)
~, **жду́щий** one shot multivibrator
мультипле́кс 1. multiplex
2. multiplex aeroprojector (photo)
мультипле́ксер multiplexer
мультицикло́н multicyclone (type of centrifugal pre-air-cleaner)
муме́талл mumetal
мунди́р tunic, parade tunic
~, **закры́тый** parade tunic with high stand-up collar
мундшту́к nozzle, mouthpiece
муть, ма́сляная oil sludge
му́фта clutch, coupling, sleeve, connecting piece
~, **быстроде́йствующая** quick-acting coupling
~ **включе́ния** 1. throw-out bearing (clutch)
2. release sleeve, withdrawal collar
~, **враща́ющаяся** fermeture nut (MG)
~, **вту́лочная** box coupling, box sleeve, box muff
~, **выжимна́я** clutch collar, withdrawal collar
~, **га́зовая** gas block (SA)
~, **гидравли́ческая** fluid clutch, hydraulic coupling
~, **двухди́сковая** twin-plate clutch, double-plate clutch
~, **ди́сковая** plate clutch, flexible disc coupling
~, **жёсткая** solid coupling
~, **запира́ющая** barrel nut (MG)
~ **затво́ра** bolt sleeve
~, **зубча́тая** gear coupling
~, **ка́бельная** cable jointing device (includes junction boxes, sleeves, potheads etc)
~, **ко́нусная** mating sleeve (friction fuse)
~, **ко́нусная фрикцио́нная** cone clutch (veh)
~, **концева́я ка́бельная** 1. terminal box (Brit)
2. cable pothead (US)
~, **кулачко́вая** dog tooth clutch
~, **ле́нточная** band coupling
~, **многоди́сковая** multi-plate clutch
~, **натяжна́я** slack adjuster, turnbuckle, screw shackle
~, **невозвра́тная** free-wheel clutch
~, **неразъёмная** fixed coupling

му́фта, ограничи́тельная muff coupling (gunsight)
~, одноди́сковая single-plate clutch
~ окуля́ра focusing ring
~ О́льдгема Oldham's coupling
~, поводко́вая flange coupling
~, подви́жная flexible coupling
~ распредели́тельного ва́ла camshaft bearing
~ расцепле́ния clutch coupling
~ с резьбо́й threaded sleeve
~, самовыра́внивающаяся self-aligning coupling
~, самосинхронизи́рующаяся self-synchronising clutch
~ свобо́дного хо́да freewheeling clutch
~, синхронизи́рующая synchronising clutch
~, скользя́щая sliding coupling, slip coupling
~ со сма́зкой fluid clutch
~, соедини́тельная 1. flange coupling 2. barrel nut, recoil spring sleeve (SA), barrel lock (SA) 3. screw plug, bolt cap, bolt sleeve
~ ствола́ barrel sleeve
~, стержнева́я pin coupling, bar coupling, needle coupling, rod coupling
~, стяжна́я 1. turnbuckle, straining screw 2. turnbuckle barrel
~, суха́я dry clutch
~ сцепле́ния driving coupling, clutch coupling
~, таре́льчатая plate clutch
~ тре́ния friction clutch
~, углова́я angle joint
~, универса́льная universal joint
~ управле́ния driving clutch
~, упру́гая flexible coupling
~, фрикцио́нная friction clutch
~, фрикцио́нная ди́сковая disc clutch
~, фрикцио́нная ко́нусная cone friction clutch
~, центробе́жная centrifugal coupling
~, шарни́рная articulated coupling, joint coupling
~, шарова́я ball socket
~, шо́мпольная cleaning rod sleeve (detachable handle)
~, эласти́чная flexible coupling
му́фта-казённик breech ring
му́шка foresight (SA)
му́шка в сбо́ре foresight assembly
~, кольцева́я ring foresight
~, кру́пная 1. "too much foresight" (incorrect aim) 2. high point of aim (SA) 3. course sight
~, ме́лкая 1. "too little foresight" (incorrect aim) 2. fire sight
~, ножеви́дная blade foresight
~, откры́тая open foresight
~ "пля́шет" unsteady aim
~, прямоуго́льная blade foresight
~, ро́вная correct aim for elevation (tip of foresight level with shoulders of backsight) (SA)
~, треуго́льная barleycorn foresight
~, ша́риковая bead foresight
му́шкель mallet
муштра́ excessive drilling (obs)
мыс cape (topo)
мытьё па́лубным спо́собом washing down, sluicing with water
мышь, лету́чая 1. bat (rodent) 2. hurricane lamp
мышья́к arsenic
мя́коть, порохова́я mealed black powder, black blasting powder
мя́со, обезво́женное dehydrated meat
~, пу́шечное "cannon fodder" (sl)
~, су́шенное dried meat
мясору́бка mincing machine
мяте́ж mutiny, revolt, riot
мяте́жник mutineer

Н

на бли́зкой диста́нции at close range
на боево́м взво́де cocked, in cocked position (SA)
"На Грудь!" "Port Arms!"
на гужево́й тя́ге animal-drawn (tpt)
"На Карау́л!" "Present Arms!"
на катка́х on wheels, wheeled
"На Ме́сте!" "Mark Time!"
на ме́сте at the halt (drill)
"На Ме́сте Ша́гом Марш!" "Mark Time!"
на о́щупь by sense of touch, by feel

"На Пе́рвый и Второ́й-Рассчита́йсь!" "Number Off in Ones and Twos!"
на плаву́ afloat
на плеча́х проти́вника on the heels of the enemy
"На Плечо́!" "Slope Arms!"
на получетвере́ньках monkey run (Sov version)
"На Пост!" "To Your Duty!" (sentry posting)
на предохрани́теле on safe (wpn)
на предохрани́тельном взво́де safety catch on
на расстро́енных ко́нтурах staggered tuning
"На Реме́нь!" "Sling Arms!"
"На Ру́ку!" first movement of "On Guard!"
на сре́дней диста́нции at medium range
на ствол per weapon
"На (Тако́й-то предме́т)-К Бо́ю!" "On (such and such an object) Form Extended Line!"
на тра́верзе abeam (nav, aeron)
на ходу́ on the move
на цы́почках on tiptoe
на четвере́ньках on all fours (crawl)
набе́г raid, foray
на́бережная quay, pier, wharf, embankment
наби́вка, асбе́стовая asbestos packing, asbestos lagging
~, водонепроница́емая watertight packing
~, воздухонепроница́емая airtight packing
~, са́льниковая stuffing box packing
~, теплоизоли́рующая lagging
~, фе́тровая felt packing
наби́тый сма́зкой packed with grease
наблюда́емый observed
наблюда́тель 1. observer 2. spotter (arty)
~, вое́нный military observer
~, возду́шный 1. air sentry 2. air observer
~, возду́шный артиллери́йский artillery air observer, artillery observation post observer
~ за во́здухом air defence look-out
~ за та́нками проти́вника anti-tank observer
~, назе́мный ground observer
наблюда́тель, передово́й артиллери́йский forward observation officer, FOO (arty)
~, противота́нковый anti-tank observer
~, хими́ческий chemical observer
наблюда́тельный observation, observer, look-out
наблюда́тель-связно́й observer-runner (pl level)
наблюда́ть to observe, to watch, to keep a lookout
наблюде́ние observation, surveillance, lookout, spotting
~, визуа́льное visual observation
~, возду́шное air observation, observation from the air
~, двусторо́ннее bilateral observation
~ за во́здухом air observation
~ за мо́рем, берегово́е coast watch
~ за по́лем бо́я battlefield surveillance
~ за разры́вами spotting (arty)
~ за стрельбо́й observation of fire
~ зна́ков разры́вов observation of bursts
~, круглосу́точное twenty-four-hour observation, round-the- clock observation
~, кругово́е all-round observation
~, назе́мное observation from the ground
~, непреры́вное continuous observation, uninterrupted observation
~ огня́ по спо́собу часо́в observation of fire by clock code
~ оповеще́ния и свя́зи, возду́шное air observation warning & communications
~, радиолокацио́нное radar surveillance, radar watch
~, слухово́е listening watch
~, сопряжённое cross observation (Brit), conduct of fire by combined observations (US)
~, сплошно́е complete coverage (radar)
~, ство́рное axial observation, observation along line battery-target (arty)
~, телевизио́нное observation by television
~, хими́ческое gas reconnaissance, CW reconnaissance
наблюде́ния, полевы́е field observations (surv)

набо́йка 1. cleat, kneeblock, needle (timber construction)
2. packing, stuffing, lining, facing, fettling
3. printed cloth
4. heeltap
набо́р 1. set, kit, collection
2. recruitment
~ высоты́ climb (aeron)
~ зубча́тых колёс set of gears, train of gears
~ инструме́нтов tool set
~ сопротивле́ний bank of resistors
~ средств для ана́лиза воды́ water testing kit
~ средств для отбо́ра проб ОВ chemical agent sampling kit
~ средств опознава́ния ОВ chemical agent analyser kit
~ шестерён gear train
набуха́ние swelling
нава́рка hard facing (welding)
наведе́нец 1. guidance officer (GW)
2. interception controller (GCI)
наведе́ние 1. control, guidance (GW)
2. laying (gunn)
~, автомати́ческое automatic laying (AA gunn), guidance (GW)
~, акти́вное active guidance (GW)
~, акти́вное радиолокацио́нное active radar homing
~, астронавигацио́нное stellar guidance (GW)
~, бли́жнее short-range guidance (GW)
~ в упреждённую то́чку constant bearing course guidance (GW)
~, вертика́льное vertical laying, laying for elevation (arty, gunn)
~, визуа́льное visual guidance (GW)
~ вручну́ю manual guidance (GW)
~, горизонта́льное 1. laying for line (gunn)
2. laying for deflection (US)
~, инерциа́льное inertial guidance (GW)
~, интерферометри́ческое радиолокацио́нное interferometer radar guidance
~, инфракра́сное infra-red guidance, IR guidance (GW)
~, кома́ндное command guidance (GW)
~, коне́чное terminal guidance (GW)
~, лучево́е beam-rider guidance, beam-riding guidance
наведе́ние на всей траекто́рии полёта all-the-way guidance (GW)
~ на сре́днем уча́стке траекто́рии mid-course guidance
~ на цель homing on a target, guidance on a target (GW)
~, нача́льное launching guidance
~, непосре́дственное direct control, direct guidance (GW)
~ отде́льной раке́ты individual missile control
~ по инфракра́сному излуче́нию infra-red guidance, IR guidance, heat-seeking guidance (GW)
~ по криво́й пого́ни pure pursuit-course guidance
~ по криво́й пого́ни, кома́ндное command pursuit-course guidance (GW)
~ по криво́й пого́ни пресле́дования pure pursuit-course guidance
~ по криво́й пого́ни с постоя́нным упрежде́нием deviated pursuit-course guidance, constant navigation guidance, fixed lead navigation guidance
~ по ла́зерному лучу́ laser (beam) guidance (GW)
~ по ли́нии визи́рования line-of-sight guidance (GW)
~ по лучу́ beam-riding guidance (GW)
~ по ме́тоду паралле́льного сближе́ния constant bearing guidance
~ по прово́дам cable guidance, wire guidance (GW)
~ по ра́дио radio guidance (GW)
~ по радиолокацио́нному лучу́ radar-beam guidance (GW)
~ по радиолучу́ beam-riding guidance (GW)
~ по "соба́чьей" криво́й pure pursuit-course guidance
~ по теплово́му излуче́нию infra-red guidance, IR guidance, heat-seeking guidance (GW)
~, полуавтомати́ческое semi-automatic guidance (GW)
~, пропорциона́льное proportional guidance (GW)
~, простра́нственное three-dimensional guidance (GW)
~, радиолокацио́нное radar guidance
~ с визуа́льным слеже́нием direct guidance (GW)

наведе́ние с опти́ческим сопровожде́нием optical-tracking guidance (GW)
~, сме́шанное combined guidance
~, телевизио́нное television guidance, TV guidance (GW)
наве́с open-sided shed, awning
навести́ (см наводи́ть)
навести́сь (см наводи́ться)
наве́тренный windward
наве́шенный hinged, hinge mounted
навига́ция navigation
~, пропорциона́льная proportional navigation
нави́нчивать/навинти́ть to screw on
навлека́ть/навле́чь to bring, to draw; **навлека́ть/навле́чь ого́нь** to draw fire
навле́чь (см навле́кать)
наводи́мый по провода́м wire-guided
наводи́ть/навести́ to direct, to bring to aim, to lay, to control, to guide (GW); **наводи́ть/навести́ в буссо́ль** to lay on the director (gunn); **наводи́ть/навести́ мост** to bridge, to lay a bridge, to launch a bridge; **наводи́ть/навести́ по а́зимуту** to position (track) in azimuth; **наводи́ть/навести́ с упрежде́нием** to aim off; **наводи́ть/навести́ снаря́д** to aim a missile
наводи́ться/навести́сь to home (GW); **наводи́ться/навести́сь на цель** to run home (GW)
наво́дка 1. laying (gunn)
2. aiming (SA)
~, автомати́ческая дистанцио́нная remote laying
~, вертика́льная laying for elevation
~, горизонта́льная 1. laying for line (Brit)
2. laying for deflection (US)
3. training (nav gunn)
~, гру́бая approximate aiming
~ мо́ста bridging, bridge laying, bridge launching
~, непрямáя indirect laying
~, нераздéльная simultaneous laying
~ ору́дий, полуавтомати́ческая semiautomatic laying (gunn)
~ ору́дий, радиолокацио́нная gun-laying by radar
~ ору́дия gun laying
~, перекрёстная crosstalk (comms)
~ под цель aiming down
наво́дка, полуавтомати́ческая semi-automatic laying
~, полуавтомати́ческая semi-automatic laying
~ по́сле ка́ждого вы́стрела re-laying after each shot
~, прице́льная visual aiming (gunn)
~, пряма́я direct laying
~, разде́льная two-stage laying (gunn)
~, то́чная fine laying (gunn)
~, то́чная горизонта́льная slow-motion traversing
~ че́рез кана́л ствола́ bore sighting
наводне́ние flooding
наво́дчик 1. layer (arty)
2. gunner (tk)
3. no. 1 (MG)
4. no. 3 (mor)
~, ба́шенный turret gunner (tk)
~, вертика́льный layer for elevation (arty)
~, горизонта́льный gun trainer
~, запасно́й reserve layer (arty)
~, оруди́йный gun layer
~ по а́зимуту layer for line
~ по углу́ ме́ста це́ли layer for elevation
наво́дчик-наблюда́тель projector controller (searchlight crew)
наво́дчик-опера́тор gunner-operator
на́вык skill acquired by experience
~, такти́ческий tactical skill acquired by experience
нага́р fouling (gun barrel)
~, порохово́й propellant fouling
нагарообразова́ние deposit formation, coking up, carbon deposit
на́гель spike, pin, treenail, trenail
~, деревя́нный wooden pin, dowel, peg
нагла́зник eye guard
~, рези́новый rubber eyepiece
нагля́дность, изобрази́тельная instruction by means of pictorial aids
~, предме́тная instruction by means of showing objects
нагля́дный 1. visual
2. obvious
нагнести́ (см нагнета́ть)
нагнета́ние 1. forcing
2. charging, injection, delivery
нагнета́тель 1. air blower
2. supercharger (ICE)
~, ротацио́нный rotary blower

нагнета́тель, центробе́жный centrifugal blower, centrifugal supercharger
нагнета́ть/нагнести́ to pump, to force, to pressurize, to supercharge
наго́рье upland, highland, elevation
нагото́ве at the ready
награ́да, де́нежная monetary award
~, посме́ртная posthumous award
~, прави́тельственная 1. decoration (award)
2. government award
награди́ть (см награжда́ть)
награжда́ть/награди́ть to award, to decorate
награжде́ние award
нагре́в heating, overheating
нагрева́ние ствола́ overheating of barrel (ord)
нагрева́тель heater, heating apparatus
нагревосто́йкий heat-resistant
нагревосто́йкость heat-resistance
нагру́дник, спаса́тельный life-jacket
нагру́зка load
~, акти́вная resistive load (elec)
~, ано́дная anode load
~, боева́я operational load
~, бо́мбовая bomb load
~, допусти́мая permissible dose, safe working load
~, инерцио́нная inertial load
~, ко́мплексная impedance load
~, максима́льно допусти́мая maximum permissible load
~, мгнове́нная instantaneous load
~ на колесо́ wheel load
~ на крюке́, тя́говая draw-bar load
~ на опо́ру bearing load (br)
~ на ось axial load, thrust load
~ на пого́нную длину́ linear load
~, наибо́льшая допуска́емая bearable load, maximum permissible load
~, номина́льная rated load
~, оптима́льная optimal load
~, осева́я end thrust, axial thrust
~, пи́ковая peak load
~, пого́нная 1. load per unit length
2. uniform loading (eng)
~, поле́зная pay load
~, попере́чная weight to calibre ratio (ammo)
~, преде́льная load capacity, breaking load (elec)
~, рабо́чая operating load
нагру́зка, разруша́ющая failure load
~, распределённая distributed load
~, расчётная rated load
~, симметри́чная balanced load
~, согласо́ванная matched load, balanced load (elec)
~, терми́ческая thermal load
~, тормозна́я brake load
~, уда́рная shock load
~, уде́льная load per unit
надба́вка increment
надви́жка пролёта launching of a span (br)
надду́б supercharging (ICE)
надду́в ди́зельного дви́гателя diesel supercharging
надёжность dependability, reliability
~, гаранти́рованная inherent reliability
~, механи́ческая mechanical reliability
~, о́бщая cumulative reliability
~ систе́мы system reliability
~, функциона́льная functional reliability
~, эксплуатацио́нная operational reliability
надели́ть (см наделя́ть)
наделя́ть/надели́ть to allot, to give, to provide; **наделя́ть/надели́ть пра́вом** to delegate authority, to authorize
надзе́мный elevated, over-head, above-ground
надзира́тель overseer
на́долбы blocks, pillars, posts, block obstacle, guard rail posts, dragons' teeth (fd eng)
~ "зу́бы драко́на", противота́нковые dragons' teeth
~, подво́дные underwater stake obstacles
~, противота́нковые anti-tank obstacles (dragons' teeth)
на́дпись, трафаре́тная stencilling
надре́з notch
надсмо́трщик, лине́йный lineman (sigs)
надстро́йка superstructure (br)
надувно́й pneumatic, inflatable
наду́льник 1. barrel mouthpiece
2. muzzle attachment
3. muzzle towing eye
наду́льник-пламегаси́тель flash eliminator, flash hider
нажа́тие, второ́е second pressure (on trigger)
~ до пе́рвого упо́ра first pressure

нажа́тие, пе́рвое first pressure
нажда́к emery
нажи́м pressure, stress, thrust
нажи́мный pressure, thrust
назва́ние designation, title, name
назе́мный ground
назнача́ть/назна́чить to allocate, to appoint, to assign, to detail; **назнача́ть/назна́чить прива́л** to designate a halt (on march)
назначе́ние 1. appointment, allocation
2. role, function, purpose
3. destination
~, осо́бое special purpose, special role
~, прямо́е proper function, primary function
~, такти́ческое tactical purpose
назна́чить (см назнача́ть)
наименова́ние nomenclature, designation
~, ко́довое code name
~, усло́вное code name, nickname
на́искось по укло́ну aslant the slope
найло́новый nylon
наказа́ние punishment
наказу́емость liability to punishment
нака́л 1. red heat, white heat, incandescence
2. heater filament circuit
3. momentum (attack)
нака́пливание 1. accumulation, storing
2. build-up, massing (tac)
нака́пливать/накопи́ть to accumulate, to amass; **нака́пливать/накопи́ть да́нные** to store data
нака́т 1. timber decking, timber platform, timber roof
2. counterrecoil, run-out (ord)
накати́ться (см нака́тываться)
нака́тник 1. roof log, room beam
2. recuperator (ord)
~, возду́шно-гидравли́ческий hydro-pneumatic recuperator
~, возду́шный pneumatic recuperator (ord)
~, гидропневмати́ческий hydropneumatic recuperator
~, пневмати́ческий pneumatic recuperator
~, пружи́нный spring recuperator
нака́тываться/накати́ться to run out (ord)
накача́ть (см нака́чывать)
нака́чивание ши́ны inflation of tyre
нака́чивать/накача́ть to pump
нака́чка "pep talk" (sl)
наки́дка cape
~, защи́тная protective cape, anti-gas cape
~, противоипри́тная anti-mustard-gas cape
наки́дка-подсти́л groundsheet cape
накипеочисти́тель scaler (veh radiators)
на́кипь scum, scale, encrustation, sinter, deposit, sediment
накла́дка 1. cover-plate, strap, gusset, cover bar
2. covering
3. overlapping
~ ба́мпера bumper guard
~ ди́ска сцепле́ния, фрикцио́нная clutch facing
~, ду́льная muzzle protector, muzzle guard
~, ли́пкая adhesive patch (for punctures)
~, скулова́я gusset (plate)
~, ство́льная hand guard (SA)
~, стыкова́я fish plate
~ сцепле́ния clutch lining
~ то́рмоза brake lining
~, тормозна́я brake lining
~, фрикцио́нная friction lining
накладна́я delivery note, receipt, bill of lading
накла́дывать/наложи́ть to put in/on/over, to lay on/in/over, to superimpose; **накла́дывать/наложи́ть повя́зку** to bandage
наклёп cold hardening, cold working
накло́н 1. tilt, slope, inclination
2. pitch (tech)
~ бронево́го листа́ angle of armour plate
~ о́си цапф cant of trunnions (ord)
~ шкво́рня наза́д castor (veh suspension)
наклоне́ние, вертика́льное range setting (arty, gunn)
~, горизонта́льное deflection setting (gunn)
~, магни́тное magnetic inclination, dip
накова́ленка anvil (ammo)
накова́льня anvil
~ ка́псюльной вту́лки primer anvil (ammo)
наколе́нник knee pad

наконéчник 1. tip
2. nozzle
3. terminal (elec)
4. ferrule
~, баллисти́ческий ballistic cap (ammo)
~, бронебóйный armour-piercing cap
~ вáла фрикциóна clutch shaft spigot
~ лéнты cartridge belt tag
~, пóлюсный pole piece, pole shoe (elec)
~, распы́ливающий atomizing nozzle
~ стéржня клáпана valve cap
~, шаровóй ball end, joint end
накопи́тель 1. tank circuit, storage circuit (elec)
2. accumulator, store (computer)
~, ёмкостный storage capacitor
~ модуля́тор delay line modulator circuit
~ на магни́тной лéнте magnetic tape store
~ на магни́тном барабáне magnetic drum store
~ на магни́тных ди́сках/кáртах magnetic disc/card store
накопи́ть (см накáпливать)
накоплéние 1. build-up (tac)
2. stockpiling
накрени́ться (см крени́ться)
накрывáть/накры́ть to cover; **накрывáть/накры́ть огнём** to bring down fire on; **накрывáть/накры́ть цель** to straddle a target (arty)
накры́тие straddle (nav gunn)
~, недолётное straddle short (nav gunn)
~, перелётное straddle over (gunn)
~ цéли straddling of the target (nav gunn)
~, чи́стое equal straddle, balanced straddle (nav gunn)
налагáть/наложи́ть to impose, to inflict; **налагáть/наложи́ть взыскáние** to award a punishment
налáдить (см налáживать)
налáдка adjustment
налáживание alignment, adjustment, tuning up, setting up
налáживать/налáдить to adjust, to tune up, to set up, to put right
"Налéво!" "Left Turn!"
"Налéво Равня́йсь!" "Left Dress!"
налёт 1. raid
2. deposit, film, residue, coating, veneer
~ авиáции air raid, air attack
налёт, возду́шный air attack, air raid
~, звёздный 1. star raid, spray raid (simultaneous attack by several ac from different directions)
2. star barrage, star strike (arty) (gunfire to confuse sound ranging)
~, масси́рованный attack in mass formation
~, огневóй artillery strike
налицó present
нали́чие presence, availability
~ боеприпáсов ammunition availability
~ для слу́жбы availability for duty
~, факти́ческое stock on hand, quantity on hand
нали́чность availability
~, теку́щая current availability, current stock level
нали́чный on hand, available, present
налóбник headrest, eye-shield, (rubber) face piece
налóг, подохóдный income tax
наложéние 1. superposing (sigs)
2. retreading (tyre)
~ взыскáния award of punishment
наложи́ть (см наклáдывать)
намагни́чивание magnetizing
намáтывание winding
намéрение intention, intent, purpose
намёт палáтки flysheet
намётка 1. clamp
2. cap square (ord)
наморáживание ice reinforcing
наму́шник foresight protector
нанесéние 1. infliction, delivery
2. drifting, heaping (snow)
3. drawing, plotting
~ делéний graduation
~ на кáрту mapping
нанести́ (см наноси́ть)
нанóс alluvium
~, снéжный snowdrift
наноси́ть/нанести́ to inflict, to deliver, to mark (map etc), to drift; **наноси́ть/нанести́ большóй урóн** to inflict heavy losses; **наноси́ть/нанести́ глáвный удáр** to deliver the main thrust; **наноси́ть/нанести́ на кáрту** to plot on the map; **наноси́ть/нанести́ поражéние** to defeat, to inflict defeat; **наноси́ть/нанести́ потéри** to cause casualties, to inflict losses;

наноси́ть/нанести́ уда́р to deliver an attack, to deliver a blow
напада́ть/напа́сть to attack; **напада́ть/напа́сть из заса́ди** to attack from ambush
напада́ющий aggressive, attacking
нападе́ние 1. assault, attack, offensive, strike
2. aggression
~, **а́томное** nuclear attack, atomic attack
~, **биологи́ческое** biological attack (NBC)
~, **внеза́пное** surprise attack
~, **возду́шное** air attack
~, **да́льнее огнево́е** defensive fire task in depth, DF task in depth
~, **неожи́данное** surprise attack
~, **огнево́е** artillery strike
~, **ожида́емое возду́шное** impending air attack
~ **с во́здуха** air attack
~, **хими́ческое** chemical attack, CW attack
~, **я́дерное** nuclear attack
напа́лм napalm
напа́лмовый napalm
напа́рник one of a two-man team
напа́сть (см напада́ть)
напёрсток запа́ла primer tube (gren)
напи́льник file (tool)
~, **вы́пуклый** half round file
напла́вить to build up (with metal)
напле́чник butt strap (SA)
наполне́ние filling
~ **зали́вкой** filling by pouring (ammo)
~ **прессова́нием** filling by pressing (ammo)
~ **шнекова́нием** screw filling (ammo)
наполни́тель filler, filling material
~, **ине́ртный** inert filler
напо́р 1. pressure, thrust
2. head (of liquid)
~, **по́лный** total head (pump)
~, **по́лный рабо́чий** total head (pump)
~, **рабо́чий** operating head (pump)
~, **скоростно́й** 1. impact pressure, dynamic pressure
2. ram, ram effect
напра́вить (см направля́ть)
направле́ние 1. direction, axis, line
2. order, directive
3. sector
4. trend, tendency
~ **ве́тра** wind direction
направле́ние, вспомога́тельное secondary sector, secondary direction
~ **гла́вного уда́ра** main axis of advance, axis of main thrust
~, **гла́вное** main axis, main direction
~ **движе́ния** traffic flow
~, **дополни́тельное** secondary direction, secondary sector
~, **испра́вленное** corrected deflection
~, **и́стинное** true direction
~ **на цель** line of fire
~, **наибо́лее угрожа́емое** most threatened sector
~ **наступле́ния** direction of attack, direction of advance
~ **наступле́ния, основно́е** axis of advance
~, **обхо́дное** alternative routing, diversion
~, **операцио́нное** 1. sector of operations
2. axis of operations
~, **ориенти́рное** bearing to aiming point (arty)
~, **осево́е** axial direction (ballistics)
~, **основно́е** 1. main sector
2. direction of main attack
3. centre of arc (gunn)
4. zero line
~ **подво́за** axis of supply
~ **подхо́да** line of approach
~ **полёта** heading (GW)
~, **пристре́лянное** reduced bearing (arty)
~ **пу́ска** launch direction (msl)
~, **реша́ющее** decisive factor
~ **свя́зи** communications link
~ **свя́зи, осево́е** main artery (comms)
~ **стрельбы́** line of fire
~ **стрельбы́, гла́вное** main axis of arc of fire
~, **танкодосту́пное** possible direction of tank approach
~, **танкоопа́сное** tank approach, tank run, likely line of tank approach
~ **тече́ния** set (current)
~ **хо́да** traverse line (surv)
напра́вленность directivity (radio)
~, **о́страя** high directivity (radio)
направля́ть/напра́вить to direct, to send, to beam (radio); **направля́ть/напра́вить в цель** to lay (gunn); **направля́ть/напра́вить движе́ние тра́нспорта по друго́му пути́** to divert traffic

направля́ющая 1. guide, projector (RL)
2. positioning arc (ord)
3. track-pin retaining cam (tk)
~ **баланси́ра теле́жки** bogie arm guide (tk)
~ **гу́сеничной це́пи** track guide
~ **затво́ра** bolt guide
~ **кла́пана** valve stem guide
~ **пусково́й устано́вки** launching rail, elevating arm, elevating frame, erecting arm (GW)
~ **реакти́вной устано́вки** rocket launcher guides, rocket launcher projector
~, **ре́льсовая** launching rail, guide rail (GW)
~, **со́товая** open frame projector (RL)
~ **теле́жки** bogie guide (RL)
~ **толка́теля** 1. tappet guide (ICE) (Brit)
2. valve lifter guide (ICE) (US)
~, **тру́бчатая** launching tube (GW)
~, **тру́бчатая пускова́я** launch tube (GW)
~, **цилиндри́ческая** barrel type projector (RL)
направля́ющий 1. guide (drill)
2. guide, guiding, leading, directing
~, **тру́бчатый** launching tube (GW)
"Напра́во!" "Right Turn!"
напряже́ние 1. voltage
2. stress, tension, strain
~ **авиа́ции** number of missions flown in 24 hours (AF)
~, **боево́е** number of sorties flown in twenty four hours (AF)
~, **вну́тренее** internal stress
~, **входно́е** input voltage
~, **высо́кое** high voltage, high tension
~, **действи́тельное** true stress
~, **допуска́емое** working stress
~, **допусти́мое** working stress
~, **запира́ющее** cut-off voltage
~, **заря́дное** charging voltage
~, **индукти́рованное** induced voltage
~, **лине́йное** line voltage
~ **на щётках** brush voltage
~, **ни́зкое** low tension, low voltage
~ **огня́** intensity of fire
~, **опо́рное** reference voltage
~ **отсе́чки** cut-off voltage
~, **пита́ющее** supply voltage
~ **поджи́га** striking voltage
~, **подключённое** cut-in voltage
~ **поко́я** quiescent voltage
напряже́ние, постоя́нное constant voltage
~, **псофометри́ческое** psophometric voltage
~ **развёртки** time-base voltage, sweep voltage
~, **разруша́ющее** rupture stress
~, **растя́гивающее** tension stress
~, **се́точное** grid voltage, grid potential (radio) (elec)
~ **смеще́ния** grid bias (elec)
~, **теплово́е** thermal stress
~, **ускоря́ющее** accelerating voltage (radar)
~, **этало́нное** reference voltage
напряжённость tension, strain
~ **магни́тного по́ля** magnetic induction flux density
~ **огня́** rate of fire
~ **по́ля** field strength (elec etc)
~ **электри́ческого по́ля** electric field strength
напыле́ние spraying
нарабо́тка, за́данная operating time, mission time
~ **на отка́з, сре́дняя** mean time between failures
нараста́ние 1. build-up (tac)
2. growth, increase
нараста́ющий cumulative, increasing, mounting
нарасти́ть (см нара́щивать)
нара́щивание 1. build-up (tac)
2. splicing (wood)
нара́щивать/нарасти́ть to build up, to intensify, to splice (wood)
наре́з 1. groove, cut
2. rifling
3. thread (screw)
наре́зать (см нареза́ть)
нареза́ть/наре́зать to cut, to rifle, to thread (screw); **нареза́ть/наре́зать по ка́рте** to allocate on the map (ground, sectors etc)
наре́зка 1. groove, cut
2. rifling
3. thread (screw)
4. allocation (ground)
~, **винтова́я** screw thread
~, **ле́вая** left-hand thread
~, **пра́вая** right-hand thread
~, **прогресси́вная** gaining twist
~, **се́кторная** interrupted thread

нарéзка, трапецievи́дная buttress screw-thread
нарезнóй 1. grooved
2. rifled
3. threaded (screw)
нарéзы, винтовы́е rifling
нарóчный courier, special messenger
нарýчники handcuffs
нарушáть/нарýшить to break, to violate, to offend against, : **нарушáть/нарýшить коммуникáции** to sever communications; **нарушáть/нарýшить прицéл** to disturb the aim
нарушéние violation, breach
~ **дисципли́ны** breach of discipline
~ **инстрýкции** infringement of rules
~ **контáкта** breaking contact
~ **обя́занностей** dereliction of duty, failure to perform duty
~ **прáвил движéния** driving offence
~ **прися́ги** violation of oath of allegiance
наруши́тель violater, offender
нарýшить (см нарушáть)
нáры bunk
~, **двухъя́русные** double-tier bunk
~, **двухэтáжные** double-tier bunk
наря́д 1. indent
2. detail
~ **вне óчереди** detail out of turn, extra duty
~, **внýтренний** detail for regimental duties
~, **гарнизóнный** garrison duty, garrison fatigue detail
~ **дежýрств** duty roster
~, **караýльный** guard detail
~ **на рабóты** fatigues detail
~ **на ремóнт** workshop indent
~ **на слýжбу** duty roster
~, **сýточный** twenty-four-hour duty, twenty-four-hour detail, daily duty detail
наря́д-закáз production order
насáдка 1. capsill, crossbeam
2. cap, capping, attachment
3. nozzle
~, **дефлéкторная** deflector cap
~, **дýльная** muzzle adapter
~, **светомаскирóвочная** black-out mask (for vehicle headlights)
~ **фáры, маскирóвочная** headlamp mask

насаждéние 1. planting
2. plantation
3. infiltration
~ **агентýры** infiltration of agents
наседáть/насéсть to press (tac)
наседáющий pressing (tac)
населéние population
~, **мéстное** local population
~, **сочýвствующее** sympathetic population, friendly population
насéсть (см наседáть)
насéчка serrature (gunn)
нáсмерть "to the death"
насóс pump
~, **баллáстный** ballast pump
~, **бензи́новый** petrol pump
~, **водооткáчивающий** bilge pump
~, **водянóй** 1. water pump
2. water circulating pump
~, **воздýшный** 1. air pump
2. priming pump
~ **впры́ска** injection pump
~, **гидравли́ческий** hydraulic pump
~ **гидравли́ческого привóда поворóта бáшни** power traverse pump
~ **горю́чего** fuel pump
~, **диафрáгменный** diaphragm pump
~, **зали́вочный** priming pump
~, **клáпанный тóпливный** valve-type fuel pump (diesel ICE)
~, **коловрáтный** rotary pump
~, **маслозакáчивающий** oil circulating pump
~, **маслоотсáсывающий** oil scavenging pump
~, **мáсляный** oil pump
~, **мáсляный турбозубчáтый** turbine lubricating gear pump
~, **мембрáнный** diaphragm pump
~, **нагнетáющий** pressure pump, pressurizing pump
~, **нали́вочный** priming pump
~, **откáчивающий** scavenging pump
~, **перекáчивающий** transfer pump
~, **питáтельный** feed pump
~, **пожáрный** fire pump
~, **продýвочный** blower
~, **ротациóнный** rotary pump
~, **ротациóнный червя́чный** screw rotary pump
~, **ручнóй воздýшный** hand air pump
~, **ручнóй поршневóй** stirrup pump

насо́с, то́пливный 1. fuel pump
2. fuel injection pump (diesel ICE)
~, топливоподка́чивающий 1. fuel supply pump (ICE)
2. fuel transfer pump (diesel ICE)
~, трю́мный bilge pump
~, ускори́тельный boost pump (ICE)
~, центробе́жный centrifugal pump
~, шестерёнчатый gear pump, gear-type pump
~, шта́нговый insert pump, rod-operated plunger pump
насо́с-форсу́нка combined pump-injector, unit injector (diesel ICE)
на́спех ad hoc
наставле́ние manual, instructions, regulations, pamphlet
~ по слу́жбе в полевы́х усло́виях Manual of Administration in the Field
~ по эксплуата́ции instructions for use, operating instructions
насти́л decking, flooring, duckboard, trenchboard
~, ве́рхний wearing surface (br)
~, вре́менный false floor
~, двойно́й перекрёстный wooden latticed wearing surface (br)
~, доща́тый slab road, plank road, slab road surface, duckboard
~, жердево́й corduroy road
~, защи́тный wearing surface (br)
~, коле́йный tread decking, treadway
~ мо́ста bridge decking
~, ни́жний попере́чный decking (br)
~, перено́сный 1. portable tracking
2. portable covering surface
~, рабо́чий decking, wearing surface (br)
~, решётчатый duckboard
~, торцо́вый wood-block flooring
~, тротуа́рный footwalk plank (br)
~, щитово́й wooden plank flooring
насти́льность траекто́рии flatness of the trajectory
настра́ивать/настро́ить to tune, to tune in, to adjust
настро́енный tuned, resonant (elec)
настро́ить (см настра́ивать)
настро́йка 1. tuning
2. adjustment
~, автомати́ческая automatic tuning
~, гру́бая coarse tuning
~, магни́тная permeability tuning
~, пла́вная continuous tuning
настро́йка, то́чная fine tuning (radio)
~, электро́нная electronic tuning
наступа́тельный aggressive, offensve, attacking
наступа́ть/наступи́ть to tread, to step, to come, to ensue; **наступа́ть** to attack, to make an offensive, to advance; **наступа́ть в пе́рвом эшело́не** to lead the assault, to advance in the first wave; **наступа́ть скачка́ми** to advance in bounds
наступи́ть (см наступа́ть)
наступле́ние 1. attack, offensive
2. advance
~, авиацио́нное 1. offensive air support
2. air offensive
~, артиллери́йское fire support plan for the assault, artillery support for the attack
~, безостано́вочное continuous offensive, sustained attack
~ во фланг flank attack
~, возду́шное air offensive
~, генера́льное general offensive
~ кру́пными си́лами major offensive
~, ло́жное diversionary attack
~, ночно́е 1. night attack
2. night advance
~, отвлека́ющее diversionary attack
~ рассве́та first light
~ с ограни́ченной це́лью attack with a limited objective
~ с плацда́рма 1. breakout of the beachhead
2. attack from a prepared position
~ темноты́ nightfall
~, фронта́льное frontal attack
~ эшело́нами attack in waves
насы́пка, порохова́я gunpowder filling (fuse)
на́сыпь embankment
~, железнодоро́жная railway embankment
~, земляна́я earth embankment, earth bank, earth mound
насыще́ние saturation
насы́щенность degree of saturation
натека́ние residual charge
на́тиск onslaught, attack
наткну́ться (см натыка́ться)
натр, е́дкий caustic soda
натрениро́ванность 1. state of training
2. physical fitness

натренировáть to train
нáтрий sodium
~, **азотнокúслый** sodium nitrate
~, **сернúстый** sodium sulphide (CW)
натыкáться/наткнýться to run (against), to stumble (across), to come (upon); **натыкáться/наткнýться на засáду** to run into ambush
натяжéние tightening, shrinkage
~ **гýсеницы** track tension (tk)
наýка, воéнная science of war
наýшники earphones
нáфта naphtha
находúться to be; **находúться в готóвности** to be in a state of readiness; **находúться в засáде** to lie in ambush; **находúться под ружьём** to be under arms
нахóдчивость 1. readiness
2. resourcefulness
находя́щийся в самовóльной отлýчке absent without leave
~ **в строю́** available for duty
нацéливать(ся)/нацéлить(ся) to aim
нацéлить(ся) (см нацéливать(ся))
начáло боевóго путú starting point of bombing/dropping run (by aircraft)
~ **движéния снаря́да** shot start (ballistics)
~ **нарéзов** commencement of rifling
начáльник 1. superior, head
2. chief (service)
3. officer in command, officer in charge
4. commander (other than fmn, unit or sub-unit)
~ **автомобúльной слýжбы полкá** regimental MT officer
~ **артиллéрии дивúзии** 1. divisional artillery commander
2. Commander Royal Artillery (Brit)
~ **артиллерúйского снабжéния полкá** regimental artillery supply officer
~ **артиллерúйско-техничéского снабжéния** ordnance officer
~, **ближáйший** immediate superior
~ **вещевóго снабжéния** clothing and equipment supply officer
~, **вышестоя́щий** superior commander
~ **гарнизóна** garrison commander
~ **караýла** guard commander
~ **клýба полкá** officer in charge of regimental club
начáльник медицúнского снабжéния officer in charge of medical supplies
~ **медицúнской слýжбы áрмии** 1. senior medical officer of army
2. Director Medical Services (Brit)
~, **непосрéдственный** next superior commander, immediate superior
~ **обóзно-вещевóго снабжéния полкá** regimental clothing and equipment supply officer
~ **оперативного отдéла** head of the operational section
~ **отдéла** departmental head
~ **оцеплéния** officer in charge of cordoning party (at firing range)
~ **пожáрной слýжбы** fire prevention officer
~ **полигóна** range officer, range warden
~ **продовóльственно-фурáжного снабжéния полкá** regimental rations and forage supply officer
~, **прямóй** direct superior, immediate superior
~ **свя́зи** senior signals officer (in unit)
~ **свя́зи áрмии** 1. army chief signals officer
2. Chief Signals Officer army (Brit)
~ **свя́зи дивúзии** 1. officer commanding divisional signals
2. Commander Royal Signals (Brit)
~ **свя́зи кóрпуса** 1. officer commanding corps signals
2. Corps Commander Royal Signals (Brit)
~ **свя́зи полкá** regimental signals officer, RSO
~ **свя́зи штáба фрóнта** 1. officer commanding signals of front
2. Commander, Army Group Signals (Brit)
~ **телегрáфной стáнции** telegraph supervisor
~ **ýзла свя́зи** officer in charge of signals office
~ **управлéния свя́зи** officer i/c point-to-point line communications (arty bn)
~ **учáстка переправы** commander of the crossing sector (river crossings)
~ **финáнсового довóльствия полкá** regimental pay officer
~ **химúческой слýжбы полкá** chief regimental chemical warfare officer
~ **штáба** chief of staff

начартди́в 1. commander of divisional artillery
2. CRA (Brit)
начартко́р 1. commander of corps artillery
2. Corps Commander Royal Artillery (Brit)
нача́ть (см начина́ть)
начеку́ on the alert
начерта́ние outline, trace
~ **зме́йкой** curved trace (trench)
~ **изло́мами** zig-zag trace (trench)
~, **кремалье́рное** square trace (trench)
~, **криволине́йное** curved trace, winding trace (trench)
~, **ло́манное** zig-zag trace
~ **транше́и, ло́манное** zig-zag trace of trench
~ **усту́пами** traversed trace (trench)
начерти́ть (см черти́ть)
начина́ть/нача́ть to begin, to start; **начина́ть/нача́ть штурм** to mount an assault
начштабри́г brigade chief of staff
наши́вка chevron, stripe
нащу́пать (см нащу́пывать)
нащу́пывать/нащу́пать to feed, to probe (tac), to prod (mine)
"не заме́чен" "Not Observed!" (arty)
не подлежи́т оглаше́нию restricted (security classification)
~ **подлежи́т ремо́нту на ме́сте** beyond local repair
~ **приго́дный к употребле́нию** unserviceable, u/s
~ **яви́вшийся на слу́жбу** absent from duty
неблагополу́чие, санита́рное insanitary conditions
небоеспосо́бный unfit for action, not battleworthy, disabled
небью́щийся shatterproof, unbreakable
невесо́мость weightlessness, zero gravity
невзведённый uncocked, unarmed (wpn, fuze)
невзры́в safe (fuse setting)
невзрыва́емость non-explosiveness
невибри́рующий vibrationless
невмене́ние freedom from responsibility
невмеша́тельство non-intervention
невозврати́мый 1. expendable
2. irreparable, irrevocable
невооружённый unarmed
невоспламеня́емый non-burning
невоспламеня́ющийся non-inflammable
невыполне́ние прика́за failure to carry out an order
невя́зка, лине́йная closure error (surv)
~, **углова́я** angular error of closure (surv)
~ **хо́да, углова́я** angular error of closure (surv)
негабари́тность bulkiness, oversize
негабари́тный out of gauge, off-gauge, excessive in size
него́дность unserviceability
него́дный 1. unserviceable, beyond repair, unsuitable
2. unfit (pers)
неде́йствующий inoperative
недисциплини́рованность lack of discipline
недогру́зка underloading, light loading
недода́ча underissue
недозво́ленный unauthorized, forbidden, unlawful
недока́т failure to run out (ord)
недокры́тие failure to close (breech)
недолёт short (gunn)
недоли́в incomplete filling
недоноси́тельство misprision (law)
недооце́нивать/недооцени́ть to underestimate
недооцени́ть (см недооце́нивать)
недооце́нка underestimation
недоста́ток 1. deficiency, shortage
2. defect, shortcoming
недоста́ча deficiency, lack, shortage
недосту́пность inaccessibility
недосту́пный inaccessible
недоукомплекто́ванный understrength
недохо́д до пере́днего положе́ния затво́ра incomplete closure of breach (ord)
недочёт 1. fault, defect
2. deficiency
незамини́рованный unmined, mine-free
незапотева́ющий anti-misting
незаражённый uncontaminated
незащищённый unprotected
неизоли́рованный bare, non-insulated, naked
неискажённый distortionless, undistorted
неисполне́ние прика́за failure to carry out an order

неиспра́вность 1. defect, fault, malfunction
2. carelessness
неиспра́вный unserviceable, out of order
нейрохирурги́ческий neuro-surgical
нейтрализа́ция neutralization
нейтрализова́ть to neutralize
нейтралите́т neutrality
нейтро́н neutron
~, бы́стрый fast neutron
~ деле́ния fission neutron
~, запа́здывающий delayed neutron
~, мгнове́нный prompt neutron
~, ме́дленный slow neutron
~, свобо́дный free neutron
~, теплово́й thermal neutron
некомпле́кт deficiency, shortage
некомпле́ктный deficient, incomplete
ненаде́жный unreliable, insecure
нена́стный inclement (met)
ненастро́енный untuned, non-resonant (elec)
необлицо́ванный unlined
необнару́женный undetected
необороня́емый undefended
необрабо́танный raw, coarse, rough, crude, untreated
необу́ченный untrained
неожи́данность surprise
неоконча́тельно снаряжённый filled, but unfused
неокопа́вшийся unentrenched, in the open, not dug in
неопо́знанный unidentified
неоса́ленный unlubricated (ammo)
неотраже́ние ги́льзы failure to eject a cartridge
непе́ль jointing clamp
непе́р napier, neper (elec, acoustics)
неплo̕тность 1. leakiness
2. looseness, unsoundness
непло́тный 1. leaky, leaking
2. loose
непобеди́мость invincibility
неповинове́ние insurbordination, disobedience
~, созна́тельное wilful defiance
неповоро́тливый unmanoeuvrable, awkward, clumsy
непого́да bad weather
непода́ча feed stoppage (wpn)
неподгото́вленный unprepared, untrained
неподо́бранность mismatch (elec)
неподтверждённый unconfirmed
неподчине́ние insubordination
непола́дка 1. failure, breakdown
2. disturbance, trouble
непосре́дственный close, immediate, direct
непотопля́емость unsinkability, ability to keep afloat
непреодоли́мость оборо́ны impenetrability of defence, impregnability of defence
неприго́дный unserviceable, unfit, unsuitable, ineligible
непристу́пный 1. impregnable, unassailable
2. inaccessible
неприце́льный unaimed
неприя́тель enemy
неприя́тельский enemy
непробива́емость impenetrability
непробива́емый impenetrable
~ оско́лками splinterproof
непроводни́к non-conductor
непрое́зжий impassable for vehicles
непрозра́чность opacity, non-transparency
непрозра́чный opaque, non-transparent
непрока́лывающийся puncture proof
непромока́емый impermeable, waterproof
непроходи́мый impassable
~ вброд unfordable
непро́чный unstable, insecure, unreliable fragile, precarious
неравноме́рность, вну́тренняя temporary deviation (control eng)
небради́вый lazy, careless
неразбо́рчивый unreadable, illegible, indecipherable
неразры́в blind (shell)
нераствори́мый insoluble
нерасторжи́мый indissolvable, inserverable
нервнобольно́й psychopathic case
нервю́ра rib (tech)
нереши́тельный indecisive, irresolute, undecided, hesitating
нержаве́ющий corrosion-proof, corrosion resistant, stainless
неро́вность ме́стности accidents of the ground
несамохо́дный towed, not self-propelled

несвоевре́менный ill-timed
несекре́тный unclassified, not secret
нескользя́щий skid-proof, antiskid
неснаряжённый 1. unarmed, unfused, unfilled, inert (ammo)
2. unloaded, empty
несогласо́ванность 1. mismatch (elec)
2. lack of coordination, lack of agreement
несоо́сность misalignment
несоотве́тствие 1. mismatch (elec)
2. disparity, discrepancy
нестанда́ртный irregular, non-standard
нести́/понести́ to bear, to sustain, to suffer
нестроеви́к non-combatant
нестроево́й non-combatant
несу́щая carrier (elec)
~, звукова́я voice carrier
~, когере́нтная coherent carrier
~ ла́зера laser carrier
~, некогере́нтная non-coherent carrier
~ опти́ческого диапазо́на optical-frequency carrier
~, радиочасто́тная radio-frequency carrier
~, шумова́я noise carrier
несу́щий supporting, carrying, bearing (mech, eng)
нета́бельный unauthorised, non-standard
нетабли́чный abnormal, non-standard
нетокси́чный non-toxic
нето́чность inaccuracy
нето́чный inaccurate, inexact
не́тто net weight
не́тчик absentee, leave-breaker (coll)
неувольне́ние из каза́рмы confinement to barracks
неуда́ча failure
неуда́чный unsuccessful
неукомплекто́ванный under-strength
неуравнове́шенный unbalanced
неустанови́вшийся 1. irregular
2. transient (elec)
3. unsteady
неутверждённый unconfirmed
неуязви́мость invulnerability
неуязви́мый invulnerable
нефтеба́за petroleum installation, tank farm
нефтеочи́стка petroleum refining
нефтепererабо́тка petroleum refining
нефтепрово́д oil pipeline
нефтепроду́кты petroleum, petroleum products
нефтетрубопрово́д oil pipeline
нефтехрани́лище petroleum installation, tank farm
нехва́тка shortage, deficiency
нечисто́ты sewage
нешта́тный 1. above establishment, surplus to establishment, nonorganic, unestablished, supernumerary
2. non-authorised, non-standard
нея́вка failure to appear, absenteeism
нея́дерный non-nuclear
неядови́тый non-toxic, clean (bomb)
нивели́р level, levelling instrument
~, техни́ческий engineer's level (surv)
нивели́рование levelling
~, барометри́ческое barometric levelling (surv)
~, геометри́ческое trigonometrical levelling
нивели́ровать to level
нивелиро́вка levelling, alignment
~, продо́льная line survey
низи́на low-lying ground
~, то́пкая swampy lowland
низково́дный low-level floating (br)
низкосо́ртный low grade
низкочасто́тный low frequency
ни́зменность plain less than 200 metres above sea level
"Ника́к Нет!" "No!", "No, Sir!" (formal negative answer to a superior)
ни́келевый nickel
никелирова́ние nickel plating
ни́кель nickel
ни́ппель 1. nipple
2. cap chamber (ammo)
~, перехо́дный nipple adapter
~, сма́зочный grease nipple
ни́ти, дальноме́рные stadia hairs
нитра́т nitrate
~ аммо́ния ammonium nitrate
нитробензо́л nitrobenzene
нитроглико́ль glycol dinitrate
нитроглицери́н nitroglycerine
нитрожелати́н blasting gelatine
нитроипри́т nitrogen-mustard gas
нитроклетча́тка nitrocellulose (expl)
нитрокрахма́л nitrostarch (expl)
нитромета́н nitromethane
нитропарафи́ны nitroparaffins
нитротолуо́л nitrotoluene

нитроцеллюло́за nitrocellulose
нить 1. strand, thread
2. filament (elec)
~, **вертика́льная** vertical hairline (instr)
~, **вольфра́мовая** tungsten filament
~, **горизонта́льная** horizontal hairline (instr)
~ **нака́ла** valve filament (elec)
~, **прице́льная** crosshair, cross-wire (instr)
~, **путево́дная** path marker tape, lead-in tape
~ **се́тки** line of graticules
~ **се́тки, горизонта́льная** horizontal line of graticules
нить-указа́тель sight graticule line
ни́ша niche, recess
~ **для боеприпа́сов** ammunition recess, ammunition niche
~, **патро́нная** ammunition niche, ammunition recess
~, **подбру́стверная** trench shelter
новичо́к new recruit
новобра́нец new recruit
нога́ 1. leg, foot
2. raker (br)
3. stand, brace
~, **ко́зловая** trestle leg (br)
~ **со́шки** bipod leg
~ **трено́ги** tripod leg
но́ги миномёта, опо́рные mortar bipod legs
нож knife
~ **отва́ла** cutting edge, leading edge (blade)
~, **фи́нский** hunting knife, trench knife, Finnish knife
но́жницы scissors, wire cutters
но́жны scabbard
~, **штыковы́е** bayonet scabbard
ножо́вка hacksaw
номенклату́ра nomenclature, designation, tallies
но́мер, а́томный atomic number
~ **ви́да колеба́ний** mode number (elec)
~ **военнослу́жащего, ли́чный** service number
~ **дета́ли** part number
~, **заво́дский** manufacturer's serial number
~ **изда́ния** edition number
~, **исходя́щий** originator's reference number (sigs)
но́мер, ли́чный service number, personal number
~, **номенклату́рный** stock number
~ **о́бласти колеба́ний** mode number (elec)
~, **опознава́тельный** identification number
~, **ору́ди́йный** gun number (pers)
~ **па́ртии боеприпа́сов** ammunition lot number
~ **полево́й по́чты** field post office number
~, **поря́дковый** serial number
~ **тра́нспорта** military train number, transport consignment number
~ **це́ли** target number
номеронабира́тель automatic telephone dial
номина́л rating
~, **преры́вистый** intermittent duty rating
номогра́мма calculating chart, nomogram
но́ниус vernier
"нора́, ли́сья" "fox's lair" (shelter trench), fox hole
нора́, мыши́ная "mousehole" (snow shelter)
но́рия bucket conveyor
но́рма боеприпа́сов normal scale of ammunition
~ **боеприпа́сов, ору́ди́йная** first line scale of ammunition
~ **вре́мени** time allowance (for doing a job)
~, **допусти́мая** maximum permissible dose
~, **ме́стная** local scale
~ **потребле́ния** consumption rate
~, **разря́дная** qualifying standard
~ **расхо́да** expenditure rate
~ **расхо́да горю́чего** fuel ration, fuel allowance
~ **снабже́ния** 1. scale of provision
2. level of supply
3. allowance
~ **снабже́ния боеприпа́сами** ammunition allowance
~ **снабже́ния, нали́чная** available supply rate
~ **содержа́ния запа́са** level of stock, stock holding level, reserves level
~, **су́точная** daily allowance, daily scale, day's quota, per diem allowance

но́рма, техни́ческая 1. technical standard
2. standard of performance
~, устано́вленная scale laid down, prescribed scale
нормализа́ция normalizing (steel)
нормати́в 1. quota, norm, standard
2. equipment establishment
нормирова́ние rating, rate setting, work quota setting
нормиро́ванный rationed scale
нормиро́вка 1. rationing, scaling
2. scheduling
но́рмы, разря́дные qualifying standards
нос 1. nose
2. bow (vessel)
носи́лки stretcher
~, басма́ческие "basmach" stretcher, mattress-type stretcher
но́скость wear life
носово́й nose, forward, front, bow
носо́к прикла́да toe of butt (SA)
~ снаря́да tip of shell
ночёвка 1. night halt
2. night shelter
ночле́г night halt, overnight bivouac
нукло́н nucleon
нуль, абсолю́тный absolute zero
нумера́ция numbering, numerical designation
нута́ция nutation, uneven flight (ballistics)
нутроме́р inside calliper
нырну́ть (см ныря́ть)
ныря́ло control plunger
ныря́ние dive
ныря́ть/нырну́ть to dive, to duck

О

обва́л landslide, fall
обвалова́ние diking
~ из мешко́в с земле́й sandbag walling
обво́д 1. outline, outlining
2. bypass
3. mesh, loop (elec)
~, вне́шний оборони́тельный perimeter defence
~, оборони́тельный defensive perimeter
обвола́кивать/обволо́чь to blanket (nucl), to envelope (gen)
обволо́чь (см обвола́кивать)
обго́н overtaking
"ОБГО́Н ЗАПРЕЩЁН" "NO OVERTAKING"
обгона́ть (см обгоня́ть)
обгоня́ть/обгона́ть to overtake
обгора́ние scorching
~ кла́пана valve burning
~ конта́ктов прерыва́теля pitting of contact breaker points
обде́лка 1. jacketing, lining
2. timbering (mining)
3. processing
4. lining (tunnel)
~ из волни́стого желёза steel tubing (mining)
обедне́ние то́плива fuel depletion
обезво́женный dehydrated
обезвре́дить (см обезвре́живать)
обезвре́женный neutralised, deactivated, defused
обезвре́живание 1. deactivation, neutralisation disarming
2. cleansing, decontamination (CW)
~ боеприпа́сов deactivation of ammunition, neutralisation of ammunition
~ взрывоопа́сных предме́тов (explosive) ordnance disposal
~ мин disarming of mines, neutralisation of mines
~ невзорва́вшихся бомб neutralisation of unexploded bombs
обезвре́живать/обезвре́дить to disarm, to neutralise, to render harmless; **обезвре́живать/обезвре́дить ми́ну** to neutralise a mine, to deactivate a mine
обезжи́риватель degreasing agent
обезжи́ривать/обезжи́рить to degrease, to extract the fat
обезжи́рить (см обезжи́ривать)
обеззара́живание purification, disinfection, decontamination
~ воды́ water purification
обеззара́живать/обеззара́зить to disinfect, to decontaminate, to purify
обеззара́зить (см обеззара́живать)
обезопа́сить (от) to secure against (perf)
обезору́живать/обезору́жить to disarm
обезору́жить (см обезору́живать)
оберто́н overtone, harmonic

обеспе́чение 1. maintenance, provision, supply and replenishment
2. protection, security
3. support, backing
4. guaranteeing, assuring
~, авиацио́нное 1. air support
2. air cover, screen
~, артиллери́йское artillery support
~, боево́е 1. protection, security in the field
2. first line logistic support
~ войск, лече́бное medical facilities, medical treatment
~ (войск), непосре́дственное инжене́рное direct engineer support
~ войск, санита́рно- профилакти́ческое preventive medical treatment
~, инжене́рное engineer support
~, лече́бное provision of medical services
~, лече́бно-эвакуацио́нное provision of medical and evacuation services
~, материа́льное maintenance, supply, logistic support, provision of material needs, logistics, "stores required" (paragraph in orders)
~, материа́льно-техни́ческое logistic support, technical maintenance, technical support, service support, logistics, provision of material needs
~ медици́нским иму́ществом supply and replenishment of medical equipment and stores
~, медици́нское 1. medical support, medical supplies
2. provision of medical supplies
~ ме́стности, инжене́рное eng operations, preparation of the ground
~, непосре́дственное close support
~, огнево́е fire support, fire cover
~ огнём covering fire, fire cover, fire support
~, первонача́льное initial maintenance
~ пита́ния provisioning
~, пла́новое routine maintenance
~ полётов flight servicing (AF)
~, противота́нковое anti-tank protection, anti-tank support, anti-tank security
~, противохими́ческое chemical security, anti-gas protection
~, противоэпидеми́ческое disease control
обеспе́чение рабо́чей си́лой labour supply
~, санита́рное 1. medical supplies
2. provision of medical supplies, supply of medical stores and equipment
~, санита́рно-профилакти́ческое provision of medical and hygiene services
~ слу́жбами ты́ла administrative support
~, та́нковое armoured support
~, техни́ческое 1. technical stores backing
2. technical maintenance, support
~ фла́нгов flank protection
обеспе́ченность supply situation, supply state
~ боеприпа́сами ammunition availability
~ горю́чим MT fuel state
~ запа́сами availability of stocks
~, материа́льная maintenance position, maintenance stocks
~ материа́льными сре́дствами stock position, supply situation
~ продово́льствием ration situation, ration state
обеспе́чивать/обеспе́чить to provide, to maintain, to supply, to afford, to ensure, to secure, to guarantee; **обеспе́чивать/обеспе́чить непреры́вное движе́ние тра́нспорта** to keep traffic moving; **обеспе́чивать/обеспе́чить прикры́тие** to afford protection, to provide cover
обеспе́чиваться/обеспе́читься to have; **обеспе́чиваться в пе́рвую о́чередь** to have first priority (stores etc); **обеспе́чиваться в после́днюю о́чередь** to have very low priority (stores etc)
обеспе́чить (см обеспе́чивать)
обеспе́читься (см обеспе́чиваться)
обеча́йка shell, drum, cowling
обжа́ть (см обжима́ть)
обжи́м 1. squeezing, pressing
2. crimper, crimpers
~ ду́льца tapering (ammo)
обжима́ние crimping
обжима́ть/обжа́ть to squeeze, to press, to crimp
обжи́мка 1. squeezing, pressing
2. crimper
~, клепа́льная rivet header

обзо́р 1. field of view
2. observation
3. scan, scanning (radar)
4. review, summary
~, боково́й sideways-looking (radar)
~, кони́ческий conical scanning
~, кругово́й 1. circular scanning, all-round scanning
2. all-round observation
~, по́лный circular scanning (radar)
~, радиолокацио́нный radar scan
~, се́кторный sector scan, sector surveillance, off-centering (radar)
обзо́рность visibility
~, за́дняя rear(ward) visibility
~, кругова́я all-round visibility
~, плоха́я poor visibility
~, хоро́шая unobstructed visibility
обката́ть (см обка́тывать)
обка́тка 1. rolling (roads)
2. running-in (engine, veh etc)
3. trial run, test run
обка́тывать/обката́ть to roll, to run in, to test out (machinery)
облака́, высо́ко-кучевы́е alto-cumulus clouds
~, высо́ко-слои́стые alto-stratus clouds
~, дождевы́е rain clouds, nimbus clouds
~, кучево́-дождевы́е cumulo-nimbus cloud
~, кучевы́е cumulus cloud
~, разо́рванно-кучевы́е fracto-cumulus clouds
~, разо́рванно-слои́стые fracto-stratus clouds
~, слои́сто-дождевы́е nimbo-stratus clouds
~, слои́сто-кучевы́е strato-cumulus clouds
~, слои́стые stratus clouds
о́блако, аэрозо́льное aerosol cloud
~, грибови́дное mushroom cloud (nucl
~, дымово́е smoke cloud
~, пе́ристое cirrus cloud
~ простра́нственного заря́да space charge cloud (electron)
~, радиоакти́вное radioactive cloud
~, электро́нное space charge cloud
о́бласть 1. area, region
2. oblast' (Sov admin sub-division)
3. field, sphere
~ высо́кого давле́ния region of high pressure (ball)
о́бласть ни́зкого давле́ния region of low pressure (ball)
~, операцио́нная operations area
~, перехо́дная threshold region (GW)
~ разреже́ния area of low pressure (expl)
~, светя́щаяся fireball (nucl)
~ сжа́тия area of high pressure
~ спе́ктра, ви́димая visible spectrum
~, тенева́я shadow region (radio)
о́блачность cloudiness, cloud cover (met)
~, ни́зкая low ceiling
обледене́ние icing-up, icing
облицо́вка facing, lining, lagging, sheeting
обли́ческий oblique
обло́мки debris, fragments
облуча́тель feed (antenna)
облуче́ние 1. irradiation (nucl)
2. illumination (radar)
~ в раскры́ве aperture illumination (radar)
~, прямо́е direct irradiation, direct radiation
~, теплово́е thermal exposure
обма́н deception
обману́ть (см обма́нывать)
обма́нывать/обману́ть to deceive
обмеле́ть (см меле́ть)
обме́н 1. traffic (radio, teleg etc)
2. exchange, interchange
~, операти́вный operational traffic (comms)
~, радиотелефо́нный radio-telephone traffic
обме́р measurement
обмора́живание frost-bite
обмороже́ние frost-bite
обморо́женный frostbitten
обмо́тка 1. winding (elec)
2. taping (elec insulation)
~ возбужде́ния field winding
~ генера́тора, по́люсная generator field coil
~, короткоза́мкнутая squirrel-cage winding
~, перви́чная primary winding (elec motor etc)
~, по́люсная field winding, field coil
~, управля́ющая control winding (elec)
~, шабло́нная former winding (elec)
~, шунтова́я shunt winding (elec)
обмо́тки puttees
обмундирова́ние uniform, clothing, outfit

обмундирова́ние, зи́мнее winter uniform
~ **и снаряже́ние** clothing and equipment
~, **ле́тнее** summer uniform
~ **парашюти́стов** jump suit
~, **повседне́вное** everyday uniform
~, **повседне́вно-полево́е** everyday field uniform
~, **полево́е** field uniform
~, **тропи́ческое** tropical uniform, tropical kit
обмундирова́ть (см обмундиро́вывать)
обмундиро́вывать/обмундирова́ть to issue with clothing, to clothe, to dress, to fit out
обмундирожа́ние, пара́дно-выходно́е parade walking-out uniform
обмы́вка 1. personal cleansing (NBC) 2. washing-down
обнаруже́ние location, detection, discovery, spotting
~, **акусти́ческое** acoustic detection
~, **бли́жнее** short range detection
~, **да́льнее** 1. early warning (radar) 2. long-range surveillance, long-range detection, long-range acquisition
~ **и запро́с, автомати́ческое** automatic error correction system, automatic ARQ system
~ **инфракра́сного излуче́ния** infra-red radiation detection
~ **мин** detection of mines, mine detecting
~ **по реакти́вной струе́** efflux detection
~, **ра́ннее** early warning, surveillance, long range warning
~ **у́мысла** evidence of intent (legal)
~, **хими́ческое** chemical detection
~ **це́ли** target detection
обнару́живать/обнару́жить to discover, to detect, to spot, to betray, to reveal (position etc), to locate (faults)
обнаружи́тель, инфракра́сный metascope, IR sensor
обнару́жить (см обнару́живать)
обновле́ние renovation, renewal
~ **запа́са** stock rotation
обобща́ть/обобщи́ть to generalise, to summarize, to crystalize; **обобща́ть/обобщи́ть зая́вки** to consolidate indents
обобще́ние 1. consolidation 2. collation, synthesis
обобщи́ть (см обобща́ть)
обогаще́ние enriching (fuel etc)
обогре́в heating(-up), warming(-up)
обогрева́ние warming up, warming, heating
обогрева́тель 1. heating apparatus 2. heater
~ **ветрово́го стекла́** hot-air windshield defroster
~ **каби́ны** cab heater
~ **прице́ла** sight heater (gunn)
о́бод колеса́ wheel tread, wheel rim
~, **колёсный** wheel rim
~, **наса́женный** built-up rim
~, **разъёмный** split wheel rim, divided-type wheel rim
ободо́к, срезно́й shear collar (ammo)
обо́з transport, train (admin)
~, **вещево́й** baggage train
~, **вью́чный** pack train
~, **корпусно́й** corps train
~, **полево́й** field train
~, **ремо́нтный** maintenance train, maintenance column
обозначе́ние мин marking of mines
~ **образца́** model type designation (wpn)
~, **усло́вное** conventional sign
обозначе́ния, усло́вные legend, key (maps, diagrams), conventional signs
обо́зник transport soldier
обо́йма 1. cartridge clip, charger, clipful of ammunition 2. ring, bracket clip, band
~, **автома́тная** magazine
~ **ката́ния** ball-bearing race
~ **миномёта** cradle (mor)
~, **патро́нная** cartridge clip
~, **па́чечная** block clip, en bloc clip
~, **пружи́нная** spring clip
~ **са́льника** gland housing
~, **снаряжённая** filled charger
~ **шарикоподши́пника** ball race, bearing cage
~ **экстра́ктора** extractor collar
обойти́ (см обходи́ть)
оболо́чка 1. jacket, case, casing, envelope, covering 2. tamper (nucl)
~ **бо́мбы** bomb case
~, **бума́жная** 1. paper shell, paper cover 2. waxed paper covering (eg on explosive charge in landmine)
~, **металли́ческая** metal envelope
~ **ми́ны** mine case
~ **пу́ли** bullet jacket
~, **скрепля́ющая** outer tube (gun barrel)

оболо́чка, сли́зистая mucous membrane
~, электро́нная electron shell (nucl)
обора́чиваемость turnover rate
оборо́на defence, defences
~, авиадеса́нтная active air defence
~, авиазени́тная active air defence
~, акти́вная aggressive defence
~, акти́вная противовозду́шная active air defence
~, акти́вная противоракéтная active anti-missile defence
~, берегова́я coast defence
~, войскова́я противовозду́шная field air defence
~, глубо́кая defence in depth
~, гражда́нская civil defence
~, долговре́менная fortified defence system, permanent defence system
~, жёсткая stout defence, stiff resistance, rigid defence
~, затяжна́я protracted defence
~, зени́тная anti-aircraft defence
~ и охра́на ты́ла defence and protection of the rear
~, ко́свенная противоракéтная indirect anti-missile defence
~, кругова́я all-round defence
~, кругова́я противовозду́шная all-round AA defence
~, кругова́я противоракéтная all-round anti-missile defence
~, манёвренная mobile defence
~, ме́стная противовозду́шная civil defence, local air defence
~, ми́нная countermining
~, моби́льная mobile defence
~ на заблаговре́менно обору́дованных пози́циях deliberate defence
~ на поспе́шно обору́дованной ме́стности hasty defence
~ побере́жья, противодеса́нтная defence of defensive coastal area
~, подви́жная mobile defence
~, подгото́вленная deliberate defence
~, позицио́нная 1. static defence, positional defence 2. static defences
~, поспе́шно за́нятая hasty defence
~ проти́вника, подгото́вленная prepared enemy positions
~, противоартиллери́йская defence against artillery
оборо́на, противовозду́шная air defence, AA defence
~, противодеса́нтная 1. defence against amphibious assault landing 2. defence against airborne assault landing
~, противоло́дочная anti-submarine defence
~, противоми́нная defence against mines (naut)
~, противопожа́рная fire cover, fire fighting, defence against fires
~, противоракéтная anti-missile defence
~, противота́нковая anti-tank defence
~, противохими́ческая anti-gas defence, defence against gas
~, про́чная stable defence
~, упо́рная stubborn defence
оборони́тельный defensive
оборони́ть (см обороня́ть)
оборо́нный pertaining to defence
обороноспосо́бность defence capabilities
обороня́ть/оборони́ть to defend
обороня́ющийся defender
оборо́т 1. revolution, turn 2. turnover (stores etc) 3. back, reverse
оборо́ты в мину́ты revs per minute
~ в секу́нды revs per second
~ дви́гателя, номина́льные rated engine speed
~, холосты́е idling revs
обору́дование 1. equipment, plant 2. engineer preparation (of ground etc)
~, авари́йное emergency equipment
~ автомати́ческой обрабо́тки да́нных automatic data processing equipment
~, бортово́е on-board equipment
~, бульдо́зерное dozer attachment, dozer blade
~, вое́нно-инжене́рное engineering equipment, engineer plant (mil)
~, вози́мое on-vehicle equipment
~, вспомога́тельное ancillary equipment, accessories
~, дезактивацио́нное decontamination equipment
~, дезинсекцио́нное facilities for delousing
~ для вы́дачи горю́чего fuel dispensing facilities

обору́дование для техни́ческого обслу́живания и теку́щего ремо́нта maintenance and repair equipment
~ **для хране́ния горю́чего** fuel storage facilities
~, **доро́жно-строи́тельное** road-building equipment
~, **душево́е** shower facilities
~, **запасно́е** reserve equipment
~, **запра́вочное** fuelling equipment, refuelling equipment
~, **землеро́йно-тра́нспортное** earth-moving equipment
~, **измери́тельное** instrumentation
~, **инжене́рное** engineer equipment
~, **контро́льная** monitoring equipment
~, **контро́льно-измери́тельное** testing and measuring equipment
~, **контро́льно-испыта́тельное** test equipment
~, **ла́герное** camp equipment, camp stores
~ **ме́стности** preparation of the ground (by engineers)
~ **ме́стности, инжене́рное** 1. field works 2. construction of field works
~, **навесно́е** 1. bulldozer attachment, engineer attachment (tk, tractor) 2. detachable equipment, attachment
~, **навигацио́нное** navigation equipment
~, **назе́мное** ground support equipment, ground equipment (GW)
~, **обслу́живающее** servicing equipment
~ **огнево́й пози́ции** setting-up of a fire position, preparation of a fire position
~, **перено́сное** portable equipment
~, **перено́сное стре́льбищное** portable small arms target eqpt
~, **подбёмно-тра́нспортное** weight handling equipment, stores handling equipment
~, **подъёмное** 1. erecting equipment (GW) 2. lifting equipment
~ **полиго́на** range facilities
~, **прибо́рное** instrumentation
~, **противопожа́рное** fire-fighting equipment
~, **пусково́е** launching equipment (msl)
~, **радиопередаю́щее** radio-transmitting equipment
~, **радиоэлектро́нное** radio and electronic equipment, radio and electronic installations
обору́дование, резе́рвное stand-by equipment, back-up equipment
~, **ремо́нтно-восстанови́тельное** recovery equipment
~, **ремо́нтное** repair equipment
~ **свя́зи** communication equipment
~, **складско́е** storage equipment
~, **сло́жное** sophisticated equipment
~, **спаса́тельное** rescue gear, survival equipment
~, **стерилизацио́нное** sterilization equipment
~, **такела́жное** rigging equipment
~ **та́нка, минноразгради́тельное** tank mine-sweeping equipment
~, **устано́вленное** installed equipment
~, **устаре́лое** out-of-date equipment, obsolete equipment
~, **фортификацио́нное** field defences
~, **холоди́льное** refrigerating equipment
обору́дованный equipped, fitted out, prepared for use (ground etc)
обору́довать to equip, to fit out, to prepare for use (ground etc)
обостре́ние aggravation, sharpening
обо́чина доро́ги road shoulder
обраба́тывать/обрабо́тать to develop (photo), to process, to tool, to machine (metall), to cultivate, to till
обрабо́тать (см обраба́тывать)
обрабо́тка 1. machining, tooling 2. processing, developing, treatment
~, **артиллери́йская** artillery preparation, softening up
~ **гру́зов** freight handling
~ **да́нных** data processing
~ **да́нных, автомати́ческая** automatic data processing
~, **дезинфекцио́нно-дезинсекцио́нная** disinfestation
~ **донесе́ний** 1. processing of messages 2. collation and interpretation of information
~ **земли́** tillage, cultivation
~, **идиологи́ческая** brain washing
~, **огнезащи́тная** fireproofing
~, **полити́ческая** political indoctrination
~, **по́лная санита́рная** full personal cleansing (NBC)
~, **санита́рная** personal cleansing (NBC)
~, **санита́рно-хими́ческая** personal cleansing (CW)

обрабо́тка, специа́льная special treatment, decontamination
~, терми́ческая heat treatment (metall)
~, части́чная санита́рная partial personal cleansing, preliminary personal cleansing (NBC)
образе́ц model, type, specimen, sample
~, боево́й operational version
~, испыта́тельный test sample
~, о́пытный experimental model
~, эксперимента́льный test model
~, этало́нный standard sample
образова́ние 1. education 2. formation, creation
~, вое́нное military education
~ нага́ра на свече́ зажига́ния spark plug carbon fouling
~ на́кипи scale formation
~ пар pair formation (nucl)
обрати́ться (см обраща́ться)
обраща́ться/обрати́ться to apply (to), to address, to accost, to turn (into), to treat, to handle, to manage; **обраща́ться/обрати́ться с боеприпа́сами** to handle ammunition
обраще́ние, непра́вильное improper use, improper handling
~ с боеприпа́сами ammunition handling
~ с пулемётом handling of a machine gun
обре́з sawn-off gun, sawn-off rifle
~ доро́ги road verge, soft shoulder
обру́шивать/обру́шить to demolish, to destroy; **обру́шивать/обру́шить ого́нь** to bring down fire (upon), to plaster with fire
обру́шиваться/обру́шиться to fall upon, to attack with great force, to collapse, to fall down
обру́шить (см обру́шивать)
обру́шиться (см обру́шиваться)
обры́в 1. break (in wire etc) 2. precipice
обры́вистый steep
обсерва́ция observation
обсле́дование investigation, reconnaissance
~, дозиметри́ческое radiac survey, radiation survey
обсле́довать to inspect, to examine, to explore
обслу́живание servicing, maintenance
обслу́живание автомоби́лей, техни́ческое car servicing
~, ба́нно-пра́чечное provision of bath and laundry facilities
~, боево́е combat maintenance
~ боево́й те́хники на по́ле бо́я battlefield maintenance
~ в полевы́х усло́виях, техни́ческое field maintenance
~ в эксплуата́ции, техни́ческое operation service, field service
~, ежедне́вное техни́ческое last-parade servicing, daily maintenance, daily technical servicing
~, еженеде́льное weekly maintenance
~ и ремо́нт, техни́ческое maintenance and repair
~, медици́нское medical attention
~ на остано́вках at-halt servicing
~, номерно́е техни́ческое mileage servicing
~, пла́новое routine maintenance, scheduled maintenance
~ по гра́фику scheduled maintenance
~, полево́е техни́ческое technical servicing in the field
~, профилакти́ческое preventive maintenance
~, сезо́нное техни́ческое bi-annual servicing
~ си́лами экипа́жа crew maintenance, crew servicing
~, теку́щее техни́ческое routine maintenance
~, техни́ческое servicing, maintenance, technical servicing
обслу́живать/обслу́жить to service, to maintain, to operate, to handle
обслу́жить (см обслу́живать)
обстано́вка situation
~, авари́йная emergency situation
~, боева́я 1. tactical situation 2. active service conditions, combat conditions
~, возду́шная air situation
~, вы́годная advantageous situation
~, исхо́дная 1. initial situation 2. opening narrative
~, конкре́тная actual situation
~, метеорологи́ческая meteorological situation, weather conditions
~, мише́нная layout of training targets
~, назе́мная ground situation

обстано́вка, операти́вная operational situation
~ по ты́лу situation in rear
~, похо́дно-боева́я field conditions
~, такти́ческая situation, tactical situation
~, тылова́я situation in rear
~, хими́ческая chemical situation
~, я́дерная nuclear conditions, atomic environment
обстоя́тельства, отягча́ющие aggravating circumstances
~, смягча́ющие mitigating circumstances
обстре́л 1. bombardment, shelling, gunfire 2. field of fire
~, артиллери́йский artillery concentration, artillery shelling, artillery quick concentration
~, кругово́й 1. all-round field of fire 2. all-round traverse
~, миномётный mortar fire
~ пло́щади area fire
~, продо́льный enfilade fire
~ хими́ческими снаря́дами chemical bombardment
обстре́ливать/обстреля́ть to bring fire to bear, to attack by fire; **обстре́ливать/обстреля́ть миномётным огнём** to mortar, to bring mortar fire to bear
обстре́ливаться/обстреля́ться to become battle-hardened, to become accustomed to fire
обстре́лянный battle-hardened, seasoned (tps)
обстреля́ть (см обстре́ливать)
обстреля́ться (см обстре́ливаться)
обта́чивать/обточи́ть to turn (lathe), to grind, to smooth
обтека́емый streamlined
обтека́ние flow
обтека́тель streamlined casing, fairing, cowling
~, анте́нный radome
~, аэродинами́ческий aero-dynamic fairing
~, опти́ческий IR-dome (GW)
~ противога́за respirator deflector tube
~, радиолокацио́нный radome
~, хвостово́й tail fairing
обти́рка wiping down, cleaning
обточи́ть (см обта́чивать)
обто́чка 1. turning, rounding off (lathe) 2. machining, facing, cleaning off
~, получистова́я semi-finished turning
~, чернова́я rough turning (eng)
~, чистова́я finish turning (eng)
обтюра́тор obturator
обтюра́ция obturation
~ ги́льзой case obturation
о́бувь footwear
~, во́йлочная felt footwear
~, ми́нно-защи́тная anti-mine boots
обу́гливание carbonization, charring
обустро́йства, железнодоро́жные rail facilities, railway installations
обуча́емый trainee
обуче́ние indoctrination, training, instruction
~ веде́нию радиацио́нной разве́дки radiac reconnaissance training
~, вое́нное military training
~, нача́льное basic training
~, одино́чное individual training
~ подрывно́му де́лу demolition training
~ стрельбе́ gunnery training
~, строево́е drill instruction (Sov term includes elementary battle drill)
~, шере́ножное squad drill
обу́ченность state of training, amount of training
обуше́к, вертлю́жный swivel eye
обхва́т close envelopment (tac)
~, наспи́нно-плечево́й upper suspension straps (para)
~, поясно́й waist strap (para)
обхо́д 1. closed traverse (survey) 2. outflanking, turning movement, wide envelopment (tac)
обходи́ть/обойти́ to bypass, to envelope, to outflank
обходно́й outflanking, bypassing, bypass, alternate
обходя́щий flanking, enveloping
обшива́ть/обши́ть to sheath, to line
обши́вка plating, lagging, sheeting, lining
~, бортова́я side plate
~, бронева́я armour plating
~ из лёгкого спла́ва light-alloy skin
~ из листово́й ста́ли steel plating
~ из стекловолокна́ fibreglass sheathing
~, теплонепроница́емая heat-proof lagging

обши́вка тормозно́й коло́дки brake shoe lining
~, фрикцио́нная brake lining, friction lining
обши́тый броне́й armour-plated
обши́ть (см обшива́ть)
обшла́г cuff
общевойсково́й combined-arms, all-arms
общела́герный all-camp
обще́ние, ли́чное personal contact
о́бщество, вое́нно-нау́чное military research group (at unit level)
о́бщий general, overall
объедине́ние, операти́вное large strategic formation, force, grouping (Sov front, army, fleet, flotilla etc)
объединённый combined, joint, centralized, united, pooled
объедини́ть (см объединя́ть)
объединя́ть/объедини́ть to join, to unite, to pool, to combine; **объединя́ть/объедини́ть зая́вки** to consolidate requisitions
объе́зд diversion, bypass
~, запа́сный emergency by-pass
объе́кт 1. objective, target, aim
2. object, thing
3. construction, project
~ ата́ки assault objective
~ бомбомета́ния bombing objective, bombing target
~, ва́жный key point
~, ло́жный decoy installation
~ нападе́ния objective
~, примо́рский coastal installation
объекти́в objective lens (optics)
~, нормальноуго́льный normal angle lens, standard lens
~, узкоуго́льный telephoto lens
~, широкоуго́льный wide-angle lens
объём ка́меры сжа́тия clearance volume (ICE)
~ ка́моры chamber capacity (gunn)
~ перево́зок volume of traffic
~, рабо́чий swept volume, displacement volume (ICE)
~, уде́льный specific volume
~ цили́ндра, по́лный total volume of cylinder (ICE)
объёмистый bulky, voluminous, capacious
объёмный 1. three-dimensional
2. bulky, voluminous, capacious (coll)
объяви́ть (см объявля́ть)
объявле́ние войны́ declaration of war
объявля́ть/объяви́ть to declare, to announce, to publish; **объявля́ть/объяви́ть вы́говор** to award a reprimand; **объявля́ть/объяви́ть прива́л** to call a halt; **объявля́ть/объяви́ть стро́гий вы́говор** to award a severe reprimand; **объявля́ть/объяви́ть трево́гу** to raise the alarm, to alert
"обыкнове́нная" "normal" (Sov message precedence)
о́быск search
обя́занность duty, obligation
~, во́инская military duty
~, всео́бщая во́инская general liability to military service
обяза́тельный obligatory, compulsory
обяза́тельство obligation, commitment
ОВ, мышьяко́вистое arsenical gas
~, мышьякосодержа́щее arsenical gas
~, туманообра́зное droplet-type gas
овладева́ть/овладе́ть to seize, to capture, to master, to become proficient
овладе́ние 1. seizure
2. mastery
овладе́ть (см овладева́ть)
овра́г ravine
овчи́на sheepskin
огиба́ние релье́фа terrain following (low flying ac)
огиба́ющая envelope (electron)
~, модуляцио́нная modulation envelope
огло́хнуть (см гло́хнуть)
огнево́й fire, firing
огнегася́щий fire-extinguishing
огнезащи́тный fire-protective
огнемёт flame-thrower
~, механизи́рованный vehicle-mounted flame-thrower
~, носи́мый portable flame thrower
~, ра́нцевый manpack flame-thrower
~, та́нковый tank flame-thrower
~, фуга́сный static flame-thrower, emplaced flame-thrower
огнемета́ние flame-throwing
огнемётчик flame-thrower operator
"ОГНЕОПА́СНО" "INFLAMMABLE"
огнеопа́сность inflammability
огнеопа́сный flammable, inflammable
огнеприпа́сы ammunition

огнесмéсь incendiary fuel, burning mixture
огнестóйкий fire-resistant, fire-resisting
огнетушúтель fire extinguisher
~, **аэрозóльный** aerosol fire extinguisher
~, **жúдкостный** wetting agent fire extinguisher
~, **пéнный** foam fire extinguisher
~, **порошкóвый** powder-type fire extinguisher
~, **сухóй порошкóвый** dry powder fire extinguisher
~, **углекислóтный** carbon-dioxide fire extinguisher
огнеупóрность heat resistance
огнеупóрный flame-proof, refractory, fire-proof, fire-resistant
огнú, опознавáтельные identification lights
оголóвник capsill
"Огóнь!" "Fire!"
огóнь 1. fire
2. light
~, **автоматúческий** automatic fire
~, **артиллерúйский** artillery fire
~, **бéглый** 1. rapid fire (SA)
2. gun fire (arty)
~, **беспокóящий** harassing fire
~, **беспорядочный** random fire
~, **блúжний** close defensive fire
~, **бортовóй** navigation light (ac)
~ **в тóчку** pin-point fire
~ **в упóр** point-blank fire
~, **винтóвочный** rifle fire
~, **внезáпный** surprise fire
~, **возврáтный** retaliatory fire
~, **губúтельный** withering fire
~, **дáльний** long-range fire
~, **действúтельный** effective fire
~, **длúтельный** sustained rate of fire
~, **дрýжный** co-ordinated fire
~, **заградúтельный** defensive fire, DF
~ **зáлпами** 1. volley fire (SA)
2. salvo fire (arty)
~, **зáлповый** 1. volley fire (SA)
2. salvo fire (arty, nav)
~, **зенúтный** anti-aircraft fire, Flak
~, **зенúтный заградúтельный** anti-aircraft barrage
~ **и манёвр** fire and movement
~ **из стрелковóго орýжия** small-arms fire
~, **измáтывающий** harassing fire
огóнь, имитациóнный simulated fire
~, **интенсúвный** intensive fire
~, **картéчный** case fire, canister fire
~, **кинжáльный** surprise short-range concentrated fire
~, **комбинúрованный** combined fire
~, **контрбатарéйный** counterbattery fire
~, **контрзаградúтельный** counter-barrage
~, **косоприцéльный** oblique fire
~, **кýчный** closely grouped fire, fire of good consistency
~, **массúрованный** concentrated fire, massed fire
~, **мéдленный** slow fire
~, **мéткий** accurate fire
~, **методúческий** battery fire (arty)
~, **миномётный** mortar fire
~, **мóщный** heavy fire
~ **на блúжних пóдступах** close-in defensive fire
~ **на большýю дáльность** long-range fire
~ **на воспрещéние** interdiction fire
~ **на дáльние дистáнции** long-range fire
~ **на задымлéние** smoke fire (arty)
~ **на изнурéние** harassing fire
~ **на окаймлéние** box barrage fire
~ **на ослеплéние** smoke fire
~ **на подавлéние** neutralization fire
~ **на поражéние** fire for effect
~ **на разрушéние** destruction fire
~ **на уничтожéние** annihilation fire, destruction fire
~ **на ходý** fire on the move
~, **наблюдáемый** observed fire
~, **навéсный** high trajectory fire
~, **навигациóнный** navigation light
~, **нарастáющий** progressive fire
~, **настúльный** flat trajectory fire
~, **нащýпывающий** probing fire
~, **неплáновый** engagement of opportunity targets
~, **неподвúжный заградúтельный** fixed defensive fire, stationary barrage
~ **непосрéдственной поддéржки** close-support fire
~, **непрерывный** sustained fire
~, **неприцéльный** unaimed fire, random fire
~ **непрямóй навóдкой** indirect fire
~ **ограждéния** obstruction light

огóнь, одинóчный single round fire, single shot fire
~, окаймля́ющий заградúтельный box barrage
~, орудúйный cannonade, gun fire
~, оружéйно-пулемётный small-arms fire
~, ответный return fire
~, ответный заградúтельный counter bombardment
~ отдéльными óчередями батарéи battery fire
~ отдéльными óчередями взвóда troop fire
~, отсéчный interdictory fire, interdiction fire
~, перекрёстный cross-fire
~, плáновый programme shoot, pre-arranged shoot, pre-arranged fire
~, плáновый заградúтельный prearranged barrage
~, плóтный dense fire
~ по вы́зову fire on call
~ по мéсту верóятного нахождéния цéли prophylactic fire
~ по ненаблюдáемым цéлям unobserved fire
~ по отдéльным цéлям engagement of separate targets
~ по плóщади area fire
~, подвúжный заградúтельный creeping defensive fire
~, полуавтоматúческий semi-automatic fire
~, посáдочный landing light (ac)
~, прикрывáющий covering fire, protective fire
~, пристрéлочный ranging fire, correction fire
~, прицéльный aimed fire
~, продолжúтельный sustained fire
~, продóльный enfilade, flanking fire
~, противопехóтный anti-personnel fire
~, противотáнковый anti-tank fire
~, прочёсывающий sweeping fire
~ прямóй навóдкой direct fire, fire over open sights
~, пулемётно-автомáтный automatic small-arms fire
~, рéдкий одинóчный slow single-shot fire
~, решúтельный aggressive fire
огóнь, ружéйно-пулемётный small-arms fire
~, ружéйный rifle fire
~ с закры́той позúции indirect fire
~ с искýсственным рассéиванием distributed fire
~ с корóтких остановóк fire from the short halt (AFV gunn)
~ с манёвром fire and movement
~ с мéста stationary fire (AFV gunn)
~ с остановóк fire from the halt
~ с ПУÁЗО AA fire using FCE
~ с хóду fire on the move (AFV gunn)
~, слáбый desultory fire
~, сосредотóченный concentrated fire
~, сплошнóй massed fire
~ стрелковóго орýжия small-arms fire
~, тóповый masthead light
~, тóчный accurate fire
~, ураѓанный hail of fire, drum fire, very heavy fire
~, флангóвый flanking fire, flank fire
~, фланкúрующий flanking fire, flank fire
~, фронтáльный frontal fire
~, чáстый rapid fire
~, чáстый одинóчный rapid single-shot fire
~ чéрез гóлову overhead fire
~, шквáльный massed intense fire
оградúть (см ограждáть)
ограждáть/оградúть to enclose, to fence in, to guard, to bar, to block
ограждéние 1. cartridge case deflector, deflector assembly
2. fence, barrier
ограничéние limiting, curtailing
~, двухсторóннее positive and negative peak limiting
~ на диóде diode limiting
~ на триóде triode limiting
~, сéточное grid current limiting (elec)
ограничéнный limited, bound
ограничивать/ограничить to border, to bound, to limit, to curtail
ограничúтель 1. limiter (elec)
2. positioning stop (MG)
3. stop (mech)
4. boundary marker
~, амплитýдный amplitude limiter
~ балансúра road wheel arm stop
~ боевóй пружúны firing spring retainer

ограничи́тель вертика́льной наво́дки elevation stop
~ **оборо́тов** engine speed governor
~ **поворо́та рукоя́тки** LBM limiting stop, breech mechanism lever limiting stop (ord)
~ **предохрани́теля** safety stop
~ **рукоя́тки, пере́дний** front catch bracket (QF ord)
~ **торсио́на** torsion bar retainer
~ **угла́ возвыше́ния** elevation limit stop (ord)
~ **це́пи** chain skid
~ **шепта́ла** sear stop
ограни́чить (см ограни́чивать)
оде́жда 1. clothing
2. revetment
3. surfacing (road)
~, **бето́нная** concrete road surface
~, **дерно́вая** turf revetment
~, **доро́жная** top dressing, road surfacing
~, **доща́тая** cut timber revetment
~, **жердева́я** log revetment, stake revetment
~, **защи́тная** protective clothing, anti-gas clothing
~ **из мешко́в с земле́й** sandbag revetment
~, **импрегни́рованная** impregnated clothing
~, **маскиро́вочная** camouflage clothing
~, **огнезащи́тная** flame-proof suit, protective clothing (fire)
~, **плетнёвая** brushwood revetment
~, **рабо́чая** working clothes, working dress
~ **с электроподогре́вом** electrically-heated clothing
~, **специа́льная** special-purpose clothing
~, **стёганая ва́тная** quilted clothing
~, **фаши́нная** fascine revetment
~, **фо́рменная** uniform
~ **хво́ростом** brushwood revetment
~, **шта́тская** civilian clothes
~, **щебёночная** gravel surface (rd)
одержа́ть (см оде́рживать)
оде́рживать/одержа́ть to gain; **оде́рживать/одержа́ть побе́ду над проти́вником** to beat the enemy
одерно́ванный turfed
оде́тый по фо́рме properly dressed
одея́ло blanket
однобо́ртный single-breasted
одновибра́тор one-shot oscillator, Kipp oscillator
однозаря́дный single-shot (weapon)
однокаска́дный single-stage (electron)
однополо́сный single side-band
однополча́нин soldier serving in same regiment, soldier from same unit
одноро́дность homogeneity, uniformity
одноство́льный single-barrelled
одноступе́нчатый single-stage
однофа́зный monophase
одноцве́тный monochrome, monochromatic
одношере́ножный single rank
одноэшело́нный of the same echelon
ожива́ло ogive
ожива́льный ogival
ожида́ние, математи́ческое mathematical expectation
~ **числа́ попада́ний, математи́ческое** expected rate of hits
ожо́г 1. burn
2. shot cutting edge of target
~ **второ́й сте́пени** second degree burn
~ **пе́рвой сте́пени** first degree burn
~, **светово́й** flash burn
~, **теплово́й** flash burn
~ **тре́тьей сте́пени** third degree burn
о́зеро lake
озо́н ozone
озонопа́уза ozonopause (atmospheric layer)
оказа́ть (см ока́зывать)
ока́зывать/оказа́ть to render, to show; **ока́зывать/оказа́ть медици́нскую по́мощь (кому)** to treat (someone) (med); **ока́зывать/оказа́ть пе́рвую по́мощь** to render first aid; **ока́зывать/оказа́ть сопротивле́ние** to offer resistance, to put up resistance
окайми́ть (см окаймля́ть)
окаймле́ние boxing-in
~, **огнево́е** box barrage
~, **трёхсторо́ннее огнево́е** open box barrage
окаймля́ть/окайми́ть to box in, to cut off (tac); **окаймля́ть/окайми́ть огнём** to box in by fire
ока́пывание digging-in, entrenching
ока́пывать(ся)/окопа́ть(ся) to dig in, to entrench
окарау́ливание guarding

окисле́ние oxidization
окисли́тель oxidant, oxidizing agent
окисли́ть (см окисля́ть)
окисля́емость oxidation susceptibility
окисля́ть/окисли́ть to oxidize
о́кись oxide
~ **азо́та** nitrogen monoxide
~ **углеро́да** carbon monoxide
окклю́зия 1. occlusion (physics, met)
2. sweating (gunn)
оккупа́нт occupying power, occupying army
оккупа́ция occupation (of territory)
оккупи́ровать to occupy (territory)
окла́д rate of pay
~, **дверно́й** cap and post sett (mining, eng)
~, **основно́й** basic pay
~ **по во́инскому зва́нию** pay for rank
~ **по до́лжности** pay for appointment
~, **прива́рочный** subsistence allowance
о́клик 1. challenge (sentry)
2. hail, call
оклика́ть/окли́кнуть to hail, to call, to challenge (sentry)
окли́кнуть (см оклика́ть)
окно́ 1. window
2. opening
3. port
~, **ветрово́е** windscreen
~, **вса́сывающее** air intake
~, **выводно́е** ejection opening
~, **выпускно́е** exhaust port
~, **выхлопно́е** exhaust port
~ **для выбра́сывания стре́ляной ги́льзы** empty case ejection opening
~ **для отраже́ния гильз** ejection slot (SA)
~ **для прохо́да гильз** ejection slot (SA)
~, **откидно́е ветрово́е** folding windscreen, folding windshield
~, **перепускно́е** transfer port
~ **пода́чи** feed opening
~ **приёмника** belt guide, belt opening (SA)
~, **приёмное** 1. magazine opening
2. feed block mouth (SA)
~, **проду́вочное** scavenging port
~, **смотрово́е** observation port, peep-hole
~ **ство́льной коро́бки** 1. breech opening, breech recess
2. magazine opening
~ **цили́ндра** cylinder port
око́вка заты́льника прикла́да butt plate
околозвуково́й transonic
око́лыш band (cap)
оконча́ние рабо́ты дви́гателя propulsion cut-off, all-burnt (GW)
оконча́тельно-снаряжённый filled and fused (ammo etc)
око́п trench, weapon pit, emplacement, foxhole
~, **артиллери́йский** artillery emplacement
~ **для боевы́х маши́н** AFV pit
~ **для миномёта** mortar pit
~ **для противота́нкового ору́дия** anti-tank gun emplacement
~ **для ручно́го пулемёта** light machine-gun emplacement
~ **для стрельбы́ лёжа, одино́чный** individual shell scrape
~ **для та́нка** tank pit
~, **ло́жный** dummy trench
~, **миномётный** mortar pit
~ **на отделе́ние, стрелко́вый** infantry section trench
~, **нано́сный** breastwork, surface trench
~ **насыпно́го ти́па** firebay formed by breastwork and parados
~ **непо́лного про́филя для огнево́го сре́дства** shallow weapon pit
~, **одино́чный стрелко́вый** foxhole, weapon slit
~, **оруди́йный** gun pit
~, **основно́й** primary trench
~, **па́рный** two-man trench
~, **пове́рхностный** breastwork
~ **по́лного про́филя** full-depth trench
~, **противота́нковый** anti-tank gun pit
~ **с круговы́м обстре́лом** all-round field-of-fire emplacement, all-round fire trench
~, **стрелко́вый** fire trench
~, **щелево́й** slit trench
окопа́вшийся entrenched, dug-in
око́панный dug-in
окопа́ть(ся) (см ока́пывать(ся))
око́пный trench
окопокопа́тель trench digger, trencher
~, **взрывно́й** explosive foxhole digger
око́п-укры́тие trench with overhead cover
око́пчик shallow weapon pit
око́п-щель foxhole, weapon slit
окочене́ть (см кочене́ть)

окóшко (little) window, aperture
окрáина outskirts
окрáска 1. paint
2. painting
3. colour
~, антикоррозиóнная rust-resisting paint
~, деформи́рующая dazzle painting
~, защи́тная camouflage paint
~, имити́рующая imitative painting (cam)
~, камуфля́жная camouflage paint
~ краскодýвкой spray painting
~, маскирóвочная camouflage net
~, мáтовая маскирóвочная matt camouflage painting
окрáшивание 1. painting
2. colouring
~, защи́тное camouflage painting
~, маскирóвочное camouflage painting
óкруг district, area
~, воéнный military district (Sov administrative military division in USSR similar to Command in UK)
окружáть/окружи́ть to surround, to encircle, to blockade, to besiege
окружáющий surrounding, encircling
окружéние encirclement, blockade, envelopment, investment
окружённый surrounded, encircled
окружи́ть (см окружáть)
окружнóй 1. (military) district (Sov)
2. Command (Brit)
окрýжность, дели́тельная pitch circle
октáнт octant
окуля́р eyepiece (optics)
окýривание fumigation
"олéнь, бегýщий" "running deer" (moving target)
óлово tin (metal)
оловя́нный tin
ом ohm
омеднéние coppering, copper fouling
оммéтр ohmmeter
~, мáлый small ohmmeter
ондуля́тор undulator (teleg)
опáлубка decking, planking, timbering, shuttering, casing
опáсности в полёте flight hazards
опáсность danger, hazard
~ взры́ва explosion hazard, explosion risk
опáсность вну́треннего облучéния internal irradiation hazard
~ воспламенéния ignition hazard
~ занóса skid hazard
~ пожáра fire hazard, fire risk
~, пожáрная fire risk
~, радиоакти́вная radiation hazard
~, тáнковая tank threat
операти́вно-стратеги́ческий strategic, tactical (depending on scale of ops)
операти́вность efficiency, proficiency
операти́вно-такти́ческий strategic, tactical
операти́вный 1. operational, operative, operating
2. strategical, tactical (depending on scale of ops)
оперáтор-слухáч hand-morse operator
операциóнная operating theatre (med)
операциóнный operations, operational
оперáция operation
~, авиадесáнтная airborne assault operation
~, армéйская army operation
~, боевáя tactical operation, large scale operations (conducted by army or higher formation)
~, воздýшно-десáнтная airborne assault operation
~ вычисли́тельной маши́ны computer operation
~, десáнтная assault landing operation (amphibious, tank-mounted, airborne or air-transported)
~, крýпная large-scale operation, major operation
~, морскáя десáнтная amphibious assault operation
~, наступáтельная offensive operation
~, оборони́тельная defensive operation
~, прихóдная receipt transaction
~, прихóдо-расхóдная receipt and issue transaction
~, рабóчая working operation
~, расхóдная issue transaction
~, совмéстная combined operation, joint operation
~, фронтовáя army group operation
опережéние overshoot (servo)
~ зажигáния spark advance, ignition advance (ICE)
~ зажигáния, автомати́ческое automatic ignition advance (ICE)

опере́ние 1. fin assembly (msl)
2. tail unit (ac)
3. bodywork (veh)
~ ми́ны mortar bomb tail vanes
~, скла́дывающееся folding fin (msl)
~, хвостово́е 1. control fins, fin assembly (msl)
2. tail unit (ac)
оперённый finned, fin stabilized
опери́ливание the fitting of handrails to a bridge
опи́ливать/опили́ть to file (metall etc)
опили́ть (см опи́ливать)
опило́вка filing (metall)
о́пись inventory
~ гру́за loading list
~ ору́жия weapons inventory
опла́та расхо́дов reimbursement of expenses
~ сто́имости reimbursement
оплести́ (см оплета́ть)
оплета́ть/оплести́ to braid, to entangle; **оплета́ть/оплести́ колю́чей про́волокой** to entangle with barbed wire
оплётка entanglement, braid
~, ме́дная copper braid
опло́т stronghold
оповести́ть (см оповеща́ть)
оповеща́ть/оповести́ть to notify, to warn
оповеще́ние notification, warning
~, возду́шное air-raid warning
~ и наведе́ние control and reporting (air defence)
опознава́ние recognition, identification
~, визуа́льное visual identification
~, радиолокацио́нное radar IFF, radar recognition system
~ самолётов aircraft recognition
~ "свой-чужо́й" IFF, identification friend or foe
~ це́ли target identification
опознава́ть/опозна́ть to identify
опозна́ние identification, recognition
~ "свой-чужо́й" IFF, identification friend or foe
опозна́ть (см опознава́ть)
ополче́ние militia, home guard
опо́ра 1. support, bracket, bearer
2. pylon
3. pier
~ баланси́ра, попере́чная hull transom (naut)

опо́ра, ба́шенная double-bent pier (br)
~ ба́шни, шарни́рная turret race, turret ring
~, берегова́я abutment (br)
~, бокова́я side extension girder (ord etc)
~, враща́ющаяся swivelling mounting, pivoting support
~, выносна́я outrigger
~ дви́гателя engine mounting
~, жёсткая 1. fixed pier, fixed support (br)
2. fixed bearing (br)
~, за́дняя end support
~, катко́вая roller bearing
~, клёточная crib pier (br)
~, козлова́я trestle pier (br)
~, концева́я abutment (br)
~, концева́я жёсткая abutment (br)
~, металли́ческая 1. metal support
2. pylon (for overhead wires)
~ мо́ста, промежу́точная bridge pier
~ на ро́ликах roller bearing
~, неподви́жная fixed support
~, однора́мная single bent pier (br)
~ опра́вки arm (support) bracket
~ о́си axle bearing
~, па́рная ра́мная two-tier trestle (br)
~, пере́дняя front support
~, плову́чая floating pier (br)
~, пло́ская ра́мная single bent pier
~, плотова́я floating pier (br)
~ подве́ски suspension support
~, подви́жная movable support, rocker base, oscillating base, suspension support
~ поплавка́ карбюра́тора carburettor float fulcrum
~, призмати́ческая knife-edge bearing
~, промежу́точная intermediate support, pier (br)
~, простра́нственная ра́мная double bent pier (br)
~, ра́мная 1. box crib trestle (br)
2. single bent pier (br)
~ рессо́ры теле́жки bogie spring block
~, ря́жевая box crib pier (br)
~, сва́йная piled pier (br)
~, сва́йная берегова́я piled abutment (br)
~ тормозно́й ле́нты brake band anchor clip
~, упру́гая resilient mounting
~, ца́пфовая trunnion support

опо́ра, четвертьволнова́я quarter-wave stub (electron)
~, ша́риковая ball race, ball-bearing support
~, шарова́я ball and socket bearing
опо́ра-пило́н tower (br)
опо́рный supporting, bearing
опорожне́ние emptying, evacuation, discharging, discharge, draining, dumping, deflation
опра́ва mounting, case, holder
"Опра́виться!" "Stand Easy!"
определе́ние definition, determination, identification, calculation, estimation, computation
~ да́льности, глазоме́рное range estimation by eye, judging distance (visual)
~ ключа́ key setting (cipher)
~ неиспра́вностей fault finding
~ радиодевиа́ции DF calibration
~ расстоя́ния judging distance
~ расстоя́ния с по́мощью постоя́нной ба́зы sub-tensing (distance measurement)
~ то́чки (своего́) стоя́ния position finding, position fixing
определённый definite, fixed, certain, defined, determined
определи́тель detector, determinant, discriminator
~ паро́в ОВ chemical detector kit
определи́ть (см определя́ть)
определя́ть/определи́ть to define, to determine; **определя́ть/определи́ть а́зимут** to take a bearing; **определя́ть/определи́ть маршру́т по́езда** to route a train; **определя́ть/определи́ть радиодевиа́цию радиопеленга́тора** to calibrate a DF set; **определя́ть/определи́ть расстоя́ние** to determine distance; **определя́ть/определи́ть расстоя́ние на глаз** to estimate distance, to judge distance
опресни́тель distiller, water distilling plant, desalination equipment
опрессо́вка pressure testing
опроки́дывание overturning
опроки́дыватель tipper
опроки́дывать/опроки́нуть to overrun (tac), to upset, to overturn, to capsize
опроки́нуть (см опроки́дывать)
опро́с interrogation
опро́счик interrogator (radar IFF)
о́птика 1. optical instruments 2. optics
оптима́льный optimum
опти́ческий optical
опуска́ние dipping, sinking, lowering, descent, depression
~ гру́нта subsidence
~ ду́льной ча́сти ствола́ muzzle droop (gunn)
опуска́ть/опусти́ть to lower, to sink, to let down, to pull down, to post (a letter), to omit, to turn down; **опуска́ть/опусти́ть домкра́том** to jack down
опусти́ть (см опуска́ть)
опу́шка edge (of a forest)
о́пыт, боево́й battle experience
~, фронтово́й battle experience
опьяне́ние intoxication
орби́та orbit
~, за́мкнутая closed orbit, stationary orbit
~ электро́на electron orbit
о́рган element, agency, establishment, organ
~ боепита́ния ammunition supply installation
~, дово́льствующий supply agency, providing authority
~, запо́рный shutting-off device
~, исполни́тельный 1. executive agency, executive organ 2. actuating unit, actuator, drive motor
~, кара́тельный penal establishment
~, разве́дывательный intelligence agency
~, рулево́й control surfaces (GW)
~, сле́дственный investigating body, investigating agency
~ снабже́ния supply organization, supply agency
~ ты́ла administrative echelon
~ управле́ния 1. control agency, command echelon 2. administrative echelon
организа́ция organization
~ обеспе́чения морско́го деса́нта, берегова́я beach organization (amph ops)
~ огня́ fire plan
организо́ванный 1. organized 2. orderly, disciplined

о́рганы, кварти́рно-эксплуатацио́нные barrack services
~, стабилиза́ции stabilizing controls
~ управле́ния 1. HQ elements
2. controls, control surfaces
~ управле́ния, хвостовы́е tail control surfaces, rear control surfaces (GW)
оргстекло́ (органи́ческое стекло́) plexiglass
о́рден order (decoration)
орденоно́сец recipient of a decoration
ордина́рец orderly, runner
ордина́та 1. ordinate
2. easting
ориента́ция orientation
~, углова́я attitude (GW)
ориенти́р landmark, outstanding point, reference point
~, ви́димый visible ground feature, visible ground mark
~, визуа́льно наблюда́емый visual reference point
~, дополни́тельный supplementary reference point
~, исхо́дный datum point
~, контро́льный reference landmark, check reference point
~, ме́стный landmark
~, назе́мный prominent ground object
~, основно́й base reference point, base point
~, пристре́лочный registration point (gunn)
~, радиолокацио́нный radar reference point
~, светово́й light marker (fired by arty), marker light
~, светя́щийся luminous reference point
ориенти́рование orientation
~, взаи́мное passing the line (arty, surv)
~ взаи́мным визи́рованием orientation and lining up (AA gunn)
~, вне́шнее relative orientation (air photo)
~, вну́треннее absolute orientation (air photo)
~ ка́рты map orientation, orientation of a map
~ ка́рты по ко́мпасу map orientation by the aid of a compass
~ на ме́стности position finding
~ на удалённой то́чке orientation and lining-up by distant object method (arty)
ориенти́рование ору́дий lining-up of guns (arty)
~ по аэросни́мкам orientation by the aid of aerial photographs
~ ПУА́ЗО AAFCE orientation
~, углово́е attitude (GW)
ориенти́ровать to orient, to orientate, to set up (instr), to line up (instr), to set (map); **ориенти́ровать в обстано́вке** to explain the situation, to "put in the picture"; **ориенти́ровать в основно́м направле́нии** to lay at zero degrees (arty); **ориенти́ровать ка́рту** to set a map
ориенти́роваться to get one's bearings
ориентиро́вочный 1. position-finding
2. tentative, rough, approximate
орке́стр band (musical)
~, духово́й brass, brass band
~, сво́дный combined band, massed bands
ортико́н orthicon (electron)
ортогона́льный orthogonal
ортодро́мия Great Circle (navigation)
ортоско́п 1. orthoscope
2. aim corrector, (mirror) aiming device (gunn)
ору́дие gun, piece, ordnance
~, автомати́ческое automatic gun
~, артиллери́йское artillery gun
~, ба́шенное turret gun
~, безотка́тное recoilless gun
~, бикали́берное tapered-bore gun
~, букси́руемое towed gun
~ в сбо́ре gun assembly
~, газодинами́ческое high-low pressure gun
~ ги́льзового заряжа́ния quick-firing gun, QF gun
~, гладкостcombо́льное smoothbore weapon
~, го́рно-вью́чное pack gun
~, го́рное mountain gun
~, дальнобо́йное long-range gun
~, дежу́рное on-call gun
~, дина́мо-реакти́вное recoilless gun
~, длинноство́льное long-barrelled gun
~ для салю́тов saluting gun
~, дульнозаря́дное muzzle-loading gun
~, жёсткое recoilless gun
~, зени́тное anti-aircraft gun
~, зени́тно-противота́нковое dual-purpose gun (AA and A Tk)
~, индивидуа́льное personal weapon/s

орудие, картузное breech loading gun, BL gun
~, кочующее roving gun
~, крепостное fortress gun
~ крупного калибра large-calibre gun
~, многоствольное multibarrelled gun
~, многоцелевое multi-purpose gun
~ навесного огня high-trajectory weapon
~, нарезное rifled gun
~ настильного огня flat-trajectory gun
~ непосредственной поддержки close support gun
~, номинальное range table gun
~, осадное siege gun
~, основное 1. pivot gun (surv) 2. standard gun, reference piece (calibration) (US)
~, патронное QF gun
~, полевое field gun
~, полуавтоматическое semi-automatic gun
~, пристрелочное ranging gun (especially at datum point ranging)
~, прицепное противотанковое towed anti-tank gun
~, противотанковое anti-tank gun
~, противотанковое безоткатное recoilless anti-tank gun
~, разборное pack gun
"орудие разряжено" "gun clear"
орудие с высокой начальной скоростью снаряда high-velocity gun
~ с небольшой начальной скоростью снаряда low-powered gun
~, самодвижущееся auxiliary-propelled gun
~, самопроизвольно-стреляющее runaway gun (automatic weapon)
~, самоходное self-propelled gun
~, скорострельное rapid-fire gun, quick-firing gun
~ сопровождения close support gun
~ среднего калибра medium-calibre gun
~, стационарное fixed gun
~, учебное drill gun, training piece
~, штурмовое assault gun
~, ядерное артиллерийское nuclear artillery gun
орудийный gun
орудия tools
оружейник armourer, gunsmith

оружие weapon, weapons, arm, arms
~, авиационное aircraft weapon system
~, автоматическое automatic weapons, automatic weapon
~, автоматическое стрелковое 1. automatic small arms 2. self-loading small arms
~, атомное atomic weapons, atomic weapon
~, бактериологическое bacteriological weapons, biological weapons, BW weapons
~, безоткатное recoilless weapon
~, бикалиберное tapered-bore weapon
~, биологическое biological weapons
~ ближнего боя short-range weapon
~ большой мощности heavy yield weapon (nucl)
~, бортовое onboard armament (ship, vehicle, ac etc)
~, бронебойное armour-piercing weapon
~, водородное hydrogen weapon, hydrogen weapons
~, вспомогательное secondary armament
~, групповое section weapons, section weapon
~, дальнобойное long-range weapon
~ для стрельбы по площади area-fire weapon, area-fire weapons
~ для стрельбы прямой наводкой direct fire weapon
~ для стрельбы с плеча shoulder-fired weapon
~, дульнозарядное muzzle loader, muzzle-loading weapon
~, заряжающееся с казённой части breech-loader, breech-loading weapon
~, зенитное скорострельное anti-aircraft quick-firing weapon(s)
~, зенитное управляемое реактивное anti-aircraft guided missiles, surface-to-air guided missiles
~, индивидуальное personal weapon/s
"Оружие к Смотру!" "For Inspection Port Arms!"
оружие, карманное pocket weapon
~, коллективное team weapon(s), group weapon(s)
~ колющего и рубящего действия cut and thrust weapon
~, короткоствольное short-barrelled weapon

ору́жие, косми́ческое space weapon
~, **ли́чное** personal weapons
~ **ма́лых кали́бров** small arms
~ **массово́го пораже́ния** weapons of mass destruction
~, **многомегато́нное я́дерное** multimegaton nuclear weapon
~ **nаве́сного огня́** high-trajectory weapon(s)
~ **насти́льного огня́** low trajectory weapon, flat trajectory weapon
~, **наступа́тельное** strike weapon, offensive weapon
~, **неавтомати́ческое стрелко́вое** hand-operated small arms, non-automatic small arms
~ **непосре́дственной подде́ржки** close support weapon
~, **носи́мое на поясно́м ремне́** side arm
~, **оборони́тельное** defensive weapon
~, **обы́чное** conventional weapons
~, **огнестре́льное** fire arm(s), fire weapons
~, **огнестре́льное противота́нковое** anti-tank fire weapon(s)
~, **однозаря́дное** single-shot weapon
~ **однора́зового де́йствия** single-shot weapon
~, **основно́е** main armament
~, **пехо́тное** infantry weapons
~, **противота́нковое реакти́вное** anti-tank rocket launcher
~, **радиологи́ческое** radiological weapon (NW)
~ **ра́зового де́йствия** one-shot weapon
~, **разрезно́е** sectionalized weapon, cut-away demonstration weapon
~, **раке́тное** 1. missiles, rockets 2. rocket weapons
~, **раке́тно-я́дерное** nuclear missiles, nuclear weapons
~, **реакти́вное** 1. rockets, missiles 2. rocket launchers, rocket weapons
~, **ручно́е** hand-held weapon
~ **с кони́ческим ствол́ом** tapered bore weapon
~, **самозаря́дное стрелко́вое** self-loading small-arms, semi-automatic small-arms
~, **самопроизво́льно стреля́ющее** runaway gun
~, **самостре́льное стрелко́вое** small-arms
ору́жие, скорострéльное rapid-fire weapon, quick firer
~, **сна́йперское** sniper's sniping weapon
~ **со свобо́дным затво́ром, автомати́ческое** blowback action weapon
"Ору́жие Соста́вь!" "Pile Arms!"
ору́жие, такти́ческое я́дерное tactical nuclear weapon, battlefield nuclear weapon
~, **термоя́дерное** thermonuclear weapons
~, **тяжёлое** heavy weapons
~, **управля́емое** guided weapon
~ **устрашне́ния** deterrent weapon
~, **ХБР** NBC weapon(s)
~, **хими́ческое** gas weapon, chemical warfare weapons, CW weapons
~, **холо́дное** silent weapons (knife etc), "cold steel"
~, **я́дерное** nuclear weapons
оса́да siege
осади́ть (см осажда́ть)
оса́дка draught (ship etc)
оса́дки precipitation
~, **ли́вневые** heavy shower (met)
~, **радиоакти́вные** radioactive fallout
осадкообразова́ние sludge formation
оса́док residue, sediment, deposit, precipitate, sludge
осажда́ть/осади́ть to lay siege, to besiege, to beleaguer
осажде́ние precipitation, settling, deposition, sedimentation, precipitate, deposit, sediment
оса́ливание greasing
осапёривание engineer modification
осва́ивать/осво́ить to master, to assimilate, to cope with
осведоми́тель informer
осведомле́ние information
освеже́ние запа́сов stock turnover, stock renewal
освети́тель, инфракра́сный infra-red searchlight, IR searchlight
освети́ть (см освеща́ть)
освеща́ть/освети́ть to illuminate; **освеща́ть/освети́ть цель** to light up a target, to illuminate a target
освеще́ние illumination
~, **вну́треннее** interior lighting
~, **люминесце́нтное** fluorescent strip lighting

освеще́ние ме́стности, иску́сственное artificial moonlight, artificial illumination of terrain
~, непрямо́е 1. artificial moonlight 2. indirect lighting
~, светомаскиро́вочное dimmed driving lights, blackout lighting (veh)
~ це́лей на по́ле бо́я battlefield illumination
освещённость intensity of illumination
освободи́ть (см освобожда́ть)
освобожда́ть/освободи́ть to release, to clear, to liberate, to free
освобожде́ние 1. liberation, release 2. elimination
освое́ние mastering, coping, assimilation
осво́ить (см осва́ивать)
оседа́ние 1. precipitation 2. settling, subsiding 3. fall-out (action)
оседла́ть доро́гу to straddle a road, to be astride a road
~ (см седла́ть)
осе́чка misfire
о́си ракеты missile axes
оско́лки debris (nucl)
оско́лок splinter, fragment
~ снаря́да shell fragment, shell splinter
~, убо́йный lethal fragment (ammo)
ослаби́тель, поршнево́й piston attenuator
ослабле́ние 1. weakening, abatement, reduction, decrease, slacking, loosening 2. attenuation (radio)
ослепи́ть (см ослепля́ть)
ослепле́ние blindness, blinding
ослепля́ть/ослепи́ть to blind
ослуша́ние insubordination
осма́ливание лыж base-waxing of skis
осмо́тр examination, inspection
~ в пути́, контро́льный halt parade servicing (veh)
~, вне́шний visual inspection
~, всесторо́нний complete inspection
~, вы́борочный spot-check inspection
~, ежего́дный annual inspection
~, ежедне́вный daily inspection
~, еженеде́льный weekly inspection
~, контро́льный 1. parade servicing (veh) 2. general inspection
~, медици́нский medical examination, medical inspection
~ ме́стности searching the ground
~ на вши́вость inspection for vermin
осмо́тр, нару́жный visual examination, cursory examination, superficial examination
~, периоди́ческий periodic inspection
~, пла́новый scheduled inspection, formal inspection
~, по́лный complete inspection
~, предвари́тельный preliminary inspection
~, предпусково́й pre-launch check (missile)
~, приёмо-сда́точный acceptance inspection, "In" inspection
~, профилакти́ческий routine inspection
~, теку́щий routine inspection
~, техни́ческий maintenance inspection, technical inspection
~, у́тренний morning inspection parade
~, эпизоди́ческий casual inspection
осмо́трщик боеприпа́сов ammunition examiner
оснасти́ть (см оснаща́ть)
оснаща́ть/оснасти́ть to equip, to fit out
оснаще́ние 1. equipment 2. equipping, provision
оснащённость state of being equipped or provided with equipment
оснащённый equipped, fitted out
осно́ва base, basis
основа́ние 1. authority, grounds 2. base, foundation
~ вертлю́га cradle plate (ord)
~ визи́ра азиму́та, подви́жное lateral foresight adjusting plate (AA ring sight)
~ визи́ра угла́ возвыше́ния, подвижно́е vertical foresight adjusting plate (AA ring sight)
~ головно́й ча́сти снаря́да shoulder of shell
~ заты́льника rear cross-piece body (MMG)
~ звена́ гу́сеницы track plate
~, кольцево́е base ring
~ му́шки foresight bracket, foresight block
~, неподви́жное 1. base plate (ord) 2. fixed base
~ ору́дия, неподви́жное gun base plate
~ расхо́да authority for issue
~ траекто́рии base of trajectory
~ тра́па track plate
~, штырево́е central pivot (ord)
осо́бенность feature, particularity

осо́бенность, конструкти́вная design feature
"осо́бо ва́жная" "especially urgent" (Sov msg precedence)
о́спа, натура́льная smallpox
оста́вить (см оставля́ть)
оставле́ние поста́ abandonment of post
~ **ча́сти, самово́льное** prolonged absence without leave
оставля́ть/оста́вить to leave, to abandon, to give up
остана́вливать/останови́ть to stop, to bring to a stop; **остана́вливать/останови́ть продвиже́ние** to call a halt
остано́в stop, catch, lock, detent
~, **автомати́ческий** automatic stop
~ **затво́ра** bolt catch, hold-open catch
~ **патро́на** cartridge stop
~ **поворо́та ба́шни** turret stop
останови́ть (см остана́вливать)
~ **(см остана́вливать)**
остано́вка halt, stop
~, **вы́нужденная** forced halt
~ **дви́гателя** engine shut-down
~, **коро́ткая** short halt
~ **по расписа́нию** regular stop, scheduled stop
~, **э́кстренная** immediate halt
оста́ток remainder, rest, balance
~, **нали́чный** balance on hand
~, **нулево́й** zero balance
~, **переходя́щий** carry over, previous balance
о́стов framework
~ **безвру́бочной констру́кции** jointless framework
~ **запа́ла** igniter body
~ **затво́ра** stripped breechblock (SA)
~ **револьве́ра** revolver frame
усто́йчивость stability, equilibrium
о́стров island
острово́к оборо́ны pocket of resistance
острогу́бцы cutting nippers
остродефици́тный in very short supply
острозара́зный highly contagious, highly infectious
остроконе́чный pointed
осцилло́граф oscillograph
~, **многошле́йфовый** multi-channel oscillograph
осыпа́ние гру́нта crumbling of soil
о́сыпь scree

ось 1. axle, axis pin, pin, spindle 2. axis
~ **абсци́сс** X-axis
~ **анте́нны, электри́ческая** lobe axis (radar)
~ **баланси́ра опо́рного катка́** road wheel arm spindle
~, **боева́я** axle, axle-tree (ord)
~ **ва́ла** shaft axis
~, **веду́щая** driving axle, leading axle
~, **вертика́льная** 1. vertical axis 2. yaw axis (GW)
~ **вертлю́га** cradle pintle
~ **взво́да** cocking lever axis pin
~ **взво́да уда́рника** outer cocking lever (ord)
~, **визи́рная** sight axis
~ **визи́рования** scan axis (radar)
~ **враще́ния** rotation axis, spin axis, axis of rotation
~ **выбра́сывателя** extractor stay
~, **динами́ческая** axis of precession (projectile)
~ **доро́ги** centre line of road
~ **дро́ссельной засло́нки** throttle spindle
~ **защёлки прикла́да** butt latch pin
~ **кана́ла ствола́** axis of the bore
~, **кача́ющаяся** oscillating axle, pivoted axle
~ **колеи́** tread axis
~, **коле́нчатая** cranked axle (veh)
~, **колёсная** wheel axle
~, **координа́тная** grid axis
~, **коро́ткая неподви́жная** stub axle
~ **кре́на** roll axis (GW)
~ **кривоши́па** crank actuating shaft (ord)
~ **ла́пок экстра́ктора** extractor spindle
~ **мо́ста** centre line of bridge
~ **направля́ющего колеса́ гу́сеницы** idler wheel spindle
~ **натяжно́го колеса́ гу́сеницы** idler wheel spindle
~, **неподви́жная** dead axle
~, **опти́ческая** optical axis
~ **ордина́т** Y-axis
~ **перемеще́ния** axis of displacement
~ **перепра́вы** axis of crossing
~ **поворо́та** steering axis
~, **подви́жная** live axle
~ **подво́за** axial road, axis of supply
~, **по́лая** hollow axle
~, **поля́рная** polar axis

ось, попере́чная transversal axis, lateral axis
~ **прице́ла** axis of sight
~, **продо́льная** longitudinal axis
~, **промежу́точная** intermediate axle
~, **разгру́женная** fully-floating axle
~, **разрезна́я** divided axle
~ **ра́мы ору́дия** gun carrier hinge lug (ord)
~ **рассе́ивания** axis of dispersion
~ **рассе́ивания, горизонта́льная** lateral axis of zone of dispersion
~ **рассе́ивания, ма́лая** lateral axis of zone of dispersion (gunn)
~ **рыска́ния** yaw axis (GW)
~ **рычага́, поворо́тная** axle arm pivot shaft
~ **свя́зи** main artery (comms)
~ **спусково́го крючка́** trigger axis pin
~ **спусково́го рычага́** sear pin, firing lever axis pin
~ **сто́пора ствола́** pivot of breech ring control arm (ord)
~ **тангажа́** pitch axis (GW)
~ **теле́жки** bogie axle, stub axle
~ **тормозно́й коло́дки** brake shoe fulcrum pin
~, **тру́бчатая** tube axle
~ **управле́ния** control axis (GW)
~ **цапф** trunnion axis (gunn)
~ **цили́ндра** cylinder axis (revolver)
~ **чувстви́тельности** gyro axis
~ **шарни́ра тормозно́й коло́дки** brake shoe link pin
~, **электри́ческая** electrical axis
~ **э́ллипса, больша́я** major axis of ellipse
~ **э́ллипса, ма́лая** minor axis of ellipse (maths)
~ **э́ллипса рассе́ивания, больша́я** longitudinal axis of zone of dispersion (gunn)
~ **э́ллипса рассе́ивания, горизонта́льная** lateral axis of zone of dispersion
~ **э́ллипса рассе́ивания, ма́лая** lateral axis of zone of dispersion
ота́пливаемый heated
отби́в the parry (bayonet)
отбива́ть/отби́ть to repulse, to beat off, to repel, to parry
отбира́ть/отобра́ть to take away, to choose, to select; **отбира́ть/отобра́ть про́бу** to take a sample
отби́ть (см отбива́ть)
о́тблеск shine, reflection
"Отбо́й!" "Stand Down!", "All Clear!"
отбо́й 1. retreat (sunset)
2. ring off, sign-off (teleph, radio)
3. stand down
4. all clear (after air alert)
~ **трево́ги** stand down, all clear
отбо́р take-off, picking, selection
~ **мо́щности** power take-off
отбо́рный choice, picked, crack, elite, selected
отбрако́вка sorting, grading, rejection
~ **неиспра́вных патро́нов** sorting of defective cartridges
отбра́сывание затво́ра, прямолине́йное straight blowback
отбра́сывать/отбро́сить to repel, to throw back
отбро́сить (см отбра́сывать)
отва́л 1. dump (mining)
2. bulldozer blade
~, **плу́жный** ploughshare
отверну́ть (см отвёртывать)
отве́рстие aperture, opening, hole, port
~ **в броне́ та́нка для стрельбы́ из ли́чного ору́жия** pistol port, port-hole (AFV)
~ **вентиля́тора, забо́рное** ventilator air inlet
~, **визи́рное** aiming aperture
~, **водовыпускно́е** drain hole
~, **воздухозабо́рное** air inlet
~, **впускно́е** inlet port (ICE)
~, **входно́е** access port, access hole
~, **выпускно́е** exhaust port (ICE)
~, **выходно́е** outlet port, outlet hole
~, **га́зовое** gas port (SA)
~ **для боево́го штыря́** pivot housing (ord)
~ **для наполне́ния** filler hole, opening
~ **для прохо́да бойка́** firing pin-hole
~ **для свечи́ зажига́ния** spark-plug hole
~ **для сли́ва ма́сла** oil drain hole
~ **для стрельбы́** firing port
~, **запа́льное** 1. tube flash channel (fuse)
2. bush firing hole (ord)
~, **затра́вочное** 1. flash channel (fuse)
2. flash hole, fire hole (ammo)
~ **кла́пана, проходно́е** valve port
~, **контро́льное** inspection hole
~, **маслоуказа́тельное** oil test point
~, **наливно́е** filler hole
~, **нарезно́е** threaded socket

отве́рстие, огнепереда́точное gas vent (mortar bomb)
~, относи́тельное relative aperture (optics)
~ парашю́та, по́люсное parachute vent
~, переда́точное time-ring connecting channel (fuse)
~, перепускно́е transfer port
~, перфори́рованное punched hole
~, рассверлённое bore
~, расто́ченное bored hole
~, сливно́е drain hole
~, смотрово́е inspection door, inspection hole,
~ ствола́, га́зовое gas port
~, форсу́ночное injection port
отве́рстия, выходны́е emission holes (smoke canister)
отверте́ть (см отвёртывать)
отвёртка screwdriver
~, крестообра́зная cross-point screwdriver, Phillip's screwdriver
~ с трещо́ткой ratchet screwdriver
отвёртывать/отверте́ть, отверну́ть to turn off, screw off
отве́с plumb, plummet
отвести́ (см отводи́ть)
отве́т, коди́рованный coded response
ответви́тель, напра́вленный directional coupler
~, трёхдецибе́льный 3-decibel coupler
отве́тственность responsibility
отве́тчик transponder, responder (IFF)
~, самолётный aircraft transponder (IFF)
отво́д 1. withdrawal
2. cut-off (pipe or circuit)
~, переливно́й overflow trap
отводи́ть/отвести́ to withdraw, to pull back, to pull out
отво́дка прице́ла sight quick-release collar
~ фрикцио́на clutch shaft fork
отвоева́ть (см отвоёвывать)
отвоёвывать/отвоева́ть to recapture
отгоризонти́ровать to level
отдава́ть/отда́ть to issue (orders etc), to give, to give away, to recoil (SA); **отдава́ть/отда́ть честь** to salute
отда́ние че́сти saluting, paying compliments
о́тданный 1. issued (orders etc)
2. given (away)
отда́ть (см отдава́ть)
отда́ча 1. output
2. recoil, "kick" (SA)
3. issue
~ анте́нны radiation efficiency, antenna efficiency
~ затво́ра blowback (SA)
~ затво́ра, заме́дленная retarded blow-back, delayed blow-back
~, свобо́дная blow-back
отде́л department, branch
~, адъюта́нтский A branch
~, вое́нно-медици́нский military medical department
~, вое́нно-фина́нсовый military finance department
~, второ́й second department (int)
~ диви́зии, полити́ческий divisional political department
~ ка́дров personnel department
~. материа́льно-техни́ческого снабже́ния department of material and technical supply
~, осо́бый special branch
~, пе́рвый ops department (Sov HQ)
~ перево́зок freight transport department
~, полити́ческий political department
~ пути́ track department (rly)
~, разве́дывательный intelligence department, intelligence branch
~ свя́зи communications department
~, секре́тно-шифрова́льный secret cypher department (comms)
~ снабже́ния supply branch
~, строево́й organisational department, administrative branch, personnel and supply records department
~ сухопу́тных войск, администрати́вно- хозя́йственный administrative and supply department of land forces (Sov)
~ ты́ла rear establishments department
~, фина́нсовый finance section
~, шифрова́льный cipher office
~ шта́ба полка́ branch of regimental HQ
~, юриди́ческий legal section
отделе́ние 1. section, detachment (unit)
2. section, department, office
3. compartment, room
~, аккумуля́торное battery section, battery compartment
~ боево́го пита́ния ammunition section

отделéние, боевóе fighting compartment (tk)
~**, боевóе зарядное** warhead (torpedo)
~ **боеуклáдок** ammunition stowage compartment
~**, вторóе** second section (int)
~ **движéния** operating division
~**, дегазациóнное** gas cleansing compartment (CW)
~ **кáдров** personnel section
~**, кузнéчно-мéдницкое** forging and brazing section
~**, мéдико-санитáрное** medical section
~ **медицúнского снабжéния** medical supplies section
~**, миномётное** mortar detachment
~**, мóечное** washroom
~**, мотóрное** engine compartment (tk etc)
~ **навóдчика** gunner's compartment
~ **обезврéживания неразорвáвщихся бомб** bomb disposal squad
~ **от самолёта** exit (of a paratroop)
~**, охраняющее** protective section, protective detachment, security section
~**, пéрвое** ops section (Sov HQ)
~**, планшéтное** plotter detachment
~**, подвúжное** mobile detachment
~**, полевóе почтóвое** field post office
~ **проводнóй связи** line troop (sig)
~**, пулемётное** machine-gun section
~ **ремóнта электрооборýдования** electrical equipment repair section
~**, санитáрное** medical detachment
~**, сапёрное** sapper section, pioneer section
~ **связи** signals section
~ **связи с пехóтой** infantry liaison section
~**, силовóе** engine compartment
~ **силовóй передáчи** transmission compartment, gear box compartment (tk)
~ **силовóй установки** engine compartment (tk)
~ **склáда, передовóе** forward stores detachment
~**, слесáрно-механúческое** fitters' section
~**, станковопулемётное** medium machine-gun section, MMG section
~**, столярно-обóйное** coach-builders' section
~**, стрелкóвое** rifle section

отделéние, топовычислúтельное survey computing section (arty)
~**, трансмиссиóнное** gearbox compartment, transmission compartment (tk)
~ **управлéния** 1. driving compartment (tk)
2. HQ section
~**, учётное** account section
~**, учётно-отчётное** accounting section
~ **химúческой развéдки** chemical reconnaissance section
~**, хозяйственное** administrative section
~ **хранéния** storage section, storage warehouse, storekeeping section, warehouse section
~**, штáтное головнóе** organic forward detachment (of base supply depot)
отделúть (см отделять)
отдéльный independent, separate, detached
отделять/отделúть to detach, to separate
óтдых rest
óтжиг annealing
óтзыв reply to challenge
откáз 1. failure, breakdown
2. refusal, denial
~ **взрыва** failure to explode
~ **двúгателя** engine failure
~ **заряда** misfire due to charge failure
~**, конструкциóнный** design error failure
~ **от дóлжности** resignation
~**, перемежáющийся** intermittent failure
~**, пóлный** complete failure, total failure
~ **тóрмоза** brake failure
~**, частúчный** partial failure
откáлывание бронú armour scabbing
откáт recoil (ord)
~**, длúнный** long recoil
~**, корóткий** short recoil
~**, непóлный** insufficient recoil
~**, нормáльный** normal recoil
~**, перемéнный** controlled recoil
~**, свобóдный** free recoil
~**, скользящий** sliding recoil
~**, заторможённый** retarded recoil
~**, увелúченный** over recoil
откатúться (см откáтываться)
откáтываться/откатúться to recoil (ord), to fall back (tac)
откáчивание pumping out, evacuating, exhausting

отка́чка pumping out, evacuating, exhausting
откидно́й hinged, folding
"Отки́нуть Штык!" "Fold Back Bayonets!"
отклоне́ние deflection, deviation, error
~, абсолю́тное absolute deviation
~, баллисти́ческое ballistic deviation
~, боково́е lateral deflection, deflection spotting correction (arty)
~, вероя́тное probable error
~, вероя́тное боково́е lateral probable error (arty)
~, вертика́льное vertical error, deflection
~, горизонта́льное horizontal error, deflection
~, изме́ренное measured deflection
~, ка́жущееся apparent error
~, кругово́е вероя́тное circular error probable (CEP)
~, магни́тное magnetic deflection
~ по боково́му направле́нию, среди́нное mean deviation in line
~ по высоте́, среди́нное mean deviation in height
~ по да́льности range error, range deviation
~ по да́льности, вероя́тное range probable error (arty)
~ по да́льности, среди́нное mean deviation in range
~, среди́нное probable error, mean deviation (in fall of shot) (gunn)
~, сре́днее mean deviation
~, сре́днее вероя́тное mean probable error
~ температу́ры, баллисти́ческое ballistic temperature
~ то́чки паде́ния снаря́да от це́нтра це́ли absolute deviation
~, углово́е angle off
~ це́нтра рассе́ивания mean point of impact error
~, часто́тное frequency excursion, frequency deviation
отклони́ть (см отклоня́ть)
отклоня́ть/отклони́ть to deflect
отключа́ть/отключи́ть to disconnect, to switch off, to cut out, to cut off
откомандирова́ть (см откомандиро́вывать)
откомандиро́вывать/откомандирова́ть to send on a mission, to detach (mil)
отко́с 1. scarp, slope
2. batter (br)
~, верхово́й upper batter (br)
~, вну́тренний inner batter (br)
~, нару́жный outer batter (br)
~, низово́й lower batter (fd wks)
открыва́ние затво́ра breech opening
открыва́ть/откры́ть to open, to clear, to uncover, to reveal, to discover; **открыва́ть/откры́ть бе́рег** to make landfall (amph ops); **открыва́ть/откры́ть ого́нь** to open fire; **открыва́ть/откры́ть ого́нь по це́ли** to engage a target
откры́тый open, exposed
откры́ть (см открыва́ть)
отла́дка adjustment
отли́в ebb tide
отли́вка founding, casting
~ та́нковой ба́шни tank turret casting
отли́чие 1. difference
2. decoration (mil)
отли́чник "excellent" soldier, "outstanding" soldier
отли́чный 1. different
2. "excellent", "outstanding" (Sov honorary title awarded for exceptional work in military and political training)
отложе́ние deposit, deposition, sediment
отлу́чка absence
~, самово́льная AWOL, absence without leave
о́тмель sandbank, shoal
отме́на 1. abolition
2. cancellation, annulment
~ за́пуска cancellation of launch (missile)
отмени́ть (см отменя́ть)
отменя́ть/отмени́ть to abolish, to abrogate, to repeal, to revoke, to rescind, to countermand; **отменя́ть/отмени́ть прика́з** to rescind an order
отме́тить (см отмеча́ть)
отме́тка 1. mark, notation
2. reading (instr)
3. pip, marker (radar)
4. spot height
~ высоты́ bench mark, spot height
~ горизонта́ли contour height
~, калибро́вочная marker pip
~ масшта́ба да́льности range marker, range pip
~, масшта́бная calibration pip (radar)

отмéтка нормáльного ýровня full-level mark
~, световáя blip (on screen)
~ ýровня height mark
~ цéли target blip (radar)
~, числовáя spot height
отмéтчик needle, hand, pointer
~ врéмени timer, time marker
отмечáние стрельбы́ witness point procedure (gunn)
отмечáть/отмéтить to register, to record, to mark; **отмечáть/отмéтить попадáние** to register a hit
отмобилизовáние complete mobilization, completion of mobilization
отмобилизовáть to mobilize fully
отмы́вка рельéфа shading (map)
отнóс боковы́м вéтром side effect of wind (ball)
относúтельность relativity
отношéние горю́чего к вóздуху fuel-air ratio (carb)
~, изотóпное isotope ratio
~, передáточное gear ratio
~ сигнáл-помéха signal-to-noise ratio
~ сигнáл-шум signal-to-noise ratio
~ тя́ги к вéсу thrust-to-weight ratio
~ хóда пóршня к диáметру цилúндра stroke to bore relationship (ICE)
отображáть/отобразúть to reflect, to represent; **отображáть/отобразúть дáнные** to display data, to indicate data, to show data
отображéние mapping, depiction, representation, display
~ информáции information display
отобразúть (см отображáть)
отобрáть (см отбирáть)
отождествлéние identification
отойтú (см отходúть)
"Отомкнýть Штык!" "Unfix Bayonets!"
отоплéние heating, warming, firing
оторвáвшийся straggler
оторвáться (см отрывáться)
отпечáток print, impression
~, контáктный contact print (photo)
~ шúны tyre tread print
отпирáние unlocking
отпóр repulse
отпотевáние sweating (metall)
отправúтель 1. originator, sender, initiator (comms)
2. consignor, dispatcher, shipper (cargo)
отпрáвить (см отправля́ть)
отпрáвка sending, shipment, dispatch, consignment
отправлéние sending, departure (trains, ships), exercise (of duties)
отправля́ть/отпрáвить to forward, to dispatch, to exercise, to perform (duties, functions)
óтпуск 1. leave (mil)
2. tempering (metall)
3. issue
~, бессрóчный indefinite leave of absence
~ докумéнта duplicate copy
~ по семéйным обстоя́тельствам compassionate leave
отпускáть/отпустúть to release, to issue, to allocate; **отпускáть/отпустúть ремéнь** to loosen slings
отпускнúк person on leave
отпускны́е 1. leave allowances
2. personnel due for leave
3. personnel absent on leave
отпустúть (см отпускáть)
отрабáтывать/отработáть to process (eg data), to work out, to finish off
отработáть (см отрабáтывать)
отрабóтка 1. processing (data)
2. working off, working out
~ взаимодéйствия мéжду тáнками и пехóтой tank-infantry cooperation
~ волны́ netting (radio)
~ приёмов с орýжием arms drill
отравлéние poisoning
отравля́ющий toxic, poisonous, poisoning
отражáтель 1. parasite (antenna)
2. reflector
3. ejector, cartridge deflector
4. reflecting mirror (sight)
5. tamper (nucl)
~ взрывнóй волны́ blast deflector
~, дезориентúрующий confusion reflector (ECM)
~, дипóльный dipole reflector
~ нейтрóнов neutron reflector
~, параболúческий parabolic reflector
~ пóршня piston reflector
~ потóка вóздуха draught deflector
~, уголкóвый corner reflector
~ фáкела jet deflector (msl)
~ фáры headlamp reflector
отражáть/отразúть to repel, to repulse, to beat off (tac), to reflect (optics etc);

отража́ть/отрази́ть ата́ку to repel an assault; отража́ть/отрази́ть контрата́ку to break up a counter attack
отраже́ние 1. image, reflection 2. repulse, repelling, beating off 3. ejection (gunn)
~ **гильз** cartridge ejection
~, **диффузио́нное** diffuse reflection
~ **от ме́стности** terrain echo, ground clutter (radar)
~ **от назе́мных предме́тов** ground clutter (radar)
~ **от су́ши** terrain echo (radar)
~, **рассе́янное** scatter propagation, scattering (radio)
~, **тропосфе́рное рассе́янное** tropospheric scattering (radio)
отрази́ть (см отража́ть)
отре́зать (см отреза́ть)
отреза́ть/отре́зать to cut off
отре́зок section, length, cut, piece, fragment, segment
~ **трубы́** pipe section
~ **четвертьволново́й ли́нии** quarter-wave section (electron)
отремонти́ровать to repair, to refit, to overhaul (perf only)
отрица́тельный negative
отро́г spur (topo)
~, **бари́ческий** ridge of high pressure
отры́в 1. break-away, abruption, separation, disengagement 2. unstick, take-off (ac)
отрыва́ть/отры́ть to dig out
отрыва́ться/оторва́ться to break-away, to break contact, to break off; **не отрыва́ться/не оторва́ться от та́нков** to keep pace with the armour
отры́вка trenching
~, **механизи́рованная** mechanical trenching
отры́ть (см отрыва́ть)
отря́д detachment
~, **авиацио́нный** detachment (AF)
~, **боково́й** flank guard
~, **броново́й** armoured detachment
~, **вертолётный** helicopter unit
~, **головно́й** vanguard
~, **деса́нтный** amphibious group (naval forces)
~ **загражде́ний, подви́жный** mobile obstacle construction detachment
отря́д, истреби́тельный paramilitary fighting detachment
~, **кара́тельный** punitive detachment
~ **ликвида́ции после́дствий** decontamination detatchment
~, **лы́жный** ski troop, ski detachment
~ **обезвре́живания взрывоопа́сных предме́тов** explosive disposal detachment
~ **обеспе́чения движе́ния** movement support detachment
~ **оборо́ны предпо́лья** covering force
~, **отде́льный** independent detachment
~, **партиза́нский** partisan detachment, guerilla detachment
~ **пе́рвой медици́нской по́мощи** first aid detachment
~, **передово́й** 1. covering force, screen 2. forward detachment, advanced guard
~, **передово́й деса́нтный** assault pathfinder force (para)
~, **подви́жный передово́й** advanced guard mobile troops
~, **полиго́нный** range detachment, range party
~ **прикры́тия** protective detachment
~, **противодеса́нтный** anti-assault-landing detachment
~, **разве́дывательный** offensive reconnaissance force (coy or bn group)
~, **санита́рный** 1. sanitary detachment 2. medical detachment
~, **сторожево́й** bivouac guard detachment
~, **ты́льный** rear main guard
~, **уче́бный** instructional detachment, training detachment
~, **фотограмметри́ческий** photogrammetry detachment
~, **штурмово́й** assault detachment
отсе́к боево́й ча́сти warhead section, warhead compartment
~, **бытово́й** living compartment (of spacecraft)
~, **водонепроница́емый** watertight compartment
~ **ка́ртера** crankcase section
~, **кормово́й** stern compartment
~, **носово́й** nose section (msl), forward compartment
~, **прибо́рный** instrument compartment
~ **управле́ния** control compartment (tk)
отсека́тель патро́на link stripper (ammo)

отсекáть/отсéчь to separate, to cut off
отселéние мéстных жи́телей evacuation of local population
отсечéние separation
отсéчка 1. feed cylinder (AA ord) 2. cut-off
отсéчка-отражáтель combined cut-off ejector and feeder guide (SA)
отсéчный cut-off, shut-off, trip (of valves), interdictory (fire)
отсéчь (см отсекáть)
отскóк rearward bound, recoil, rebound
отслоéние протéктора loosening of tread, tread separation (from tyre)
отсрóчка deferment
отставáние lagging behind, straggling
"Отстáвить!" "As You Were!"; **отстáвить от слýжбы** to dismiss, to discharge, to retire
отстáвить (см отставля́ть)
отстáвка retirement
отставля́ть/отстáвить to set aside, to put aside, to dismiss
отстáвший straggler
отстóй sediment, residue, sludge
отстóйник sump
отстрáивать/отстрóить to off-tune, to tune out, to build up
отстрéл shoot
~, прóбный proof firing (gunn)
отстрóить (см отстрáивать)
отступáть/отступи́ть to withdraw, to retreat
отступи́ть (см отступáть)
отступлéние withdrawal, retreat
~, вы́нужденное forced withdrawal
~, плáновое planned withdrawal
~, такти́ческое tactical withdrawal
~ частоты́, допусти́мое frequency tolerance
отсýтствие absence
отсýтствовать to be absent
отсýтствующий 1. absent 2. absentee
отсчёт 1. reading (instr) 2. bearing 3. time interval
~, нулевóй zero reading
~, основнóй zero bearing (arty)
~ радиокóмпаса radio compass reading
оттáивание defrosting
оттáлкивание repulsion
оттáлкивать/оттолкнýть to repel
óттепель thaw
оттесни́ть (см оттесня́ть)
оттесня́ть/оттесни́ть to drive back, to force back
оттолкнýть (см оттáлкивать)
оттормáживание release of brake
оттушёвка tinting, shading-in (map)
оття́гивать/оттянýть to divert, to draw off (tac), to delay, to procrastinate; **оття́гивать/оттянýть врéмя** to gain time
оття́жка stay, anchor wire
~, áнкерная anchor guy rope, stay
~, боковáя guy for lateral stability, anti-sway bracing (suspension bridge)
~, обрáтная anchorage cable (br)
оттянýть (см оття́гивать)
отфильтровáть (см отфильтрóвывать)
отфильтрóвывание filtering off
отфильтрóвывать/отфильтровáть to filter out (elec)
отхóд 1. retirement, retreat, withdrawal 2. departure (trains) 3. sailing (ships)
отходи́ть/отойти́ to retire, to withdraw, to fall back, to move away, to depart, to pull out, to deviate from, to digress from
отходя́щий retreating, withdrawing
отцепи́ть (см отцепля́ть)
отцепля́ть/отцепи́ть to uncouple, to unhook
отчёт account, report, consolidated return, summary
отчётность accounting, records
отчёты, свóдные comprehensive returns
отчуждéние сóбственности seizure of property, sequestration
отъёмный detachable
отыскáние tracking down, locating, location
~ ви́лки bracketing fire
~ ви́лки дáльности range bracketing
~ неисправности fault finding (tech)
~ цéлей target locating
отыскáть (см оты́скивать)
~ (см оты́скивать)
оты́скивать/отыскáть to find, to run to earth, to look for, to search; **оты́скивать/отыскáть боковýю ви́лку** to bracket for line
офицéр officer

офице́р администрати́вно-хозя́йственной слу́жбы A/Q officer, administration officer
~, аккредито́ванный accredited officer
~ батаре́и, ста́рший gun position officer, GPO (arty)
~, дежу́рный duty officer, orderly officer
~ запа́са reserve officer
~, ка́дровый 1. regular officer 2. cadre officer (Sov)
~, мла́дший junior officer (junior Lt to Capt in Sov army)
~ наведе́ния interception controller (radar)
~ по вооруже́нию armament officer
~ по снабже́нию supply officer
~ свя́зи liaison officer
~ свя́зи авиа́ции air force liaison officer
~, ста́рший 1. senior officer 2. gun position officer (arty)
~, штабно́й staff officer
офице́р-артиллери́ст artillery officer
офице́р-квартирье́р billeting officer
офице́р-миномётчик mortar officer
офице́р-наведе́нец guidance officer (SAM site)
офице́р-сослужи́вец brother officer
офице́р-специали́ст по обезвре́живанию неразорва́вшихся бомб bomb disposal officer
офице́р-танки́ст tank officer
офице́р-хозя́йственник Q officer
официа́льно при́нятый authorized
офо́рмить (оформля́ть)
оформле́ние 1. drawing-up 2. mounting, appearance, design
оформля́ть/офо́рмить to draw up, to make official
охва́т envelopment, outflanking
~, ножно́й leg strap (parachute)
охлади́тель cooler, refrigerant, coolant
охлади́ть (см охлажда́ть)
охлажда́ть/охлади́ть to cool, to refrigerate
охлажде́ние cooling, refrigeration
~, вне́шнее external cooling
~, вну́треннее internal cooling
~, водяно́е water cooling
~ водяно́й руба́шкой waterjacket cooling
~, возду́шное air cooling
~, жи́дкостное liquid cooling
~, закры́тое closed circuit cooling
охлажде́ние, иску́сственное air conditioning
~, по́ристое sweat cooling
~, принуди́тельное forced cooling
~, регенерати́вное regenerative cooling, closed-cycle cooling
~, эффузио́нное sweat cooling (arty)
охо́та hunting
~ на лис "fox hunting" (Sov DF competition)
охра́на protection, guard
~, берегова́я coast watch
~, вольнонаёмная civilian guard (at mil establishment - Sov equivalent to Ministry of Defence constabulary)
~, назе́мная ground protection, ground cover
~, пограни́чная border guard, frontier guard
~, пожа́рная fire service
~, противопожа́рная fire-fighting service
охране́ние protection, security
~, бли́жнее local protection, local security
~, боево́е battle outposts
~, боево́е подвижно́е mobile search security guard (van guard company in AB forces)
~, боево́е похо́дное march security
~, боково́е flank protection, flank guard, flank security
~, головно́е advance guard, frontal security
~, кругово́е all-round protection
~ на о́тдыхе protection at rest
~, непосре́дственное immediate protection, local protection
~, передово́е advanced screen
~, похо́дное protection on the move, march protection
~, сторожево́е outposts, protection at the halt, bivouac guard
~, тылово́е rear guard
~, ты́льное rear guard
~, фланго́вое flank guard, flank protection
~, фронта́льное advanced guard
охрани́ть (см охраня́ть)
охра́нный restricted (area)
охраня́ть/охрани́ть to guard, to protect
охраня́ющий protective, protecting, guarding

оце́нивать/оцени́ть to appreciate, to assess, to evaluate, to estimate; **оце́нивать/оцени́ть да́нные** to evaluate data, to estimate data; **оце́нивать/оцени́ть обстано́вку** to appreciate the situation, to estimate the situation; **оце́нивать/оцени́ть цель** to select a target

оцени́ть (см оце́нивать)

оце́нка appreciation, estimation, estimate, evaluation, assessment, appraisal, valuation

~ **ме́стности** ground appreciation

~ **обстано́вки** appreciation of the situation

~ **обстано́вки в инжене́рном отноше́нии** engineer appreciation

~ **разры́вов** artillery spotting

~ **техни́ческих ка́честв** engineering evaluation

~ **ХБР-обстано́вки** NBC appreciation

оцепле́ние полиго́на range guard

оцинко́ванный galvanized

оцинкова́ть (см оцинко́вывать)

оцинко́вывать/оцинкова́ть to galvanise

оцифро́вка numbering

оча́г бактериологи́ческого пораже́ния centre of bacteriological contamination (the area affected by enemy bacteriological attack)

~ **зараже́ния** centre of infection (area subjected to bacteriological attack)

~ **инфе́кции** centre of infection

~ **пораже́ния, хими́ческий** centre of chemical contamination

~ **сопротивле́ния** centre of resistance, pocket of resistance

очерёдность 1. priority, order of priority 2. precedence (comms)

~ **выполне́ния огневы́х зада́ч** priority of fire

о́чередь 1. ripple (missile)
2. burst (SA)
3. rounds, salvo (arty)
4. sequence
5. line, queue

~, **батаре́йный** battery fire, salvo fire (arty)

~, **втора́я** second precedence (message precedence designator)

~, **дли́нная** long burst

~, **контро́льная** ranging salvo

~, **коро́ткая** short burst

о́чередь методи́ческого огня́ rounds of battery fire (arty)

~, **пе́рвая** first degree of precedence (comms)

~, **поража́ющая** burst of fire for effect

~, **после́дняя** lowest degree of precedence (comms)

~, **пристре́лочная** 1. ranging burst (SA) 2. ranging salvo, ranging rounds (arty)

очерта́ние form, outline, contour, silhouette

очисти́тель cleaner, cleanser, purifier, rectifier, treater

очи́стить (см очища́ть)

очи́стка 1. mopping up (tac) 2. cleaning

~ **воды́** water purification

~, **дробестру́йная** blast cleaning, shot blasting

~, **пескостру́йная** sand blasting

~, **хими́ческая** chemical purification

~ **щётками** brushing, sweeping

очища́ть/очи́стить to purify, to clean, to mop up, to clear (tac)

очки́ 1. eyepieces
2. spectacles
3. points (in games etc)

~, **защи́тные** eye protectors, goggles

~, **незапотева́ющие** anti-dim eyepieces

~, **предохрани́тельные** safety goggles

~, **сне́жные** snow goggles

очко́ взрыва́теля fuse hole (ammo)

~ **ка́псюльной вту́лки** primer hole (ammo)

оши́бка error, mistake

~ **буссо́ли, индивидуа́льная** individual compass error

~ **в расчёте** miscalculation

~, **вероя́тная** probable error

~, **вероя́тная кругова́я** circular error probable (CEP)

~ **взрыва́теля, постоя́нная** constant fuse error

~, **допусти́мая** permissible error

~ **за́пуска, углова́я** angular launching error (GW)

~ **из-за нето́чной устано́вки прибо́ра** installation error

~, **инструмента́льная** instrument error

~ **ко́мпаса** compass error

~ **наведе́ния, углова́я** target error angle (GW)

~ **направле́ния** line error, error in line

оши́бка, номина́льная nominal error
~, **ночна́я** night effect (radio)
~, **относи́тельная** relative error
~, **параллакти́ческая** parallax error
~, **перехо́дная динами́ческая** step acceleration error (servo)
~, **постоя́нная** systematic error
~, **случа́йная** random error, accidental error
~ **сопровожде́ния** tracking error
~, **сре́динная** probable error, mean error (ball)
~, **сре́дняя арифмети́ческая** mean deviation (maths, gunn)
~, **сре́дняя квадрати́ческая** root mean square deviation (gunn)
~, **стати́ческая** steady state error, velocity error (servo)
~, **сумма́рная** cumulative error
~, **сумма́рная вероя́тная** overall probable error
~, **углова́я** angular error
~, **установи́вшаяся динами́ческая** constant acceleration error (servo)
оьране́ние, боево́е подвижно́е mobile march security guard (van guard company in AB forces)
оюк brushwood pyramid signpost

П

павильо́н, стрелко́вый indoor rifle range
па́водковый flood
па́водок high water, flood
па́вший в бою́ killed in action
паде́ние fall, drop
~ **давле́ния** pressure drop, pressure loss
~ **напряже́ния** voltage drop
~, **свобо́дное** free fall
~ **снаря́да** fall of shot
паёк ration
~, **высококалори́йный** high calorie value ration
~, **индивидуа́льный** individual ration pack
~, **неприкоснове́нный** emergency rations
~, **полево́й** field ration, field service ration
паёк, расфасо́ванный pack ration
~, **су́точный** daily ration, 24-hour ration
~, **сухо́й** dry provisions, dry rations
~, **штурмово́й** assault ration
пажи́лина wheelguide, riband
паз slot, groove, channel
~ **и вы́ступ** tongue and groove
~, **попере́чный** cross slot
~, **сма́зочный** oil groove
~, **спира́льный** spiral groove
~, **Т-обра́зный** T-cut
~ **уда́рника, кольцево́й** ball-retaining striker groove (gren)
~, **фикси́рующий** locating slot
~, **шпо́ночный** key bed, keyway
па́йка soldering, brazing
~ **лату́нью** brass brazing
~ **мя́гким припо́ем** soft soldering
~, **твёрдая** brass brazing
~ **твёрдым припо́ем** hard soldering, brazing
пакга́уз warehouse, storehouse
пакеля́ж stone pitching (road eng)
паке́т, взрывно́й thunder flash
~, **дегазацио́нный силикаге́левый** silica-gel decontamination packet
~, **дегази́рующий** personal CW decontamination kit
~, **заказно́й** registered package
~, **индивидуа́льный** first aid packet
~, **индивидуа́льный дегазацио́нный** personal CW decontamination kit
~, **индивидуа́льный перевя́зочный** first field dressing
~, **индивидуа́льный противохими́ческий** personal CW decontamination kit
~ **мо́ста** bridge composite roadbearer
~ **направля́ющих** cluster of rocket launchers, multi-round rocket launcher
~, **подрывно́й** satchel charge
~, **санита́рный** first field dressing kit
~, **служе́бный** official despatch
~, **хими́ческий** personal anti-gas equipment
па́кля oakum
пакт pact
~ **о ненападе́нии** non-aggression pact
пала́та ward (hospital etc)
пала́тка tent
~, **операцио́нная** surgical tent
~, **шестигра́нная** hexagonal tent
~, **штабна́я** office tent
пале́тка map case

па́лец 1. digit, finger, toe
2. pin
~, балансиро́вочный balancing plug (GW)
~, вставно́й male pin
~, гу́сеничный track pin (tk)
~ звена́ гу́сеницы track pin
~ карда́на universal-joint pin
~ колесоотбо́йного бру́са riband pin
~, опо́рный anchor pin
~, пла́вающий поршнево́й floating gudgeon pin
~ ползуна́ приёмника belt feed pawl
~, поршнево́й gudgeon pin, piston pin, wristpin
~ прицепно́го шатуна́ articulated rod knuckle pin (V-type ICE)
~, рессо́рный spring shackle pin
~ тормозно́й коло́дки, опо́рный brake anchor pin
~ тормозно́й коло́дки, шарни́рный brake shoe hinge pin
~ тра́ка grouser bar, track pin
~ шарни́ра joint pin, link pin, swivel pin
~ шатуна́ gudgeon pin
~ щёткодержа́теля brush holder arm
палиса́д 1. paling, fence
2. stockade
пали́ть to fire (a weapon) (coll), to burn
па́лка, бараба́нная drumstick
~, лы́жная ski stick
~, трениро́вочная dummy rifle stick, training stick
па́луба deck
~, кормова́я after-deck
~, носова́я fore-deck
палудри́н paludrine (anti-malaria drug)
пальба́ gunfire (coll)
пальцеотбо́йник track pin controller
па́мятка instruction booklet, instruction pamphlet
~ деса́нтнику standard operating procedures for assault personnel
~ по обраще́нию с ору́жием weapon handling guide
па́мять memory, storage computer
~, операти́вная working memory
~, стира́емая erasable memory
па́ндус fixed ramp
пане́ль board, panel
~ автома́тов защи́ты се́ти circuit-breaker panel
пане́ль, звукопоглоща́ющая sound-absorbing panel
~ коммута́ции switchboard panel, switching panel
~, ла́мповая valve holder
~, пере́дняя 1. dashboard (veh)
2. instrument panel (tk)
~ переключе́ний switchboard panel, switching panel
~, прибо́рная instrument panel, instrument console
~, распредели́тельная distribution panel
~ с гнёздами jack panel
~, съёмная access panel, detachable panel
~, термоизоляцио́нная insulating panel
па́ника panic
паникёр panic-monger
панора́ма 1. panorama
2. dial sight, panoramic sight
~, артиллери́йская artillery panorama
панора́мный panoramic
панто́граф pantograph
па́нцирь 1. armour, coat of mail
2. shell, test (armadillo, tortoise etc)
папа́ха "Papakha" (Caucasian high fur hat worn with winter uniform by full colonels and above)
папиро́са 1. cigarette
2. hammer, striker head (ammo)
па́пка file, folder
па́пка-планше́т map case, map folder
пар 1. steam
2. vapour
~ высо́кого давле́ния high-pressure steam
~ ОВ toxic vapour
па́ра, зубча́тая gear wheel and pinion
~ сил couple (mech)
~ сил, выпрямля́ющая righting couple (physics)
~, скру́ченная twisted pair (electron)
~, червя́чная worm and worm-wheel
~, шарова́я ball and socket joint
пара́бола безопа́сности parabola of safety (ballistics)
параболи́ческий parabolic
парава́н paravane
пара́д parade, review
парази́т parasite (med)
парали́ч paralysis
паралла́кс parallax
~ ба́зы subtense angle

параллáкс, вертикáльный vertical displacement
~, **горизонтáльный** horizontal displacement
параллелогрáм parallelogram
~, **визи́рный** parallel motion gear (course and speed sights)
паралле́ль parallel
парáметр parameter, characteristic
~, **баллисти́ческий** ballistic parameter
~, **курсовóй** 1. course profile (GW)
2. ground range (gunn)
3. plan range (nav gunn)
парáметры, основны́е конструкти́вные basic design data
~ **рабóты** working conditions
~, **тáктико-техни́ческие** performance data (eqpt)
~, **эксплуатациóнные** technical data, operating data (eqpt)
парапе́т parapet
парафи́н paraffin wax
~, **жи́дкий** medicinal paraffin, liquid paraffin
парашю́т 1. parachute
2. cargo tray
~, **авари́йный** emergency parachute
~ **автомати́ческого де́йствия** automatic parachute
~, **вытяжнóй** extractor parachute, auxiliary parachute, pilot parachute
~, **глáвный** main parachute
~, **грузовóй** supply dropping parachute
~, **запаснóй** reserve parachute
~, **кольцевóй** annular parachute
~, **ле́нточный** ribbon parachute
~, **многокýпольный** parachute cluster (for heavy dropping)
~, **нагрýдный** chest-type parachute, chest-pack parachute
~, **наспи́нный** back-type parachute, back-pack parachute
~ **свобóдного де́йствия** free parachute
~, **стабилизи́рующий** stabilizing parachute
~, **тормознóй** brake parachute
~, **трениро́вочный** practice parachute, training parachute
~, **управля́емый** guided parachute
парашюти́зм parachute activity
парашю́тик little parachute
~, **вытяжнóй** extractor parachute
парашюти́рование parachuting, pancaking (ac)
парашюти́ровать to parachute, to pancake (ac)
парашюти́ст parachutist
парашюти́ст-десáнтник paratrooper
парашюти́ст-инстру́ктор parachute instructor
парашюти́ст-лы́жник para-skier
парашю́тный parachute
пари́ровать to parry, to ward off, to avert
парите́т, я́дерный nuclear parity
парк park, store, depot, fleet, complement, workshop (mil unit), yard, pool
~, **автобронетáнковый** vehicle establishment (of unit or formation)
~, **автóбусный** bus fleet
~, **автомоби́льный** 1. vehicle park, MT park
2. vehicle pool
~, **артиллери́йский** artillery ordnance park
~ **боевы́х маши́н** AFV park
~, **вагóнный** rolling stock (rly)
~, **грузовóй** freight yard, goods yard
~, **деревя́нный мостовóй** wooden bridging park
~ **десáнтно-вы́садочных средств, свóдный** amphibious boat pool
~, **инвентáрный** inventoried stock
~, **лёгкий переправóчный** light bridging train
~, **маневрóвый** shunting yard (rly)
~, **мостовóй** bridging park
~ **основнóй сортирóвки** classification yard (rly)
~, **переправóчный** bridging train
~, **полевóй** field maintenance park (for servicing and repair of vehicles and equipment)
~, **понтóнный** pontoon park, bridge train
~, **постоя́нный** base maintenance park (with permanent installations for servicing and repair work)
~ **прибы́тия** receiving yard (rly)
~ **приём** receiving yard (rly)
~, **приёмный** receiving yard, reception siding (rly)
~, **приёмо-отпрáвочный** reception and departure sidings (rly)
~ **путе́й** sidings (rly)
~ **ремóнта** repair area, workshop area

парк сбо́ра маши́н, передово́й advanced vehicle reception park, forward vehicle depot
~, сортиро́вочный sorting sidings (rly)
~, сортиро́вочный механизи́рованный mechanised marshalling yard
~, та́нковый tank park
~ техни́ческого обслу́живания servicing area
~ тра́нспортных маши́н motor park
~ тягаче́й tractor park
~, уче́бно-боево́й field training area
~ эвакуи́руемых маши́н returned vehice park
парк-стоя́нка parking area
парламентёр truce envoy
парово́з steam locomotive
парово́й steam
парога́з steam (produced in HTP (high test peroxide) turbine)
парогазогенера́тор 1. combustion chamber (high test peroxide engine)
2. steam generator
пародина́мо steam dynamo
паро́ль 1. password
2. authentication signal
~, переговóрный authentication signal
паро́м ferry
~, автомоби́льно-пассажи́рский passenger-car ferry
~, большегру́зный моторизо́ванный heavy motor ferry
~, кана́тный cable ferry
~, мостово́й pontoon bridge section
~, мото́рный motor ferry
~ на возду́шной поду́шке hover-ferry
~ на подво́дных кры́льях hydrofoil ferry
~ на понто́нах, перевозно́й pontoon ferry
~, речно́й river ferry
~, самохо́дный motor ferry, powered ferry, motorized ferry, self-propelled ferry
~, сваебо́йно-обстро́ечный pontoon-mounted piling rig
~, спаса́тельный life raft
~, та́нковый tank ferry
~, тяжёлый heavy ferry
~, цепно́й chain ferry
паро́м-амфи́бия amphibious ferry
пароморе́йс ferry trip
партбюро́ (communist) party bureau
партиза́н partisan, guerilla
партиза́нский partisan, guerilla
па́ртия 1. party, group
2. lot, batch (ammo)
3. consignment, set
4. The Communist Party
~ беста́рного гру́за bulk consignment
~ боеприпа́сов ammunition lot, ammunition batch
~ ваго́нов waggon batch
~ горю́чего consignment of fuel
~ гру́за consignment, shipment
~ гру́за для ча́сти unit consignment
~ гру́за на ваго́н car load (rly)
~ заря́дов по́роха powder lot
~ по́роха lot of propellant
~ по́роха, образцо́вая type powder lot
~, рекогносциро́вочная reconnaissance party
партко́м communist party committee (Sov)
парто́рг communist party organiser (Sov)
паруси́на canvas
паруси́нка collapsible boat
па́ры fumes
па́спорт 1. identity card
2. log (tech)
~ автомоби́ля motor log book, motor certificate
~ дви́гателя engine's log book
~, заводско́й manufacturer's rating plate, factory certificate
~, заграни́чный passport
~, техни́ческий log book (veh etc), equipment record
пассати́жи combination pliers
пасси́в liabilities (accounting)
пассиви́рование anti-corrosion coating (of metals)
пасси́вный passive
па́ста paste, compound
~, вулканизи́рующая vulcanizing compound
~, герметизи́рующая sealing paste, sealing compound
~, прити́рочная abrasive paste, grinding paste
патоге́нный pathogenic
патрио́т patriot
патро́н 1. cartridge, round
2. chuck (lathe)
3. socket
~ без закра́ины rimless cartridge

патро́н, боево́й live round
~, боево́й охолощённый drill round
~ боково́го огня́ rim-fire cartridge
~, винто́вочный rifle cartridge, rifle round
~, вы́шибной propelling cartridge
~, гопкали́товый hopcalite cartridge
~, динами́тный a stick of dynamite, dynamite cartridge
~ для винто́вочной грана́ты grenade cartridge
~, до́сланный chambered round
~, дробово́й shot-gun cartridge, canister round
~, зажига́тельный incendiary charge
~, заклини́вшийся jammed round
~, запа́льный 1. detonator (mine)
2. combustion torch, starting cartridge (diesel)
~, имитацио́нный artillery shellburst simulator
~, малокали́берный 1. small-calibre round (arty)
2. miniature round (SA)
~, освети́тельный illuminating cartridge, flare cartridge
~ освеще́ния lamp holder
~, охо́тничный shot-gun cartridge
~, переснаряжённый refilled cartridge
~, пистоле́тный pistol cartridge
~, подрывно́й blasting cartridge, explosive cartridge
~, прове́рочный dummy cartridge, dummy round (for test purposes)
~ противога́за respirator canister
~, раке́тный signal cartridge
~, реакти́вный освети́тельный rocket flare cartridge
~, регенерати́вный regenerative canister (breathing eqpt)
~ ручно́го снаряже́ния hand-loaded cartridge
~ с бесшля́пной ги́льзой, име́ющий опо́рный бу́ртик belted rimless cartridge
~ с бронебо́йной пу́лей armour-piercing round
~ с ги́льзой без закра́ины rimless cartridge
~ с зажига́тельной пу́лей incendiary round
~ с цилиндри́ческой ги́льзой straight-case cartridge
патро́н, самоцентру́ющий self-centring chuck (lathe)
~, сигна́льный signal cartridge
~, станда́ртный standard cartridge
~, унита́рный quick-firing (QF) fixed round
~, уче́бно-контро́льный drill round
~, уче́бный drill round, drill cartridge (SA)
~, хвостово́й primary cartridge (mortar)
~, холосто́й blank cartridge
~ шпи́лечного воспламене́ния pin-fire cartridge
~, электри́ческий lamp holder
патро́н-воспламени́тель, динами́тный dynamite cap
патро́нник chamber (ord, SA)
патронта́ш ammunition pouch, bandolier
~, нагру́дный bandolier
патро́ны с трасси́рующими пу́лями tracer ammunition
па́трубок 1. gas passage
2. branch pipe, connecting pipe, overflow pipe
~, выхлопно́й short exhaust outlet pipe
~ ра́мы затво́ра carrier pintle (ord)
патрули́рование patrolling (non-tactical)
патрули́ровать to patrol (non-tac)
патру́ль patrol (non-tac, police etc)
~, вну́тренний regimental police patrol
~, гарнизо́нный garrison patrol, town patrol
~, коменда́нтский town patrol, garrison patrol (Sov equivalent of mil police)
~, офице́рский patrol commanded by an officer
~, сержа́нтский patrol commanded by an NCO
патру́льный prowler sentry, prowler guard
па́уза 1. pause, interval
2. paralysis (elec)
3. resting time (radar etc)
~, операти́вная pause in operations, lull in operations
па́хота ploughed land
па́чка pack, packet, package, bundle, file
па́шня ploughland, arable land
пая́льник soldering iron
пая́ть to solder, to braze
педа́ль акселера́тора accelerator pedal, throttle pedal

педáль взрывáтеля, нажи́мная igniter plunger
~ **выключéния** cut-out pedal
~ **гáза** accelerator pedal, throttle pedal
~**, нажи́мная** pressure plunger (min)
~ **ножнóго спýска** firing pedal (ord)
~ **переключéния передáч** gear shift pedal
~ **подáчи тóплива** accelerator pedal
~**, спусковáя** trigger pedal
~ **сцеплéния** clutch pedal
~**, тормознáя** brake pedal
~ **фрикциóна** clutch pedal
пек pitch (tar)
пекáрня bakery
пéкарь baker
пéленг bearing
~**, боевóй** attack vector (AD)
~**, визуáльный** visual bearing
~**, звуковóй** sound bearing
~**, и́стинный** true bearing
~**, кóмпасный** compass bearing
~**, магни́тный** magnetic bearing
~**, обрáтный** reciprocal bearing
~ **ориенти́ра, и́стинный** true bearing of a reference point
~ **ориенти́ра, магни́тный** magnetic bearing of a reference point (from aircraft)
~ **радиостáнции** radio bearing
~ **упреждённой тóчки** collision heading, collision point bearing (GW)
пеленгáтор direction finder, DF set
~**, автомати́ческий** automatic DF set
~**, зени́тный тепловóй** AA heat radiation detector
~**, инфракрáсный** infra-red DF set
~ **кóмпаса** compass azimuth circle
~**, тепловóй** thermal detector
пеленгáция taking of a bearing, direction finding, DF
~**, автомати́ческая** automatic direction finding, automatic DF
пеленгировáние direction finding, DF
пеленги́ровать to take a DF bearing
пеленговáть to take a DF bearing
пéна foam, suds, scum, head, lather
~**, противопожáрная** fire fighting foam
пенáл 1. small container
2. pencil box
пенёк chuck, batten
~ **ствола́** barrel shoulder (ord)
пéнистость foaminess, frothiness
пéнистый foamy, frothy
пеницилли́н penicillin
пеногенерáтор foam generator
пенообразовáние foaming, frothing, foam production
пенопла́ст foamed plastics, aerated plastics, cellular plastics
~**, жёсткий** rigid foam plastics
~**, эласти́чный** flexible foam plastics
пенополиуретáн polyurethane foam
пéнсия pension
~ **за вы́слугу лет** service pension
~ **по инжали́дности** disability pension
~ **по слýчаю смéрти военнослýжащего** family pension
пентагри́д heptode, pentagrid
пентóд pentode
~ **с перемéнной крутизнóй** variable-mu pentode
пентоли́т pentolite (expl)
пентри́т penthrite, PETN (expl)
пень tree stump
пéня fine
пéпел ash
перви́чный primary
первоклáссный first-rate
первоочереднóй top-priority
первосóртный high grade, first-rate
переаттестáция reclassification
перебази́рование re-deployment, transfer from one base to another
перебази́ровать to re-deploy, to transfer from one base to another
перебéжка 1. bound, rush
2. defection
перебéжчик defector
перебóй 1. unevenness, irregularity, stoppage, delay, interruption
2. miss (engine)
~ **в зажигáнии** misfiring (ICE)
~ **в снабжéнии** breakdown in supplies, breakdown in supply system
перебóрка 1. bulkhead, partition
2. refit (coll)
перебрáсывать/перебрóсить to switch, to transfer, to throw over
перебрóсить (см перебрáсывать)
перебрóска 1. transfer, switch
2. switch over (instr)
~ **войск по вóздуху** air trooping, airlift
~ **по вóздуху** airlift
~ **средств ты́ла** delivery of logistical support

переброска через реку river crossing (by troops)
перевал pass (topo)
~, горный mountain pass
перевалка trans-shipment, transloading
перевес superiority, preponderance
~ дульной части орудия muzzle heavy
~, численный numerical superiority
перевод 1. transfer 2. interpretation, translation
~, денежный money order
~, почтовый postal order
переводчик 1. interpreter, translator 2. change-over lever, change lever, change-over switch
~ для автоматического огня automatic fire selector lever
перевоз, канатный cable ferry
перевозимый transportable
~ по воздуху 1. air portable (eqpt) 2. air transportable (pers)
перевозка transportation, haulage, carriage, lift, shipment
~, автомобильная motor transportation, carriage by MT
~ боеприпасов ammunition movement
~, водная water transportation
~, воздушная airlift, air transportation, air movement
~ войск movement of troops
~ войск, воздушная air trooping, airlift
~ грузов, бестарная bulk movements of materials
~ грузов наливом bulk lifting, bulk transportation
~, железнодорожная rail travel, carriage by rail
~ на короткое расстояние short haul
~ опасных грузов, водная transfer of dangerous cargoes by river and canal
~ от корабля к берегу ship-to-shore transportation
~, сквозная through movement
~, срочная 1. urgent shipment 2. urgent move (personnel)
перевозки 1. (air) lift 2. (troop) movements
~, бестарные bulk deliveries
~, военные military traffic
~, воздушные airlift, air movement
~, воинские troop movements
~, встречные reverse movements
~, грузовые freight movements
перевозки, дальние long-distance haulage, long-distance traffic
~, контейнерные container traffic
~, короткопробежные short-distance traffic, short-haul traffic
~, междугородные inter-city traffic
~, оперативно-стратегические воздушные strategic airlift
~, оперативно-тактические tactical transportation
~, плановые routine movements
~, снабженческие supply lift
~, транзитные through shipments
переворот revolution, overthrow, coup
перевязать (см перевязывать)
перевязка bandage, dressing
~, стерильная sterile dressing, antiseptic dressing
перевязочная dressing room, dressing ward (med)
перевязывать/перевязать to bandage
перевязь 1. shoulder belt, Sam Browne 2. sling (med)
перегиб bend, twist
перегнать (см перегонять)
перегон stage, lap, leg
перегонка distillation, distilling
~ самолётов ferrying of aircraft
перегонять/перегнать to overtake, to outdistance, to drive out, to distill
перегородка 1. bulkhead, partition 2. septum, iris (electron)
~, герметичная air-tight partition
~, ёмкостная capacitive iris (waveguide)
~, огнестойкая fire-proof bulkhead
~, отражательная baffle plate
~, противопожарная fire wall fire bulkhead
перегрев overheating
~, местный hot spot (nucl)
перегревание overheating
перегружать/перегрузить to overload, to transfer a load, to trans-ship
перегруз overload, excess load
перегрузить (см перегружать)
перегрузка 1. overloading 2. transfer of load, trans-shipment
~, длительная sustained overloading
~, допустимая permissible overload
~, максимальная maximum overload, extreme overload

перегрузка, предельная 1. acceleration limit (aeron)
2. maximum permissible overload
перегруппировать(ся) (см перегруппировывать(ся))
перегруппировка regrouping
перегруппировывать(ся)/ перегруппировать(ся) to regroup
передавать/передать to transmit, to hand over; **передавать/передать данные** to transmit data, to transfer data, to send data, to furnish data, to deliver data; **передавать/передать заявки в** to refer a request to; **передавать/передать ключом** to send morse by key, to key; **передавать/ передать открыто** to transmit in clear; **передавать/передать права** to delegate authority
передаточный transmission, transmitting
передатчик transmitter
~, аварийный emergency transmitter, radio transmitter
~ активных радиопомех active jammer
~, длинноволновый long-wave transmitter
~ заградительных помех barrage jammer, barrage transmitting jammer
~, искровой spark gap transmitter (radio)
~ искусственных помех jamming transmitter
~ команд command transmitter (GW)
~, коротковолновый short-wave transmitter
~, лазерный laser transmitter
~, ламповый valve transmitter, valve sender
~, мешающий jamming transmitter, jammer
~, направленный directional transmitter
~, оптический optical transmitter
~ ответчика responding transmitter, answering transmitter
~ помех, бортовой on-board jammer, airborne jammer
~ помех, магнетронный magnetron jammer
~ помех прицельного типа spot jammer
~ помех, радиолокационный radar jammer
передатчик, радиозондовый radiosonde transmitter
~ радиолокационных помех radar jammer
~, радионавигационный navigational transmitter
~ радиопомех jamming transmitter
~, резервный auxiliary transmitter, stand-by transmitter
~, ретрансляционный relay transmitter
~ с временным разделением каналов transmitter with time division of channels
~ с двумя боковыми полосами частот double-side-band transmitter
~ с кварцевой стабилизацией crystal-stabilised transmitter
~ с одной боковой полосой частот single-side-band transmitter
~ с фазовой модуляцией phase-modulated transmitter
~ с частично подавленной боковой полосой vestigial side-band transmitter, asymmetrical-side-band transmitter
~ с частотным разделением каналов transmitter with frequency division of channels
~ с частотой, заданной кварцем crystal-controlled transmitter
~, самолётный карцинотронный aircraft carcinotron transmitter
~ синхронных радиопомех synchronous jammer
~, телевизионный television transmitter
~, телеметрический telemetering transmitter
~, фототелеграфный facsimile transmitter
передатчик-ответчик transponder, transmitter-responder
передать (см передавать)
передача 1. hand over, transfer
2. transmission, broadcast
3. transmission, drive
4. gear
~, автоматическая automatic transmission
~, бесступенчатая infinitely variable transmission, stepless gear
~, бесшумная silent gear, quiet gear
~, бортовая final drive (tracked vehicle)
~ в заднем мосте rear-axle drive

передáча, винтовáя screw gear
~, **вторáя** second gear
~, **высóкая** high gear, top gear
~, **вы́сшая** high gear, top gear
~, **геликоидáльная** helical gear
~ **ги́бкой свя́зью** flexible gearing
~, **гидравли́ческая** hydraulic transmission
~, **гидродинами́ческая** hydrodynamic transmission
~, **гидромехани́ческая** hydromechanical transmission
~, **гидромехани́ческая силовáя** hydromechanical transmission
~, **гипóидная** hypoid gear
~, **глáвная** final drive (wheeled veh)
~, **групповáя** group transmission, group working (sigs)
~ **дáнных** data transmission
~ **двумя́ боковы́ми полосáми** double side-band transmission
~ **дискрéтной информáции** digital transmission
~, **дифференциáльная** differential gear, differential drive
~, **замедля́ющая** (speed) reducing gear
~, **зубчáтая** gear train, gearing
~ **изображéния** picture transmission
~, **и́мпульсная** pulse transmission
~ **информáции** data transmission
~, **кардáнная** cardan drive, universal-joint drive
~, **клиноремённая** V-belt drive, texrope drive
~ **ключóм** keying, hand morse transmission
~ **ключóм и приём на слух** morse transmission and reception
~, **колёсная** wheel gearing
~, **конéчная** final drive
~, **кони́ческая** bevel drive
~, **кривоши́пная** crank gear(ing)
~, **кривоши́пно-шату́нная** crank gear(ing)
~, **лёгкая синхрóнная** indicator magslip transmission (FCE)
~, **микрофóнная** microphone transmission, voice transmission, R/T transmission
~, **многоканáльная** multi-channel transmission, multiplex transmission

передáча, многокрáтная multiplex transmission, multi-channel transmission
~, **многоступéнчатая** multiple gearing
~ **на боковóй полосé** side-band transmission
~ **на две пáры колёс** four-wheel drive
~ **на двух боковы́х полосáх** double-side-band transmission
~ **на зáдние колёса** rear-wheel drive
~ **на однóй боковóй полосé** single-side-band transmission
~ **на перéдние колёса** front-wheel drive
~ **на четы́ре колесá** four-wheel drive
~, **напрáвленная** 1. beaming (radar) 2. beamed transmission (radio) 3. directional propagation (radio)
~, **непосрéдственная** direct drive (mech)
~, **ни́зшая** low gear
~ **однóй боковóй полосóй** single-side-band transmission
~, **однополóсная** single side-band transmission
~ **от полуóси к колесу́** wheel-hub drive
~, **откры́тая** plain-language transmission
~, **пéрвая** first gear
~, **планетáрная** epicyclic gear, planetary gear
~, **повышáющая** step-up gear, overdrive
~ **подотчётного иму́щества** transfer of equipment on charge
~, **понижáющая** step-down gear
~, **поперéчная** cross drive
~, **промежу́точная** intermediate drive, intermediate gearing
~, **прямáя** direct drive
~, **реду́кторная** reduction gear
~, **рéечная** rack-and-gear drive
~, **рéечная рулевáя** rack and pinion steering gear
~, **ремённая** belt drive
~ **с больши́м передáточным числóм** high-ratio gear
~ **с врéменным уплотнéнием канáлов** time multiplex transmission, transmission with time division of channels
~ **с постоя́нным зацеплéнием** constant mesh transmission
~ **с синхронизáтором** synchromesh transmission

передáча с уравнúтелем скóростей synchromesh transmission
~ **с частúчно подáвленной боковóй полосóй** vestigial side-band transmission, asymmetrical-side-band transmission
~ **с частóтным уплотнéнием канáлов** frequency multiplex transmission, transmission with frequency division of channels
~, **силовáя** power transmission, power drive
~, **силовáя синхрóнная** power magslip transmission (FCE)
~, **синхрóнная** synchro transmission, synchro data transmission (FCE)
~ **со свобóдным хóдом** free wheel
~ **трéнием** friction drive
~, **трéтья** third gear
~, **ускоря́ющая** overdrive
~, **фототелегрáфная** facsimile transmission
~, **фрикциóнная** friction gearing
~, **цепнáя** chain drive
~, **цилиндрúческая зубчáтая** spur gearing
~, **червя́чная** worm gear
~, **четвёртая** fourth gear
~, **шестерéнчатая** pinion drive
~, **электрúческая синхрóнная** synchro transmission, magslip transmission
передвижéние в рост the walk (method of movement)
~ **вне дорóг** cross-country movement
~ **войск** troop movement/s
~ **перекáтами** leap-frogging
~ **переплóзанием** crawling (method of movement)
~ **ползкóм** crawling (method of movement)
~ **скачкáми** movement by bounds
~ **согнýвшись** crouching (method of movement)
~, **тактúческое** tactical movement
передемпфúрование overdamping, oversmoothing
передéржка overexposure (photo)
передислокáция redeployment
передовóй 1. forward, advanced, leading (tac)
2. leading (gen)
передóк limber (ord)
передóк, двухколёсный two-wheeled limber (arty)
переды́шка breathing space, lull
переéзд 1. level crossing
2. journey
3. permanent move
~, **железнодорóжный** level crossing
переезжáть/переéхать to cross, to (re)move (to); **переезжáть/переéхать вброд** to ford
переéхать (см переезжáть)
пережóг burnout
перезарядúть (см перезаряжáть)
перезаря́дка 1. recharging (elec)
2. overcharge (elec)
перезаряжáние re-loading (ammo)
перезаряжáть/перезарядúть to recharge (elec), to reload (weapon)
перейтú (см переходúть)
перекáт 1. shoals (river)
2. leap-frog (tac)
перекáчка pumping, transfer, dispensing
~ **горю́чего** fuel transfer
~ **по трубопровóду** pipeline transfer
перекисловодорóдный hydrogen peroxide
пéрекись peroxide
~ **азóта** nitrogen peroxide
~ **водорóда** hydrogen peroxide
~ **маргáнца** manganese dioxide
переклáдина 1. horizontal bar (gym)
2. transom
3. cap, capsill, top sill
~, **кóзловая** trestle transom
переклáдка transom
переклёпка re-riveting
перeклúчка roll-call, call-over
переключáтель change-over switch
~, **антéнный** 1. transmit-receive switch, TR switch
2. anti TR switch, ATR switch
3. transmitter-blocker cell, TB cell (pulse radar)
4. antenna change over switch (switch, cell, tube, box are synonymous in this context)
~ **бáшни, глáвный** turret master switch
~ **блúжнего свéта фар, ножнóй** foot-operated dip switch
~, **включáемый излучéнием** light-activated switch
~, **возбуждáемый свéтом** light-activated switch

переключа́тель, гла́вный main switch, master switch
~ **да́льности** range switch
~ **диапазо́нов** frequency-range selector switch
~ **затво́ра** breech mechanism change-over lever
~ **кана́лов** channel selector
~ **кана́лов, дистанцио́нный** remote channel selector (DME)
~ **кана́лов свя́зи** channel selector switch
~**, кно́почный** push-button switch
~**, концево́й** limit switch
~**, многоконта́ктный** multi-contact switch
~ **переда́чи-приёма** ATR switch, transmit-receive switch, TR switch
~**, перекидно́й** throw-over switch
~**, поворо́тный** rotary switch
~ **по́стов управле́ния** control point change-over switch (elec)
~**, рыча́жный** lever switch
~ **све́та фар** headlamp dipper switch
~**, солено́идный** solenoid switch
~ **указа́теля поворо́та** direction indicator switch, trafficator switch
~**, электромагни́тный** electromagnetic switch
~**, электро́нный** electronic switch
переключа́ть/переключи́ть to switch over; **переключа́ть/переключи́ть переда́чу** to change gear; **переключа́ть/переключи́ть ско́рость** to change gear
переключе́ние 1. switching (elec)
2. shifting (mech)
~ **анте́нны** lobe switching (radar)
~ **переда́ч** gear shifting change
~ **переда́ч без включе́ния сцепле́ния** clutchless gear change
~ **переда́ч рычаго́м на рулево́й коло́нке** steering-column gear change
~ **рычаго́м, располо́женным в полу́** floor gear change
~**, синхронизи́рованное** synchromeshing (gears)
~**, электро́нное** gating (electron)
переключи́ть (см переключа́ть)
переко́с warping
перекре́стие graticule, cross-wires (instr)
перекрёсток crossroads
~ **с кругов́ым движе́нием** roundabout
перекре́стье graticule, cross-wires (instr)
перекрыва́ть/перекры́ть to overlap, to cover mutually, to dim (headlights), to exceed
перекры́тие 1. overhead cover, overhead protection
2. overlap, overlapping
~ **кла́панов** valve overlap (ICE)
~**, пло́ское** overhead cover at ground level, ground level cover
~**, попере́чное** lateral overlap (air photo)
~**, продо́льное** line overlap (topo)
перекры́ть (см перекрыва́ть)
переле́сок copse
перелёт 1. flight
2. over (arty gunn)
~**, внутриарме́йский** intra-army flight, flight within the area of an army
~**, межарме́йский** inter-army flight (AF)
~**, обра́тный** flyback (CRT)
перелива́ние кро́ви blood transfusion
перело́м 1. break
2. turning point
переме́нная variable (maths)
~**, зави́симая** dependent variable
~**, ко́мплексная** complex variable
~**, случа́йная** random variable
переме́нный 1. variable
2. alternating (elec)
переме́нчивый changeable (weather)
перемести́ть (см перемеща́ть)
перемести́ться (см перемеща́ться)
переме́шивание mixing, stirring, agitation
перемеща́ть/перемести́ть to transpose, to move, to shift
перемеща́ться/перемести́ться to move location, to change position, to shift
перемеще́ние change of position, displacement, shift
~ **це́ли, углово́е** target angular travel, angular displacement of target
переми́рие armistice, truce
перемы́чка 1. connector, bonding strip, bonding jumper
2. bulkhead, partition
3. coffer-dam
~**, грудна́я** chest crossing straps (para)
перенацели́ть to redirect, change the direction of (tac)
перенести́ (переноси́ть)
перено́с 1. carried forward (book-keeping)
2. transfer, carry-over
~ **огня́** switching of fire

перено́с огня́ спо́собом коэффицие́нта "К" "K" transfer (US) (gunn)
переноси́ть/перенести́ to carry/transfer from one place to another; **переноси́ть/перенести́ ого́нь** to switch fire (gunn)
перено́ска на носи́лках litter carry
перено́сный portable
переобору́дование 1. refit 2. re-equipping
переотклоне́ние overshooting, overswing (electron)
переоце́нка over-estimation
перепа́д jump, drop, fall (temp, pressure etc)
~ давле́ний differential pressure, pressure differential
~ напо́ра тече́ния flow gradient
~ температу́р temperature drop
перепа́лка exchange of fire, fusillade, skirmish
перепи́ска, служе́бная official correspondence
перепи́счик transcriber
~ на маши́нке typist
переподчине́ние артилле́рии regrouping of artillery command
переподчини́ть (см переподчиня́ть)
переподчиня́ть/переподчини́ть to resubordinate
перепо́лзание crawling, crawl
~ на боку́ side crawl
~ на получетвере́ньках form of crawl (similar to Brit monkey run)
перепо́нка 1. septum (electron) 2. membrane
~, бараба́нная eardrum
перепра́ва crossing (usually water obstacle)
~ вброд fording
~, деса́нтная assault crossing (of a water obstacle)
~, запа́сная alternate crossing
~, ледяна́я ice bridging
~, ло́жная feint crossing
~, мостова́я bridge crossing
~ на паро́мах crossing by ferries
~ на пла́вающих сре́дствах ferrying
~ на плота́х raft ferry, rafting
~ на понто́нах pontooning
~, односторо́нняя one-way crossing
~, паро́мная ferry crossing
~, понто́нная pontoon crossing, pontoon bridge crossing
перепра́ва, регули́руемая controlled crossing
~ с бо́ем forced crossing, opposed crossing
~ че́рез реку́ river crossing
перепра́виться (см переправля́ться)
переправля́ться/перепра́виться to swim across, to sail across; **переправля́ться/перепра́виться вброд** to ford; **переправля́ться/перепра́виться че́рез реку́ по дну** to wade a river (using a schnorkel)
перепра́вочно-деса́нтный assault-crossing
перепра́вочно-мостово́й bridging
перепу́ск by-pass
перерабо́тка re-treatment, re-processing
перераспределе́ние re-distribution, re-allotment
~ ве́са weight transfer
перерасхо́д excessive consumption
~ то́плива excessive fuel consumption
переры́в break, stoppage
~ в подво́зе break in supply
пересече́ние intersection
~ доро́г crossroads
пересечённость ruggedness (of terrain)
переско́к skip effect (radio)
переснаряди́ть (см переснаряжа́ть)
переснаряжа́ть/переснаряди́ть to replenish, to recharge, to re-equip
переснаряже́ние replenishment, recharging, re-equipping
перестволе́ние relining (ord)
перестра́ивать/перестро́ить to re-form, to re-build, to re-tune (radio)
перестрахо́вка over-cautiousness
перестре́лка exchange of fire
~, артиллери́йская artillery engagement, exchange of artillery fire
перестрое́ние change of formation
перестро́ить (см перестра́ивать)
перестро́йка 1. re-tuning (radio) 2. reconstruction 3. reorganization
пересчёт conversion
переуспокое́ние overdamping, oversmoothing
перехва́т interception (ac, GW, radio etc)
~, управля́емый с земли́ ground controlled interception, GCI
перехвати́ть (см перехва́тывать)
перехва́тчик interceptor

перехва́тывать/перехвати́ть to intercept
перехо́д 1. conversion, change-over, transition
2. march, passage
3. distance covered in a given unit of time
4. crossing, cross-over
~, вынужденный stimulated transition, induced transition (elec)
~, коле́йный trackway
~ мо́рем sea passage
~, пехо́тный day's march
~, подзе́мный underground passage
~, рабо́чий signal transition (elec)
~, спонта́нный spontaneous transition (elec)
~, су́точный day's march
~, электро́нно-ды́рочный positive-negative junction, P-N junction (of a semi-conductor)
переходи́мый вброд fordable
переходи́ть/перейти́ to get across, to get over, to take, to assume; **переходи́ть/перейти́ вброд** to wade; **переходи́ть/перейти́ в наступле́ние** to go over to the offensive; **переходи́ть/перейти́ к оборо́не** to go over to the defensive; **переходи́ть/перейти́ на пораже́ние** to switch to fire for effect
перехо́дник adapter
перехо́дный transient, transitional
пе́речень list, schedule
~ шта́тной армату́ры list of standard fittings
перечисле́ние 1. enumeration
2. transfer
~ фо́ндов transfer of funds
переша́гивание сбо́ку side-stepping (ski)
перешéек isthmus
перигéй perigee
периге́лий perihelion (astron)
пери́ла handrail
~ мо́ста bridge railing
пери́од безотка́зной рабо́ты reliable life, trouble-free life
~ вы́броски duration of drop (airborne ops)
~ испыта́ний test period
~, ла́герный camp season
~, меже́нный average water level (river etc)
~, межремо́нтный interval between repairs, mileage intervals
пери́од нака́пливания build-up phase
~ норма́льной эксплуата́ции normal operational period
~ обеспе́чения ата́ки period of planned support (tac)
~ облуче́ния duration of irradiation
~ обно́ски breaking-in stage, breaking-in period
~ обраще́ния orbital period (ESV)
~ освое́ния assimilation period. breaking-in period
~ повторе́ния и́мпульсов pulse repetition period
~, подготови́тельный preparatory stage
~ полураспа́да half life (nucl)
~ последе́йствия га́зов period of secondary action of gases (ball)
~ прирабо́тки running-in period
~ прогре́ва warm-up period
~ радиоакти́вного распа́да radioactive-decay period
~ развёртки scan period
~ разогре́ва warm-up time
~ сближе́ния advance-to-contact phase
~ сгора́ния combustion period
~ сканди́рования scan period
~ скры́того де́йствия latent period (med)
~ техни́ческого обслу́живания maintenance period
~, угрожа́емый period of threat (preceding outbreak of war)
периоди́ческий routine, periodic, recurring
периоди́чность frequency, rate of recurrence
~ отбо́ра проб frequency of sampling
~ переда́чи донесе́ний message transmission rate
~ техни́ческого обслу́живания maintenance rate
перисko̒п periscope
~ ба́шенного стрелка́ turret gunner's periscope (tk)
~, бинокуля́рный periscopic binoculars
~, зени́тный skysearch periscope
~, инфракра́сный infra-red periscope, IR periscope
~, ма́чтовый mast periscope
~, око́пный trench periscope
~, транше́йный trench periscope
~, убира́ющийся retractable periscope
перископи́ческий periscopic

перискóп-развéдчик scout's periscope
пёрка flat drill
перманганáт permanganate
~ **кáлия** potassium permanganate
~ **кáльция** calcium permanganate
~ **нáтрия** sodium permanganate
перó 1. pen
2. feather
3. fin
~ **мúны** mortar bomb vane
~ **стабилизáтора** mortar bomb vane
перрóн platform
персонáл personnel
~, **обслýживающий** 1. maintenance personnel
2. ground servicing staff (AF)
~, **техни́ческий** technical personnel
~, **эксплуатациóнный** maintenance staff
перспекти́вный long-term, perspective, advanced
пертурбáция perturbation
перфокáрта perforated card
перфолéнта perforated tape
перфорáтор perforator
~, **вычисли́тельный** calculating punch (punched cards)
перхлорáт аммóния ammonium perchlorate
~ **нáтрия** sodium perchlorate
перчáтки gloves
~, **двупáлые** mittens
~, **защи́тные** protective gloves
~, **огнезащи́тные** anti-flash gloves
пески́, бархáнные inland sand dunes
~, **бугри́стые** sand knolls
~, **зыбýчие** shifting sands, quicksands
~, **незакреплённые** soft sand, "sand sea"
пескодýвка sand blasting
песóк sand
~, **зыбýчий** quicksand
~, **сыпýчий** quicksand
песчáный sandy
петáрда black-powder pellet (fuse)
~, **порохова́я** powder magazine (fuse)
петеэ́ровец anti-tank rifleman
петли́ца 1. gorget patch
2. frog (sword)
пéтля 1. loop, noose
2. snare
~, **букси́рная** towing loop
~, **верёвочная** rope loop
~, **гистерези́сная** hysteresis loop (elec)
~, **грузоподъёмная** hoisting sling
пéтля, двойнáя clove hitch
~, **замóчная** locking collar
~, **носкóвая** foot binding (snowshoe)
~ **свя́зи** coupling loop (electron)
пехóта infantry
~, **воздýшно-десáнтная** airborne infantry
~, **крылáтая** airborne infantry
~, **морскáя** marine infantry
~, **моторизóванная** motorized infantry
~, **окопáвшаяся** entrenched infantry
~, **посáженная на тáнки** tank-borne infantry
пехоти́нец infantryman
пехóтный infantry
печáть seal, stamp
~, **гéрбовая** official seal, official stamp
~ **чáсти** unit stamp
пéчка, палáточная tent stove
печь 1. stove, incinerator, furnace
2. heater
~, **борóвная** stove with horizontal flue
~, **имитациóнная** gas simulator apparatus
~, **мусоросжигáющая** refuse incinerator
~, **похóдная разбóрная** collapsible tent stove
~, **электри́ческая** electric furnace
печь-мангáлка brazier
пéшее по тáнковому dismounted tank training
пéший foot, dismounted
пéшим поря́дком on foot, dismounted
пéшня heavy crowbar
пещéра cave, cavern
~, **снéжная** snow shelter
пикáп pick-up (veh)
пикéт 1. bench mark (survey)
2. "piquetage" point (arty)
пикетáж мéстности piquetage of the zone (arty)
пики́рование diving (aeron)
пики́ровать to dive (ac, missiles)
пикофарáда picofarad (elec)
пикрáт аммóния ammonium picrate (expl)
пикратóл picratol (expl)
пилá saw
~, **ди́сковая** circular saw
~, **крýглая** circular saw
~, **лéнточная** band saw
~, **лучкóвая** bow saw

пила́, мото́рная power saw
~, попере́чная cross-cut saw
~, продо́льная rip saw, pit saw
~, ручна́я portable saw
~, цепна́я chain saw
пиломатериа́лы sawn timber
пило́н tower (br), pylon
пилообра́зный saw-tooth shaped
пило́тка field service cap, forage cap
пи́льщик sawyer
пинце́т tweezers
пипе́тка для взя́тия про́бы электроли́та battery electrolyte measurer
пирами́да 1. pyramid
2. rifle rack
~, винто́вочная rifle stack
~, руже́йная rifle rack
пирами́ды, железобето́нные dragon's teeth
пирозамо́к separation charge
пирозапа́л powder igniter
пирокла́пан propellant valve
пироксили́н gun cotton
пироли́з pyrolysis
пиро́метр pyrometer
~, цветопоглоща́ющий colour extinction pyrometer
пирона́фт illuminants
пиропатро́н pyrotechnic cartridge
~, пусково́й pyrotechnic igniter
пиропро́вод ignition wire
пиросвеча́ igniter cartridge
пиростá́ртер cartridge starter
пироте́хник 1. pyrotechnics expert
2. member of bomb disposal squad
пироте́хника pyrotechnics
пиротехни́ческий pyrotechnics
пирс pier
~ на понто́нах pontoon pier
~, сбо́рно-разбо́рный понто́нный sectional floating causeway
пи́сарь clerk
~, дежу́рный duty clerk
~ подразделе́ния orderly room clerk
~ ро́ты company clerk
пи́сарь-делопроизводи́тель administrative clerk
пистоле́т 1. pistol
2. nozzle, gun
~, автомати́ческий automatic pistol
~, га́зовый gas-operated metal spray gun, gas-operated spray gun
пистоле́т, раке́тный Very pistol, signal pistol, flare gun
~, сигна́льный signal pistol
~, целево́й target pistol
пистоле́т-краскораспыли́тель paint spray gun
пистоле́т-пулемёт sub-machine gun, machine carbine, machine pistol
пистоле́т-раке́тница signal pistol
пистоле́т-распыли́тель paint spray gun
писто́н 1. percussion cap
2. hollow rivet
3. eyelet
пита́ние 1. supply
2. feeding, messing, catering
3. fare, sustenance, nourishment
4. power supply
~, авари́йное emergency supply
~, автомати́ческое automatic feed
~, автоно́мное independent power supply
~, боево́е 1. supply in the field
2. ammunition supply (arty)
~ боеприпа́сами ammunition supply
~ войск при бива́чном расположе́нии bivouac feeding
~, ле́нточное belt feed (ammo)
~, магази́нное magazine feed (ammo)
~ от вне́шнего исто́чника эне́ргии external power supply
~ от се́ти mains supply
~, паралле́льное shunt feeding (elec)
~, после́довательное series feeding (elec)
пита́ния прибо́ра управле́ния зени́тным огнём артиллери́йским зени́тным огнём artillery director power plant
пита́тель feeder, feeding device
пита́ть to supply, to feed, to nourish
пла́вать/плыть to float, to swim, to sail (ship)
пла́вающий amphibious, floating
пла́вка melting, fusion, fusing
пла́вни water meadows, reed swamps
пла́вники fins, flippers (swimming)
пла́вный 1. continuous
2. smooth, even
плавпbriча́л floating jetty, pontoon jetty
плавсре́дства water crossing equipment, flotation equipment
плаву́честь buoyancy
плаву́чий 1. buoyant
2. floating

пла́зма plasma
~ **кро́ви** blood plasma
плакиро́ванный clad, coated, plated
пламегаси́тель flash reducer, flash eliminator
2. flash inhibitor, flash reducing charge (ammo)
пламеотража́тель blast shield, flame shield, flame deflector
пла́мя flame, flash
~, **ду́льное** muzzle flash
~, **засоплово́е** back-blast
~, **обра́тное** back-blast, back-flash
план plan, programme, schedule
~ **боевы́х де́йствий** tactical plan
~ **бо́я** tactical plan
~ **возду́шных перево́зок войск** air movement table
~ **восстановле́ния иму́щества** reclamation programme
~ **выгрузки эшело́нов** 1. detraining table
2. debussing table
3. deplaning table
~ **вы́садки деса́нта** landing plan
~, **генера́льный** general plan, overall plan
~, **грузово́й** loading chart, stowage plan
~ **движе́ния тра́нспорта** traffic circulation plan, vehicle movement schedule
~ **загражде́ний** obstacle plan
~, **за́дний** background
~ **инжене́рного обеспе́чения** engineer operational plan
~ **маскиро́вки** camouflage plan
~ **мини́рования** mine plan
~ **ми́нного по́ля** minefield plan
~, **мобилизацио́нный** scheme of mobilization, mob scheme
~ **накопле́ния запа́сов** supply build-up programme
~ **наступле́ния** plan of attack
~ **оборо́ны** defence plan
~ **огнево́й подде́ржки с во́здуха** air support plan
~ **огня́** fire plan
~, **операти́вный** operational plan
~ **организа́ции движе́ния** traffic flow chart
~, **основно́й** master plan
~ **ПВО** air defence plan
~ **перево́зок** movements schedule
~, **пере́дний** foreground
план по ты́лу administrative plan
~ **погру́зки** 1. embarkation and tonnage schedule
2. entraining schedule
3. embussing schedule
4. emplaning schedule
~ **подрывны́х рабо́т** demolition plan
~ **примене́ния хими́ческого ору́жия** chemical operation plan
~ **примене́ния я́дерного ору́жия** nuclear fire plan
~, **произво́дственный** production plan
~ **противота́нковой оборо́ны** anti-tank defensive schedule
~ **рабо́т** work schedule, work plan
~ **рабо́ты, су́точный** daily work target
~ **рабо́ты ты́ла** administrative plan
~ **разве́дки** intelligence plan
~ **разруше́ний** demolition plan
~ **рассредото́чения** dispersal plan
~ **расхожде́ния** dispersal plan (group target, AA)
~ **ремо́нта** repair forecast
~ **свя́зи** signals plan
~ **снабже́ния** supply plan
~, **темати́ческий** syllabus
~, **углово́й** range deflection fan (gunn) (for transference of data between OPs or GPs)
планёр 1. glider
2. airframe
планери́зм gliding
планери́ст glider pilot
планета́рный epicyclic, planetary
пла́н-зая́вка estimate
планиме́тр planimeter
плани́рование 1. planning
2. gliding
~ **загражде́ний** obstacle planning
~, **предвари́тельное** preliminary planning, advance planning
~ **рабо́ты ты́ла** administrative planning
плани́ровать/сплани́ровать to glide (aeron), to plan; **плани́ровать/сплани́ровать потре́бности** to frame requirements
планиро́вка layout, planning
пла́нка 1. plank, board, bar, plate, strip
2. clamp, strap
~, **защи́тная** protective strap
~, **конта́ктная** contact disc (min)
~, **монта́жная** mounting strip
~, **нажи́мная** pressure board (min)

пла́нка, направля́ющая carriage guide (ord)
~ **ни́жней анта́бки** butt-swivel plate (SA)
~, **о́рденская** medal ribbon bar
~, **откидна́я прице́льная** folding sight leaf
~, **попере́чная** diagonal brace
~ **прице́ла** sight leaf (SA)
~, **прице́льная** backsight leaf
~ **с о́рденскими ле́нтами** medal ribbon bar
~, **соедини́тельная** bolt coupling unit
~, **сре́дняя** centre batten (min)
~, **сто́порная** stop plate
~, **упо́рная** bearing ring (ord)
пла́н-конспе́кт проведе́ния заня́тия lesson plan
пла́новый 1. scheduled, systematic, planned
2. plan
планоме́рность ordered process
планоме́рный planned, according to plan, deliberate
планша́йба face chuck, face plate
планше́т 1. artillery board
2. plane table
3. map-board
~, **автомати́ческий** automatic plotting board
~, **автомати́ческий радиолокацио́нный** automatic radar plot
~, **звукометри́ческий** sound ranging plotting board
~, **огнево́й** artillery board, fire chart
~, **радиолокацио́нный** radar plotting board
~, **радиофици́рованный** combined plotting board and radio set (arty obsn)
~, **электро́нный** electronic data display, electronic plotting board
планшети́ст radar plotter, target plotter
планше́т-построи́тель field plotter (arty)
~ **попра́вок да́льности** range adjustment board, range correction board
планши́р gunwhale, gunnel
планши́рь gunwhale, gunnel
пласт stratum, layer, bed, seam
пласти́на 1. plate
2. flitch, split log
~, **биметалли́ческая** bi-metallic plate
~, **вертика́льно-отклоня́ющая** Y-plate (CRT)
пласти́на, горизонта́льно-отклоня́ющая X-plate (CRT)
~ **колле́ктора** commutator segment
~, **отклоня́ющая** deflector plate (electron)
~ **радиа́тора** radiator fin
~ **распредели́тельного ва́ла, упо́рная** camshaft thrust plate
пласти́нка 1. plate
2. record (disc)
~, **кварцева́я** quartz plate
пласти́нчатый lamellar, laminated, sheet
пласти́ны, направля́ющие deflector plates (electron)
пласти́т plastic explosive
пластифика́тор softener, plasticizing agent
пласти́ческий plastic
пласти́чность 1. ductility
2. plasticity
пластма́сса plastic material
~, **слои́стая** laminated plastic
пла́та plate, board, card, sheet, slice (elec)
~ **за хране́ние** storage charge
~, **кре́мниевая** silicon slice
~, **монта́жная** wiring board
~, **объединя́ющая** integrating board
~, **печа́тная** printed-circuit card
~ **с логи́ческой схе́мой** logic card
~ **с несоединёнными микрокомпоне́нтами, станда́ртная** master slice
~ **с печа́тным монтажо́м** printed-circuit card
~, **транзи́сторная** transistor slice
платинотро́н platinotron (elec)
плато́ plateau, table-land
платфо́рма 1. platform
2. flat wagon, flat (rly)
3. pallet
~, **бокова́я** side-loading platform (rly)
~, **выдвижна́я** roll-out platform
~, **высо́кая** raised platform
~, **гиростабилизи́рованная** gyro-stabilised platform
~, **грузова́я** cargo bed
~, **двухо́сная** four-wheeled flat wagon (rly)
~, **железнодоро́жная** 1. flat, flat car, flat wagon (rly)
2. railway platform
~, **комбини́рованная** combined end and side loading platform (rly)
~ **ору́дия, поворо́тная** turn-table (wpn)

платфо́рма, островна́я island platform (rly)
~, парашю́тная heavy-drop platform (paradrop)
~, сбо́рно-разбо́рная temporary platform
~, торцо́вая end-loading platform
~, четырёхо́сная flat bogie wagon
пла́тье, шта́тское civilian clothes
плафо́н roof lamp
плац barrack square, parade ground
~, строево́й drill square, parade ground
~, уче́бный 1. training area
2. drill square, parade ground
плацда́рм bridgehead, springboard, lodgement area (tac), airhead
~ возду́шного деса́нта airhead
~ вы́садки морско́го деса́нта beach-head
~ десанти́рования airhead
~, деса́нтный 1. beachhead
2. airhead
~, инжене́рный springboard for attack provided with engineer works
~, исхо́дный springboard for attack
~, предмо́стный bridgehead
~, прибре́жный bridgehead, beach-head
пла́шка die, screw-thread die
плашко́ут pontoon, barge
~, вы́садочно-вы́грузочный landing scow
~, вы́садочный landing barge
~, деса́нтно-вы́садочный landing craft mechanized, LCM
~, пехо́тно-вы́садочный landing craft assault, LCA
~, та́нко-деса́нтный landing craft utility, LCU
плащ, брезе́нтовый tarpaulin cloak, heavy-duty rain-cape (usually with hood and sometimes provided for sentries)
пла́щ-наки́дка rain-cape
пла́щ-пала́тка rain-cape, ground sheet (mil)
пла́щ-пальто́ cold weather greatcoat/cape
плексигла́с plexiglass
плен captivity
плене́ние capture
плени́ть (см пленя́ть)
плёнка 1. film
2. tape (sound recording)
~, кату́шечная roll film
плёнка, незапотева́ющая anti-dim disc, anti-dim eyepiece insert (respirator)
~, окси́дная oxide film, blue finish (metall)
~, панхромати́ческая panchromatic film
~, противозапотева́ющая anti-dim disc, anti-dim eyepiece insert (respirator)
~, фосфа́то-ла́ковая parkerized and lacquered coating
~, фотографи́ческая photographic film
пле́нный prisoner
~, контро́льный prisoner taken for identification
пленя́ть/плени́ть to take prisoner
плес reach, stretch of water
плете́нь brushwood hurdle, wattle hurdle
плетнёвый brushwood, wattle
плечо́ вперёд forming (drill)
плита́ 1. plate, slab
2. cooking stove
~ бортово́й брони́ side plate
~, бронева́я armour plate
~, ве́рхняя пере́дняя накло́нная бронева́я glacis plate
~, железобето́нная 1. reinforced concrete slab (br)
2. reinforced concrete panel
3. pre-fabricated concrete panel
~, кухо́нная cooking range
~ миномёта, опо́рная mortar base plate
~, нажи́мная pressure plate (min)
~, опо́рная base plate
~, фунда́ментная base ring (AA ord)
пли́тка tile, table, plate
~, боева́я recoil shield (arty)
плове́ц, боево́й frogman, underwater swimmer
плову́честь buoyancy
пло́мба 1. seal (on door, packing case etc)
2. tooth stopping, filling
пломбиро́вка sealing, filling
плоского́рье plateau, table-land
плоскогу́бцы flat-nosed pliers
~, комбини́рованные combination pliers
плоскодо́нка flat-bottomed boat
пло́скости, крестообра́зные несу́щие cruciform lifting surfaces
плоскостно́й two-dimensional
пло́скость plane (geom), surface
~ ба́зы base plane
~ броса́ния plane of departure (gunn)
~, вертика́льная vertical plane

плóскость, горизонтáльная horizontal plane
~, картúнная image plane
~ крéна roll plane (GW)
~, несýщая lifting surface (aeron)
~, нулевáя datum plane
~, опóрная 1. datum plane 2. bearing surface
~ оптúческого изображéния focal plane
~, поперéчная roll plane (GW)
~ прицéливания sighting plane
~ рыскáния yaw plane
~ стрельбы́ plane of departure, plane of fire (gunn)
~ управлéния control surface (GW)
~, фокáльная focal plane
~ цéли vertical plane of sight, plane of position (gunn)
плот raft
~, бревéнчатый log raft, timber float
~, надувнóй спасáтельный inflatable life raft
плóтик small raft
~, надувнóй inflatable raft
плотúна dam, dyke
~, водослúвная overflow weir
~, водохранúлищная storage dam
~, дéрево-землянáя earth-and-timber dam
~, землянáя earth dam
~, ряжевáя crib dam (br)
~, фашúнно-хворостянáя fascine-brushwood dam
плóтник carpenter
плотномéр density meter
плóтность 1. density, specific gravity 2. tightness, compactness
~, артиллерúйская gun density, artillery density (number of guns, including mortars and rocket launchers per kilometre of front)
~, бóмбовая bomb density
~ вóздуха, баллистúческая ballistic density, ballistic air density
~ вóздуха, нормáльная standard air density
~ движéния traffic density
~ заграждéний obstacle density (number of obstacles, including mines, per sq km)
~ заря́да, объёмная charge density (elec)
~ заряжáния propellant density
плóтность минúрования minefield density (number of mines either per hectare or per linear kilometre of minefield)
~ мúнных заграждéний mine density (number of mines per kilometre of front)
~ оборóны degree of concentration of forces and resources in defence
~ огня́ weight of fire, density of fire
~ огня́ артиллéрии weight of artillery fire (number of shells per unit area per minute)
~ по мóщности, спектрáльная power density function, power spectrum
~ подавлéния weight of fire necessary to neutralize a target
~ поражéния shell density (number of shells, including mortar bombs, per unit area per minute)
~, противотáнковая anti-tank weapon density (number of anti-tank weapons per linear kilometre of front in defence)
~, тактúческая tactical density
~ тáнков tank density (number of tanks and self-propelled guns per kilometre of front)
~, тáнковая tank density, number of tanks per kilometre of frontage (Sov)
~ частúц density of particles
~ электролúта specific gravity of electrolyte
~, я́дерная nuclear density
площáдка platform, stage, stand, area, ground, space, helicopter pad
~, взлётно-посáдочная landing strip, runway
~, выноснáя пулемётная forwardly projecting machine-gun emplacement
~, дегазациóнная 1. CW decontamination centre 2. CW decontamination area
~, дегазациóнно-дезактивациóнная decontamination area
~ для вы́дачи issue area
~ для вы́садки десáнта с вертолётов helicopter landing site
~ для контрóльного ýровня clinometer plane
~ для противотáнковых рýжей anti-tank weapon platform
~ для пулемёта machine-gun pit
~ для ремóнтных рабóт repair area
~ для сбóрки assembly area

площа́дка, запра́вочная fuelling area, refuelling area
~, контро́льная clinometer plane (gunn)
~, мо́ечная wash area (veh)
~, монта́жная construction platform
~ обслу́живания service platform
~, огнева́я firing area (GW)
~ опробова́ния proving site
~, оруди́йная gun platform
~, погру́зочная loading area
~, подъёмная cargo tray
~ подъёмника lifting platform
~, поса́дочная 1. airfield, landing ground 2. landing site (airborne ops)
~ приземле́ния landing zone
~ приземле́ния вертолётов landing point (helicopters)
~, пулемётная machine-gun emplacement
~, пускова́я 1. launching pad (GW) 2. firing area (GW)
~, разгру́зочная unloading platform, discharging platform
~, сбо́рочная assembly site
~, ста́ртовая 1. launching pad (GW) 2. firing area, launching site (GW)
~, строева́я drill square, parade ground
~, строи́тельная construction site, building site
~, тормозна́я brake platform (rly)
~ управле́ния control area (GW)
площа́дка-эстака́да trestle platform
пло́щадь 1. area 2. square (eg "Red Square")
~ вхо́да intake area
~, де́йствующая effective area (antenna)
~, опо́рная area of bearing surface
~ опо́ры area of bearing service
~ отраже́ния, эквивале́нтная equivalent echoing area (radar)
~ паде́ния снаря́дов beaten zone (ball)
~ поглоще́ния, эффекти́вная effective absorption area (electron)
~, поража́емая 1. beaten zone (ball) 2. casuality producing area (nucl)
~, поража́емая уда́рной волно́й blast damage area (nucl)
~ пораже́ния area of destruction
~ пораже́ния оско́лками fragment pattern (gunn)
~ по́ршня piston area
~ приёмной анте́нны, эффекти́вная effective absorbing area (radar)
пло́щадь прилега́ния contact area
~, просмо́тренная scanned area, swept area (radar)
~ разлёта оско́лков area of burst
~ рассе́ивания zone of dispersion (ball)
~ сече́ния sectional area
~ скла́да, поле́зная effective storage area
~ та́нка, лобова́я frontal area of tank
~ це́ли, уязви́мая vulnerable area of target
~ части́чной дегаза́ции оде́жды site partial decontamination (clothes)
~ части́чной дегаза́ции тра́нспорта site partial decontamination (transport)
~, эффекти́вная отража́ющая equivalent echoing area (radar)
плуг plough
~ лафе́та trail spade (ord)
~, ми́нный mine plough
плужо́к plough (of plough mine layer)
плу́нжер plunger
плуто́ний plutonium
плуто́новый plutonium
плыть (см пла́вать)
плюс 1. plus 2. plus, over (arty)
"Плюс!" "Over!" (arty)
плюс батаре́и positive battery terminal
пляж beach
пнева́ние tree stump barrier
пневма́тик tyre
пневмати́ческий pneumatic
пневмосисте́ма pneumatic system
пневмощито́к compressed air panel
по а́зимуту in azimuth
по бездоро́жью 1. across country 2. along bad or impassable roads (due to weather or surface conditions)
по вы́зову on call, at call
по инста́нции through normal channels
по кварти́рам in billets, in quarters
"По Маши́нам!" "Mount!"
"По Места́м!" "Mount!", "Take Post!"
по поря́дку номеро́в numerically, in numerical order
"По Поря́дку-Рассчита́йсь!" "Squad, Number!" (order)
по разделе́ниям by numbers (drill)
по спосо́бности on one's own initiative, independently
по тре́бованию on call, on demand
по фро́нту laterally

по це́ли on target
по часово́й стре́лке clockwise
по частя́м in detail, piecemeal
по-батаре́йно by battery (arty)
побе́г desertion, escape, flight
побегу́шка распредели́теля rotor arm (veh)
побе́да victory
победи́тель victor, winner
победи́ть (см побежда́ть)
побе́дный victorious
победоно́сный victorious, triumphant
побежда́ть/победи́ть to defeat, to conquer, to vanquish
побере́жье shore, coastland, beaches
~, необору́дованное open beach
поби́ть (см бить)
побо́чный secondary, side, collateral
побуди́тель stimulus
побу́дка reveille
побужде́ние 1. stimulation, prompting 2. motive, inducement, incentive
побы́вка short home leave
поваго́нно by wagon load (rly)
по́вар cook
по́вар-инстру́ктор instructor-cook
поведе́ние manner, bearing, conduct
~, во́инское military conduct, soldierliness
пове́рка 1. audit, auditing, checking 2. verification, roll-call, muster, check
~, вече́рняя evening roll call
поверну́ть (см повора́чивать)
пове́рхности, рабо́чие working surfaces
пове́рхность surface
~, аэродинами́ческая airfoil, aerodynamic surface
~, глисси́рующая hydrofoil, planing surface
~, грани́чная boundary surface, interface
~ ката́ния rolling contact surface
~ кла́пана, опо́рная valve face
~, накло́нная sloping surface
~, необрабо́танная unfinished surface
~, несу́щая lifting surface (aeron)
~, обрабо́танная machined surface
~, опо́рная bearing area, bearing surface
~, отража́ющая echoing area (radar)
~, рабо́чая operating surface, working surface
~, рулева́я control surface (missile)
~ скольже́ния rubbing surface, slipping surface
пове́рхность, скользя́щая running surface (ski)
~, ско́шенная camway (brake shoe)
~, сфери́ческая surface of sphere, spherical surface
~ торможе́ния brake surface
~, торцо́вая plane surface
~ тре́ния rubbing surface, friction surface
~, тру́щаяся friction surface, bearing surface
~ уплотне́ния wearing surface
~ управле́ния control surface
~, у́ровенная datum level, reference level, datum plane, reference plane
~, эффекти́вная отража́ющая equivalent echoing area (radar)
пове́стка 1. registration notice 2. evening roll-call (bugle call)
~ дня agenda
повзво́дно 1. by platoons 2. by troops (arty)
пови́нность duty, obligation, service
~, во́инская compulsory military service, conscription
~, всео́бщая во́инская compulsory military service
повинове́ние obedience, subordination
по́вод occasion, cause, ground,
пово́дка warping, distortion, buckling
поводо́к catch plate, guiding piece, leading-in piece (eng)
~ возвра́тного механи́зма return spring rod (SA)
пово́зка 1. wheeled vehicle 2. carriage (arty)
~, лафе́тная platform wagon (ord)
~, санита́рная medical vehicle
пово́зочный animal transport driver
поволо́к selector block (tk)
повора́чивать/поверну́ть to turn, to swing, to change; **повора́чивать/поверну́ть ба́шню** to traverse the turret; **повора́чивать/поверну́ть дви́гатель** to crank the engine
поворо́т 1. turn, traverse 2. bend, turning, curve (road etc)
~ ба́шни turret rotation, turret traverse
~ в сто́рону "укло́на" downhill turn (ski)
~ "ве́ером" stationary step-around turn (ski)
~ ве́тра wind shift

поворо́т, зигзагообра́зный "Z" bend
~, кругово́й all-round traverse, 360 degree traverse
~ круго́м 180 degree turn (aeron)
~ "на паралле́льных лы́жах" из "полуплу́га" stem-christiana turn (ski)
~ ору́дия gun traverse
~ "пересту́пом" stationary step turn (ski)
~, пла́вный steady turn, sweeping turn
~ "плу́гом" snowplough turn (ski)
~ "полуплу́гом" half-snowplough turn (ski)
~ "прыжко́м" stationary jump turn (ski)
~ "че́рез но́гу круго́м" stationary kick turn (ski)
поворо́тливость 1. manoeuvrability, agility, ease of handling 2. turning radius
поворо́тный rotary, turning, traversing, revolving
поворо́ты в движе́нии turnings on the march (drill)
~ на ме́сте turnings on the march (drill)
повреди́ть (см поврежда́ть)
поврежда́ть/повреди́ть to damage, to injure
поврежде́ние damage
~, легко́ устрани́мое superficial damage
~, неисправи́мое irreparable damage
~, опа́сное critical damage
~, радиацио́нное radiation damage
~, уста́лостное fatigue failure (metall)
повреждённый damaged, broken-down, impaired, defective, faulty
повста́нец rebel, insurgent
повтори́тель, като́дный cathode follower
повтори́ть (см повторя́ть)
повторя́ть/повтори́ть to repeat, to duplicate; **повторя́ть/повтори́ть текст** to read back (comms)
повы́сить (см повыша́ть)
повыша́ть/повы́сить to develop, to heighten, to raise, to promote, to improve, to increase
повыше́ние improvement, rise, advancement, increase, promotion
повы́шенной проходи́мости cross-country, of improved cross-country performance (referring to vehs)
повя́зка 1. bandage 2. armband
~, ги́псовая plaster cast
~, ма́рлевая gauze bandage
повя́зка, нарука́вная armband
~, перви́чная field dressing
~, тра́урная mourning armband
"Погаси́ть Огни́!" "Lights Out!"
"Погаси́ть Свет!" "Lights Out!"
погаси́ть (см гаси́ть)
погаше́ние по́езда breaking up of a train (rly)
по́гиб ствола́ limited barrel wear
погиба́ть/поги́бнуть to perish, to be killed
поги́бнуть в бою́ to fall in battle, to be killed in battle
~ (погиба́ть)
поги́бший в бою́ killed in action
~ на по́ле бо́я killed in action
поглоти́тель 1. killer (elec) 2. absorbent
~, хими́ческий chemical absorber, chemical absorbent
поглоти́ть (см поглоща́ть)
поглоща́ть/поглоти́ть to attenuate (elec), to absorb, to swallow, to consume, to devour
поглоще́ние absorption
~ га́мма-излуче́ния gamma-ray absorption
~, избира́тельное resonance absorption
~, молекуля́рное molecular absorption
пого́да, тума́нная foggy weather
погодосто́йкий weather-proof
погодосто́йкость weather resistance
поголо́вно to a man, without exception
пого́н 1. shoulder-board, shoulder strap 2. race (mech)
~ ба́шни turret race
~ ба́шни, ве́рхний upper rotating ring, moving race, upper turret race, moving turret race
~ ба́шни, ни́жний lower turret race, stationary turret race
~, зубча́тый cog race
~, нашивно́й sewn-on shoulder board
~, съёмный detachable shoulder-board
~, шарово́й ball race
пого́нный linear
пого́нщик 1. animal transport driver 2. transport worker (sl)
пого́ня pursuit
пограни́чник frontier guard
пограни́чный border, frontier
погранотря́д border detachment

погранохрáна frontier guard, border guard
пóгреб cellar, vault, magazine
~, артиллерийский magazine, ammunition stores (nav)
~, патрóнный magazine, ammunition store
~, пороховóй powder magazine (nav)
погребéние burial
погребóк для боеприпáсов ammunition shelter
погрéшность error
~, допусти́мая permissible error
~ прицéла, суммáрная absolute error (arty)
~, суммáрная absolute error
~, четвертнáя quadrantal error (radio)
погрохрáна border guard, frontier guard
погружéние свай pile driving
погрузи́ться (см грузи́ться)
погрýзка 1. loading (stores) 2. embussing, emplaning, entraining (pers)
~ бронетранспортёров на ходý loading-on of APCs afloat (amph ops)
~ вручнýю hand loading (vehs etc)
~ десáнта embarkation of assault personnel
~, крáновая crane loading
~, механизи́рованная mechanised loading
~, прóбная 1. stowage test 2. embarkation dry run
~ с земли́ ground-level loading
~ сбóку side loading
~ сзáди rear loading (veh)
~, такти́ческая tactical stowage
~ тéхники loading of equipment and war stores
~ эвакуи́руемого имýщества backloading (eqpt)
~ эшелóна entrainment (rly)
погрýзочный loading
погрýзчик lift truck
~, ви́лочный fork-lift truck
~, ковшóвый bucket elevator loader
~, конвéйерный conveyer loader
~, лéнточный belt loader
под арéстом under arrest
"Под Знáмя Слýшай на Караýл!" "To the Colours Present Arms!"
под распи́ску against receipt
под ружьём under arms
"Пóд руку!" "Secure Arms!"
подавáльщик mess waiter
подавáтель 1. magazine platform (SA) 2. feed lever, feed arm
подавáть/подáть to give, to offer, to serve (food), to supply (stores etc), to feed (wpn); **подавáть/подáть заявку** to submit an indent, to put in an indent, **подавáть/подáть заявку на боеприпáсы** to indent for ammunition
подави́тель 1. damper, silencer (mech) 2. killer (elec)
подави́ть (см подавля́ть)
подавлéние 1. suppression (elec etc) 2. neutralizing (mil) 3. jamming (radar etc)
~ несýщей частоты́ carrier wave suppression
~ огнём neutralization by fire
~, электрóнное jamming
подáвленный neutralized, suppressed, overwhelmed
подавля́ть/подави́ть to neutralize, to suppress, to overwhelm
подавля́ющий overwhelming
подáть (см подавáть)
подáча 1. supply, bringing up 2. feed (weapon, lathe etc)
~ вóздуха air supply
~ горю́чего fuel feed
~ горю́чего, комбини́рованная gravity and exhaust pressure fuel feed
~ горю́чего насóсом fuel pump fuel feed
~ горю́чего под давлéнием pressure fuel feed
~ горю́чего под давлéнием отрабóтавших гáзов exhaust pressure fuel feed
~ горю́чего под давлéнием сжáтого вóздуха air pressure fuel feed
~ горю́чего разрежéнием vacuum fuel feed
~ горю́чего самотёком gravity fuel feed
~, лéнточная belt feed
~ патрóнов cartridge feed
~ патрóнов, беззвеньевáя linkless ammunition feed
~ патрóнов, звеньевáя linked ammunition feed
~ патрóнов лéнтой belt cartridge feed

подáча самотёком gravity feed, flow by gravity
~ тóплива, баллóнная gas-pressure propellant-feed system (GW)
~ тóплива, вытеснúтельная pressurised propellant feed system (GW)
~ тóплива, насóсная turbopump propellant feed system (GW)
~, фитúльная capillary feed
подаю́щий feed
подбáбок bearing block
подбáжка stringer (br)
подбáлка bolster, corbel (br)
подбивáть/подбúть to knock out of action; **подбивáть/подбúть танк** to knock out a tank
подбúвка packing, tamping, lining, padding
подбирáть/подобрáть to select, to pick out, to pick up; **подбирáть/подобрáть живóт** to pull in the stomach (drill)
подбúтый knocked out (of action)
подбúть (см подбивáть)
подбóр selection, assortment
подбoрóдник chin strap
подбру́стверный trench
подвáл basement, cellar
подвáльный basement, cellar
подвéдомственность jurisdiction, competence
подвéдомственный dependent, under the jurisdiction of, in the charge of
подвергáть/подвéргнуть to expose to, to subject to, to inflict
подвергáться/подвéргнуться to undergo; **подвергáться/подвéргнуться обстрéлу** to be exposed to fire, to undergo fire
подвéргнуть (см подвергáть)
подвéргнуться (см подвергáться)
подвéс suspension, hanger, suspender
~, кардáнный gimbal joint
подвéска 1. suspension, suspension system
2. sling (br)
~, балансúрная bogie suspension
~, гúбкая flexible suspension
~, гидропневматúческая hydropneumatic suspension, oleo-pneumatic suspension
~, жёсткая rigid suspension
~, завúсимая conventional suspension, rigid-axle type suspension
подвéска зáдних колёс, независимая independent rear suspension
~, индивидуáльная independent suspension
~, лю́лечная swing link
~ мóста suspender (br)
~ на шаровы́х шарнúрах ball joint suspension
~, независимая independent suspension (veh)
~, плáвающая floating suspension
~, пневматúческая air suspension
~, пневмогидравлúческая air-oil suspension
~, пружúнная coil spring suspension
~, рессóрная spring suspension
~, рычáжная wishbone suspension
~ с качáющимися рычагáми suspension by swinging arms
~, слóжная балансúрная compound twin-bogie suspension (tk)
~ со стержневы́ми рессóрами torsion bar suspension
~, стержневáя torsion-bar suspension
~ телéжки bogie suspension
~, торсиóнная torsion-bar suspension
подвестú (см подводúть)
подвéшивание suspension
пóдвиг act of gallantry, feat
подвúжность mobility
~, боевáя tactical mobility
~ вне дорóг off-road mobility
~ на пересечённой мéстности cross-country mobility
~, тактúческая tactical mobility, battlefield mobility
подвóд вóздуха air supply
~ мóщности power input
~ тóплива fuel feed, fuel supply
подводúть/подвестú to bring, to lead up to, to supply; **подводúть/подвестú балáнс** to balance an account
подвóдка admission, supply, feed (tech)
подвóз supply, bringing up of supplies, transport, delivery
~ материáльных средств в боевы́х услóвиях tactical resupply
~, плáновый routine transportation
подворотничóк detachable collar lining, under-collar
подголóвник, регулúруемый adjustable head rest

подго́нка 1. matching (elec)
2. fitting, adjustment
~, небольша́я minor adjustment
подгоня́ть/подогна́ть to urge on, to drive, to hurry, to adjust, to fit, to match
подгора́ние burning
подготови́тельный preparatory
подгото́вка 1. preparation
2. training
~, авиацио́нная air preparation
~, артиллери́йская artillery preparation, preparatory bombardment, artillery softening up
~ ата́ки preparation for attack
~, баллисти́ческая calculation of ballistic range correction
~, боева́я training, military training, combat training
~ бронета́нковых войск, боева́я tank training
~, вневойскова́я pre-conscription military training
~ води́телей та́нков tank driver training
~, вое́нная military training
~, возду́шно-деса́нтная airborne training
~ войск, такти́ческая tactical training
~ да́нных collection and analysis of planning data
~, допризы́вная pre-conscription military training
~, инжене́рная field engineering training. engineer training
~ исхо́дных да́нных preparation of initial data (gunn)
~ исхо́дных да́нных, глазоме́рная preparation of initial data by eye
~ исхо́дных да́нных, по́лная full preparation of initial data (arty)
~ исхо́дных да́нных, сокращённая abbreviated preparation of initial data (arty)
~, команди́рская officer training
~, методи́ческая 1. instructional technique training
2. methodical training
~ наступле́ния, авиацио́нная air preparation
~, нача́льная initial training
~, нача́льная одино́чная basic training
~, огнева́я 1. preparatory fire
2. weapon training
подгото́вка, одино́чная individual training
~, одино́чная строева́я individual drill
~, одино́чная такти́ческая individual tactical training
~, оконча́тельная calculation for range and line (gunn)
~, основна́я basic training
~, парашю́тная parachute training
~, первонача́льная basic training
~, повы́шенная advanced training
~, полити́ческая political training
~, предвари́тельная calculation of initial errors (gunn)
~, предста́ртовая pre-launch preparation (GW)
~, противохими́ческая anti-gas training, CW training
~, специа́льно-строева́я special-training drill (armoured units)
~, стрелко́вая weapon training
~, строева́я drill training (Sov term includes elementary battle training)
~, такти́ческая tactical training, combat training
~, техни́ческая technical training
~ ты́ла administrative preparation
~, физи́ческая physical training
~, хими́ческая CW training
подгото́вленность preparedness, state of training
подгру́ппа sub-group
поддежу́рный alternate, reserve (personnel)
поддержа́ние maintenance, upkeep
поддержа́ть (см подде́рживать)
подде́рживать/поддержа́ть to support, to back up, to maintain; **подде́рживать/поддержа́ть запа́с** to keep up stocks; **подде́рживать/поддержа́ть огнём** to cover with fire
подде́рживающий supporting, backing up, maintaining
подде́ржка support
~, авиацио́нная air support
~, артиллери́йская artillery support
~, артиллери́йская корабе́льная naval gunfire support (amph ops)
~, взаи́мная огнева́я mutual fire support
~, возду́шная air support
~ войск, авиацио́нная close air support
~, непла́новая огнева́я impromptu covering fire

поддержка, непосредственная close support
~, непосредственная огневая close fire support
~, общая general support
~, огневая fire support
~, танковая tank support
~, ядерная nuclear support, nuclear backup, atomic support, atomic backup
поддиапазон sub-band (radio)
поддон 1. sump (veh etc)
2. sabot (ammo)
3. pallet (air supply)
4. base disc (ammo)
~ картера crankcase pan
~, масляный oil pan
~ орудия firing platform
~ подкалиберного снаряда sabot (APDS)
поддон-волокуша toboggan-type pallet
поддув supercharging
подействовать (см действовать)
поджиг striking. triggering (elec)
поджигатель incendiary
~ войны warmonger
поджим выбрасывателя extractor retaining lug
подзаряд boosting charge, recharge, recharging
~, непрерывный trickle charge (elec)
подземный underground, subterranean
подив divisional political department
подивизионно by divisions (tac)
подкалиберный sub-calibre
подкачка priming (pump)
~, ручная hand priming
подкладка 1. footsole (mining)
2. mudsill (br)
3. backing, lining
~, дощатая footsole (mining)
~ железно-дорожного вагона bolster (rly)
~, фетровая felt pad
подклинивание chocking (veh)
подклинивать/подклинить to wedge up
подклинить (см подклинивать)
подключать/подключить to connect, to switch in, to tap (line of comms)
подключаться/подключиться to cut in (engine)
подключение tapping (lines of comms)
подключить (см подключать)
подключиться (см подключаться)
подковообразный horseshoe-shaped
подкоп sap, undermining
подкос strut, raker, angle brace
подкрадываться/подкрасться to creep up to, to steal up to
подкраска touchup (paint)
подкрасться (см подкрадываться)
подкрепить (см подкреплять)
подкрепление 1. reinforcement, bracing (tech)
2. reinforcement (mil)
3. confirmation, corroboration
подкреплять/подкрепить to support, to confirm, to reinforce
подлежать to be subject to, to be liable to
подлежащий subject to, liable
подлесок undergrowth
подлёт hover (helicopters)
подлинник original
подлинный original, authentic, genuine
подлодка submarine
подложка sublayer, backing, substratum, substrate
подлокотник elbow rest
подмодулятор pre-modulator
подмости "самолёт" improvised horizontal piling frame (br)
подмостки scaffolding, platform, stage
поднадзорный person under surveillance
поднесущий sub-carrier (elec)
поднимать/поднять to lift, to raise; **поднимать/поднять домкратом** to jack up; **поднимать/поднять по тревоге** to put on the alert, to alert; **поднимать/поднять тревогу** to give the alarm, to raise the alarm
подножие foot, base (topo)
подножка footboard, running board
подносчик ammunition handler, ammunition number
~ боеприпасов ammunition bearer, handler
~ патронов ammunition number, carrying number, ammunition carrier
поднятие флага hoisting the flag
поднять (см поднимать)
подобный similar
подоболочка sub-shell
подобрать (см подбирать)
подогнать (см подгонять)
подогрев warming-up, heating, pre-heating
подогреватель heater, preheater

подогрева́тель во́здуха air preheater
~ **во́здуха, поступа́ющего в карбюра́тор** carburettor air intake preheater
~ **дви́гателя** engine preheater
~ **пи́щи** hot plate
подозрева́ть/заподо́зрить to suspect
подозре́ние suspicion
подозри́тельный suspicious
подойти́ (см подходи́ть)
подо́пытный experimental, tentative
подорва́ть(ся) (см подрыва́ть(ся))
подотчётность accountability
подотчётный accountable
подо́шва 1. foot (topo)
2. sole (footwear)
подпи́лок file, rasp
подписа́ть (см подпи́сывать)
по́дписи на ка́рте map lettering
подпи́сывать/подписа́ть to sign, to add (to), to subscribe for; **подпи́сывать/подписа́ть акт** to sign a statement
по́дпись на горизонта́лях contour height figures
подполза́ть/подползти́ to creep up, to crawl up
подползти́ (см подполза́ть)
подполко́вник lieutenant-colonel (Sov Army, Air and Naval Land Forces)
подпо́лье 1. cellar
2. underground resistance movement
подпо́льный underground, resistance
подпо́ра strut, prop
подпо́рка bearer, prop, support
~, **вре́менная** temporary support
подпо́чва subsoil, substratum
подпо́чвенный subsurface
подпружинённый spring-loaded, sprung
подпры́гивание ствола́ при вы́стреле barrel jump (gunn)
подпуска́ть/подпусти́ть to allow to approach
подпусти́ть (см подпуска́ть)
подпя́тник bearing socket, axial bearing
~, **шарово́й** ball socket
подра́внивание 1. adjusting
2. dressing (drill)
подразделе́ние 1. sub-unit (Sov battalion and below)
2. element (tac)
~, **ба́нно-дезинфекцио́нное** sterilisation and bath sub-unit
подразделе́ние вое́нно-полево́й по́чты army postal sub-unit
~ **второ́го эшело́на** second echelon sub-unit
~, **дежу́рное** duty sub-unit
~, **инжене́рно- мостострои́тельное** engineer bridge building sub-unit
~, **инжене́рно-доро́жное** engineer road construction sub-unit
~, **инжене́рно-позицио́нное** engineer works service sub-unit
~, **инжене́рно-сапёрное** sapper sub-unit
~, **картографи́ческое** mapping sub-unit
~, **курса́нтское** 1. trainee sub-unit, training sub-unit
2. NCO's cadre
~, **лине́йное** 1. line sub-unit (inf)
2. line detachment (sigs)
~, **лине́йное разве́дывательное** line reconnaissance sub-unit
~, **манёвренное** manoeuvre element
~, **маскиро́вочное** camouflage sub-unit (eng sub-unit with camouflage duties)
~ **обезвре́живания неразорва́вшихся бомб** bomb disposal sub-unit
~, **огнево́е** fire sub-unit
~ **охра́ны объе́кта, подлежа́щего разруше́нию** demolition guard
~ **охра́ны объе́кта, подлежа́щего подрыва́нию** demolition guard
~, **парашю́тно-деса́нтное** airborne sub-unit
~, **передово́е** forward sub-unit
~, **подде́рживающее** support element
~, **подру́чное** stand-by sub-unit
~ **полево́го водоснабже́ния** field water supply sub-unit
~, **понто́нно-мостово́е** pontoon bridge sub-unit
~, **при́данное** attached sub-unit
~, **пулемётное** machine-gun team
~, **разве́дывательное** 1. reconnaissance sub-unit
2. intelligence sub-unit
~ **размини́рования** mine-clearing sub-unit
~, **ремо́нтное** repair element
~, **сапёрное** sapper sub-unit
~, **сапёрно-подрывно́е** engineer demolition sub-unit
~ **сбо́ра иму́щества** salvage collecting sub-unit
~ **сме́шанного соста́ва** mixed sub-unit

подразделение, соседнее adjacent sub-unit
~, спасательное rescue section
~, специальное specialist sub-unit
~, стрелковое infantry sub-unit
~, строевое line sub-unit (inf)
~, топографическое survey sub-unit
~, тыловое 1. rear sub-unit
2. administrative sub-unit
~ усиления reinforcing sub-unit
~, учебное training sub-unit, training installation
~ химической защиты CW defence sub-unit
~, штатное разведывательное
1. organic intelligence sub-unit
2. organic reconnaissance sub-unit
~, эвакуационное recovery sub-unit
~, эвакуационно-ремонтное recovery sub-unit
подрайон subsector
подрамник 1. sub-frame
2. bearer
~ кузова body sub-frame
подрессоренный sprung
подрессоривание springing, spring suspension
~, независимое independent suspension
подрессорник spring actuated lever
подручный at hand, to hand, improvised, stand-by
подрыв 1. demolition, blasting
2. undermining
3. detonation, destruction
4. subversion
5. injury
~, аварийный emergency destruction (missiles)
подрывание 1. cut (dml)
2. demolition
~ мостов bridge demoltion
~ туннелей tunnel blasting
подрывать/подрыть to undermine, to sap
подрывать(ся)/подорвать(ся) to blow, to explode, to subvert; **подрывать/подорвать боевую часть** to detonate a warhead; **подрываться/подорваться** to be blown up, to blow up (expl)
подрывник demolition sapper
подрывной 1. demolition, blasting
2. subversive
подрыть (см подрывать)
подрядчик contractor
подсветка illumination
~, подкузовная rear-axle light, convoy light
~ целей target illumination
подсектор subsector
подслушать (см подслушивать)
подслушивание 1. eavesdropping
2. interception (sigs)
~ и перехват сообщений monitoring and interception of communications
подслушивать/подслушать to eavesdrop, to listen in, to monitor, to intercept
подснаряд sub-projectile
подсобный auxiliary, secondary
подставка 1. support, prop, rest, stand
2. trestle
3. bracket
4. plate-rack
подстанция sub-station
подстил folding sheet, groundsheet
подстилка 1. sheet, ground sheet
2. litter (in stables etc)
подстраивать/подстроить to tune in
подстраиваться/подстроиться to take up, to adopt; **подстраиваться/подстроиться в боевой порядок** to take up battle formation
подстрекательство persuading, instigation, incitement
подстреливать/подстрелить to wound by a shot, to shoot down
подстрелить (см подстреливать)
подстроить (см подстраивать)
подстроиться (см подстраиваться)
подстройка, глинобитная clay construction, earthern construction
~ частоты frequency tuning, frequency control
~ частоты, автоматическая automatic frequency control
~, электронная electronic tuning
подступ approach
~, близкий tactical approach
~ к мосту bridge approach
~, скрытый defiladed route, covered line of approach
~, укрытый covered approach
подступать/подступить to approach
подступить (см подступать)
подсудность jurisdiction, competence (law)
подтвердить (подтверждать)

подтверждáть/подтвердѝть to confirm, to corroborate; **подтверждáть/подтвердѝть получéние** to acknowledge receipt
подтверждéние confirmation, acknowledgement, corroboration
~ **приёма радиогрáммы** acknowledgement of receipt (radio comms)
подтекáние leak
подтовáрник chock, wedge, slat
подтя́гивание 1. moving up, bringing up (tac)
2. pulling up, tightening up
3. hauling, warping
подтя́гивать/подтянýть to bring up, to move up, to tighten up, to draw up
подтя́нутый spruce, neat, smart looking
подтянýть (см подтя́гивать)
подýшка 1. pillow, cushion, pad
2. spoil placed on top of fieldworks
~, **асбéстовая** asbestos pad
~, **бетонѝрованная** concrete pad
~, **кольцевáя** ring-shaped pad (obturator)
~ **обтюрáтора** obturator pad
~ **обтюрáтора, асбéстовая** asbestos obturating pad (ord)
~, **опóрная** cradle (ord)
~, **противомозóльная** anti-callous pad
~, **съёмная** detachable pad, detachable cushion
~, **узловáя** joint block, junction block (eng, br)
подфáрник parking light, side lamp
подхвáт, скоростнóй high-speed buffeting (aeron)
~, **срывнóй** low-speed (pre-stall) buffeting (aeron)
подхóд approach
~ **к бéрегу** beach approach route, beach entrance (amph ops)
~, **твóрческий** creative approach, imaginative approach
подходѝть/подойтѝ to come up (to), to approach, to fit, to suit, to approach (a question); **подходѝть/подойтѝ на дистáнцию действѝтельного огня́** to get within effective range
подцáпфенник trunnion bearing (ord)
подцéпщик soldier employed in rigging aircraft for dropping loads
подчинéние subordination
подчинённость subordination
подчинённый 1. subordinate
2. under command
подчинѝть (см подчиня́ть)
подчинѝться (см подчиня́ться)
подчиня́ть/подчинѝть to subordinate, to put under the command of; **подчиня́ть/подчинѝть когó-нибýдь своéй вóле** to impose one's will upon someone
подчиня́ться/подчинѝться to be subordinate to, to submit, to obey, to come under the command of
подшабрóвка scraping (tech)
подшивáть/подшѝть to sew, to hem, to line, to sole; **подшивáть/подшѝть к дéлу** to file (docs)
подшѝпник bearing (tech)
~ **большóй голóвки шатунá** big-end bearing
~, **брóнзовый** bronze bearing
~, **бýксовый** axle bearing
~ **вáла** shaft bearing
~ **включéния фрикциóна** clutch bearing
~, **внéшний** outer bearing
~, **внýтренний** inner bearing
~, **выжимнóй** release bearing (clutch)
~ **выключéния сцеплéния** clutch bearing, clutch release bearing
~ **выключéния фрикциóна** clutch release bearing
~, **глáвный** main bearing
~, **головнóй** small-end bearing (ICE)
~ **дифференциáла** differential bearing
~, **игóльчатый** needle bearing
~, **игóльчатый рóликовый** needle roller bearing
~ **кардáнного шарнѝра** universal-joint bearing
~ **качéния** rolling contact bearing
~ **качéния, бессепарáторный** cageless bearing
~ **качéния, однорядный** single-row anti-friction bearing
~ **колéнчатого вáла, креннóй** crankshaft main bearing
~ **колесá** wheel bearing
~, **конѝческий** tapered bearing
~, **концевóй** end bearing
~, **креннóй** main bearing
~, **моты́левый** big-end bearing
~, **нарýжный** outer bearing
~, **не трéбующий смáзки** lubricated-for-life bearing
~, **неразъёмный** bush bearing

подши́пник, опо́рный supporting bearing
~, осево́й axle bearing
~ перви́чного ва́ла primary shaft bearing
~, пла́вающий floating bearing
~, подде́рживающий bearing
~ промежу́точного ва́ла lay-shaft bearing
~, промежу́точный intermediate bearing
~, радиа́льный radial bearing, journal bearing
~, радиа́льный ша́риковый radial ball bearing
~ распредели́тельного ва́ла camshaft bearing
~, регули́руемый adjustable bearing
~, ро́ликовый roller bearing
~ с кони́ческими ро́ликами tapered roller bearing
~, самосма́зывающийся self-lubricating bearing
~, самоустана́вливающийся self-aligning bearing
~ скольже́ния sliding bearing, sleeve bearing
~, скользя́щий plain bearing
~, сфери́ческий spherical race bearing
~, упо́рный 1. thrust block, thrust bearing
2. axial bearing
~ фрикцио́на, выжимно́й clutch release bearing
~, ша́риковый ball bearing
~, шату́нный big-end bearing
~, шкворнево́й king-pin bearing
подши́ть (см подшива́ть)
подшле́мник helmet liner, cap comforter
~, шерстяно́й balaclava helmet
подъе́зд 1. entrance, porch
2. approach
подъём 1. rise, ascent, upgrade (topo)
2. reveille
3. ascent (GW)
4. lifting, hoisting, raising
~, большо́й steep gradient
~ "ёлочкой" herring-bone ascent (ski)
~ "зигза́гом" zig-zag ascent (ski)
~ ка́рты colour featuring of a map
~ кулачка́ lift of cam, cam rise
~ "ле́сенкой" side-stepping ascent (ski)
~ "полуёлочкой" semi-herringbone ascent (ski)
подъём, преде́льный maximum slope
~, прямо́й straight ascent
подъёмник 1. erector (GW)
2. jack, hoist
~ для автомоби́лей vehicle lift
~, цепно́й chain hoist
~, электри́ческий electric stacker
подъёмный lifting, hoisting, elevating
поеди́нок duel, single combat
по́езд 1. train
2. gun plus prime mover (arty)
~, автомоби́льный 1. road train, truck train (MT)
2. tractor-trailer unit
~, бронево́й (бронепо́езд) armoured train
~, вое́нно-санита́рный ambulance train
~, во́инский troop train
~, грузово́й freight train
~, почто́вый mail train
~ прице́пов road train
~, санита́рный ambulance train
~, са́нный sled train
~, сбо́рный pick-up goods train
~, ско́рый express train
~, э́кстренный special train
пое́здка, полева́я 1. battlefield tour (by a group of officers to study individual tactical and military historical problems or a whole theatre of operations)
2. tactical exercise without troops, TEWT (the Sov term includes CPXs)
пожа́р fire, conflagration
~, верхово́й forest fire
~, верхово́й лесно́й forest fire, scrub fire (in which complete trees burn)
~, низово́й scrub fire
пожа́рник fireman
пожа́рный 1. fireman
2. fire, fire-fighting
пожаробезопа́сность fire safety
пожароопа́сность fire risk, fire danger
пожаротуше́ние fire fighting
позёмок 1. drifting snow
2. a winter wind
позити́в positive
позитро́н positron
пози́ции, исхо́дные forming-up positions
позицио́нный static, positional
пози́ция position
~, артиллери́йская огнева́я artillery gun position
~ батаре́и, огнева́я battery gun position

пози́ция боево́го охране́ния forward edge of the battle area, FEBA
~, **взво́дная** platoon locality
~, **вре́менная** temporary position
~, **вре́менная огнева́я** temporary gun position
~, **вспомога́тельная** secondary position
~, **вспомога́тельная огнева́я** secondary fire position
~, **вы́годная** advantageous position
~, **выжида́тельная** 1. assembly area, assembly position (tk)
2. waiting area (inf)
3. "hide" (GW tac)
~, **деревя́нно-земляна́я огнева́я** earth and timber fire position
~ **для зени́тного огнево́го сре́дства** anti-aircraft emplacement
~ **для пристре́лки, огнева́я** fire registration position
~ **для противота́нковой пу́шки, огнева́я** anti-tank gun emplacement
~ **для стрельбы́ прямо́й наво́дкой, огнева́я** direct-fire position
~, **закры́тая огнева́я** 1. covered fire position (SA)
2. indirect fire position (arty)
~, **запасна́я** 1. alternative position
2. third line (Sov offensive posn)
~, **запасна́я огнева́я** alternative gun position, alternative fire position
~, **исхо́дная** starting point, forming-up place, FUP
~, **кома́ндная** dominating position, commanding position
~, **ло́жная огнева́я** dummy fire position
~ **на обра́тном ска́те, огнева́я** reverse slope fire position
~ **наведе́ния** guidance area, guidance and control area (GW)
~, **наси́женная** long-occupied position
~, **ночна́я** night position (vacated during day time)
~, **оборони́тельная** defensive position
~, **обороня́емая** defended position
~, **обору́дованная** organized position, developed position
~, **обору́дованная в инжене́рном отноше́нии** dug-in defensive position
~, **огнева́я** 1. fire position (tac)
2. firing point (range)
3. gun position (tac)
~, **основна́я** main position

пози́ция, основна́я огнева́я main fire position
~, **откры́тая** 1. exposed position
2. direct fire position (arty)
~, **откры́тая огнева́я** 1. exposed gun position
2. direct fire position (arty)
~, **отсе́чная** switch position, alternative position
~, **передова́я** forward post, forward position
~, **подгото́вленная** prepared position
~, **полузакры́тая огнева́я** semi-concealed gun position (guns hidden unless firing)
~, **поспе́шно обору́дованная** hastily organised position, hasty position
~, **предмо́стная** forward bridgehead position
~ **пристре́лочного ору́дия, огнева́я** fire registration position
~, **промежу́точная** 1. intermediate position (tac)
2. second defensive position (Sov organised defence)
~ **пусково́й устано́вки** launcher site (GW)
~ **с круговы́м обстре́лом, обору́дованная огнева́я** all-round field-of-fire position
~, **скры́тая огнева́я** 1. indirect fire position (arty)
2. concealed fire position
3. concealed gun position
~, **ста́ртовая** launching area, launcher area, launcher position (GW)
~, **та́нковая** tank position
~, **техни́ческая** missile service area, missile assembly and test area
~, **укреплённая** entrenched position, fortified position
~, **укры́тая** covered position
~ **управле́ния** guidance and control area
познако́мить (см знако́мить)
позывно́й call, callsign
~, **лине́йный** link callsign (comms)
~, **несменя́емый** fixed callsign
~, **циркуля́рный** collective callsign, group callsign (comms)
по́имка detection, picking-up (radar)
по́иск 1. search, quest, hunt
2. raid (mil)
3. sweep, scan, search (radar)

пóиск, автомати́ческий automatic search
~, визуáльный visual search
~, дневнóй daylight fighting patrol (to capture prisoner(s) for interrogation), daylight raid, daylight "snatch"
~ информáции information retrieval
~, круговóй all-round scanning, circular scanning, circular sweep
~, радиолокациóнный radar scan, radar sweep, radar search
~, ручнóй manual search
~, сéкторный sector scanning
~ цéли target location
пóйма flood plain, water meadow (topo)
покáз demonstration, showing, display
~, нагля́дный demonstration
показáние 1. reading (instr)
2. statement
~ спидóметра 1. speedometer reading
2. mileometer reading
показáтель 1. index, coefficient
2. feature, characteristic
3. indicator
~, чи́сленный numerical indicator, numerical designator
показáтельный 1. demonstration, model
2. Exemplary, Model (Sov honorary title awarded for exceptional work in military and political training)
показнóй demonstration, show
покáзчик 1. "enemy" (on exercise)
2. demonstrator
3. marker (butt party)
покарáть (см карáть)
покáтость incline, slope, declivity
покачáть (см качáть)
покóвка forging
поколебáться (см колебáться)
покрóв, расти́тельный vegetation, vegetative cover
~, снéжный snow cover
покрывáть/покры́ть to cover, to roof, to overlay; **покрывáть/покры́ть бронёй** to armour
покры́тие overhead covering, cover, covering, coverage
~, антикоррозиóнное 1. rust preventive coating
2. underseal (vehicle)
~, антиотражáтельное anti-reflection coating
~, асфальтобетóнное bituminous concrete surface
покры́тие, биту́мное bituminous coating
~, боевóе overhead protective cover
~, броневóе armour plating
~, декорати́вное decorative coating
~, дорóжное 1. road surface
2. track expedients
~, защи́тное protective covering, overhead cover
~, зеркáльное mirror coating
~, колéйное wheel treadway
~, колéйное щитовóе treadway
~ крáской painting
~, лáковое lacquer coat, varnish coat
~ ли́нзы, осветля́ющее lens coating, optics coating
~, маскирóвочное camouflage cover
~, перви́чное primer coat
~, поглощáющее 1. absorbent covering
2. anti-radar coating
~, противокоррозиóнное anti-corrosion coating
~, противооскóлочное splinterproof cover
~, разбóрно-перенóсное road mat (wheel track of steel or timber (eng))
~, сбóрное дорóжное prefabricated trackway
покры́ть (см покрывáть)
покры́шка 1. casing, cover
2. outer casing (tyre)
~, кóжаная leather cover, leather cap
~ объекти́ва, кóжаная leather cap for object glass (arty)
покушéние на преступлéние attempt to commit a crime
пóле 1. field
2. background
3. land (rifled barrel)
~, береговóе ми́нное beach minefield
~, беспокóящее ми́нное nuisance minefield
~ большóй плóтности, ми́нное high-density minefield
~ бóя battle area, battlefield, field of battle
~ бóя, глáвное main battle area
~ ви́димости field of vision
~, вспáханное ploughed field
~ заграждéний obstacle in depth, obstacle belt
~, защи́тное ми́нное protective minefield

по́ле зре́ния 1. field of view
2. field (optics)
~ **излуче́ния** radiation field
~ **изображе́ния** useful field (optics)
~, **имитацио́нное** battle simulation area
~ **инду́кции** induction field (elec)
~, **конта́ктное** bank (teleph)
~, **лётное** landing ground (ac)
~, **ло́жное ми́нное** phoney minefield, dummy minefield
~, **магни́тное** magnetic field
~ **ма́лой пло́тности, ми́нное** low-density minefield
~, **мише́нное** target range
~ **наблюде́ния** field of observation
~ **наре́за** land (rifling)
~ **неви́димости** dead ground
~ **обзо́ра** field of view
~ **облуче́ния** irradiation field
~, **оборони́тельное ми́нное** defensive minefield
~ **обстре́ла** field of fire
~, **отде́льное ми́нное** mine belt
~, **противодеса́нтное ми́нное** anti-airborne minefield
~, **противопехо́тное ми́нное** anti-personnel minefield
~, **противота́нковое ми́нное** anti-tank minefield
~, **радиолокацио́нное** radar coverage
~ **рассе́яния** stray field (elec)
~ **с хими́ческими ми́нами, ми́нное** chemical minefield
~, **сме́шанное ми́нное** mixed minefield
~ **сраже́ния** battlefield
~, **такти́ческое** battle range, tactical training area
~, **такти́ческое ми́нное** tactical minefield
~, **управля́емое ми́нное** controlled minefield
~, **учёбное** training ground
~, **фи́рновое** glacier field
~, **электромагни́тное** electromagnetic field
полево́й field
поле́зный useful, profitable, beneficial, helpful
полёт flight
~, **акти́вный** powered flight
~, **аэродро́мный** local flight (up to 50km from base without landing)
~, **боево́й** operational flight
полёт, бре́ющий contour flying, ground-level flying, hedge-hopping
~, **неуправля́емый** free-flight, unguided flight
~, **пасси́вный** coasting (GW)
~ **по потолка́м** cruise-climb flight
~, **разве́дывательный** reconnaissance flight
~, **свобо́дный** free flight
~, **тра́нспортно-связно́й** transport and communications flight
~, **управля́емый** controlled flight
по́лзание по-пласту́нски leopard crawl
ползко́м crawling, by crawling
ползу́н 1. slideblock, slide bar, slidepawl
2. rebound slide (revolver)
3. selector rod (gear)
4. feedslide (MG)
~ **амортиза́тора отда́чи** recoil slide
~ **затво́ра** retracting slide, retractor slide
ползуно́к slipper, slide, sliding contact (elec)
~, **оруди́йный** gun slide
ползу́честь creep (metall)
поли́вка ОВ CW spraying
полиго́н 1. training ground, proving ground
2. range (firing)
~, **артиллери́йский** artillery range
~, **винто́вочный** rifle range
~, **инжене́рный** engineer training and testing ground
~, **испыта́тельный** proving ground, test range
~, **о́пытный та́нковый** tank proving ground
~ **поса́дки** landing area (of re-entering earth satellite)
~ **приземле́ния** landing area (of re-entering earth satellite)
~, **раке́тный** rocket firing range
~, **та́нковый** tank training ground, tank firing range
~, **хими́ческий** chemical proving ground
поликли́ника dispensary, sick quarters
полирова́ние polishing, buffing, burnishing
полиспа́ст load pulley block, block and tackle, pulley tackle
поли́тико-просвети́тельный politico-educational
политотде́л political department
политрабо́тник political worker

политрýк political commissar (obs)
поли́ция, воéнная military police (in non-Soviet forces)
полк regiment
~, артиллери́йский artillery regiment
~, бронекавалери́йский armoured regiment
~, вертолётный helicopter regiment
~, гáубичный howitzer regiment
~, горнострелкóвый mountain rifle regiment
~, зени́тно-артиллери́йский anti-aircraft regiment
~, зени́тный anti-aircraft regiment
~, инженéрный engineer regiment
~, истреби́тельно-противотáнковый артиллери́йский tank-destroyer artillery regiment
~, истреби́тельный противотáнковый артиллери́йский tank-destroyer artillery regiment
~, кавелери́йский cavalry regiment
~ крýпного кали́бра, зени́тный артиллери́йский heavy anti-aircraft regiment, HAA regiment
~ лёгкой зени́тной артиллéрии LAA regiment
~, миномётный mortar regiment
~, мотострелкóвый motor rifle regiment
~, полевóй инженéрный field engineer regiment
~, посáдочно-десáнтный артиллери́йский airborne artillery regiment
~, пýшечный артиллери́йский conventional artillery regiment, gun artillery regiment
~, самохóдно-артиллери́йский 1. SU artillery regiment (Sov)
2. self-propelled artillery regiment
~, свóдный composite regiment
~ свя́зи signal regiment
~, стрелкóвый rifle regiment
~, тáнковый tank regiment
~, тяжёлый артиллери́йский heavy artillery regiment
~, тяжёлый тáнковый heavy tank regiment
~, тяжёлый танкосамохóдный heavy tank/ self-propelled artillery regiment
пóлка berth, rack
~, кругова́я holding-down ring (ord)
полквы́лет regimental sortie
полкóвник colonel (Sov Army, Air and Naval Land Forces; equivalent to group captain RAF)
полковóдец military leader, general (not mil rank)
полковóй regimental
полнокрóвный 1. full-blooded
2. up to strength (coll)
полномóчие authority, powers
полнотéльный solid (ammo)
пóлный overall, full, complete
пол-оборóта half-turn
"Пoлоборóта Налéво!" "Left Incline!"
"Полоборóта Напрáво!" "Right Incline!"
полови́на лáмпы, вторáя second half of a dual valve (electron)
половинéние halving
полови́нить to split; **полови́нить ви́лку** to split the bracket (arty); **полови́нить ви́лку дáльности** to split the range bracket; **полови́нить ширóкую ви́лку** to break down the long bracket
половóдье 1. spring high water (in river etc)
2. period of spring floods, flood-time
полóгий sloping, inclined
положéние 1. situation, position, condition
2. state, status
3. regulations, statute
~, безрazли́чное position of neutral equilibrium
~, боевóе 1. firing position (gunn)
2. armed position (fuze)
3. protection position (respirators anti-gas)
~, взведённое retracted position
~ включéния "on" position, engaged position
~, воéнное martial law
~ выключéния "off" position, released position
~ для стрельбы́ firing position
~ для стрельбы́ лёжа lying firing position
~ для стрельбы́ с колéна kneeling fire position
~ для стрельбы́ с упóра firing position with weapon supported
~ для стрельбы́ си́дя sitting firing position
~ для стрельбы́ стóя standing firing position
~, исхóдное forming-up place, FUP

положéние, кáжущееся apparent position, apparent situation
~, мúрное peace footing
~ наготóве 1. ready position (wpn) 2. alert position (respirator)
~, невзведённое unarmed condition
~, нейтрáльное neutral (gear)
~, нерабóчее "off" position
~ неустóйчивого равновéсия state of unstable equilibrium
~, нулевóе secured position (arty, ord)
~ о дéнежном довóльствии allowance regulations
~, осáдное state of siege
~ пóлного запáздывания fully retarded position (ICE ignition)
~ пóлного опережéния fully advanced position (ICE ignition)
~ пóршня в вéрхней мёртвой тóчке top-centre piston position (ICE)
~, похóдное 1. travelling position 2. slung position (respirator)
~ противогáза на пóясе belt position of respirator
~ противогáза, похóдное slung position of respirator
~, рабóчее working position, operating position
~ равновéсия state of equilibrium
~ рулéй, нейтрáльное streamline position of control surfaces, normal position of control surfaces, mid-position of control surfaces
~ с боеприпáсами ammunition situation
~, сбалансúрованное trim position (aeron)
~, служéбное status
~ "смúрно" position of attention
~ стволá пóсле накáта gun runout position
"положéние, угрожáемое" "threat of air attacks" (an air-raid warning condition in which air attack seems likely)
положéние, угрожáемое threatening situation (preceding outbreak of war)
~, упреждённое future position
~ устáва regulation laid down in the manual
~ цéли, кáжущееся apparent target position
~ цéнтра вращéния, оптимáльное optimum pivot position
~, чрезвычáйное state of emergency
положéния, основнúе doctrine, fundamental tenets
полóжено по штáту authorized on establishment
положúтельный positive
"Положúть Орýжие!" "Ground Arms!"
пóлоз runner (tech)
~, поддéрживающий track skid (tk)
полозкú люльки, направля́ющие cradle slipper guides (ord)
~, направля́ющие slipper guides (ord)
~ стволá lower cradle slides (ord)
полозóк 1. runner (pontoon) 2. key guide (eng) 3. cradle guide (ord)
~ ног сóшки bipod shoe
полóзья sledge runners, skids
полóмка breakage, breakdown
полосá strip, band, zone, belt, area, stripe
~ безопáсности safety strip
~ безопáсности, концевáя runway overshoot area
~, бетонúрованная concrete strip
~ боевы́х дéйствий zone of contact, operations
~, боковáя side-band (radio)
~, боковáя сердцевúнная seventy percent breadth zone, (dispersion, gunn)
~ бригáды brigade area of responsibility
~ веро́ятного поражéния zone of probable damage
~, взлётно-посáдочная airstrip, landing strip, runway
~, взлётно-посáдочная грунтовáя airstrip (hard mud or earth)
~, вторáя оборонúтельная second defensive zone
~, глáвная main runway
~ движéния traffic lane
~ движéния огневóго вáла barrage lane
~ действúтельного поражéния effective beaten zone
~ дéйствия 1. area of responsibility (defence) 2. zone of advance (attack) 3. zone of operations (general)
~, дополнúтельная огневáя secondary fire sector
~, заградúтельная 1. fire break 2. obstacle zone, obstacle belt
~, заграждáемая suppressed band (electron)

полосá заграждéний obstacle zone, obstacle belt
~ **заграждéния** attenuation band
~ **задержáния** attenuation band
~ **кóрпуса** corps area of responsibility
~ **лýчшей половúны попадáний** fifty percent zone (gunn)
~, **межфронтовáя** no-man's land
~, **металлúческая** metal strip runway, pierced steel planking runway, PSP runway
~, **мúнная** mine strip
~ **наступлéния** zone of advance
~, **нейтрáльная** no-man's land, neutral zone
~, **ничéйная** no man's land
~ **обеспéчения** forward defence area security zone
~, **оборонúтельная** defensive zone
~ **оборóны** defended area, defensive zone
~ **оборóны, вторáя** second defensive zone
~ **оборóны, глáвная** main defensive zone
~ **оборóны, дивизиóнная** divisional defence area
~ **оборóны, пéрвая** first defensive zone, main defensive zone
~ **оборóны, трéтья** third defensive zone
~ **обстрéла** zone of fire
~ **огневóго вáла** barrage lane (arty)
~ **огня́** arc of fire
~ **огня́, дополнúтельная** secondary arc of fire
~, **основнáя огневáя** main fire support sector (arty)
~ **отвóда** 1. lane of land on which road is built
2. site selected for road
3. overall width of road (incl footpaths and verges)
~ **охранéния** area of responsibility
~ **переправы на льду** swept lane of ice crossing
~ **по боковóму направлéнию, сердцевúнная** percentage breadth zone (of dispersion of shot) (ball) (percentage depends on weapon - eg for tank gunn figure is seventy percent)
~ **по высотé, сердцевúнная** percentage height zone (of dispersion of shot) (gunn)

полосá по дáльности, сердцевúнная percentage length zone (of dispersion of shot) (gunn)
~ **поражéния** danger zone
~, **посáдочная** landing strip, landing runway (ac)
~ **препя́тствий** 1. assault course
2. obstacle zone
~ **препя́тствий, учéбная** obstacle course
~, **прибрéжная** coastal area, coastal strip
~, **прифронтовáя** combat zone, forward edge of the battle area, FEBA
~ **прóволочных заграждéний** barbed-wire obstacle belt
~ **продвижéния** zone of advance
~ **пропускáния** pass band, band-width
~ **противотáнковых препя́тствий** anti-tank obstacle zone
~ **развéдки** zone of observation
~, **разделúтельная** traffic line (white line on road)
~ **разрушéния** zone of destruction
~, **сердцевúнная** effective beaten zone, seventy percent zone (ball)
~, **сердцевúнная боковáя** seventy percent breadth zone
~ **сопротивлéния, глáвная** main area of resistance
~ **сплошнóго огня́** band of fire
~, **тыловáя** rear area, communication zone
~, **укреплённая** fortified zone, prepared position
~ **частóт** frequency interval, frequency band
~ **частóт, вéрхняя** upper side band (radio)
~ **частóт, нúжняя** lower sideband
~, **штурмовáя** assault course
полóсно-пропускáющий band-pass
пóлость hollow, cavity, chamber, housing (tech)
полóтнище 1. visual signalling panel
2. gore (parachute)
~, **авиасигнáльное** air-ground liaison panel
~, **опознавáтельное** marking panel, identification panel
~ **палáтки** tent section, bivouac sheet
~ **Пóпхема** Popham panel (ship or ac visual signalling panel)
~, **сигнáльное** air-ground signalling panel

полотно́ 1. bed (rly, rd, etc)
2. cloth, linen
~, гу́сеничное track (veh)
~, доро́жное road bed, carriageway and shoulders
~, железнодоро́жное railway track, rail bed
~, земляно́е sub-grade (road)
~, мостово́е roadway of bridge
полуавтома́т 1. semi-automatic weapon
2. semi-automatic lathe
полуавтома́тика 1. semi-automatic mechanism
2. semi-automatic breech mechanism
полуавтомати́ческий semi-automatic
полуваго́н open wagon (rly)
полувёрстка scale 1:21,000 map
полувзведённый half-cocked
полувзво́д half cock
полуволново́й half-wave
полуглиссёр hydroplane
полугу́сеничный half-track
полу́да tinning
полукапони́р single caponier (fd wks)
полукилометро́вка scale 1:50,000 map
полукольцо́ half-ring
~, светя́щееся luminous semi-circle
полукомпле́кт one terminal set (teleg, teleph)
полукру́г 1. semi-circle
2. cluster (of mines)
полуму́фта half-coupling
полуокруже́ние semi-encirclement, semi-blockade, semi-envelopment
полуо́стров peninsula
полуо́сь 1. half-shaft
2. stub axle
~, полуразгру́женная semi-floating half-shaft
~, разгру́женная по́лностью fully floating half-shaft
полупереxо́д half of a day's march
полупе́тля half-loop
полуподши́пник arch bearing
полупонто́н pontoon section, pontoon unit
~, носово́й end section of sectional pontoon
~, сре́дний pontoon centre section
полуприце́п semi-trailer
~, двухо́сный tandem-axle semi-trailer
~, низкора́мный low bed semi-trailer
полуприце́п-рефрижера́тор refrigerated semi-trailer
полупроводни́к semi-conductor, transistor
~, селе́новый selenium semi-conductor
полупроводнико́вый transistorized
полураспа́д half-life (nucl)
полускоба́ toe iron (ski)
полусто́йкий semipersistant
полу́торка one and a half tonner (veh) (coll)
полуфабрика́т semi-prepared product
полуфигу́ра half-figure (gunn)
получа́тель 1. recipient, receiver, drawing unit
2. destination, addressee, consignee
получа́ть/получи́ть to receive, to draw, to procure; **получа́ть/получи́ть боеприпа́сы** to draw ammunition; **получа́ть/получи́ть поврежде́ние** to suffer damage, to sustain damage
получе́ние 1. receipt, receiving
2. procurement
получи́стый semi-finished
получи́ть (см получа́ть)
полушле́м, вя́заный woollen helmet (balaclava type)
полушу́бок three-quarter length sheep-skin coat
полшага́ half a pace
по́лы пала́тки tent fly, tent door
по́лый hollow
полынья́ clear water (in ice)
по́люс pole
~, географи́ческий true pole, geographic pole
~, отрица́тельный negative terminal (elec)
~, положи́тельный positive terminal (elec)
поля́на natural clearing, glade (in forest)
поляриза́ция polarization
~ анте́нны antenna polarization
~, кругова́я circular polarization
~, лине́йная linear polarization
поля́рность polarity (elec)
поля́рный 1. polar
2. arctic
поме́тка note, notation, mark
поме́ха 1. interference, noise, jamming
2. hindrance, handicap
~, акти́вная electronic jamming (ECM)
~, взаи́мная mutual interference
~, вне́шняя external interference (elec)

поме́ха, заградительная barrage jamming, non-selective defensive jamming (GW, ECM)
~, зерка́льная image interference
~, и́мпульсная pulse jamming (ECM)
~, иску́сственная artificial interference, jamming (radio, radar)
~, наведённая induced noise
~, негати́вная passive ECM (radar)
~, немодули́рованная continuous wave jamming, C.W. jamming
~, непреры́вная continuous jamming (ECM)
~ от сле́дующего за сосе́дним кана́ла second-channel interference
~ от сосе́днего кана́ла adjacent-channel interference
~, пасси́вная reflection ECM, passive ECM
~, прице́льная spot jamming, selective jamming (ECM)
~, промы́шленная man-made noise, man-made static, "QRM"
~, симметри́чная image interference (electron)
поме́хи interference, noise, jamming
~, акти́вные active jamming, electronic jamming (ECM)
~, амплиту́дно-модули́рованные amplitude-modulated noise, jamming, AM noise, jamming
~, аппарату́рные set noise (electron)
~, атмосфе́рные atmospheric noise, atmospherics
~, взаи́мные mutual interference
~, загради́тельные barrage jamming
~, зерка́льные image interference
~, иску́сственные artificial interference, jamming
~, косми́ческие cosmic noise, galactic noise
~, ме́стные clutter (radar)
~, модули́рованные акти́вные modulated jamming
~, непреры́вные continuous noise
~ от дождя́ rain clutter (radar)
~ от дробово́го эффе́кта shot noise
~ от морски́х волн sea clutter (radar)
~ от фо́на background interference
~, полуприце́льные semi-selective jamming (ECM)
~, прице́льные selective jamming
поме́хи, промы́шленные industrial interference, man-made interference
~ радиолокацио́нным сре́дствам electronic radar countermeasures
~ радиосвя́зи 1. jamming 2. interference
~, случа́йные random noise, accidental jamming
~, теплов́ые thermal noise
~, узкополо́сные narrow-band interference
~, умы́шленные jamming
~, часто́тно-модули́рованные frequency-modulated noise, jamming, FM noise, jamming
~, электри́ческие electrical noise
помехозащищённость noiseproof feature, immunity to jamming, resistance to jamming
помехозащищённый anti-jamming, jam-free, interference free
помехоусто́йчивость noiseproof feature, immunity to jamming, resistance to jamming
помехоусто́йчивый anti-jamming, jam-free, interference suppression device
помеще́ние building, premises, room, space
~, жи́лое living space, accommodation, living quarters
~, карау́льное guardroom
помкомвзво́да 2 i/c platoon
помо́ст superstructure
помо́щник 1. assistant, helper 2. No. 2, second-in-command, deputy
~ гранатомётчика No. 2 (infantry rocket launcher crew)
~ команди́ра second-in-command, 2 i/c
~ команди́ра батаре́и battery captain (arty), battery 2 i/c
~ команди́ра перепра́вы assistant crossing control officer (or landing officer)
~ наво́дчика No. 2 (MG)
~ нача́льника шта́ба deputy chief of staff
~ нача́льника шта́ба, пе́рвый first assistant to chief of staff
по́мощь, безотлага́тельная враче́бная immediate medical treatment
~, ветерина́рная veterinary aid
~, лече́бная medical assistance, medical treatment

по́мощь, медици́нская medical aid
~, неотло́жная медици́нская emergency medical treatment
~, неотло́жная хирурги́ческая emergency surgery
~, пе́рвая first aid
~, пе́рвая медици́нская first aid
~, сро́чная хирурги́ческая emergency surgery
~, техни́ческая technical assistance, first-aid repairs
по́мпа pump
~, заливна́я priming pump
~, ма́сляная oil pump
~, подка́чивающая accelerator pump, auxiliary pump
~ приеми́стости carburettor pump
~, то́пливная fuel pump, fuel injection pump
~, топливоподка́чивающая fuel pump
помпа́ж 1. pumping
2. surging, surge
3. hunting
понести́ поте́ри to suffer losses, to sustain losses
~ (см нести́)
понижа́ть/пони́зить to lower, to reduce, to demote; **понижа́ть в до́лжности** to demote to a lower appointment; **понижа́ть в зва́нии** to reduce in rank
пони́зить (см понижа́ть)
понто́н pontoon
~, брезе́нтовый canvas pontoon
~, закры́тый decked-in pontoon
~, кормово́й stern pontoon section (br)
~, коро́бчатый стально́й rectangular steel pontoon
~, кра́йний bow section pontoon
~, надувно́й air pontoon, pneumatic pontoon
~, норма́льный standard pontoon
~, носово́й end section of sectional pontoon
~, ордина́рный bipartite pontoon, standard pontoon
~, полу́торный tripartite pontoon
~, самохо́дный self-propelled pontoon
~, складно́й folding pontoon
~, сре́дний centre-section pontoon
~, фане́рный plywood pontoon
понтонёр pontoneer
понто́нный pontoon
поодино́чке one by one, singly
поору́ди́йно gun by gun, by guns (arty)
поощре́ние commendation
попада́ние hit (projectile)
~ в броню́ под угло́м к но́рмам angled strike on armour (ref: normal)
~ по норма́ли right-angled hit (at normal)
~ при ма́лом угле́ встре́чи glancing impact
~, прямо́е direct hit
~ с рикоше́та ricochet hit
попада́ть/попа́сть to hit, to strike (ball), to fall, to get into; **попада́ть/попа́сть в заса́ду** to fall into an ambush; **попада́ть/попа́сть в цель** to hit the target; **попада́ть/попа́сть под ого́нь** to come under fire
попа́сть (см попада́ть)
попере́чина 1. cross girder (br)
2. cross member (chassis)
попере́чный transverse, cross-
поплаво́к float
~, капко́вый kapok float
~, надувно́й pneumatic float
пополне́ние 1. replenishment (stores)
2. replacement, reinforcement (pers)
~ боеприпа́сов replenishment of ammunition
~ ли́чного соста́ва personnel replacement
~, ма́ршевое draft of reinforcements
~ расхо́да replenishment
~ расхо́да горю́чего fuel resupply
~ расхо́да, пла́новое automatic replacement of issues. routine replenishment
~ то́плива refuelling
пополни́тель ору́дия recoil filling tank (ord)
пополни́ть (см пополня́ть)
пополня́ть/пополни́ть to replenish (stores), to replace, to reinforce (personnel)
попра́вить (см поправля́ть)
попра́вка correction
~ а́зимута azimuth correction
~, азимута́льная azimuth correction
~, аэродинами́ческая pressure error correction (ac)
~, баллисти́ческая ballistic correction (gunn)

поправка, боковая 1. correction for line, line correction (arty)
2. lateral deflection (SA)
~ **в дальности** range correction
~ **в направлении** correction for line (gunn)
~ **в направлении на деривацию** deflection correction for drift
~ **взрывателя** fuse correction
~, **гравитационная** gravity correction
~ **дальномера** range indicator correction
~ **дальности, исчисленная** range correction of the moment
~ **дальности, метеорологическая** range correction of the moment (arty)
~ **дальности на изменение давления воздуха** range correction for barometric pressure
~ **дальности на изменение начальной скорости** MV correction
~ **дальности на изменение температуры воздуха** range correction for ballistic air temperature
~ **дальности на изменение температуры заряда** range correction for charge temperature
~ **дальности на изменение веса снаряда** correction to range for non-standard weight of projectile
~ **дальности на продольный ветер** range correction for head or following component of equivalent constant wind (Sov nav gunn)
~ **дня** correction of the day (Sov nav gunn)
~, **дополнительная** closing angle (air navig)
~ **за провешивание** correction for alignment
~, **инструментальная** instrument correction, correction for instrument errors
~ **к показаниям компаса** compass correction
~, **курсовая** course correction
~, **метеорологическая** meteorological correction
~ **на боковой ветер** 1. correction for cross component of equivalent constant wind, line C of M (gunn)
2. aim-off for wind (SA)
~ **на боковой ветер и деривацию** wind and drift correction

поправка на веер special correction
~ **на ветер** equivalent constant wind correction
~ **на вращение Земли** correction for rotation of the Earth
~ **на высоту** altitude correction
~ **на давление воздуха** correction for barometric pressure
~ **на движение цели** target speed correction, lead correction
~ **на девиацию** correction for deviation
~ **на деривацию** correction for drift
~ **на износ** correction for wear
~ **на инерцию вращения** correction for rotational inertia
~ **на интерференцию** interference correction
~ **на колпачок взрывателя** projectile correction for non-standard fuse, "cap on" correction
~ **на кривизну канала ствола** barrel bend correction
~ **на массу откатных частей** correction for recoil
~ **на метеорологические условия** correction of the moment
~ **на наклон боевой оси** correction for tilt
~ **на наклон орудийных цапф** correction for trunnion tilt
~ **на начальную скорость** MV correction
~ **на неокрашенность** projectile correction for unfinished (unpainted) shell (Sov)
~ **на отдачу** correction for recoil
~ **на отклонение температуры воздуха от нормальной** ballistic air temperature correction
~ **на отступление веса снаряда от нормальной** non-standard weight of projectile correction
~ **на отступление плотности воздуха от нормальной** correction for barometric pressure
~ **на отступление температуры заряда от нормальной** correction for charge temperature
~ **на параллакс** parallax correction
~ **на постоянную ошибку** correction for constant error
~ **на превышение орудий** height difference correction

попра́вка на разнобо́й MV correction (gunn)
~ **на ра́зность высо́т** correction for altitude difference
~ **на ра́зность дерива́ции** non-standard drift correction
~ **на ско́рость** speed compensation
~ **на смеще́ние** correction for displacement, target offset
~ **на снос** drift correction
~ **на снос ве́тром** windage correction
~ **на сопротивле́ние кана́ла ствола́** correction for bore resistance
~ **на упрежде́ние** lead correction
~ **на усту́п ору́дий** correction for range difference (gunn)
~ **на ход це́ли** 1. aim-off (arty)
2. target movement correction, allowance for target movement, target speed correction
~ **накло́нной да́льности** slant-range correction factor
~ **направле́ния** deflection, lateral correction
~ **направле́ния, метеорологи́ческая** line correction of moment (arty)
~ **направле́ния на боково́й ве́тер** line correction for cross component of equivalent constant wind
~ **направле́ния на дерива́цию** line correction for drift
~ **по результа́там наблюде́ния** observation correction, spotting correction
~, **пристре́лочная** ranging correction
~, **пристре́лянная** correction obtained by datum point ranging
~, **расчётная** computed correction
~ **с учётом да́нных упрежде́ния** correction for lead
~, **сумма́рная** total correction
~ **угла́ прице́ливания** elevation correction
~ **угла́ прице́ливания на у́гол ме́ста це́ли** elevation correction for target position
~, **углова́я** angular correction
~ **угломе́ра, оконча́тельная** final deflection (gunn)
~, **эмпири́ческая** empirical correction
поправля́ть/попра́вить to repair, to mend, to correct
попра́вочник, артиллери́йский computer graph (arty, gunn)
~ **ПУАО** FCE present range corrector
попусти́тельство connivance
По́пхем Popham air-ground panel (visual signalling)
поража́емость 1. degree of vulnerability (to fire)
2. destructiveness, lethality
поража́емый 1. vulnerable
2. dangerous
поража́ть/порази́ть to strike, to hit, to defeat, to rout, to destroy; **поража́ть/порази́ть штыко́м** to bayonet
пораже́нец defeatist
пораже́ние 1. defeat
2. destruction, damage, injury
3. hit
~, **лучево́е** radiation injury
~, **ма́ссовое** mass destruction
~ **от проника́ющей радиа́ции** radiation injury
~ **по частя́м** defeat in detail
~, **си́льное** severe damage
~, **такти́ческое** tactical reverse, tactical defeat
~ **та́нка** tank kill
~, **терми́ческое** thermal radiation damage
~ **це́ли** target kill
поражённый 1. casualty
2. hit, struck
3. destroyed
4. contaminated
5. defeated
пораже́нство defeatism
порази́ть (см поража́ть)
пориене́ние wound
поржа́веть (см ржа́веть)
по́ристость porosity
по́ристый porous
поро́г 1. rapids (river)
2. threshold
~ **реше́ния, адапти́вный** self-adjusting threshold (radio)
~ **чувстви́тельности** threshold of sensitivity
поро́да 1. rock, layer, stratum
2. spoil, muck (mining)
~, **водопроница́емая** permeable stratum
~, **коренна́я** bed rock
~, **мелова́я** chalk layer (topo)
поро́жный empty

порожня́к empty vehicle
~, обра́тный returning "empties" (vehs etc)
по́рознь separately, apart, in detail
поро́к flaw, defect, blemish, imperfection, fault
по́росль coppice, young forest
поро́тно by companies
по́рох powder, powder propellant
~, безды́мный smokeless propellant
~, бе́лый white gunpowder
~, беспла́менный flashless propellant, flashless powder
~, бу́рый cocoa powder
~, быстрогоря́щий quick-burning propellant
~, гигроскопи́ческий hygroscopic powder
~, горя́щий hot propellant
~, дегресси́вный degressive propellant
~, детони́рующий fulminating powder
~, ди́сковый circular cross-section powder grain
~, ды́мный gunpowder (Brit), black powder (US)
~, зернёный chopped powder, grain powder
~, кана́льный тру́бчатый perforated tube propellant
~, кольцево́й single perforated tube propellant
~, коротконаре́занный тру́бчатый short-length single perforated powder grain
~, крупнозерни́стый pebble, coarse-grained powder, large- grain powder
~, куби́ческий cube grain propellant, cube powder
~, ле́нточный strip propellant, strip powder grain, strip grain propellant
~, медленногоря́щий slow-burning powder
~, мелкозерни́стый fine-grained powder
~, многокана́льный multi-perforated grain powder
~, многоосно́вный multibase powder
~, мо́лотый mealed powder
~, негигроскопи́ческий non-hygroscopic powder
~, нитроглицери́новый double base powder (expl), nitroglycerine propellant
~, нитрогуаниди́новый nitro-guanidine propellant
по́рох, однокана́льный тру́бчатый single-perforated tube propellant
~, одноосно́вный single-base solid propellant
~, одноро́дный homogeneous propellant
~, пикри́новый picrate
~, пироксили́новый single-base powder, nitro-cellulose smokeless powder
~, пласти́нчатый powder grain, square cross-section grain propellant
~, полиро́ванный glazed powder
~, полубезды́мный semi-smokeless powder
~, порошкообра́зный mealed powder
~, призмати́ческий prismatic grain powder (expl)
~, прогресси́вный progressive powder
~, прукто́вый cord propellant
~, раке́тный solid propellant (missile)
~, реакти́вный solid propellant (missile)
~ со сфери́чески шестигра́нными зёрнами spherical-hexagonal grained powder
~, сфери́ческий spherically grained propellant, powder
~, тру́бочный time fuze powder
~, тру́бчатый single-perforated tube propellant
~, флегматизи́рованный moderated propellant
~, холо́дный беспла́менный cool flashless propellant
~, цилиндри́ческий solid cylindrical grain propellant
~, чёрный gunpowder (Brit), black powder (US)
~, шлифо́ванный moulded powder
порошкообра́зный powdered, in powdered form (expl)
порошо́к powder, dust
~, индика́торный detector powder (CW)
~, нажда́чный emery dust
~, у́гольный carbon powder
порт port, dock
~ вы́грузки port of disembarkation
~ погру́зки морско́го деса́нта mounting base (amph ops)
портати́вность portability
портати́вный portable
портпле́д equipment holdall
портупе́я sword-belt, waist-belt
поруче́ние mission, assignment

поручéние, персонáльное personal assignment, personal mission
пóручень handrail, lifting handle
пóрция batch, lot, portion
пóрча damage, deterioration
пóршень piston
~ **амортизáтора** damping piston, shock-absorber piston
~, **вращáющийся** rotary piston
~, **встрéчный** opposing piston
~, **гáзовый** piston, piston stem (SA)
~, **демпфи́рующий** damping piston
~ **затвóра** breech screw, screw breech block
~ **компенсáтора** recoil buffer piston (ord)
~ **накáтника** recuperator piston, counterrecoil piston
~ **насóса-ускори́теля в карбюрáторе** carburettor accelerator pump piston
~, **охлаждáемый мáслом** oil-cooled piston
~, **плáвающий** floating piston
~ **противооткáтного устрóйства** recuperator piston, buffer piston (ord)
~, **пустотéлый** hollow piston
~ **с вóгнутым дни́щем** concave-head piston, dished-head piston
~ **с плóским дни́щем** flat-headed piston
~ **с разрезнóй ю́бкой** split-skirt piston
~ **с цéльной ю́бкой** full-skirt piston
~ **стéпени сжáтия** variable compression-ratio piston
~, **ступéнчатый** stepped thread breech screw (ord)
~ **тóрмоза откáта** recoil piston
~ **тормознóго цили́ндра** brake cylinder piston
пóрыв, наступáтельный aggressive spirit, offensive spirit
поря́док order, sequence, method, formation, routine, procedure
~, **боевóй** i. battle formation, deployment, disposition, formation
2. order of battle, battle order
3. disposition (gunn, mor) (Sov term includes CP, OP, main and reserve fire positions and ammunition niches)
~ **в оди́н эшелóн, боевóй** line formation
~ **в предви́дении бóя, похóдный** tactical column, tactical march formation, approach march formation
поря́док ведéния огня́ fire plan, sequence of firing
~ **взаимодéйствия** arrangement for cooperation
~ **вы́садки** disembarkation schedule
~, **глубокó эшелони́рованный боевóй** column formation
~ **движéния** order of march
~ **движéния на мáрше** order of march
~ **дéйствия** technique, mode of operation, sequence of operations
~ **демонтáжа** dismantling procedure
~ **десáнта, похóдный** cruising disposition of ships in amphibious group
~, **дисциплинáрный** schedule of summary punishments
~ **дня** agenda
~ **дóпуска** procedure for admittance
~ **зажигáния** firing order, firing sequence (engine)
~ **закреплéния** sequence of reorganisation
~, **линéйный** linear formation
~ **мáрша** order of march, march order
~, **мáршевый** order of march
~ **обеспéчения** priority list
~ **огня́** nature of fire (gunn)
~ **освещéния цéлей** sequence for target illumination
~ **отбóра** 1. selection system for call-up
2. sequence for selection
~ **подчинённости** chain of command
~ **построéния** order of forming-up
~, **похóдный** march formation
~, **предбоевóй** approach march formation
~ **провéрки** test procedure
~ **произвóдства подрывны́х рабóт** demolition procedure
~ **произвóдства разрушéний** demolition procedure
~ **рабóты** i. procedure
2. firing order, firing sequence
~ **рабóты по рáдио** radio procedure, radio operating procedure
~ **радиообмéна** radio procedure
~ **разбóрки** dismantling procedure
~, **развёрнутый** deployed formation
~, **расчленённый** deployed formation
~ **рóмбом, боевóй** diamond formation
~ **сбóрки** assembly order, assembly procedure

поря́док снабже́ния system of supply
~ **угло́м вперёд** 1. arrow-head formation
2. one-up formation (tk)
3. wedge formation (AFV)
~ **угло́м наза́д, боево́й** inverted wedge formation
~, **уста́вный** procedure laid down by the regulations
~ **учёта и отчётности** accounting procedure
~, **ша́хматный** staggered formation, chessboard formation (tac)
посади́ть (см сажа́ть)
поса́дка 1. landing (ac)
2. embussing, entraining, embarking
3. planting
4. fit, seating
~, **авари́йная** emergency landing
~, **вы́нужденная** forced landing
~, **глуха́я** force fit
~ **на автомоби́ль** embussing
~ **на кора́бль** embarkation
~ **на по́езд** entraining
~ **на самолёт** emplaning
~ **на су́дно** embarkation
~, **подви́жная** clearance fit, running fit (eng)
~ **пу́ли в ги́льзу** seating of bullet in cartridge case
~, **скользя́щая** sliding fit
поса́дочно-деса́нтный air-landed, air-landing
поса́дочный 1. landing
2. embarkation, mounting
3. planting
поса́женный на шпо́нке keyed
посе́в sowing
посёлок village, settlement
посереди́не in the middle, half-way along
посети́ть (см посеща́ть)
посеща́ть/посети́ть to call, to visit, to attend; **посеща́ть/посети́ть курс** to attend a course
посла́нец 1. envoy, messenger
2. man on draft (coll)
послевое́нный post-war
последи́ть (см следи́ть)
после́довательно 1. in series (elec)
2. in succession, in turn
после́довательно-соединённый connected in series
после́довательность 1. consecutiveness, sequence, succession
2. consistency
3. (wave) train, (pulse) train
последова́тельность затя́жки tightening sequence
после́довательность и́мпульсов pulse train (electron)
после́дствие after-effect, consequence
послесвече́ние afterglow, persistence (CRT)
послужи́ть (см служи́ть)
послуша́ние obedience
посме́ртный posthumous
посо́бие 1. allowance, grant, aid (financial)
2. textbook, manual, study aid
~, **возвра́тное** recoverable grant
~, **выходно́е** service gratuity, terminal gratuity
~, **де́нежное** cash allowance, monetary allowance
~, **единовре́менное** 1. one-time grant
2. terminal grant
~, **инвали́дное** disability benefits
~, **нагля́дное** visual aid
~, **учéбное** training aid
посо́бия, нагля́дные visual aids
посо́бник accomplice, accessory before the fact
поспеши́ть (см спеши́ть)
поспосо́бствовать (см спосо́бствовать)
посре́дник umpire (on exercise)
пост post
~ **акусти́ческого обнаруже́ния самолётов** air listening post
~, **боево́й** action station
~, **боево́й информацио́нный** operations room
~, **водоме́рный** water gauge post
~ **возду́шной свя́зи** air force communications post
~, **дальноме́рный** range-finder director (nav)
~, **диспе́тчерский** 1. air traffic control post
2. control room
~ **дистанцио́нного управле́ния** remote control post
~, **дозиметри́ческий** radiac monitoring station
~, **звуково́й** sound ranging post, microphone base

пост, звукометри́ческий sound ranging post
~, зени́тно-артиллери́йский метеорологи́ческий anti-aircraft artillery meteorological post
~, инжене́рный наблюда́тельный engineer observation post
~ инструмента́льной разве́дки sound ranging post
~, контро́льно-распредели́тельный check and distribution point
~, контро́льный control post, check point
~, медици́нский medical station
~, метеорологи́ческий weather station
~ ми́нного наблюде́ния mine watching post
~, наблюда́тельный look-out position, look-out post, observation post, OP
~, отде́льный сторожево́й outlying sentry post
~ охране́ния sentry post
~ ПВО air defence reporting post
~ подслу́шивания listening post
~, пожа́рный fire picquet
~ предупрежде́ния sound ranging advanced post
~ радиолокацио́нной наво́дки radar director
~, радиолокацио́нный radar site, radar station
~ регули́рования check point, control point
~ регули́рования движе́ния traffic control point
~ регули́рования движе́ния, подвижно́й traffic patrol
~ санита́рно-хими́ческой обрабо́тки CW cleansing station
~, светометри́ческий flash-spotting OP (arty)
~, сигна́льный signal box
~, сна́йперский sniper's post
~ управле́ния 1. command post GW 2. controlling point (elec)
~, хими́ческий наблюда́тельный CW observation post
~ хими́ческого наблюде́ния CW observation post
~, центра́льный кома́ндный control centre
поста́вить на предохрани́тель to apply the safety catch
~ (см поставля́ть)
поста́вить (см ста́вить)
поста́вка delivery, supply
поста́вленный на ремо́нт laid up for a repair
поставля́ть/поста́вить to supply
поста́вщик supplier
постано́вка 1. erection, raising 2. putting, placing, setting
~ дымово́й заве́сы smoke-screen laying, smoke-screening
~ зада́ч allocation of tasks
~ зада́чи briefing
~ зада́чи в полёте in-flight briefing (airborne ops)
~ мин mine-laying
~ на боево́й взвод cocking the striker (gren)
~ на дли́тельное хране́ние moth-balling
~ на ме́сто replacing
постановле́ние decree, resolution
~ вое́нного суда́ court-martial order
постоя́нная constant
~, баллисти́ческая ballistic constant
~ вре́мени time constant
~ вре́мени, ма́лая short-time constant
~, гравитацио́нная universal gravitational constant
~ дифракцио́нной решётки grating factor (antenna)
~, навигацио́нная navigation constant, navigational correction (GW)
~ распа́да decay factor, decay constant (nucl)
~ распростране́ния propagation coefficient
~, сосредото́ченная lumped constant (electron)
~, фа́зовая phase-constant, phase counter
пострада́вший victim
~ в бою́ battle casualty
построе́ние 1. forming-up, parade 2. formation 3. construction
~, боево́е combat formation
~, лине́йное linear formation
~ оборо́ны defensive lay-out
~, одноэшело́нное one-echelon formation
~, операти́вное tactical order of battle
~ паралле́льного ве́ера paralleling (arty)
~ полосы́ оборо́ны defensive lay out

построе́ние це́лей, боево́е target formation
~, эшелони́рованное echelon formation
построи́тель 1. plotter (AAFCE) 2. computer (obs)
постро́ить (см стро́ить)
постро́иться (см стро́иться)
постро́йка construction, erection, building
~ доро́г road building
~ мо́ста bridging
~, надво́рная outbuilding
постро́йки, оборони́тельные defensive works, defence works, defences
~, полевы́е оборони́тельные field defences
поступа́тельно-возвра́тный reciprocating
поступа́ть/поступи́ть to act, to do; **поступа́ть/поступи́ть в подчине́ние** to come under the command, to become subordinate to; **поступа́ть/поступи́ть в снабже́ние** to come into service; **поступа́ть/поступи́ть на вое́нную слу́жбу** to enlist, to join up (mil); **поступа́ть/поступи́ть на вооруже́ние** to enter service
поступи́ть (см поступа́ть)
поступле́ние 1. intake (stores) 2. entrance, admission 3. arrival, start 4. enlistment
посту́пок conduct, action, deed
посу́да 1. crockery 2. container
посудомо́йка washing-up machine
посы́лка 1. impulse, pulse 2. parcel 3. premise
~, бесто́ковая space pulse (t/p)
посы́льный runner, messenger, orderly
~, ко́нный mounted orderly
~, пе́ший runner (messenger)
посяга́тельство encroachment
потайно́й 1. secret, hidden 2. countersunk, flush
потенциа́л 1. potential 2. potential voltage (elec)
~ зажига́ния striking voltage, striking potential
~ ионизáции ionization potential
~, ионизацио́нный ionization potential
~ ко́рпуса earth potential
потенциа́л, я́дерный nuclear potential
потенцио́метр potentiometer
~ обра́тной свя́зи feedback potentiometer
поте́ри casualties, losses
~, безвозвра́тные 1. irrecoverable losses 2. equipment beyond economical repair 3. killed and missing personnel
~, боевы́е casualties by enemy action
~, боевы́е санита́рные wounded personnel (Sov term includes shell shock, burns and frost bite cases)
~ в бою́ battle casualties
~ в диэле́ктрике dielectric losses
~ в та́нках tank losses
~, вое́нные casualties, war losses
~, возврати́мые recoverable losses (includes repairable equipment and curable sick and wounded personnel)
~ горю́чего, о́бщие gross leakage (fuel)
~, дифракцио́нные diffraction losses
~, диэлектри́ческие dielectric losses
~ на поглоще́ние absorption losses
~ на рассе́яние scattering losses
~, небоевы́е санита́рные losses through sickness
~, неопра́вданные unjustified losses
~, о́бщие 1. gross leakage (fuel) 2. overall losses
~, оми́ческие resistive losses, ohmic losses (elec)
~ преобразова́ния conversion loss
~, тяжёлые heavy losses, heavy casualties
поте́рна gallery, tunnel (fortifications)
потерпе́ть (см терпе́ть)
поте́ря loss, waste
потеря́ть (см теря́ть)
пото́к 1. flow, current, flux 2. stream
~, адиабати́ческий adiabatic flow
~, возду́шный air blast
~, га́зовый gas flow
~ гру́за material flow
~ движе́ния traffic stream
~, двухме́рный two-dimensional flow
~, дозвуково́й subsonic flow
~, конвекцио́нный convection current
~, магни́тный magnetic flux
~, набега́ющий approach flow
~ нейтро́нов neutron flux
~, сверхзвуково́й supersonic flow
~, светово́й heat flash (nucl)

потóк, тепловóй heat-transfer
~, удéльный тепловóй heat transfer rate
потолóк ceiling (aeron)
~, максимáльный maximum vertical range (AA ord)
~ полёта flight ceiling (ac)
~ стрельбы́ ceiling (AA gunn)
потóчный line, assembly line
потреби́тель consumer, user
потреби́ть (см потребля́ть)
потреблéние consumption
~, сýточное daily consumption
~ тóплива fuel consumption
~, удéльное specific consumption
потребля́емый expendable
потребля́ть/потреби́ть to consume, to use
потрéбности needs, requirements
~ войск, повседнéвные day-to-day requirements
~, коли́чественные quantitative requirements
~, первоочерёдные first priority requirements
~, перспекти́вные плáновые long-term planning requirements
~, плáновые forecast requirements
~, повседнéвные day-to-day requirements
~ текýщего довóльствия day-to-day requirements
потрéбность requirement, requisite, need
~ в боеприпáсах ammunition demand
~ в горю́чем petrol requirement
~ в рабóчей си́ле labour requirements
~, перспекти́вная long-term requirement
~, плáновая estimated requirement
~, текýщая current demand
потрéбный requisite
похвалá commendation
похóд campaign, march
похóдный march, marching, mobile, field, field-type
похóды, альпи́йские mountaineer operations, mountain campaign
пóчва soil
~, тáлая thawed soil
почвозацéп 1. grouser (track) 2. spud (tk track)
пóчести honours
~, боевы́е battle honours
~, вóинские military honours
почи́н initiative
почи́нка repair, repairing, mending
почи́стить (см чи́стить)
пóчта, воéнная forces postal system
~, воéнно-полевáя field post office, FPO
~, полевáя field post office
"Пошёл!" "Jump!"
поэшелóнно by echelons
пóяс belt
~, вéрхний top chord (br)
~, горизонтáльный band of latitude
~, ни́жний bottom chord (br)
~, патрóнный cartridge belt
~ препя́тствий obstacle belt
~, спасáтельный lifebelt
~, стыковóй jointing strip
~, умéренный temperate zone
~, часовóй time zone
поясóк band, rib
~, ведýщий driving band (ammo)
~, центри́рующий centring band (ammo)
поясóк-противоомедни́тель decoppering band
правá, води́тельская driving licence
~ води́теля driving licence
"Правéе!" "Bear Right !"
прáвила rules, regulation
~ безопáсности safety regulations
~, водохóдные rule of the road (on water)
~ движéния rule of the road, movement regulations
~ движéния по автомагистрáльным дорóгам highway regulations
~ затемнéния blackout regulations
~ и нóрмы code of practice
~ пожáрной безопáсности fire prevention rules
~ пóльзования instructions for use
~ радиообмéна radio procedure
~, санитáрно-гигиени́ческие hygiene and sanitary regulations
~ стрельбы́ firing drills
~ тéхники безопáсности safety precautions, regulations, accident-prevention rules
~ ýличного движéния traffic regulations
~ хранéния storage precautions
~ эксплуатáции operating rules
прави́ло handspike, hand lever
прáвило rule, regulation
~, эмпири́ческое rule of thumb
прави́тельство government
прáво authority, right, law

пра́во, вое́нное military law
~, вое́нно-уголо́вное military penal law
~ прое́зда right of way
~, уголо́вное criminal law
правопоря́док law and order
правофланго́вый right-flank, right-wing
пра́вый 1. right(-hand)
2. right-wing (political)
3. right, correct
4. righteous, just
пра́ктика, боева́я battle experience, combat experience
практика́нт trainee
пра́ктикум practical training
пра́порщик warrant officer
пра́чечная laundry
пребыва́ние в плену́ period spent in captivity
~ на посту́ on sentry-go
превосхо́дство superiority
~ в во́здухе air superiority
~ в си́лах superiority of forces
~ в та́нках tank superiority
~, огнево́е fire superiority
~, подавля́ющее overwhelming superiority
~, я́дерное nuclear superiority
преврати́ть (см превраща́ть)
превраща́ть/преврати́ть to transform, to convert
превраще́ние transformation, conversion
превыше́ние 1. exceeding, excess
2. height difference (arty)
~, взаи́мное difference in height between two points (topo)
~, относи́тельное relative height
~ траекто́рии height of trajectory (ball)
~ траекто́рии над гре́бнем укры́тия, наиме́ньшее minimum crest clearance
прегра́да obstacle, barrier
~, во́дная water obstacle, wet gap
~, есте́ственная natural obstacle
~, огнева́я curtain of fire
прегради́ть (см прегражда́ть)
прегражда́ть/прегради́ть to obstruct, to block
преда́тельство treason
предбоево́й 1. approach mark
2. pre-combat
предви́дение anticipation, foresight
предвое́нный pre-war
предго́рье foot-hill
преддетона́ция predetonation, pre-ignition (ICE)
преде́л bound, limit
~, взрывоопа́сный limit of sensitiveness to explode
~ ви́димости limit of visibility
~ выно́сливости fatigue limit (metall)
~, досяга́емости extreme range, limit of maximum range
~ обнаруже́ния detection threshold
~ огневы́х возмо́жностей limit of effective range
~ пропорциона́льности elastic limit (metall)
~ про́чности tensile limit (metall)
~ про́чности при сжа́тии compressive strength (metall)
~ ско́рости limiting speed
~ сцепле́ния limit of adhesion
~ теку́чести 1. proof stress
2. yield point (metall)
~ упру́гого сопротивле́ния ствола́ maximum safe barrel pressure (ord)
~ упру́гости elastic limit (metall)
~ уста́лости fatigue limit (metall)
~ утомля́емости fatigue limit (metall)
преде́лы, габари́тные clearance limits
предизбира́тель preselector
предизбира́тельность preselection
предка́мера pre-combustion chamber
предлага́ть/предложи́ть to offer, to propose, to suggest; **предлага́ть/предложи́ть бой** to offer battle
предложи́ть (см предлога́ть)
предме́т 1. object, article, item
2. subject, theme
3. feature (topo, radar)
~, есте́ственный ме́стный natural terrain feature
~, заме́тный landmark
~, испо́льзуемый в оборо́не, есте́ственный ме́стный natural defensive feature
~, контингенти́рованный rationed item
~, ме́стный local feature (topo, radar)
~ снабже́ния, дефици́тный item in short supply
~ снабже́ния, нета́бельный unauthorised item
~ снабже́ния, нехо́дкий slow-moving supply item
~ снабже́ния, хо́дкий fast moving items, fast moving stores

предме́ты, ме́стные local features (topo, radar)
предоста́вить (см предоставля́ть)
предоставле́ние allocation, assignment
предоставля́ть/предоста́вить to allot, to make available, to allow, to grant
предостереже́ние, штормово́е gale warning
предосторо́жность safeguard, precaution
предотвраще́ние prevention, averting, warding off
~ ава́рий accident prevention
предохрани́тель 1. safety catch, safety switch
2. electrical fuse, cut out
~, автомати́ческий 1. grip safety catch (SA)
2. automatic safety device
~ дистанцио́нного уда́рника time pellet safety casing (fuse)
~, ла́пчатый stirrup spring (fuse)
~ му́шки foresight protector
~, неавтомати́ческий manual safety catch
~, патро́нный cartridge fuse (elec)
~, пла́вкий electrical fuse
~, про́волочный пла́вкий wire fuse (elec)
~ спусково́го механи́зма trigger lock
"Предохрани́тель Ставь!" "Apply Safety Catches!"
предохрани́тель уда́рника stirrup spring (T and P fuse)
~, центробе́жный centrifugal safety lock
~, эле́ктро-хими́ческий electrochemical safety element
предписа́ние written order, written instruction
~, командиро́вочное travel order
предписа́ния, погру́зочные loading instructions
предпо́лье 1. security zone, forward defence area
2. no-man's land
предпосы́лка pre-requisite
предпринима́ть/предприня́ть to undertake; **предпринима́ть/предприня́ть ата́ку** to mount an attack; **предпринима́ть/предприня́ть наступле́ние** to mount an assault, to mount an offensive
предприня́ть (см предпринима́ть)
предприя́тие, стациона́рное ремо́нтное static repair workshop
предпрыжко́вая pre-jump practice (parachute)
предсказа́ние prediction, forecast
представи́тель representative
~, вое́нный military representative
предста́вить (см представля́ть)
представля́ть/предста́вить to present, to offer; **представля́ть/предста́вить зая́вку** to submit an inventory
предупреди́тельный preventive, safety, warning
предупрежде́ние 1. warning, notice
2. prevention
~, заблаговре́менное ample warning, timely warning
~, штормово́е gale warning
предше́ственник precursor, predecessor, fore-runner
предъионизáтор keep-alive electrode
преждевре́менный premature
преиму́щество advantage
прекрати́ть (см прекраща́ть)
прекраща́ть/прекрати́ть to stop, to cease, to discontinue; **прекраща́ть/прекрати́ть ого́нь** to stop firing, to cease firing; **прекраща́ть/прерати́ть сраже́ние** to break off the battle
прекраще́ние огня́ cessation of fire, cease-fire
~ слеже́ния за це́лью track rejection (radar)
преломле́ние refraction
премирова́ние granting of cash awards, granting of bonus payments
преобразова́ние transformation, re-organisation, conversion
~ высоты́ altitude conversion
~, группово́е group translation (sigs)
~, индивидуа́льное channel translation (comms)
~, интегра́льное integral transform (math)
~ координа́т co-ordinate conversion
~, сверхгруппово́е super-group translation (teleph, teleg, radar)
~ све́та, когере́нтное coherent light conversion
преобразова́тель converter (elec)
~ а́зимута azimuth data converter
~, анало́го-цифрово́й analog-to-digital converter

преобразова́тель, баллисти́ческий ballistic converter
~, вибрацио́нный vibrator converter
~ высоты́ altitude converter
~ да́льности range converter
~ изображе́ния, инфракра́сный infrared image convertor
~ импеда́нса impedance transformer
~, каска́дный cascade converter
~, масшта́бный scale converter
~, одноя́корный rotary converter
~, умфо́рмерный rotary converter (elec)
~, фотоэлектри́ческий photo-emissive converter, iconoscope, storage-type TV camera tube
~, цифроанало́говый digital-analog converter
~ частоты́ frequency converter
~, электро́нно-опти́ческий image converter, photo-electric conversion unit
~, электроопти́ческий electrical image converter
преобразова́ть (см преобразо́вывать)
преобразо́вывать/преобразова́ть to reform, to reorganize, to process (data), to transform, to convert
преодолева́ть/преодоле́ть to cross, to negotiate, to surmount, to overcome; **преодолева́ть/преодоле́ть вброд** to wade, to ford; **преодолева́ть/преодоле́ть ми́нное по́ле** to cross a minefield; **преодолева́ть/преодоле́ть препя́тствие** to negotiate an obstacle
преодоле́ние passage, crossing, negotiation
~ во́дной прегра́ды crossing a water obstacle, negotiation of a water obstacle
~ ми́нного по́ля minefield breaching, minefield crossing
~ препя́тствий obstacle negotiation
преодоле́ть (см преодолева́ть)
препара́т preparation
~, противорадиацио́нный anti-radiation drug, anti-radiation pill
~, сульфами́дный sulphonamide
препара́ты, сульфами́дные sulphonamides
~, сульфанилами́дные sulphanilamide drugs
преподава́тель instructor
препя́тствие obstacle
~, берегово́е beach obstacle
~, вертика́льное vertical obstacle
препя́тствие, во́дное water obstacle
~, есте́ственное natural obstacle
~, есте́ственное противота́нковое natural anti-tank obstacle
~, иску́сственное artificial obstacle
~, иску́сственное противота́нковое artificial anti-tank obstacle
~, ло́жное dummy obstacle
~, малозаме́тное 1. low wire entanglement
2. low trip-wire obstacle
~, мини́рованное mined obstacle
~, ми́нное mine obstacle
~ на доро́ге road obstruction
~, перено́сное movable obstacle
~, пла́вающее floating obstacle
~, подво́дное underwater obstacle
~, про́волочное wire obstacle
~, противопехо́тное infantry obstacle
~, противота́нковое anti-tank obstacle
~, такти́ческое tactical obstacle
~, фортификацио́нное field fortifications
~, электризо́ванное electrified obstacle
препя́тствовать to hinder, to impede, to obstruct
прерва́ть (см прерыва́ть)
прерыва́тель contact-breaker, circuit breaker, interrupter, chopper
~, конта́ктный contact breaker
~ постоя́нного то́ка d.c. chopper (elec)
~ пото́ка spoiler (GW)
прерыва́тель-распредели́тель distributor and contact-breaker unit
прерыва́ть/прерва́ть to interrupt, to break off
преры́вистый interrupted, intermittent, non-continuous
пресле́дование 1. pursuit
2. persecution, victimization
~, паралле́льное parallel pursuit (tac), pursuit in parallel formation
~, фронта́льное direct pursuit
пресле́довать to pursue, to chase
пресс, ки́пный baler, bundling machine
~, сма́зочный grease gun
пре́сс-маслёнка grease nipple
преступле́ние crime
~, вое́нное war crime
~, во́инское military offence (mil law)
~, госуда́рственное crime against the state
престу́пник criminal

престу́пник, вое́нный war criminal
престу́пность criminal nature, criminality
прете́нзия complaint, claim
прецессия precession
при нали́чии subject to availability
прибега́ть/прибе́гнуть to resort (to), to have recourse (to), to fall back (upon); **прибега́ть/прибе́гнуть к ору́жию** to resort to arms
прибе́гнуть (см прибега́ть)
приближа́ться/прибли́зиться to approach, to approximate
приближе́ние approach
прибли́зиться (см приближа́ться)
прибо́йник rammer (ord)
прибо́р 1. instrument, device, apparatus
2. set (of articles)
3. furniture (rifle)
~ **автомати́ческой наво́дки** automatic gun laying equipment
~, **азимута́льный** bearing drive, bearing control, bearing indicator (arty)
~, **артиллери́йский** artillery instrument
~, **букси́рный** towing pintle
~ **вертика́льного наведе́ния, принима́ющий** elevation receiver (AA gunn, FCE)
~, **визи́рный** view-finder (photo)
~, **выливно́й** spraying device
~, **выливно́й авиацио́нный** aircraft spray (CW)
~, **газосигнализа́торный** gas alarm
~, **геодези́ческий** surveying instrument
~, **гироскопи́ческий** gyroscopic instrument
~ **горизонта́льного наведе́ния, принима́ющий** bearing receiver (AA gunn)
~ **дальноме́рного гра́фика** mean range calculator (gunn)
~, **дегазацио́нно-душево́й** decontamination shower unit
~, **дегазацио́нный** decontamination apparatus
~ **дистанцио́нного управле́ния огнём** remote-control firing device
~ **для автомати́ческой сма́зки** automatic oiler
~ **для дегаза́ции ме́стности** area CW decontaminating equipment
~ **для извлече́ния стре́ляных гильз** clearing plug (SA)
прибо́р для ночно́й стрельбы́ night-firing device
~ **для осмо́тра кана́ла ствола́, опти́ческий** bore inspection telescope
~ **для прице́ливания в темноте́** night sight
~ **для прове́рки зажига́ния** ignition tester
~ **для прове́рки нали́чия паро́в алкого́ля в дыха́нии води́теля** breathaliser
~ **для прове́рки свече́й зажига́ния** spark-plug tester
~ **для прове́рки устано́вки прице́ла** sight testing gauge
~ **для слеже́ния, ла́зерный** tracking laser
~, **дозиметри́ческий** radiation-monitoring instrument
~, **дозиро́вочный** dose-measuring device (fluids)
~, **дымово́й** smoke generator
~, **ёмкостный реша́ющий** capacitance resolver (electron)
~ **"за́йчик"** spotlight projector
~ **заме́ра ка́меры** chamber wear measuring device
~, **измери́тельный** measuring instrument
~, **кислоро́дный изоли́рующий** oxygen breathing set
~, **кома́ндный** 1. command box (wire-controlled GW)
2. control device
~, **контро́льный** monitoring device
~, **кре́шерный** interior pressure gauge, crusher gauge
~, **магнитоэлектри́ческий** moving coil instrument
~, **метеорологи́ческий измери́тельный** meteorological measuring instrument
~, **метеорологи́ческий** meteorological instrument
~ **наблюде́ния** vision device
~ **наблюде́ния команди́ра, перископи́ческий** commander's periscope
~ **наблюде́ния наво́дчика** gunner's periscope
~ **наблюде́ния наво́дчика, телескопи́ческий** gunner's telescope
~ **наблюде́ния, назе́мный** ground surveillance device
~ **наведе́ния** guidance unit (msl)

прибóр, навигациóнный navigation instrument
~ навóдчика, смотровóй перископи́ческий gunner's periscope
~ навóдчика, смотровóй телескопи́ческий gunner's telescope
~ ночнóго ви́дения infra-red device, night vision device
~, обогревáтельный heating unit
~, огнетуши́тельный fire extinguisher
~ опти́ческого сопровождéния optical tracking device
~, перенóсный portable device
~, перископи́ческий indirect vision device
~, покáзывающий indicating instrument
~, полупроводникóвый semi-conductor device
~, предохрани́тельный safety device
~, принимáющий receiver, data receiver
~ провéрки форсýнок injector testing device
~, провéрочный test set
~, рáнцевый дегазациóнный pack decontaminant sprayer (CW)
~, регистри́рующий recording device, recorder
~ с зáписью на лéнте, самопи́шущий tape recorder
~, самопи́шущий recording instrument, graphic instrument
~, светосигнáльный signal lamp, heliograph equipment
~, светочувстви́тельный photo-sensitive device
~, смотровóй observation instrument
~ снаряжéния arming device
~ со счётчиком Гéйгера, дозиметри́ческий Geiger counter
~, стрелкóво-тренирóвочный musketry training device
~, счётно-решáющий computer
~, счётный calculating device
~, тепловизиóнный thermal imaging device
~, тóчный precision instrument
~ тренирóвки расчёта ПУÁЗО AAFCE simulator
~, тя́говый towing pintle
~, угломéрный azimuth gear (mor)
~ управлéния control instrument
прибóр управлéния артиллери́йским зени́тным огнём (ПУÁЗО) anti-aircraft fire control equipment, AAFCE
~ управлéния артиллери́йским огнём gun data computer (arty)
~ управлéния огнём 1. fire control instrument, FCI, fire control equipment, FCE (arty)
2. gun control equipment, GCE (tk)
~, учéбный дозиметри́ческий radiac training device
~, фоточувстви́тельный photo-sensitive device
~ хими́ческой развéдки chemical agent detector kit
~ хими́ческой развéдки, войскóвóй CW field reconnaissance instrument
~ целеуказáния, принимáющий target indicator receiver (FCE)
~ центрáльной навóдки, даю́щий director-setting transmitter
~, центрáльный (ПУÁЗО) predictor (AAFCE)
~, электровáкуумный electric vacuum device (lighting), vacuum tube (electron)
~, электроизмери́тельный electrical measuring instrument
~, электрóнный electronic device, electronic instrument
~, эталóнный calibration instrument
прибóр-индикáтор áзимута bearing indicator (arty)
прибори́ст 1. operator
2. instrument mechanic
прибóр-предупреди́тель advanced post control unit (sound ranging)
прибóры, контрóльно-измери́тельные testing and measuring instruments
~, опти́ческие optical instruments
прибрéжный coastal, riverside
"При́был по Вы́зову!" "Reporting!" (set phrase for reporting arrival when called for by a superior)
привáл halt
~, большóй long halt on the march (two-four hours)
~, мáлый short halt on the march (20-30 minutes)
привáренный welded on
привáрка welding-on
привáрок staple ration
приведéние в дéйствие starting, actuating, triggering

приведе́ние к норма́льному бо́ю zeroing (wpns)
~ **ми́ны в безопа́сное состоя́ние** disarming a mine
~ **ми́ны в боево́е положе́ние** arming a mine
приве́рженцы большо́й огнево́й мо́щи adherents of heavy firepower school (tank theory)
~ **тяжёлой брони́** adherents of heavy armour school (tank theory)
привести́ (см приводи́ть)
приве́тствие, во́инское salute
~, **тройно́е** three cheers
приви́вка inoculation, vaccination, injection
~, **предохрани́тельная** prophylactic vaccination, preventive inoculation, prophylactic inoculation
~, **противостолбня́чная** anti-tetanus injection
~, **профилакти́ческая** immunization, prophylactic inoculation, preventive inoculation
привлека́ть/привле́чь to attract, to call upon, to enlist (the services of)
привле́чь (см привлека́ть)
приво́д 1. drive
2. switchgear
~ **агрега́тов** accessory drive
~, **аккумуля́торный** accumulator drive
~ **вентиля́тора** fan drive
~, **вспомога́тельный** accessory drive, auxiliary drive
~, **гидравли́ческий** hydraulic drive
~, **гла́вный** main drive
~, **двухступе́нчатый** double-step drive
~ **диста́нции** range transmitting gear (AA sight)
~, **дистанцио́нный** remote drive
~, **дифференциа́льный** differential (gear) drive
~, **кла́панный** valve actuating gear
~ **колёс** wheel drive
~, **кулачко́вый** cam gear
~ **механи́зма поворо́та ба́шни** power traverse drive
~, **механи́ческий** mechanical drive
~, **механи́ческий тормозно́й** mechanical brake linkage
~ **на две о́си** four-wheel drive
~ **на за́дние колёса** rear-wheel drive
~ **на пере́дние колёса** front-wheel drive
приво́д на четы́ре колеса́ four-wheel drive
~ **от кулачко́вого ва́ла** camshaft drive
~ **от маховика́** flywheel drive
~ **от распредели́тельного ва́ла** camshaft drive
~, **пневмати́ческий** air-operated drive
~ **постоя́нных оборо́тов** constant-speed drive
~, **реверси́вный** reversible drive
~, **ремённый** belt drive
~, **ручно́й** manual operation
~ **с двумя́ веду́щими о́сями** four-wheel drive
~, **силово́й** mechanical drive
~, **силово́й автомати́ческий** servo mechanism
~ **ско́рости** target speed transmitting gear (AA sight)
~, **следя́щий силово́й** servomechanism, power-operated follow-up control system
~ **спидо́метра** speedometer drive
~ **тахо́метра** tachometer drive
~, **тормозно́й** brake linkage
~, **трансмиссио́нный** transmission drive
~ **управле́ния коро́бкой переда́ч** gearbox control linkage
~ **управле́ния тормоза́ми** brake controls, brake control linkage
приводи́ть/привести́ to lead, to bring, to cause; **приводи́ть/привести́ в боеву́ю гото́вность** to bring to combat readiness; **приводи́ть/привести́ в де́йствие** to actuate, to set in motion; **приводи́ть/привести́ в него́дность** to put out of commission; **приводи́ть/привести́ взрыва́тель в де́йствие** to trigger a fuse; **приводи́ть/привести́ к прися́ге** to attest; **приводи́ть/привести́ ми́ну в безопа́сное состоя́ние установ́кой предохрани́теля** to disarm a mine; **приводи́ть/привести́ ми́ну в боево́е положе́ние** to arm a mine; **приводи́ть/привести́ ору́дие к норма́льному бо́ю** to zero a gun
приводно́й drive, driving, actuating (tech)
привяза́ть (см привя́зывать)
привя́зка survey, surveying
~ **по ка́рте** grid-sheet survey
~, **топографи́ческая** 1. topographical survey, topographical surveying
2. fixation (surv)

привя́зчик, топографи́ческий topographical survey vehicle
привя́зывать/привяза́ть to fix by survey, to tie in, to control by survey
пригна́ть (см пригоня́ть)
пригово́р sentence, verdict
приго́дность suitability, fitness
~, боева́я operational suitability
~, такти́ческая tactical suitability
~, физи́ческая physical fitness
приго́дный fit, applicable, serviceable
приго́нка 1. fitting, adjustment, fit
2. bedding-in (mech)
~, пло́тная close fitting
~ снаряже́ния fit of equipment
пригоня́ть/пригна́ть to fit, to fit on, to adjust, to drive home, to bring home (trans)
пригора́ние sticking, burning, scorching
~ поршневы́х коле́ц seizing of piston rings
"Пригото́виться!" "get ready to jump!" (para)
пригранй́чный frontier, border
придава́ть/прида́ть to place under the command of, to attach, to impart, to communicate
прида́ние 1. attachment
2. giving, imparting, communication
~ водонепроница́емости water-proofing
при́данный attached, placed under command of
прида́ток attachment
прида́ть (см придава́ть)
прида́ча placing under command
"Приём!" "Over!" (R/T procedure)
приём 1. movement, position (drill)
2. reception, receiving
3. method, procedure
4. admission (hospital etc)
~, амбула́торный sick parade
~, дежу́рный listening watch (comms)
~, машинопи́сный taking down of radio messages direct on to typewriter
~ на слух C.W. reception, morse reception (by pers)
~, однопра́вильный unilateral reception (DF)
~ по разделе́ниям drill by numbers
~, руже́йный rifle drill, arms drill, rifle exercise
приём с промежу́точной пода́чей то́плива double de-clutching (down)
~, сдво́енный diversity reception (comms)
~, слухово́й C.W. reception, morse reception
~, строево́й drill movement
приеми́стость pick up (of an engine)
приёмка receipt, acceptance, taking on
~, техни́ческая acceptance inspection
приёмник 1. radio receiver
2. feed block
3. ammunition tray
~, батаре́йный battery-operated radio receiver
~ вертика́льной наво́дки elevation receiver (gunn)
~, войсково́й recruit reception centre
~, вре́мя-и́мпульсный pulse-time electrical receiver
~, всеволново́й all-wave receiver
~, гетероди́нный heterodyne receiver
~ горизонта́льной наво́дки bearing receiver (gunn)
~, дете́кторный crystal receiver
~ излуче́ния, дио́дный diode detector
~, инфракра́сный infra-red receiver
~, ква́нтовый quantum detector
~ кома́нд command receiver
~ кома́ндных сигна́лов command receiver
~ магази́на magazine feed-piece opening
~ на транзи́сторах transistorised radio
~ опти́ческого диапазо́на optical receiver
~, панора́мный panoramic receiver
~, пеленга́торный direction-finding receiver
~, по́исковый radar search receiver (ECM)
~, портати́вный radio receiver
~ прямо́го ви́дения, инфракра́сный direct vision infra-red receiver
~ прямо́го усиле́ния straight receiver (radio)
~ пулемёта MG feed block body
~, радиолокацио́нный radar receiver
~ систе́мы наведе́ния guidance receiver (GW)
~ систе́мы самонаведе́ния homing receiver (GW)

приёмник, телевизио́нный television receiver, television set
~ **теплово́го излуче́ния** infra-red detector, IR detector, thermal detector
~, **терми́сторный** thermistor detector
~, **фотоэлектри́ческий** photoelectric detector
приёмник-демодуля́тор receiver-demodulator
приёмник-дешифра́тор receiver-decoder
приёмоотве́тчик, радиолокацио́нный transponder
приёмопереда́тчик transceiver transceiver
приём-сда́ча handover-takeover
приёмы произво́дства разруше́ний demolition techniques
~ **с ору́жием** arms drill
~ **стрельбы́** firing drill
~, **техни́ческие** techniques
прижа́тый 1. pinned-down 2. low-altitude
прижа́ть (см прижима́ть)
прижима́ть/прижа́ть to press, to clasp; **прижима́ть/прижа́ть к земле́** to pin down (tac)
призва́ть (см призыва́ть)
приземле́ние 1. landing (ac or paratroops) 2. dropping flat (of troops)
~, **то́чное** spot landing (ac)
приземли́ться (см приземля́ться)
приземля́ться/приземли́ться to land, to touch down
при́зма prism
~, **головна́я** head prism
~ **До́ве** Dove prism
~, **измери́тельная** wedge prism
~ **коллима́тора** collimator prism
~, **крышеобра́зная** Amici prism, roof prism (optics)
~, **обора́чивающая** erecting prism (optics)
~, **отража́тельная** reflecting prism
~, **повора́чивающая** deflecting prism
~, **прямоуго́льная** right-angle prism
~, **пятиуго́льная** pentaprism
~, **ромби́ческая** rhomboidal prism
~, **смотрова́я** viewing prism
~ **смотрово́го прибо́ра** direct vision block
при́зма-клин wedge prism
при́зма-отража́тель reflecting prism
при́зменный prismatic, prism
признава́ть/призна́ть to recognise, to acknowledge, to pronounce
при́знаки, вне́шние external characteristics
~, **демаски́рующие** revealing signs, give-away signs, identification features
призна́ние, факти́ческое factual recognition, recognition de facto
призна́ть него́дным к испо́льзованию to sentence as unserviceable, to condemn as unserviceable
~ **(см признава́ть)**
призы́в на вое́нную слу́жбу conscription, call-up
призыва́ть/призва́ть to call, to summon, to urge; **призыва́ть/призва́ть к ору́жию** to call to arms; **призыва́ть/призва́ть на вое́нную слу́жбу** to conscript, to call up
призывни́к conscript
прика́з order, command
~, **боево́й** operation order
~ **на возду́шную перево́зку** air movement order
~ **на перево́зку** traffic order, movement order
~ **на произво́дство взры́ва** demolition order
~ **на произво́дство подрывны́х рабо́т** demolition order
~ **на произво́дство разруше́ний** demolition order
~, **операти́вный** operation order
~, **пи́сьменный** written order
~ **по строево́й ча́сти** general order
~ **по ты́лу** administrative order
~ **по ча́сти** unit order
~, **секре́тный** classified order
~, **строево́й** routine order
~, **у́стный** verbal order
приказа́ние order, command, instruction
~ **по свя́зи** signal operation order
~, **слове́сное** verbal order, verbal command
приказа́ть (см прика́зывать)
прика́з-инстру́кция, постоя́нный standing order
прика́зывать/приказа́ть to order, to command, to give orders
прикла́д rifle butt, rifle stock
~, **винто́вочный** butt end, rifle butt
~, **откидно́й** collapsible butt
~, **отъёмный** removable stock (SA)

прикла́дка holding (SA)
прико́лыш tent pin
~, пала́точный tent peg
прикомандирова́ние attachment
прикомандиро́ванный attached
прикомандирова́ть to attach (perf only)
прикрепле́ние attachment, fastening
прикрыва́ть/прикры́ть to cover, to protect, to screen; **прикрыва́ть/прикры́ть огнём** to cover by fire, to provide covering fire; **прикрыва́ть/прикры́ть отхо́д** to cover the withdrawal; **прикрыва́ть/прикры́ть продвиже́ние** to cover the advance
прикры́тие cover, covering action, covering forces
~ авиа́цией air cover
~, авиацио́нное air cover, air umbrella
~, артиллери́йское artillery cover
~, боево́е covering force
~, есте́ственное natural cover
~, зени́тное anti-aircraft cover
~, истреби́тельное fighter cover
~ истреби́тельной авиа́цией fighter cover
~, непосре́дственное 1. immediate cover 2. cover using materials at hand
~ огнём fire cover
~ огнём артилле́рии artillery cover
~, пехо́тное infantry escort, infantry covering force
~, щитово́е gun shield
прикры́ть (см прикрыва́ть)
прилега́ющий adjoining, adjacent, neighbouring
прили́в 1. lug, boss 2. flood tide, incoming tide
~ ствола́ barrel lug
~, ца́пфенный trunnion boss
прили́во-отли́вный tidal
прилипа́ние adhesion
приложе́ние appendix, enclosure, annexure
~ по ты́лу logistics annexure
"приложи́сь!" "Take Aim!"
прима́нка decoy, bait
примене́ние use, application
~, боево́е tactical use
~ к ме́стности use of ground
~ ХБР-ору́жия, боево́е NBC warfare operations
~ я́дерного ору́жия, неограни́ченное unrestricted use of nuclear weapons
примени́ть (см применя́ть)
применя́ть/примени́ть to apply, to employ, to use; **применя́ть ору́жие** to use arms (imp only)
приме́рный apparent, approximate
при́месь, антикоррозио́нная rust inhibitor
"Примкну́ть Штык!" "Fix Bayonets!"
примо́рский coastal, maritime
принадле́жности, посте́льные bedding
~, такела́жные tackle
принадле́жность 1. accessories 2. vehicle kit 3. belonging (to), membership (in)
~, оруди́йная gun accessories
принима́ть/приня́ть to take, to accept, to receive; **принима́ть/приня́ть боево́й поря́док** to deploy for action, to deploy into battle formation, to adopt battle formation; **принима́ть/приня́ть на вооруже́ние** to take into service, to adopt (mil); **принима́ть/приня́ть поря́док** to adopt a formation; **принима́ть/приня́ть пост** to take over a post; **принима́ть/приня́ть реше́ние** to make a decision, to make a plan; **принима́ть/приня́ть строеву́ю сто́йку** to come to attention
принужде́ние force, coercion
при́нцип principle
~ боево́го примене́ния мин mine policy
приня́ть (см принима́ть)
приобрести́ (см преобрета́ть)
приобрета́ть/приобрести́ to acquire
приобрете́ние acquisition
припа́сы stores, supplies, munitions
~ и запа́сы, вое́нные munitions
~, огнестре́льные ammunition
~, съестны́е food, rations
припи́лка filing
припо́й solder
~, лату́нный brass brazing alloy
~, мя́гкий soft solder
~, оловя́нно-свинцо́вый lead-tin solder
~, сере́бряный silver solder
~, твёрдый brazing metal, hard solder
при́пуск allowance, surplus
прираба́тывать/прирабо́тать to run in
прирабо́танный run-in, worn-in
прирабо́тать (см прираба́тывать)
прирабо́тка running-in (mech)
~, горя́чая running in (mech)
приса́дка additive

присáдка, антикоррозиóнная anti-corrosion additive, anti-rust agent
~, антиокислúтельная anti-oxidising agent
~ к мáслу oil additive
~ к ракéтному тóпливу propellant additive (msl)
~, мóющая detergent additive, dispersing additive
~ прóтив осадкообразовáния anti-deposit additive, anti-sludging agent
~, противоизнóсная wear-preventive additive, anti-wear additive
~, противокоррозиóнная anti-corrosion additive, anti-rust agent
~, противообледенúтельная anti-icing additive
присáсывание choking (engine)
присвáивать/присвóить to award (decoration etc), to grant (rank); **присвáивать/присвóить звáние** to promote
присвоéние 1. award
2. appropriation
~ воéнного имýщества stealing public property, misappropriation of service property
~ звáния promotion
~ звáния, внеочереднóе accelerated promotion
~ звáния, вы́борочное promotion by selection, selective promotion
~ звáния, досрóчное accelerated promotion
присвóить (см присвáивать)
присоединúть (см присоединя́ть)
присоединя́ть/присоединúть to join, to connect, to annexe (territory)
приспособлéние attachment, gear, equipment, device
~, блокирóвочное interlock (elec)
~, блóковое block and tackle
~, буксúрное towing attachment, towing gear
~, взрывнóе firing gear
~, визúрное view-finder, sighting device
~ для взрывáния actuating mechanism, explosive actuating device
~ для переéзда машúной глубóких брóдов schnorkel device
~ для провéрки углóв установки передних колёс wheel tracking alignment gauge
приспособлéние для разминúрования mine-clearing device
~ для снаряжéния металлúческих лент link-loading machine (ammo)
~ для стрельбы́ холосты́ми боеприпáсами blank firing attachment
~ для установки aligning device
~, запáльное firing device, ignition device, priming assembly
~, измерúтельное metering device
~, нагля́дное visual aid
~, нажúмное pressure piece
~, натяжнóе 1. track tensioner (tracked vehicle, tk)
2. tensioning device
~ неизвлекáемости anti-handling device, activating device activator
~, ночнóе прицéльное night sight
~, прицéльное sights, aiming equipment, sight gear
~, смáзывающее lubricator
~, фиксúрующее jig
~, холостóе blank firing attachment
~, центрúрующее centring device
приспособлéния, перегрýзочные trans-shipment facilities
приспосóбленность для перевóзки по вóздуху air transportability
приспосóбленный adapted
~ для перевóзки по вóздуху airportable, air transportable
приспущённый at half-mast (flag)
приставка attachment
~, ретрансляциóнная re-broadcast apparatus, relay attachment
~, тонáльная voice frequency apparatus
прúстань quay, pier, landing stage
~, плавýчая floating landing stage
~ снабжéния navigation head
пристёгиваться/пристегнýться to strap oneself in (to an aircraft)
пристегнýться (см пристёгиваться)
пристрéливать/пристрелúть to shoot down, to kill
to range, to register, to zero (SA gunn)
пристрелúть (см пристрéливать)
пристрéлка ranging, registration
~ батарéй, одноврéменная collective ranging
~ воздýшного репéра air burst ranging
~ дáльности ranging
~ зáлпами salvo ranging

пристрéлка, контрóльная check fire, verification fire
~ **направлéния** shooting in for line, range for line
~ **очередны́ми зáлпами** ranging by single step spotting
~ **по измéренным отклонéниям** ranging by cross observation, short base registration
~ **по наблюдéнию знáков разры́вов** ranging to establish a bracket
~ **по ненаблюдáемым цéлям** blind fire
~ **по планшéту** ranging from the target map
~ **по репéру** registration fire
~ **приближéнием к цéли** creeping method of ranging
~ **репéра** registration fire
~ **с дальномéром** ranging by range-finder
~ **устýпом** laddering (gunn) (nav)
пристрéлочный spotting, ranging
пристреля́ть (см пристрéливать)
присýтствовать to attend, to be present
прися́га oath
~, **воéнная** oath of allegiance, military oath
~, **служéбная** oath of allegiance
притаи́ться to lurk, to hide, to lie low (perf only)
притерéть (см притирáть)
притирáть/притерéть to grind in (valves)
прити́рка grinding in (valves), lapping
притóк tributary
притормози́ть to brake slightly
приты́к butt joint
притяжéние, гравитациóнное gravitational attraction
приучáть/приучи́ть to train, to break in
приучи́ть (см приучáть)
прихвáт jam, jamming, seizure
прихвáтывание поршневы́х колéц seizure of piston rings
~ **тóрмоза** brake drag
прихóд 1. receipts, income 2. arrival
"Прицéл!" "Range...!" (fire order)
прицéл 1. sight 2. backsight (rifle) 3. sight setting
~, **автомати́ческий** 1. automatic sight 2. course and speed sight
прицéл, автомати́ческий зени́тный course and speed sight (ord)
~, **боковóй** side sights (training aid)
~, **боковóй телескопи́ческий** side-mounted telescope
~, **вéкторный** vector sight
~, **винтóвочный** rifle sight
~, **винтóвочный перископи́ческий** sniperscope
~, **выдвижнóй** bar sight
~, **гироскопи́ческий** gyro (gun)sight
~ **двухкрáтного увеличéния** sight with a magnification factor of two
~, **диоптри́ческий** aperture sight, peep sight
~, **дистанциóнно-управля́емый** centrally controlled sight (AD gunn)
~ **для ведéния огня́ прямóй навóдкой, тáнковый** tank direct fire sight
~ **для мáлых дистáнций, щиткóвый** short-range leaf sight
~ **для прямóй навóдки, опти́ческий** telescope sight
~ **для стрельбы́ прямóй навóдкой** direct-fire sight
~, **дневнóй** day sight
~ **дня** fired correction of the moment (ball)
~, **дуговóй** arc sight
~, **зави́симый** sight mounted on recoiling parts
~, **зени́тный** anti-aircraft sight
~, **зени́тный раккýрсный** predictor sight (AA gunn)
~, **инфракрáсный** infra-red sight
~, **исхóдный** initial elevation (gunn)
~, **колéнчатый телескопи́ческий** elbow-type telescopic sight
~, **коллимáторный** collimator sight
~, **колóдочный** ramp backsight
~, **кольцевóй** ring sight, annular sight
~, **кольцевóй раккýрсный** cartwheel sight (AA ord)
~, **лáзерный** laser sight
~, **лéвый** trainer's sight
~, **назéмный** ground target sight
~, **наимéньший** crest angle (Brit), minimum elevation (US) (gunn)
~, **незави́симый** non-recoiling sight
~, **нормализóванный** standard gun sight
~, **ночнóй** night sight
~, **опти́ческий** telescopic sight

прицéл, оптúческий кольцевóй optical ring sight
~, орудúйный gunner's quadrant, gun sight
~, откиднóй folding sight
~, откры́тый open sight
~, панорáмный panoramic sight
~, передвижнóй adjustable sight
~, перископúческий periscopic sight
~, постоя́нный fixed sight, battle sight, non-adjustable sight
~, радиолокациóнный radar sight, aiming radar
~, радиолокациóнный пу́шечный gun-laying radar
~, рáмочный tangent sight
~ с диóптром aperture sight
~ с завúсимой лúнией прицéливания sight with rocking bar sight system
~ с независимой лúнией прицéливания sight with independent line of sight system
~ с осветúтельным устрóйством illuminated sight
~ с подвижнóй сéткой, оптúческий adjustable graticule sight
~ с подсвéтом illuminated sight
~ с полузавúсимой лúнией прицéливания sight with semi-independent line of sight system
~ с рáмкой, откиднóй leaf sight
~, светя́щийся luminous sight
~, сéкторный ramp sight (SA)
~, снáйперский sniper's telescopic sight
~ со счётно-решáющим устрóйством computing sight
~, стрелкóвый радиолокациóнный radar ranging gyro gunsight (ac)
~, тáнгенсный tangent sight
~, теку́щий present range (gunn)
~, телескопúческий telescopic sight
~, телескопúческий шарнúрный articulated telescopic sight
~, учéбный training sight
~, цáпфенный trunnion sight
~, шарнúрный телескопúческий articulated telescope sight
прицéливание laying, sighting, aiming
~ по цéнтру вúдимой чáсти цéли aiming at centre of observed mass
прицéливаться/прицéлиться to sight, to range
прицéлиться (см прицéливаться)
прицéл-квадрáнт quadrant sight
прицéл-пострóитель computing sight
прицéльный 1. sightsetter
2. aiming, aimed, sighting
3. sight, backsight
прицéп trailer
~, безрáмный frameless trailer
~, двухколёсный two-wheeled trailer
~, двухóсный two-axle, twin axle trailer
~, кры́тый closed-top trailer
~, многоколёсный multiwheel trailer
~, многоóсный multi-axle trailer
~, низкорáмный low-bed trailer, drop-frame trailer
~, одноóсный single-axle trailer
~, откры́тый open-top trailer
~, плáвающий amphibious trailer
~, пожáрный fire trailer
~ с пожáрным насóсом fire-pump trailer
~, седéльный semi-trailer
~, топливозапрáвочный fuel trailer
прицéп-бáня bath-unit trailer
прицéп-водоцистéрна water trailer
прицéп-рефрижерáтор refrigerated trailer
прицéп-рóспуск pole trailer
прицéп-салóн trailer-saloon (aircrew rest veh)
прицéп-самосвáл dump trailer, tilt-deck trailer
~ с боковóй разгру́зкой side-dump trailer
~ с зáдней разгру́зкой rear-dump trailer
прицéп-фургóн van-type (body) trailer
~ для перевóзки лошадéй horse box trailer
прицéп-цистéрна tanker trailer
причáл 1. mooring, mooring place
2. mooring rope, mooring line
причáливать/причáлить to moor
причáлить (см причáливать)
причинúть (см причиня́ть)
причиня́ть/причинúть to cause, to occasion; **причиня́ть/причинúть потéри** to inflict losses
пришвартовáть (см пришвартóвывать)
пришвартóвывать/пришвартовáть to moor, to make fast
проанализúровать обстанóвку to make an appreciation of the situation
~ (см анализúровать)

прóба 1. trail, test
2. sample
~, **порохóвая** proof firing
пробéг 1. run, distance, haul (road)
2. run-down (voltage etc)
~, **безостанóвочный** non-stop run
~ **в мúлях** road mileage
~ **в мúлях, межремóнтный** mileage between repairs
~ **в мúлях, срéдний** average mileage
~ **в мúлях, суммáрный** total mileage
~ **в мúлях, сýточный** daily mileage
~, **вúдимый** visual range (nucl)
~, **годовóй** annual mileage
~, **длúтельный** duration running
~ **за перúод вы́бега** distance run during free-running (coasting) period
~ **за перúод ускорéния** distance run during accelerating period
~ **за тормознóй перúод** distance run during braking period
~, **испытáтельный** test run
~, **контрóльный** trial run, test run
~, **межремóнтный** time between repairs
~ **по дорóге** road running
~, **показáтельный** demonstration run
~, **прóбный** trial run
~ **самолёта** 1. landing run
2. landing distance
~, **среднесýточный** average daily mileage
~, **сýточный** daily run
~, **холостóй** dead mileage
пробéл gap, interval
пробивáемость 1. penetrability
2. piercing performance, penetrative performance
пробивáние 1. piercing, penetration, punching
2. breakdown, rupture (elec)
~ **бронú** penetration of armour
пробивáть/пробúть to penetrate, to pierce; **пробивáть/пробúть брешь** to effect a breach; **пробивáть/пробúть броню́** to pierce armour, to penetrate armour
пробирáться/пробрáться to make one's way through
пробúть (см пробивáть)
прóбка 1. bung, stopper, cork, plug
2. traffic jam (sl)
3. plug gauge
4. electric fuse
прóбка, броневáя armour plug
~ **взрывáтеля** fuse cap
~, **воздýшная** 1. air lock
2. air plug
~ **выпускнóго отвéрстия** drain plug
~ **горловúны радиáтора** radiator filler cap
~, **дорóжная** traffic jam, hold-up, road congestion
~, **дýльная** muzzle plug, tampion
~ **заливнóго отвéрстия** filler plug
~ **зарядного отвéрстия** filling hole plug (gren)
~ **наливнóго отвéрстия** filler plug
~, **направля́ющая** positioning stop (breach block)
~, **нарезнáя** threaded plug
~, **нúжняя** base plug (gren)
~, **паровáя** vapour lock
~ **пароотвóда** steam plug
~, **пластмáссовая** plastic plug
~ **с резьбóй** screw plug, thread plug
~, **сливнáя** drain plug
~, **спускнáя** drain plug
~, **стержневáя** core plug
~, **трáнспортная** traffic congestion, traffic jam
~, **установóчная** arming plug (mine)
~, **холостáя** filling plug, transit plug (gren)
пробкóвый cork
прóбник proof plug (gunn)
прóбный tentative, experimental, trial, test
пробóина 1. hole, gap
2. bullet hole, shot hole
пробóй 1. sparkover, breakdown of insulation
2. puncture
пробóйник 1. punch, drift (tool)
2. rammer (arty)
пробрáться (см пробирáться)
пробуксóвка spin(ning), slip(ping), slippage
~ **колёс** wheel spin
~ **сцеплéния** clutch slip
~ **фрикциóна** clutch slip
пробурúть (см бурúть)
провáл 1. failure
2. gap (topo)
~ **чувствúтельности** dead spot (radio, radar)
провáр fusion (welding)

проведе́ние 1. construction, building, laying, installing
3. piloting
2. conducting, carrying out, execution
прове́рить (см проверя́ть)
прове́рка inspection, check, audit, testing
~ **боево́й подгото́вки** state of training check, state of training inspection
~ **бо́я** checking the zero, accuracy test (of a gun)
~ **в эксплуата́ции** field testing
~, **вы́борочная** spot check, random check
~ **компре́ссии** compression check (ICE)
~, **контро́льная** practice inspection
~ **на вы́борку** random inspection, sampling inspection, spot check
~ **на вы́держку** random inspection
~ **на отсу́тствие коро́ткого замыка́ния** short-circuit test (elec)
~ **на проры́в** continuity check (elec wire)
~ **натяже́ния гу́сеницы** checking track tension
~, **непла́новая** snap inspection, snap check
~ **нулевы́х обстано́вок прице́льных приспособле́ний** zeroing the sights
~, **обра́тная** read-back (comms)
~ **отчётности** examination of accounts, auditing
~ **по при́нципу да-нет** "go-no-go" test (GW)
~ **под давле́нием** pressure check
~ **приеми́стости** performance checking (veh)
~ **прице́льной ли́нии** bore sighting (arty)
~, **пропускна́я регла́ментная** check-out (GW)
~ **техни́ческого состоя́ния** serviceability check
~ **физи́ческой подгото́вки к бо́ю** physical proficiency test, battle efficiency test
проверя́ть/прове́рить to check, to verify, to test, to inspect ; **проверя́ть/прове́рить градуиро́вку** to calibrate (instr); **проверя́ть/прове́рить нали́чность** to check stocks; **проверя́ть/прове́рить укомплекто́ванность** to check for deficiencies
прове́с dip, sag, slack (of wire)
провести́ (см проводи́ть)
прове́тривание ventilation
прове́тривать/прове́трить to air, to ventilate
прове́трить (см прове́тривать)
прове́шенный signposted, staked
прове́шивание signposting, marking with stakes
провиса́ние droop
провиса́ющий sagging, drooping
про́вод wire, cable, lead, conductor
~, **бензи́новый** petrol pipe, petrol line
~, **возду́шный** open wire, overhead wire
~, **двухжи́льный** twin-core conductor, two-core conductor
~ **зажига́ния** ignition cable
~, **магистра́льный** main cable
~, **норма́льный** standard cable
~, **обво́дный** bypass conduit, bypass canal
~ **под напряже́нием** live wire
~, **сапёрный** firing wire, demolition cable
~, **скры́тый телефо́нный** underground telephone cable
~, **сталеалюми́ниевый** steel-cored aluminium conductor
~, **то́лстый** heavy gauge wire, thick cable
~ **то́плива** fuel pipe
~ **управле́ния** guidance wire (GW)
~, **экрани́рованный** screened cable
провода́, свечны́е plug leads
проводи́мость 1. transmittance (nucl)
2. admittance, conductance, susceptance (elec)
3. conductivity
~, **акти́вная** conductance (elec)
~, **входна́я акти́вная** input conductance
~, **входна́я ко́мплексная** input admittance
~, **ды́рочная** hole conductivity, P-type conductivity (semi-conductor)
~ **изоля́ции** leakance (elec)
~, **ко́мплексная** admittance (elec)
~, **магни́тная** permeance (magnetic)
~ **обра́тного де́йствия** short-circuit feedback admittance
~, **проходна́я** short-circuit transfer admittance
~, **реакти́вная** susceptance (elec)
~, **теплова́я** thermal conduction

проводи́ть/провести́ to take, to lead, to build, to conduct, to carry out, to spend (time), to cheat, to run (over), to pass (over), to book, to draw (lines etc), to install, to pilot; **проводи́ть/провести́ вы́садку возду́шного деса́нта** to launch an airborne assault
прово́дка 1. wiring (elec)
2. tracking by radar operators
3. plotting (of an ac track)
проводни́к 1. guide, conductor, guard
2. conductor (elec)
провожа́тый guide, escort
про́волока wire
~ **в двух ко́льях** double-row barbed wire obstacle
~ **в четырёх ко́льях** four-row barbed wire obstacle
~ **взрыва́теля, натяжна́я** firing wire, demolition cable
~ **"внабро́с"** hastily-erected wire obstacle
~, **гла́дкая** plain wire
~, **колю́чая** barbed wire
~, **Константа́новая** Constantan wire, Eureka wire
~, **контрова́я** safety wire, securing wire
~, **натяжна́я** trip wire (mine)
~, **роя́льная** piano wire
~, **сигна́льная** trip-wire, trip-flare wire
~, **уде́рживающая** restraining wire (fuze)
~, **холоднотя́нутая** cold-drawn wire
проволо́чка delay, procrastination
про́волочка fine wire
~, **Константа́новая** Constantan wire, Eureka wire
~, **плати́но-ири́диевая** iridio-platinum wire
про́волочный wire
прогазо́вка engine acceleration when double declutching (down)
прога́р burnthrough, burnout (missiles)
проги́б 1. bending, warping, buckling
2. bend, arc
~, **весово́й** arch (of ski)
прогно́з 1. forecast
2. prognosis (med)
~ **выпаде́ния радиоакти́вных оса́дков** fallout prediction
~ **зараже́ния** contamination forecast
~ **пого́ды** weather forecast
прогнози́рование forecasting, prediction
прого́н 1. roadbearer, stringer, stretch (br)
2. bearer, baulk, beam
3. run, trip
~ **мо́ста** bridge joist
~, **сло́жный** composite beam
~, **составно́й** made-up beam
прогора́ние burning-out (ICE valves, pistons)
програ́мма programme, syllabus
~ **боево́й подгото́вки** training curriculum
~ **операти́вно-такти́ческой подгото́вки** tactical programme
~ **раке́ты** rocket programme
~, **хронциро́ванная** timed programme
программи́рование programming
~, **автомати́ческое** automatic programming, machine programming
~ **для вычисли́тельной маши́ны** computer programming
~, **лине́йное** linear programming
~, **математи́ческое** mathematical programming
~, **нелине́йное** non-linear programming
~, **полуавтомати́ческое** semi-automatic programming
програ́ммный pre-programmed, programmed, scheduled
прогре́в warming-up
прогрева́ть/прогре́ть to warm up, to warm thoroughly, to heat, : **прогрева́ть/прогре́ть дви́гатель** to warm up the engine
прогре́ть (см прогрева́ть)
прогрохоти́ть (см грохоти́ть)
продвига́ться/продви́нуться to advance, to move, to push on/further/ forward; **продвига́ться/продви́нуться вперёд** to advance; **продвига́ться/продви́нуться перебе́жками** to advance by rushes; **продвига́ться/продви́нуться скачка́ми** to advance by bounds
продвиже́ние advance, progress
~ **кру́пными си́лами** advance in force
~ **от укры́тия к укры́тию** advance from cover to cover
~ **по слу́жбе** advancement, promotion
продви́нуться (см продвига́ться)
проде́лать (см проде́лывать)
проде́лывание прохо́дов breaching (of obstacles)

проде́лывать/проде́лать to do, to perform; **проде́лывать/проде́лать прохо́д** to effect a breach; **проде́лывать/проде́лать прохо́д в ми́нном по́ле** to clear a passage through a minefield; **проде́лывать/проде́лать прохо́д в препя́тствии** to breach an obstacle
продешифри́ровать to analyse; **продешифри́ровать плёнку** to analyse a film
продово́льственный food, supply, ration
продово́льствие 1. provisions, food supplies
2. rations (US)
3. supplies (Brit)
~, скоропо́ртящееся perishable foodstuffs
продолжи́тельность duration, time, period
~ вое́нной слу́жбы length of military service
~ запра́вки filling time
~ заря́дки charging time (elec)
~ заряже́ния loading time (gunn)
~ огнево́го налёта time-on-target
~ полёта по ине́рции coast time (missile)
~ ремо́нта repair time
~ сбо́рки assembly time
~ эксплуата́ции in-service time
~ экспози́ции length of exposure
продо́льный longitudinal
продува́ние venting
продува́ть/проду́ть to scavenge (tech), to blow through, to blow out
проду́вка 1. scavenging (tech)
2. blowing through
проду́кт, побо́чный by-product
проду́кты, газообра́зные gaseous products
~ деле́ния fission products (nucl)
~, заморо́женные frozen food, frozen goods
~, пищевы́е food, foodstuffs
~, расфасо́ванные packaged goods
~ сгора́ния combustion products
~ сгора́ния, выбра́сываемые с выхлопны́ми га́зами exhaust products (ICE)
~, скоропо́ртящиеся perishable goods
проду́ть (см продува́ть)
продфура́ж rations and forage
"ПРОЕ́ЗД ВОСПРЕЩЁН" "NO ENTRY"
прое́кт project, projected design
~ проекти́рования 1. designing, draft, planning, draw up plans
2. projection
~, рабо́чий working project
~, техни́ческий technical project
~, эски́зный draft project
прое́ктно-констру́кторский drawing and designing, project designing
прое́ктно-сме́тный relating to estimates for a project
прое́ктный 1. projecting, designing
2. drafting
прое́кция projection (math, carto), view, elevation
~, азимута́льная azimuthal projection
~, бокова́я side projection, side view
~ ве́ктора vector projection
~, вертика́льная elevation, vertical projection, front view
~, видоизменённая поликони́ческая international map projection
~, вне́шняя external perspective projection, Clarke's projection (geom)
~ Га́усса transverse Mercator projection, Gauss conformal projection
~, гноми́ческая gnomonic projection
~, изометри́ческая isometric projection
~, картографи́ческая cartographic projection, map projection
~, квадра́тная цилиндри́ческая simple cylindrical projection
~, кони́ческая conical projection
~ Мерка́тора Mercator's projection, cylindrical orthomorphic projection
~, мерка́торская Mercator's projection (map)
~, норма́льная конфо́рмная кони́ческая Lambert conformal conic projection
~, норма́льная поля́рная zenithal equidistant projection
~, норма́льная прямá́я zenithal equidistant projection
~, орфографи́ческая orthographic projection
~, перспекти́вная perspective projection
~ пло́скости plane projection
~, поликони́ческая polyconical projection
~, поля́рная polar projection
~, попере́чная transverse projection

прое́кция, попере́чная экваториа́льная equatorial zenithal projection
~, **проста́я цилиндри́ческая** simple cylindrical projection
~, **равновели́кая азимута́льная** zenithal equal-area projection (carto)
~, **равновели́кая цилиндри́ческая** cylindrical equal-area projection (carto)
~, **равнопромежу́точная кони́ческая** simple conical projection with one standard parallel
~, **равноуго́льная кони́ческая** conical orthomorphic projection
~, **стереографи́ческая азимута́льная** stereographic projection
~, **фронта́льная** front projection
~, **центра́льная** gnomonic projection, central projection
проём aperture, opening
проже́ктор searchlight
~, **воспроизводя́щий** reading gun, viewing gun (electron)
~ **залива́ющего све́та** floodlight
~, **запи́сывающий** writing gun (electron)
~, **зени́тный** anti-aircraft searchlight
~, **инфракра́сный** searchlight with infra-red filter
~, **ксено́новый** xenon searchlight
~, **о́блачный** cloud searching
~, **обы́чный** white light searchlight
~, **стира́ющий** erasing gun (electron charactron tube etc)
~, **та́нковый** tank-mounted searchlight
~, **электро́нный** electron gun
проже́ктор-иска́тель spotlight
прожектори́ст searchlight operator
прозра́чность атмосфе́ры air transparency, atmospheric visibility
произвести́ (см производи́ть)
производи́тельность output, performance, productivity
~ **артилле́рии, огнева́я** rate of artillery fire
~ **насо́са** pump capacity
~, **пла́новая** production target
производи́ть/произвести́ to make, to carry out, to execute, to promote (to, to the rank of); **производи́ть/произвести́ взрыв** to let off explosion, to execute demolition; **производи́ть/произвести́ запро́с** to challenge, to interrogate (radar IFF); **производи́ть/произвести́ испыта́ния** to carry out trials; **производи́ть/произвести́ прице́ливание** to take aim; **производи́ть/произвести́ разве́дку** to reconnoitre, to make a reconnaissance; **производи́ть/произвести́ реви́зию** to audit; **производи́ть/произвести́ сострéл ору́дий** to calibrate (gunn)
произво́дство 1. production 2. execution
~ **в офице́ры** commissioning as an officer
~, **вое́нное** war production, military production
~ **вы́стрела** firing of a shot
~, **ма́ссовое** mass production, gross production, high-volume production
~, **мелкосери́йное** small-scale production
~ **разруше́ний** demolition work
~, **сери́йное** batch production, series production, line production
проинструкти́ровать to brief thoroughly
проинформи́ровать (см информи́ровать)
происше́ствие event, incident
~, **лётное** aircraft accident, flying accident
~, **чрезвыча́йное** 1. unusual occurrence 2. serious accident
прока́лывать/проколо́ть to puncture, to pierce, to perforate
прока́т rolled metal (section or sheet)
прока́тка rolling (metall)
прокла́дка 1. gasket, packing, shim, washer, spacer, pad 2. laying (cable, pipes, roads)
~, **асбе́стовая** asbestos gasket, asbestos packing
~, **бензосто́йкая** petrol-proof gasket
~, **бу́ферная** buffer pad
~, **воскова́я** wax pad
~, **герметизи́рующая** gasket
~ **голо́вки бло́ка** cylinder head gasket
~ **голо́вки бло́ка цили́ндров** cylinder head gasket
~, **зазо́рная** shim
~, **изоляцио́нная** sealing ring (mine), sealing washer
~ **ка́беля** cable laying
~, **ко́жаная** leather washer
~ **коло́нного пути́** trail blazing
~ **кры́шки** cover joint
~ **кры́шки бло́ка** cylinder head gasket

прокла́дка кры́шки подши́пника bearing cap shim
~ лы́жни laying a ski-trail
~, мета́лло-асбе́стовая metal asbestos gasket
~ полуо́си axle shaft gasket
~, регулиро́вочная shim (made of metal or paper)
~, рези́новая rubber gasket, rubber seating
~, слюдяна́я mica spacer
~, теплоизоли́рующая thermal insulation
~ то́рмоза, амортизи́рующая brake lining
~, фе́тровая felt packing, felt gasket
~ цили́ндра, уплотня́ющая cylinder packing
прокла́дывать/проложи́ть to lay, to build, to construct; **прокла́дывать/проложи́ть доро́гу** to lay a road; **прокла́дывать/проложи́ть ка́бель** to lay cable
проко́л puncture, blow-out
проколо́ть (см прока́лывать)
проконтроли́ровать (см контроли́ровать)
прокорректи́ровать (см корректи́ровать)
прокурату́ра, вое́нная Judge Advocate's Office
~ вооружённых сил, Гла́вная вое́нная Judge Advocate's Office of the Armed Forces
прокуро́р procurator, prosecutor
~, вое́нный judge advocate
пролёт 1. span (br)
2. flypast, overflight (ac)
3. air distance
4. flight
~, а́нкерный anchor arm, anchor span
~, берегово́й abutment span, shore bay (br)
~, козло́вый trestle span
~, мостово́й bridge span
~, перекрыва́емый gap crossable (by AFV)
~, подвесно́й suspended span (br)
~, расчётный effective span (br)
проли́в strait, sound
проложе́ние, горизонта́льное horizontal distance, horizontal range
проло́м hole (in wall), breach, gap, fracture
прома́сленный oiled
прома́сливание oiling, oil impregnating, treating with oil
про́мах miss (SA, arty)
прома́хиваться/промахну́ться to miss, to miss the mark
промахну́ться (см прома́хиваться)
промежу́ток space, interval, distance, gap
~ вре́мени time interval
~ вре́мени ме́жду вспы́шкой и зву́ком вы́стрела flash to bang time
~, искрово́й spark gap, sparking plug gap
~ ме́жду поса́дками landing interval (helicopters)
~, мёртвый dead interval (missile)
~, необстре́ливаемый gap not covered by fire
проме́р measuring-off, taping (surv)
промерза́ть/промёрзнуть to freeze solid
промёрзнуть (см промерза́ть)
проме́ривать/проме́рить to measure, to survey
проме́рить (см проме́ривать) to make a mistake in measurement (perf only)
проме́рщик tapeman (surv)
проми́лле a thousandth part
промота́ние вое́нного иму́щества fraudulent misapplication of military property
промыва́ние flushing, washing-out
промы́вка flushing, washing-out
промы́шленность industry
~, вое́нная war industry, armaments industry
проника́ние penetration (of armour etc)
проника́ть/прони́кнуть to penetrate
проникнове́ние penetration (tac)
прони́кнуть (см проника́ть)
проница́емость 1. permeability, penetrability
2. permittivity (electron)
~ ла́мпы inverse valve amplification factor (electron)
~, магни́тная magnetic permeability
пропа́вший бе́з вести missing
пропада́ть/пропа́сть to be missing, to be lost, to fade (radio)
пропа́н propane
пропа́сть (см пропада́ть)
пропиле́н propylene, propene
пропи́танный impregnated
пропита́ть (см пропи́тывать)

пропи́тка impregnation, impregnating
~, огнезащи́тная fire-proofing impregnation
пропи́тывать/пропита́ть to impregnate, to soak, to saturate
пропорциона́льный proportional, proportionate
про́пуск 1. non-attendance, lapse, omission, blank, gap
2. pass, permit, password
3. admission
~ зажига́ния misfire (ICE)
~, постоя́нный permanent pass
~, ра́зовый one-time pass
пропуска́тель acceptor (electron)
пропуска́ть/пропусти́ть to pass, to let pass, to let in, to let out, to omit, to miss; **пропуска́ть/пропусти́ть во́ду** to leak
пропускни́к, санита́рный medical check point, decontamination centre, field cleansing and disinfestation centre
пропускно́й permit, pass, traffic
пропусти́ть (см пропуска́ть)
прора́б foreman
прорва́ть (см прорыва́ть)
прорва́ться (см прорыва́ться)
прорези́ненный rubberized
прорези́нивать/прорези́нить to rubberize
прорези́нить (см прорези́нивать)
про́резь slot, notch, slit, slotted hole
~ предохрани́теля safety slot
~ прице́ла "U" of the backsight, backsight notch
~, прице́льная "U" of the backsight, backsight notch
про́рубь ice hole
проры́в 1. breakthrough, breach (tac)
2. break, gap
~ га́за в ка́ртер compression blow-by (ICE)
~ га́зов gas escape
~ на не́скольких уча́стках multiple penetration
~ пороховы́х га́зов blow back
~, такти́ческий tactical breakthrough
прорыва́ть/прорва́ть to break through, to pierce; **прорыва́ть/прорва́ть блока́ду** to run the blockade; **прорыва́ть/прорва́ть оборо́ну** to breach the defences
прорыва́ться/прорва́ться to break through, to pierce
проса́чивание 1. infiltration (tac)
2. percolation, soakage
3. leakage, oozing
проса́чиваться/просочи́ться to infiltrate, to percolate, to leak, to ooze
просве́рливать/просверли́ть to bore, to drill, to pierce, to perforate
просверли́ть (см просве́рливать)
просве́т 1. clearance, gap
2. bay, aperture, opening
3. central stripe on officer's shoulder board (Sov)
~, доро́жный ground clearance
просветле́ние опти́ческих линз lens blooming
про́сек, противопожа́рный fire lane, fire break
про́сека opening, cutting (in forest)
просёлок country road
проско́к overshoot, overswing
просле́дование passage
просло́йка layer, stratum, interface
~, возду́шная air seal
прослу́шивание, контро́льное monitoring (electron)
просмо́тр observation, field of observation
просочи́ться (см проса́чиваться)
проста́вка intermediate piece, transition piece, adapter
просто́й 1. plain, ordinary, simple
2. non-running time, idle time (machines)
3. demurrage
простота́ в эксплуата́ции simplicity of operation
~ управле́ния ease of control
простра́нственный three-dimensional, spatial, space
простра́нство space
~, баллисти́ческое ballistic area
~, возду́шное air space
~ группирова́ния drift space (electron)
~, излётное danger area (range) (spent bullet area in front of fire position)
~, мёртвое dead ground, dead space, dead zone, dead area
~, ме́стное поража́емое true dangerous space (ball)
~, необстре́ливаемое dead zone, dead area
~, непоража́емое dead ground, dead zone, zone immune from fire (ball)

простра́нство, откры́тое open space
~, поража́емое danger area, beaten zone (ball)
~ преобразова́ния drift space (electron)
~, прикры́тое defiladed zone, enfiladed zone, zone covered by fire
~, прице́льное поража́емое theoretical dangerous space
простре́л effective fire cover
простре́ливать/прострели́ть to rake with fire, to cover by fire
прострели́ть (см простре́ливать)
просту́пок offence, breach of regulations
~, во́инский military misdemeanour (mil law)
~, дисциплина́рный breach of discipline, disciplinary offence
просчёт miscalculation
проте́з artificial limb
проте́зы, зубны́е false teeth
протеи́н protein
протека́ть/проте́чь to flow past
проте́ктор 1. protector, protective covering
2. tread (tyre)
~, изно́шенный worn tread
~, самозатя́гивающийся self-sealing tube
проте́чь (см протека́ть)
проти́вник enemy, opponent, adversary
~, возду́шный 1. enemy air forces
2. enemy aircraft
~, назе́мный enemy ground forces
~, обозна́ченный skeleton enemy (trg)
противобо́рствующий opposing
противове́с counterweight, counterpoise
противовозду́шный air defence, AA
противовспе́ниватель anti-foam agent
противовы́стрел countershot
противога́з respirator, gas-mask
~, гражда́нский civil defence respirator, civilian- pattern respirator
~, изоли́рующий breathing apparatus
~, у́гольный charcoal respirator, respirator using activated charcoal
~, учёбный training mask
~, фильтру́ющий respirator
противога́зовый anti-gas
противогангрено́зный anti-gangrene
противодавле́ние back-pressure
противоде́йствие opposition, counter-measure, counteraction
противоде́йствие, гидроакусти́ческое sonar countermeasure
~, радиолокацио́нное 1. radar countermeasures
2. electronic countermeasures
противоде́йствия, радиоэлектро́нные electronic countermeasures, ECM
противоде́йствовать to counteract, to counter, to oppose, to resist, to react
противодеса́нтный anti-landing, anti-airborne, anti- assault-landing
противодетонацио́нный anti-knock (ICE)
противозано́сный antiskid
противозени́тный anti-aircraft
противоизно́сный wear-preventive
противомедни́тель decoppering agent, decoppering charge (ammo)
противоме́ра countermeasure
противоми́нный anti-mine
противообледени́тель de-icing device, defroster
противоокисли́тель anti-oxidant
противомедни́тель decoppering agent, decoppering charge (ammo)
противооско́лочный splinter-proof
противоотка́тный anti-recoil, counter-recoil
противопенообразова́тель anti-foamer, anti-foaming agent
противопехо́тный anti-personnel
противопожа́рный fire-fighting
противополо́жный opposite, contrary
противопоме́ховый anti-jamming
противопу́льный anti-small-arms
противорадиацио́нный anti-radiation
противораке́та anti-missile missile
противораке́тный anti-missile
противосамолётный anti-aircraft
противоснаря́дный shell-proof
противостари́тель anti-deteriorant, preservative
противостолбня́чный anti-tetanus
противостоя́ть to withstand, to resist, to hold out against
противостоя́щий opposing
противотанки́ст anti-tank gunner
противота́нковый anti-tank
противото́к reverse flow, backward flow, counterflow, countercurrent
противоуго́н 1. anchor
2. anti-creep device
противоукло́н opposite elevation

противохими́ческий anti-chemical, anti-gas, anti-CW
противоэпидеми́ческий anti-epidemic
противоя́дие antidote
прoти́рка pull-through
~, шо́мпольная scourer
прото́к, во́дный water channel, water duct, canal
протоко́л 1. protocol
2. proceedings, minutes of a meeting
прото́н proton
прото́чка, кольцева́я circular groove (ammo)
протравни́к vent bit (ord)
протра́ленный cleared (minefield)
протра́ливать/протра́лить to clear (minefield)
протра́лить (см протра́ливать)
протя́гивание broaching (eng)
протяжённость expansion, extension, spread, range, length
протя́жка broach (tool)
проу́шина lug, ear, eye
профила́ктика prevention
~, пожа́рная fire prevention
профилакти́ческий prophylactic, preventive
профилакто́рий 1. health centre
2. preventive maintenance workshop
~, похо́дный mobile preventive maintenance workshop
профили́ровать to grade, to level; **профили́ровать доро́гу** to grade a road, to level a road
про́филь 1. section (map reading)
2. cross section, profile
3. sphere, field
~ бандажа́ tyre profile
~, изно́шенный worn profile
~, изыска́тельский survey profile
~ кулачка́ cam profile
~ ме́стности relief section
~, по́лный full section (carto)
~, попере́чный end elevation
~, продо́льный sectional elevation
~ проте́ктора tread contour
~, сокращённый hasty section (topo, carto)
~ ши́ны tyre section
прохо́д gap, breach, gangway, lane, passage
~ в ми́нном по́ле minefield gap, minefield lane
прохо́д, го́рный mountain pass
~, коле́йный track lane, treadway lane (through a minefield)
~, обозна́ченный marked lane (through an obstacle)
~, первонача́льный initial lane
~ по пересечённой бездоро́жью cross-country ability, cross-country capability
~, пожа́рный fire aisle, fire-fighting aisle (in a store)
~, у́зкий defile
проходи́мость 1. cross-country performance
2. going (state of terrain)
3. trafficability, passability
~, высо́кая high mobility, high cross-country performance (of veh)
~ по бездоро́жью cross-country capability, cross-country performance
~ по пересечённой ме́стности cross-country ability, cross-country capability
~, повы́шенная cross-country capability, cross-country performance
~, проходна́я short-circuit feedback admittance (elec)
проходи́мый passable, practicable
прохожде́ние transit, passage
~ торже́ственным ма́ршем ceremonial march past
проце́сс бо́я flow of battle, course of battle
~ взрыва́ния firing chain
~ образова́ния па́ры pair production (electron)
~, перехо́дный transient (electron)
~, рабо́чий operating procedure
прочéрчивание plotting (on graphs etc)
прочеса́ть (см прочёсывать)
прочёска 1. screening (security)
2. sweep, search
прочёсывание scanning, combing (ground etc)
прочёсывать/прочеса́ть to sweep, to comb, to search
прочи́стка scavenging, purging
про́чность endurance, durability, strength (of materials)
~ на изло́м breaking strength
~ на разры́в tensile strength
~ на сжа́тие compression strength
~ при сдви́ге shear strength
~, уда́рная impact strength
прощу́пать (см прощу́пывать)

прощу́пывание prodding (mine detection)
прощу́пывать/прощу́пать to prod (mine detection)
прояви́тель developer (photo)
проявле́ние, фотографи́ческое photographic development
пружи́на spring, coil spring
~ **амортиза́тора** shock-absorber spring
~, **боева́я** 1. main spring (striker, ord) 2. striker spring (gren) 3. hammer spring (SA)
~ **бойка́** striker spring
~ **бу́фера** buffer spring
~ **взво́да** sear spring
~, **взводя́щая** arming spring (of fuze)
~, **винтова́я** helical spring, coil spring
~, **возвра́тная** 1. recoil spring, return spring 2. fusee spring (MG) 3. main spring (pistol)
~, **возвра́тно-боева́я** return spring (MG)
~, **вспомога́тельная** secondary spring
~ **гне́тка** extractor spring
~, **двупёрая боева́я** V-type main spring
~, **ди́сковая** disc spring
~, **заводна́я** clock spring
~, **закрыва́ющая** closing spring (breech)
~ **замедли́теля** retard spring
~, **за́мкнутая** flat spring
~ **запира́ющего механи́зма** breech closing spring
~ **затво́ра** bolt (driving) spring
~, **кла́панная** valve spring
~, **кольцева́я** ring spring
~ **компенса́тора** compensator spring
~, **контрпредохрани́тельная** creep spring (ammo)
~ **ложево́го кольца́** band spring
~, **нажи́мная** pressure spring, thrust spring
~ **нажи́мной кры́шки** pressure plate spring (mine)
~ **нака́тника** recuperator spring (ord)
~, **натяжна́я** recoil spring
~, **ни́жняя возвра́тная** barrel plunger spring (ord)
~, **опо́рная** support spring
~ **отсе́чки** cut-off spring
~, **оттяжна́я** retaining spring
~, **пласти́нчатая** laminated leaf spring, leaf spring
~ **подава́теля** magazine spring
~, **подаю́щая** lifting spring
пружи́на подве́ски suspension spring
~ **ползуна́ приёмника** belt feed pawl spring
~, **предохрани́тельная** detent spring (fuse)
~ **предохрани́теля подвижны́х ре́ек** feed rod spring
~, **рессо́рная** leaf spring
~ **са́льника** stuffing box spring
~ **соба́чки то́рмоза** brake pawl spring
~, **спира́льная** spiral spring
~, **спускова́я** trigger spring
~ **спусково́го крючка́** trigger spring
~ **спусково́го рычага́** sear spring
~ **спусково́й тя́ги** trigger spring (SA)
~, **сто́порная** retaining spring
~, **таре́льчатая** plate spring, disc spring
~ **то́рмоза отка́та** recoil spring
~ **то́рмоза, оттяжна́я** brake return spring
~ **тормозно́й коло́дки, возвра́тная** brake shoe return spring
~, **торсио́нная** torsion spring
~ **уда́рника** striker spring
~, **уравнове́шивающая** relief spring, balancing spring
~ **фикса́тора** retaining spring
~ **фрикцио́на, выжимна́я** clutch release spring
~ **фрикцио́на, нажи́мная** clutch pressure spring
~ **шепта́ла** sear spring
~ **щёточного конта́кта** brush contact spring
пружи́на-контрпредохрани́тель creep spring (fuze)
пружине́ние springing, spring action
пружи́нка, боева́я main spring (striker, ord)
пружи́нный spring, spring-actuated
прут, сва́рочный welding rod
пруто́к rod
пры́гать/пры́гнуть to jump, to leap, to bound; **пры́гать/пры́гнуть с парашю́том** to drop by parachute
пры́гнуть (см пры́гать)
прыжо́к, группово́й stick jump (para)
~, **затяжно́й** delayed jump (para)
~, **пристре́лочный** test jump (before starting trainees' jumping practice) (para)
~ **с парашю́том** parachute jump

прыжо́к со стабилиза́ции delayed drop, free-fall jump (para)
~, учéбный парашю́тный practice parachute jump
пря́жа yarn
~, асбéстовая asbestos yarn
~, стекловолокни́стая glass fibre yarn
пря́жка buckle, fastener; bar to decoration (Brit)
~, Д-обра́зная D-ring
~, передвижна́я adjustable buckle
пряма́я straight line
"Пря́мо!" "Forward!" (drill)
прямо́й 1. straight, direct 2. immediate, foremost 3. through, straightforward
~ наво́дки over open sights
прямоуго́льник rectangle
прямоуго́льный rectangular
психохими́ческий psychochemical
псофо́метр psophometer
пуа́з poise
ПУА́ЗО AAFCE
~, моностати́ческий central station computer (AAFCE)
пуансо́н punch, stamp
пу́говка курка́ cocking piece knob
пуд pood (36 lb; 16.38 kg)
пудово́й pood (weighing 36 lb)
пу́дра powder
пузырёк bubble, blister
пулемёт machine-gun
~, ба́шенный turret machine gun, turret mounted machine gun (tk)
~ в спо́нсоне sponson machine-gun
~ Го́чкиса Hotchkiss gun
~, еди́ный general purpose machine-gun, GPMG
~, зени́тный anti-aircraft machine-gun
~ ко́мплексной устано́вки coaxial machine-gun
~, кочу́ющий roving machine-gun
~, крупнокали́берный heavy machine-gun, HMG
~, курсово́й hull machine-gun (tk)
~, лёгкий light machine-gun, LMG
~, лобово́й hull machine-gun (tk)
~ Лью́иса, ручно́й Lewis automatic rifle
~ Ма́ксима, станко́вый Maxim machine-gun
~, многоство́льный multi-barrelled machine-gun
~, пристре́лочный ranging MG
пулемёт, ро́тный company machine-gun
~, ручно́й light machine-gun, LMG
~ с тяжёлым стволо́м heavy-barrelled machine-gun
~, синхронизи́рованный synchronised machine-gun
~ со свобо́дным затво́ром blowback operated machine-gun
~, спа́ренный coaxial machine-gun
~, станко́вый medium machine-gun, MMG
~ та́нка, лобово́й forward hull machine-gun (tk)
~, та́нковый tank machine-gun
~, туре́льный turret machine-gun
~, универса́льный general-purpose machine-gun (GPMG)
пулемётный machine-gun
пулемётчик machine-gunner
~, дежу́рный duty machine-gunner (sentry)
~, лобово́й hull-gunner (tk)
пуленепроница́емый bullet-proof
пулепоглоти́тель backstop (on firing range)
пулесто́йкий bullet-proof
пулеула́вливатель bullet stop, bullet catcher, bullet trap (on rifle range)
пулеулови́тель bullet trap (rifle range)
пульвериза́тор atomizer, spray gun
пульса́ция ripple, pulsation
пульси́рующий pulsating
пульт desk, panel, console
~ дежу́рного operator's desk
~ для прове́рки испра́вности подрывны́х маши́нок demolition exploder test set
~ для прове́рки напряже́ния подрывны́х маши́нок demolition exploder voltage tester
~ опера́тора operator's desk
~ управле́ния control panel, control console, control desk
~ управле́ния, выносно́й remote-position control panel
~ управле́ния, кно́почный thumb-button control box (GW)
~ управле́ния миноиска́теля mine detector tuning panel
~ управле́ния огнём fire control console
~ управле́ния радиокла́ссов universal wireless training set

пульт управле́ния, центра́льный central panel, control panel
~**, центра́льный** central console
~ **энергопита́ния** power distribution panel
пу́ля bullet
~**, безоболо́ченная** solid bullet (without envelope or without jacket)
~**, бронебо́йная** armour-piercing bullet
~**, бронебо́йно-зажига́тельная** armour-piercing incendiary bullet
~**, веретенообра́зная** boattailed bullet
~**, винто́вочная** rifle bullet
~ **"дум-дум"** dumdum bullet
~**, зажига́тельная** incendiary bullet
~ **на излёте** spent bullet
~**, нарезна́я** grooved bullet
~**, оболо́ченная** jacketed bullet
~**, остроконе́чная** pointed bullet
~**, плакиро́ванная** plated bullet
~**, полубронебо́йная** semi-armoured piercing bullet
~**, пристре́лочно-зажига́тельная** ranging-adjustment incendiary bullet
~**, разрывна́я** explosive bullet
~ **с металли́ческой оболо́чкой** metal-cased bullet
~ **с мя́гкой голо́вкой** soft-nosed bullet
~ **с полусфери́ческой голо́вкой** round-nosed bullet
~ **со сре́занной голо́вкой** hollow-nosed bullet
~ **со стально́й оболо́чкой** steel-jacketed bullet
~**, трасси́рующая** tracer bullet
~**, тупоно́сая** blunt-nosed bullet
~**, шальна́я** stray bullet
~**, шрапне́льная** shrapnel bullet
пункт 1. point, post
2. item, paragraph (comms)
~ **авари́йного ремо́нта** first aid repair area (for eqpt)
~ **авари́йных маши́н, сбо́рный** vehicle casualty collecting point
~ **автотра́нспорта, коне́чно-вы́грузочный** roadhead, truckhead
~**, артиллери́йский наблюда́тельный** artillery observation post
~**, артиллери́йский подвижно́й наблюда́тельный** mobile artillery observation post
~**, батальо́нный запра́вочный** battalion fuelling point
пункт, батальо́нный патро́нный battalion ammunition point
~ **батаре́и, кома́ндный** battery command post
~**, батаре́йный команди́рский** battery command post (includes CP control room, FC & TC radar (AD))
~**, бензозапра́вочный** petrol filling point
~**, берегово́й контро́льный** bank control (water crossing)
~**, берегово́й опо́рный** beach stronghold
~ **боево́го пита́ния** ammunition point
~ **боепита́ния** ammunition point
~ **боепита́ния, обме́нный** ammunition refilling point, ammunition point
~ **боеприпа́сов, передово́й** forward ammunition point
~**, боково́й** cross observation post (gunn, arty)
~**, боково́й наблюда́тельный** flank observation post, flank OP
~**, брониро́ванный наблюда́тельный** armoured observation post
~ **взво́да, опо́рный** platoon strong point
~**, взво́дный патро́нный** platoon ammunition point
~ **во́дного тра́нспорта, коне́чно-вы́грузочный** waterhead, navigation head
~**, во́дный** water point
~**, водоразбо́рный** mains water supply point
~**, водораспредели́тельный** water distribution point, water point
~ **водоснабже́ния** water-supply point
~**, вое́нно-продово́льственный** rations delivery point
~ **возду́шного тра́нспорта, коне́чно-вы́грузочный** airhead, air terminal
~ **возду́шной свя́зи** air liaison centre
~**, возду́шный комма́ндный** airborne command post
~**, возду́шный наблюда́тельный** air observation post
~**, вре́менный медици́нский** temporary medical post
~**, вре́менный наблюда́тельный** temporary observation post
~**, вспомога́тельный наблюда́тельный** auxiliary observation post
~**, вспомога́тельный эвакуацио́нный** auxiliary evacuation point

пункт вы́броски, запасно́й alternative drop zone (para)
~ вы́броски, основно́й main drop zone (para)
~, выжида́тельный assembly area
~ вы́садки деса́нта assault landing point (either airborne or seaborne)
~, вычисли́тельный computer station
~, геодези́ческий trig point, trigonometrical point
~, головно́й ро́тный патро́нный advance company ammunition point
~ движе́ния, исхо́дный initial point
~, дегазацио́нно-обмы́вочный CW decontamination centre
~, дегазацио́нный CW decontamination centre
~, деж́урный телефо́нный manned telephone
~, дивизио́нный медици́нский 1. divisional medical station (Sov) 2. casualty clearing station (Brit)
~, дивизио́нный обме́нный divisional exchange point
~, диспе́тчерский 1. air traffic control point 2. control room
~, диспе́тчерский контро́льный dispatchers' control point (MT)
~ донесе́ний, сбо́рный staff message centre
~, доро́жно-ремо́нтный road repair point
~, доро́жный спра́вочный roadside information post
~ доста́вки боеприпа́сов ammunition delivery point
~, запасно́й наблюда́тельный alternative OP, reserve OP
~, запа́сный кома́ндный alternative command post
~ запра́вки ГСМ POL point, refuelling point
~, запра́вочный petrol point, refuelling point
~, исхо́дный starting point, start point
~, команди́рский command post, main observation post
~, команди́рский наблюда́тельный main observation post
~, кома́ндно-диспе́тчерский air traffic control tower
пункт, кома́ндно-наблюда́тельный 1. command and observation post 2. commander's observation post (sub units)
~, кома́ндный command post
~, коне́чно-вы́грузочный 1. road head, truckhead 2. railhead 3. airhead, air terminal 4. waterhead, navigation head
~, коне́чный termination, terminal
~, конте́йнерный container depot
~, контро́льно-испыта́тельный check and test point
~, контро́льно-пропускно́й traffic control point, check point, main gate check point
~, контро́льно-распредели́тельный check and distribution point (NBC)
~, контро́льно-техни́ческий technical control point
~, корпусно́й обме́нный corps supply refilling point
~, корректиро́вочный observation post (arty)
~, ло́жный наблюда́тельный dummy OP
~ маршру́та, исхо́дный point of departure (ac)
~ маршру́та, коне́чный flight destination point
~ маршру́та, поворо́тный turning point
~, медици́нский medical aid post
~ медици́нской по́мощи medical aid post
~ медици́нской по́мощи, баталь́онный 1. battalion medical post 2. regimental aid post (Brit)
~ мо́йки и запра́вки washing and refuelling point
~ мостово́й перепра́вы bridge site, bridge-crossing point
~ на та́нке, подви́жный наблюда́тельный tank observation post
~, наблюда́тельный observation post, OP
~ наблюде́ния и свя́зи, головно́й forward observation and reporting post
~ наведе́ния, кома́ндный ground controlled interception station, GCI station
~ назначе́ния destination
~, населённый built-up area, inhabited locality

пункт обеззара́живания оде́жды clothing decontamination point
~, **обме́нный** exchange point, supply refilling point
~, **обмы́вочно-дегазацио́нный** cleansing station (NBC)
~ **обогре́ва** warming post
~ **обрабо́тки** 1. processing point 2. computing centre (arty)
~, **опо́рный** 1. defended post 2. strong point
~, **опо́рный противота́нковый** anti-tank strong point
~, **основно́й наблюда́тельный** principal observation post, main OP
~ **остано́вки** halt point
~ **отправле́ния** point of departure
~, **пайко́вый** ration point
~ **паро́мной перепра́вы** raft site, ferrying site
~, **патро́нный** ammunition point (SA)
~ **перви́чной санита́рной обрабо́тки** emergency decontamination centre
~, **перева́лочный** exchange point, transfer point, trans-shipment point, stores loading point, relay point
~, **перевя́зочный** field dressing station
~, **перегру́зочный** reloading point, trans-shipment point
~, **переда́точный** transfer point, intermediate point
~ **переда́чи информа́ции** information transmission point
~, **передово́й артиллери́йский наблюда́тельный** forward artillery OP, forward artillery observation post
~, **передово́й комма́ндный** forward command post
~, **передово́й медици́нский** 1. forward aid post 2. regimental aid post, RAP (Brit)
~, **передово́й наблюда́тельный** forward observation post
~, **передово́й перевя́зочный** advanced dressing station
~, **передово́й распредели́тельный** advance distribution point
~, **передово́й ремо́нтно-восстанови́тельный** forward repair point
~ **перезаря́дки** refilling point
~ **перепра́вы** crossing place
~ **перепра́вы вброд** fording site
пункт перепра́вы вплавь swimming site (tk crossing)
~ **перепра́вы по льду** ice crossing site
~ **повреждённых маши́н, сбо́рный** damaged vehicle collection point, equipment collecting point
~, **пограни́чный** frontier point
~ **погру́зки** 1. loading point (equipment) 2. embarkation area (personnel)
~, **подви́жный артиллери́йский наблюда́тельный** mobile artillery OP
~, **подви́жный кома́ндный** mobile command post, command post vehicle
~ **подслу́шивания** listening post
~ **полево́го водоснабже́ния** field water point
~, **полково́й медици́нский** regimental aid post, RAP
~ **помы́вки** bath point, bath station
~ **поса́дки** point of embarkation
~ **постоя́нной дислока́ции** 1. permanent disposition 2. garrison location base
~ **прибы́тия** point of arrival
~ **приёма автобронета́нковой те́хники** vehicle reception depot
~, **приёмный** reception centre
~, **продово́льственный** ration supply point
~ **про́пуска, контро́льный** traffic control point
~ **про́пуска на перепра́ву, контро́льный** crossing traffic control point (river crossings)
~, **противота́нковый опо́рный** anti-tank strong point
~ **разда́чи карт** map distribution point
~, **разде́льный** block post (rly)
~, **разли́вочный** refuelling point
~, **райо́нный диспе́тчерский** area control centre (ACC)
~ **распределе́ния боеприпа́сов** ammunition distribution point, ammunition point
~ **распределе́ния горю́чего** fuel distribution point
~, **распредели́тельный** bulk breaking point, distributing point, reduction point (supplies)
~ **регули́рования движе́ния** traffic control centre, traffic control point (TCP)
~, **ремо́нтно-восстанови́тельный** recovery station

пункт, ро́тный боево́й распредели́тельный company battle distribution point
~, ро́тный патро́нный company ammunition point
~, ро́тный разда́точный company distribution point (for distributing cooked food)
~ ро́ты, опо́рный company locality, company strong point
~ санита́рной обрабо́тки cleansing station (NBC)
~ сбо́ра rallying point, assembly point, RV
~ сбо́ра донесе́ний message centre
~ сбо́ра донесе́ний, передово́й forward message centre
~ сбо́ра, запасно́й alternative rallying point
~ сбо́ра и эвакуа́ции маши́н vehicle recovery post
~ сбо́ра иму́щества, передово́й advance salvage point, equipment collecting point
~ сбо́ра по трево́ге alarm post
~, сбо́рный collecting point, assembly point, rallying point, rendezvous, RV
~ сбра́сывания горю́чего fuel dropping point
~ систе́мы поса́дки, диспе́тчерский airfield GCA control point
~ снабже́ния боеприпа́сами ammunition supply point
~ снабже́ния горю́чим petrol point, fuel point
~ снабже́ния, передово́й advanced supply point, forward stores depot
~ снабже́ния, полково́й regimental supply point
~ сопряжённого наблюде́ния cross-observation post
~ сосредото́чения деса́нтных средств boat marshalling area, marshalling area (assault crossings)
~ специа́льной обрабо́тки special decontamination centre, cleansing station (NBC)
~, такти́ческий tactical locality, tactical point
~ техни́ческого контро́ля technical check point
пункт техни́ческого наблюде́ния technical inspection post (Sov) (duties similar to those of forward repair and recovery element (Brit))
~ техни́ческого обслу́живания technical servicing point
~ техни́ческого обслу́живания и ремо́нта technical servicing and repair point, service and repair bay
~, транзи́тный shipping point
~, триангуляцио́нный triangulation point, trig point
~, тригонометри́ческий trigonometrical point, trig point
~ трубопрово́да, коне́чный pipeline terminal point
~ трубопрово́да, нача́льный pipeline start point
~, укреплённый strong point
~ управле́ния control point, command post
~ управле́ния, вспомога́тельный auxiliary control point
~ управле́ния, вы́несенный remote control point
~ управле́ния за́пуском launcher control post LCP
~ управле́ния, инжене́рный technical servicing control point
~ управле́ния огнём fire control post
~ управле́ния огнём зени́тной артилле́рии AA artillery operations centre
~ управле́ния, подвижно́й mobile command post
~ управле́ния сре́дствами ПВО air defence operations centre
~, хими́ческий наблюда́тельный CW observation post
~ хозя́йственного дово́льствия ration supply point (issues cooked food in bulk)
~, эвакуацио́нный backloading point, clearing station
пункти́р dotted line
пупиниза́ция coil loading (elec)
пурга́ snowstorm, blizzard
пуск 1. start (inscription on instrument panel or starter button)
2. launch (missile)
~ в ход starting up, starting (veh, m/c)
~ встре́чного огня́ counterfiring (fire-fighting)

пуск раке́ты missile launching
~ **с земли́** ground launch (msl)
~, **ядовитоды́мный** release of toxic smoke
пусто́тный vacuum, hollow
пусты́рь waste ground
путёвка 1. work ticket (MT)
2. travel and accommodation warrant
3. movement order, travel permit
путево́й travel, course, route, trip
путеме́р odometer
путепрово́д flyover, overpass, viaduct
путепрокла́дчик route clearer
~ **БАТ** route clearance dozer
путеразруши́тель track-wrecking machine
пути, вну́тренние во́дные inland waterways
пути́, коммуникацио́нные lines of communication
~ **подво́за и эвакуа́ции** supply and evacuation routes, "up-and- down" routes
~ **сообще́ния** lines of communication
~ **сообще́ния, во́дные** navigable waterways
пу́ти сообще́ния, вое́нные military lines of communication
путь path, route, track, way, journey
~, **бесстыково́й** continuous welded rail (rly)
~, **боево́й** campaign record (of a unit etc)
~, **боково́й** siding (rly)
~ **вероя́тного подхо́да** likely approach
~, **во́дный** inland waterway
~, **вытяжно́й** sorting track (rly)
~, **га́зовый** gas passage (ord, SA)
~, **гла́вный** main line (rly)
~ **движе́ния** line of march
~, **двойно́й** double track
~, **дыха́тельный** respiration tract
~, **железнодоро́жный** railway track
~, **запа́сный** 1. alternate route
2. side track (rly)
~, **коле́йный** track-way
~, **коло́нный** cross-country road
~, **кру́жный** detour
~ **манёвра** tactical route
Путь, Мле́чный Milky Way
путь, моноре́льсовый monorail (track)
~ **обхо́да** bypass
~, **обхо́дный** bypass
путь, объездно́й indirect route, roundabout route, detour
~ **отхо́да** line of withdrawal
~, **погру́зочный** loading siding (rly)
~ **подво́за** supply route
~ **подво́за, основно́й** main supply road, main supply route (MSR)
~ **подво́за, полково́й** regimental supply route
~ **подхо́да** approach
~, **подъездно́й** 1. access road, approach road
2. railway siding, spur track
~, **подъездно́й железнодоро́жный** spur track, railway siding
~, **подъездно́й коло́нный** access trail
~, **попере́чный** lateral road
~, **приёмо-отпра́вочный** reception and departure siding (rly)
~, **ре́льсовый** track, railway line
~, **ре́льсовый подъездно́й** rail access
~, **рока́дный** lateral road, lateral line of communication
~, **сме́жный** adjacent route
~ **снабже́ния, гла́вный** axis of supply, main supply route
~ **снаря́да в кана́ле ствола́, по́лный** total shot travel (ball)
~ **сообще́ния** communication route
~ **торможе́ния** stopping distance, braking distance
~ **эвакуа́ции** evacuation route
~ **эвакуа́ции, полково́й** regimental evacuation point
~, **электрофици́рованный** electrified track
пу́чность antinode
~ **ско́рости** velocity antinode
~ **то́ка** current antinode
пучо́к 1. bundle, bunch
2. beam
~, **дополни́тельный** incremental charge (ammo)
~, **запи́сывающий** writing beam (electron)
~ **заря́да** bundle (of cordite in a charge)
~, **остронапра́вленный** pencil beam
~, **стира́ющий** erasing beam (electron)
~, **счи́тывающий** reading beam, viewing beam (electron)
~ **траекто́рий** dispersion cone (ball)
~, **фикси́рующий** holding beam (electron)

пу́шечный gun
пу́шка gun, cannon
~, **автомати́ческая** 1. automatic cannon 2. automatic gun
~, **автомати́ческая зени́тная** automatic anti-aircraft gun
~, **арме́йская** army gun (122-155 mm, Sov)
~, **а́томная** atomic cannon
~, **батальо́нная** battalion gun (45-76mm, Sov)
~, **бикали́берная** tapered-bore gun
~ **большо́й мо́щности** long-range gun
~, **букси́руемая** towed gun
~, **гладкоство́льная** smooth-bore gun
~, **го́рная** mountain gun
~, **го́рно-вью́чная** mountain gun, pack gun
~, **деса́нтная** light assault gun (airborne and amph ops)
~, **дивизио́нная** divisional gun (75-88 mm, Sov)
~, **длинноство́льная** long-barrel gun
~, **зени́тная** anti-aircraft gun
~, **корпусна́я** corps gun (100-115 mm, Sov)
~, **ла́зерная** laser(-beam) gun
~, **линемета́тельная** line-throwing gun
~, **малокали́берная** small calibre gun
~, **нарезна́я** rifled gun
~, **полева́я** field gun
~, **полкова́я** regimental gun
~, **противота́нковая** anti-tank gun
~ **с автома́тикой, испо́льзующей си́лу отда́чи, автомати́ческая** recoil-operated gun
~ **с высо́кой нача́льной ско́ростью снаря́да, та́нковая** high-velocity tank gun
~, **самодви́жущаяся** auxiliary-propelled gun
~, **самохо́дная** self-propelled gun
~, **самохо́дная противота́нковая** self-propelled anti-tank gun
~, **сверхдальнобо́йная** super-long-range gun
~, **скорostре́льная** rapid-fire gun, quick-fire gun
~, **спа́ренная** twin-barrelled gun
~, **среднекали́берная** medium gun
~, **универса́льная** dual-purpose gun, dual-role gun
~, **шта́тная** standard gun
пу́шка, электро́нная electron gun
пыж wad, swab (gunn)
~, **во́йлочный** felt wad (ammo)
~, **восково́й** wax wad
~, **карто́нный** compressed paper wad (ammo)
~, **контро́льный** check swab (arty)
~, **мы́льный** soap wad
~, **оса́ленный** grease wad
пыжева́ние swabbing out (gunn)
пыжива́ние swabbing out (gunn)
пылело́вка, центробе́жная centrifugal air filter
пылеме́р dust counter (instr)
пыль, радиоакти́вная radioactive dust
пьезоэлектри́ческий piezo-electric
пя́льцы, закладны́е collar sett (mining)
пята́ 1. vertical journal (eng) 2. pivot bearing 3. heel
~ **о́си** axle journal
~, **шарова́я** ball joint, socket joint
пята́к 1. number 5 2. 5-kopeck piece
пятибо́рье pentathlon
~, **офице́рское** officers' pentathlon
пятивёрстка scale 1 : 210,000 map
пятикилометро́вка scale 1 : 500,000 map
пятикле́ммный five-terminal
пятило́пастный five-bladed
пятио́кись азо́та nitrogen pentoxide
пя́тка heel end, heel
~ **прикла́да** heel of the butt (SA)
пятно́ spot, stain, blemish
~, **като́дное** cathode spot
~, **светово́е** light spot
~, **тёмное** dark spot (television)

Р

рабо́та 1. work, task, job 2. performance 3. operation, working, running (of machines)
~, **безотка́зная** reliable service
~, **беспере́бо́йная** trouble-free running, trouble-free operation
~, **боева́я** fighting performance, operational efficiency

рабо́та в боевы́х усло́виях battle performance
~, **дли́тельная** continuous running
~, **ду́плексная** duplex operation (comms)
~, **закры́тая** classified work
~, **землеро́йная** excavation
~, **землеро́йно-тра́нспортная** earth-moving
~, **канцеля́рская** office work, clerical duties
~, **монта́жно-демонта́жная** assembling and dismantling
~ **на бе́дной сме́си** weak running (ICE)
~ **на бога́той сме́си** rich running (ICE)
~ **на холосто́м ходу́** idling
~, **нау́чно-иссле́довательская** scientific research
~, **непреры́вная** continuous running
~ **по снабже́нию** supply operations
~ **по теку́щему содержа́нию** maintenance work
~, **поша́говая** step-by-step operation
~, **ремо́нтная** repair work
~, **сверхуро́чная** overtime
~, **синфа́зная** in-phase operation (elec)
~, **сла́женная** well co-ordinated work, teamwork
~, **спаса́тельная** rescue operations
~ **тыла** 1. rear services, administrative services
2. work of rear area
~, **учебно-воспита́тельная** training
~, **хозя́йственная** fatigues, fatigue duties
рабо́тать (imp only) to work, to operate, to run (machine); **рабо́тать ключо́м** to transmit by telegraphy, to send morse to key (comms); **рабо́тать микрофо́ном** to transmit by voice (comms)
рабо́тник worker
~, **медици́нский** medical orderly
~, **нау́чный** scientific worker
работоспосо́бность 1. efficiency
2. fitness for work, working capacity
рабо́ты, авари́йно-спаса́тельные salvage and rescue operations
~, **авари́йные** emergency rescue and recovery
~, **взрывны́е** demolitions, demolition work
~, **демонта́жные** dismantling
рабо́ты, доро́жно-мостовы́е road and bridge construction and repairs, road and bridge works
~, **доро́жно-строи́тельные** road construction work
~, **доро́жные** road works
~, **землеро́йно-тра́нспортные** earth-moving
~, **земляны́е** earthworks
~, **инжене́рные** engineer works, engineer tasks
~, **контрми́нные** counter-mining
~, **пиротехни́ческие** fireworks
~ **по размини́рованию** mine clearing activity
~, **подзе́мно-ми́нные** underground mining, tunnelling (mil eng)
~, **подрывны́е** demolition work
~, **регла́ментные** periodic technical servicing (ac, radar)
~, **регулиро́вочные** checking and adjustment
~, **ремо́нтно-строи́тельные** construction and repair work
~, **ремо́нтные** repair work
~, **сапёрные** engineer tasks (fd wks), field engineering
~, **спаса́тельные** rescue work
~, **стано́чные** machining
~, **транше́йные** trenching operations
~, **фортификацио́нные** protective works, field works (fd eng)
рабо́чий 1. worker, workman, operator
2. work
3. operating, actuating, working
равне́ние 1. dressing (drill), alignment
2. levelling, equalization
"Равне́ние на Среди́ну!" "Eyes Centre!" (Sov)
"Равне́ние Нале́во!" "Eyes Left!"
"Равне́ние Напра́во!" "Eyes Right!"
равни́на lowland, plain (topo)
равни́нный flat, unbroken (topo)
равнове́сие equilibrium, balance
~, **адиабати́ческое** adiabatic equilibrium
~, **аэродинами́ческое** aerodynamic balance
~, **безразли́чное** neutral equilibrium, neutral stability
~, **бинаура́льное** binaural balance
~, **гравитацио́нное** gravitation equilibrium

равнове́сие, динами́ческое dynamic equilibrium
~ **моме́нтов** balance of couples, moment balance
~, **неусто́йчивое** unstable equilibrium
~, **относи́тельное** relative equilibrium
~, **полити́ческое** balance of power, political equilibrium
~ **пути́** stability in yaw (GW)
~ **сил** balance of power
~, **температу́рное** temperature balance
~, **теплово́е** thermal balance
~, **термодинами́ческое** thermodynamic equilibrium
~, **усто́йчивое** stable equilibrium
равноде́йствующая resultant (maths)
равноде́нствие equinox
равноме́рность uniformity
равноме́рный even, uniform
равносигна́льный equisignal
равночасто́тный equifrequent
ра́вный equal, similar, alike
"Равня́йсь!" "Right Dress!"
равня́ться/сравня́ться to align, to equal, to dress (drill), to keep up with (tac); **равня́ться/сравня́ться в заты́лок** to cover off from front to rear
рад rad (a unit of absorbed radiation dose; 100 ergs per gram of the tissue)
рада́р radar
радиа́н radian
радиа́тор 1. radiator (veh) 2. cooling jacket (SA)
~, **боково́й** side radiator
~ **в сбо́ре, водяно́й** water radiator assembly
~, **водяно́й** water radiator
~, **возду́шный** air cooler
~, **змеевико́вый** coiled radiator
~, **кольцево́й** annular radiator
~, **ма́сляный** oil cooler, oil radiator
~ **отопле́ния** heating element
~, **пласти́нчатый** film coil radiator
~, **ребри́сто-тру́бчатый** fin and tube radiator
~, **ребри́стый** finned radiator
~, **со́товый** honeycomb radiator, cellular radiator
~, **тру́бчатый** tube radiator
~, **яче́истый** cellular radiator, honeycomb radiator
радиацио́нно-хими́ческий radiation-chemical
радиацио́нный radiac, radiation
радиа́ция, втори́чная secondary radiation
~, **ионизи́рующая** ionizing radiation
~, **косми́ческая** cosmic radiation
~, **нача́льная** immediate nuclear radiation, initial nuclear radiation
~, **нача́льная я́дерная** immediate nuclear radiation, initial nuclear radiation
~, **оста́точная** residual radiation
~, **оста́точная я́дерная** residual nuclear radiation
~, **первонача́льная** initial radiation
~, **проника́ющая** 1. penetrating radiation, hard radiation 2. immediate nuclear radiation, initial nuclear radiation
~, **рассе́янная** scattered radiation
~, **рассе́янная проника́ющая** scattered radiation
~, **сумма́рная** total radiation
~, **фотосинтети́ческая акти́вная** photosynthetically active radiation
~, **я́дерная** nuclear radiation
ра́дий radium
ра́дио radio, wireless
радиоавтомоби́ль radio truck
радиоакти́вно-заражённый contaminated with radioactivity
радиоакти́вность radioactivity
~, **атмосфе́рная** atmospheric radioactivity
~ **во́здуха** airborne radioactivity
~, **иску́сственная** induced radioactivity
~, **наведённая** induced radioactivity
~, **оста́точная** residual radioactivity
радиоакти́вный radioactive
радиоальтиме́тр radio altimeter
радиоаппара́т radio set
~, **вью́чный** pack set (radio)
ра́дио-аппарату́ра radio equipment
радиоастроно́мия radio astronomy
радиоба́шня radio tower
радиоблокно́т message pad
радиобюро́ radio bureau (sigs)
радиовеща́ние broadcasting
радиовзво́д wireless platoon
радиовзрыва́тель 1. radar fuse, radar proximity fuse, radar pulse fuse 2. radio fuse, radio proximity fuse, radio-actuated proximity fuze, VT fuse
~, **акти́вный** active radio fuse

радиовзрыва́тель, до́плеровский Doppler-reflection fuse
~, часто́тно-индукцио́нный electronic frequency influence fuse
радиовойна́ electronic warfare
радиоволна́ radio wave
радиовы́зов radio call
радиовысотоме́р 1. radio altimeter 2. height-finding radar
радиовы́шка radio tower
"Радиогазе́та" "forces magazine" radio programme
радиоглуши́тель jamming station
радиогонио́метр radio goniometer (DF)
радиогоризо́нт 1. radio horizon 2. radar horizon
радиогра́мма 1. radio telegram 2. radiogram
радиогру́ппа radio net
радиодальноме́р radio range-finder
~, зени́тный anti-aircraft range finder (radar)
радиода́нные radio operating data
радиодевиа́ция radio deviation (DF)
~, полукругова́я semi-circular error (DF)
~, постоя́нная constant deviation (radio DF)
~, четвертна́я quadrantal error (DF)
радиодежу́рство radio watch
радиодезинформа́ция radio deception
радиодермати́т radiodermatitis
радиоди́ктор radio announcer
радиодисципли́на radio discipline
радиодонесе́ние radio message
радиозапро́счик interrogator (radar IFF)
радиозасе́чка radio fix (DF)
радиозо́нд radio sonde
радиоизлуче́ние radio emission
радиоиндика́тор диста́нции про́маха radio miss-distance indicator (AD arty)
радиокана́л radio channel
радиоко́д radio code
~, штормово́й storm warning radio code
радиокома́нда radio command
радиокорреспонде́нция radio traffic
радиола́мпа valve, vacuum tube (electron)
радиоли́ния microwave link (comms)
~ управле́ния, кома́ндная command wire-guidance (missiles)
радиолока́тор radar set, radar
~ визи́рования це́ли target-tracking radar
радиолока́тор, высотоме́рный height-finding radar
~, контрминомётный mortar-locating radar
~ наведе́ния command radar, guidance radar
~, обзо́рный surveillance radar
~ обнаруже́ния surveillance radar
~ обнаруже́ния облако́в cloud warning radar
~ обнаруже́ния це́ли acquisition radar
~, по́исковый scan radar, sweep radar, search radar
~ с автомати́ческим слеже́нием radar with automatic tracking system, automatic tracking radar
~ с непреры́вной генера́цией continuous wave radar, C.W. radar
~ слеже́ния за це́лью target-tracking radar
~ сопровожде́ния tracking radar
~, телевизио́нный television radar (TV camera focused on radar screen)
радиолока́торщик radar operator
радиолокацио́нный radar
радиолока́ция radar, radiolocation
~, акти́вная active radiolocation
~, до́плеровская Doppler radar system
~, монои́мпульсная monopulse radiolocation
~, пасси́вная passive radar
~, трёхкоордина́тная volumetric radar
радиолу́ч 1. radio beam 2. radar beam
~, поса́дочный radio landing beam
радиомаскиро́вка 1. signals security and deception 2. communications security
радиома́стер radiotechnician
радиома́чта radio mast
радиомая́к radio beacon
~, визуа́льно-звуково́й visual-aural radio range, VAR (ac)
~, всенапра́вленный omnirange, omnidirectional radio beacon
~, глисса́дный glide-path transmitter (airfield radio beacon)
~, курсово́й runway localizer transmitter (airfield ILS)
~, ма́ркерный radio marker beacon
~, равносигна́льный radio range, radio range station (navig beacon)

радиомая́к-отве́тчик responder beacon, transponder
радиомегафо́н powered megaphone, loud-speaker
радиоместоопределе́ние radio fixing (DF)
радио́метр 1. radiometer 2. radiac meter (nucl)
радиоме́трия 1. measurement of radiation 2. measurement of degree of radio-activity contamination
радиомеха́ник radio technician
радиомеша́ние jamming
радиомолча́ние 1. radio silence 2. electronic silence
радионаблюде́ние monitoring, radio monitoring, radio interception
радионаведе́ние electronic guidance (GW)
радионавига́ция radio navigation
радионаправле́ние radio link (comms)
радиообме́н radio traffic, communication by radio
радиообору́дование radio equipment
радиоокно́ radio spectrum
радиоопера́тор radio operator
радиоотделе́ние radio detachment
радиопаро́ль authenticator (comms)
радиопе́ленг radio bearing, radio DF bearing
~, и́стинный true radio bearing
радиопеленга́тор 1. radar direction finder 2. radio DF station, radio range station
~, земно́й ground DF station
~, коротково́лновый short wave radio DF station
радиопеленга́ция radio direction finding DF, radio DF, radio goniometry
радиопеленгова́ние DF, radio direction finding, radio DF, radio goniometry
радиопереговóры radio traffic
радиопереда́тчик radio transmitter
~, авари́йный emergency transmitter, radio distress transmitter
~, кома́ндный command transmitter (GW)
~ напра́вленного де́йствия beam transmitter
радиопереда́ча radio transmission
радиопередви́жка mobile radio station
радиоперехва́т radio interception
радиопого́да radio-wave propagation conditions
радиоподслу́шивание radio monitoring, radio interception
радиопозывно́й radio callsign
радиополиго́н signals training area
радиополуко́мпас radio compass
радиопоме́ха 1. interference (radio) 2. jamming (radio)
~, акти́вная active jamming
~, загради́тельная barrage jamming (ECM)
~, немодули́рованная continuous wave jamming, C.W. jamming
радиоприём receiving, reading, taking (sigs)
~, слухово́й C.W. reception, morse reception
радиоприёмник radio receiver
~, автомоби́льный car radio receiver
~, дежу́рный receiver on listening watch (comms)
~, разве́дывательный intercept receiver, search receiver
~, супергетероди́нный superheterodyne receiver, superhet
радиоприце́л 1. radio sight 2. radar sight
радиопроже́ктор radar-directed searchlight, radar- controlled searchlight
радиопрозра́чность radio permittivity
радиопротиводе́йствие electronic countermeasures, ECM
~, акти́вное active electronic countermeasures
~, пасси́вное passive electronic countermeasures, passive ECM
радиоразве́дка signals intelligence (SIGINT, incl COMINT and ELINT)
радиореле́йный radio-relay
радиоретрансля́ция radio-relay, radio-relaying
радиосвя́зь radio communication
~, двусторо́нняя two-way radio communication
~, ду́плексная duplex radio communication
~, ионосфе́рная ionospheric radio communication
~, многокана́льная multi-channel radio communication
~, односторо́нняя one-way radio communication

радиосвя́зь, си́мплексная simplex radio communication
~, тропосфе́рная troposphere (radio) communication
радиосе́ть 1. antenna system
2. radio net
~, авари́йная emergency radio net
~, арме́йская army command net
~, дивизио́нная divisional command net (comms)
~, звёздная star (radio comms)
~, кольцева́я star with lateral working (radio)
~, кома́ндная command net
~ кома́ндования command net
~, комбини́рованная joint net, combined net (radio)
~, ко́рпусная corps command net (comms)
~, ло́жная deception radio network
~ начартди́ва divisional artillery commander's net
~, свобо́дная free net (comms)
~ ты́ла диви́зии divisional administration net
~ целеуказа́ния target indication net
радиосигна́л radio signal
~, напра́вленный directional radio signal
~, отражённый radio echo
радиосле́жка traffic analysis (comms)
радиосре́дство means of radio communications
радиоста́нция 1. radio transceiver, radio set
2. radio station
~, автомоби́льная radio car, vehicular radio set
~, ведо́мая slave radio station
~, веду́щая master radio station
~, вое́нная combat radio (set)
~, контро́льная control station (radio)
~, миниатю́рная miniature radio (set)
~, назе́мная ground-based radio station
~, пеленга́торная direction-finding radio station
~, перено́сная portable transmitter-receiver, portable radio set
~, подви́жная radio communication vehicle
~, подслу́шивающая intercept station (sigs)
радиоста́нция, портати́вная portable transmitter-receiver, portable radio set
~, портати́вная ду́плексная walkie-talkie
~, приёмно-передаю́щая transceiver, transmitter-receiver
~, промежу́точная relaying station, intermediate station (radio)
~, ра́нцевая man-portable radio set, portable transceiver, man-pack radio set
~, распоряди́тельная control (radio net), control station (sigs)
~, ретрансляцио́нная radio-relay station
~, та́нковая tank radio
~, трансляцио́нная relay station, relaying station
~ тропосфе́рной свя́зи tropospheric radio relay station
~, устано́вленная на маши́не vehicular radio set
~, шле́мная helmet radio
радиотелегра́мма radio telegram, radio message
радиотелегра́ф 1. radio telegraph
2. radio telegraph equipment
3. radio telegraph office
радиотелеграфи́ст radio telegraphist
радиотелегра́фия radio telegraphy
радиотелегра́фный radio-telegraph
радиотелеизмере́ние radio telemetry
радиотелеметри́я radio telemetry
радиотелемеха́ника radio telecontrol, radio telemechanics
радиотелепри́нтер radio teleprinter
радиотелеуправле́ние remote control by radio, radio control
радиотелефо́н radio telephone
радиотелефони́ст radio telephonist, radio telephone operator
радиотелефони́я radio telephony
радиотелефо́нный radio-telephone
радиоте́нь screened zone (radio)
радиоте́хник radio technician
радиоте́хника 1. science of radio, radio technology
2. electronics
3. radio engineering
радиотехни́ческий 1. electronic
2. radio
3. radio engineering
радиото́чка public loudspeaker, room speaker
радиотрансля́ция rebroadcast (comms)

радиотрена́ж radio training
радиоуглеро́д radiocarbon
радиоу́зел 1. radio centre 2. broadcasting centre
~, приёмный central radio receiving room
радиоуправле́ние radio control
радиоуправля́емый radio-controlled
радиоустано́вка wireless set, radio set
радиоустро́йство radio equipment
~ для слепо́й поса́дки instrument landing system
радиофици́рованный radio-equipped
радиофици́ровать to equip with radio (imp only)
радиофототелегра́ф facsimile radio-telegraphy
радиофуга́с radio-fired mine
радиоце́нтр radio communications centre
радиочастота́ radio frequency, RF
радиочасто́тный radio frequency, RF
радиошу́мы radio interference, radio noise
радиощу́п electronic mine detector
радиоэкспеди́ция signals despatch office (radio)
радиоэлектро́ника radio electronics, electronics
радиоэлектро́нный electronic
радиоэ́хо radio echo
ради́ровать to radio (imp only)
ради́ст radio operator
ради́ст-заряжа́ющий operator/loader (tank crew)
ради́стка woman radio operator
ради́ст-кодиро́вщик radio operator who plots meteorological information
ради́ст-эсти́ст radio and keyboard operator
ра́диус действи́тельного пораже́ния effective radius of beaten zone (gunn)
~ де́йствия 1. sensitive range (proximity fuze) 2. radius of action
~ де́йствия на пересечённой ме́стности cross-country range
~ де́йствия, ограни́ченный restricted radius of action
~ закругле́ния bending radius
~ зо́ны разруше́ния radius of destruction
~ криво́й curve radius
~ кривоши́па crank throw
ра́диус, ларморо́вский radius of gyration (nucl)
~, метацентри́ческий metacentric radius
~ ожива́льной ча́сти radius of ogive (ammo)
~ очерта́ния головно́й ча́сти снаря́да calibre radius head, CRH (ammo)
~ поворо́та radius of turn
~ поража́ющего де́йствия уда́рной волны́ взры́ва blast damage radius
~ пораже́ния, преде́льный acceptable miss distance (GW)
~ разлёта оско́лков effective fragmentation radius
~ разруше́ния radius of destruction
~ разры́ва bursting radius
~ сплошно́го пораже́ния lethal radius
~ убо́йного де́йствия effective casualty radius
~ убо́йного де́йствия оско́лков effective fragmentation radius
ра́диус-ве́ктор radius vector (surv)
разаррети́рование uncaging (of gyroscopes)
разаррети́ровать to uncage (tech); **разаррети́ровать гироско́п** to uncage a gyroscope
разассигнова́ние appropriation, allocation
разбави́тель diluent (chem)
разбала́нс unbalance
разбе́г take-off run
разбега́ться/разбежа́ться(разбе́гаться (разг)) to scatter, to disperse (tps) (intrans)
разбива́ть/разби́ть to defeat, to rout, to smash, to break, to pitch (tents)
разби́вка laying-out, lay-out, siting
~ деса́нта по самолётным гру́ппам allocation of assault personnel to aircraft, allocation of "chalk" numbers (airborne ops)
~ диапазо́на allotment of frequencies, allocation of frequencies
~ ла́геря laying out of a camp
разбира́емость ease of dismantling
разбира́ть/разобра́ть to strip, to dismantle, to sort out, to investigate (charge etc); **разбира́ть/разобра́ть винто́вки из ко́зел** to unpile arms; **разбира́ть/разобра́ть ору́жие** to strip a weapon

разби́ть (см разбива́ть)
разбо́йничий pirate, marauding
разбо́р 1. investigation, trial
2. summing-up, critique (after an exercise)
3. analysis
~ **стрельбы́** analysis of firing
разбо́рка 1. dismantling, stripping
2. sorting
~ **в боевы́х усло́виях** field stripping
~, **непо́лная** partial stripping
~, **по́лная** complete stripping
разбо́рный 1. collapsible, dismantleable, dismantling
2. dismountable
3. sectional
разбо́рчивость 1. legibility
2. readability
разбра́сывать/разброса́ть to scatter
разбро́с scatter, dispersion
~ **частоты́** frequency splitting (electron)
разбро́санность dispersion
разбры́згивание splashing, spraying
разва́л break up, disintegration
~ **колёс** wheel camber (veh)
разва́лины 1. debris
2. blowdown
разве́данный reconnoitred
разве́дать (см разве́дывать)
разведгру́ппа reconnaissance party
развеddáнные reconnaissance information, intelligence data
разведдозо́р reconnaissance patrol
разведдонесе́ние reconnaissance report
разве́дка 1. reconnaissance
2. intelligence
3. patrol, reconnaissance group
~, **авиацио́нная** air reconnaissance
~, **агенту́рная** clandestine intelligence
~, **артиллери́йская** artillery reconnaissance and observation
~, **артиллери́йская инструмента́льная** artillery reconnaissance and observation
~, **бактериологи́ческая** biological reconnaissance, BW recce
~ **бе́рега** beach reconnaissance (amph ops)
~, **бли́жняя** close reconnaissance, reconnaissance for protection
~, **боева́я** battle reconnaissance
~ **бо́ем** 1. fighting patrol
2. reconnaissance in force
разве́дка, бронета́нковая 1. armoured reconnaissance
2. armoured reconnaissance group
~, **бурова́я** reconnaissance using trial bore-holes (tunneling)
~, **визуа́льная** visual reconnaissance
~, **возду́шная** air reconnaissance
~, **возду́шная радиацио́нная** aerial radiological survey
~, **возду́шная телевизио́нная** aerial television reconnaissance
~ **возду́шного проти́вника** air defence reconnaissance, air defence surveillance
~, **войскова́я** operational reconnaissance
~, **войскова́я инжене́рная** combat engineer reconnaissance
~, **войскова́я хими́ческая** operational chemical reconnaissance
~, **гидрографи́ческая** hydrographic reconnaissance
~, **глубо́кая** long-range reconnaissance
~, **да́льняя** long-range reconnaissance, distant reconnaissance, deep reconnaissance
~, **дозиметри́ческая** radiac reconnaissance
~, **заса́дная** the gaining of information by ambush
~, **звукова́я** sound ranging (arty obsn)
~, **звукосветометри́ческая** sound ranging and flash spotting (arty obsn)
~, **инжене́рная** engineer reconnaissance, engineer intelligence
~ **инфракра́сными сре́дствами** infra-red reconnaissance, IR reconnaissance, infra-red battlefield surveillance
~ **исто́чников воды́** reconnaissance for water
~, **команди́рская** commander's personal reconnaissance
~, **ли́чная** personal reconnaissance
~ **маршру́та** route reconnaissance
~, **маршру́тная** surface reconnaissance, route reconnaissance
~, **медици́нская** medical and biological warfare reconnaissance
~ **ме́стности** reconnaissance of the ground, area reconnaissance
~, **метеорологи́ческая** meteorological reconnaissance
~ **мин** mine detection
~ **ми́нных поле́й** minefield reconnaissance

разве́дка, моторизо́ванная motorized reconnaissance
~ **на во́ду** water supply intelligence
~ **наблюде́нием** visual reconnaissance
~, **назе́мная** ground reconnaissance
~, **оконча́тельная** pre-orders reconnaissance
~, **операти́вная** operational reconnaissance, operational intelligence
~, **опти́ческая** 1. visual reconnaissance, visual observation
2. flash spotting, flash ranging (arty)
~, **пе́шая** reconnaissance on foot
~ **пого́ды** weather reconnaissance
~ **подслу́шиванием** gaining information by use of listening posts
~ **по́иском** gaining information by use of fighting patrol
~, **предвари́тельная** preliminary reconnaissance
~ **пути́** route reconnaissance, routing survey
~, **радиацио́нная** radiac reconnaissance
~, **радиоакти́вная** radiac monitoring
~, **радиолокацио́нная** radar observation, radar battlefield surveillance
~, **радиотехни́ческая** electronic intelligence, electronic reconnaissance, ELINT
~ **радиоэлектро́нных средств** electronic reconnaissance
~ **родо́в войск** special-to-arm reconnaissance
~, **санита́рная** medical reconnaissance
~, **си́льная** reconnaissance in force (tac)
~, **совме́стная** joint reconnaissance
~, **специа́льная** special reconnaissance
~ **сре́дствами свя́зи** signals intelligence
~, **стратеги́ческая** strategic reconnaissance
~, **такти́ческая** tactical reconnaissance
~, **телевизио́нная** television reconnaissance
~, **техни́ческая** technical reconnaissance, technical intelligence
~ **тылово́го райо́на** service area reconnaissance
~, **фотограмметри́ческая** photogrammetric reconnaissance
~, **фотографи́ческая** photographic reconnaissance, photographic intelligence

разве́дка, хими́ческая chemical reconnaissance, chemical intelligence
~ **ши́лом** narrow-front reconnaissance (tks)
разведо́рган reconnaissance unit, intelligence agency
разведподразделе́ние reconnaissance sub-unit
разведро́та reconnaissance company
разведсво́дка intelligence summary
разведтеодоли́т artillery survey theodolite
разведча́сть reconnaissance unit
разве́дчик 1. scout, member of reconnaissance patrol
2. forward observer (arty)
3. observation aircraft
4. selector (Baudot receiver)
~, **артиллери́йский** artillery forward observer
~ **пого́ды** weather reconnaissance aircraft
разве́дчик-мотоцикли́ст motorcycle scout
разве́дчик-наблюда́тель scout observer, forward observation post observer
разве́дчик-парашюти́ст reconnaissance paratrooper
разве́дчик-слуха́ч member of listening patrol
разве́дывание carrying out of reconnaissance, reconnoitring
разве́дывательно-диверсио́нный commando-type reconnaissance
разве́дывательный 1. reconnaissance, scout
2. intelligence
разве́дывать/разве́дать to reconnoitre
разверну́ть (см развёртывать)
развёртка 1. sweep (radar display)
2. scanning (television camera)
3. trace (CRT)
4. reamer
5. development (blueprint)
~, **азимута́льная** azimuth sweep (radar PPI display)
~ **бегу́щим лучо́м** flying-spot scanning, light-spot scanning, indirect scanning
~, **бы́страя** fast scanning
~, **вертика́льная** vertical scanning
~, **временна́я** time base (radar display)
~, **горизонта́льная** horizontal scanning

развёртка, ждущая ratchet time-base, triggered sweep (radar)
~, кадровая frame scanning, vertical scanning
~, кольцевая circular sweep (radar display)
~, коническая taper reamer
~, линейная linear sweep (radar display)
~, линейная временная linear time-base (radar display)
~, нелинейная non-linear time base (radar display)
~, перемежающаяся interlaced scanning (television)
~, последовательная 1. sequential scanning (television, Brit)
2. progressive scanning (television, US)
~, прогрессивная 1. sequential scanning (television, Brit)
2. progressive scanning (television, US)
~, прямая linear sweep
~, прямолинейная rectilinear sweep (radar display)
~, прямоугольная rectangular scanning (radar)
~, радиальная radial sweep (radar display)
~, радиально-круговая PPI sweep (radar display)
~, спиральная spiral scanning
~, строчная line scanning (television)
~, ступенчатая step scanning
~, телевизионная television scanning
~, трёхмерная three-dimensional scanning
~, чересстрочная staggered scanning, interlaced scanning
~, экспоненциальная exponential timebase, exponential sweep (radar display)
~, электронная electronic scanning
развёртывание 1. scanning (by radar beam)
2. deployment
3. development
4. unfolding, unrolling, unfurling
5. setting-up, erection
6. reaming (tech)
~, винтовое helical scanning (radar)
~, коническое conical scanning (radar)
~, круговое circular scanning (radar)
~, спирально-коническое helical scan, spiral scanning (radar)
развёртыватель scanner (electron)
развёртывать/развернуть to unroll, to unfold, to unfurl, to deploy, to develop, to establish, to set up, to erect, to square (shoulders), to turn out (feet), to ream (tech), to scan (radar), to expand (a bn into a regt etc)
развёртываться в боевой порядок to deploy into battle formation
развести (см разводить)
разветвление 1. bifurcation
2. road fork, road junction
~, т-образное hybrid junction, T-junction (electron)
развивать/развить to develop; **развивать/развить наступление** to press home an attack; **развивать/развить ход** to get under way (nav)
развилина 1. bifurcation
2. road fork, road junction
развилка 1. bifurcation
2. road fork, road junction
развинтить (см развинчивать)
развинчивать/развинтить to screw out, to unscrew, to screw off
развитие development, evolution
~ прорыва exploitation of the breakthrough
~ успеха exploitation (tac)
развить (см развивать)
развод караулов guard mounting
~ суточного наряда daily fatigues parade
разводить/развести to take, to conduct to, to part, to separate, to dilute; **разводить/развести караулы** to mount the guard; **разводить/развести на посты** to post reliefs; **разводить/развести часовых** to post the sentries
разводной separating, opening
разводящий NCO marching reliefs, NCO i/c reliefs
разворот, боевой chandelle (flt manoeuvre) (aeron)
~ в ласточкином хвосте three point turn (veh)
развязать (см развязывать)
развязывание 1. untying, unfastening, uncoupling
2. decoupling (electron)
развязывать/развязать to untie, to undo, to uncouple; **развязывать/развязать войну** to unleash war

разгадáть (см разгáдывать)
разгáдывать/разгадáть to unravel, to guess, to discover; **разгáдывать/разгадáть цифр** to break a cypher
разгáр 1. heat, height, climax, full swing 2. wear, erosion, scoring (of bore)
~ канáла стволá erosion of bore
~, кольцевóй annular erosion
~, нúточный cord wear
разгибáтель ferrule (fuze)
разглашéние воéнной тáйны breach of security, divulgence of military secrets
разговóры, перехóдные cross-talk (radio)
разгóн acceleration
разгóнка тóплива distillation of fuel
разгонять/разогнáть to accelerate, to rev up, to race (engine)
разгрaдúть (см разграждáть)
разграждáть/разградúть to clear (mines), to remove (obstacles), to breach
разграждéние obstacle clearing
~ препятствий removal of obstacles, obstacle clearing
разгранúчивать/разгранúчить to delimit, to mark the limit
разгранúчить (см разгранúчивать)
разгранлúния boundary (between formations or units)
разгрóм 1. rout 2. destruction, devastation
разгромúть (см громúть) to rout, to destroy, to devastate (perf only); **разгромúть нáголову** to put to rout
разгружáть/разгрузúть to unload, to offload
разгрузúть (см разгружáть)
разгрýзка 1. unloading, discharge 2. relief 3. handling
~, первоочереднáя priority unloading
~, сýточная daily unloading average
разгруппирóвка de-bunching (elec)
раздáвливание overrunning, overwhelming (tac)
раздáтчик distributor, distributing agent
раздвижнóй telescopic, adjustable, expanding, extensible, collapsible, sliding
раздвоéние bifurcation, branching, forking
раздевáльня undressing room, undressing tent
раздéл 1. partition, division, separation 2. section (document) 3. break-sign (sigs)
разделéние separation, division, distribution, allocation
~ во врéмени time division (sigs)
~ канáлов, временнóе time-division multiplexing
~ канáлов, частóтное frequency-division multiplexing
~ огня allocation of fire tasks
разделённый на тысячные graduated in mils
раздражáющий 1. caustic, corrosive 2. irritant (CW)
раздробúть (см дробúть)
раздýтие канáла barrel bulge (ord)
раздýтость канáла barrel bulging (ord)
разжáлование demotion, reduction in rank
разжáловать to degrade (perf only); **разжáловать в рядовые** to reduce to the ranks
раззенкóвка rebore
разлагáть/разложúть to decompose, to analyse; **разлагáть на составляющие чáсти** to resolve into components
разлёт убóйных оскóлков fragmentation patterns, spread (ball)
разлúв 1. flood, overflow 2. distribution (liquid)
~ горючего в мéлкую тáру breakdown of fuel into containers, bulk-breaking of fuel
разливáться/разлúться to overflow
разлúться (см разливáться)
различúтель discriminator (electron)
~, фáзовый phase discriminator
разложéние 1. resolution 2. dissociation, decomposition, disintegration, decay
~ вéкторов resolution of vectors
~ изображéния resolution of image (TV)
~, химúческое chemical decomposition
разложúть (см разлагáть)
размагнúтить (см размагнúчивать)
размагнúчивать/размагнúтить to demagnetize
размáтывание unwinding, unreeling, paying out, uncoiling
размáтывать/размотáть to unwind, to unreel, to pay out

размáх 1. swing, sweep
2. scale, scope
3. throw (mech)
4. span, spread (wings)
~ **оперáции** scope of the operation (Sov term defines frontage and depth)
размедни́тель decoppering agent (ammo)
размéр size, dimension, rate, amount, scale, extent, level (of supply etc)
~, **внéшний** overall dimension
~, **внýтренний** internal dimension
~, **габари́тный** overall dimension, overall size
~ **запáса** level of supply, stock level, supply level
~, **нарýжный** external dimension
~, **номинáльный** nominal size
~, **нормáльный** standard size
~, **основнóй** basic dimension
~ **пéнсии** rate of pension
~ **под ключ** width across flats (of nut or bolt)
~, **пóлный** overall size
~, **стандáртный** standard size
~ **цéли** target size, target dimension
размéры, габари́тные clearance limits, overall dimensions
~ **цéли, приведённые** corrected target dimensions (area in which a shot can be considered a direct hit)
размести́ть (см размещáть)
размéтка 1. marking, plotting, marking out
2. layout
~ **вагóнов** marking off (rly)
размещáть/размести́ть to locate, to arrange, to place, to distribute, to billet, to quarter; **размещáть в палáтках** to accommodate under canvas
размещéние accommodation
~ **по кварти́рам** billeting
~, **факти́ческое** physical layout
~ **частóт** frequency allocation
разминёр member of gapping team (mine clearance)
размини́рование mine clearing, mine lifting
~ **вручнýю** hand lifting of mines, hand removal of mines
~ **маршрýта** route mine clearance
~, **сплошнóе** area mine clearance
~ **учáстка (мéстности)** minefield clearance, minefield lifting
размини́рованный cleared of mines
размини́ровать to clear mines (imp only)
разморáживание defrosting
размотáть (см размáтывать)
размóтка unwinding, unreeling, paying out
размыкáние 1. breaking, disconnection (elec)
2. opening out (tac)
~, **бы́строе** quick release (mech)
размыкáтель circuit breaker, trip (elec)
размыкáть/разомкнýть to disconnect, to break
размягчéние softening
разнаря́дка 1. order, warrant
2. allocation
~ **материáльных средств** distribution order
разнесённость, прострáнственная space diversity (electron)
разнобóй 1. dislocation, lack of co-ordination
2. ranging difference (arty)
~ **орýдий** spread (gunn)
разнобóйность dispersion, spread (gunn)
разнокали́берный different calibre
разнообрáзие навóдки inherent human error in laying (gunn)
разнóс 1. separation
2. racing (of engine)
~ **по частотé** frequency separation (electron)
~, **частóтный** frequency diversity (electron)
рáзность difference, variety, diversity
~ **времён** time interval (sound ranging)
~ **высóт** altitude difference
~ **дáльностей** range difference
разóбранный disassembled, dismantled, taken apart
разобрáть (см разбирáть)
разобщéние disengaging, releasing
разобщи́тель 1. disconnector
2. disconnector rod (pistol)
разогнáть (см разгоня́ть)
разогревáть/разогрéть to warm up; **разогревáть/разогрéть двигатели** to warm up engines
разогрéтый warmed up, heated
разогрéть (см разогревáть)
"Разойди́сь!" "Fall Out!", "Dismiss!"
"Разойти́сь!" "Fall Out!", "Dismiss!"
разойти́сь (см расходи́ться)

"Разомкнись!" "Open Out!"
разомкнуть (см размыкать)
разорвавшийся exploded
разорвать (см разрывать)
~ **фронт** to break through the front
разоружать/разоружить 1. to disarm
2. to dismantle
разоружение disarmament
~ **мины** rendering mine safe
разоружить (см разоружать)
разрабатывать/разработать to cultivate, to exploit (mining), to work out; **разрабатывать/разработать план** to formulate a plan
разработать (см разрабатывать)
разработка 1. development
2. solving, working out
3. processing
4. cultivation
~, **контрольная** practice, practice runthrough
разражаться/разразиться to break out, to burst out
разразиться (см разражаться)
разрегулировка misalignment, maladjustment
разрежение 1. vacuum
2. tenuity (physics)
3. rarefaction
разрез 1. section (drawing)
2. cut, slit
~, **боковой** lateral section
~, **вертикально поперечный** cross-section
~, **полный** full section
~, **продольный** longitudinal section
~, **сложный** off-set section
~, **частичный** broken section (eng drawing)
разрешение authority, permission, permit, clearance
~ **на нанесение ядерного удара** atomic release
~ **на отпуск материальных средств** authority for issue
~ **на право перевозки** authority for move
~ **пролёта** flight clearance
разрешённый authorized
разрозненный 1. uncoordinated
2. isolated, separate
разруха devastation, ruin
разрушать/разрушить to demolish, to blow up, to break down
разрушение demolition, damage, destruction
~ **моста** bridge demolition
~ **объектов оперативно-стратегического значения** strategic demolition
~ **основных объектов** key demolitions
~ **от усталости** fatigue failure (metall)
~, **подводное** underwater demolition
~, **полное** complete disintegration, total destruction
разрушительный destructive
разрушить (см разрушать)
разрыв 1. burst, explosion
2. gap, breach, rupture
~ **вблизи цели** near-miss
~, **воздушный** air burst
~, **высокий** high burst
~ **гильзы, поперечный** separated case (QF fixed ammo)
~, **действительный** effective burst
~ **до взведения взрывателя** premature detonation
~ **на поверхности земли** ground burst
~, **наземный** surface burst, ground burst
~, **наивыгоднейший** effective burst
~ **ниже цели** air burst below target (AD)
~, **низкий** low burst
~, **нормальный** normal burst
~, **перелётный воздушный** over (air burst)
~, **преждевременный** "premature" (gunn), premature detonation
~ **при ударе** impact burst
~, **противопожарный** firebreak, fire prevention lane, ride
~, **рикошетный** ricochet burst
разрывать/разорвать to tear, to break, to disrupt, to blow up (intrans)
разрыхление disintegration, dilation, distension, expansion, loosening(-up)
разряд 1. class, category, grade
2. discharge (elec, wpn)
~, **грозовой** lightning discharge, lightning
~, **дуговой** arc discharge (elec)
~, **коронный** corona discharge (elec)
~, **первый по стрельбе** first-class shot (category)
~, **тлеющий** glow discharge
разрядить (см разряжать)

разря́дник 1. discharge tube (radar)
2. spark gap
3. lightning conductor
4. sportsman with official rating, listed sportsman
~, враща́ющийся искрово́й rotary spark gap (radar)
~, га́зовый spark gap tube (radar)
~, грозово́й lightning arrester, lightning conductor
~, искрово́й spark gap (elec device)
~ магази́на 1. clip release (ord)
2. magazine catch (SA)
~, резона́нсный rhumbatron (electron)
"Разряжа́й!" "Unload!"
разряжа́ние 1. unloading, unpriming
2. discharging (elec)
разряжа́ть/разряди́ть to unload, to clear (wpn), to unprime (gren), to disarm (mine), to discharge (elec)
разъеда́ние corrosion, pitting
разъедине́ние separation, interruption, disconnection, disengagement, releasing, uncoupling
разъедини́тель disconnecting switch, isolating switch
разъедини́ть (см разъединя́ть)
разъединя́ть/разъедини́ть to split up, to isolate, to disconnect, to uncouple
разъе́зд 1. small cavalry detachment
2. lay-by
3. (passing) siding (rly)
4. departure
~, боево́й mounted patrol
разъём, быстроде́йствующий quick-release plug, snatch plug
~, ште́псельный 1. plug connection
2. umbilical plug (GW)
разъёмный split
райо́н 1. area, region, zone
2. rajon (Sov administrative region)
~, арме́йский army area
~, артиллери́йский противота́нковый artillery anti-tank zone
~ артиллери́йских пози́ций gun position
~ а́томного уда́ра atomic strike area
~, аэродро́мный airfield complex
~, бива́чный bivouac area
~ боевы́х де́йствий 1. operational zone, theatre of operations
2. battle area, combat zone
райо́н бо́я 1. battle area, operational zone
2. battle ground
~, взво́дный platoon locality
~, войсково́й divisional area of responsibility (Sov)
~, войсково́й тылово́й communications zone
~ вы́броски dropping zone, DZ, dropping area
~ вы́броски деса́нта dropping zone, DZ, dropping area
~ вы́грузки unloading area, debussing area, de-training area
~, вы́грузочный unloading area, debussing area, de-training area
~, выжида́тельный assembly area, concentration area
~ выпаде́ния радиоакти́вных оса́дков fallout area (NBC)
~ вы́садки landing area, landing ground
~ вы́садки морско́го деса́нта seaborne assault landing area
~ де́йствий area of operations
~ де́йствий а́рмии army area
~ деса́нта assault landing zone, dropping zone, DZ
~ десанти́рования assault landing area, dropping zone, DZ
~ деса́нтных опера́ций intervention area (amph ops)
~, дивизио́нный тылово́й divisional rear area
~ для наступле́ния, исхо́дный forming-up place, FUP
~ загражде́ний delaying area
~ запра́вки горю́чим refuelling area
~ зараже́ния contaminated area
~, исхо́дный forming-up place, FUP
~ коммуника́ций Lines of Communication area
~, малонаселённый sparsely populated area
~ обеспе́чения maintenance area, support area
~ обеспе́чения, арме́йский army maintenance area
~ обеспе́чения, берегово́й beach maintenance area (amph ops)
~ обеспе́чения коне́чно-вы́грузочного пу́нкта возду́шного тра́нспорта airhead maintenance area
~ обеспе́чения, корпусно́й corps maintenance area

рай́он обеспéчения, передовóй forward maintenance area
~, **обеспéчиваемый** supported area
~, **оборони́тельный** defensive area
~ **оборóны** defended area
~ **оборóны, батальóнный** battalion defence area, battalion area of responsibility
~ **оборóны, основнóй** vital ground, main defended area
~ **оборóны, рóтный** company defensive area, company area of responsibility
~ **оборóны, укреплённый** organised tactical locality
~, **обстрéливаемый** beaten zone (ball)
~ **огневы́х пози́ций** deployment area (arty)
~ **ожидáния** waiting area, transit area
~ **óтдыха** rest area
~ **охранéния** area of responsibility
~ **падéния снаря́дов** impact area
~ **переправ́ы** river crossing area
~ **переправ́ы, исхóдный** assault crossing forming-up place
~ **погру́зки** 1. embussing area, embarkation area (personnel) 2. loading area (equipment)
~, **погру́зочный** loading area
~, **подчинённый бáзовый** base subordinate area
~, **позициóнный** missile deployment area, missile launching area (GW)
~ **посáдки** embarkation area
~, **посáдочный** embarkation area
~, **прибрéжный** coastal strip
~ **приземлéния десáнта** landing zone, LZ (airborne ops)
~, **противотáнковый** tank-killing area
~, **противотáнковый укреплённый** tank-proof area, tank proof locality
~ **разрушéний** demolished area
~ **расквартировáния войск** accommodation area
~ **расположéния** location, unit lines
~ **рассредотóчения** area of dispersion
~, **рóтный** company area, company lines
~ **сбóра** assembly area, rallying area
~ **сбóра десáнта** assault force rendezvous, RV, assault landing rallying area
~, **сбóрный** assembly area, rendezvous, RV
~ **снабжéния** supply area

рай́он сосредотóчения concentration area
~ **стáртовых пози́ций** launching area (missiles)
~ **стоя́нки тáнков** harbour area (tk)
~, **танконедосту́пный** tank-proof locality
~, **тыловóй** rear area
~, **укреплённый** fortified area, fortified region, fortified locality
~ **формировáния автоколóнн** convoy marshalling area
~ **формировáния десáнта** lowering position (amph ops)
~ **цéли** target area

рак, лучевóй radiation cancer

ракéта 1. signal cartridge 2. rocket, missile
~, **авиациóнная** air launched missile
~, **артиллери́йская** artillery rocket
~, **баллисти́ческая** ballistic missile
~, **бéлая** white signal rocket
~, **бескры́лая** wingless missile
~ **бли́жнего дéйствия** tactical missile, short-range missile
~, **высóтная** high-altitude missile
~, **гроздевáя сигнáльная** cluster, cluster signal cartridge, multi-star signal cartridge
~ **дáльнего дéйствия** long-range rocket, long-range missile
~, **двухступéнчатая** two-stage missile
~, **деревя́нная лóжная** wooden dummy rocket
~, **дымовáя сигнáльная** smoke flare
~, **жи́дкостная** liquid-propellant rocket, liquid propellant missile
~ **заграждéния** air defence barrage rocket (releasing a hawser on a parachute)
~, **звуковáя** maroon (audible signal rocket)
~, **зени́тная** anti-aircraft missile, surface-to-air missile
~, **зени́тная неуправля́емая** anti-aircraft free-flight rocket
~, **зени́тная управля́емая** anti-aircraft guided missile, surface-to-air guided missile
~ **клáсса "вóздух-вóздух"** air-to-air missile
~ **клáсса "вóздух-земля́"** air-to-surface missile

ракéта клáсса "землñ-вóздух" surface-to-air missile
~ **клáсса "землñ-землñ"** ground-to-ground missile
~, **крылáтая** cruise missile
~, **крылáтая** cruise missile
~, **крылáтая зенúтная** winged anti-aircraft missile
~, **линеметáтельная** line-throwing rocket, life-saving rocket
~, **лóжная деревñнная** wooden dummy missile
~, **межконтинентáльная** inter-continental rocket, inter-continental missile
~, **межконтинентáльная баллистúческая** inter-continental ballistic missile ICBM
~, **метеорологúческая** meteorological rocket
~, **многозвёздная сигнáльная** multi-star signal rocket, cluster signal rocket
~, **многоступéнчатая** multi-stage rocket
~, **многоцелевáя управлñемая** multi-purpose guided missile
~, **неоперённая** finless rocket
~, **неуправлñемая** free-flight rocket, unguided rocket, free rocket
~, **низковысóтная** low-altitude missile
~, **одноступéнчатая** single-stage rocket
~, **оперённая** fin-stabilized projectile, fin- stabilized rocket, fin-stabilized missile
~, **осветúтельная** illuminating rocket
~, **оскóлочно-зажигáтельная** fragmentation incendiary rocket
~, **пороховáя** solid fuel rocket, solid propellant rocket
~, **противовертолётная** anti-helicopter missile
~, **противорадиолокациóнная** anti-radar missile
~, **противотáнковая** anti-tank missile
~, **радиоуправлñемая** radio-controlled missile
~ **с жúдкостным двúгателем** liquid-propellant rocket
~ **с лáзерной систéмой наведéния** laser-guided missile
~ **с химúческой боевóй чáстью** rocket with chemical warhead
~ **с ñдерной боевóй чáстью** missile with nuclear warhead
ракéта, самолётная баллистúческая air-launched ballistic missile
~, **самонаводñщаяся** homing missile
~, **сверхдáльняя баллистúческая** intercontinental ballistic missile, ICBM
~, **сигнáльная** signal flare rocket
~ **со склáдывающимся оперéнием** folding-fin missile
~, **составнáя** multi-stage rocket
~ **срéдней дáльности** intermediate-range missile
~, **стабилизúруемая вращéнием** spin-stabilized missile
~, **стабилизúруемая по крéну** roll-stabilized missile
~, **стáртовая** jet-assisted take-off unit (ac), JATO unit, booster rocket
~, **стратегúческая** strategic missile
~, **тактúческая** tactical missile
~, **твердотóпливная** solid-propellant missile
~, **телеуправлñемая** remote-controlled rocket, remote- controlled missile, guided missile, radio-controlled rocket
~, **трёхступéнчатая** three-stage missile
~, **управлñемая** guided missile, guided rocket
~, **управлñемая по проводáм** wire-controlled missile, wire-guided missile
~, **цветнáя** coloured flare, coloured rocket, coloured signal light
ракéта-ловýшка decoy missile
ракéта-мишéнь target-missile
ракéта-перехвáтчик anti-missile missile
ракéтница signal pistol, Very pistol, flare pistol
ракéтно-авианóсный carrier-borne missile-armed
ракéтно-техничéский missile
ракéтно-ñдерный nuclear missile
ракéтный rocket(-propelled), missile
ракетонóсец rocket carrier
ракетоплáн rocket aircraft
ракéтчик 1. member of launcher detachment, launcher number
2. battle-field illuminant operator
ракýрс 1. foreshortening ratio, angle of approach (Sov air gunn)
2. aspect (GW)
рáма 1. frame
2. sett (mining)
3. breech block carrier (ord)
4. chassis

ра́ма, бреве́нчатая timber frame, log frame (mining)
~, брусча́тая squared timber frame, frame of square timber (min)
~, гу́сеничная track frame (veh)
~, дверна́я door frame
~, доща́тая case (min)
~, жёсткая rigid frame
~, затво́рная 1. breechblock carrier (ord) 2. piston extension (SA)
~ каре́тки подве́ски bogie frame (tk)
~ кача́ния rocking frame
~, кача́ющаяся 1. rocking frame (vehicle movement- simulator used for tank-firing training) 2. rocking trestle (for launching spans, br)
~, лесопи́льная frame saw, gang saw
~, мише́нная target frame
~, монта́жная mounting bracket
~, мото́рная engine bed, subframe, underframe
~, опо́рная 1. strutting frame 2. sett (mining)
~ ору́дия, поворо́тная gun turntable
~, пласти́нная half-round timber frame
~, поворо́тная turntable (ord)
~, подко́сная "A" frame
~, подмото́рная engine sub-frame, auxiliary frame
~, по́исковая search coil (mine detector)
~ поршнево́го затво́ра breech screw
~, распо́рная spreader frame
~, сварна́я welded frame
~, спускова́я butt slide (SA)
~, ство́льная barrel extension (ord)
~ теле́жки подве́ски bogie frame (tk)
~, торцо́вая end frame (eng)
~, хребтова́я backbone frame (veh)
~ хребтово́й фо́рмы backbone frame
~ ша́сси main frame, chassis
ра́мка 1. frame 2. loop antenna, frame antenna 3. backsight leaf 4. body (pistol)
~ автомати́ческого пистоле́та automatic pistol receiver
~ боево́й пружи́ны mainspring housing
~, враща́ющаяся rotating frame antenna
~, враща́ющаяся радиопеленга́торная rotating frame radio direction finder
~, иска́тельная search-head, search coil unit (mine detector)
ра́мка, по́исковая search coil (mine detector)
~, прице́льная backsight leaf
~ револьве́ра revolver frame
ра́мпа ramp
ра́на wound
~, гно́йная septic wound
~, ко́лотая punctured wound
~, рва́ная lacerated wound
~, сквозна́я perforated wound
~, шрапне́льная shrapnel wound
ране́ние wound, wounding
~, смерте́льное fatal wound
ра́неный 1. wounded man 2. wounded
~, лежа́чий lying case (class of wounded)
~, носи́лочный stretcher case (class of wounded)
~, сидя́чий sitting case (class of wounded)
~ сре́дней тя́жести mild case (class of wounded)
~, ходя́чий walking wounded
~, че́люстно-лицево́й maxillo-facial case (class of wounded)
ра́нец 1. haversack, knapsack, pack 2. parachute pack
~, заплéчный haversack
ра́нец-рюкза́к haversack, rucksack, large pack
ра́нить to wound
ра́нцевый pack
ра́порт 1. report 2. application (request)
~ о предоставле́нии о́тпуска application for leave
~ по кома́нде application submitted through normal channels
рапортова́ть to report verbally
расквартирова́ние accommodation, billeting, quartering
расквартирова́ть (см расквартиро́вывать)
расквартиро́вывать/расквартирова́ть to billet, to accommodate
раскла́дка apportionment, allotment, distribution
~ проду́ктов 1. bill of fare 2. diet sheet 3. ration schedule
~, стандáртная standard diet-sheet
раскла́дчик minelayer

раскла́дчик, механи́ческий ми́нный mechanical mine layer, mine-laying vehicle
~, ми́нный minelayer
раскоди́ровать to decode
расконсерва́ция taking equipment out of reserve storage
раско́с strut, brace, truss
раскрути́ть (см раскру́чивать)
раскру́тка acceleration (gas turbine engine)
раскру́чивать/раскрути́ть to untwist, to untwine, to develop full power (engines)
раскры́в aperture (antenna)
раскрыва́ние парашю́та development of parachute
раскры́тие opening(-out), uncovering, disclosing, exposing
~ парашю́та parachute deployment
раскула́чивание маши́н cannibalization
распа́д 1. disintegration, breaking up
2. decomposition
3. decay (nucl)
~, а́томный atomic disintegration
~, взрывно́й explosive disintegration
~ по не́скольким путя́м multiple disintegration, branched disintegration (nucl)
~, радиоакти́вный radioactive decay
~, разветвлённый multiple disintegration (nucl)
распада́ться/распа́сться to disintegrate, to break up, to decay, to decompose
распа́дов в секу́нду disintegrations per second (nucl)
распа́ивать/распая́ть to unsolder
распакова́ть (см распако́вывать)
распако́вывать/распакова́ть to uncrate, to uncase, to unpack
распа́сться (см распада́ться)
распатрони́ровать to break into component parts (imp only)
распая́ть (см распа́ивать)
распира́ющий коло́дки brake expander
расписа́ние time-table, schedule
~, боево́е 1. order of battle
2. watch and station bill (nav)
~ заня́тий training schedule
~ посто́в time and post roll (guard)
~ сил проти́вника, боево́е enemy order of battle
распи́ска receipt
расплю́щивание поле́й наре́зов в кана́ле ствола́ flattening of lands (ball)
распознава́ние identification, recognition
распознава́ть/распозна́ть to identify, to recognise, to distinguish
распозна́ть (см распознава́ть)
располага́ть/разположи́ть to place, to position, to situate, to arrange, to have available, to have at one's disposal
располага́ться/расположи́ться to be situated,to be stationed; **располага́ться/расположи́ться ла́герем** to set up camp
расположе́ние 1. layout, distribution, location, situation
2. disposition, position, dispositions
~, бива́чное bivouacking
~, боево́е combat dispositions
~, каза́рменное barrack accommodation
~, кварти́рно-бива́чное close billeting
~, кварти́рное barrack equipment
~, ла́герное field location
~ на ме́стности ground layout
~ на о́тдыхе disposition (tps) in rest area or bivouac
~, оборони́тельное defensive dispositions
~ по ка́рте disposition on the map
~ проти́вника enemy's dispositions
~, ря́дное in line (ICE)
~ цили́ндров, ря́дное cylinders in line
располо́женный located, situated
расположи́ть (см располага́ть)
расположи́ться (см располага́ться)
распо́рка strut, cross-bar, stretcher, spreader
распо́рки для лыж ski rack
распоряди́тельность administrative ability
распоряди́тельный 1. active, capable, efficient (pers)
2. administrative
распоря́док routine, schedule, order
~ дня daily routine
распоряже́ние order, instruction, ordinance
~, боево́е operational instructions
~, пи́сьменное written instruction
~ по свя́зи signal instruction
~, предвари́тельное warning order
~, у́стное verbal instruction, oral instruction
распределе́ние distribution, allocation

распределéние артиллéрии, боевóе allotment of artillery
~ артиллерийских средств allotment of artillery effort
~ вéса weight distribution
~ волн allocation of wavelengths
~ Гáусса Gaussian distribution
~ зарядов charge distribution
~ огня distribution of fire
~ ответственности allocation of responsibility
~ ошибок error distribution
~ попаданий pattern, shot group
~ продовольствия breakdown of rations
~ разрывов по высотé dispersion in height (gunn)
~ целей allocation of targets
~ частóт frequency allocation
распределитель distributor (ICE, teleg etc)
~ зажигания distributor (ICE)
распределить (см распределять)
распределять/распределить to allocate, to suballocate, to breakdown, to dispense, to distribute, to allot
распредщит switchboard
распространéние 1. propagation, dissemination, spreading
2. fanning out (after break-through) (tac)
~, аномáльное anomalous propagation (electron)
~ радиовóлн propagation of radio waves
~, тропосферное tropospheric propagation (radio)
распространённость 1. prevalence
2. abundance (nucl)
распространитель слухов rumourmonger
распространить (см распространять)
распространять/распространить to spread, to disseminate
распускáть/распустить to let loose, to allow discipline to slacken, to dismiss (drill), to disband, to dissolve
распустить (см распускáть)
распутица season of bad roads (spring and autumn)
распутье 1. road fork, cross-roads
2. parting of the ways
распылéние 1. atomization, atomizing
2. impingement (of jet)
~ горючего fuel atomization
распылённость splitting up, dispersal
распыливание 1. atomization, atomizing, pulverisation
2. injection (diesel)
~, воздушное blast injection (diesel)
распылитель 1. atomizer
2. spray gun
3. jet (carb)
4. spray tip (diesel injector)
~ форсунки injection nozzle
распылить (см распылять)
распылять/распылить to spray, to atomise, to pulverise, to disperse, to split up
рассверливание boring out
рассвéт first light, dawn
рассéивание dispersion, scattering
~, баллистическое ballistic dispersion
~, боковóе dispersion in line
~ в дáльности dispersion in range
~ выстрелов shot pattern
~, естéственное natural dispersion
~ оскóлков fragmentation pattern
~ отдéльных снарядов round-to-round variation
~ по высотé dispersion in height
~ по дáльности dispersion in range
~ по направлéнию dispersion in line
~, пóлное total dispersion
~ разрывов dispersion of bursts
~ снарядов dispersion of shot
~, среднеарифметическое arithmetic mean of dispersion
~, угловóе angular dispersion
рассéиватель diffuser (optics)
рассéивать(ся)/рассеять(ся) to disperse/ to dissipate
рассекáть/рассéчь to split, to cleave
рассекрéтить (см рассекрéчивать)
~ (см рассекрéчивать)
рассекрéчивать/рассекрéтить to declassify
рассéчь (см рассекáть)
рассéяние scatter, dissipation
~, анóдное anode dissipation (electron)
~ излучéния radiation scatter
~ назáд back-scatter, back-scattering (nucl)
~, обрáтное back scattering, back scatter (nucl)
рассéять(ся) (см рассéивать(ся))
расслéдование inquiry, investigation
~, административное administrative enquiry

рассмотре́ние прете́нзий examination of claims
рассогласова́ние error, response lag, mismatch (servo, electron)
рассогласо́ванный mismatched
расспро́с questioning, interrogation
рассредото́чение 1. dispersion, dispersal
2. distribution (electron)
~ **во́йск** troop dispersion, troop deployment
~ **деса́нтных сил на перехо́де** dispersal of amphibious group on passage (amph ops)
~, **такти́ческое** battlefield dispersion
рассредото́чивать/рассредото́чить to disperse, to dissipate
рассредото́чить (см рассредото́чивать)
расста́вить (см расставля́ть)
расставля́ть/расста́вить to place, to arrange, to post (sentry)
расстано́вка 1. spacing, arrangement, order
2. employment (of personnel)
расстоя́ние distance, range, way, spacing, interval
~, **безопа́сное** safe distance
~, **ви́димое** apparent distance
~, **вы́раженное во вре́мени** time distance
~, **горизонта́льное** map range
~ **до це́ли** target range, distance to the target
~ **изображе́ния** image distance
~ **ме́жду кумуляти́вным заря́дом и пове́рхностью подрыва́емого объе́кта** stand-off distance (HEAT ammo)
~ **ме́жду ли́ниями координа́тной се́тки** grid interval
~ **ме́жду це́нтрами окуля́ров** interpupillary distance
~ **по ка́рте** map distance
~ **скачка́** skip distance (radio)
~, **тормозно́е** braking distance
~, **углово́е** angular distance
~, **фо́кусное** focal length (optics)
расстра́ивать/расстро́ить to disorder, to upset, to put out of order/alignment/ tune; **пасстра́ивать/расстро́ить ата́ку** to break up the attack
расстре́ливать/расстреля́ть to subject to heavy fire, to execute by shooting, to use up (ammo)
расстреля́ть (см расстре́ливать)
расстро́ить (см расстра́ивать)
расстро́йка detuning (electron)
~ **по частоте́** frequency separation
расстро́йство 1. disorder, disarrangement
2. maladjustment
"Рассчита́йсь!" "Number!"
рассы́паться (см рассыпа́ться)
рассыпа́ться/рассы́паться to crumble, to fall to pieces, to spill, to scatter; **рассыпа́ться/рассы́паться в цепь** to form extended line
раста́скивание dragging away, removal
раста́чивать/расточи́ть to bore, to bore out, to chisel out
растая́ть (см та́ять)
раство́р 1. flare (horn)
2. aperture (antenna)
3. solution (chemical)
~, **во́дный** aqueous solution
~, **дегази́рующий** CW decontaminant liquid
~, **дезинфекцио́нный** disinfectant solution
~, **дезинфици́рующий** disinfection solution
~, **индика́торный** gas detecting solution
~, **незамерза́ющий** anti-freeze solution
~, **рези́новый** rubber solution
~, **цеме́нтный** cement mortar
растворе́ние dissolving
раствори́мый soluble
раствори́тель solvent
~, **лету́чий** volatile solvent
~, **труднолету́чий** non-volatile solvent
расторо́пность alertness
расточи́ть (см раста́чивать)
расто́чка boring (metal, wood etc)
~ **цили́ндров** cylinder boring
растр raster, raster trace, grid lines (CRT, radar)
растра́та вое́нного иму́щества fraudulent conversion of military property
растру́б бойни́цы mouth of an embrasure
~ **га́зового по́ршня** piston cup
~ **кожуха́** bell-housing extension
растру́бный bell-shaped
растя́гивание диапазо́на band spread (radio)
растя́гивать/растяну́ть to stretch, to elongate

растяже́ние tension
растя́жка 1. stretching, extending
2. guy wire
~, про́волочная wire guy-rope, tension wire
~, тросова́я rope guy, fibre rope guy, cable stay
~ фро́нта extending of front, extension of front
растяну́ть (см растя́гивать)
расфасо́вка sorting and packaging
расфокуси́рование defocussing
расформирова́ние disbanding, breaking up
расформирова́ть (см расформиро́вывать)
расформиро́вывать/расформирова́ть to disband, to break up
расхище́ние pillage, embezzlement
расхо́д 1. expenditure, expense, outlay
2. consumption
~ бензи́на petrol consumption
~ боеприпа́сов ammunition consumption, ammunition expenditure
~ ВВ consumption of explosives
~ горю́чего fuel consumption
~ горю́чего, уде́льный specific fuel consumption
~ и прихо́д income and expenditure, debit and credit
~, километро́вый consumption per kilometre
~, ма́ссовый mass flow
~ на киломе́тр consumption per kilometre
~, номина́льный rated consumption
~, пла́новый forecast expenditure, normal expenditure
~, по́лный total rate of flow
~, постоя́нный constant flow
~, сре́дний mean consumption
~ то́плива, уде́льный specific fuel consumption
~, уде́льный specific consumption
~, часово́й consumption per hour, hourly consumption
расходи́ться/разойти́сь to break ranks, to break up, to diverge, to differ
расхо́дование боеприпа́сов expenditure of ammunition
расхо́довать/израсхо́довать to expend, use up
расходоме́р flowmeter, flow(-rate) indicator, flow gauge
~ то́плива fuel-metering device
расхо́дуемый expendable
расхо́ды, накладны́е overhead expenses
расхожде́ние discrepancy, deviation, divergence
~ часто́т frequency distortion (elec)
расцве́тка, усло́вная colour code
расцепи́ть (см расцепля́ть)
расцепле́ние disengagement, uncoupling
"Расцепля́й!" "Unhook!" (arty)
расцепля́ть/расцепи́ть to uncouple, to unhook
расча́лка bracing, bracing cable
расчёт 1. team, crew, party, detail, detachment
2. calculation, estimate
3. allotment, detailing, delegation
4. payment
~, аэродинами́ческий aerodynamic analysis
~ боево́го сре́дства weapon crew
~, боево́й allotment of battle tasks (as sentries, snipers, observers etc)
"Расчёт в Укры́тие!" "Detachment Take Cover!" (arty)
расчёт ли́чного соста́ва personnel distribution
~, миномётный mortar crew, mortar team
~, нали́чный payment in cash, cash payment
~, непо́лный skeleton crew
~, огнево́й gun detachment
~, огнемётный flame team
~, оруди́йный 1. gun detachment (Brit)
2. gun section (US)
~, пулемётный machine-gun crew
~ сил и средств estimate of manpower and equipment
~, сна́йперский миномётный sniper mortar crew
~, ста́ртовый launcher detachment (missile)
~, строево́й 1. drill team
2. team drill
расчётный 1. team, crew
2. estimated
3. rated
4. account
расчи́стка 1. clearing (topo)
2. clearing in minefield

расчи́стка обстре́ла clearing the field of fire
расчи́танный computed, calculated
расчлене́ние dispersal, breaking up, splitting up
расчлени́ть (см расчленя́ть)
расчленя́ть/расчлени́ть to split, to disperse, to dismember
расшире́ние expansion
~, адиабати́ческое adiabatic expansion
расшири́тель expander, spreader, widener
~ и́мпульсов pulse widener (electron)
~ поршнево́го кольца́ compressor inner ring, expander ring
расши́рить (см расширя́ть)
расширя́ть/расши́рить to enlarge, to widen, to expand; **расширя́ть/расши́рить плацда́рм** to enlarge a bridgehead
расшифрова́ние 1. interpretation (int, photo)
2. deciphering
расшифрова́ть (см расшифро́вывать)
расшифро́вка 1. interpretation (int, photo)
2. deciphering
расшифро́вывать/расшифрова́ть to interpret (int, photo), to decipher, to identify (en eqpt)
расще́лина cleft, crevice
расщепле́ние 1. fission (nucl)
2. spallation (nucl)
3. decomposition (chem)
~ ядра́ nuclear disintegration
рацио́н diet, ration, fare
рационализа́тор rationalizer
рационализа́ция rationalization
рациона́льный rational, judicious, efficient
ра́ция 1. radio station
2. radio set
ра́шпиль rasp (mech)
реаге́нт reagent, chemical reactant
реаги́рование reaction, response
~, ги́бкое flexible response (strategic)
реакти́в reagent, chemical agent, reactant
реакти́вный 1. rocket
2. jet, jet propelled
3. reaction, reactive
реа́ктор reactor
~, а́томный nuclear reactor
~, кипя́щий atomic boiler
реа́ктор, термоя́дерный thermonuclear reactor, fission reactor
~, я́дерный nuclear reactor
реа́ктор-конве́ртер-размножи́тель breeder-converter (nucl)
реа́ктор-размножи́тель breeder (nucl)
реа́кция, биохими́ческая biochemical reaction
~, взрывна́я explosive reaction
~ деле́ния fission reaction (nucl)
~ деле́ния, управля́емая цепна́я я́дерная controlled fission chain reaction
~ нагру́зки, часто́тная pulling figure (electron)
~ опо́ры back pressure
~ по ускоре́нию acceleration response (control systems)
~, самоподде́рживающая(ся) self-sustaining reaction (nucl)
~ си́нтеза thermonuclear reaction, fusion reaction
~ соедине́ния thermonuclear reaction, fusion reaction
~, термоя́дерная thermonuclear reaction
~, хими́ческая chemical reaction
~, цепна́я chain reaction
ребо́рда 1. wheel flange
2. collar rim (of pulley)
рёбра жёсткости webs (mor baseplate)
~ лы́жи, боковы́е edges (skis)
~ лю́льки, направля́ющие cradle guides (ord)
~, направля́ющие cradle guides (ord)
ребри́стый finned
ребро́ rib, fin
~, вну́треннее inner edge (of ski)
~ жёсткости reinforcing rib, stiffening rib, strengthening rib
~, защи́тное fender plate (naut)
~, направля́ющее spline (eng)
~, стабилизи́рующее stabilizing fin
ре́верс reversing gear, reverse
реверси́рование reversing, reversal
реви́зия audit, inspection
револьве́р revolver
~, беску́рковый hammerless revolver
~, несамовзво́дный single-action revolver
~ с открыва́ющимся ство́лом hinge-action revolver, hinged frame-type revolver
~, самовзво́дный double-action revolver

револьве́р, шестизаря́дный six-chambered revolver
реву́н 1. siren
2. shell fitted with sound signalling device
регенера́ция 1. reclamation, reclaiming
2. regeneration (control systems)
3. purification (air)
регенери́руемость recoverability, reclaimability
реги́стр вычисли́тельной маши́ны computer register
~ **кома́нд** register (computer)
регистра́тор recording device, recorder
регистра́ция registration, recording
регла́мент 1. regulations
2. periodic technical servicing
3. time limit
4. adjustment of centres
регули́рование adjustment, regulation, control
~ **вла́жности** humidity control
~ **движе́ния** traffic control, movement control
~ **мо́щности** power control
~, **неусто́йчивое** increased oscillation control
~, **позицио́нное** position control (GW)
~ **рабо́ты дви́гателя на оборо́тах холосто́го хо́да** idle control, slow-running control (ICE)
~ **ско́рости** speed control
регули́ровать to adjust, to regulate, to control; **регули́ровать дви́гатель** to tune up the engine
регулиро́вка 1. adjustment
2. control
~ **винто́м** screw adjustment
~ **гро́мкости, автомати́ческая** automatic gain control, automatic volume control, AVC
~, **гру́бая** coarse adjustment
~ **зажига́ния** ignition adjustment, spark adjustment
~ **зазо́ра в свече́ (зажига́ния)** spark gap adjustment (ICE)
~ **зазо́ра ме́жду пере́дним сре́зом затво́ра и дном ги́льзы** head-space adjustment (SA)
~ **карбюра́тора** carburettor adjustment
~, **ме́лкая** minor adjustment, fine adjustment
регулиро́вка моме́нта зажига́ния ignition timing, spark timing (ICE)
~, **ручна́я** finger-tip adjustment
~ **сцепле́ния** clutch adjustment
~, **то́нкая** delicate adjustment
~ **тре́тьей щёткой** third-brush control (elec)
~ **усиле́ния** gain control
~ **усиле́ния, автомати́ческая мгнове́нная** instantaneous automatic gain control
~ **усиле́ния, автомати́ческая** automatic gain control
~ **усиле́ния, временна́я** sensitivity time control, STC, swept gain (radar)
~ **усиле́ния, ручна́я** manual gain control (electron)
~ **усиле́ния с заде́ржкой, автомати́ческая** delayed automatic volume control, delayed automatic gain control
~ **чувстви́тельности, автомати́ческая** automatic gain control, automatic volume control, automatic sensitivity control
регулиро́вочный regulating, adjusting
регулиро́вщик pointsman, traffic controller
регули́руемый adjustable, controlled
регули́рующийся adjustable
регуля́тор 1. regulator
2. gas regulator (SA)
3. governor (mech)
~ **автомати́ческого опереже́ния зажига́ния** automatic advance and retard (ICE)
~, **всережи́мный** variable speed governor
~, **га́зовый** gas regulator (SA)
~ **давле́ния** pressure regulator
~ **давле́ния ма́сла** oil-pressure regulator
~, **максима́льный** maximum speed governor
~ **ма́лых оборо́тов** slow speed governor
~, **микрометри́ческий** micrometer screw
~ **направля́ющего колеса́ гу́сеницы** track idler adjuster (AFV)
~ **напряже́ния** voltage regulator
~ **напряже́ния, вибрацио́нный** vibrator-type voltage regulator
~ **натяжно́го колеса́ гу́сеницы** track idler adjuster (AFV)
~ **отка́та** cut-off gear

регуля́тор пода́чи во́здуха air governor
~ **расхо́да то́плива** fuel flow controller, fuel flow governor, fuel control unit
~ **с у́гольным сто́лбиком** carbon-pile regulator (elec)
~ **ско́рости** governor (mech)
~ **ско́рости, центробе́жный** centrifugal (speed) governor
~ **усиле́ния** gain control (electron)
~, **центробе́жный** centrifugal governor
~ **я́ркости се́тки прице́ла** sight reticle dimmer control
реда́н step (in keel of hydroplane or seaplane float)
редколе́сье sparse wood
редколле́гия editorial board (unit newspapers)
реду́ктор 1. reduction gear
2. reduction valve
~, **бортово́й** final drive gear (tracked vehicle)
~ **давле́ния во́здуха** air pressure
~ **поворо́тного механи́зма** traverse gear-box
~, **промежу́точный** intermediate reduction gear
рее́стр register, list, inventory
ре́ечник rodman (surv)
ре́ечный rodman (surv)
"Ре́же Ход!" "Slacken the Pace!"
режи́м 1. method, system, practice, mode
2. schedule, routine
3. conditions of work
4. phase
5. level
6. rate, speed
7. regime
~ **автомати́ческого сопровожде́ния це́ли** auto-tracking mode (radar)
~ **горизонта́льного полёта** straight and level flight, horizontal flight phase
~, **дежу́рный** listening watch (comms)
~ **де́йствий** routine
~ **дли́тельного хране́ния** long-term storage (of missiles)
~, **дли́тельный** continuous duty rating
~ **дня** daily routine
~ **зависа́ния** hovering (helicopters)
~ **кругово́го обзо́ра** circular scanning mode (radar)
~ **ма́лого га́за** slow-running mode
~, **ма́ршевый** sustained flight (missile)
режи́м, микрофо́нный voice working, radio-telephony, R/T (radio comms)
~, **неустанови́вшийся** transient effect
~, **неусто́йчивый** unstable state
~ **номина́ла** maximum continuous power
~ **огиба́ния** terrain following (low flying ac)
~ **огня́** 1. maximum permissible rate of fire
2. method of fire
~ **освеще́ния, незатемнённый** lights unrestricted (with blackout mask) (Brit)
~, **питьево́й** water discipline
~ **поведе́ния** routine
~ **прице́ливания** mode of aiming
~ **промежу́точной гото́вности** intermediate readiness (of missiles)
~ **рабо́ты** method of operating, method of working
~ **рабо́ты, норма́льный** standard duty (mech)
~ **рабо́ты, переме́нный** varying duty (mech)
~ **рабо́ты, постоя́нный** constant duty (mech)
~ **рабо́ты, пусково́й** starting duty (mech)
~ **рабо́ты, установи́вшийся** stable running (mech)
~, **рабо́чий** normal operating conditions
~ **се́кторного по́иска** sector scan (radar)
~, **ста́ртовый** boost (mode of operation of rocket motor), booster phase (missile)
~, **телегра́фный** telegraph working, telegraph mode
~, **температу́рный** working temperature range
~ **управле́ния огнём** fire control mode
~ **хране́ния** storage (of missiles)
~, **экономи́ческий** economical cruising speed
резе́рв reserve, reserves
~, **артиллери́йско-противота́нковый** anti-tank artillery reserve
~, **ближа́йший** immediate reserve
~ **боеприпа́сов** ammunition reserve
~ **Верхо́вного Главнокома́ндования** Supreme High Command Reserve
~ **гла́вного кома́ндования** GHQ reserve
~, **общевойсково́й** general reserve (of formation)
~, **о́бщий** general reserve
~, **операти́вный** tactical reserve

резерв, продовольственный food stocks
~, противодесантный anti-airborne reserve, anti-assault landing reserve
~, противотанковый anti-tank reserve
~, стратегический strategic reserve
~, тактический tactical reserve
резервист reservist
резервный reserve, alternative
резервуар storage tank, container
~, бензиновый petrol tank
~, бескольевый тканевый collapsible fabric static tank
~, вспомогательный auxiliary tank
~, главный main reservoir
~, заглублённый underground (storage) tank
~, запасной reserve tank
~, надземный above-ground (storage) tank
~, нефтяной oil storage tank
~, подземный buried tank, underground storage tank
~ радиатора, верхний top radiator tank, header tank (veh)
~ радиатора, нижний bottom radiator tank (veh)
~ распределительный distribution tank
~, резинотканевый rubberized-fabric tank
~, сборно-разборный sectional storage tank, bolted storage tank
~, топливный fuel tank
~, уравновешивающий balance tank
резервуар-цистерна storage tank (water)
резерж, подвижный mobile reserve
резец cutter (eng)
резидент 1. stationary agent (int)
2. Resident
резина vulcanised rubber
~, бензостойкая petrol-resistant rubber
~, вулканизированная vulcanized rubber
~, маслостойкая oil-resistant rubber
~, пенистая foam rubber
~, сырая uncured rubber
резиновый rubber
резино-тканевый (made of) rubber(ized)-fabric
резистор resistor
~, регулируемый variable resistor
резка, соломенная chopped straw
резкость sharpness, definition of image (opt)

резонанс resonance (electron)
~ напряжений series resonance (electron)
~, параллельный parallel resonance (electron)
~, последовательный series resonance (electron)
~ токов parallel resonance (electron)
резонансный resonance, resonant (electron)
резонатор ringing circuit, resonator (electron)
~ магнетрона magnetron cavity, magnetron resonator (electron)
~, объёмный cavity resonator (electron)
~, полый cavity resonator (electron)
резонатор-группирователь buncher cavity (electron)
резонатор-улавливатель catcher cavity (electron)
результат испытания test data
результаты огня fire effect
резьба thread (screw)
~ для соединения со ствольной коробкой barrel-locking thread(s)
~, левая left-handed thread
~, правая right-handed thread
~, сорванная stripped thread
~, трапециодальная acme screw-thread
~, упорная buttress thread
резьбовой threaded, thread
резьбомер thread gauge
рейд 1. raid
2. roads, roadstead (nav)
~, внутренний inner harbour area, inner roadstead
рейка 1. rod, batten
2. rule, ruler, straight-edge
3. arm, pole crosspiece (line comms)
4. rack (eng)
5. surveyor's pole
~, водомерная depth gauge
~ дальномера, выверочная stadia rod
~, дальномерная stadia rod (surv)
~ для обучения прицеливанию aiming post (arty)
~, зубчатая gear rack
~, измерительная yardstick, gaugestick, dipstick, measuring rod, sounding rod
~, неподвижная stop pawl holder (ord)
~, подвижная feed pawl holder (QF ord)
~ поршня piston rack

ре́йка с вы́резами, вертика́льная slotted batten (line comms)
~ **с деле́ниями** graduated rule
~, **снегоме́рная** snow scale, snow stake
~, **топографи́ческая** surveyor's pole
ре́йки, неподви́жные stop-pawl holders (ord)
~, **подви́жные** feed-pawl holders (ord)
ре́йнджер ranger (US)
рейс 1. trip, journey, crossing (one way) 2. flight (river crossing)
~, **обра́тный** return trip
~, **пе́рвый** maiden voyage
ре́йсо-расчёт wave, boat-load, ferry-load (tps)
ре́йс-расчёт wave, boat-load, ferry-load (tps)
рейсши́на T-square
река́ river
~, **сре́дняя** medium-width river (width 60-300m) (Sov)
~, **судохо́дная** navigable river
~, **у́зкая** narrow river (width up to 60m) (Sov)
~, **широ́кая** wide river (width more than 300m) (Sov)
реквизи́ровать to commandeer, to requisition
реклама́ция claim, complaint, protest
рекогносци́ровать to reconnoitre
рекогносциро́вка 1. commander's reconnaissance 2. reconnaissance, reconnoitring (of ground)
~, **инжене́рная** engineer reconnaissance
~, **команди́рская** commander's personal reconnaissance
~ **маршру́та** route reconnaissance
~, **фотографи́ческая** photographic reconnaissance
рекогносциро́вочный reconnaissance, reconnoitring
реконсерва́ция reprocessing for storage
реко́рдер recorder, recording head (sound recording)
ректифика́ция rectification, redistillation (chem)
рекупера́тор recuperator (gunn)
релакса́тор relaxation oscillator (electron)
релакса́ция relaxation
реле́ relay (elec)
~, **биметалли́ческое** bimetallic relay
~ **вре́мени** timer, time-delay relay (elec)
реле́, встре́чное built-in relay
~, **встро́енное** built-in relay
~, **гальванометри́ческое** galvanometer-type relay
~, **минима́льное** reserve current cut-out relay
~ **обра́тного то́ка** cut-out (veh)
~, **отключа́ющее** cut-out relay
~ **переме́нного то́ка** AC relay (elec)
~, **поляризо́ванное** polarised relay (elec)
~, **пусково́е** starting relay (elec)
~, **электромагни́тное** electromagnetic relay
реле́йщик member of radio-relay unit
реле́-регуля́тор 1. current and voltage regulator unit 2. cut-out and voltage regulator
релье́ф relief, contour
~ **дна** bottom contour (underwater)
~ **ме́стности** relief of the ground, configuration of the ground
рельс rail
~, **го́рный** sprag (veh)
~, **заземлённый** earthed rail
~, **направля́ющий** guide rail
~, **предохрани́тельный** safety rail
~, **токопроводя́щий** live rail, contact rail (rly)
реля́ция recommendation (for award etc)
реме́нь strap, belt, sling
~ **безопа́сности** safety belt
~ **вентиля́тора** fan belt
~, **вентиля́торный** fan belt
~, **винто́вочный** rifle sling
~, **напле́чный** shoulder strap (eqpt)
~, **носко́вый** toe-strap (ski)
"Реме́нь Отпусти́ть!" "Loosen Slings!"
реме́нь, перено́сный sling
~, **плечево́й** shoulder strap
~, **подборо́дочный** chin strap
"Реме́нь Подтяну́ть!" "Tighten Slings!"
реме́нь, поясно́й waist belt, belt
~, **привязно́й** seat belt
~, **пя́точный** heel strap (ski)
~, **руже́йный** carrying sling, carrying strap, rifle sling
~, **увя́зочный** lash rope
ремешо́к с переключа́телем микрофо́на, ше́йный microphone switch box and neck strap
ремзаво́д repair workshop
ремзи́п repair spare parts kit
ремни́, напле́чные shoulder-straps

ремо́нт 1. repair, repairs
2. renovation
3. refit (nav)
4. overhaul, maintenance
~, авари́йный emergency repairs
~, агрега́тный unit repair
~, ба́зовый base repair
~ боеприпа́сов renovation of ammunition
~ в ба́зовой мастерско́й base repair
~ в боевы́х усло́виях combat maintenance, combat repair
~ в полевы́х усло́виях field maintenance, field repair
~, внепла́новый unscheduled repair
~, войсково́й first line repair
~, вре́менный temporary repair, emergency repair
~ второ́го эшело́на second line repair, second echelon maintenance
~, заво́дский factory repair, major repair
~, капита́льный 1. complete overhaul, major overhaul
2. base repairs (Brit)
~, мастерско́й workshop repair
~, ме́лкий minor repair
~, модернизацио́нно-восстанови́тельный modernization and reconstruction repair
~, непла́новый сро́чный emergency repair
~, о́бщий general repair
~, первичный initial repair
~ пе́рвого эшело́на first echelon maintenance, first line repair
~, пла́ново-предупреди́тельный preventive maintenance
~, пла́ново-профилакти́ческий routine maintenance, preventive maintenance
~, пла́новый periodic servicing
~, полево́й field repair
~, полково́й unit repairs, unit maintenance
~, профилакти́ческий preventive maintenance
~, сре́дний medium repair, field repair, unit repair
~, сро́чный emergency repair, urgent repair
~, теку́щий unit repairs, running repairs
~, трудоёмкий major repair
~, эксплуатацио́нный running repairs, maintenance
ремо́нт, эшелони́рованный line repair
ремонтёр, аэродро́мный airfield runway repair equipment
ремонти́ровать to repair, to remount (of cavalry)
ремо́нтно-восстанови́тельный repair, recovery, reclamation
ремо́нтный 1. repair
2. maintenance
ремонтопригóдность maintainability, repairability
ремфо́нд repair pool, spare parts pool
рента́бельность profitability, profitableness
рентге́н roentgen (unit of X-ray radiation)
рентге́новский 1. X-ray
2. ghosted, cut-away (drawing, model etc)
рентгено́лог radiologist
рентгеноло́гия radiology
рентгенолюминесце́нция roentgeno-luminescence
рентгеноме́тр roentgen meter, dose-rate meter, radiac survey meter
рентгеноскопи́я radio examination, X-ray examination
рентге́н-эквивале́нт, биологи́ческий man-roentgen equivalent (NBC)
реорганиза́ция reorganization
реорганизова́ть to reorganise
реоста́т rheostat
~ высоты́ height control (ac gunsight)
~ подсве́тки прице́ла sight dimmer rheostat
~, пусково́й starting rheostat
~ ско́рости speed control (ac gunsight)
репе́р 1. registration point (gunn)
2. check point (US) (gunn)
3. bench mark (surv)
4. datum point
~, возду́шный air registration point (gunn)
~, возду́шный звуково́й air registration point (sound ranging)
~, действи́тельный 1. actual registration point (Sov term denotes easily visible or registered point, co-ordinates of which are known)
2. check point (US) (gunn)
3. datum point
~ для сostре́ла ору́дий calibration point

репе́р, звуково́й 1. registration point (sound ranging)
2. sound ranging check point (US)
~, контро́льный witness target
~, назе́мный ground registration point
~, пристре́лочный registration point
~, фикти́вный imaginary registration point (Sov term denotes mpb of a group of bursts, co-ordinates of which are determined by resection)
~, фикти́вный возду́шный imaginary air registration point (gunn)
~, фикти́вный назе́мный imaginary ground registration point (gunn)
репети́ция, генера́льная dress rehearsal
репре́ссия reprisal, punitive measure
репроду́кция reproduction (photo, carto)
ресиве́р receiver
респира́тор respirator
рессо́ра spring (mech)
~, бу́ферная buffer spring
~, вита́я coil spring
~, гидропневмати́ческая air spring, pneumatic spring
~, за́дняя rear spring
~, листова́я leaf spring
~, пере́дняя front spring
~, пневмати́ческая air spring, pneumatic spring
~, полуэллипти́ческая semi-elliptical spring
~, пружи́нная coil spring
~ с переме́нной жёсткостью variable-rate spring
~, спира́льная coil spring
~, стержнева́я torsion bar, torsion shaft
~ теле́жки bogie spring
~, че́тверть-эллипти́ческая quarter-elliptical spring
~, шарни́рно-закреплённая articulated spring
~, эллипти́ческая elliptical spring
ресу́рс 1. resource
2. "life", expected life (mech)
3. overhaul life, life left till overhaul required
~, безопа́сный safe life (expl etc)
~ дви́гателя engine life
~, лётно-боево́й combat flight capability
~, устано́вленный overhaul life (engines)
~ ши́ны tyre life
~, эксплуатацио́нный working life
ресу́рсы, материа́льные resources, material resources
~, ме́стные local resources
ретрансли́ровать to relay, to re-transmit, to re-broadcast
ретрансля́тор repeater (comms)
ретрансляцио́нный relay, radio-relay, re-broadcast
ретрансля́ция relay, radio-relay, re-broadcast
~, ручна́я manual relaying (RR)
ретрансми́ттер retransmitter
ре́тушь, светова́я dodging (photo)
рефле́ктор 1. reflector
2. tamper (nucl)
~, параболи́ческий parabolic reflector (radar)
~, уголко́вый corner reflector (radio)
рефра́кция refraction
~, берегова́я coastal refraction (radio)
~, норма́льная атмосфе́рная standard atmospheric refraction (electron)
реце́нзия review, analysis
реце́пт 1. prescription
2. recipe
рециркуля́ция recirculation
ре́чка river
речно́й river
реша́ющий decisive, conclusive
реше́ние команди́ра commander's plan
~ команди́ра по ты́лу administrative decision
~, при́нятое plan (tac)
решётка 1. grating, grille
2. array (antenna)
~, лине́йная анте́нная linear array (antenna)
~, остронапра́вленная pencil beam array
~, попере́чно-излуча́ющая broad-side array (aerial display)
~ радиа́тора, защи́тная radiator guard
~, сверхнапра́вленная super-directive array (antenna)
~ свя́зи communications net (used only of non- Soviet nets)
~, тесо́вая crate
~, фазиро́ванная phased array
реши́тельный 1. decisive
2. determined (attack, assault etc)
ржа rust (obs)
ржа́веть/поржа́веть to rust, to corrode
ржаве́ющий rusty, corrosive

ржа́вление rusting, corrosion
ржа́вчина rust
ри́гель cross-bar, brace
рикоше́т ricochet
рикошети́ровать to ricochet
ри́ска 1. graduation line
2. mark, groove, channel
~, **контро́льная** reference mark
~, **осева́я** centre mark
~, **перекрёстная** crosswire (opt)
~ **указа́теля** index line, index mark
~, **устано́вочная** graticule, setting mark, index mark
~ **хому́тика** slider setting mark (SA sight)
рису́нок бегово́й ча́сти tread pattern (tyre)
~ **покры́шки, проте́кторный** tread pattern (tyre)
~ **проте́ктора** tread pattern (tyre)
рифлёный grooved, channelled
робури́т roburite (expl)
ров moat, large ditch
~, **крепостно́й** moat
~, **односторо́нний** one-way tank ditch
~, **противота́нковый** anti-tank ditch, tank ditch
~, **противотра́нспортный** anti-vehicular ditch
ро́вик foxhole, weapon pit
~, **отхо́жий** latrine trench
~, **полево́й** field latrine
~, **противограна́тный** bomb trap (in trench)
~, **противоскоростно́й** vehicle hindrance scarp
ро́вик-гранатоулови́тель bomb trap (in trench)
рога́тка knife rest (fd eng obstacle)
рого́жа 1. matting, bast matting
2. hopsack (textile)
рогу́лька forestry barbed wire picket
род авиа́ции air force command (ie bomber, fighter, coastal etc)
~ **войск** arm of service
~ **войск, боево́й** fighting arm
~ **ме́стности** nature of terrain, land form
~ **сил, обеспе́чивающий** supporting arm (of the service)
рожки́ спусково́го крючка́ trigger horns (SA)
ро́за permanent echoes, pattern of permanent echoes (radar)
розе́тка socket outlet (elec)
~, **бортова́я** external electrical supply socket
~, **ште́псельная** socket outlet (elec)
ро́зыск search, inquiry
Ро́зыск, Уголо́вный Criminal Investigation Department
ро́йма reamer (machine tool)
рока́да lateral line of communication, line of communication parallel to front (rly, road or earth road)
~, **артиллери́йская** lateral road for artillery
рокиро́вка lateral move (tac)
ро́лик 1. roller
2. bobbin, porcelain cleat (line comms)
~, **кулачко́вый** cam roller
~, **опо́рный** 1. return roller
2. jockey wheel (track assembly)
~ **подвижны́х ре́ек** feed rod roller (ord)
~, **подде́рживающий** track support roller (tk)
~ **толка́теля кла́пана** cam follower roller
роликоподши́пник roller bearing
~, **иго́льчатый** needle roller bearing
~, **кони́ческий** tapered roller bearing
рольга́нг roller conveyor, roller track, roller platform
ромб 1. rhombus
2. diamond formation
3. diamond-shaped badge of rank (obs)
ро́мбик mark on aircraft gunsight graticule
ро́спуски pole trailer
ро́ссыпь валу́нов boulder-strewn ground
ро́та company
~, **автомоби́льная** MT company
~, **автопроже́кторная** motorised searchlight company
~, **автосанита́рная** motorized medical company
~, **автотра́нспортная** MT company, transport company
~ **в соста́ве аванга́рда** advance-guard company
~, **вое́нно-строи́тельная** military construction company (Sov)
~, **деса́нтная перепра́вочная** assault (river) crossing company
~, **доро́жная** road-construction company
~, **зени́тная** air defence company, anti-aircraft company

ро́та, инжене́рно-доро́жная engineer road construction and repair company
~, инжене́рно-мостострои́тельная engineer bridge-building company
~, инжене́рно-ремо́нтная engineer maintenance company
~, инжене́рно-сапёрная sapper company
~, ка́бельно-шестова́я telephone construction company
~, курса́нтская 1. cadet company 2. trainee company, training company (at military schools or on courses)
~, лине́йно-строи́тельная line-laying company (sigs)
~, ме́дико-санита́рная medical company
~, миномётная mortar company
~ морско́й пехо́ты marine infantry company
~ обезвре́живания неразорва́вшихся бомб bomb disposal company
~, отде́льная авторемо́нтная independent MT repair company
~, парашю́тно-деса́нтная parachute company
~, парашю́тно-стрелко́вая parachute rifle company
~, па́рковая 1. maintenance company 2. equipment holding company
~ перево́зок по во́здуху air dispatch company
~, перепра́вочно-деса́нтная assault crossing company (Sov eng sub-unit)
~, пехо́тная infantry company
~ подво́за боеприпа́сов ammunition supply company
~ подде́ржки support company
~, понто́нная pontoon company
~, понто́нно-мостова́я pontoon bridge company
~, порто́вая port company, port squadron
~, при́данная attached company
~, пулемётная machine-gun company
~, путева́я ремо́нтная railway maintenance company, road maintenance rifle company
~, радиолокацио́нная radar company
~, радиотехни́ческая radar operating company
ро́та, разве́дывательная 1. reconnaissance company 2. reconnaissance party
~ распределе́ния горю́чего fuel distribution company
~, ремо́нтная repair company
~, самока́тная cyclist company
~, сапёрная engineer company, field engineer company
~, сво́дная composite company
~ свя́зи, лине́йная line company (sigs)
~, стрелко́вая rifle company
~, стрелко́во-парашю́тная parachute rifle company
~, та́нковая tank company
~, та́нко-деса́нтная tank-borne infantry company
~, телегра́фная telegraph company (mil)
~, телегра́фно-строи́тельная telegraph line-laying company
~, телегра́фно-эксплуатацио́нная telegraph line-maintenance company
~, телефо́нно-ка́бельная telephone-cable company (mil)
~, тра́нспортная transport company
~, трубопрово́дная pipeline company
~ тяжёлого-ору́жия support company
~, учёбная training company
~ хими́ческой защи́ты chemical defence company
~, штабна́я staff company, HQ company
~, штурмова́я assault company
~, эвакуацио́нная recovery company
ро́тный company
ротолёт guided missile with annular wing (AAM or SAM)
ро́тор 1. rotor 2. bucket wheel (trench digger)
ро́ща small wood, grove
ртуть mercury
~, грему́чая mercry fulminate
руба́нок plane (tool)
руба́ха, водола́зная diving dress
руба́шка 1. shirt 2. jacket (mech)
~ бло́ка water jacket
~, водяна́я water jacket
~, нагрева́тельная heating jacket
~, охлажда́ющая cooling jacket
~ охлажде́ния cooling jacket
~, свинцо́вая lead sleeve (bullet)
~, свобо́дная loose liner, inner "A" tube (ord)

руба́шка, теплозащи́тная heat-insulating jacket, thermal- insulating jacket
~, фо́рменная service shirt
рубе́ж bound, line (tac)
~ ата́ки start line (tac)
~ а́томной безопа́сности troop safety line, risk distance line (NBC)
~ безопа́сного бомбомета́ния bomb line
~ безопа́сного удале́ния troop safety line, risk distance line (NBC)
~ безопа́сного удале́ния огня́ minimum range line (arty)
~ боево́го охране́ния battle outpost line (tac)
~ бомбомета́ния bomb line (forward of which aircraft may attack targets without danger to own ground forces)
~ вво́да в бой start line of second wave (tac)
~ веде́ния огня́ firing line (tac)
~, во́дный 1. water obstacle
2. river line, canal line, water's edge on defender's side
~, гла́вный оборони́тельный main line of resistance, main line of defence
~ для ата́ки, исхо́дный start line, jump-off line
~ для броска́ в ата́ку assault line (tac)
~ для стрельбы́ forward edge of the battle area, FEBA
~ загради́тельного огня́ defensive fire line
~ загради́тельного огня́, основно́й main defensive fire line
~, исхо́дный start line, line of departure (tac)
~, коне́чный limit of exploitation (tac)
~, назна́ченный report line (tac)
~ обнаруже́ния radar range perimeter
~ обнаруже́ния, бли́жний inner limit of detection (radar)
~, оборони́тельный line of defence, defensive line (tac)
~ огнево́го ва́ла, пе́рвый opening line of barrage (arty)
~ огнево́го ва́ла, после́дний final line, final depth line (arty)
~, огнево́й firing point (SA range)
~ огня́ line (arty barrage)
~ откры́тия огня́ limit of opening fire (tac)
~, пе́рвый основно́й opening line (arty)
~ перехва́та line of interception (AF)
рубе́ж, преде́льный 1. final defence line (mobile defence)
2. limit of exploitation (attack)
~ прекраще́ния огня́ cease-fire line (arty)
~ препя́тствий obstacle line
~, промежу́точный intermediate position (tac)
~, промежу́точный уравни́тельный phase line, report line (tac)
~, противота́нковый anti-tank gun line (tac)
~ радиолокацио́нного обнаруже́ния radar range perimeter
~ развёртывания deployment line (tac)
~ размыка́ния break-off line (AF)
~ регули́рования report line, phase line (tac)
~, регули́рующий phase line, report line (tac)
~ ро́спуска formation break-off line (AF)
~, танконедосту́пный tank barrier
~, тылово́й rear boundary
~, укреплённый fortified line
~, уравни́тельный phase line, report line (tac)
~ я́дерной безопа́сности nuclear safety line
рубero ́ид Ruberoid (lightweight roof-sheeting material) (trade name)
"Руби́льник!" 1. "Switch on!"
2. "Switch Off!"
руби́льник knife-switch, contact breaker, cut-out (elec)
~, перекидно́й double-throw switch
ру́бка 1. driving compartment
2. felling, hewing, chopping
~ в ла́пу dovetailing (jointing of corners of log walls)
~ в ча́шку notching (semi-circular jointing of corners of log walls)
рубль 1. rouble
2. one hundred (gunn)
ружболва́нка rifle billet (gunsmithing)
ружейно-пулемётный small arms
руже́йный rifle
ружпа́рк company arms store
ружьё 1. rifle
2. shotgun
3. hand-held rocket launcher
~, духово́е air gun, air rifle

ружьё, противота́нковое 1. anti-tank rifle
2. anti-tank rocket launcher
~, противота́нковое реакти́вное anti-tank rocket launcher
~, реакти́вное rocket launcher, bazooka
~ с гла́дким ство́лом smooth-bore gun (SA)
рука́в 1. sleeve
2. hose, flexible tube
3. branch (river)
~, бензозапра́вочный petrol hose
~, брониро́ванный armoured hose, reinforced hose
~, вса́сывающий intake hose
~, запра́вочный filling hose
~, нагнета́тельный pumping hose (water supply)
~, напо́рный pressure hose
~, оплетённый braided hose
~, пожа́рный fire hose
~, рези́новый rubber hose
~, резинотканьёвый rubberized fabric hose
~, тормозно́й flexible brake hose pipe
рукави́цы gauntlets
"Ру́ки Вверх!" "Hands Up!"
руководи́тель supervisor
~ заня́тий 1. director of an exercise
2. instructor, chief instructor
~ перепра́вы, техни́ческий river crossing senior engineer officer
~ полётов officer in charge of flying
~ стрельб range officer, officer in charge of firing exercises
руководи́ть to direct, to control, to supervise (imp only)
руково́дство 1. control
2. leadership
3. instructions, guide
~ по ремо́нту repair manual
~ по стрельбе́ из та́нка tank gunnery instruction
~ по эксплуата́ции operation manual, operating instructions
~ слу́жбы service manual
~, техни́ческое 1. technical manual
2. technical guidance
~, централизо́ванное centralized control
рукоприкла́дство striking (with the fist or hand) (mil law)
рукоя́тка 1. handle, hand-grip
2. lever
3. grip, stock (pistol)
~ взво́да cocking handle
~ для отво́да це́нтра за́дней ба́бки tailstock locking lever (veh)
~ для перено́ски carrying handle (SA)
~ домкра́та jack handle
~, заводна́я starting handle
~ заряжа́ния cocking handle (SA)
~ затво́ра 1. cocking handle (SA)
2. breech mechanism lever, LBM (ord)
~ затво́рной ра́мы cocking handle (SA)
~ механи́зма горизонти́рования cross-level adjustment cap (mor)
~ перезаряжа́ния 1. breech mechanism lever
2. cocking handle (SA)
~ пистоле́та pistol grip
~, пистоле́тная pistol grip
~, поворо́тная turning handle
~ поворо́тного механи́зма traversing handle
~ подъёмного механи́зма elevating gear operating handle
~ предохрани́теля safety lever
~ прикла́да butt handle (SA)
~ пулемёта machine-gun carrying handle
~, пускова́я starting handle, starter crank
~ ручно́й пода́чи то́плива hand throttle
~ рычага́ lever handle
~, спускова́я firing key, firing handle
~ ствола́ carrying handle (MG)
~ сто́пора хо́да, переводна́я axle-tree retaining bolt locking pin (ord)
~, сто́порная stop lever
~, тормозна́я brake arm
~ управле́ния 1. control handle
2. joystick (GW controller)
рулево́й 1. steersman, coxswain
2. steering, rudder
руле́ние taxiing (ac)
руле́тка measuring tape
ру́ли, возду́шные control fins (GW)
руло́н roll, reel (of paper, tape, etc)
руль 1. steering wheel
2. rudder
3. fin, vane (aeron)
4. handlebars
~, аэродинами́ческий aerodynamic control surface, vane

руль, возду́шный air vane, control surface
~, га́зовый jet vane (GW)
~, газодинами́ческий jet vane (GW)
~, газостру́йный jet vane (GW)
~, носово́й nose vane (GW)
~ управле́ния steering rudder
румб 1. bearing, compass point 2. rhumb (navig)
~, ко́мпасный compass point
~, четвертно́й semi-cardinal point, inter-cardinal point
румбатро́н rhumbatron (electron)
ру́мбы ко́мпаса, гла́вные cardinal points of the compass
ру́мпель tiller, rudder
ру́пор megaphone
~, бикони́ческий biconical horn (electron)
ру́сло channel, river bed
~, меже́нное permanent channel (river etc)
руче́й 1. stream, burn, brook 2. groove
ру́чка handle, haft, knob
~, букси́рный towing yoke
~ для перено́ски carrying handle (SA)
~, заводна́я crank handle, starting handle
~, ко́жаная leather carrying handle
~ перезаряжа́ния cocking handle (SA)
~, подъёмная hoisting eye, lifting eye
~ рукоя́тки LBM handle, handle of breech mechanism lever (ord)
~ рукоя́тки затво́ра breech-block handle
~, складна́я folding handle
~ то́рмоза brake handle
~ управле́ния steering handle
~ червячка́ угломе́ра azimuth-gear release handle
ру́чки заты́льника double spade grips (MG)
ручни́к hand hammer
~, трафаре́тный stencil brush
ручно́й 1. hand, hand-held, hand-operated, manual 2. tame
ры́вки pulling (electron)
рым ring, ring bolt, lifting eye
~, швварто́рный mooring ring
ры́мы, подъёмные lifting eyes
рыска́ние yaw (aeron, naut)
рыхли́тель, доро́жный road scarifier
рыча́г 1. lever, arm 2. crank
~ автомати́ческого взведе́ния automatic cocking lever (ord)
~ амортиза́тора shock-absorber lever
~ боево́го взво́да striker lever (gren)
~ взведе́ния cocking lever
~ взво́да cocking lever
~ включе́ния starting lever, engaging lever, coupling lever
~ выключе́ния releasing lever, disconnecting lever, disengaging lever
~ гу́сеницы, натяжно́й track adjusting lever
~, двупле́чий double-arm lever (arty), bell crank
~ дро́ссельной засло́нки throttle control lever
~, закрыва́ющий breech closing lever
~ избира́теля selector lever (gearbox)
~, кла́вишный key lever (teleg)
~, кла́панный valve rocker (engines)
~, коле́нчатый crank handle (MG)
~, конта́ктно-поворо́тный tilt rod (mine)
~, контро́льный control arm (ord)
~, коромы́словый rocker arm
~ кулачко́вого коромы́сла rocker arm (eng)
~, кулачко́вый кача́ющийся rocker arm (eng)
~ кули́сы change speed lever, gear lever (tk)
~ му́фты coupling lever
~, ножно́й foot lever, foot pedal
~ опо́рного катка́ bogie wheel axle arm (tk)
~ освобожде́ния то́рмоза brake release lever
~ остано́ва затво́ра bolt catch lever, breechblock stop lever
~ параллелогра́мма, ве́рхний upper sight operating arm (ord)
~ параллелогра́мма, ни́жний lower sight operating arm (ord)
~ педа́ли pedal lever
~, перекидно́й rocker arm
~ переключе́ния переда́ч gear lever
~ поворо́тной ца́пфы steering arm (veh)
~ пода́чи feed lever, feed arm
~, подаю́щий feed lever, feed arm
~ подве́ски suspension arm

рычáг подъёмного механи́зма elevating and depressing lever, elevating bar
~ ползунá приёмника feed-slide lever (gunn)
~, пружи́нный spring lever
~, регули́рующий regulating lever
~ рулевóго управлéния tiller, steering lever (tk)
~ ручнóго взведéния hand cocking lever
~ ручнóго тóрмоза hand brake lever
~, спусковóй 1. trip bar (teleph)
2. sear bar (trigger mechanism)
3. striker lever (gren)
4. release lever (mine)
5. firing lever (MG)
~, стóпорный locking lever
~ стоя́ночного тóрмоза parking brake lever
~ тóрмоза brake lever
~ удáрного механи́зма striker actuating lever (ord)
~ управлéния control lever
~ управлéния гáзом throttle arm
~ управлéния, лéвый Left Hand steering lever (tk)
~ управлéния поворóтом бáшни, команди́рский commander's turret traverse control lever
~ управлéния, прáвый Right Hand steering lever (tk)
~, фасóнный curved lever
~ фрикциóна clutch lever
~ экстрáктора extractor rod (revolver)
рычажóк 1. (small) lever
2. tumbler
3. trigger
~ предохрани́теля safety lever (SA)
рябь ripple (gen and elec)
ряд 1. row, line, series
2. order, sequence
3. file (pers)
~, глухóй blank file
~ мин mine row
~, ми́нный mine row
~, непóлный blank file
~, планетáрный epicyclic gear
~, радиоакти́вный decay series (nucl)
~, урáновый uranium series, uranium decay series
рядовóй 1. private, private soldier
2. airman
3. seaman, rating
4. ordinary, common
"Ря́ды Вздвой!" "Form Fours!"
ряж crib (eng)
ря́ж-колóдец hollow crib (eng)

С

с воздýшным охлаждéнием air-cooled
с катóдной свя́зью cathode coupled
с колéна from the kneeling position, in a kneeling position
с корóткой остано́вки from the short halt
с лéнточной подáчей belt-fed (MG)
с лéнточным питáнием belt fed
с магази́нным питáнием magazine-fed
с мéста 1. from the halt
2. from a fixed fire position
с настрóенной сéткой tuned grid (oscillator etc)
с настрóенным анóдом tuned anode
с остано́вок from the halt
с отражáтелем tamped (nucl)
с отсоединённой сéткой floating, floating grid (oscillator etc)
с перемéнной крутизнóй variable-mu (electron)
с сня́той фáской bevelled
с хóда from the line of march (tac)
с хóду on the move (tk gunn)
сáбля sabre
саботáж sabotage
саботáжник saboteur
сабули́т sabulite (expl)
савáнна savannah (topo)
сади́ться/сесть to land (ac), to go flat (battery), to set (sun), to shrink (of cloth), to sit down; **сади́ться/сесть на мель** to run aground, to get into difficulties
сáжа soot, carbon black
сажáть/посади́ть to seat, to land (ac); **сажáть/посади́ть под арéст** to put under arrest
сáжень, морскáя fathom (naut)
~, шестифýтовая fathom (naut)
сáйлентблок Silentbloc (bushing)

сала́зки 1. small sled
2. slides
~, мише́нные target sled
~, реакти́вные rocket sled, test sled
са́ло 1. fat, lard, suet, tallow
2. new thin ice, slush
~, пу́шечное gun grease
сало́н saloon, cabin (ac etc)
салфе́тка 1. swab (CW decontamination)
2. serviette
3. filter cloth
са́льник 1. packing, seal, grommet, oil seal
2. stuffing-box, gland
~ ка́ртера маховика́ bell-housing oil seal
~, лабири́нтный labyrinth packing, labyrinth gland, labyrinth sealing
~ полуо́си axle shaft oil seal
~, самоуплотня́ющийся self-sealing stuffing box
~ ствола́ barrel packing
салю́т salute (guns etc)
самби́ст unarmed combat specialist
са́мбо unarmed combat
самовзво́дный self-cocking
самовозбужде́ние self-excitation (elec)
самовозгора́ние spontaneous combustion
самово́льный 1. arbitrary, unauthorized
2. insubordinate, self-willed
самовоспламене́ние spontaneous combustion, self-ignition
самовоспламеня́ющийся hypergolic, spontaneously inflammable
самовса́сывающий self-priming
самовыключе́ние дви́гателя engine failure
самовыта́скивание self-recovery
самовыта́скиватель self-recovery expedient (a form of winch for self-recovery)
самого́н moonshine, hooch, illegally distilled spirit
самодви́жущийся self-moving, self-propelled
самодегаза́ция weathering (CW agent)
самоде́йствующий automatic, self-acting, self-operating
самоде́льный improvised, home-made
самоде́ятельность individual initiative, individual enterprise, self-initiated activity
самозаку́поривающийся self-sealing
самозамыка́ющийся self-locking
самозаря́дный self-loading, semi-automatic (wpn)
самозатя́гивающийся self-sealing
самозащи́та self-defence
~ без ору́жия unarmed combat
самоинду́кция self-induction
самока́т bicycle
самока́тчик cyclist
самокрепле́ние autofrettage (eng)
самолёт 1. aircraft, aeroplane
2. improvised horizontal piling frame (sl, eng)
~, беспило́тный pilotless aircraft, drone
~, боево́й operational aircraft
~, винтово́й propeller-driven aircraft
~, вое́нно-тра́нспортный military transport aircraft
~, па́лубный deck-landing aircraft
~, поршнево́й piston-engined aircraft
~, радиоуправля́емый radio-controlled aircraft, drone
~, разве́дывательный reconnaissance aircraft
~, раке́тный rocket aircraft
~, реакти́вный jet-propelled aircraft
~ с вертика́льным взлётом и поса́дкой VTOL aircraft
~ с коро́тким взлётом и поса́дкой STOL aircraft
~, санита́рный ambulance plane, air ambulance
~, сверхзвуково́й supersonic aircraft
~ свя́зи liaison aircraft
~, тра́нспортно-деса́нтный troop-carrying aircraft, troop carrier
самолёт-амфи́бия amphibious aircraft
самолёт-запра́вщик tanker aircraft
самолёт-зондиро́вщик meteorological reconnaissance aircraft
самолёт-истреби́тель fighter aircraft
самолёт-корректиро́вщик spotter plane, artillery observation aircraft
~, артиллери́йский air observation post, air OP
самолёт-ма́тка mother-plane, parent plane (from which missile is launched)
самолёт-мише́нь target drone, pilotless target aircraft
самолёт-носи́тель carrier aircraft, host aircraft
самолёто-вы́лет aircraft sortie

самолёт-освети́тель pathfinder
самолёт-разве́дчик reconnaissance aircraft
самолёт-ракетоно́сец missile-carrying aircraft
самолёт-ретрансля́тор relay aircraft (comms)
самолёт-снаря́д cruise missile
~, межконтинента́льный intercontinental cruise missile
самолёт-цель target aircraft, drone
самолёт-штурмови́к ground-attack aircraft
"Самолёты!" "Plane!", "Aircraft!" (warning of air attack)
самоликвида́тор self-destruction device
~, центробе́жный centrifugal self-destruction element (GW)
самоликвида́ция self-destruction
самонаведе́ние homing, homing guidance
~, акти́вное active homing guidance
~, акти́вное радиолокацио́нное active radar homing
~, акусти́ческое acoustic homing
~, инфракра́сное infra-red homing, IR homing
~ на коне́чном уча́стке terminal homing
~ на цель target homing
~, опти́ческое optical homing
~, пасси́вное passive homing guidance
~ по ме́тоду паралле́льного сближе́ния constant bearing homing
~ по электромагни́тному излуче́нию electromagnetic homing
~, полуакти́вное semi-active homing guidance
~, радиолокацио́нное radar homing
~ с по́мощью инфракра́сных средств infra-red homing, IR homing
~ с по́мощью ла́зерных средств laser homing
~ с по́мощью опти́ческих средств optical homing
~ с по́мощью радиолокацио́нных средств radar homing
~, теплово́е heat(-seeking) homing
самонаводя́щийся homing, seeker
самооборо́на 1. self-defence 2. local defence (arty)
самоока́пывание digging in
самоопроки́дывающийся self-dumping, self-tipping (veh)
самопи́сец recording instrument, graphic instrument, automatic pen recorder
~, ветрово́й wind recorder
~ давле́ния pressure recorder
~, цифрово́й digital recorder
самопи́шущий self-recording
самоподгото́вка self-tuition
самоподде́рживающий(ся) self-sustaining
самопо́мощь self-help
самопроизво́льный spontaneous
самопрослу́шивание self listening, monitoring of own signal transmission (sigs)
самопу́ск self-starter
~, возду́шный air starter
саморазря́д self-discharge (elec)
саморегистри́рующий self-recording
саморегули́рующийся self-adjusting
саморемо́нт self-maintenance
самосва́л dump truck, tipping truck
~ с боково́й разгру́зкой side-dump truck, side-dump tipper
~ с гидравли́ческим подъёмником hydraulic dump truck, hydraulic dump tipper
самоскле́ивающийся self-sealing
самоскрепле́ние self-hooping, auto-frettage (gun production)
самосма́зывающийся self-oiling, self-lubricating
самостре́л malingerer guilty of self-wounding
самотёк gravity feed, gravity drainage
самотёком by gravity
самоуплотне́ние self-sealing
самоускоре́ние auto-acceleration
самоустана́вливающийся self-adjusting, self-aligning
самохо́дка self-propelled gun, SU gun (Sov), assault gun
самохо́дный self-propelled
самохо́дчик SU gunner (Sov)
санато́рий sanatorium
санба́т medical battalion
са́ни sledge
сан-инстру́ктор medical instructor, NCO i/c medical party in action (Sov)
санита́р medical orderly
~, ро́тный company medical orderly

санита́р-дезинфе́ктор disinfection orderly
санитари́я sanitation
~, полева́я field hygiene
санита́рка nurse
санита́рно-гигиени́ческий sanitary and hygiene
санита́р-носи́льщик stretcher-bearer
санита́рный sanitary, medical
са́нкция sanction
санотделе́ние medical detachment
санпропускни́к medical check point, decontamination centre, field cleansing and disinfestation centre
санро́та medical company
санча́сть medical unit
са́па sap (fd works)
~, двойна́я перекидна́я single sap with double parapet (eng)
~, одино́чная перекидна́я single sap with one parapet (fd eng)
сапёр sapper, combat engineer
сапёр-минёр sapper mining specialist
сапёр-подрывни́к sapper in demolition party
сапёр-разве́дчик member of sapper recce party
сапёры бронета́нковых войск armoured engineers
~, корпусны́е corps engineer troops
сапёры-штурмовики́ assault engineers
сапоги́ boots
~, ке́рзовые canvas-top boots
~, рези́новые rubber knee-boots, gum boots
сапо́жник cobbler
сапожо́к ра́мы piston post (SA)
сапроточа́с sapper company man-hour
сапу́н air vent, breather
сара́й barn, shed
сарппогра́мма flight recorder read-out (ac)
сателли́т 1. satellite
2. planetary bevel wheel
~ планета́рной переда́чи planetary pinion
сбаланси́ровать (см баланси́ровать
сбереже́ние care (eqpt etc)
~ ору́жия, плохо́е abuse of weapon, misuse of weapon
сберка́сса, полева́я field savings bank
сбива́ние пе́ленга meaconing (MEAsuring and CONfusing of navigation signals) (ECM)
сбива́ть/сбить to put together, to knock together, to shoot down (an aerial target)
сбива́ться/сби́ться to be knocked out, to go out of tune, to go off, to go astray, to get lost, to go out of adjustment
сби́тый shot down, destroyed (aerial target)
сбить (см сбива́ть)
сби́ться (см сбива́ться)
сближа́ться/сбли́зиться to approach, to close in on
сближе́ние approach, closing-in
~ меридиа́нов convergence of meridians (carto)
~ с проти́вником advance to contact (tac)
сбли́зиться (см сближа́ться)
сблоки́ровать (см блоки́ровать)
сбой break (comms), interval, pause
сбо́ку 1. on the flank, on the side
2. from the flank, from the side
сбор 1. assembly, rally
2. course of instruction
3. assembling, collecting, rallying
4. signal to assemble
5. salvage
~, ежего́дный ла́герный annual camp
~, ла́герный 1. camp guard
2. camp strength (total compliment of camp including elements from miscellaneous formations)
~, ли́хтерный lighterage
~ по трево́ге turn out, muster on alarm, stand-to
~, уче́бно-методи́ческий training methods meeting, training methods course
~, уче́бный training course, training assembly
сбо́рка assembly, assembling (wpn)
~, конои́дная cam assembly
~ паро́ма raft assembly
сбо́рник collector, receiver, collection tank
сбо́рно-разбо́рный prefabricated and demountable, sectional
сбо́рный 1. mixed, combined, miscellaneous
2. scratch, ad hoc
3. prefabricated

сбóрочный 1. assembly, erecting
2. unitized
сбóрщик assembler
сбрáсываемый jettisonable
сбрáсывание 1. dropping, discarding, jettisoning
2. release (from ac)
~, **аварийное** emergency drop
~, **беспарашю́тное** free dropping
~ **вы́мпела** message drop
~ **грýзов, беспарашю́тное** free-dropping of supplies (airborne ops)
~ **мин** minelaying (by ac)
~, **парашю́тное** air dropping by parachute
~ **парашю́тным спóсобом** supply dropping (para)
сбрáсыватель bomb release mechanism
сбрáсывать/сбрóсить to drop, to release, to jettison; **сбрáсывать/сбрóсить бóмбы** to drop bombs; **сбрáсывать/сбрóсить на парашю́тах** to drop by parachute
сброс 1. fault (geol), steep break in glacier ice
2. discarding
3. zeroing, resetting (of instruments etc)
4. drop/release (bombing)
сбрóсить (см сбрáсывать)
сваедержáтель pile head bracket
свáйка 1. stake
2. marline-spike
свáйный piling, piled, on piles
свáливание canting
свáливать/свали́ть to tip, to cant
свали́ть (см свáливать)
свáлка 1. heap
2. fight, brawl
~, **мýсорная** refuse dump
свáривать/свари́ть 1. to weld
2. to boil, to cook
свари́ть (см свáривать)
свáрка welding
~, **автогéнная** oxy-acetylene welding, gas welding
~, **автомати́ческая** automatic welding
~, **áтомно-водорóдная** atomic-hydrogen arc welding
~, **ацетилéновая** acetylene welding
~, **ацетилéно-кислорóдная** oxy-acetylene welding
~ **внакрóй** lap weld
~ **внахлёстку** lap weld
свáрка встык flash welding
~, **гáзовая** gas welding
~, **газопрéссовая** oxy-acetylene welding
~ **давлéнием** pressure welding
~, **дуговáя** arc welding
~, **контáктная** pressure contact welding, electric resistance welding
~ **плавлéнием** fusion welding
~, **ручнáя кузнéчная** blacksmith's welding
~, **стыковáя** butt-welding
~, **торцóвая** butt welding
~, **тóчечная** spot welding
~, **электродуговáя** electric arc welding
свáрочный welding
свáрщик welder
свáя pile
~, **деревя́нная** timber pile
~, **железобетóнная** reinforced-concrete pile
~, **креннáя** bearing pile
~ **мостовóй опóры** bridge pile
~, **откóсная** raker pile (br)
~, **причáльная** mooring pile
сведéние 1. contraction, cramp (med)
2. settling (of an account)
свéдение information, data, report
~ **о дорóгах** road state report
~ **о нали́чии боеприпáсов** ammunition state
~ **о нали́чности** availability report
~, **развéдывательное** reconnaissance information, intelligence information
свéдения, агентýрные information supplied by agents
~, **достовéрные** reliable information
~ **о нали́чии боеприпáсов** ammunition returns
~ **о состоя́нии стволóв** gun tube report
~ **о состоя́нии ты́ла** logistic state
~, **развéдывательные** reconnaissance information, intelligence information
свéрить (см сверя́ть)
свéрка collation
~, **автомати́ческая** automatic correlation
~ **часóв** synchronization of watches
сверлéние drilling, boring
~, **кольцевóе** trepanning
~, **мáсляное** oil way (in machine, in engine)
сверли́ть to drill, to bore (imp only)
сверлó drill, drilling bit

сверло́, алма́зное diamond drill
~, америка́нское twist drill, helical drill
~, перово́е twisted flat drill
~, спира́льное twist drill
сверну́ть (см свёртывать)
сверну́ться (см свёртываться)
свёртка кружка́ в колпачо́к cupping (ammo manufacture)
свёрток 1. bundle, roll
2. turn-off (minor road from major)
свёртывание 1. closing up (tac)
2. packing up, rolling up
3. coagulation
4. curtailment, temporary closure
~ боево́го поря́дка reorganization
~ в коло́нны forming columns
свёртывать/сверну́ть to turn aside, to roll up, to close down (radio), to coagulate, to curdle; **свёртывать/сверну́ть ла́герь** to break camp; **свёртывать/сверну́ть пала́тки** to strike tents; **свёртывать/сверну́ть папиро́су** to roll a cigarette; **свёртывать/сверну́ть с доро́ги** to turn off the road
свёртываться/сверну́ться to close up (tac), to curl up, to roll up
сверхдавле́ние superpressure
сверхда́льний very long range, super long range
сверхзатуха́ние overdamping
сверхзвуково́й supersonic
сверхкомпле́ктный supernumerary, above establishment
сверхмо́щний superpower
сверхнормати́вный above norm, above permitted standard
сверхору́жие superweapon
сверхпреломле́ние super-refraction, trapping (radio)
сверхпроводи́мость superconductivity (elec)
сверхрефра́кция super-refraction
сверхслу́жащий extended-service man
сверхсро́чник extended-service man, regular (Sov)
сверхсрочнослу́жащий extended-service man, regular (Sov)
сверхсро́чный 1. extended, longer-lasting
2. urgent, pressing
сверхтяжёлый superheavy
сверхшта́тный surplus to establishment, above establishment
сверя́ть/све́рить to compare, to collate, to verify
свес ба́шни turret overhang
свести́ (см своди́ть)
свет 1. light
2. world
~, залива́ющий flood-light
~, затемнённый dimmed light
~, защи́тный khaki
~, иску́сственный лу́нный artificial moonlight
~, про́блесковый flashing light
~, ультрафиоле́товый ultra-violet light
~ фар head lamp light
свети́ло, небе́сное heavenly body, celestial body
свети́льник lamp, lamp fitting
~, люминесце́нтный luminescent lamp, fluorescent lamp
светогра́мма light signal
светоко́пия photostat
~, си́няя blue-print
светомаскиро́вка black-out, blacking out (cam)
светоме́трия flash ranging (arty)
светонепроница́емый light-proof
светопеча́тание phototype process
светопла́н situation display (electron)
светопо́ст flash-spotting OP (arty)
светорассе́яние light diffusion, light dispersion
светосигнализа́ция light signalling
светоси́ла 1. focal ratio (lens)
2. luminosity, candlepower
светососта́в luminous paint, luminescent compound
светосхе́ма illuminated chart
светофи́лтр, инфракра́сный infra-red filter, IR filter
светофи́льтр light filter
~, маскиро́вочный blackout filter
светофо́р 1. traffic light
2. signal light, luminous signal, light-signalling device
~ для регули́рования движе́ния traffic light
светочувстви́тельность sensitivity to light, photosensitivity
светочувстви́тельный light-sensitive
светя́щийся luminous, luminescent

свечá 1. candle
2. candlepower
3. sparking plug
4. suppository (med)
~, **дымовáя** smoke candle
~ **зажигáния** sparking plug
~ **зажигáния, экранúрованная** screened spark plug
~, **запáльная** spark plug
~ **накáла** heater plug, glow plug (diesel)
~, **неразбóрная** integral sparking plug
~, **разбóрная** removable sparking plug
~, **химúческая** chemical candle
~, **ядовúтая дымовáя** toxic smoke candle
свечéние glow, luminescence, radiance, phosphorescence
свидéтель witness
свидéтельство 1. evidence
2. certificate
~, **инспéкторское** inspection certificate (stores etc)
~, **приписнóе** draft registration slip (conscription)
~, **проходнóе** certificate of military service
свинéц lead (metal)
свинúна pork
свинцевáние канáла стволá bore leading
свинцóвый lead, leaden
свистóк whistle, blast
свищ fistula (med)
свободнонесýщий cantilever
свобóдный 1. free, available, loose
2. independent, separate
свод 1. code (law etc)
2. arch, roof, vault
~, **горя́щий** web (expl)
~ **давлéния** pressure arch (min)
~, **двухфлáжный** two-flag code (comms)
~, **естéственный** natural arch (mining)
~, **сигнáльный** signal code
~, **снéжный** snow arch
сводúть/свестú to bring together, to amalgamate, to bring down, to remove, to fell (trees); **сводúть/свестú в бригáды** to brigade
свóдка summary, resume, report
~, **артиллерúйская метеорологúческая** artillery meteor telegram
~ **весовы́х дáнных** weights report (GW)
свóдка, метеорологúческая meteorological report, weather report
~ **о налúчии боеприпáсов** ammunition report
~, **оперативная** situation report, SITREP
~ **по ты́лу** logistics report
~, **развéдывательная** intelligence summary, INTSUM
~ **свя́зи** signals location statement and communications state
свóдный combined, composite
своеврéменность opportuneness, good timing, timeliness
своúм хóдом under (its) own power
"свой" friendly (IFF)
свóйства, баллистúческие ballistic properties, ballistics
~, **боевы́е** military characteristics, combat characteristics
~ **мéстности, тактúческие** tactical features of the ground
~ **орýжия** characteristics of (a) weapon
~, **смáзочные** lubricating properties, lubricating qualities
~ **тáнка, ходовы́е** ride characteristics of a tank
свóйство property, characteristic, attribute
~, **антидетонациóнное** anti-knock property, anti-knock value (liquid fuels)
~, **баллистúческое** ballistic property
"свой-чужóй" IFF, Identification Friend or Foe
"свой-чужцú", опознавáтельное устрóйство IFF system
свя́занный 1. bound, tied, connected (with), coupled
2. bonded (nucl)
свя́зи, вéрхние продóльные top lateral bracing (br)
~, **нúжние продóльные** bottom lateral bracing
~, **подвóдные** underwater bracing (br)
связúст signaller
свя́зка 1. cluster
2. strapping (magnetron)
3. string (interconnected gren A tk obstacle)
свя́зки внахлёстку echelon strapping (magnetron)
~, **двойны́е кольцевы́е** double ring strapping (electron)

свя́зки, одино́чные кольцевы́е single ring strapping (electron)
~, **ступе́нчатые** echelon strapping (magnetron)
связно́й orderly, messenger
~, **пе́ший** runner (messenger)
связь 1. communications, intercommunications
2. co-operation, liaison
3. contact
4. coupling, connection (elec)
5. cross member, tie (eg trail of arty piece)
~, **авари́йная** emergency communication
~, **автотрансформа́торная** auto-inductive coupling, bottom end inductive coupling (electron)
~, **акусти́ческая** communication by sound, sound signalling (whistle etc)
~, **бесперебо́йная** continuous communications, unbroken communications
~, **беспроводна́я** radio communications, wireless communications
~ **в бли́жней инфракра́сной о́бласти спе́ктра** near infra-red communication
~ **в да́льней инфракра́сной о́бласти спе́ктра** far infra-red communication
~ **в инфракра́сном диапазо́не** infra-red communication, IR communication
~, **взаи́мная** interconnection
~ **взаимоде́йствия** co-operation communications, liaison communications
~ **взаимоде́йствия с авиа́цией** air liaison
~, **визуа́льная** visual contact
~, **вне́шняя** external communications
~, **вне́шняя ёмкостная** top-end capacitive coupling (elec)
~, **вну́тренняя** intercommunication, intercom
~, **вну́тренняя ёмкостная** direct capacitive coupling, bottom-end capacitive coupling (elec)
~, **внутрита́нковая** tank intercom
~, **вое́нно-почто́вая** forces postal service
~, **возду́шная** air liaison
~, **гальвани́ческая** resistance coupling (elec)
~, **да́льняя** long-range communications, long-distance communications
связь, двухсторо́нняя two-way communications
~, **двухсторо́нняя группова́я телефо́нная** conference call (teleph)
~, **дистанцио́нная** telecommunication, long-range communication
~, **ду́плексная** duplex communication
~, **ёмкостная** capacitance coupling (elec)
~, **ёмкостно-оми́ческая** capacitive ohmic coupling, resistance-capacitance coupling, R-C coupling (elec)
~, **железнодоро́жная** railway communication
~, **жёсткая обра́тная** follow up direct feedback (electron)
~, **жива́я** communication by live agent (eg runner)
~ **за счёт тропосфе́рного рассе́яния** tropospheric scatter communication
~, **загоризо́нтная** over-the-horizon communication
~, **звукова́я** sound signalling
~, **зри́тельная** visual communications
~, **изодро́мная обра́тная** proportional feedback (electron)
~, **индукти́вная** inductive coupling (elec)
~ **кле́ром** plain language communications, communications in clear
~, **кома́ндная** command communication
~ **кома́ндования** command communications
~, **коротково́лновая** short wave communications
~ **ко́рпуса** corps signals
~, **крити́ческая** critical coupling (elec)
~, **курье́рская** courier service
~, **ла́зерная** laser communication
~, **локтева́я** close contact (tac)
~ **ме́жду пехо́той и та́нками** infantry-tank communication
~, **метео́рная** meteor burst propagation (radio)
~, **многокана́льная** multi-channel connection (comms)
~, **многокра́тная** multiplex communications, multi-channel transmission (comms)
~ **на УКВ** microwave communications
~, **назе́мная** ground communication
~, **обра́тная** feedback (electron, servo)
~, **огнева́я** co-ordination of fire, mutual fire support plan

связь, односторо́нняя one-way communication
~ **оповеще́ния** air and NBC warning service net
~, **опти́ческая** optical communication
~, **парази́тная** parasitic coupling (elec)
~, **перекрёстная** crosstalk
~, **пе́шая** communication by runner
~ **по направле́нию** point-to-point communications
~ **по ускоре́нию, стабилизи́рующая обра́тная** acceleration stabilizing feed-back, rate -stabilizing feed-back (servomechanism)
свя́зь, подви́жная 1. signals despatch service, SDS
2. message carrying service
~ **подви́жными сре́дствами** communication by message carrying
связь, попере́чная lateral bracing (br)
~, **постоя́нная** fixed communications
~, **почто́вая** postal service
~, **проводна́я** line telecommunications
~, **продо́льная** lateral bracing (br)
~, **радиореле́йная** radio relay communications
~, **радиотелефо́нная** radio telephone communication
~, **регенерати́вная обра́тная** regeneration, regenerative feed-back, positive feed-back (electron, servomechanism)
~, **речева́я** speech communication
~, **селе́кторная** selective circuit communication (teleph)
~, **си́льная** close coupling (elec)
~, **си́мплексная** one-way communication, simplex communication
свя́зь, скоростна́я обра́тная velocity feedback
связь, сла́бая loose coupling (elec)
~ **слу́жбы оповеще́ния** air and NBC warning service net
~, **служе́бная** engineering communication, (radio-relay, RR, between service engineers and operators)
~ **со штырём** saddle pivot bracket (ord)
~, **спу́тниковая** satellite communications
~, **такти́ческая** tactical liaison (between units during battle)
связь та́нков с пехо́той tank-infantry communication
~, **телегра́фная** telegraphic communications
~, **телефо́нная** telephonic communications, voice communications
~, **те́сная** close liaison
~, **трансформа́торная** transformer coupling, inductive coupling (elec)
~, **тропосфе́рная** troposphere (scatter) communication
~ **тыла** administrative communications
~ **управле́ния огнём** fire-control communication
~, **фельдъе́герская** courier service communications
~, **фототелегра́фная** facsimile communication, telephoto communication
~, **цифрова́я** digital communication
~, **электри́ческая** telecommunications
~, **электро́нная** electron coupling
свяще́нник chaplain, padre
сгла́живание smoothing (electron)
сго́вор 1. agreement
2. collusion, conspiracy
сгора́емый combustible
сгора́ние combustion
~, **безды́мное** smokeless combustion
~, **вну́треннее** internal combustion
~, **затяжно́е** delayed combustion
~, **каталити́ческое** catalytic ignition
~, **непо́лное** incomplete combustion
~, **неравноме́рное** non-uniform combustion
~, **по́зднее** afterburning
~, **по́лное** complete combustion
~, **после́дующее** after-burning
~, **предохрани́тельное** blowout of a fuse
~, **равноме́рное** uniform combustion
сгружа́ть/сгрузи́ть to off-load
сгрузи́ть (см сгружа́ть)
сгруппирова́ться (см группирова́ться)
сгусти́тель горю́чего fuel thickener
сгу́стки ма́сла oil sludge
~, **ма́сляные** oil sludge
сгу́сток 1. clot (of blood etc)
2. bunch, cluster
~, **электро́нный** electron bunch (RR)
~ **электро́нов** electron bunch (RR)
сгуще́ние 1. thickening, condensation, concentration
2. coagulation, clotting

сдава́ться/сда́ться to capitulate, to surrender
"Сдать пост!" "Sentries-Pass!" (order for relief of sentry)
сда́ться (см сдава́ться)
сда́ча 1. delivery, hand over, turn in 2. surrender
~, безогово́рочная unconditional surrender
~ поста́ relief of post (sentry)
сдва́ивать/сдвои́ть to double
сдвиг 1. slip, shift, displacement 2. shear, shearing 3. fault, dislocation (geol)
~, доплеро́вский Doppler shift, Doppler drift
~ щёток brush lead (elec)
сдвои́ть (см сцва́ивать)
сде́лать (см де́лать)
сдержа́ть (см сде́рживать)
сде́рживание containment
сде́рживать/сдержа́ть to restrain, to hold back, to pull up; **сде́рживать/сдержа́ть бо́ем** to contain by battle
се́вер North
~, и́стинный True North
~ координа́тной се́тки grid north
~, магни́тный magnetic north
~ се́тки Grid North
се́верный north, northern
се́веро-восто́к northeast
се́веро-восто́чный northeast, northeastern
се́веро-за́пад northwest
се́веро-за́падный northwest, northwestern
седла́ть/оседла́ть to saddle
седло́ saddle, seat
~, вью́чное pack saddle
~ кла́пана valve seat, valve insert
~, кла́панное valve seating, valve seat
седлови́на col, saddle (topo)
сейсмофо́н seismophone
секре́т 1. secret 2. listening post
секре́тно secret (gen, also security grading)
секре́тность secrecy
секста́нт sextant
се́ктор arc, area, sector, segment
~ ви́димости arc of visibility
~ выключа́теля 1. cut-out arc (Sov mor) 2. taboo setting (gunn)
се́ктор, гла́дкий interruption (screw)
~, дополни́тельный alternative arc of fire, secondary arc of fire
~, зубча́тый toothed quadrant
~ наблюде́ния area of observation
~ обзо́ра angular field of view
~ обслу́живания servicing area
~ обстре́ла 1. arc of fire (wpn) 2. arc of responsibility (AFV)
~ обстре́ла, дополни́тельный subsidiary arc of fire, alternative arc of fire, secondary arc of fire
~ обстре́ла, основно́й primary arc of fire
~ огня́ arc of fire
~ ПВО air defence sector
~ подъёмного механи́зма, зубча́тый elevating arc (AA ord)
~, подъёмный elevating arc (gunn)
~ по́иска search sector
~ разве́дки sector of reconnaissance
~, распредели́тельный distributor segment (magneto)
~ рычага́ то́рмоза brake lever quadrant
~ со шка́лой arc and scale (ord)
~ стрельбы́ arc of fire
~ стрельбы́, основно́й primary arc of fire
~ устано́вки elevating arc (ord)
~, червя́чный steering worm sector (veh)
секундоме́р stop-watch
секу́нды, счи́танные a few seconds
секу́щая secant (math)
се́кция, нагнета́тельная delivery section (pump)
~, отка́чивающая suction section (pump)
~, согласу́ющая matching section (electron)
селекти́вность selectivity
селе́ктор selector
~ по да́льности range discrimination unit (missile)
селе́кция 1. discrimination 2. selection (radar)
~ дви́жущихся целе́й moving target indication, MTI, selection of moving targets (radar)
~, ка́чественная quality selection
~, ко́довая code selection (GW)
селе́н selenium (chem)
селе́ние settlement, village
сели́тра saltpetre
~, аммони́йная ammonium nitrate

селитра, калиевая potassium nitrate
~, чилийская Chile saltpetre (sodium nitrate)
село village
сельсин, бесконтактный contactless synchro (a specialized form of synchro developed in USSR) (elec)
сельсин-датчик synchro transmitter, data synchro, self- synchronous transmitter, selsyn transmitter
сельсин-приёмник synchro receiver, selsyn receiver
сельсин-трансформатор synchro transformer, selsyn transformer
семафор semaphore, arm signal
семейство кривых family of curves (ball)
~, радиоактивное decay series (nucl)
~ траекторий family of trajectories (ball)
семинар course (of study), seminar
сенситометр sensitometer
сепаратор separator
~ подшипника качения bearing cage (veh)
~, центробежный centrifugal separator
~, электромагнитный electromagnetic mass separator
сепсис sepsis (med)
сера sulphur
серводвигатель servomotor, servo
сервоклапан servovalve
сервомеханизм servomechanism, servo unit, servo system
сервомотор servomotor, servo
~, гидравлический hydraulic servomotor
~, масляный hydraulic servo motor
сервопривод servo, servo link
~ органов управления servo control
~ тормозной системы brake servo
сервосистема servosystem
~, устойчивая stable servosystem
сервотормоз servo-assisted brake, power brake
сервоуправление servocontrol
сервоусилитель servo amplifier
сервоустройство servo device
сердечник core
~, бронебойный armour-piercing core
~, карбидо-вольфрамовый tungsten carbide core (ammo)
~, настроечный tuning slug (elec)
~ пули bullet slug
~, расслоенный laminated core (elec)
сердечник снаряда shell core
~, стальной steel core
~, стальной бронебойный hard steel core (as in APDS)
сердцевина, пороховая gunpowder core (safety fuse)
~ рассеивания 50% zone (of dispersion of shot) (gunn)
серебро silver
середина толщины свода centre of the web (expl)
середнячок average performer (soldier who has obtained average results)
серёжка shackle, fastening
сержант sergeant
~, взводный platoon sergeant
~, младший junior sergeant (Sov roughly equivalent to Brit Cpl or Lance Sgt)
~, старший senior sergeant (roughly equivalent to Brit Colour Sgt, Staff Sgt)
сержант-химик sergeant-chemical troops
серийность 1. message precedence (system of)
2. series production
серия 1. series
2. stick (bombs)
3. degree of precedence (message)
4. ripple (series launching of missiles)
~ Б routine (Sov message precedence)
~ беглого огня series of volleys
~ выстрелов group of rounds
~ Г operational priority (Sov message precedence)
~, зачётная scoring group
~ импульсов pulse train (electron)
~ К priority (Sov message precedence)
~, контрольная controlled series (adjusting mean point of impact)(gunn)
сероводород hydrogen sulphide
серьга shackle, link
~ боевой пружины stirrup spring (fuze)
~, буксирная tow-yoke, towing shackle
~ подвески suspension shackle
сестра, медицинская nurse, nursing sister
сесть (см садиться)
сетка 1. net
2. grid (carto, electron)
3. reticle, graticule (optical instr)
~, азимутальная azimuth grid
~, антидинатронная suppressor grid (electron)

се́тка, армату́рная compression bars (reinforced concrete)
~, артиллери́йская координа́тная fire-control grid (gunn)
~, географи́ческая 1. graticule system 2. geographical grid
~ гипе́рбол hyperbolic net (navigation)
~, грузова́я cargo net
~ дальноме́ра range scale
~, дальноме́рная range finding scale
~ для целеуказа́ния target grid (arty)
~, заземлённая earthed grid, grounded grid (electron)
~, защи́тная suppressor grid (valve)
~, картографи́ческая map graticule (topo)
~, квадра́тная картографи́ческая rectangular cartographic grid, graticule system of practically square rectangles
~, кольчу́жная wire mesh
~, координа́тная grid system
~, маляри́йная mosquito net
~, маскиро́вочная camouflage net
~ маслофи́льтра strainer (in oil filter)
~ на вое́нно-топографи́ческих ка́ртах, координа́тная military grid (carto)
~, неподви́жная fixed graticule (instr)
~, носи́мая personal camouflage net
~, обрамлённая ring sight, frame sight (ac gunsight)
~, о́бщая earthed grid, common grid (electron)
~, опти́ческая optical reticle
~, подви́жная moving graticule (gunsight etc)
~ прице́ла, неподви́жная fixed graticule
~, про́волочная 1. wire net, wire netting 2. gauze
~, противодинатро́нная suppressor grid (elec)
~, противомоски́тная mosquito net, mosquito netting
~ рассе́ивания по да́льности dispersion pattern in range
~ рассе́ивания по направле́нию dispersion pattern in deflection
~, угломе́рная graticule
~, управля́ющая control grid (valve)
~, усло́вная координа́тная arbitrary grid (surv)
~, фильтрова́льная filtering gauze, filtering strainer
~, экра́нная screen grid (electron valve)

се́точный grid
сеть 1. net, network, circuit 2. system 3. netting
~, автомати́ческая телефо́нная automatic telephone circuit
~, артиллери́йская опо́рная artillery control network (surv, established from datum networks of benchmarks)
~, бортова́я on-board electrical supply system
~, взрывна́я 1. firing circuit 2. demolition circuit
~, взрывна́я паралле́льная parallel firing circuit
~, взрывна́я паралле́льно-после́довательная series-parallel firing circuit
~, взрывна́я после́довательная series firing circuit
~, геодези́ческая geodetic network, control network (surv)
~, геодези́ческая опо́рная geodetic network, control network (surv)
~, гидрографи́ческая hydrographic pattern
~, горизонта́льная опо́рная horizontal control network (surv)
~, детонацио́нная detonating circuit
~, доро́жная road and rail network
~, еди́ная координа́тная common grid system (topo)
~, железнодоро́жная rail network
~, канализацио́нная sewerage system, sewage system
~, магистра́льная main circuit, ring main
~, маляри́йная mosquito net
~, маскиро́вочная camouflage net
~ МБ magneto exchange, magneto switchboard exchange
~, ме́стная артиллери́йская опо́рная local artillery control network (surv) (established from theatre maps and astro obsn)
~ на побере́жье, радиолокацио́нная coastal radar screen
~ но́мер 1 для стрелка́ face veil (cam)
~ но́мер 4, маскиро́вочная 10 x 2.5 metre camouflage net
~, огнепрово́дная firing circuit
~ оповеще́ния warning network (for air, NBC etc)

сеть, опóрная control network (surv, datum)
~, опóрная геодези́ческая geodetic network, control network (surv)
~, опóрная топографи́ческая control network (surv)
~, освети́тельная lighting system
~, питáющая power network (elec)
~, подрывнáя demolition circuit
~, прóволочная wire net, wire netting, wire entanglement
~, радиолокациóнная radar network, radar system
~, рыболóвная fishing-net
~, сапёрная перенóсная concertina wire obstacle
~ свя́зи, воéнная military communications net
~ свя́зи, дивизиóнная divisional communications net
~, стационáрная fixed net (radio)
~, тáбельная маскирóвочная standard camouflage net, regulation camouflage net
~, тáнковая tank (camouflage) net
~ управлéния огнём fire-control net
~, участкóвая circuit branches (elec and fuse)
~, хлопчатобумáжная cotton net (cam)
~, шатрóвая double apron fence (wire obstacle)
~, электри́ческая electrical comment
~, электровзрывнáя electrical demolition circuit
сечéние section, cross-section
~, вы́деленное detail section (drawing)
~, выходнóе exit area (jet/rocket motor nozzle)
~, поперéчное cross-section
~, расчётное reference plane
~ соплá, проходнóе nozzle area
сжáтие 1. compression
2. cavitation (expl)
~, адиабати́ческое adiabatic compression
~, изотерми́ческое isothermal compression
~, изэнтропи́ческое isentropic compression
сжáтый 1. compressed
2. compacted, contracted
3. condensed, concise
сжать (см сжимáть)
сжечь (см сжигáть)
сжигáние 1. combustion, burning
2. consumption (of fuel)
сжигáть/сжечь to burn, to burn up, to consume, to burn off, to set on fire
сжиди́ть (см сжижáть)
сжижáть/сжиди́ть to liquefy, to condense, to compress
сжижéние 1. liquefaction
2. condensation
сжимáемость compressibility
сжимáтель-расшири́тель compandor (teleph)
сжимáть/сжать to compress
сигнáл 1. call, signal
2. horn (on veh)
~, авари́йный distress signal
~, акусти́ческий audible signal
~, амплиту́дно-модули́рованный amplitude-modulated signal
~, аналóговый analogue signal
~ бди́тельности warning signal
~ бéдствия distress signal
~, блуждáющий ghost signal (radar)
~, боевóй alarm signal
~ боевóй готóвности stand-to signal
~, ви́димый visual signal
~ включéния указáтеля поворóта direction indicator tell-tale
~ "Возду́шная Тревóга" "red alert!" (air raid warning)
~ врéмени time-signal
~, входнóй input signal
~ вы́зова call signal
~, вызывнóй calling signal (teleph)
~ выключéния дви́гателя shutoff signal (GW)
~, выходнóй output signal
~ гироскóпа gyro signal
~ готóвности ready signal
~, дискрéтный discrete signal
~ "зáнято" busy signal, engaged signal
~ запрóса interrogating signal, interrogator (IFF)
~, звуковóй audio-signal, sound signal
~, зелёный line clear signal
~, зри́тельный visual signal
~, зу́ммерный tone signal, buzzer signal
~, и́мпульсный pulse signal
~, исполни́тельный executive signal
~, контрóльный control signal, pilot signal

сигна́л, корректи́рующий corrective command (GW)
~, кра́сный 1. red strip indicating "primed" (gren) (Sov) 2. red signal
~, ло́жный confusion signal (ECM)
~, маскиру́ющий masking signal (radar, radio)
~, мига́ющий flashing signal
~, минима́льный обнару́живаемый minimum detectable signal
~, модули́рованный modulated signal
~, модули́рованный по фа́зе phase-modulated signal
~, модули́рующий modulating signal
~, настро́ечный tuning call, netting call
~ настро́йки tuning call, netting call, tuning signal
~ оповеще́ния alarm, warning signal
~ опознава́ния recognition signal, identification signal
~, опознава́тельный recognition signal, identification signal
~, опознава́тельный светово́й identification lamp
~, опо́рный reference signal
~, опти́ческий optical signal, visual signal
~ "Отбо́й" clearing signal (teleph)
~ "Отбо́й ВТ" (возду́шная трево́га) "All Clear" (air raid signal)
~, отве́тный answer signal, answering signal
~, отражённый echo, reflected signal (radar)
~ оши́бки error signal
~, паро́льный authentication signal
~, пиротехни́ческий pyrotechnic signal
~, пове́рхностный ground wave signal
~ поворо́та turn signal
~ подтвержде́ния acknowledgement
~, позывно́й callsign signal
~, постоя́нный fixed signal
~, предвари́тельный 1. warning signal 2. distant signal (rly)
~, про́блесковый intermittent signal
~ "путь свобо́ден" proceed signal
~, различи́мый recognizable signal
~ самонаведе́ния homing signal
~, светово́й light signal, indicating light
~, синхронизи́рующий synchronizing signal
~, та́ктовый cadence signal (teleg)
сигна́л, тона́льный audio signal
~ трево́ги alert, alarm signal
~ управле́ния 1. command signal (GW) 2. tactical control code signal
~, усло́вный pre-arranged signal
~ фо́на background signal
~, ча́стотно-модули́рованный frequency-modulated signal
~, электри́ческий electrical impulse
сигна́л-генера́тор signal generator
сигнализа́тор 1. signalling device 2. indicator
~ давле́ния ма́сла oil pressure warning device
сигнализа́ция signalling
~, автомати́ческая automatic signalling
~, звукова́я sound signalling
~, опти́ческая visual signalling
~, охра́нная detection and alarm system
~, пожа́рная fire warning
~, светова́я flashing, light signalling
~, семафо́рная mechanically-worked semaphore signalling
сигнализи́ровать to signal (imp and perf)
сигнали́ст 1. bugler, signal-bugler 2. signaller
сигнали́ст-бараба́нщик bugler-drummer
сигнали́ст-семафо́рщик semaphore signaller
сигна́л-кома́нда command signal (msl)
сигна́лы взаимоде́йствия co-ordination signals (pre-arranged)
~, постоя́нные служе́бные operating signals (codes for standard phrases)
сигна́льщик signaller
marshaller (ac)
сигна́л-э́хо echo (radar)
сиде́ние seat
~, катапульти́руемое ejection seat
сиде́нье seat
~ води́теля driver's seat
~ команди́ра commander's seat
~ наво́дчика gunner's seat
~, регули́руемое adjustable seat
сиде́ть to sit, to be, to fit; **сиде́ть на ме́ли** to be in a difficult situation, to be in difficulties (imp only)
си́ла force, power, strength
~ аэродинами́ческого сопротивле́ния drag force (aeron)
~ ве́тра wind force, wind strength
~, вне́шняя external force
~ воды́, подъёмная buoyancy of water

си́ла, возвраща́ющая restoring force
~, возмуща́ющая aerodynamic disturbance force
~, гравитацио́нная gravitational force
~, дви́гательная motive power
~, дви́жущая propelling force, motive force, motive power
~, де́йствующая effective force
~, динами́ческая подъёмная dynamic lift (aeron)
~, жива́я 1. kinetic energy 2. personnel, troops, manpower
~ земно́го притяже́ния earth's pull
~ земно́го тяготе́ния earth's pull
~ ине́рции inertial force, set-back force
~, кориоли́совая force due to Coriolis acceleration (GW)
~, коэрцити́вная coercive force, coercivity
~, куло́новская Coulomb force (elec)
~ лобово́го сопротивле́ния drag (aeron)
~, лошади́ная horse power
~, магне́то-дви́жущая magnetomotive force
~, магнитодви́жущая magnetomotive force
~ ма́гнуса Magnus force (GW)
~, мгнове́нная instantaneous force
~, мета́тельная projectile force
~, напра́вленная вниз downward force
~, огнева́я fire power
~ огня́ volume of fire, firepower
~, осева́я axial force
~, остана́вливающая stopping power (ball)
~ отда́чи recoil force
~, подъёмная 1. lift (aeron) 2. carrying capacity, lifting capacity (wagons, cranes, etc)
~, по́лная аэродинами́ческая over-all aerodynamic force
~, попере́чная аэродинами́ческая cross velocity force
~ притяже́ния 1. gravity 2. force of attraction, attraction force
~, пробивна́я penetrative power
~, рабо́чая manpower, working-strength
~, равноде́йствующая resultant force, equivalent force
~, разреша́ющая resolving power (optics)
~, разруши́тельная destructive force
си́ла, реакти́вная 1. jet power, jet thrust 2. reaction force
~ сигна́ла signal strength
~ сно́са cross-velocity force
~ сопротивле́ния во́здуха air resistance
~ сопротивле́ния руле́й control-surface drag
~ сцепле́ния cohesion
~ та́нка, уда́рная striking power of a tank
~, термоэлектродви́жущая thermoelectromotive force
~ то́ка, безопа́сная largest safe current (for testing detonators, mil eng, dml)
~ то́ка, гаранти́йная minimum current, fusion test current (to ensure detonation)
~ то́ка, де́йствующая root mean square current, RMS current
~, толка́ющая propelling force
~ торможе́ния brake power
~ тре́ния friction force
~ тя́ги 1. tractive force, traction 2. thrust (rocket motor etc)
~ тя́ги винта́ thrust of propeller
~ тя́ги, максима́льная gross tractive effort
~ тя́ги по сцепле́нию adhesive force
~ тяготе́ния attraction force
~ тя́жести force of gravity
~, тя́нущая pull
~, убо́йная lethal power
~, уда́рная striking power, hitting power
~, центробе́жная centrifugal force
~, центростреми́тельная centripetal force
~, электродви́жущая electromotive force, EMF
~, я́дерная nuclear force
силикаге́ль silica gel
силико́н silicone
силки́, про́волочные trip-wire entanglement
сило́к snare (for animals, or used in groups as personnel traps)
силуэ́т 1. silhouette 2. figure target
си́лы forces
~ авангáрда, гла́вные main guard (tac)
~, амфиби́йные amphibious forces
~, бронета́нковые armoured forces
~, вое́нно-возду́шные air force
~, вооружённые armed forces
~, гла́вные main body (tac)

си́лы, деса́нтные landing force, landing forces
~, духо́вные moral fibre, morale
~ и сре́дства resources, men and equipment
~, кру́пные бронета́нковые massed armour
~, парашю́тно-деса́нтные parachute forces
~, превосхо́дные superior forces
~, превосходя́щие superior forces
~ прикры́тия covering troops
~, разноро́дные all-arms force, all-arms forces
сильфо́н concertina connection (mech)
си́менс mho (elec)
синдро́м syndrome
си́нтез synthesis
синтети́ческий synthetic
синусо́ида sine wave
синусоида́льный sinusoidal
синхрогенера́тор synchronizing signal generator
синхрои́мпульс synchronizing pulse
синхрониза́тор synchronizer, lock unit
~, акусти́ческий acoustic resonance device
синхрониза́ция synchronization, timing
синхронизи́рованный synchronized, tuned
синхрофазотро́н proton synchrotron (nucl)
синхроциклотро́н synchro-cyclotron (nucl)
си́нька blueprint
сире́на siren
~, электри́ческая electric siren
систе́ма system
~, авари́йная emergency system
~ авари́йного сли́ва то́плива fuel jettisoning system
~ автоматизи́рованного контро́ля automatic monitoring system
~, автомати́ческая бортова́я automatic flight control system
~, автомати́ческая телефо́нная automatic telephone system
~ автомати́ческого контро́ля automatic control system
~ автомати́ческого регули́рования servomechanism, closed-loop control system, closed-sequence control system, closed-cycle control system
систе́ма автомати́ческого сопровожде́ния automatic tracking system (FCE)
~ автомати́ческого управле́ния automatic control system
~ автомати́ческого управле́ния огнём automatic fire control system
~ автомати́ческой обрабо́тки да́нных automatic data processing system
~ автоно́много наведе́ния pre-programmed guidance system, pre-set guidance system, self-contained guidance system, autonomous guidance system (GW)
~ автоно́много управле́ния self-contained guidance system, pre-set guidance system (GW)
~, анте́нная antenna system
~, аритми́чная start-stop system (teleg)
~, астати́ческая "ideal" synchronous system (data transmission, servo)
~, боева́я weapons system
~ веду́щего луча́ two-beam radar track guidance system (GW)
~ вентиля́ции ventilation system
~, взрывна́я firing system
~, визи́рная sighting system
~ вну́тренней свя́зи intercommunication system
~, водоотли́вная drainage system, bilge system
~ возду́шного отопле́ния и охлажде́ния air conditioning system
~ вооруже́ния armament system
~ воспламене́ния ignition system
~ воспламене́ния, терми́ческая thermal ignition system (GW)
~ впры́ска injection system (fuel)
~ впры́ска, одностру́йная solid jet injection system
~, всепого́дная all-weather system
~, встро́енная built-in system
~, вычисли́тельная computer system
~ генера́тора, щёточная generator brush assembly
~, гиперболи́ческая навигацио́нная hyperbolic navigation system (Loran)
~ да́льнего обнаруже́ния баллисти́ческих раке́т ballistic missile early warning system, (BMEWS)
~, дальноме́рная circular navigation system (radio- navigation)

систéма, дальномéрно-рáзностная range difference navigation system (radio-navigation)
~, **двои́чная** binary system
~, **десяти́чная** decimal system
~, **дифференциáльная дýплексная** differential duplex system
~, **долговрéменная оборони́тельная** fortified defence system, permanent defence system
~ **достáвки** delivery system
~ **достáвки я́дерных боеприпáсов к цéли** nuclear weapons delivery system
~ **жёсткого автомати́ческого управлéния** open-loop control system, open-cycle control system
~ **жизнеобеспéчения, автонóмная** self-contained life-support system
~ **загражде́ний** obstacle system, obstacle belt
~ **зажигáния** ignition system
~ **зажигáния бесконтáктная** contact breakerless ignition system
~ **зажигáния от магнéто** magneto ignition system
~ **зажигáния, терми́ческая** thermal ignition system (GW)
~ **зажигáния, электрóнная** electronic ignition system
~ **зáмкнутого ци́кла, следя́щая** closed-loop servo system
~ **зáпуска** launching system
~ **имитациóнных радиопомéх** electronic countermeasures deception equipment
~, **инерциáльная навигациóнная** inertial navigation system
~ **инженéрных загражде́ний** obstacle system, obstacle belt
~, **интегрáльная электрóнная** integrated electronic system
~, **инфракрáсная** infra-red system, IR system
~ **классификáции мостóв** bridge classification system
~, **комáндная** command (guidance) system (GW)
~ **кондициони́рования вóздуха** air-conditioning system
~ **координáт** system of co-ordinates
~ **координáт, анóдная** Ia/Va co-ordinates (electron)
~ **координáт, биполя́рная** whole circle system (surv)

систéма координáт, земнáя terrestrial reference system (GW)
~ **координáт, кони́ческая** conical co-ordinate system
~ **координáт, неподви́жная** fixed system of
~ **координáт, óбщая** common co-ordinate system (missile and target)
~ **координáт, прямоугóльная** rectangular grid system (carto)
~ **координáт, сéточная** Ig/Vg co-ordinates (elec)
~ **координáт, цилиндри́ческая** cylindrical co-ordinate system
~ **косми́ческого орýжия** space weapon system
~, **лáзерная** laser system
~, **ли́нзовая оборáчивающая** lens erecting system
~, **многоканáльная** multi-channel system, diversity system (comms)
~, **многокаскáдная** multi-stage system (electron)
~, **многокомпонéнтная** multi-component system
~, **многокýпольная парашю́тная** parachute cluster (used in heavy dropping)
~ **наблюдéния** observation system, surveillance system
~ **наблюдéния, глобáльная** global observations system
~ **наблюдéния, телевизиóнная** television pick-up system
~ **наведéния** guidance system
~ **наведéния, автонóмная** pre-programmed guidance system, pre-set guidance system, self-contained guidance system, autonomous guidance system (GW)
~ **наведéния, акти́вная тепловáя** active heat-seeking guidance system
~ **наведéния, астроинерциáльная** inertial celestial guidance system (GW)
~ **наведéния, астроинерциóнная** inertial-celestial guidance system (GW)
~ **наведéния, астронавигациóнная** celestial navigation guidance system (GW)
~ **наведéния, бортовáя** on-board guidance system (GW)
~ **наведéния вводá в луч** gathering beam system of guidance (GW)

систе́ма наведе́ния, вне́шняя кома́ндная external command guidance system (GW)
~ **наведе́ния, всепого́дная** all-weather guidance system
~ **наведе́ния, гиперболи́ческая** hyperbolic guidance system (GW)
~ **наведе́ния, инерциа́льная** inertial guidance system (GW)
~ **наведе́ния, инерцио́нная** all-inertial guidance system, inertia guidance system (GW)
~ **наведе́ния, инерцио́нно-астронавигацио́нная** inertial-celestial guidance system (GW)
~ **наведе́ния, инерцио́нно-гравитацио́нная** inertial-gravitational guidance system (GW)
~ **наведе́ния, инфракра́сная** infra-red guidance system, heat-seeking guidance system
~ **наведе́ния, кома́ндная** command guidance system (GW)
~ **наведе́ния, ко́мплексная** combined guidance system (GW)
~ **наведе́ния, кругова́я** circular navigation guidance system (GW)
~ **наведе́ния, магнитометри́ческая** magnetic guidance system
~ **наведе́ния, монои́мпульсная** monopulse guidance system (GW)
~ **наведе́ния, назе́мная** ground control guidance system (GW)
~ **наведе́ния по лучу́** beam-riding guidance system (GW)
~ **наведе́ния по лучу́, двухлучева́я** two-beam radar beam-riding guidance system (GW)
~ **наведе́ния по лучу́, однолучева́я** single-beam radar beam-riding command guidance system (GW)
~ **наведе́ния по провода́м** wire command guidance system (GW)
~ **наведе́ния по радиолучу́** beam-riding guidance system (GW)
~ **наведе́ния, радиоинерцио́нная** radio-inertial navigation system, radio-inertial guidance system (GW)
~ **наведе́ния, радиокома́ндная** radio command guidance system
~ **наведе́ния, радиолокацио́нная** radar guidance system (GW)

систе́ма наведе́ния, радионавигацио́нная radionavigation guidance system (GW)
~ **наведе́ния, телевизио́нная** television radio-command guidance system (GW)
~ **наведе́ния, теплова́я** infra-red guidance system (GW)
~ **наведе́ния, электро́нная** electronic guidance system (GW)
~ **навига́ции, инерцио́нная** inertial navigation system (GW)
~ **навига́ции, кругова́я дальноме́рная** circular navigation system (GW)
~ **навига́ции/наведе́ния, дальноме́рно - ра́зностная** range difference navigation/guidance system (GW)
~, **назе́мная навигацио́нная** ground navigation system
~ **накопле́ния да́нных** data storage system
~ **обнаруже́ния** warning system, detection system, target acquisition system
~ **обнаруже́ния, вспомога́тельная** ancillary detection system
~, **обора́чивающая** erector assembly, erecting system (optics)
~ **оборо́ны** defensive system, defences
~ **обрабо́тки да́нных** data handling system, data processing system
"систе́ма обуче́ния, потого́нная" "slave-driving method of training" (sl)
систе́ма, огнева́я fire plan
~ **огня́** fire plan
~ **операти́вного управле́ния маши́но-тра́кторным па́рком** automatic control system for tractor and vehicle parks
~ **опознава́ния, радиолокацио́нная** radar recognition system (IFF)
~ **опознава́ния це́ли** target identification system
~, **о́птико-электро́нная** opto-electronic system
~, **опти́ческая визи́рная** optical sighting system
~, **опти́ческая кома́ндная** optical command system (GW)
~ **ору́жия** weapon system
~ **освеще́ния** illumination set, lighting set
~ **отсчёта** reference frame (GW)
~ **охлажде́ния** cooling system

систе́ма охлажде́ния, герметизи́рующая sealed cooling system
~ охлажде́ния, за́мкнутая closed circuit cooling system
~ охлажде́ния, термосифо́нная thermosyphon cooling system
~ переда́ч, синхро́нно-следя́щая self-synchronous transmission system
~ переда́чи да́нных data transmission system
~ переме́нного то́ка a.c. system (elec)
~ пита́ния fuel system
~, пневмати́ческая тормозна́я air brake system
~ поворо́та ба́шни turret traversing system
~ пода́чи feed system
~ подве́ски suspension system
~, подвесна́я harness system (para)
~, подви́жная moving parts (SA)
~ подры́ва destruct system
~, пожа́рная спри́нклерная fire sprinkler system
~ по́иска и сопровожде́ния це́ли search-and-tracking system
~, по́лностью автоматизи́рованная fully automatic system
~ По́рро, обора́чивающая Porro erecting system (optics)
~ поса́дки самолётов по прибо́рам instrument landing system, ILS
~ постоя́нного то́ка d.c. system (elec)
~ препя́тствий obstacle system
~ приво́да actuating system
~ приво́да, электрогидравли́ческая electrohydraulic drive system, electrohydraulic servo mechanism
~ приорите́тов priority system
~, прице́льная sighting system
~ прово́дки wiring system
~ продо́льного управле́ния pitch control system
~ противовозду́шной оборо́ны AA defence plan
~ противовозду́шной оборо́ны, еди́ная integrated air defence system
~ противовозду́шной оборо́ны (ПВО), еди́ная integrated air defence
~, противопожа́рная fire-prevention system
~, противота́нковая anti-tank defence plan
систе́ма противота́нкового огня́ anti-tank fire plan
~ противота́нковой оборо́ны anti-tank defence plan
~, прямоуго́льная координа́тная rectangular co-ordinate grid system
~, рабо́тающая на при́нципе испо́льзования отда́чи recoil-operated system
~, радиолокацио́нная radar system
~, радиолокацио́нная и́мпульсная pulse-modulation radar system
~, радиолокацио́нная навигацио́нная radar navigation system
~ радиопротиводе́йствия electronic countermeasures system, ECM system
~, радиореле́йная radio-relay system (comms)
~ радиосвя́зи, однополо́сная single side-band system of radio-communication
~, радиотелеметри́ческая telemetry system, radiotelemetry system
~, радиотехни́ческая дальноме́рная radar range measuring equipment
~ радиоуправле́ния radio-command system
~, ра́зностно-дальноме́рная навигацио́нная range difference navigation system, hyperbolic navigation system, LORAN system
~, разо́мкнутая open-loop system, open-cycle system (elec)
~, распредели́тельная distributing system
~, резе́рвная stand-by system
~ рулево́го управле́ния steering system
~ рычаго́в system of levers
~ с автомати́ческой коммута́цией automatic switching system
~ с включе́нием в де́йствие при возникнове́нии неиспра́вности fail-safe system
~ с высо́кой разреша́ющей спосо́бностью high-resolution system
~ с дли́нным хо́дом ствола́ long recoil system (ord)
~ с коро́тким хо́дом ствола́ short recoil system (ord)
~ с отво́дом порохов́ых га́зов gas-operated system
~ с отда́чей затво́ра blow-back action

систе́ма с радиолокацио́нным управле́нием radar control system (GW)
~ **с управле́нием по провода́м** wire control system (GW)
~ **с управле́нием по ра́дио** radio control system (GW)
~ **с часто́тной модуля́цией** frequency-modulation system
~ **самонаведе́ния** homing guidance system (GW)
~ **самонаведе́ния, акти́вная** active homing guidance system
~ **самонаведе́ния, акусти́ческая** acoustic passive homing system (GW)
~ **самонаведе́ния, звукопеленга́торная** acoustic passive homing guidance system (GW)
~ **самонаведе́ния, инфракра́сная** heat-seeking homing guidance system, heat homing guidance system, infra-red homing guidance system, infra-red passive homing guidance system (GW)
~ **самонаведе́ния на цель** target homing guidance system (GW)
~ **самонаведе́ния, опти́ческая** light sensitive passive homing guidance system, optical passive homing guidance system (GW)
~ **самонаведе́ния, пасси́вная** passive homing guidance system (GW)
~ **самонаведе́ния, полуакти́вная** semi-active homing guidance system (GW)
~ **самонаведе́ния, радиолокацио́нная** radar homing guidance system (GW)
~ **самонаведе́ния, теплова́я** heat-seeking (infra-red) passive homing guidance system (GW)
~ **самонаведе́ния, электромагни́тная** electromagnetic homing guidance system (GW)
~, **самонастра́ивающаяся** self-adjusting system
~ **сбо́ра да́нных** information-gathering system
~ **сбро́са** jettison system
~ **свя́зи** communications system
~ **свя́зи, порайо́нная** area type communications system
~ **сдво́енного приёма** diversity reception system
~ **сигнализа́ции** signalling system

систе́ма синхро́нных свя́зей synchro data transmission system
~ **скачко́в уплотне́ния** system of shock waves (aeron)
~, **следя́щая** servomechanism
~ **сма́зки** lubrication system
~ **сма́зки под давле́нием** pressure lubricating system, force-feed lubrication system
~ **сма́зки, принуди́тельная** pressure lubricating system
~ **сма́зки разбры́згиванием** splash lubrication system
~ **сма́зки с мо́крым ка́ртером** wet-sump lubrication system (ICE)
~ **сма́зки с сухи́м ка́ртером** dry-sump lubrication system (ICE)
~ **сма́зки ша́сси, централизо́ванная** centralized chassis lubrication system
~, **сме́шанная** hybrid system
~ **снабже́ния запа́сными ча́стями** spare parts supply system
~ **спасе́ния** recovery system (ESVs etc)
~ **стабилиза́ции пу́шки** gun stabilization system (tk)
~, **телеметри́ческая** telemetering system
~ **телеуправле́ния** command guidance system, remote control system, command control system
~ **телеуправле́ния, радионавигацио́нная** radio navigation guidance system (GW)
~ **телеуправле́ния, дальноме́рная навигацио́нная** circular navigation guidance system, range-finding navigation guidance system
~ **телеуправле́ния, кома́ндная** command guidance system (GW)
~ **телеуправле́ния по лучу́** beam-riding guidance system
~ **телеуправле́ния по тропе́** beam-riding guidance system
~ **телеуправле́ния, угломе́рно-дальноме́рная навигацио́нная** bearing-distance radio navigation guidance system
~ **теплоизоля́ции** heat insulating system
~, **то́пливная** fuel system
~, **топливоокисли́тельная** fuel oxidizer system
~ **топливопода́чи** fuel-feed system
~, **тормозна́я** brake system
~, **тормозя́щая** braking system

систéма транспортирóвки порóды surface disposal of spoil (mining)
~ **траншéй** trench system
~ **трубопровóдов** piping
~ **укреплéния мéстности** fortification system
~ **управлéния** guidance system, control system
~ **управлéния, автомати́ческая** automatic control system
~ **управлéния, автонóмная** pre-programmed guidance system, self-contained guidance system
~ **управлéния, астроинерциáльная** inertial celestial guidance system (GW)
~ **управлéния вéктором тя́ги** thrust vector control system
~ **управлéния, гироскопи́ческая** inertial guidance system
~ **управлéния, дистанциóнная** telecontrol system, remote-control system
~ **управлéния и наведéния** control and guidance system
~ **управлéния, инерциáльная доплерóвская** inertial Doppler guidance system
~ **управлéния, инерциáльная** inertial guidance system
~ **управлéния, инерциóнная** inertial guidance system
~ **управлéния, комáндная** command guidance system (GW)
~ **управлéния, кóмплексная** combined guidance system (GW)
~ **управлéния, назéмная** ground control guidance system
~ **управлéния огнём** fire-control system
~ **управлéния орýжием** weapon-control system
~ **управлéния по крéну** roll control system
~ **управлéния по кýрсу** yaw control system
~ **управлéния по углý рыскáния** yaw control system
~ **управлéния, полуакти́вная** semi-active guidance system
~ **управлéния, радиоинерциóнная** radio-inertial guidance system
~ **управлéния, радиокомáндная** radio command guidance system (msl)
систéма управлéния стрельбóй fire control system
~ **управлéния углóм тангажá** pitch control system
~, **устóйчивая** stable system
~ **целеуказáния** target designating system
~ **централизóванной подкáчки шин** central tyre inflation system
~, **циркуляциóнная смáзочная** circulating lubricating system
~ **шифрóванной передáчи рéчи** scrambled speech system
~ **эвакуáции** evacuation chain
~ **экрани́рованных проводóв зажигáния** screened ignition wiring
~, **электрогидравли́ческая** electro-hydraulic system
~, **электрóнная** electronic system
~, **электрóнно-опти́ческая** electron optics system
ситуáция, авари́йная emergency situation
сифóн siphon
скáла 1. rock, cliff
2. scale (graduated)
скáлывание spallation (nucl)
скамéйка bench
скани́рование scanning
~, **автомати́ческое** automatic scan, automatic scanning (radar)
~, **винтовóе** helical scanning (radar)
~, **кони́ческое** conical scanning (radar)
~, **кругoвóе** circular scanning
~, **прямолинéйное** rectilinear scanning (radar)
~, **прямоугóльное** rectangular scanning
скáпливать/скопи́ть to amass, to build up
скат 1. slope
2. wheel (of veh)
3. complete tyre (inner tube and outer cover)
4. shoulder (ammo)
~, **бóгнутый** concave slope
~, **волни́стый** undulating slope
~, **встрéчный** forward slope
~, **вы́пуклый** convex slope
~ **ги́льзы** case chamfer, case shoulder (ammo)
~, **обрáтный** reverse slope
~ **патрóнника, кони́ческий** chamber cone (ord)

скат патро́нника, пере́дний bullet seat (ammo)
~, пере́дний forward slope
~, прямо́й uniform slope
~, ро́вный uniform slope
~, ступе́нчатый stepped slope
ска́тка greatcoat roll (coat rolled up for carrying across the shoulders)
скафа́ндр 1. pressurized suit 2. diving suit 3. space suit
~, водола́зный diving suit, diving equipment
"Скачо́к!" "Add!" (an OP order in airburst ranging) (Sov)
скачо́к bound
~ уплотне́ния compression wave, shock wave, pressure shock, shock drag (aeron)
~ уплотне́ния, входно́й inlet shock wave (aeron)
~ уплотне́ния, головно́й leading edge shock wave (aeron), bow shock wave (GW)
~ уплотне́ния, косо́й inclined shock wave, angle shock wave (aeron)
~ уплотне́ния, ме́стный local shock, compressibility burble (aeron)
~ уплотне́ния на вхо́де inlet shock wave
~ уплотне́ния, отсоединённый detached shock wave
~ уплотне́ния, присоединённый attached shock wave
~ уплотне́ния, прямо́й normal shock wave (ball)
ска́шивание chamfering, bevelling
сква́жина 1. crack, slit 2. borehole
~, артезиа́нская artesian borehole
~, бурова́я borehole
сквозни́к passageway
~, коле́нчатый double entry passageway (eng fortification)
сквозня́к draught
скиатро́н skiatron, dark-trace tube (electron)
скин-эффе́кт skin-effect (elec)
скипида́р turpentine
склад depot, store
~ автобронета́нковой те́хники vehicle ordnance depot
~, артиллери́йский artillery supply depot
склад, ба́зовый base depot
~ боеприпа́сов ammunition store, ammunition depot
~ боеприпа́сов, полево́й field ammunition depot
~ бронета́нковой те́хники tank depot
~, вещево́й clothing store
~, вое́нный military supply depot, military store
~, во́инский military supply depot, military store
~ вооруже́ния munition depot, arms depot
~, выда́ющий issuing depot
~, гарнизо́нный garrison store, garrison depot
~, гла́вный артиллери́йский main artillery store
~, головно́й артиллери́йский forward artillery store
~ горю́чего fuel storage depot, POL store
~ горю́чего, передово́й forward fuel depot
~ ГСМ POL depot, fuel storage depot
~, дивизио́нный divisional supply dump
~, дово́льствующий issuing depot, parent depot
~ инжене́рного иму́щества engineer supply depot, engineer store
~, квартирме́йстерский quartermaster stores (non-Sov)
~ кро́ви, передово́й advanced blood bank (med)
~, окружно́й 1. (military) district depot 2. Command depot
~ ору́жия armoury
~, передово́й forward depot, forward store
~, передово́й ме́дико-санита́рный forward medical depot
~, передово́й полево́й advance dump, forward stores depot
~, подзе́мный underground store
~, полево́й field supply depot, field dump, field store
~, полево́й арме́йский army field depot, rear stores depot
~, полково́й regimental dump, regimental store
~, продово́льственно-фура́жный rations and forage depot
~, продово́льственный supply depot
~ снабже́ния supply depot

склад специа́льного ору́жия special weapons dump
~, топографи́ческий survey depot
~, центра́льный меди́ци́нский и санита́рно-хозя́йственный medical base depot
~ эвакуи́руемых маши́н returned vehicle park
складúрование stockpiling, storage
складúровать to store
скла́дка 1. dumping
2. fold (geol)
3. crease, wrinkle, fold
~ ме́стности fold in the ground
складно́й collapsible, folding
скла́дывание stacking
~ ка́рты "гармо́никой" "concertina" map folding
скла́дывать/сложи́ть to put, to place, to fold, to make up, to surrender; **скла́дывать/сложи́ть в запа́с** to return to stock; **скла́дывать/сложи́ть ору́жие** to lay down arms, to capitulate
скле́ивание adhesion, glueing
скле́ивать/склеи́ть to glue, to stick together
склеи́ть (см скле́ивать)
склон slope
~, круто́й steep slope
~, поло́гий gentle slope
~, фи́рновый glacier slope
склоне́ние 1. depression (arty)
2. declination, deflection, variation
3. slope, incline, declivity
~, магни́тное magnetic declination, magnetic deviation
~ магни́тной стре́лки magnetic variation, magnetic declination
~ небе́сного свети́ла declination of heavenly body (astron)
скло́нность tendency
скоба́ shackle, staple, clip, bracket, dog, catch
~, букси́рная towing shackle
~ крепле́ния toe-irons (ski binding)
~, подъёмная lifting lug
~, предохрани́тельная 1. safety catch, fire-and-safe unit safety catch (SA)
~, прямáя staple, cramp, dog (carpentry)
~, спускова́я 1. trip plate, plate of plate withdrawal igniter (min)
2. trigger guard (SA)
~, сто́порная stop plate
скоба́, строи́тельная dog (mech)
ско́бка 1. bracket
2. staple
скова́ть (см ско́вывать)
ско́вывание containing (enemy, tac)
ско́вывать/скова́ть to contain, to immobilise, to hold, to pin down (tac), to chain, to fetter, to lock (of frost or ice),
сколо́ченный 1. cohesive, well-knit
2. hammered together
скольже́ние sliding, slipping, skidding
скоммути́ровать (см коммути́ровать)
сконцентри́рованный concentrated
скопи́ть (см ска́пливать)
скопле́ние massing, accumulation
скорова́рка pressure cooker
скороподъёмность rate of climb
скоропортя́щийся perishable
скоростре́льность rate of fire
~, боева́я actual rate of fire, normal rate of applied fire (SA)
~, высо́кая high rate of fire
~, максима́льная maximum rate of fire
~, ма́лая slow cyclic rate of fire
~, ни́зкая low rate of fire
~, практи́ческая maximum useable rate of fire, practicable rate of applied fire (SA)
~, сре́дняя mean rate of fire
~, теорети́ческая cyclic rate of fire
скоростре́льный quick-firing, rapid-firing, QF
ско́рость 1. speed, velocity, rate
2. gear
~, баллисти́ческая ballistic speed
~ в конце́ акти́вного уча́стка траекто́рии burnout velocity
~, вертика́льная vertical velocity
~ ве́тра 1. wind speed
2. wind velocity
~, взлётная take-off speed
~ возмуще́ния disturbance velocity
~, возраста́ющая acceleration
~ враще́ния rate of revolution, spin rate, rotation rate
~ враще́ния анте́нны antenna scan rate, speed of antenna rotation
~ враще́ния, углова́я turn rate, rate of rotation (GW)
~ встре́чи striking velocity, impact velocity (projectile)
~, втора́я косми́ческая escape velocity (missiles, ESV)

ско́рость, гиперзвукова́я hypersonic speed
~ **горе́ния** burning rate
~ **горе́ния, лине́йная** rate of recession (ball, solid fuel, expl)
~, **горизонта́льная** horizontal speed, level speed
~, **группова́я** group velocity (electron)
~ **движе́ния** running speed
~ **движе́ния по бездоро́жью** speed across country, cross-country speed
~ **движе́ния по гра́фику** schedule speed
~, **дозвукова́я** subsonic speed
~, **ду́льная** initial velocity, muzzle velocity (MV)
~ **зву́ка** speed of sound
~, **звукова́я** speed of sound
~ **измене́ния по вре́мени** time rate of change
~ **истече́ния** exhaust velocity (GW)
~, **и́стинная** true speed, actual velocity
~, **и́стинная возду́шная** true airspeed
~, **коне́чная** remaining velocity, terminal velocity (ball, projectile)
~, **крити́ческая** critical speed
~, **ма́ршевая** average rate of march
~, **маршру́тная** journey speed
~ **на впу́ске** inlet velocity
~, **нача́льная** muzzle velocity
~, **нача́льная углова́я** initial angular velocity
~, **ограни́ченная** restricted speed
~, **околозвукова́я** transonic speed
~, **оконча́тельная** remaining velocity, striking velocity (ballistics)
~, **окру́жная** annular velocity, peripheral speed
~, **операти́вная** handling speed (comms)
~ **оско́лков, нача́льная** initial fragment velocity
~, **оста́точная** remaining velocity
~ **отда́чи** recoil velocity
~ **отка́та** recoil velocity
~, **относи́тельная** relative velocity
~, **пе́рвая косми́ческая** orbital velocity (ESV)
~ **переда́чи, преде́льная** limiting transmission speed
~ **(передвиже́ния) по пересе́ченной ме́стности** cross-country speed
~, **переме́нная** variable speed
~ **переме́шивания** rate of mixing (msl fuels)

ско́рость пики́рования diving speed
~ **полёта** flight speed
~ **полёта, крити́ческая** critical flight speed, transonic speed
~ **по́ршня, сре́дняя** mean piston speed (ICE)
~, **поступа́тельная** forward speed
~, **практи́ческая** operational speed
~, **преде́льная** 1. terminal velocity (falling body) 2. maximum speed
~ **при встре́че** striking velocity, impact velocity (projectile)
~, **приме́рная** apparent speed, approximate speed
~ **пробива́ния** penetration velocity
~ **протя́жки плёнки** film travel speed
~, **путева́я** ground speed
~, **разрешённая** authorized speed
~ **распростране́ния** velocity of propagation (electron)
~, **расчётная** design speed
~, **результи́рующая** resultant velocity
~ **сближе́ния** closing rate, approach speed
~, **сверхвысо́кая нача́льная** super high muzzle velocity
~, **сверхзвукова́я** supersonic speed
~ **сгора́ния** rate of combustion
~ **снаря́да** missile velocity
~ **сраба́тывания** speed of response
~, **сре́дняя** average speed
~, **сре́дняя эксплуатацио́нная** cruising speed
~, **табли́чная нача́льная** standard muzzle velocity, standard MV (gunn)
~ **теплообме́на** rate of heat exchange
~ **тече́ния** current velocity, speed of the current, rate of flow
~, **тра́нспортная** road speed
~ **убега́ния** escape velocity (ESV)
~, **углова́я** angular velocity
~, **уда́рная** impact velocity
~, **фа́зовая** phase velocity (elec)
~ **фла́ттера** flutter speed (aeron)
скороте́чный 1. fast flowing, fast moving 2. transient
скос 1. slant, bevel, slope 2. tilt (air photo)
~, **бронево́й** sloped armour, glacis plate (tk)
~, **отжи́мный** camway (brake & shoe)
~ **пото́ка** downwash

скоси́ть (см коси́ть)
скот, мясно́й beef cattle
скотобо́йня, полева́я field butchery
ско́шенный bevelled
скребо́к scraper (tool)
скре́пер scraper (earth-moving eqpt)
скрепи́ть (см скрепля́ть)
скрепле́ние 1. fastening, clamping, tightening
2. joint
3. hooping (ord)
~ **ко́льцами** hoop strengthening (ord)
~ **му́фтами** tube strengthening (ord)
~ **про́волокой** wire strengthening (ord)
скрепля́ть/скрепи́ть to secure, to make fast; **скрепля́ть/скрепи́ть по́дписью** to countersign, to validate
скреще́ние 1. junction, meeting
2. crossing, interlacing, intermixing
скре́щивание доро́г cross-roads
~ **прово́дов** transposition (line comms)
скрип squeak, squeal
скру́чивание twist, torsion, coiling action
скру́чивающий twisting, torsional
скрыва́ть/скрыть to conceal, to hide, to suppress (information)
скры́тие concealment
скры́тность concealment, secrecy, security
~ **свя́зи** communications security
скры́тый hidden, concealed, suppressed
скрыть (см скрыва́ть)
ску́ченность bunching (tac)
ску́ченный bunched, crowded
ску́чивание 1. crowding, boxing up
2. bunching (tac)
слабина́ play, slack, sag (cable etc)
слага́емое component, addend
слага́ющая component (math)
~ **баллисти́ческого ве́тра** wind component
сла́женность teamwork, co-ordination
~**, боева́я** teamwork in action
сла́женный co-ordinated, concerted, harmonious
сла́живание arrangement, agreement
слань 1. dunnage, flat dunnage
2. bearing platform (supporting eqpt)
сле́ги ramp, skids
след 1. track, trail, spoor, scent
2. pattern (nucl fallout)
~**, дымово́й** smoke trail
след инве́рсии contrail, vapour trail, condensation trail (GW, ac)
~**, инверсио́нный** contrail, vapour trail, condensation trail (GW, ac)
~**, конденсацио́нный** vapour trail, condensation trail, contrail (GW, ac)
следи́ть/последи́ть to follow, to track, to keep watch over, to leave tracks; **следи́ть/последи́ть за да́нными** to monitor information
сле́дование 1. following
2. observation (rules etc)
сле́дователь investigating officer
~**, вое́нный** investigating officer
сле́дствие 1. investigation
2. consequence, result
следы́, лы́жные ski tracks
слеже́ние tracking (radar)
~**, автомати́ческое** automatic tracking
~ **за це́лью** target tracking
~**, опти́ческое** optical tracking
~**, полумеханизи́рованное** aided tracking (radar)
"Слеза́й!" "Dismount!"
слезотече́ние lachrymation
слепота́, сне́жная snow blindness
~**, цветова́я** colour blindness, daltonism
слеса́рный fitting
сле́сарский fitter's
сле́сарь fitter, mechanic, locksmith
сле́сарь-монта́жник fitter-mechanic
слив draining, discharge, decanting
~ **горю́чего** fuel dumping
сливно́й drain, draining
сли́ток ingot, rod, bar, pig (metal)
слича́ть/сличи́ть to compare
сличи́тель упрежде́ний lead computer
сличи́ть (см слича́ть)
слия́ние blending, confluence
слобода́ large village, settlement
сло́во, ко́довое code word, codeword
~**, кома́ндное** word of command
сложи́ть (см скла́дывать)
сло́жный sophisticated
слой layer, sheet, stratum, lamella
~ **атмосфе́ры** atmospheric layer
~**, ве́рхний по́чвенный** topsoil
~**, водоно́сный** water-bearing stratum
~**, гидроизоляцио́нный** impervious layer, impervious stratum
~ **Д** D layer (radio)
~ **двукра́тного ослабле́ния** half-value thickness (NW)

слой Е E layer, Kennelly-Heaviside layer (radio)
~, изоляцио́нный insulation layer
~, ионизи́рованный ionized layer
~ К K shell (nucl)
~ Л L shell (nucl) usually written with latin L
~, ламина́рный laminar layer
~ М M shell (nucl)
~ ма́сла film of oil
~, пове́рхностный 1. superficial layer, surface layer
2. topsoil
~, пограни́чный boundary layer (ball, aeron)
~ полови́нного ослабле́ния half-value thickness (NW)
~ полови́нного поглоще́ния half-value thickness (NW)
~, распредели́тельный distributing layer (fieldworks eng)
~, сма́зывающий lubricating film
~, теплоизоляцио́нный heat-insulating layer
~, теплопоглоща́ющий heat-sink layer
~ Ф F layer, Appleton layer (radio) usually written with latin F
слома́ть (см лома́ть)
сломи́ть to break; **сломи́ть сопротивле́ние** to overcome opposition, to break down resistance
слу́жба service, duty
~, авари́йная breakdown service
~, авари́йно-восстанови́тельная emergency repair service
~, авари́йно-ремо́нтная breakdown service
~, авари́йно-спаса́тельная salvage and rescue service (nav)
~, авиадиспе́тчерская air traffic control service
~, автотра́кторная 1. MT, MT system or organisation
2. MT duties
~, администрати́вная administrative services, administrative service
~, артиллери́йская фотограмметри́ческая artillery photogrammetry service
~ артиллери́йского снабже́ния ordnance service (Sov) (concerned with supply of all wpns and ammo)
слу́жба, аэродро́мная airfield maintenance and supply service
~, аэродро́мно-диспе́тчерская air traffic local and approach control
~ в де́йствующей а́рмии active service, field service
~, ветерина́рная veterinary service
~ вещево́го снабже́ния clothing supply service
~ ВНОС (возду́шного наблюде́ния, оповеще́ния и свя́зи) air warning service, air observation warning and communication service
~, вну́тренняя administration within the unit
~, вое́нная military service
~, вое́нно-медици́нская military medical service
~, вое́нно-почто́вая forces postal service
~, вое́нно-санита́рная medical service
~, вое́нно-топографи́ческая military survey service
~, вое́нно-тра́нспортная military transportation service
~, вое́нно-хими́ческая chemical warfare service
~, вое́нно-юриди́ческая army legal service, army legal services
~ вое́нных сообще́ний (ВОСО) military railway transport service
~, во́инская military service
~, гарнизо́нная garrison service, garrison duty
~, действи́тельная active service, colour service
~, дово́льственная supply services
~, доро́жно-коменда́нтская provost traffic-control service
~, инжене́рная engineer service
~, инжене́рно-техни́ческая technical engineering branch
~, интенда́нтская Q service, supply service
~, карау́льная guard duty, guard service
~, картографи́ческая map service
~, квартирме́йстерская quartermaster corps (non-Sov)
~ кварти́рного дово́льствия barrack services
~, кварти́рно-эксплуатацио́нная barrack services

слу́жба, коменда́нтская 1. provost service, traffic control service 2. guards and defence of HQ 3. billeting and equipment of HQ
~, контро́льно-спра́вочная signals office supervision and enquiries service
~ материа́льно-техни́ческого обслу́живания material and technical supply service
~, ме́дико-санита́рная medical service
~, медици́нская medical service
~, метеорологи́ческая meteorological service
~, непреры́вная вое́нная continuous military service, unbroken military service
~, обеспе́чивающая supporting service
~, огнева́я gun drill
~ оповеще́ния air warning service
~, парашю́тно-деса́нтная parachute troops
~, па́рковая servicing of vehicles/eqpt in barracks
~ перелива́ния кро́ви blood transfusion service
~ пого́ды weather service
~, пожа́рная fire service
~ пожа́рной охра́ны fire service
~, похоро́нная 1. graves registration service 2. burial service, memorial service
~, продово́льственная catering service
~ продово́льственного снабже́ния food supply service
~ противота́нкового оповеще́ния anti-tank warning service
~, сверхсро́чная extended service
~ свя́зи Signals (arm of the service)
~, спаса́тельная rescue service, life-saving organization
~, станцио́нно-эксплуатацио́нная radio-telegraph operating service (mil)
~, сторожева́я 1. security service 2. piquets and patrols
~, техни́ческая technical service
~, топографи́ческая topographical service
~, тра́нспортная transport service
~ управле́ния, диспе́тчерская (air-traffic) control service
~, фармацевти́ческая pharmaceutical service
~, фельдъе́герская courier service
слу́жба фина́нсового дово́льствия pay services
~, хими́ческая CW service
~ хозя́йственного дово́льствия Q services (in unit)
~ штабо́в, полева́я field service staff regulations
~, эвакоспаса́тельная rescue and recovery service
~, эвакуацио́нно-спаса́тельная rescue and recovery service (veh)
слу́жбы ты́ла rear services, administrative services
служе́бный 1. official, service 2. procedural
служи́ть/послужи́ть to serve, to be employed; **служи́ть/послужи́ть укры́тием** to afford cover
слуха́ч 1. listening sentry 2. morse operator (comms) 3. monitor (comms)
слухово́й 1. morse, C.W. (comms) 2. auditory
слу́чай, кра́йний emergency
~, несча́стный accident, misfortune
"Слу́шай Мою́ Кома́нду!" "At My Command!", "On My Order!"
"Слу́шайте Все!" "Pay Attention!"
"Слу́шаюсь!" "Sir!" (acknowledgement of order by junior)
слы́шимость audibility
слюда́ mica
сма́зать (см сма́зывать)
сма́зка 1. lubrication, oiling, greasing 2. lubricant, oil, grease
~, автомати́ческая self-lubrication, automatic lubrication
~, амуни́чная harness grease, accoutrements grease
~, антикоррозио́нная rust-preventive compound, slushing compound, rust-preventive grease, slushing grease
~, графи́тная graphite grease, graphite lubricant
~, графи́товая graphite grease, graphite lubricant
~, густа́я grease lubrication
~ для гипо́идных переда́ч hypoid lubricant
~, дозиро́ванная metered lubrication
~, жи́дкая oil, lubricating oil
~, защи́тная protective grease
~, зи́мняя winter lubricant

сма́зка, зи́мняя оруже́йная winter gun oil
~, ка́ртерная wet sump lubrication, bath lubrication (engines)
~, консисте́нтная lubricant grease, lubricating grease
~ лыж ski waxing
~ ма́сляным тума́ном oil mist lubrication
~, многофункциона́льная multi-purpose grease
~ под давле́нием, по́лная full-pressure lubrication
~ под давле́нием, части́чная partial-pressure lubrication
~, противокоррозио́нная rust-preventive compound, slushing compound, rust-preventive grease, slushing grease
~ разбры́згиванием splash lubrication
~, руже́йная rifle oil, rifle lubrication
~, свинцо́вая консисте́нтная lead grease
~, синтети́ческая synthetic grease
~, снаря́дная shell grease
~, тугопла́вкая heat-resistant grease
~, централизо́ванная centralized lubrication (system)
сма́зывание lubrication, greasing, oiling
сма́зывать/сма́зать to lubricate, to oil, to grease
смазь, зи́мняя winter lubricant
сма́тывание reeling-in
сма́тывать/смота́ть to uncoil, to unwind; **сма́тывать/смота́ть ка́бель** to recover a cable
смека́лка common sense, sharpness, keenness of wit
сме́лость bravery, courage
сме́лый daring, courageous, brave, bold
сме́на 1. relief (replacement of unit or guard)
2. exchange
3. shift (labour force)
~ вещево́го иму́щества clothing exchange
~ войск relief of troops
~, дневна́я day shift
~ карау́ла guard relief
~ ле́йнера retubing (ord)
~ ма́сла oil change
~ отли́ва turn of the tide
~, пе́рвая first relief, first watch
сме́на прили́ва turn of the tide
~, рабо́чая working shift, operating shift
~ часовы́х relief of sentries
смени́вшийся the relieved (unit, guard)
смени́ть (см сменя́ть)
сме́нный interchangeable, replaceable
сменя́емый outgoing, to be relieved (unit, guard)
сменя́ть/смени́ть to replace, to relieve
сменя́ющий incoming, relieving (unit, guard)
смерте́льный fatal, deadly, lethal
сме́ртность mortality rate
смертоно́сный lethal, deadly
смерч 1. waterspout, spout (met)
2. sandstorm
~, водяно́й waterspout
смесеобразова́ние carburation, carburetion
смеси́тель mixer
~, бала́нсный balanced mixer (electron)
~, волново́дный бала́нсный hybrid-T balanced mixer (radar, elec)
~, дио́дный diode mixer (electron)
~, криста́лловый crystal mixer (electron)
~, односе́точный single input mixer (electron)
~ с двойны́м вхо́дом double-input mixer (electron)
~ с коаксиа́льной ли́нией coaxial mixer (electron)
смесь 1. mixture, mix
2. blend
~, бе́дная lean mixture, weak mixture
~, бензи́но-возду́шная petrol-air mixture
~ Бе́ргера "Berger's mixture" (smoke agent)
~, бога́тая rich mixture, fuel-rich mixture
~, быстрогоря́щая quick-burning mixture
~, взрывна́я explosive mixture
~, воспламени́тельная igniting mixture
~, горю́чая combustible mixture, fuel mixture
~, двухкомпоне́нтная горю́чая bipropellant
~, дымова́я smoke composition (smoke-producing substance)
~ Ершо́ва "Yershov's mixture" (smoke agent)
~, зажига́тельная incendiary mixture

смесь, напáлмовая napalm mixture
~, незамерзáющая anti-freeze mixture, anti-freeze compound
~, непрореагúровавшая unreacted mixture (fuel)
~, несамовоспламеня́ющаяся non-hypergolic mixture
~, обогащённая enriched mixture (fuel)
~, огнемётная flame-thrower fuel
~, охлаждáющая coolant, cooling mixture
~, рабóчая air/fuel mixture (engines)
~, самовоспламеня́ющаяся hypergolic mixture, self-igniting mixture
~, содержáщая аммонúйную селúтру ammonium nitrate mixture
~, францу́зская trimonite (expl)
смéта estimate
~, перспектúвная long-term estimate
~, предварúтельная advance estimate, preliminary state
~, финáнсовая financial estimate
смещéние 1. bias (electron) 2. displacement, shifting 3. switch, azimuth increment (gunn)
~ за счёт катóдного тóка, автоматúческое cathode self-bias (electron)
~ за счёт сéточного тóка, автоматúческое grid leak bias (electron)
~, обрáтное back bias (electron)
~, сéточное grid bias (elec)
~, угловóе angular displacement
смещённый displaced, out of line
"Смúрно!" 1. "Attention!" 2. "March to Attention!"
смúрно quietly, peacefully, meekly
смолá resin, tar
смонтúрованный mounted, assembled
смонтúровать (см монтúровать)
смотáть (смáтывать)
смотр parade, inspection, review
~, строевóй inspection parade
смывáние washing-down (NBC decontamination)
смы́вка paint remover
смык joint, connection (wires, pipes)
смыкáние 1. closing in, closing 2. closing (on centre or flank) (Sov drill)
смыкáть/сомкну́ть to close up, to clench; **смыкáть/сомкну́ть ряды́** to close up (in depth) (Sov drill); **смыкáть/сомкну́ть строй** to close (laterally) (Sov drill)
смягчáть/смягчúть to soften, to mitigate, to ease; **смягчáть/смягчúть толчóк при падéнии** to break a fall, to absorb the landing shock
смягчéние 1. softening 2. cushioning, damping
смягчúть (см смягчáть)
смятéние confusion, commotion, panic
смять to crush (mil), to rumple, to crumple
снабдúть (см снабжáть)
снабжáемый по вóздуху air-supplied
снабжáть/снабдúть to supply; **снабжáть/снабдúть по заявкам** to supply on indent
снабжéнец member of supply services
снабжéние supply, supplies, provisions
~, артиллерúйское artillery supply
~, артиллерúйско-техúческоe ordnance supplies
~ боеприпáсами ammunition supply
~, воéнно-технúческое military technical supply
~, воéнно-хозя́йственное administrative supply, quartermaster's supplies
~ ГСМ supply of fuels and lubricants
~ по вóздуху air supply, supply by air
~, продовóльственное food supply
снáйпер sniper
снаря́д 1. projectile, missile, shell, shot 2. apparatus, equipment
~, авиациóнный баллистúческий air-launched ballistic missile, ALBM
~, агитациóнный propaganda shell, leaflet shell
~, активно-реактúвный rocket-assisted projectile
~, артиллерúйский химúческий chemical artillery shell
~, áтомный atomic shell
~, баллистúческий ballistic missile
~, бетонобóйный concrete piercing shell
~, боевóй реактúвный military rocket
~, бризáнтный high explosive shell
~, бронебóйно-зажигáтельный armour-piercing incendiary shell
~, бронебóйно-трассúрующий armour-piercing projectile with tracer
~, бронебóйный armour-piercing projectile, shot or shell

снаря́д, бронебо́йный подкали́берный armour-piercing composite shot (eg armour-piercing discarding sabot) (APDS) shot

~, бронепрожига́тельный hollow-charge shell, HESH (high explosive squash head) shell

~, бронепрожига́ющий hollow-charge shell

~ в ине́ртном снаряже́нии inert shell

~ вспомога́тельного назначе́ния round for proof and testing purposes

~, гимнасти́ческий gymnastic apparatus

~, гиперзвуково́й hypersonic projectile

"снаря́д, говоря́щий" aircraft dropped loudspeaker device

снаря́д, далеколета́ющий long-range missile

~, дальнобо́йный long-range shell

~, дистанцио́нный time shell

~ для вкладно́го ство́лика sub-calibre projectile

~ для за́пуска спу́тника ESV launcher vehicle

~ для целеуказа́ния target indication shell

~ до́нного выбра́сывания base-ejection shell

~, дымово́й smoke shell

~, зажига́тельный incendiary shell

~, зени́тный anti-aircraft shell

~, зени́тный управля́емый реакти́вный anti-aircraft guided missile, surface-to-air guided missile

~, изгото́вленный из бума́жной ма́ссы paper shot (trials, ball)

~, кали́берный full calibre projectile, full calibre non-composite shell, full charge shell

~ кла́сса "во́здух-во́здух", реакти́вный air-to-air missile, AAM

~ кла́сса "во́здух-во́здух", управля́емый air-to-air missile, AAM

~ кла́сса "во́здух-земля́", реакти́вный air-to-ground missile, AGM

~ кла́сса "во́здух-земля́", управля́емый air-to-surface missile, ASM

~ кла́сса "во́здух-земля́", "во́здух-кора́бль", реакти́вный air-to-surface missile, ASM

~ кла́сса "земля́-во́здух", управля́емый surface-to-air missile, SAM

снаря́д кла́сса "земля́-земля́", управля́емый surface-to-surface missile, SSM

~ кла́сса "пове́рхность-пове́рхность", управля́емый surface-to-surface missile, SSM

~, крыла́тый winged missile, aerodynamic missile

~ кумуляти́вного де́йствия, бронебо́йный armour-piercing hollow-charge shell

~, кумуляти́вный hollow-charge projectile, shaped-charge projectile

~, кумуляти́вный бронепрожига́ющий HEAT round

~, кумуляти́вный противота́нковый реакти́вный shaped-charge anti-tank missile

~, лафетопро́бный paper shot (mounting test round), proof shot

~, малокали́берный small-calibre shell

~, межконтинента́льный баллисти́ческий inter-continental ballistic missile, ICBM

~, морти́рный mortar bomb (obs)

~, наводи́мый по лучу́ beam-riding missile

~, надкали́берный spigot mounted projectile (operating on rifle grenade principle)

~, назе́мный управля́емый surface-to-surface missile, SSM

~, напо́лненный водо́й, испыта́тельный water shot (trials, ball)

~, невраща́ющийся non-rotating projectile

~, неокра́шенный unpainted shell (Sov without anti- corrosive paint, factor in ballistic corrections)

~ непосре́дственного пораже́ния impact missile

~, неразорва́вшийся blind shell, blind

~, нестре́ляный unfired round

~, неуправля́емый реакти́вный free-flight rocket, unguided rocket, free rocket

~, норма́льный normal shot, right angle shot (ball)

~, оконча́тельно-снаряжённый complete projectile (gunn)

~, окра́шенный painted shell (Sov with anti-corrosive paint, standard peace-time projectile)

снаря́д, оперённый finned missile, fin stabilized missile
~, оперённый реакти́вный finned missile, fin-stabilized missile
~, о́пытный proof shot (ball)
~, освети́тельный star shell, illuminating shell
~, оско́лочно-дымово́й H.E. smoke shell
~, оско́лочно-трасси́рующий tracer-fitted fragmentation shell
~, оско́лочно-фуга́сный HE fragmentation shell
~, оско́лочно-хими́ческий chemical fragmentation (bursting) shell
~, оско́лочный fragmentation shell
~ основно́го назначе́ния primary purpose shell (nav gunn)
~ осо́бого назначе́ния special-purpose shell
~, остроголо́вый sharp-nosed projectile
~, остроголо́вый бронебо́йный sharp-pointed armour-piercing projectile
~, парашю́тный parachute ground training apparatus
~, перфора́торный drilling barge (rock dredging)
~, плитопро́бный armour-proving round
~, подкали́берный composite shot, sub-calibre shot (eg APCR, APDS, APCNR shot)
~, подкали́берный бронебо́йный AP composite shot (eg APDS)
~, полноте́льный solid shot (as distinct from shell)
~, полубронебо́йный semi-armour-piercing shell
~, пороховой раке́тный solid propellant missile
~, практи́ческий practice projectile
~, пристре́лочный ranging shot, ranging shell
~, про́бный proof projectile, proof shot (ball)
~, противорадиолокацио́нный anti-radar missile
~, противота́нковый anti-tank shell, anti-tank missile
~, противота́нковый авиацио́нный реакти́вный air-to-surface anti-tank missile
~, противота́нковый кумуляти́вный hollow-charge anti-tank shell
снаря́д, противота́нковый реакти́вный anti-tank rocket, anti-tank missile
~, противота́нковый управля́емый anti-tank guided missile
~, противота́нковый управля́емый реакти́вный anti-tank guided missile
~, разрывно́й explosive shell
~, раке́тный rocket-propelled projectile
~, реакти́вный rocket missile (usually applied to anti- tank and other small missiles)
~ с баллисти́ческим наконе́чником, остроголо́вый бронебо́йный APCBC (armour-piercing capped ballistic-capped) projectile
~ с бронебо́йным наконе́чником armour-piercing capped (APC) projectile
~ с бронебо́йным наконе́чником, остроголо́вый APCBC projectile
~ с вви́нченным взрыва́телем fused shell
~ с вы́шибным дном base ejection shell
~ с головны́м взрыва́телем nose-fused shell
~ с ине́ртным снаряже́нием inert shell
~ с неконта́ктным взрыва́телем proximity-fused shell
~ с обы́чным заря́дом, бронебо́йный armour-piercing high-explosive shell
~ с отделя́ющимся поддо́ном, бронебо́йный подкали́берный APDS shot
~ с пласти́ческим взры́вчатым вещество́м HESH (high-explosive squash head) shell
~ с поддо́ном, подкали́берный бронебо́йный APDS shot
~ с по́лной гидроскопи́ческой усто́йчивостью over-stabilized projectile
~ с радиовзрыва́телем radio-fused shell
~ с разрывны́м заря́дом, бронебо́йный APHE shot
~ с расплю́щивающейся головно́й ча́стью HESH (high-explosive squash-head) shell
~, сверхзвуково́й supersonic projectile
~, скалодроби́тельный rock-breaker dredger
~, снаряжённый фо́сфором, дымово́й phosphorus smoke shell

снаря́д специа́льного назначе́ния special-purpose shell
~ специа́льного назначе́ния, артиллери́йский carrier shell
~, сплошно́й shot
~, сплошно́й бронебо́йный AP shot (solid)
~ сре́днего ра́диуса, баллисти́ческий intermediate-range ballistic missile, IRBM
~ сре́днего ра́диуса де́йствия, баллисти́ческий intermediate-range ballistic missile, IRBM
~, телеуправля́емый guided missile, command-guided missile
~, терми́тный thermite shell
~, толстосте́нный thick-walled shell
~, тонкосте́нный thin-walled shell
~, трасси́рующий tracer shell
~, трасси́рующий оско́лочно-фуга́сный flat-headed shell, flat-headed shot, blunt-nosed shell, blunt-nosed shot
~, трениро́вочный 1. training equipment, training device
2. drill round (arty)
~, тупоголо́вый blunt-nosed projectile
~, турбореакти́вный spin-stabilized missile
~ уда́рного де́йствия percussion shell
~, управля́емый controlled projectile
~, управля́емый по провода́м wire-guided missile
~, управля́емый противота́нковый anti-tank guided missile
~, управля́емый реакти́вный guided missile
~, уче́бный drill shell, drill projectile, dummy shell
~, фо́сфорный зажига́тельный phosphorus shell
~, фуга́сно-хими́ческий gas high-explosive shell
~, фуга́сный HE shell
~, шрапне́льный shrapnel shell
~, я́дерный nuclear shell
снаря́д-антираке́та anti-missile missile
снаряди́ть (см снаряжа́ть)
снаря́дный 1. ammunition number (arty)
2. shell, ammunition
снарядосто́йкий shell-proof
снаря́д-перехва́тчик раке́т anti-missile missile
снаря́ды, неоконча́тельно снаряжённые plugged ammunition (minus fuse)
~, оконча́тельно снаряжённые fused ammunition
снаряжа́ть/снаряди́ть to fill (shell etc), to arm (mine), to arm, to equip (troops), to supply, to furnish
снаряже́ние 1. filling (shell)
2. arming (mine)
3. charging (clip, magazine)
4. personal equipment
~, автоно́мное возду́шное self-contained air apparatus (diving)
~, боево́е fighting equipment
~ боеприпа́сов 1. ammunition filling (in factory)
2. selection of charge and fuse (in the field)
~ взрыва́теля fuse loading
~, вое́нное munitions
~, защи́тное protective clothing
~, ине́ртное inert filler
~, ко́жаное Sam Browne belt
~, легководола́зное frogman's equipment
~ ле́нты патро́нами belt filling (MG)
~, ли́чное personal equipment, accoutrements
~ магази́на magazine filling
~, неоконча́тельное filling (but not fusing) (ammo)
~, облегчённое похо́дное light marching order
~, оконча́тельное filling and fuzing (ammo)
~ патро́нной ле́нты belt filling, belt loading (SA)
~ патро́нов, ручно́е hand loading
~, полево́е battle order (pers eqpt)
~, по́лное похо́дное field service marching order (FSMO)
~, футля́рное block filling (shell production)
~ шнекова́нием screw filling (method) (shell)
~, штурмово́е battle order (personal eqpt)
снаряжённый 1. armed, fuzed
2. filled (ammo)
3. fully equipped, equipped, fitted out
снасть 1. equipment, tackle
2. guy, guy-rope (tent)
3. guideline, hauler

снасть, пала́точная tent guy rope
снег snow
~, обледене́лый утрамбо́ванный glazed packed snow
~, разрыхлённый loose snow, loosened snow
~ с дождём sleet
"снег", углеки́слый carbon-dioxide foam
снег, уто́птанный packed snow
~, утрамбо́ванный packed snow
снега́, ве́чные permasnow
снегоме́р snow gauge
снегоочисти́тель snowplough
~, газодинами́ческий snow blower, blower snow plough
~, двухотва́льный V-type snow-plough
~, ротацио́нный rotary snow-plough
~, ро́торный rotary snow-plough
~, та́нковый armoured snowplough
~, та́нковый универса́льный tank snowplough attachment
~, центробе́жный snow blower
~, шне́ко-ро́торный rotary-screw snow plough
снегоочи́стка snow removal
снегопа́д snowfall
снегосту́п snowshoe
снегоубо́рка snow removal, snow clearance
снегохо́д oversnow vehicle
снежура́ slush
снижа́ть/сни́зить to lower, to bring down, to reduce; **снижа́ть/сни́зить темп ата́ки** to diminish momentum of attack
снижа́ться/сни́зитьця to descend
сниже́ние descent
~ анте́нны antenna downlead
~ в во́инском зва́нии reduction in rank
~ в до́лжности reduction in appointment
~, максима́льное maximum depression (ord)
~ на парашю́те parachute descent
сни́зить (см снижа́ть)
сни́зиться (см снижа́ться)
снима́ть/снять to take off, to take away, to remove; **снима́ть/снять бе́рег** to make a beach survey (amph ops); **снима́ть/снять блока́ду** to lift (raise) blockade; **снима́ть/снять ка́бель** to reel in cable; **снима́ть/снять ми́ну** to remove a mine, to lift a mine; **снима́ть/снять оса́ду** to raise a siege; **снима́ть/снять пала́тки** to strike tents; **снима́ть/снять с передка́** to unlimber (arty); **снима́ть/снять с предохрани́теля** to take off the safety catch, to release the safety catch; **снима́ть/снять с учёта** to take off strength, to write off; **снима́ть/снять стру́жку** to tear a strip off (coll), to reprimand; **снима́ть/снять штук** to unfix a bayonet; **снима́ть/снять электри́ческие характери́стики** to take electrical readings
снима́ться/сня́ться to have ones photograph taken; **снима́ться/сня́ться с бива́ка** to break camp; **снима́ться/сня́ться с ла́геря** to break camp
сни́мок photograph
~, измери́тельный perspectogram (air photo)
~, крупномасшта́бный large-scale aerial photograph (1:4,000-1:8,000)
~, мелкомасшта́бный small-scale photograph (air) (1:20,000-1:25,000)
~, радиолокацио́нный PPI photograph (radar)
~, тенево́й shadowgraph
сноп sheaf, bunch
~ траекто́рий cone of fire, cone of dispersion (gunn)
сноровúстый dexterous, slick
сноро́вка proficiency, skill, ability
~ экипа́жа crew proficiency
снос 1. drift
2. deflection
3. demolition, tearing down (bldg)
~ ве́тром windage, wind deflection (ball)
сня́тие 1. removal, taking away
2. raising (siege)
3. unfixing (bayonet)
4. weighing (anchor)
5. unlimbering (gun)
6. dismissal (work)
7. recovering (circuit)
~ мин mine clearance
~ предохрани́тельных устро́йств arming (mine etc)
сня́тый с вооруже́ния obsolete
снять (см снима́ть)
сня́ться (см снима́ться)
соба́ка, вое́нно-служе́бная war dog
~, ездова́я harness dog
~, карау́льная guard dog, watchdog

соба́ка минорозыскно́й слу́жбы mine-detecting dog
~, посы́льная messenger dog
~, санита́рная medical dog (used by medical orderly to search for wounded)
~, сторожева́я guard dog
собаковожа́тый dog handler
соба́чка trigger, detent, trip, pawl
~ выбра́сывателя extractor pawl
~, запо́рная lock pawl
~ крюка́ towing hook attachment catch
~ неподви́жных ре́ек stop pawl (automatic loader)
~, пере́дняя rammer releasing lever (QF ord automatic loader)
~ подава́теля feed pawl
~ подвижны́х ре́ек feed pawl
~ рукоя́тки то́рмоза handbrake lever pawl
~, сто́порная retaining pawl
~, храпова́я ratchet pawl
собесе́дник 1. subscriber (teleph) 2. collocutor, interlocutor
собира́ние gathering, assembling, collecting
собира́ть/собра́ть to assemble, to erect, to put together
соблюда́ть/соблюсти́ to observe, to keep, to keep to; **соблюда́ть/соблюсти́ радиомолча́ние** to observe radio silence
соблюде́ние observation (of regulations etc)
~ пра́вил маскиро́вки camouflage discipline
соблюсти́ (см соблюда́ть)
собра́ние 1. meeting, gathering, assembly 2. collection (of works, laws etc)
со́бранный collected, gathered, mustered (troops)
собра́ть (см собира́ть)
со́бственность, казённая government property
со́бственный (one's) own, personal, proper, unladen (weight) (veh), set (noise) (radio, radar), natural, free
соверша́ть/соверши́ть to accomplish, to perform, to make, to commit, to perpetrate; **соверша́ть/соверши́ть зве́рства** to commit atrocities; **соверша́ть/соверши́ть отхо́д** to make a withdrawal
соверше́нно секре́тно top secret
соверше́нство perfection
соверше́нствование perfecting, improving
соверше́нствовать/усоверше́нствовать to perfect, to improve
соверши́ть (см соверша́ть)
Сове́т, Верхо́вный Supreme Soviet
сове́т, вое́нный 1. council of war 2. Military Council (Sov) (Group of senior officers to advise and assist a commander, found at Front (Army Group) level and above)
сове́тник adviser
~ по артилле́рии artillery adviser
совеща́ние conference, consultation
совино́вничество complicity, co-guilt
совмести́мость compatibility
~, электромагни́тная electromagnetic compatibility
совмести́тельство doubling-up (of duties, functions etc)
совмеща́ющий rate setter (AD FCE)
~ бокового́ упрежде́ния lateral rate setter (AD FCE)
~ вре́мени полёта снаря́да time drum operator (AD FCE)
~ паралла́кса displacement and wind correction setter (AD FCE)
~ угла́ возвыше́ния elevation rate setter (AD FCE)
~ упрежде́ния по да́льности range setter (AD FCE)
~ устано́вки взрыва́теля fuze setter (AD FCE)
совмеще́ние 1. combination 2. matching, comparing 3. superimposition (photo)
~, нето́чное mismatch (elec)(dial setting AD FCE)
совнархо́з Council of National Economy (Sov)
сово́к-скребо́к scoop-scraper (snow clearing implement)
совоку́пность totality, combination, sum total
совпаде́ние 1. coincidence, coinciding 2. synchronization, synchronism
совхо́з state farm, sovkhoz
совхо́зник state farmworker
согла́сие 1. consent, permission, assent 2. harmony, concord, accord

согласова́ние 1. agreement, concord
2. co-ordination
3. collimation (sight)
4. matching (electron)
5. synchronization
~ де́йствий экипа́жа crew co-ordination
~ огня́ и движе́ния combination of fire and movement
~, опти́ческое optical comparison (inertial navig)
~ прице́ла с ору́дием coordination of sight with gun sight-gun relationship
~ угло́в карда́нновых подве́сов gimbal angle matching (inertial navig)
согласо́ванность co-ordination
согласо́ванный co-ordinated, synchronized
согласова́ть (см согласо́вывать)
согласо́вывать/согласова́ть to co-ordinate, to tie in
соглаше́ние 1. agreement, understanding
2. covenant
~ о переми́рии armistice agreement
~ о прекраще́нии огня́ cease-fire agreement
со́да, кальцини́рованная soda ash, anhydrous sodium carbonate
~, углеки́слая soda ash, sodium carbonate, washing soda
соде́йствие 1. assistance, help, support
2. co-operation
~, боево́е tactical support, direct support
~, непосре́дственное direct support, direct assistance
~, огнево́е fire support
~ проти́внику aiding the enemy
~ с во́здуха, боево́е tactical air support, direct air support
соде́йствовать to co-operate, to assist, to aid, to support
содержа́ние 1. maintenance, keep, upkeep
2. contents, capacity
3. matter, substance
4. scope
~, де́нежное pay and allowances, pay
~ доро́ги road maintenance
~ на гауптва́хте confinement in the guardroom
~, теку́щее running maintenance
содержа́ть to maintain, to keep, to contain, to include, to comprise, to support, to provide for;
содержа́ть в испра́вности to maintain in good order
содержи́мое contents
~ па́мяти memory content
соедине́ние 1. formation (tac)
2. connection, union, splicing, joint
3. combination
4. compound (chem)
5. force (nav)
~, авиацио́нное air force formation
~, артиллери́йское artillery formation
~, болтово́е bolt joint
~, бронета́нковое armoured formation (mil)
~ ва́лов shaft coupling
~, взры́вчатое хими́ческое explosive compound
~, винтово́е screw joint
~, возду́шно-деса́нтное airborne assault force
~, войсково́е formation, military formation
~, волново́дное waveguide joint
~ вполде́рева halving (timber joint)
~, враща́ющееся rotary joint
~, вре́менное temporary connection
~ встык butt joint
~, вышестоя́щее higher formation (mil)
~, ги́бкое flexible joint
~, дро́ссельное choke joint
~, жёсткое rigid connection
~ звездо́й star connection (elec)
~, карда́нное universal coupling, cardan joint
~, коле́нчато-пыча́жное toggle joint
~, манёвренное task force (nav)
~, междуэлеме́нтное cell connecting bar (elec)
~, морско́е деса́нтное seaborne assault force
~, му́фтовое spigot and socket joint
~ на ма́ссу earth strap connection
~, общевойсково́е all-arms formation (mil)
~, операти́вное task force
~, передово́е деса́нтное forward assault-landing formation
~, после́довательное series connection, series connecting (elec)
~ растру́бом spigot and socket joint
~, рука́вное 1. sleeve coupling
2. hose coupling
~, сварно́е welded joint

соедине́ние, сво́дное деса́нтное операти́вное joint amphibious task force
~, скользя́щее sliding joint
~, скользя́щее шлицево́е splined slip joint
~, сме́шанное series-parallel connection, series- parallel connecting (elec)
~, стрелко́вое rifle formation (tac)
~, стыково́е butt joint
~, стыково́е сварно́е butt welding
~, торцо́вое edge joint
~ треуго́льником delta connection (elec)
~ труб, сварно́е welded pipe joint
~, углово́е gusset, corner joint
~, универса́льное шарни́рное gimbal joint, universal joint
~, фла́нцевое flange point
~, хими́ческое chemical compound
~, шарни́рное hinge joint
~, шарово́е ball joint
~ ши́пом tenon joint
~ шла́нгов hose connection, rubber tube connection, flexible tube connection, hose coupling
~, шту́церное nipple joint, nipple union
соединённый joined, combined, united
соедини́тель connector
созве́здие constellation
создава́ть/созда́ть to create, to originate, to give rise to, to found, to establish; **создава́ть/созда́ть брешь** to breach; **создава́ть/созда́ть дымову́ю заве́су** to lay a smokescreen
созда́ние 1. build-up, development 2. creation
созда́ть (см создава́ть)
сойти́ (см сходи́ть)
сократи́ть (см сокраща́ть)
сокраща́ть/сократи́ть to shorten, to curtail, to reduce, to dismiss, to cancel; **сокраща́ть/сократи́ть рабо́тное вре́мя** to reduce reaction time
сокруша́ть/сокруши́ть to overwhelm, to rout
сокруши́ть (см сокруша́ть)
солда́т soldier
~, молодо́й recruit, untrained soldier, young soldier
~, Неизве́стный Unknown Warrior
солда́т-артиллери́ст artillery man, member of gun detachment
солда́т-наблюда́тель other-rank observer
солда́т-парашюти́ст parachutist
солено́ид solenoid
солидо́л solidol (general-purpose calcium-soap- base grease)
~, жирово́й natural solidol (refined or neutral mineral oils with calcium soaps of cottonseed oil)
~, синтети́ческий synthetic solidol (mineral oils and calcium soaps of synthetic fatty acids obtained by oxidising high-molecular hydrocarbons)
~, эмульсио́нный sett solidol (made by emulsifying mineral oils of medium viscosity with lime water, using rosin acid soaps as stabilizers)
солидолонагнета́тель grease gun
~, ручно́й grease gun (hand held)
со́лион solion (elec)
солифлю́кция solifluction, soil creep
солнцестоя́ние solstice
соло́ма straw
солонова́тый brackish
солонча́к salt marsh, salt flat, sabkhat
солончако́вый 1. salt-saturated 2. brackish
соль, азотноки́слая nitrate
~, Бертоле́това Berthollet's salt, potassium chlorate
соля́рка diesel oil (coll)
"Сомкни́сь!" "Close!" (Sov drill, to close on centre or either flank)
сомкну́ть (см смыка́ть)
сообща́ть/сообщи́ть to report
сообще́ние report, message, communication
~, беспереса́дочное through service (rly)
~, междугоро́дное inter-city traffic
~, ме́стное local traffic
~, при́городное suburban service
~, прямо́е 1. direct connection 2. through traffic
сообщи́ть (см сообща́ть)
соорудить (см сооружа́ть)
сооружа́ть/сооруди́ть to erect, to construct, to build, to equip
сооруже́ние building, construction, structure, installation
~, берегово́е оборони́тельное shore fortification

сооружéние, бронирóванное огневóе armoured emplacement
~, дéрево-землянóе earth-and-timber field defence work
~, дéрево-землянóе оборони́тельное earth-and-timber strongpoint, earth-and- timber field defence works
~, дéрево-землянóе огневóе earth and timber pill-box
~ для зени́тного огневóго срéдства anti-aircraft emplacement
~, долговрéменное permanent installation
~, долговрéменное оборони́тельное permanent strong point, permanent defensive installation, pill box
~, долговрéменное огневóе (ДОС) permanent strong point, pill box
~, железобетóнное reinforced concrete structure
~, железобетóнное полевóе оборони́тельное concrete field fortification
~, закры́тое огневóе covered (weapon) emplacement, pill-box
~, землянóе earthworks
~ из готóвых элемéнтов prefabricated structure
~, инженéрное engineer construction
~, лóжное decoy installation, dummy installation
~, маскирóвочное camouflaged construction, concealed construction
~, нанóсное surface structure
~, оборони́тельное defensive installation
~, огневóе weapon emplacement
~ операти́вно-стратеги́ческого значéния, оборони́тельное strategic fortification
~, откры́тое полевóе оборони́тельное open field fortification
~, подзéмное tunnelled defences, underground works (mil eng)
~ полевóго ти́па, огневóе field emplacement
~, полевóе инженéрное pioneer work
~, портóвое port installations, harbour installations
~, фортификациóнное fortification
сооружéния граждáнской оборóны civil defence works
~, дорóжные road works
сооружéния, инженéрные fieldworks, engineer works
~, оборони́тельные defence works, defences
~, полевы́е field works
~, полевы́е оборони́тельные field defences
~, полудолговрéменные оборони́тельные provisional defences
~, речны́е river installations
~, фортификациóнные protective works
соóсность coaxiality, alignment
соóсный coaxial
соотвéтственный corresponding, appropriate, pertinent
соотношéние сил relative strengths
~ сил и средств relative strengths in men and equipment
~ чи́сленности тáнков и пехóты tank-infantry ratio
сóпка hill, mound, knoll
соплó nozzle, venturi, jet
~, вoзду́шное air intake (carb)
~, входнóе air duct, intake duct
~, забóрное air intake duct
~, калибрóванное metering nozzle
~ карбюрáтора, возду́шное caburettor air intake
~ лавáла convergent-divergent nozzle
~, реакти́вное reaction nozzle, jet nozzle, propelling nozzle, exhaust nozzle
~, соóсное axial exit nozzle (msl)
~, сходя́щееся convergent nozzle
~ форсу́нки venturi
сопостáвить (см сопоставля́ть)
сопоставлéние 1. comparison, confrontation, contrast
2. correlation (of data)
сопоставля́ть/сопостáвить to compare, to confront, to contrast, to balance, to collate
соприкосновéние contact
~, боевóе armed contact
~, непосрéдственное close contact (with enemy)
сопровождáть to accompany, to convoy, to escort, to track (radar); **сопровождáть огнём** to support by fire, to give direct support fire; **сопровождáть по áзимуту** to track in azimuth
сопровождáющий escort, convoy man

сопровожде́ние 1. accompaniment
2. escort, convoy
3. support
4. tracking (radar)
~, **автомати́ческое** automatic tracking
~, **артиллери́йское** artillery support
~, **инерцио́нное** inertial tracking
~, **огнево́е** supporting fire, close support fire, support fire
~ **освещённой це́ли** tracking of illuminated target
~ **по ско́рости** rate aided tracking
~, **полуавтомати́ческое** semi-automatic tracking, aided tracking (radar)
~, **прямо́е** direct tracking
~, **радиолокацио́нное** radar tracking
~, **ручно́е** manual tracking
~, **углово́е** angle tracking
~ **це́ли** tracking of a target
сопротивле́ние 1. resistance, opposition, drag
2. resistor (elec)
~, **акти́вное** resistance
~, **аэродинами́ческое** aerodynamic drag
~, **аэродинами́ческое лобово́е** aerodynamic drag
~, **балла́стное** ballast resistor
~ **ве́тра** wind resistance
~, **вне́шнее лобово́е** external drag
~, **вноси́мое** reflected impedance (elec)
~ **во́здуха** air resistance
~, **волново́е** 1. wave impedance, characteristic impedance, surge impedance (electron)
2. wave drag (missile, ball)
~, **вооружённое** armed resistance, armed opposition
~, **вре́дное** parasite drag, parasitic drag (aeron)
~ **вса́сывания** intake resistance, suction drag
~, **входно́е** input impedance
~, **входно́е ко́мплексное** input impedance
~, **вы́ходное по́лное** output impedance
~, **гася́щее** damping resistor, dropping resistor (elec), (electron)
~ **дви́гателя, вну́треннее лобово́е** burner drag (GW, aeron)
~ **движе́нию** resistance to motion
~, **доба́вочное** series resistor
~, **ёмкостное** capacitive reactance (elec)
~ **излуче́ния** radiation resistance (elec)
сопротивле́ние, индукти́вное inductive reactance, inductive resistance (elec)
~ **истече́нию** back pressure
~ **каче́ния** rolling resistance
~, **ко́мплексное** impedance
~, **конта́ктное** contact resistance (elec)
~ **коро́ткого замыка́ния** short-circuit impedance (elec)
~, **лобово́е** drag, frontal drag
~, **магни́тное** reluctance (elec)
~ **нагру́зки** loading resistor (elec)
~, **нагру́зочное** loading resistor
~, **о́бщее** impedance (elec)
~, **ограничи́тельное** limiting resistor
~, **оми́ческое** ohmic resistance, resistance
~, **отрица́тельное** negative resistance (elec)
~ **переме́нного то́ка, вну́треннее** dynamic resistance (elec)
~, **переме́нное** variable resistor
~, **перехо́дное** contact resistance (elec)
~ **пове́рхностного тре́ния** skin friction drag (aeron)
~, **по́лное** impedance
~ **растека́ния** spreading resistance (elec)
~, **реакти́вное** reactance (elec)
~, **резона́нсное** dynamic resistance (elec)
~ **свя́зи** reflected impedance
~ **скачка́** shock drag (aeron)
~ **скру́чиванию** torsion strength
~, **слага́ющее** component resistance (elec)
~, **уде́льное** specific resistance, resistivity (elec)
~, **упо́рное** stubborn opposition (tac)
~ **уте́чки** grid-leak resistor (elec)
~ **форси́рования** engraving resistance, resisting force (ball)
~, **характеристи́ческое** wave impedance, surge impedance, characteristic impedance (elec)
~ **холосто́го хо́да** open-circuit impedance (electron)
~, **шумово́е** equivalent noise resistance (electron)
~, **шунти́рующее** by-pass resistance, shunting resistance
~ **щётки, перехо́дное** brush contact resistance (elec)
~ **щитка́** flap drag (aeron)

сопротивля́емость 1. power of resistance 2. resistivity (elec)
~, магни́тная reluctivity (elec)
сопротивля́ться to resist, to offer resistance, to oppose
сопряга́ть/сопря́чь to gang, to couple (elec), to combine
сопряже́ние 1. junction, coupling, linking 2. ganging, coupling (elec) 3. interfacing
сопряжённый coupled
сопря́чь (см сопряга́ть)
соразме́рить (см соразмеря́ть)
соразмеря́ть/соразме́рить to adjust, to match (for size), to harmonize
сора́тник, боево́й comrade-in-arms
сорва́ть (см срыва́ть)
соревнова́ние competition, contest
сорт горю́чего fuel grade
~ ма́сла oil grade
сортиме́нт assortment
сортирова́ть to sort, to grade, to size; **сортирова́ть боеприпа́сы** to sort ammunition, to classify ammunition
сортиро́вка sorting, classification, segregation
~, перви́чная preliminary sorting
сортиро́вочная sorting ward (med), sorting department
со́ртность 1. grade, rating 2. knock rating (of aviation gasolines)
сосе́ди flank units, flank formations, adjacent units
сосредото́чение 1. concentration 2. build-up 3. marshalling
~ войск troop concentration, assembly of forces
~ деса́нтных средств boat marshalling point (river etc crossing)
~ огня́ concentration of fire
~ огня́ артиллери́йской гру́ппы artillery group concentration (group is of no standard size and is formed for special support)
~ огня́ диви́зии divisional artillery concentration
~ огня́ корпусно́й артилле́рии corps artillery concentration
~ огня́, непла́новое concentration of fire on opportunity targets
сосредото́чение огня́, обеспе́чивающее действи́тельное пораже́ние це́ли effective concentration of fire
~ огня́, пла́новое deliberate concentration of fire
~ огня́ по одно́й то́чке pin-point concentration of fire
~ огня́ полка́ regimental artillery concentration
~ огня́, после́довательное rolling barrage
~, операти́вное operational build-up
сосредото́ченность degree of concentration
сосредото́ченный concentrated, intensive, convergent
сосредото́чивать/сосредото́чить to concentrate, to fix, to focus
сосредото́чить (см сосредото́чивать)
соста́в 1. complement, composition 2. filling (projectile) 3. train (rly)
~, антикоррозио́нный anti-rust compound, anti-corrosion compound
~, боево́й fighting strength, effectives
~ взрыва́теля, дистанцио́нный powder train (in time fuse)
~, взры́вчатый explosive compound
~, води́тельский driver complement
~, вое́нно-хозя́йственный administrative personnel
~, воспламени́тельный firing composition (elec detonator)
~, воспламеня́ющий fuse composition
~, вы́сший кома́ндный Senior officers, Higher Command personnel (Maj-Generals and above in Soviet army)
~, газообра́зный взры́вчатый explosive gas
~, грему́чий fulminate, initiating compound
~, дистанцио́нный time fuse composition, powder train
~ для удале́ния кра́ски paint remover
~, дымово́й smoke composition
~, дымообразу́ющий smoke-producing composition
~, зажига́тельный incendiary composition
~, зерни́стый порохово́й granular form (ammo)
~, инжене́рно-техни́ческий technical engineer personnel

соста́в, иницииру́ющий fulminate, initiating compound
~, **ка́дровый** 1. regular personnel 2. cadre
~, **ка́псюльный** cap detonating composition, cap priming composition, primer mixture
~, **коли́чественный** numerical strength
~, **кома́ндный** 1. officers of arms (army) 2. line officers (nav) 3. officers of policy branches (AF)
~, **ко́нский** complement of horses
~ **корабля́, ли́чный** ship's company
~, **легковоспламеня́ющийся** inflammable compound
~, **лётно-подъёмный** aircrew
~, **лётный** aircrew
~, **ли́чный** personnel
~, **людско́й** personnel
~, **медленногоря́щий** slow-burning composition (ammo)
~, **мла́дший кома́ндный** junior command personnel, NCOs (Sov includes Junior Sgt to WO)
~, **нали́чный** strength, effective personnel
~, **обслу́живающий** operating personnel, servicing personnel
~, **освети́тельный** illuminating compound
~, **офице́рский** officers, officer personnel
~, **подвижно́й** rolling stock (rly)
~, **по́ездный** train (rly)
~, **полити́ческий** political staff
~, **постоя́нный** 1. permanent staff 2. permanent strength
~, **предохраня́ющий от запотева́ния стёкол** anti-dim (respirators, NBC)
~, **противозамерза́ющий** anti-freeze
~, **противокоррози́йный** anti-rust compound
~, **рядово́й** rank and file
~, **светя́щий(ся)** luminescent compound, luminous paint
~, **сержа́нтский** (the) sergeants (as a body), (the) senior NCOs
~, **спи́сочный** paper strength, listed strength
~, **сре́дний кома́ндный** Junior officers, Intermediate Command Personnel (Sov includes Junior Lt to Capt)
соста́в, ста́рший кома́ндный Field officers, Senior Command Personnel (Sov includes Major to Col)
~, **ста́рший офице́рский** field officers
~, **старши́нский** (the) warrant officers
~, **тёрочный** friction compound
~, **трасси́рующий** tracer composition
~, **уда́рный** percussion composition, cap composition
~, **хими́ческий** chemical composition
~, **чи́сленный** numerical strength
~, **щелочно́й** alkaline compound
состави́тель 1. compiler 2. originator (comms)
~ **поездо́в** yardmaster (rly)
соста́вить (см составля́ть)
составле́ние drafting (message), compiling
~ **прое́ктов** designing, projecting
составля́ть/соста́вить to draft (message), to build up (reserves), to compose, to form, to make up; **составля́ть/соста́вить акт** to draw up a statement; **составля́ть/соста́вить винто́вки в ко́зли** to pile arms; **составля́ть/соста́вить зая́вку** to make out an indent
составля́ющая component, constituent, ingredient
~, **акти́вная** in-phase component, real component, active component (elec)
~ **переме́нного то́ка** AC component
~ **постоя́нного то́ка** DC component
~ **углово́й ско́рости, горизонта́льная** azimuth component of angular velocity (AD)
составно́й sectional, composite
состоя́ние condition, state, status
~ **боево́й гото́вности** 1. state of alert 2. degree of readiness
~, **взведённое** cocked position (SA)
~, **газообра́зное** gaseous state
~ **гото́вности** state of readiness
~, **жи́дкое** fluid state, liquid state
~, **испра́вное** serviceability, serviceable condition, good working order
~, **небоеспосо́бное** disablement, unserviceable condition, non-battle-worthy condition, unfit state
~, **нетре́звое** drunken condition
~ **поко́я** state of rest
~, **работоспосо́бное** serviceable condition

состоя́ть to consist of, to be composed of, to be part of; **состоя́ть на действи́тельной слу́жбе** to be on the active list; **состоя́ть на дово́льствии** to be on ration strength; **состоя́ть на дово́льствии с котла́** to be on messing strength; **состоя́ть налицо́** to be present; **состоя́ть по та́белю** to be according to the inventory

состре́л ору́дий comparative calibration, calibration adjustment (arty)

состре́лка ве́ера adjustment of mpi of whole fire unit

состяза́ние competition

~, вое́нно-прикладно́е military medley competition (includes various forms of physical and military training)

сосу́д container, vessel

~ дю́ара Dewar flask, insulated tank

сот honeycomb

~ (скры́тая огнева́я то́чка) concealed fire position

сотру́дник collaborator, colleague

сотру́дничество collaboration

сотрясе́ние 1. vibrating, shaking
2. concussion, shock

со́ты 1. honeycomb
2. honeycomb entanglement (fd defences)
3. rigging line stowage loops (parachute)

соударе́ние collision

со́ус 1. "sauce" (decontaminant)
2. sauce, gravy

соуча́стие complicity, participation

соуча́стник participator, accomplice

сохране́ние 1. conservation, preservation
2. retention
3. care, custody

сохрани́ть (см сохраня́ть)

сохра́нность 1. safe keeping, safety, security
2. shelf life (battery)

сохраня́ть/сохрани́ть to conserve, to preserve, to maintain, to retain, to keep; **сохраня́ть/сохрани́ть темп продвиже́ния** to maintain the momentum of the advance

сочета́ние conjunction, combination

~ манёвра с огнём fire and movement

сочета́ть to combine

сочине́ние composition, essay

сочлене́ние joint, linkage

сочлене́ние бортово́й переда́чи, карда́нное final drive universal joint

~, волново́дное waveguide joint (electron)

~ гла́вной переда́чи, карда́нное final drive universal joint

~, ёмкостное capacity joint

~, карда́нное universal joint

~, самоустана́вливающееся ball joint

~, сфери́ческое ball joint

со́шка 1. prop, support
2. front support, bipod, monopod (LMG etc)
3. share (of plough)

~, откидна́я folding bipod

~, рулева́я drop arm, steering gear arm, Pitman arm (veh)

~ руля́ drop arm, steering gear arm, Pitman arm (veh)

со́шки bipod group, monopod group (LMG etc)

сошни́к spade (gun trail etc), shoe, spike

~, откидно́й folding spade (ord)

~ пулемёта bipod spike (LMG)

~, съёмный detachable spade (ord)

сою́з union, alliance

сою́зник ally

спад и́мпульса trailing edge of pulse (radar)

~ радиоакти́вности radioactive decay

спа́ивать/спая́ть to solder, to braze, to weld together

спай soldered joint, joint, seam

спа́йка 1. soldering, welding, brazing
2. "team spirit"
3. commissure (anat)

~, боева́я teamwork in action, teamwork in battle

спа́ренный duplex, coupled, double, paired, twin, dual, coaxial, twin-barrelled

спа́ривание coupling, pairing, matching

спа́ривать/спа́рить to couple, to pair

спа́рить (спа́ривать)

"Спаса́йся Кто Мо́жет!" "Every Man For Himself!"

спаса́ть/спасти́ to save, to rescue

спасе́ние rescue, salvage

спасти́ (см спаса́ть)

спа́сшийся survivor

спа́янность cohesion

спа́янный united, well-knit

спая́ть (см спа́ивать)

спектр spectrum
~ **обтека́ния** flow pattern
~ **поглоще́ния** absorption spectrum
~, **светово́й** "light" spectrum (includes infra-red, visible and ultra-violet)
~ **часто́т** frequency spectrum
~, **часто́тный** frequency spectrum
~, **я́дерный** nuclear spectrum
спектроанализа́тор spectrum analyser
спектро́граф spectrograph
спектро́метр spectrometer
спектроме́трия spectrometry
спец specialist
специали́ст specialist
~, **кла́ссный** qualified tradesman, qualified specialist (mil)
~ **по обезвре́живанию неразорва́вшихся бомб** bomb disposal expert
~ **по подрывны́м рабо́там** demolition specialist
специали́ст-боди́ст Baudot operator (teleg)
специали́ст-минёр mine-warfare specialist
специали́ст-подрывни́к demolition specialist
специа́льность 1. speciality
2. branch of service, branch of arm
~, **сме́жная** associated trade
специнструме́нты special tools (for a particular job, piece of eqpt etc)
специ́фика inherent characteristics
спецификáция specification
спецобору́дование special equipment
спецобрабо́тка special treatment, decontamination
спецо́вка special-purpose clothing
спецоде́жда, защи́тная protective clothing
спецору́жие special weapon
спецотря́д special detachment
спецскла́д special store
спецча́сть special unit
спе́шивание dismounting
спе́шиваться/спе́шиться to dismount
спеши́ть/поспеши́ть to hurry, to be fast (timepiece)
спе́шиться (см спе́шиваться)
спидо́метр speedometer
спира́ль 1. spiral
2. "spiral" (tk gunn ex)
3. helical guides (drum magazine)
спира́ль бруно́ concertina wire
~, **двойна́я** coiled-coil filament (elec)
~, **про́волочная** coil of concertina wire, concertina
спирт alcohol
~, **бути́ловый** butyl alcohol
~, **винилбути́ловый** vinyl butyl alcohol, vinyl butyl ether
~, **денатури́рованный** denatured alcohol
~, **изопропи́ловый** isopropyl alcohol
~, **мети́ловый** methyl alcohol
~, **мото́рный** alcohol fuel
~, **нашаты́рный** ammonium hydroxide
~, **твёрдый** solid alcohol
~, **техни́ческий** industrial alcohol
~, **фурфури́ловый** furfuryl alcohol
~, **фурфуро́ловый** furfural alcohol
~, **эти́ловый** ethyl alcohol
спирто́вка spirit lamp
списа́ние write-off
~ **в расхо́д** writing-off
списа́ть (см спи́сывать)
спи́сок register, list, roll, muster roll
~ **адресо́в** distribution list
~, **именно́й** nominal roll, personnel register
~, **кома́ндный** crew list
~, **послужно́й** service record (Sov officers)
спи́сывать/списа́ть to write off; **спи́сывать/списа́ть в расхо́д** to strike off charge
спи́ца spoke
спи́чка подрывника́ slow match
~, **сапёрная** slow match
~, **тле́ющая** slow match
спи́чки matches
сплав 1. alloy
2. floatage, floating (rafts)
~, **антифрикцио́нный** bearing metal, antifriction metal
~, **бе́лый** white metal
~, **лёгкий** light alloy
~, **магниевоцирко́ниевый** magnesium zirconium alloy
~, **твёрдый** cutting alloy, tungsten carbide, hard metal
сплавле́ние fusion
сплани́ровать (см плани́ровать)
сплочённость solidarity
спо́йлер spoiler (aeron)
сполза́ние drop (ball)
спо́нсон sponson (tk)

спóровый fungal
спортзáл gymnasium
спортсмéн-разря́дник trained sportsman, qualified sportsman
спóсоб ведéния огня́ fire control (gunn)
~ ведéния подрывны́х рабóт demolition technique
~ взрывáния firing method (eng, dml)
~ взрывáния, электри́ческий electrical firing method (eng, dml)
~ иниции́рования взры́ва priming method
~ круговóго визи́рования plane table intersection, radiation (surv)
~ наведéния guidance principle (GW)
~, пароаммиáчный steam ammonia process (CW decontamination)
~ перпендикуля́ров perpendicular-dropping method (surv)
~, поля́рный radiation method (surv)
~ равносигнáльной зóны 1. equisignal zone method (ILS) 2. split-beam method (radar)
~ размини́рования, взрывнóй explosive method of mineclearing
~, такти́ческий tactical method
~ целеуказáния method of target indication
спосóбность автомоби́льной дорóги, пропускнáя road capacity
~ береговóй полосы́, пропускнáя beach capacity (amph ops)
~, бронепробивнáя armour defeating performance, armour penetration, armour penetrating ability
~, высóкая разрешáющая high power of resolution
~, грузопропускнáя cargo-handling capacity
~, затемня́ющая total obscuring power (smoke)
~, излучáтельная radiation efficiency (antenna)
~, ионизи́рующая ionizing power
~, крóющая total obscuring power
~, отражáющая reflecting power
~ по áзимуту, разрешáющая azimuth resolution, bearing resolution
~ по врéмени, разрешáющая time resolution
~ по дáльности, разрешáющая range resolution (radar set)
спосóбность по угловы́м координáтам, разрешáющая angular resolution
~ по углу́ мéста, разрешáющая elevation resolution
~ по углу́, разрешáющая azimuth resolution (radar set)
~ пóрта, пропускнáя port capacity
~, пробивнáя penetrative performance, piercing performance
~, пропускнáя 1. telegraph speed (teleg), transmission speed (comms) 2. capacity (rd, rly)
~, разрешáющая 1. discrimination, resolution (radar) 2. resolving power (photo)
~, разрешáющая по áзимуту bearing discrimination (radar)
~, разрешáющая по дáльности range discrimination (radar)
~ склáда, пропускнáя depot capacity
~, тя́говая towing performance
~, угловáя разрешáющая angular resolution (electron)
~, эмиссиóнная emissive capacity
спосóбствовать/поспосóбствовать to assist, to aid, to abet, to expedite
спóсобы ведéния огня́ fire techniques, gunnery techniques
~ и приёмы мини́рования и размини́рования mine-warfare techniques
~ подрывны́х рабóт demolition techniques
~ произвóдства разрушéний demolition techniques
~ разграждéния obstacle breaching techniques
~ управлéния огнём fire direction techniques
спотыкáч low anti-personnel obstacle, tripwire
спрáва на лéво from right to left
спрáвиться (см справля́ться)
спрáвка 1. inquiry, return, memorandum 2. information 3. reference number
справля́ться/спрáвиться to consult, to cope; **справля́ться/спрáвиться с обстанóвкой** to have situation in hand
спрáвочник manual, handbook, reference book
спри́нклер sprinkler, diffuser branch pipe (of fire hose)

спринцо́вка syringe, syringeing
спры́гивание jumping down, jumping off
спрыск nozzle (of fire hose branch)
спурт spurt
спуск 1. incline, slope, descent, fall (topo)
2. release
3. downhill running (ski)
4. stirrup (pistol)
5. detent, trigger (ord)
~ **в ни́зкой сто́йке** downhill run in full crouch position (ski)
~ **в сре́дней сто́йке** downhill run in semi-crouch position (ski)
~, **круто́й** sharp decline
~ **курка́** trigger release
~, **механи́ческий** manual firing (arty)
~, **ми́нный** incline (mining)
~ **на караби́не** seat rappel descent (mountaineering)
~ **"на́искось" скло́на** downhill traverse (ski)
~, **накло́нный** ramp
~, **предохрани́тельный** safety sear
~, **предупреди́тельный** first pressure (on trigger, SA)
~, **ручно́й** hand trigger
~ **с опо́рой на коле́ни** descent on skis leaning elbows on knees
~ **с предупреди́телем** two-stage trigger
~, **туго́й** heavy trigger pull, snatch (SA)
спуска́ться/спусти́ться to descend
спусти́ться (см спуска́ться)
спу́тник 1. satellite
2. travelling companion
~ **земли́, иску́сственный** artificial earth satellite
сраба́тывание 1. operating cycle (of mechanism)
2. wearing, wear and tear
~ **взрыва́теля** fuze action
~, **непра́вильное** malfunction
~, **случа́йное** random operation
срабо́танность teamwork, harmonious co-operation
сравне́ние comparison
сравня́ться (см равня́ться)
сраже́ние battle, engagement, action
~, **встре́чное** encounter battle
~, **оборони́тельное** defensive battle
срасти́ть (см сра́щивать)
сра́щивание 1. splicing
2. joining, binding
сра́щивать/срасти́ть to joint, to splice

среда́ 1. medium
2. environment
3. Wednesday
~, **неподви́жная** quiescent atmosphere (ball, aeron)
~, **теплопередаю́щая** heat transfer agent
среднекали́берный medium-calibre
среднеквадра́тный root mean square
среднесу́точный average per day, daily average
сре́дний average, medium, mean, middle, central
сре́дства, авиацио́нно-тра́нспортные transport aircraft
~, **артиллери́йские** artillery, artillery resources
~, **бактериа́льные** bacteriological warfare agents, biological warfare agents, BW agents
~ **бактериологи́ческого нападе́ния** biological warfare agents, bacteriological warfare agents
~, **боевы́е** arms, weapons, combat resources
~ **взрыва́ния** firing accessories (dml)
~, **водола́зные** diving equipment
~, **водоотка́чивающие** water pumping equipment
~, **водоочисти́тельные** water treatment equipment
~ **дегаза́ции** decontamination equipment
~, **дежу́рные огневы́е** weapons on call, duty fire weapons
~ **дезориента́ции** deception devices
~, **де́нежные** funds, cash, moneys
~ **деса́нта, вы́грузочные** assault landing craft (cargo)
~ **деса́нта, вы́садочные** assault landing craft (personnel)
~, **деса́нтно-вы́садочные** 1. assault landing vehicles
2. landing craft
~, **деса́нтно-перепра́вочные** assault river-crossing equipment
~, **деса́нтные плаву́чие** assault landing craft, assault landing equipment
~ **дистанцио́нного управле́ния взры́вом** remote-controlled means of detonation
~, **доро́жно-восстанови́тельные** road maintenance equipment
~ **доста́вки биологи́ческого ору́жия** means of disseminating biological agents

сре́дства доста́вки я́дерного ору́жия к це́ли, такти́ческие tactical nuclear delivery means
~ защи́ты, коллекти́вные collective protective equipment
~, зени́тные air defence weapons
~ зо́ны боевы́х де́йствий, ремо́нтные combat zone repair facilities
~, имитацио́нные simulation devices, simulants
~ индивидуа́льной противохими́ческой защи́ты personal anti-gas equipment
~ индивидуа́льной хими́ческой защи́ты personal anti-gas equipment
~ инжене́рного вооруже́ния engineer equipment
~, инжене́рные engineer equipment
~ маскиро́вки, техни́ческие artificial means of concealment
~ массово́го пораже́ния weapons of mass destruction, means of mass destruction
~, материа́льные material resources, stores, munitions, stores vehicles and ammunition
~, ме́стные local resources
~ механиза́ции mechanical means, mechanical equipment
~ мини́рования и размини́рования mine-warfare equipment
~, мостострои́тельные bridge-building equipment
~, мотомеханизи́рованные mechanized forces
~ наблюде́ния observation facilities
~ обеззара́живания decontamination equipment, decontamination facilities
~ обнаруже́ния хими́ческого зараже́ния chemical contamination detection devices
~, огневы́е guns, fire weapons, firearms
~, огнемётные flame-throwing weapons
~, освети́тельные means of illumination, illuminants
~, перево́зочные transportation
~ передвиже́ния means of transport
~, перепра́вочные water crossing equipment, wet gap crossing equipment
~ пехо́ты, огневы́е infantry weapons, infantry fire resources
~ пехо́ты, противота́нковые infantry anti-tank weapons
сре́дства, пиротехни́ческие pyrotechnic equipment
~, подвижны́е transport, MT
~, подвижны́е ремо́нтные mobile repair facilities
~, подде́рживающие supporting weapons
~ подде́ржки та́нков tank-support means
~, подру́чные materials at hand, improvised means
~, подрывны́е demolition pyrotechnic equipment, blasting explosives
~ пораже́ния weapons, means of destruction
~ пораже́ния, обы́чные conventional means of destruction, conventional weapons
~ противоде́йствия радиолока́ции radar counter-measures, RCM
~, противохими́ческие anti-gas equipment, anti-gas materials
~ противохими́ческой защи́ты anti-gas equipment, chemical defence facilities
~ противохими́ческой защи́ты, индивидуа́льные personal anti-gas equipment
~ противохими́ческой защи́ты, та́бельные коллекти́вные unit protective equipment (scale issue)
~, радиолокацио́нные radar equipment
~ радиопротиводе́йствия, акти́вные active electronic counter-measures, active ECM
~ разве́дки мин mine-detecting equipment
~ размини́рования mine-clearing devices
~, реакти́вные 1. rocket weapons, missiles 2. rockets
~, ремо́нтные repair facilities
~, самохо́дные перепра́вочно-деса́нтные self-propelled assault water-crossing equipment
~, светосигна́льные light-signalling equipment
~ свя́зи communications equipment, signals equipment, means of communication, communications facilities
~ свя́зи, возду́шно-звуковы́е sound-signalling equipment

сре́дства свя́зи, гидроакусти́ческие sonic signals equipment (nav)
~ **свя́зи, подвижны́е** mobile communications facilities
~ **свя́зи, про́волочные** line communications
~ **свя́зи, резе́рвные** uncommitted communications equipment
~ **свя́зи, техни́ческие** technical means of communication (radio, line, dispatch riders etc)
~ **слу́жбы снабже́ния горю́чим, техни́ческие** POL supply servicing equipment, fuel and oil distributing equipment
~ **созда́ния радиопоме́х** jamming devices
~, **спаса́тельные** salvage appliances, survival appliances
~, **сухопу́тные тра́нспортные** land transport
~, **та́бельные** scale equipment, authorized equipment, organic equipment, weapons laid down in authorized scale, equipment laid down in authorized scale
~, **такела́жные** block and tackle gear
~, **техни́ческие** 1. technical means 2. technical equipment, hardware
~ **управле́ния огнём** fire-control facilities
~ **усиле́ния** 1. support weapons 2. means of reinforcement
~ **устраше́ния, я́дерные** nuclear deterrent
~, **хими́ческие** chemical agents
~, **шта́тные огневы́е** organic weapons
~ **эвакуа́ции повреждённых маши́н** vehicle recovery equipment
~, **эвакуацио́нные** evacuation facilities
~, **я́дерные подрывны́е** atomic demolition munitions
сре́дство 1. device, means, facility 2. weapon 3. remedy (med)
~ **автомати́ческого де́йствия, боево́е** robot weapon
~, **автомати́ческое огнево́е** automatic weapon
~, **антикоррози́йное** rust preventive
~, **а́томное боево́е** nuclear weapon
~, **бактериологи́ческое** bacteriological agent
сре́дство биологи́ческой разве́дки biological detection device
~ **бли́жнего бо́я, огнево́е** short-range weapon
~, **боево́е** weapon
~, **болеутоля́ющее** analgesic, pain reliever
~, **возбужда́ющее** stimulant
~, **возду́шно-деса́нтное боево́е** airborne assault weapon
~, **вспомога́тельное огнево́е** secondary weapon
~, **дезинфици́рующее** disinfectant
~, **деса́нтно-вы́садочное** 1. assault landing vehicle 2. assault landing craft
~ **для борьбы́ с та́нками** anti-tank resources
~ **для истребле́ния та́нков** tank killer
~ **для форси́рования водно́й прегра́ды** assault river crossing craft
~ **доста́вки** delivery system
~ **доста́вки я́дерного ору́жия** nuclear weapon delivery system
~, **дымово́е** smoke weapon
~, **дымообразу́ющее** smoke generator
~, **жаропонижа́ющее** febrifuge, antipyretic (med)
~, **зажига́тельное** incendiary weapon
~ **запра́вки** refuelling equipment
~ **защи́ты** 1. means of defence 2. protective equipment (anti-gas eqpt etc)
~ **звуково́й разве́дки, электро́нное** electronic sound locating device (arty)
~, **зени́тное огнево́е** high-angle weapon
~ **индика́ции отравля́ющих веще́ств** detector equipment (CW etc)
~ **насти́льного огня́, огнево́е** flat-trajectory weapon
~, **обезжи́ривающее** degreasing agent
~ **обнаруже́ния** detection device
~ **оборо́ны** defensive weapon, means of defence
~, **огнево́е зени́тное** AA weapon, surface-to-air weapon
~, **огнево́е противота́нковое** anti-tank weapon
~, **огнегаси́тельное** fire-extinguishing agent
~, **огнетуши́тельное** fire-extinguishing agent

сре́дство, освети́тельное illuminant, illumination device
~, отпу́гивающее насеко́мых insect repellant
~ ПВО air defence resources
~ перепра́вы, коле́йное treadway gap-crossing equipment (raft, br)
~, подде́рживающее огнево́е fire support weapon
~ подде́ржки support, means of support
~ пораже́ния (destruction) weapon
~ примене́ния а́томного ору́жия atomic weapon delivery system
~ про́тив скольже́ния anti-skid device
~, противопожа́рное fire-fighting equipment
~ радиацио́нной разве́дки radiological detection device
~ размини́рования mine-clearing device
~, раке́тное rocket weapon, missile
~ санита́рного тра́нспорта ambulance
~ свя́зи, полево́е field communications equipment, means of communication in the field
~, та́нко-деса́нтное плаву́чее tank landing craft
~, термоя́дерное thermonuclear weapon
~ усиле́ния reinforcing means, reinforcing weapon
~ устраше́ния, глоба́льное global deterrent
срез 1. cutting, cut, section
2. shear (stress)
~, ду́льный muzzle, muzzle face
~ затво́ра,пере́дний bolt face (SA)
~, казённый breech end, breech face (ord)
~ ствола́, казённый breech face (ord)
сре́зать (см сре́зывать)
сре́зывание shearing
сре́зывать/сре́зать to cut off, to shear, to fail (examination)
срок time, time limit, deadline, period
~ вы́слуги length of service (time already served)
~, гаранти́йный warranty period
~ го́дности 1. useful life, working life
2. expiration date
~, межремо́нтный mileage interval (between repairs)
~ обслу́живания maintenance period
~ рабо́ты, гаранти́йный guarantee period
срок ремо́нта repair time
~ слу́жбы 1. life (eqpt)
2. term of service
~ слу́жбы, гаранти́йный warranty life, guaranteed service life
~ слу́жбы дви́гателя engine life
~ сма́зки lubrication interval
~ хране́ния 1. warranty (period)
2. shelf life
3. life (commodity)
сро́сток splice, joint (cable)
"сро́чная" "immediate" (Sov message precedence designation)
срочнослу́жащий conscript (soldier serving normal period of compulsory military service), non- regular soldier
сро́чность, первоочерёдная top priority
сро́чный urgent, high priority, "rush"
сруб, бреве́нчатый square log frame
срыв 1. blowout (of flame)
2. breakdown, failure, stoppage
3. burbling (aerodynamics)
срыва́ть/сорва́ть to break up, to tear down, to tear off, to ruin (ofenemy plans) **срыва́ть/сорва́ть ата́ку** to break up an attack
срыва́ть/срыть to raze to the ground
срыть (см срыва́ть)
ссу́да loan
стабилиза́тор 1. stabilizer
2. tail fin, tail unit (missile, bomb)
~ взры́вчатого вещества́ explosive stabilizing agent
~, гироскопи́ческий gyroscopic stabilizer (tk gunn)
~, кольцево́й tail drum, drum-shaped tail
~ напряже́ния voltage stabilizer
~, носово́й nose fin (GW)
~, скла́дывающийся folding fin (msl)
~, хвостово́й rear fin, tail fin
стабилиза́ция stabilization
~, аэродинами́ческая aerodynamic stabilization
~ в вертика́льной пло́скости vertical stabilization
~ в горизонта́льной пло́скости horizontal stabilization
~ в двух пло́скостях horizontal and vertical stabilization
~ враще́нием spin-stabilization
~, гироскопи́ческая gyrostabilizing, gyrostabilization

стабилиза́ция, ква́рцевая crystal stabilization
~ **по горизо́нту** horizontal stabilization
~ **по кре́ну** roll stabilization
~ **по ку́рсу** yaw stabilization
~ **по тангажу́** pitch stabilization
~ **по углу́ рыска́ния** yaw stabilization
~ **ракéты** rocket stabilization, missile stabilization
~ **следя́щей систéмы** resetting (servomechanism)
~ **частоты́** frequency stabilizing
~, **чрезме́рная** overstability, overstabilization (servo)
стабилизи́рующий stabilizing
стабилотро́н/стабилитро́н "stabilotron", stable element (instrument that maintains desired orientation)
стаби́льность stability
~, **баллисти́ческая** ballistic stability
~ **горю́чего** fuel stability
~ **по́роха** propellant stability
~ **при хране́нии** storage stability
стаби́льный stable
ста́вень shutter
~, **предохрани́тельный** safety shutter (fuse)
ста́вить/поста́вить to put, to place, to apply, to put on, to stage, to bet, to raise (a question), to organise; **ста́вить/поста́вить взрыва́тель на предохрани́тель** to set a fuse at safe; **ста́вить/поста́вить грана́ту на боево́й взвод** to pull the pin from a grenade; **ста́вить/поста́вить гриф** to classify (docs etc); **ста́вить/поста́вить дымову́ю заве́су** to lay a smoke screen; **ста́вить/поста́вить зада́чу** to allot a task; **ста́вить/поста́вить ми́ны** to lay mines; **ста́вить/поста́вить на предохрани́тельный взвод** to set at safe (SA); **ста́вить/поста́вить на ремо́нт** to put in for repair; **ста́вить/поста́вить на стоя́нку** to park (veh); **ста́вить/поста́вить огнево́й вал** to put down a barrage (arty)
ста́вка 1. headquarters (of higher fmns) 2. rate of pay
ста́вни shutters
ста́вня shutter
стадио́н stadium
ста́дия phase, stage
ста́дия проекти́рования blueprint stage, design stage
стаж 1. period, length of time 2. stage, period of training, period of probation
~ **слу́жбы** length of service
стажёр 1. probationer 2. student undergoing practical training with troops
стажиро́вка 1. probationary training period 2. practical training with troops for students
стака́н 1. hollow metal cylinder, hollow metal sleeve 2. canister (bounding mine) 3. glass (drinking vessel)
~ **закрыва́ющей пружи́ны** closing spring case (ord, automatic loading)
~, **запа́льный** 1. burster tube, priming composition cap 2. gaine (nose fuse) 3. magazine cover (base fuse) 4. adaptor booster
~, **нажимно́й** pressure sleeve (arty)
~ **подши́пника** bushing sleeve
~ **снаря́да** shell body
~, **сто́порный** stop plunger, retaining catch plunger
~, **шарообра́зный зажига́тельный** spherical glass incendiary body (gren)
стака́нчик igniter cup (gren)
ста́лкиваться/столкну́ться to collide; **ста́лкиваться/столкну́ться с проти́вником** to encounter the enemy, to "bump"
сталь steel
~, **быстроре́жущая** high-speed tool steel
~, **вана́диевая** vanadium steel
~, **волни́стая** corrugated steel
~, **высокока́чественная** high quality steel
~, **двутавро́вая** I-beam, H-beam, rolled-steel beam
~ **для та́нков, (бронева́я)** tank (armour) plate
~, **жаропро́чная** creep-resisting steel
~, **жаросто́йкая** oxidation-resisting steel
~, **жароупо́рная** heat-resisting steel, oxidation- and creep-resisting steel
~, **инструмента́льная** tool steel
~, **ка́чественная** good quality steel

сталь, конструкцио́нная engineering steel
~, коры́тная channel-section steel
~, леги́рованная alloy steel
~, листова́я steel plate, sheet steel
~, листова́я жаропро́чная heat resistant sheet steel
~, лита́я бронева́я cast armour steel
~, малоуглеро́дистая mild steel
~, марте́новская open-hearth steel
~, нержаве́ющая stainless steel
~, низколеги́рованная высокопро́чная low-alloy high-tensile steel
~, оруди́йная gun-barrel steel
~, прока́тная rolled steel
~, про́фильная section steel
~, ре́льсовая rail steel
~, рессо́рная spring steel
~, специа́льная alloy steel
~, ство́льная оруди́йная gun barrel steel
~, тавро́вая T-bar (steel)
~, углеро́дистая carbon steel
~, углеро́дистая оруди́йная carbon gun steel
~, углеро́дистая пу́шечная carbon gun steel
~, фасо́нная section steel
~, хро́мистая chromium steel
~, хромокоба́льтовая cobalt-chromium steel
~, хромомолибде́новая chrome-molybdenum steel
~, хромони́келевая chrome-nickel steel
стаме́ска wood chisel
стан 1. stature, height
2. camp, halting place
3. mill (eng)
~, полево́й 1. field camp
2. tactical rendezvous
~, прока́тный rolling mill (eng)
станда́рт 1. standard, norm
2. level
стандартиза́ция standardization
стандартизи́ровать to standardize
станда́ртность conformity to standard
станда́ртный standard
стани́на 1. mount, bed, base (for machinery etc)
2. trail (ord)
~, за́дняя rear trail (ord)
~, коро́бчатая box trail (ord)
~ крестообра́зного лафе́та outrigger (of cruciform trail)(ord)
стани́на лафе́та carriage sidepiece (ord)
~, основна́я platform frame (ord)
~, откидна́я folding extension girder assembly (ord)
стани́ны, раздвижны́е split trail (ord)
станки́ст machine-gunner (No. 1 in MMG det)
"Станови́сь!" "Fall In!"
станови́ться/стать to stand, take one's stand, to stop, to become, to get, to grow; **станови́ться/стать в заты́лок** to cover off **станови́ться/стать на я́корь** to anchor
становле́ние settling-in, acclimatisation, making the grade
стано́к 1. mount, mounting (ord)
2. machine tool
3. stall (in stable)
~, балансиро́вочный balancing stand
~, букрово́й boring rig
~, ве́рхний saddle (ord)
~, двухколёсный (пулемётный) two-wheeled mount (MG)
~ для расто́чки cylinder boring machine
~ заряжа́ния loading machine
~, зени́тный anti-aircraft mounting
~ криворо́та aim tester (training aid)
~, круглошлифова́льный grinding lathe
~, мета́тельный zero-length launcher
~, ни́жний 1. carriage body (ord)
2. trail (ord)
3. saddle support (ord)
~, оруди́йный gun mounting
~, прице́льный aiming rest (SA trg eqpt)
~, пулемётный machine-gun mounting, machine-gun carrier
~, пусково́й ground equipment, launching table (missile site)
~, сверли́льный drilling machine, boring machine
~, тока́рный lathe
~, трено́жный tripod mounting
~, универса́льный universal mount
~, фре́зерный milling machine
стано́к-треноѓа tripod mounting
ста́нция 1. station, post
2. set, station (radio, radar)
3. plant, outfit
~, авари́йная emergency station
~, автодегазацио́нная bulk decontamination vehicle (clothing and equipment) (CW)

ста́нция, автозапра́вочная petrol station
~, автомати́ческая телефо́нная automatic telephone exchange
~ автомати́ческого сопровожде́ния, радиолокацио́нная automatic tracking radar
~, автомоби́льная водоочисти́тельная water purification truck
~, автомоби́льная электри́ческая truck mounted generating set
~, авторазли́вочная chemical lorry, chemical spraying truck (tanker with mechanical pump)
~, акти́вная радиолокацио́нная active radar
~, артиллери́йская метеорологи́ческая artillery meteorological station
~, артиллери́йская радиолокацио́нная artillery fire control radar
~, бензозапра́вочная petrol station
~ берегово́й оборо́ны, радиолокацио́нная coastal defence radar
~ боково́го обзо́ра, радиолокацио́нная sideways looking radar
~, бортова́я радиолокацио́нная on-board radar, airborne radar
~, ведо́мая slave station (radio navigation)
~, веду́щая master station (radio navigation)
~, водонасо́сная water pumping station
~, водоочисти́тельная water purification unit
~, вое́нно-почто́вая forces post office, FPO
~, вози́мая фи́льтровая portable water purification set
~ вы́вода самолётов на поса́дку, радиолокацио́нная approach control radar
~, вы́грузочная railhead, unloading station, detraining station
~, генера́торная generating station
~, грузова́я goods station, freight station
~ да́льнего де́йствия, радиолокацио́нная long-range radar
~ да́льнего обнаруже́ния early-warning radar
~ да́льнего обнаруже́ния, бортова́я радиолокацио́нная airborne early-warning radar
ста́нция да́льнего обнаруже́ния самолётов, радиолокацио́нная AD early-warning radar
~, дальноме́рная range-finder position
~, диспе́тчерская радиолокацио́нная ground control approach radar, GCA radar
~ для горю́чего, перека́чивающая fuel-pumping set (mounted on a truck chassis)
~, до́плеровская радиолокацио́нная Doppler radar
~, загоризо́нтная радиолокацио́нная over-the-horizon radar
~, загради́тельная barrage jammer (EW)
~, задаю́щая master station (radio navig)
~, запра́вочная fuel filling station, filling point
~, заря́дная charging station, charging plant, battery charging outfit
~ засе́чки миномётных пози́ции, радиолокацио́нная mortar-locating radar
~, звукометри́ческая sound ranging station, sound ranging apparatus
~, звукоула́вливающая sound location station
~, зени́тная артиллери́йская метеорологи́ческая anti-aircraft artillery meteorological station
~, и́мпульсная радиолокацио́нная pulse radar, pulse radar set
~, и́мпульсно-до́плеровская радиолокацио́нная pulsed Doppler radar
~, испыта́тельная test station
~ истреби́теля, радиолокацио́нная fighter radar
~ истреби́теля-перехва́тчика, радиолокацио́нная interceptor radar
~, когере́нтная радиолокацио́нная coherent radar
~, компре́ссорная compressor unit (veh)
~, коне́чная железнодоро́жная railhead, railway terminus, rail terminal
~, коне́чно-вы́грузочная железнодоро́жная railhead, railway terminus, rail terminal
~, корабе́льная радиолокацио́нная marine radar, shipborne radar
~, коротково́лновая short wave radio (station, set)

ста́нция кругово́го обзо́ра tactical control radar, PPI (station, set)
~ **кругово́го обзо́ра, радиолокацио́нная** plan position indicator, PPI
~, **ло́жная радиолокацио́нная** decoy radar station, decoy radar set
~, **междугоро́дная телефо́нная** trunk exchange (comms, teleph)
~ **метеонаблюде́ния** weather observation station
~, **метеорологи́ческая** meteorological station, weather station
~, **метеорологи́ческая радиолокацио́нная** meteorological radar
~, **меша́ющая** jamming station
~, **многоцелева́я радиолокацио́нная** multi-function radar, multiple radar
~, **монои́мпульсная радиолокацио́нная** monopulse radar
~ **наблюде́ния за артиллери́йскими разры́вами** artillery locating radar (station, set)
~ **наведе́ния** command radio (GW)
~ **наведе́ния истреби́телей, радиолокацио́нная** ground control interception radar, GCI radar
~ **наведе́ния, назе́мная** ground-controlled interception station, GCI station
~ **наведе́ния по лучу́** beam-rider guidance radar
~ **наведе́ния по лучу́, радиолокацио́нная** beam-rider guidance radar
~ **наведе́ния проже́кторов** searchlight control equipment (radar)
~ **наведе́ния раке́т** missile-guidance radar
~ **наведе́ния раке́т, радиолокацио́нная** missile-guidance radar
~, **навигацио́нная радиолокацио́нная** navigation radar
~, **назе́мная радиолокацио́нная** ground radar
~ **назе́мной артиллери́йской разве́дки (СНАР)** ground artillery reconnaissance radar station, battlefield surveillance radar
~, **насо́сная** pumping station
~, **некогере́нтная радиолокацио́нная** non-coherent radar
~, **неподавля́емая радиолокацио́нная** unjammable radar

ста́нция непреры́вного излуче́ния, радиолокацио́нная continuous-wave radar, CW radar
~ **обеззара́живания тра́нспорта** vehicle decontamination station
~, **облуча́ющая** target-track illuminator radar
~ **облуче́ния це́лей** target-illuminating radar, target-track illuminator radar
~ **облуче́ния це́лей, радиолокацио́нная** target illuminating radar, target-track illuminator radar
~ **обнаруже́ния** warning radar, surveillance radar
~ **обнаруже́ния и наведе́ния** radar reporting and control centre
~ **обнаруже́ния и сопровожде́ния це́ли, радиолокацио́нная** acquisition radar
~ **обнаруже́ния и целеуказа́ния, радиолокацио́нная** search and target designation radar
~ **обнаруже́ния, лёгкая радиолокацио́нная** small search radar
~ **обнаруже́ния, лёгкая** light warning station LWS (radar)
~ **обнаруже́ния миномётных пози́ций, радиолокацио́нная** mortar-locating radar
~ **обнаруже́ния миномётов, артиллери́йская радиолокацио́нная** artillery mortar-locating radar
~ **обнаруже́ния миномётов проти́вника** mortar-locating radar, counter-mortar radar
~ **обнаруже́ния назе́мных це́лей, радиолокацио́нная** ground-target search radar
~ **обнаруже́ния, радиолокацио́нная** warning radar
~ **обнаруже́ния самолётов, радиолокацио́нная** aircraft warning radar
~ **обрабо́тки да́нных** data processing station (computer)
~ **обслу́живания** service station
~ **о́бщего назначе́ния, радиолокацио́нная** general-purpose radar
~, **однои́мпульсная радиолокацио́нная** monopulse radar
~, **оконе́чная** terminal station
~, **оконе́чная телегра́фная** telegraph terminal

ста́нция определе́ния да́льности, радиолокацио́нная distance-measuring radar
~ **организо́ванных поме́х** jammer (EW)
~ **оруди́йной наво́дки** fire-control radar, gunlaying radar
~, **пано́рамная радиолокацио́нная** panoramic radar
~, **пасси́вная радиолокацио́нная** passive radar
~, **пеленга́торная** direction-finding station
~ **переда́чи кома́нд** command transmitter (GW)
~, **передаю́щая** transmitting station
~ **пере́днего обзо́ра, радиолокацио́нная** forward-looking radar
~, **перека́чивающая** pumping station
~, **перено́сная** portable radio (station)
~ **перехва́та, бортова́я радиолокацио́нная** airborne intercept radar
~ **пита́ния ору́дий** vehicle-mounted generator supplying power for guns
~, **подви́жная дегазацио́нная** truck-mounted decontamination equipment
~, **подви́жная заря́дная** battery-charging truck
~, **подви́жная компре́ссорная** compressor truck
~, **подви́жная радиолокацио́нная** mobile radar
~, **подрывна́я** firing point (dml)
~, **подслу́шивающая** intercept station (comms)
~ **по́иска, радиолокацио́нная** search radar
~, **полева́я почто́вая** field post office, FPO
~ **поме́х** jammer (ECM)
~, **помехоусто́йчивая радиолокацио́нная** unjammable radar
~, **приводна́я** homer (beacon)
~, **приёмная** receiving station
~, **приёмно-передаю́щая** transceiver
~, **приёмо-передаю́щая** transceiver
~, **прице́льная радиолокацио́нная** fire-control radar
~, **проводновеща́тельная** wired broadcasting station
~ **"прожзву́к"** coupled searchlight-sound ranging station

ста́нция, промежу́точная relaying station (radio)
~ **прямо́й наво́дки** fire-control radar (set)
~, **рабо́тающая по ме́тоду сравне́ния фаз, радиолокацио́нная** phase-comparison radar
~, **рада́рная** radar station
~, **радиовеща́тельная** broadcasting station
~, **радиолокацио́нная (см радиолока́тор)** radar, radar station, radar set
~ **радиолокацио́нных поме́х** jamming transmitter (radar)
~, **радиопеленга́торная** radio direction-finding station, radio DF set
~ **радиопоме́х** jamming station
~ **радиопослу́шивания** 1. radio intercept station 2. monitoring station
~, **радиопроже́кторная** radar-directed searchlight, RDSL, radar-controlled searchlight
~, **радиореле́йная** radio relay station
~, **радиотелеметри́ческая** radio-telemetering station
~ **разве́дки** early warning radar
~ **разве́дки и целеуказа́ния** tactical control radar, surveillance and acquisition radar
~ **разве́дки стреля́ющих миномётов** mortar-locating radar
~ **ра́ннего предупрежде́ния, радиолокацио́нная** early warning radar station
~ **распознава́ния це́лей, радиолокацио́нная** decoy-discrimination radar, discrimination radar
~, **распоряди́тельная** 1. regulating station (rly) 2. control (radio net)
~, **ретрансляцио́нная** relay station, rebroadcast station
~, **ручна́я** manual exchange (comms)
~, **ручна́я водола́зная** hand-pump diving station
~ **с автомати́ческим сопровожде́нием це́ли, радиолокацио́нная** automatic tracking radar
~ **с бы́строй развёрткой, радиолокацио́нная** rapid-scanning radar

ста́нция с высо́кой разреша́ющей способностью, радиолокацио́нная high-resolution radar
~ **с и́мпульсной модуля́цией, радиолокацио́нная** pulse-modulated radar
~ **с и́мпульсным режи́мом рабо́ты, радиолокацио́нная** pulse radar
~ **с неподви́жной анте́нной, радиолокацио́нная** fixed array radar
~ **с неподви́жной многоэлеме́нтной анте́нной, радиолокацио́нная** fixed array multi-element radar
~ **с непреры́вным излуче́нием** continuous wave (CW) set (radio/radar)
~ **с непреры́вным излуче́нием эне́ргии, радиолокацио́нная** continuous wave radar, C.W. radar
~ **с селе́кцией дви́жущихся це́лей, радиолокацио́нная** moving-target indication radar, fixed-target rejection radar
~ **с синтези́рованной анте́нной решёткой, радиолокацио́нная** synthetic array radar
~ **с фазиро́ванной (анте́нной) решёткой, радиолокацио́нная** phased-array radar
~ **с часто́тной модуля́цией, радиолокацио́нная** frequency-modulated radar, FM radar
~, **самолётная радиолокацио́нная** airborne radar set
~, **сантиметро́вая радиолокацио́нная** centimetre-band radar, super-high-frequency radar, SHF radar
~, **сверхдлинново́лновая радиолокацио́нная** long-wave radar
~ **свя́зи** communications centre
~ **систе́мы оборо́ны пло́щади, радиолокацио́нная** air defence radar
~ **систе́мы опознава́ния "свой-чужо́й", радиолокацио́нная** identification friend or foe radar, IFF radar
~ **слеже́ния за це́лью** target-tracking radar
~ **слеже́ния, радиолокацио́нная** tracking radar
~ **слепо́го бомбомета́ния** blind-bombing radar
~ **снабже́ния** railhead

ста́нция со скани́рующей в вертика́льной пло́скости анте́нной, радиолокацио́нная nodding radar, vertical scanning radar
~ **сопровожде́ния** tracking radar
~ **сопровожде́ния раке́т(ы), радиолокацио́нная** missile-tracking radar
~ **сопровожде́ния це́ли** target-tracking radar
~ **сопровожде́ния це́ли, радиолокацио́нная** target-tracking radar
~, **сортиро́вочная** marshalling yard (rly)
~, **стыкова́я** point of junction between railway systems, junction station (rly)
~, **такти́ческая радиолокацио́нная** tactical radar
~, **тало́нная** checking station (rly)
~, **телегра́фная** telegraph station, telegraph office
~, **телеметри́ческая** telemetering station
~, **телефо́нная** telephone exchange
~, **теплопеленга́торная** thermal bearing set, thermal DF set
~ **техни́ческого обслу́живания и ремо́нта** servicing and repair station
~ **токораспределе́ния** power distribution station
~ **то́чного сопровожде́ния** precision-tracking radar
~, **трансляцио́нная** 1. relay station 2. repeater station
~, **трёхкоордина́тная радиолокацио́нная** three-dimensional radar
~, **трёхме́рная радиолокацио́нная** three-dimensional radar
~ **трубопро́вода, коне́чная** pipeline terminal, pipe head
~, **узлова́я** railway junction, junction station
~ **управле́ния артиллери́йской стрельбо́й, радиолокацио́нная** gun-laying radar, gun-directing radar
~ **управле́ния огнём** fire-control radar
~ **управле́ния огнём, радиолокацио́нная** fire-control radar
~ **управле́ния перехва́том** ground controlled interception radar, GCI radar
~ **уче́бных поме́х** jamming trainer, ECM trainer
~, **учрежде́нская телефо́нная** private branch exchange, PBX (teleph)

стáнция формирования поездóв railway marshalling depot
~ **ЦБ** central-battery exchange (teleph)
~ **целеуказáния** acquisition radar
~ **целеуказáния, радиолокациóнная** tactical control radar, acquisition radar
~, **централизóванная заправочная** centralised refuelling station
~, **центрáльная телефóнная** telephone exchange
~ **черни́льной зáписы на шесть звуковы́х постóв** six-microphone sound-ranging pen recorder
~, **эвакуациóнная** casualty clearing station
~, **электри́ческая** 1. electric power station
2. generating set
~, **электропитáющая** power station, electric power station
стáпел для подвéсных систéм airborne-training gantry
старéние ageing, seasoning
стáрица ox-bow lake, mort lake (topo)
старослу́жащий senior soldier
старт 1. start
2. launch (missile)
~, **автомати́ческий** automatic launch (GW)
стáртер starter
~, **компрéссорный** air motor starter (diesel engines)
~, **ножнóй** kick starter
~, **педáльный** kick starter
~, **пневмати́ческий** compressed air starter
~, **электри́ческий** electric starter
~, **электроинерциóнный** inertia electric starter, non-axial starter (engines)
стáртер-генерáтор starter-generator
стартови́к launcher number (missile crew)
стáртовый starting, starter, launcher, launching, take-off, booster
старшеку́рсник senior-class student
стáрший senior, superior
~ **автомоби́ля** vehicle commander
~ **по звáнию** senior in rank
старшинá 1. sergeant-major, warrant-officer (army)
2. petty officer (navy)
3. warrant officer (air force)
~ **1-й статьи́** petty officer (Sov navy)
старшинá 2-й статьи́ leading seaman (Sov navy)
~, **батальóнный** regimental sergeant-major, RSM
~, **батарéйный** battery sergeant-major, BSM (arty)
~, **глáвный** chief petty officer (Sov navy)
~ **рóты** company sergeant-major
~ **сверхсрóчной слу́жбы** extended-service warrant officer
старшинствó seniority
стáскивание мин mine pulling (exploding of mines as they are discovered)
стáтика statics
стати́стика statistics
стати́ческий static
стáтор stator (elec)
статоскóп statoscope
статья́ расхóда item of expenditure
~, **расхóдная** expenditure account number
~ **смéты** budget item
стационáр 1. permanent out-station
2. hospital
стационáрность permanence (of location etc)
стационáрный static, fixed
ствол 1. barrel/tube (wpn)
2. trunk (tree)
~, **бикали́берный** conical barrel (arty, ord)
~, **быстросменя́емый** quick-change barrel
~, **быстросъёмный** quick-change barrel
~ **в сбóре** barrel assembly
~, **винтóвочный** rifle barrel
~, **вкладнóй** sub-calibre tube (ord)
~, **вспомогáтельный** sub-calibre tube (ord)
~, **гладкостéнный** smooth-bore barrel
~, **запáсный** spare barrel
~, **испы́танный стрельбóй** proved barrel (ball)
~, **моноблóчный** monobloc barrel
~, **нарезнóй** rifled barrel
~, **некреплённый** monobloc barrel (ord)
~, **облегчённый** light barrel
~, **опрóбованный стрельбóй** proved barrel (ball)
~ **ору́дия** gun barrel
~, **подкали́берный** sub-calibre tube (ord)

ствол, пожа́рный branch (pipe end of fire-fighting hose)
~, приставно́й sub-calibre barrel
~, ребри́стый finned barrel (MG)
~ с кожухо́м jacketed barrel
~, свобо́дный loose barrel (ord)
~, скреплённый hooped barrel, bound barrel, built-up barrel, hooped gun-body
~, скреплённый с про́волокой wire-wound gun body
~ со свобо́дной трубо́й barrel with loose tube
~ со свобо́дным ле́йнером barrel with loose liner
~, уче́бный sub-calibre tube
~, уче́бный малокали́берный sub-calibre barrel (trg)
~, хроми́рованный chromium-plated barrel
ство́лик barrel, tube
~, вкладно́й sub-calibre barrel, sub-calibre adapter
~, малокали́берный sub-calibre barrel
ствол-монобло́к monobloc-type barrel
створ мо́ста, осево́й bridge axis line
~, осево́й bridge axis line
~ я́корных ли́ний anchor line transit (br)
ство́рка flap
створоско́п alidade rule (surv)
ство́ры, светові́е alignment lights (tac) (Sov used in pairs to show direction of attack at night)
стеара́т ли́тия lithium stearate
сте́бель 1. shank, shaft
2. body (fuze)
3. pillar (gunsight)
~, дугово́й arc shaft (ord)
~ затво́ра bolt cylinder (SA)
~ прице́ла sight bracket, sight arm shank
~ прице́ла, дугово́й curved elevation bar
~ стрелы́ towbar, towing bar, drawbar
~ уда́рника striker rod, hammer arm (SA)
стёганый quilted
стёкла очко́в eyepiece discs (respirator)
стекло́, армиро́ванное wired glass
~, безоско́лочное shatter-proof glass, splinter-proof glass, safety glass
~, бронево́е bullet-proof glass, armoured glass
~, ветрово́е windscreen
стекло́, жи́дкое waterglass, sodium silicate solution
~, за́днее rear window
~, ма́товое ground glass
~, не пробива́емое пу́лями bullet-resistant glass
~, небью́щееся shatter-proof glass, safety glass
~, непоте́ющее non-dimming eye-piece (respirator)
~, органи́ческое organic glass, transparent plastic
~, пере́днее windscreen (veh)
~, покры́тое специа́льным соста́вом, опти́ческое coated optics
~, противоотбле́сковое anti-glare glass
~, пулесто́йкое bullet-resistant glass
~, слои́стое laminated glass
~ три́плекс safety glass, triplex safety glass
~, увеличи́тельное magnifying glass
стеклобло́к, многосло́йный смотрово́й laminated glass vision block
стекловолокно́ glass fibres, fibreglass, glass wool
стеклообогрева́тель de-icer, demister
стеклоочисти́тель windscreen wiper
~, пневмати́ческий suction-type windscreen wiper
~ со стеклоомыва́телем windscreen wiper with washer
стеклоподогрева́тель de-icer, demister
стеклотка́нь fibre-glass cloth
~, яче́истая fibre-glass honeycomb
стелла́ж 1. shelving, racking
2. stand
~, откры́тый open racking
~, универса́льный многоя́русный all-purpose multiple-storage rack
~, я́щичный binning
сте́лька insole
~, во́йлочная felt insole
стена́ 1. wall
2. barricade block (fd wks)
~, напо́льная side of field works facing enemy, wall facing enemy (of building, installation etc)
стенд 1. stand
2. showcase
3. test bench
~ для испыта́ния test bench, test bed
~, испыта́тельный test stand, test bed
~, контро́льный test bench

стенд, поворóтный tilt platform
~, сбóрочный assembly stand
стéнка, вертикáльная side plate of saddle, side member (ord)
~, волноотбóйная sea wall
~, забóрная abutment wall
~ канáла стволá barrel wall
~ магазúна, зáдняя rear guide of automatic loader (AA ord)
~ магазúна, передняя front guide of automatic loader (AA ord)
~, подпóрная revetment, retaining wall
~ сзáди мишéней backstop, bank (firing range)
~, швéдская wall with wall-bars (gym)
стенограмма shorthand report
стенóграф shorthand writer, shorthand typist (male)
стенографúстка shorthand writer, shorthand typist (female)
стенотипúстка shorthand typist (female)
стенотрóн stenotron (Sov electron valve)
стéол "Steol" (Sov hydraulic fluid)
~ М "Steol M" (glycerine and ethyl-alcohol mixture used as hydraulic fluid)
стéпень 1. degree, grade
2. class
3. power (maths)
~ боевóй готóвности degree of readiness
~ измельчéния degree of atomization
~ ожóга, пéрвая first degree burn
~ распылéния degree of atomization
~ секрéтности security classification
~ сжáтия compression ratio (ICE)
~ твёрдости degree of hardness
~ утрáты трудоспосóбности degree of disability
~ уязвúмости degree of vulnerability
стерадиáн steradian (maths)
стереовысотомéр stereoscopic height and range finder
стереогониóметр stereogoniometer
стереодальномéр stereoscopic rangefinder
стереодальномéр-высотомéр stereoscopic height-and-range-finder
стереóметр stereometer
стереопáра stereoscopic pair, stereopair (photo)
стереопланигрáф stereoplanigraph (photo surv)
стереоскóп stereoscope
стереоскóп, лúнзово-зеркáльный lens and mirror stereoscope
стереоскопúст anti-aircraft artillery range taker
стереоскопúчность stereoscopy
стереоснúмок stereoscopic photograph
стереосъёмка stereoscopic photography
стереотрубá battery commander's OP telescope, stereoscopic telescope, "donkey's ears"
~, артиллерúйская battery commander's OP telescope, stereoscopic telescope, "donkey's ears"
стереофотограмметрúческий stereophotogrammetric
стереофотоснúмок stereopair, anaglyph
стерженёк предохранúтеля подвúжных рéек feed rod plunger (automatic loader) (ord)
стéржень rod, shaft, bar, shank
~ бýфера, опóрный buffer stop (ord)
~ возврáтно-боевóй пружúны return spring rod
~ возврáтного механúзма return spring cap, return spring centring plunger
~ гáзового пóршня gas piston rod (SA)
~, грибовúдный mushroom head, axial vent (obturator, B L ord)
~ дистанциóнного удáрника time pellet stem (of time and percussion fuze)
~ для постанóвки звéньев гýсениц, направля́ющий track pin drift
~ для постанóвки пáльцев трáков, направля́ющий track pin drift
~ замедлúтеля, инерциóнный graze pellet stem, inertia pellet stem (ammo)
~ затя́жки лéнточного тóрмоза brake-band pull rod
~ изоля́тора insulator spindle
~, колéнчатый chamber cleaning rod (MG)
~, контáктно-поворóтный tilt rod (on anti-clearance mine)
~, маслоизмерúтельный oil dipstick
~, направля́ющий main spring guide (SA, pistol)
~, переводнóй speed change rod
~ подвéски suspension rod
~, поршневóй piston rod
~, предохранúтельный 1. arming stem (of butterfly bomb)
2. detent (fuse)

сте́ржень, прице́льный aiming post
~ **пружи́ны спусково́го рычага́, направля́ющий** sear spring post (MG) (SA)
~ **пружи́ны спусково́й тя́ги, направля́ющий** trigger spring plunger (SA)
~, **регули́рующий** control rod
~ **рукоя́тки затво́ра** LBM catch plunger (ord)
~, **скру́чивающийся** torsion bar, torsion shaft
~, **то́пливный** fuel rod (nucl)
~, **топливоизмери́тельный** fuel dipstick
~, **торсио́нный** torque bar
~, **уда́рный** striker (ammo), striker stem (fuse)
~, **ура́новый** uranium rod (nucl)
~ **экстра́ктора** extractor shank (revolver)
стерилиза́тор sterilizer
стерилиза́ция sterilization
стерилизова́ть to sterilize
стеснённый confined, limited, hampered
стесни́ть (см стесня́ть)
стесня́ть/стесни́ть to put restraint on, to hinder; **стесня́ть/стесни́ть манёвр** to restrict movement
стетоско́п 1. stethoscope (med) 2. defectoscope (eng)
сте́чкин Stechkin 9mm machine pistol (Sov)
стильб stilb (surface brightness unit: one candle per sq cm)
сти́мул stimulus, impetus
стимуля́тор stimulant
сти́рка laundering
стифна́т свинца́ lead styphnate
стихи́я the elements (collective)
стог се́на haystack
"Стой!" "Halt!"
"Стой! Кто Идёт?" "Halt! Who Goes There?"
сто́йка 1. post, leg, stanchion, side tree (eng) 2. bay, rack (comms) 3. stake, column, pillar 4. picket, revetment picket
~, **амортизацио́нная** shock absorber strut
~ **дифференциа́льных систе́м** differential system rack (comms)
сто́йка для крепле́ния ствола́ barrel boss (pistol)
~ **затво́рной ра́мы** piston post (SA)
~ **индивидуа́льного преобразова́ния** channel translation rack (comms)
~ **несу́щих и контро́льных часто́т** carrier and pilot frequency rack (comms)
~, **ни́зкая** full crouch position
~, **основна́я** position of attention (ski)
~, **пала́точная** tent pole
~, **пери́льная** handrail support
~ **прице́ла** sight shank (SA)
~ **спусково́й скобы́** slide locking projection (pistol)
~, **сре́дняя** semi-crouch position (ski)
~, **строева́я** position of attention
~, **телефо́нная** roof standard (Sov for telephone lines)
~ **тона́льного вы́зова** ringer rack, ringer bay (comms)
~, **углова́я** corner post
~ **четырёхпроводно́й коммута́ции** four-wire switching rack (comms)
~ **што́ка га́зового по́ршня** piston post (SA)
сто́йки прице́ла telescope carrier (gunsight)
сто́йкий 1. staunch, firm 2. persistent 3. durable
сто́йкость 1. steadfastness, steadiness 2. resistance (power of), resistability 3. stability (expl) 4. durability
~, **детонацио́нная** anti-knock value (fuel)
~ **к излуче́нию** radiation tolerance
~ **ОВ** persistence (of war gas)
~ **органи́зма к облуче́нию** biological radiation tolerance
~, **терми́ческая** thermal stability
~, **хими́ческая** chemical resistance
сто́йло bay (in workshop)
стол, пусково́й launching pad
~, **ста́ртовый** launching platform (missile)
столб pole, post
~ **ды́ма, грибови́дный** mushroom-shaped cloud of smoke (nucl)
~, **километро́вый** kilometre post
~ **пы́ли** column of dust
~, **телегра́фный** telegraph pole

сто́лбик 1. peg
2. small pole
~, контро́льный marker (route etc)
~, пике́тный marking peg
~, селе́новый selenium pile (electron)
столбня́к tetanus, lock-jaw
столе́шница table top
сто́лик, прокла́дочный plotting table, chart table
столкнове́ние encounter, collision, clash
~ авангáрдов first encounter (tac)
~, боево́е engagement, encounter, bump (tac)
~, вооружённое armed conflict
~, неожи́данное unexpected encounter, surprise encounter
столкну́ться (см ста́лкиваться)
столо́вая dining room, dining hall, mess-room
~, солда́тская cookhouse
столя́р joiner
стометро́вка (разг) 1. map to scale 1/10000
2. hundred-yard race
"Стоп!" "Stop!"
стопа́ 1. sole (of foot)
2. step (walk)
3. pile, ream
стопи́н quickmatch (expl)
стоп-кран 1. emergency brake (rly)
2. cut-off valve
сто́пор lock, locking device, stop, stopper, detent, catch, pawl
~ бараба́на cylinder stop (revolver)
~ ба́шни turret (traversing) lock
~ взво́да sear (of firing mech)
~ газово́й ка́моры gas-cylinder lock (SA)
~ затво́ра breech-block detent
~ о́си взво́да уда́рника check plunger (breech mech)
~ о́си ра́мы затво́ра breech-block detent
~ отсе́чек feed gate stop (automatic loader)
~, предохрани́тельный safety stop
~ регуля́тора gas regulator adjustment lug
~ ствола́ breech ring stop
~ хо́да axle-tree locking gear
~, центробе́жный centrifugal bolt (fuse)
сто́порить/засто́порить to stop, to check, to brake
сто́пор-ныря́ло setback pin
стоп-сигна́л stop signal
стопсигна́л, светово́й stop lamp
сторожи́ть to watch, to guard
сторона́, вою́ющая belligerent
~ горизо́нта cardinal point
~ зу́ба, конта́ктная driving side of tooth (gearing)
~ зу́ба, нерабо́чая coast side of tooth (eng)
~, наве́тренная windward side
~, напо́льная side of field work facing the enemy
~, обра́тная reverse side
~ погру́зки loading side
~, подве́тренная downwind, leeward side
~, рабо́чая drive side (eng)
~ стро́я, ты́льная rear (drill)
~, ты́льная rear
~ хо́да leg of traverse (surv)
сто́я standing
стоя́нка 1. halt, place of halt
2. parking area (veh)
3. hard standing (ac)
~, вертолётная helicopter pad
"СТОЯ́НКА ЗАПРЕЩЕНА́" "NO PARKING"
стоя́нка та́нков tank harbour, leaguer
стоя́ть to stand, to be situated, to be, to stop, to be at a standstill, to stay, to live (at), to stand up for, to hold (one's own) (imp only); **стоя́ть на посту́** to be on duty at a post; **стоя́ть на я́коре** to lie at anchor, to ride at anchor; **стоя́ть по строево́й сто́йке** to stand at attention
страви́ть (см стра́вливать (стравля́ть))
стра́вливать (стравля́ть)/страви́ть to let off, to release (pressure, steam etc), to veer, to pay out (rope etc)
стра́жа guard, watch
страна́ све́та cardinal point (surv)
страте́г strategist
стратеги́ческий strategic
страте́гия strategy, grand strategy
~ войны́ на истоще́ние strategy of attrition
~ контрси́лы counterforce strategy
~ непрямы́х де́йствий strategy of the indirect approach
стратосфе́ра stratosphere
страх пе́ред возме́здием fear of retaliation

страхо́вка верёвкой belaying (mountaineering)
~ че́рез крюк piton belay (mountaineering)
~ че́рез ска́льный вы́ступ rock belay (mountaineering)
стре́жень natural channel (deepest part of river)
стрела́ 1. jib, derrick, boom
2. drawbar, doubling bar
3. perch
4. arrow, pointer, indicator
5. trail (ord)
~, двутавро́вая "I" beam jig, "H" beam jig
~ кра́на crane jib
~ прове́са dip (sag of cable from horizontal)
~, рабо́тающая live boom (crane)
~ рессо́ры spring arch, spring camber
стре́лка 1. needle, pointer
2. point (rly)
3. perch
4. trail (ord)
~ ко́мпаса, магни́тная magnetic compass needle
~, ко́мпасная compass needle
~, контро́льная control point setting pointer (FCE)
~, магни́тная magnetic needle
~, оруди́йная gun reader arm (arty)
~, прице́льная sight reader arm (arty)
~ угло́в пики́рования angle of dive lever pointer (AA course and speed sight)
стрелко́вый rifle, infantry
стрелови́дность sweepback
стрело́к 1. rifleman, infantryman
2. gunner (tk, ac etc)
~, ба́шенный turret gunner (tk)
~, ме́ткий marksman
~, отли́чный marksman
~, пере́дний bow gunner (AFV)
~, сверхме́ткий expert marksman
стрело́к-команди́р маши́ны gunner-commander
стрело́к-лы́жник rifleman on skis
стрело́к-наво́дчик gun layer
стрело́к-пехоти́нец infantry rifleman
стрело́к-разря́дчик first-class shot
стре́лочник pointsman (rly)
стрельба́ fire, firing
стрельба́, автома́тная 1. fire from automatic weapons
2. sub-machine gun fire
3. automatic fire
~, артиллери́йская artillery fire
~ батаре́и battery firing
~ бе́глым огнём rapid fire
~ без подгото́вки impromptu fire
~, беспузы́рная firing without bubble (submarine torpedo firing)
~, боева́я field firing
~ боевы́ми патро́нами live firing
~, бы́страя snap shooting
~ в ти́ре indoor shooting
~ в упо́р firing at point blank range
~, возду́шная aerial gunnery
~, группова́я salvo fire
~ "жа́лом" firing with fixed firing pin (mor)
~ за́лпами salvo firing
~, за́лповая volley fire (SA)
~, зачётная classification firing, recorded shoot
~, зени́тная anti-aircraft fire
~ из винто́вки rifle shooting
~ из вкладно́го ство́лика sub-calibre firing
~ из вспомога́тельных стволо́в sub-calibre firing
~ из пистоле́та pistol firing
~ из положе́ния лёжа firing from the prone position
~ из положе́ния с коле́на firing from kneeling position
~ из положе́ния сто́я firing from standing position
~ из уче́бных стволо́в sub-calibre firing
~ из-за укры́тия firing from cover
~, имитацио́нная simulated fire
~, ине́ртная practice firing
~, исчи́сленная predicted fire
~, ка́рликовая sub-calibre fire
~, контрбатаре́йная counterbattery fire
~, контро́льная 1. check firing (weapons)
2. qualification shoot (pers)
~, ку́чная grouping (gunn)
~ лёжа firing from lying position
~ малокали́берными патро́нами, трениро́вочная small-bore practice shooting
~, ме́ткая accurate fire, marksmanship
~, морти́рная high-angle fire

стрельба́ на большу́ю да́льность long-range fire
~ **на да́льние диста́нции** long-range fire
~ **на да́льность прямо́го вы́стрела** point-blank fire
~ **на ма́лые диста́нции** short-range fire
~ **на миниатю́р-полиго́не** miniature range practice
~ **на подавле́ние** neutralizing fire
~ **на полиго́не** range firing
~ **на полиго́не, уче́бная** range practice
~ **на пораже́ние** fire for effect
~ **на разруше́ние** destruction fire (buildings, vehicles, installations)
~ **на рикоше́тах** ricochet fire
~ **на уничтоже́ние** annihilation fire (personnel, wpns)
~ **на ходу́** fire on the move
~, **навесная** high trajectory fire, high angle fire
~ **навски́дку** snap shooting
~, **насти́льная** flat trajectory firing, very low angle fire (Sov below 20 degree angle)
~, **нача́льная** range practice firing
~, **нема́я** dry firing
~, **непроизво́льная** firing by run-away gun
~ **одино́чными вы́стрелами** single shot fire, firing single rounds
~, **о́пытная** experimental firing, proof firing
~, **отвёрнутая** throw-off firing (AA gunnery practice)
~ **о́чередями** 1. firing in bursts (automatic wpn) 2. firing rounds of battery fire (arty)
~, **пла́новая** programme shoot, planned shoot, pre-arranged fire
~ **по береговы́м це́лям** naval gunfire, firing at shore targets
~ **по возду́шным це́лям** anti-aircraft fire
~ **по мише́ням** target shooting
~ **по наблюда́емой це́ли** observed fire
~ **по неви́димой це́ли** indirect fire
~ **по ненаблюда́емой це́ли** unobserved fire
~ **по одино́чным живы́м це́лям** sniping
~ **по отде́льным це́лям** precision fire
~ **по пло́щади** area fire, area shoot
~ **по провéшенному направле́нию** fire on fixed line

стрельба́ по случа́йной вы́годной це́ли opportunity fire
~ **по табли́це огня́** schedule fire
~ **"по штурмово́й"** close-target shooting (AA)
~, **показна́я** demonstration fire
~ **полево́й артилле́рии** field gunnery
~, **полиго́нная** range firing
~, **практи́ческая** practice firing, field firing
~, **предзачётная** preliminary firing
~ **при наблюде́нии с во́здуха** air-observed fire
~, **пристре́лочная** registration shoot
~, **пристре́льная** precision shoot, aimed fire
~, **прице́льная** aimed fire
~, **прове́рочная** check fire (gunn)
~ **прямо́й наво́дкой** direct fire
~, **пулева́я** rifle shooting (sport)
~, **пу́льная** gunnery practice using aiming rifle, aiming rifle firing (nav gunn)
~ **реакти́вными снаря́дами, за́лповая** rocket salvoes
~, **режи́мная** prearranged fire
~, **рикоше́тная** ricochet fire
~, **руже́йная** rifle fire
~ **с больши́м угло́м возвыше́ния** high angle fire
~ **с закры́той огнево́й пози́ции** indirect fire (gunn)
~ **с зерка́льным отворо́том** 180-degree throw-off shoot
~ **с иску́сственным рассе́иванием по да́льности** searching fire
~ **с иску́сственным рассе́иванием по высоте́** searching fire
~ **с коле́на** firing from a kneeling position
~ **с коро́тких остано́вок** fire from short halts (tk gunn)
~ **с лыж** firing on skis
~ **с наблюде́нием** observed fire
~ **с назе́мным наблюде́нием** ground-observed fire
~ **с полузакры́той огнево́й пози́ции** semi-indirect shooting
~ **с по́мощью откры́того прице́ла** open-sight firing
~ **с предвари́тельной подгото́вки да́нных** predicted fire, predicted shoot

стрельба́ с прикла́дом прижа́тым к бо́ку firing from the hip
~ с пристре́лкой registered fire
~ с упо́ром прикла́да в бедро́ fire from the hip
~ с упо́ром прикла́да в плечо́ fire from the shoulder
~ с фотопулемётом camera-gun firing
~ с хо́да firing on the move
~ с хо́ду fire on the move
~ се́риями ripple fire (GW)
~, ско́ростная rapid fire, quick firing
~, сна́йперская sniping fire
~ со ста́нцией оруди́йной наво́дки radar controlled fire
~, состяза́тельная competition shooting, match shooting
~, спа́ренная coaxial fire
~ спо́собом выжида́ния це́ли engaging a likely approach (inf)
~ спо́собом сопровожде́ния це́ли swinging traverse fire (MG)
~ "спу́ском" firing using firing mechanism (mor)
~ сто́я firing from standing position
~, трениро́вочная range practice
~, уда́рная percussion fire
~, уче́бная range practice
~, уче́бная боева́я target practice, range practice
~, уче́бно-показа́тельная fire-power demonstration shoot
~ холосты́ми патро́нами blank ammunition firing
стре́льбище range, range firing, firing range
~ для стрельбы́ боевы́ми патро́нами live firing range
~, электрифици́рованное electric target range
стре́льбы range course, range practice
~, боевы́е field firing
~, боевы́е одино́чные individual field firing, individual battle practice firing (inf trg)
~, нача́льные range practice firing
стрельбы́, одино́чные боевы́е individual field firing, individual battle practice firing
стре́льбы, практи́ческие field firing, practice firing
~, уче́бные firing training
стре́льчатый tapered, pointed, arched
стре́ляный spent, fired (ammo)
стреля́ть/вы́стрелить to shoot, to fire; **стреля́ть/вы́стрелить в упо́р** to fire point blank; **стреля́ть/вы́стрелить за́лпами** to fire volleys **стреля́ть/вы́стрелить на аво́сь** to fire at random; **стреля́ть/вы́стрелить наверняка́** to shoot to kill; **стреля́ть/вы́стрелить науга́д** to fire at random
стреля́ющий 1. firer
2. TCO, Tactical Control Officer (arty)
3. LCO, Launcher Control Officer (GW)
4. firing, shooting
стреми́тельность rapidity, swiftness, impetuosity
стреми́тельный rapid, swift, impetuous
стреми́ться to rush, to strive for, to aim at, to aspire (imp only)
стремле́ние desire, inclination, tendency, aspiration
стремя́нка 1. U-bolt, rebound clip
2. light ladder
~ крепле́ния U-bolt
~ рессо́ры spring clamp
стре́нга strand (rope)
стрептомици́н streptomycin
стри́нгер stringer (longitudinal member)
строб-и́мпульс gate pulse, strobe pulse (electron)
строби́рование gating, strobing, radar ranging
строб-ла́мпа gate valve, strobe valve
стробоско́п stroboscope
строево́й 1. operational, line, front-line,
2. drill, ceremonial
строе́ние, ве́рхнее 1. superstructure (br, rly)
2. overlay for reinforcing material (ice br)
~, пролётное superstructure (br)
~ пути́, ве́рхнее ballast section (rly)
~, сре́днее ба́лочное suspended bay, suspended span (br)
строи́тельство, гражда́нское civil engineering
~, доро́жное road construction
~, доро́жно-мостово́е road and bridge construction
стро́ить/постро́ить to build, to construct, to form up, to draw up (trans); **стро́ить/постро́ить паралле́льный ве́ер в основно́м направле́нии** to get the guns parallel on their zero lines

стрóиться/пострóиться to fall in, to form up (intrans)
строй 1. formed body
2. formation, order, line
3. regime, system
~, боевóй order of battle, ORBAT
~ в лúнию, расчленённый dispersed formation in line (Sov drill)
~, двухшерéножный two ranks, two deep (drill)
~, зáмкнутый close order, close formation
~ "клин" arrowhead formation
~, похóдный in column
~, простóй single line formation
~, развёрнутый line (fmn)
~, разóмкнутый open order formation (Sov drill formation - interval between files one pace or more)
~, рассыпнóй extended order (fmn)
~, расчленённый deployed formation
~ "ромб" diamond formation
~, сóмкнутый 1. mass formation, close formation
2. close dressing (drill)
стрóйка 1. building, construction
2. construction site
строймателиáлы building materials
строп 1. sling, hoisting sling, strop
2. rigging line (para)
~ подъёма lifting sling
~, простóй bale sling, strop
стропá shroud line, suspension line (para)
~, тросовáя 1. cable sling, rope sling
2. cable harness (para)
стропóвка lifting by sling
стрóпы, пáрные snotters (pair of)(cargo handling ropes)
струбцúнка clamp, cramp, adjustable clamp
струг 1. plane (carpentry)
2. grader (rd constr)
3. flat-bottomed riverboat
~, дорóжный blade grader
~, путевóй grader (rd constr)
стрýжка shaving, chip
стружóк, ручнóй spokeshave
структýра, организациóнно-штáтная organization and establishment
струя́ stream, flow, jet, spray
~ горю́чего fuel jet, fuel spray
~, кáпельная spray of droplets
струя́, кинжáльная narrow high-pressure jet
~, кумулятúвная hollow-charge jet
~, óгненная jet of burning fuel (flame thrower)
~, рассéивающая spray
~, реактúвная jet efflux, jet exhaust
~, спýтная wake (naut)
стýдень, гремýчий nitrogelatine
стýжа hard frost, bitter cold
стук knock, clatter, rap
~, детонациóнный knocking, pinking (ICE)
~ пóршня piston slap (ICE)
ступéнь 1. step
2. fire step
3. grade
4. stage
~, основнáя main stage (missile)
~, отрабóтавшая expended stage (GW)
~ параллéльности convergence difference
~, пушпýльная push-pull stage (electron)
~ ракéты rocket stage, missile stage
~, стрелкóвая fire step
ступéнька, стрелкóвая fire step
ступúца boss, hub
~ дúска фрикциóна clutch centre
~ колесá wheel hub
~ кривошúпа crank boss
стык 1. joint, butt, junction
2. boundary (between units, fmns etc)
~ дорóг road junction
~, свáренный welded joint
стыкóвка joining, jointing, docking (space vehs)
сты́чка skirmish, engagement
стэнд stand
стя́жка 1. tightening device, turnbuckle
2. leg stay (mor bipod)
~, винтовáя screw coupling
стя́жка-мýфта straining screw, turnbuckle
субординáция order of seniority, subordination
субподря́дчик sub-contractor
сувóровец Suvorov cadet (Suvorov mil academies in Moscow and Tula)
суглúнистый clay and sand (of soil)
суглúнок loam
сугрóб snowdrift
~, снéжный snowdrift

суд, вое́нно-полево́й field court-martial, drumhead court-martial
~, вое́нный military court, court martial
~, полево́й field court-martial
~ че́сти, офице́рский officer's court of honour (Sov, convened by Commanding Officer to try officers for minor offences)
суде́йство 1. judging
2. umpiring (ex)
су́дно, букси́рное towing vessel, tug
~, водоналивно́е water boat
~, госпита́льное hospital ship
~, деса́нтное amphibious ship
~, затону́вшее wreck, sunken vessel
~, нефтеналивно́е fuel tanker vessel (ship)
~, пехо́тно-деса́нтное infantry landing craft, LCI
~ подде́ржки, деса́нтное landing craft support
~, санита́рное hospital ship
~ сре́днего тонна́жа, деса́нтное landing ship medium
~, тамо́женное customs vessel
~, та́нко-деса́нтное tank landing ship, LST
~, торго́вое merchant ship
~, тра́нспортно-вы́садочное landing ship logistic, LSL
~, тра́нспортное деса́нтное 1. landing ship logistic, LSL
2. landing ship assault, LSA
~, уда́рное деса́нтное minor landing craft, assault landing craft
~, штурмово́е деса́нтное mechanised landing craft, LCM
су́дорога convulsion, cramp
судохо́дность navigability
судохо́дный navigable
суже́ние зрачка́ constriction of the pupil, contraction of the eye (symptom of nerve gas poisoning)
~, относи́тельное reduction of area (metall)
сукно́, шине́льное military cloth, serge
сульфури́л, хло́ристый sulphuryl chloride
су́мерки twilight
~, вече́рние evening twilight
су́мка pouch, satchel
~, грана́тная grenade pouch
~ для магази́нов magazine pouch
су́мка, лине́йная line repair bag, line repair kit, line repair satchel (sigs)
~, медици́нская войскова́я medical aid kit
~, парашю́тная parachute pack
~, патро́нная ammunition pouch
~, переме́тная saddle bag
~, полева́я map case
~, поясна́я waist pouch (carried on belt)
~, противога́зовая respirator haversack
~, противохими́ческая CW first aid kit
~, санита́рная first-aid satchel, medical pack
сумма́рный total, combined
супергетероди́н superheterodyne
суперо́ртикон image orthicon (elec)
суперрефра́кция super-refraction, trapping (radio)
су́песок sandy loam
супесча́ный crumbly and sandy (of soil)
су́песь sandy loam
су́ппорт 1. support
2. rest, carriage (lathe)
су́рик red lead
~, свинцо́вый red lead
суспе́нзия suspension
су́тки 24-hour period (starting midnight)
~, и́стинные solar day
~, лу́нные lunar day
сутода́ча day's ration
су́точные ration allowance
су́точный day's, 24-hour, diurnal, daily
суха́рик split collar (eng)
суха́рь 1. dry biscuit
2. thrust bearing, insert piece
суходо́л dry valley, dry gap, wadi
сухопа́рник steam dome (locomotive)
сухопу́тный land, ground
су́ша dry land
су́шеный dried, dehydrated
суши́лка drying room
сфе́ра 1. sphere
2. zone, realm, scope
~ компете́нции sphere of responsibility
сфери́чность sphericity
сформи́ровать (см формирова́ть)
сфотографи́ровать (см фотографи́ровать) to photograph
схва́тка 1. skirmish, encounter
2. brace, bracing
~, рукопа́шная hand-to-hand fighting, skirmish
схва́тывание setting (concrete)

схéма 1. plan, diagram, scheme
2. network, circuit
~, аналóговая analogue circuit
~, аэродинами́ческая aerodynamic configuration
~, балáнсная balanced circuit
~ бóя battle plan
~ в мóдульном исполнéнии, печáтная printed-circuit module
~, вычисли́тельная computer circuit
~ вы́явленных цéлей recorded targets
~, ги́бкая печáтная flexible printed circuit
~, гибри́дная hybrid circuit
~ движéния трáнспорта traffic circuit
~, двухтáктная push-pull circuit (electron)
~, дешифрáторная decoding circuit
~, диóдная diode circuit
~, заграждáющая rejector circuit
~ задéржки 1. delay network
2. gate stage (electron)
~, запоминáющая memory circuit, storage circuit
~ затя́жки tightening diagram (eng)
~, звездовáя star network (elec)
~, измери́тельная metering circuit
~, и́мпульсная pulse circuit
~, интегрáльная integrated circuit
~, интегри́рующая integrating circuit
~ кáбельных соединéний cable diagram
~, каскáдная cascade connection, cascade circuit
~ коммутáций circuit diagram
~, лабораторная breadboard (elec)
~, линéйная linear circuit
~, логи́ческая logic circuit
~ мáрша route plan and march formations
~, мáтричная matrix circuit
~ мéстности area sketch, site sketch
~, микроминиатю́рная micro-miniature circuit
~, микроминиатю́рная печáтная micro-printed circuit
~, микроэлектрóнная micro-electronic circuit
~ ми́нного пóля minefield pattern
~, многозвéнная ladder circuit
~, многослóйная multi-layer circuit, multi-planar circuit
~, многоступéнчатая multi-stage circuit
~, многофáзная polyphase circuit

схéма, многофункционáльная multi-function circuit
~, мóдульная modular circuit
~, молекуля́рная molecular circuit
~, моноли́тная monolithic circuit
~, монтáжная breadboard (elec)
~, мостовáя bridge circuit (elec)
~ на керáмике, печáтная ceramic printed circuit
~ на нéскольких кристáлликах multi-chip circuit
~ на однóм кристáллике single-chip circuit
~ на оснóве твёрдого тéла solid circuit
~ на полупроводникóвых прибóрах transistor circuit
~ на твёрдом тéле solid circuit
~ на тóлстых плёнках thick-film circuit
~ на тóнких плёнках thin-film circuit
~ на транзи́сторах transistorized circuit
~, напылённая thin-film circuit
~, нейтроди́нная neutrodyne circuit
~, нелинéйная non-linear circuit
~, несимметри́чная unbalanced network
~, нормáльная аэродинами́ческая normal configuration (GW)
~, обостря́ющая peaking circuit (elec)
~, óбщая 1. general plan
2. block diagram (elec)
~, объёмная three-dimensional circuit
~ огня́ fire-chart, tactical map (at arty HQ showing important tgts, zones of fire, own tps and OPs)
~, одновáльная single-shaft configuration
~, одноперекиднáя single flip-flop circuit (electron)
~, оконéчная termination (elec)
~, опроки́дывающая flip-flop circuit
~, óптико-электрóнная optic-electronic circuit
~, óпто-электрóнная optic-electronic circuit
~, ориенти́рная sketch map, map showing local features and reference points
~ ориенти́ров range card, panorama drawn at observation post
~, перекиднáя flip-flop circuit (electron)
~, перспекти́вная panoramic sketch
~, печáтная printed circuit
~ полéй неви́димости visibility chart

схе́ма, по́лностью интегра́льная fully integrated circuit
~, помехоподавля́ющая anti-clutter circuit, interference suppression circuit (radar)
~, после́довательная series circuit (elec)
~, принципа́льная 1. actual layout, actual equipment circuit
2. block diagram, basic diagram
3. functional diagram
~, пропуска́ющая gate circuit
~, противоме́стная side-tone reduction circuit, side reduction circuit (radio)
~ рабо́ты 1. flow diagram
2. flow of data chart (FCE)
~ радиосвя́зи radio diagram
~, разве́дывательная reconnaissance report sketch
~ расположе́ния мин в ми́нном по́ле minefield pattern
~ ро́тного райо́на plan of company layout
~ с дискре́тными элеме́нтами, интегра́льная partially integrated circuit
~ свя́зи signal diagram
~ сгла́живания smoothing circuit
~ синхрониза́ции synchronizing circuit
~ сма́зки lubrication diagram
~ со структу́рой мета́лл-о́кисел-кре́мний metal-oxide-silicon integrated circuit
~, со́бранная печа́тная printed-circuit assembly
~ совпаде́ний lock-on circuit (radar)
~, строби́рующая gate circuit
~, структу́рная block diagram
~, супергетероди́нная super-heterodyne circuit
~, твёрдая integrated circuit (elec)
~, толстоплёночная thick-film circuit
~, транзи́сторная transistorized circuit
~ треуго́льника delta network
~, три́ггерная trigger circuit
~ управле́ния стрельбо́й fire control system
~ уча́стков зараже́ния радиоакти́вными оса́дками fallout pattern, radiation pattern
~, фазовраща́тельная phase-switching network
~, фазосдвига́ющая phase-shifting network
схе́ма, функциона́льная functional diagram, functional circuit
~ це́лей sketch plan of targets
~, части́чно интегра́льная partially integrated circuit
~, электри́ческая circuit diagram
~ электри́ческих соедине́ний circuit diagram
схе́ма-ка́лька map overlay
сход lift-off (missile)
сходи́ть/сойти́ to go down, to come down, to descend, to get off, to come off, to pass by, to be taken (for); **сходи́ть/сойти́ на бе́рег** to land on a beach (amph ops)
схо́дня ramp, gang-plank
схожде́ние колёс wheel toe-in
~ пере́дних колёс, обра́тное wheel toe-out
сцентри́ровать to align
сце́пка coupling
~, жёсткая rigid tow bar
~, многозве́нная articulated coupling
сцепле́ние 1. coupling, linking
2. meshing, engagement (gears)
3. clutch (mech)
4. connecting (elec)
~, автомати́ческое automatic clutch
~, гидравли́ческое hydraulic clutch
~, ди́сковое disc clutch, plate clutch
~, многоди́сковое multi-plate clutch, multiple-disc clutch
~, одноди́сковое single-plate clutch, single-disc clutch
~, трёхди́сковое triple-plate clutch
~, фрикцио́нное friction coupling
~, центробе́жное centrifugal clutch
сцинтилля́ция scintillation
счал pusher-tug barge train, barge train
счёт 1. bill
2. counting
3. grid numbering system (topo)
~ очко́в scoring (shooting)
~, теку́щий current account
счетверённый quadruple
счётчик indicator, meter
~ а́льфа-части́ц alpha counter
~, ампе́р-часо́в ampere-hour meter
~ бе́та-части́ц beta counter
~ ватт-часо́в watt-hour meter
~ вре́мени timer
~, газоразря́дный gas discharge counter, gas discharge counter tube (nucl)

счётчик Ге́йгера Geiger counter
~ **ге́йгера-мю́ллера** Geiger-Muller counter
~ **излуче́ний** radiation counter
~, **и́мпульсный** pulse counter
~, **контро́льный** monitor counter
~, **ма́сляный** oil meter
~ **оборо́тов** revolution counter, tachometer
~, **погружно́й** immersion counter (nucl)
~, **складско́й** storage meter
~, **сцинтилляцио́нный** scintillation counter
~ **ускоре́ния** acceleration meter
~ **фото́нов** quantometer
~ **числа́ оборо́тов** rev counter
~ **электро́нов** electron counter
счисли́тель computer
~ **баллисти́ческих да́нных** ballistic data computer
~ **боковы́х попра́вок** azimuth-adjusting slide rule
~, **звукометри́ческий** sound-ranging calculator
~ **ку́рса це́ли** target course computer
~ **упрежде́ний** deflection computer, lead computer
счи́тывающий высоту́ height-reader (AA gunn)
съеда́ть/съесть to corrode
съезд 1. congress, conference, meeting, convention
2. descent
3. approach
4. ramp
съёмка 1. removal, taking-off
2. survey
~, **аэрофотографи́ческая** aerial photogrammetry (surv)
~, **вое́нно-топографи́ческая** 1. military topographical survey
2. ordnance survey (Brit)
~, **вы́борочная снегоме́рная** spot survey of snow conditions
~, **глазоме́рная** (field)sketching, eye survey
~, **горизонта́льная** planimetric survey
~, **инструмента́льная** instrumental survey
~, **картографи́ческая** cartographical photography, mapping photography
~, **маршру́тная** route survey
~, **ме́нзульная** plane table survey
съёмка ме́стности, радиолокацио́нная radar mapping
~, **снегоме́рная** snow survey
~, **топографи́ческая** survey, topographical survey
~, **фотографи́ческая** photographic survey
съёмник remover, puller, replacer (tool)
~, **просто́й** extractor
~, **универса́льный** general-purpose extractor tool
съёмный 1. removable, detachable, dismountable
2. interchangeable, replaceable
съёмщик мин puller (member of mine-clearance party)
~ **отсчётов** film booker (Srg)
съесть (см съеда́ть)
сы́воротка serum
сыгра́ть (см игра́ть)
сыпь rust trace (in barrel)
сырьё, стратеги́ческое strategic raw materials
~, **то́пливное** fertile material (nucl)
сюрпри́з booby trap
~, **взрывно́й** booby trap

Т

та́бель 1. list, table
2. timetable
3. equipment table, equipment schedule
~ **вое́нного вре́мени** war entitlement equipment table
~ **вооруже́ния и снаряже́ния** equipment schedule
~ **да́льностей** range table
~ **движе́ния сро́чной операти́вной документа́ции** schedule of timed operational traffic (comms)
~ **девиа́ции** deviation table (compass)
~ **донесе́ний** report priority log (comms)
~ **иму́щества** table of stores and equipment
~ **инвента́рного иму́щества** unit equipment tables
~ **ми́рного вре́мени** peace entitlement equipment table

та́бель обмундирова́ния и снаряже́ния scale of clothing and personal equipment
та́бельный 1. authorized
2. standard
3. organic
табле́тка tablet, pill (med)
~, противорадиацио́нная anti-radiation pill
табле́тки для обеззара́живания воды́ water-sterilizing tablets
табли́ца table, chart
~ автоперево́зок, пла́новая MT timetable
~, баллисти́ческая ballistic table
~ бо́я, пла́новая operational timetable
~ вероя́тности probability factor table (ball)
~ взаимоде́йствия, пла́новая timing co-ordination table
~ волн и позывны́х frequency and callsign table
~ деса́нта, пла́новая air movement table for air landing
~, ко́довая code table, enciphering table
~ норм загру́зки loading table
~ огня́ task table (arty)
~ огня́, пла́новая fire plan chart (arty)
~ перево́да мер conversion table
~ перево́зок, пла́новая troop movement table
~, перегово́рная procedure table, procedure chart
~ перепра́вы, пла́новая assault river crossing plan
~, пересчётная conversion table
~ погру́зки, пла́новая embarkation table
~ попра́вок calibration table
~ попра́вок да́льности range correction table
~ после́довательности вы́грузки schedule of sequence of unloading (amph ops)
~ прили́вов tide table
~ радиосигна́лов brevity code chart, procedure chart (comms)
~, радиосигна́льная brevity code chart, procedure chart (comms)
~ сигна́лов brevity code chart, procedure chart (comms)
~ сма́зки lubrication table
~ стрельбы́ range table
табли́ца стрельбы́ артилле́рии artillery range table
~ стрельбы́, числова́я tabular range table
~, тара́жная table of composition of the charges (arty range tables)
~ угло́в прице́ливания elevation table
~ упрежде́ний target lead table
~ устано́вок угломе́ра для сосредото́чения огня́ table of concentration corrections (gunn)
~ эквивале́нтов conversion table
табли́цы стрельбы́, го́рные mountain firing tables
табли́чка, надкрова́тная bed card (showing particulars of occupant)
~, оруди́йная gun data board
табли́чный standard, as laid down
табло́ illuminated display, illuminated board
~ боево́го соста́ва weapons availability table
~, светово́е lamp register (teleph, teleg)
~, спра́вочное information board
табуля́тор punch-card sorting machine, tabulating machine
табу́н "tabun" (non-persistent nerve gas)
таво́т grease, lubricating grease
таво́тница 1. grease gun
2. grease cup, grease nipple, compression grease cup
тавотонагнета́тель grease gun
таёжный taiga (topo)
таз 1. basin, wash basin, pan
2. pelvis
3. bush (mech)
тайга́ taiga
та́йна 1. secret
2. secrecy
~, вое́нная 1. military secret, classified military information
2. military secrecy
~, служе́бная official secret
~ телефо́нных перегово́ров line security
та́йнопись secret writing
тайфу́н typhoon
"Так Держа́ть" "Steady!" (order to maintain same course, speed and interval)
"Так То́чно!" "Yes, Sir!", "Yes!"
такела́ж loading tackle, rigging

такт 1. tact
2. stroke (mech)
~ **впу́ска** intake stroke, suction stroke (ICE)
~ **вса́сывания** intake stroke, suction stroke (ICE)
~ **вы́пуска** exhaust stroke (ICE)
~ **вы́хлопа** exhaust stroke (ICE)
~, **пе́рвый** suction stroke, induction stroke, intake stroke (ICE)
~ **рабо́чего хо́да** power stroke (ICE)
~, **рабо́чий** power stroke (ICE)
~ **расшире́ния** expansion stroke (ICE)
~ **сжа́тия** compression stroke (ICE)
~, **тре́тий** power stroke (ICE)
та́ктик tactician
та́ктика tactics
~ **бронета́нковых во́йск** tank tactics
~ **кочу́ющих ору́дий** shoot-and-scoot technique
~ **ме́лких подразделе́ний** minor tactics
~, **общевойскова́я** combined arms tactics
~, **прикладна́я** applied tactics
такти́ческий tactical
та́ли pulley blocks, lifting tackle
та́лик talik, unfrozen spot within permafrost
тало́н 1. counterfoil
2. chit, coupon
~ **эксплуата́ции** registration document (veh)
та́лреп straining screw, turnbuckle
таль 1. tackle, block and tackle, pulley block
2. jack, hoist
~, **цепна́я** chain pulley block
тальк talc(um)
та́мбур 1. air-lock (at entrance to gas-proof room)
2. reel, drum, bobbin
3. vestibule, lobby
тамо́жня customs house, customs office
тампо́н plug, pad, tampon (med)
~, **ва́тный** cotton wool wad
тампона́ж tamponade, tamponage
танга́ж pitch (aeron)
та́нгенс tangent (math)
та́нгента transmitter cut-out switch
та́ндем-приво́д tandem-drive
та́ндем-ускори́тель tandem boost (GW)
танк 1. tank (AFV)
2. tank (container)

танк - ми́нный тра́льщик mine-clearing tank
~, **аэротранспорта́бельный** air portable tank
~, **безба́шенный** turretless tank
~, **веду́щий** command tank
~, **вко́панный** dug-in tank
~, **воздушнодеса́нтный** airportable tank
~, **вооружённый зени́тными сре́дствами** anti-aircraft tank
~, **вооружённый пу́шкой** gun tank
~, **вы́веденный из стро́я** knocked-out tank
~, **головно́й** command tank
~, **дежу́рный** guard tank
~, **деса́нтный** airportable tank
~, **завя́знувший в болоте́** bogged tank
~, **завя́знувший в гря́зи** bogged tank
~, **застря́вший в болоте́** bogged tank
~, **застря́вший в гря́зи** bogged tank
~, **зени́тный** anti-aircraft tank
~, **ко́рпус кото́рого скрыт релье́фом ме́стности** hull-down tank
~, **лёгкий** light tank
~, **мостово́й** bridgelaying tank, armoured bridgelayer
~ **непосре́дственной подде́ржки пехо́ты** close-support tank
~, **обору́дованный отва́лом** tank dozer
~, **огнемётный** flame-thrower tank
~, **оснащённый сре́дствами размини́рования** mine-clearing tank
~, **основно́й боево́й** main battle tank
~, **пехо́тный** infantry tank
~, **пла́вающий** amphibious tank
~, **подби́тый** knocked-out tank, tank casualty
~, **ради́йный** radio-equipped tank
~, **разве́дывательный** reconnaissance tank
~, **резе́рвный** replacement tank
~ **с бо́йковым тра́лом** flail tank
~ **с задра́енными лю́ками** closed down tank
~ **с катко́вым тра́лом** tank with mine-clearing roller
~, **сапёрно-подрывно́й** engineer tank
~ **сопровожде́ния** supporting tank
~, **сре́дний** medium tank
~, **станда́ртный** standard tank
~, **телеуправля́емый** 1. remote-controlled tank
2. radio-controlled tank

танк, тяжёлый heavy tank
~, учéбный training tank
~, штурмовóй assault tank
~, штурмовóй инженéрный assault engineer tank
танк-амфи́бия amphibious tank
танк-бульдóзер tank dozer
танкдóзер tank dozer
тáнкер 1. fuel tanker, oil carrier
2. tanker, tank ship
танкéтка light tank, tankette (obs)
танки́ст tank crew member
танки́ст-навóдчик tank gunner
танки́ст-стрелóк tank gunner
"танкобоя́знь" 1. tank sickness
2. fear of tanks
танковóз tank carrier, tank transporter
тáнковый 1. tank
2. armoured
танкодосту́пный accessible to tanks
танкодрóм tank training area, tank park
танконедосту́пный tank-proof, inaccessible to tanks
танконóсец tank landing vessel, LCT
~, удáрный tank landing craft, LCT
танкоопáсный favourable for tanks, tankable
танкострелкóвый tank gunnery
танкостроéние tank construction, tank building
танкострои́тель tank builder, tank constructor
танкофóн tank intercom
танк-охóтник tank hunter
танкошлéм tank helmet (worn by tank crew members)
танк-развéдчик scouting tank
танк-разгради́тель mine-clearing tank
танк-трáльщик mine-clearing tank
~, цепнóй flail mine-clearing tank
тáра 1. container, crate
2. fuel can
3. outer packing (rly)
4. tare (veh)
~, гермети́ческая sealed container
~, грузовáя supply container, packaging
~ для тяжеловéсных гру́зов, парашю́тно- десáнтная heavy-duty supply-dropping container
~, кáпсюльная bomb cell container (air despatching)
~, мешóчная sack
тáра, мя́гкая flexible container, flexible packing
~, парашю́тно-десáнтная supply-dropping crate, aerial delivery crate
тарáн 1. ram (mech)
2. ramming
3. breakthrough (tac)
4. penetration force (tac)
тарáнить to ram
тарéлка клáпана valve-stem collar, valve spring washer, valve disc, valve-spring carrier
~, клáпанная valve spring washer, valve stem collar (ICE)
~, упóрная 1. spring thrust washer, spring thrust collar, spring locating cup, spring locating buffer
2. thrust plate
тарéль 1. feed guide (ord)
2. jack footplate (ord)
тари́рование calibration, calibrating
тари́ровать to calibrate
тахеóметр tacheometer (surv)
тахеомéтрия tacheometry (surv)
тахи́метр tachymeter (surv)
тахóграф tachograph
тахóметр tachometer, revolution counter
тачáнка light horse-drawn vehicle, horse-drawn MG mounting
тáчка wheelbarrow
тáяние melting, thawing
тáять/растáять to melt, to thaw
твердéние при старéнии age hardening
твёрдость по Бринéлю Brinell hardness (metall)
~ по Мосу Mohs hardness, scratch hardness, impact hardness (metall)
твёрдотя́нутый hard-drawn (metall)
твóрчество creative thinking, creative work, creation
теáтр theatre
~ воéнных дéйствий theatre of war
~ войны́ theatre of war
текст text (comms)
~, незашифрóванный clear text
~, откры́тый clear text, plain language (comms)
~, смысловóй plain language text
~, цифровóй figures text
~, шифрóванный ciphered text
тексту́ра texture, structure

теку́честь 1. flow, fluidity
2. yield (metall)
3. turnover (pers)
теку́щий 1. present (AA gunn)
2. current
телеви́дение television
телевизио́нный television, TV
телеви́зор 1. television receiver, television set
2. radar scope
телегра́мма telegram, cablegram, cable
~, проходя́щая "through" message (received at signal office for onward transmission)
~, служе́бная service message (comms)
~, шифро́ванная cryptogram
телегра́ф telegraph
~, слухово́й C.W. telegraphy
"телегра́ф, солда́тский" rumour, "bush telegraph", "the grapevine"
телеграфи́рование telegraphy
~, ду́плексное duplex telegraphy
~, многокра́тное multiplex telegraph
~, надтона́льное super-audio telegraphy
~, си́мплексное simplex telegraphy
~, тона́льное voice-frequency telegraphy
телеграфи́ст telegraphist
телеграфи́я telegraphy
~, амплиту́дная amplitude-shift telegraphy
~, одноканáльная часто́тная single-channel frequency shift telegraphy
телегра́фно-строи́тельный line construction
телегра́фный telegraph
теле́жка 1. landing gear, undercarriage
2. bogie (tracked veh)
3. bogie (rly)
4. cart, trolley
5. truck
~, ви́лочная fork lift truck
~, двухо́сная four-wheel bogie
~, ми́нная miner's truck (coal-mine etc)
~, мото́рная motor dolly
~, ручна́я wheelbarrow, hand trolley
~, технологи́ческая handling trolley (GW)
телеизмере́ние telemetry
телема́рк telemark, telemark turn (ski)
телеметри́я telemetry
телемеха́ника remote control
телепереда́тчик television transmitter
телеско́п 1. telescope
2. telescope, optical bar (range-finder)
~, автоколлимацио́нный auto-collimating telescope
~, бинокуля́рный binocular telescope
~ визуа́льного слеже́ния tracking telescope
телета́йп teleprinter
телеуправле́ние 1. remote control
2. radio control
~, радиолокацио́нное radar control (GW)
телефо́н telephone
~, автомати́ческий dial telephone
~, батаре́йный battery-operated telephone
~ вну́тренней свя́зи interphone
~, головно́й headset, headphone
~, доба́вочный telephone extension, extension number
~, клавиату́рный teleprinter
~, опти́ческий telephone using modulated light
~, ручно́й handset (teleph)
телефони́рование telephony
~, высокочасто́тное carrier telephony
телефони́ровать to telephone
телефони́ст telephonist
~ коммута́тора telephone switchboard operator
телефони́я telephony
~, опти́ческая telephony by modulated light beam
телефо́нный telephone
телефоногра́мма phonogram, telephonogram
телефо́ны 1. telephones
2. headphones, earphones
телефотогра́фия telephotography
телефотосвя́зь telephotographic communications
те́ло 1. body
2. substance, solid
3. shank, shaft (tool)
~, инерцио́нное inertia block (fuse)
~ ору́дия gun body, chase (ord)
телогре́йка padded jacket, jerkin
телохрани́тель bodyguard (one person)
те́льник undershirt, vest
тембр timbre, tone (radio)
темп tempo, rate, impetus, speed, momentum
~ ата́ки momentum of attack
~ вы́дачи да́нных data rate

темп наступле́ния 1. momentum of advance
2. momentum of attack
~ огня́ rate of fire
~ огня́, батаре́йный battery fire interval (nav)
~ огня́, оруди́йный gun-fire interval (nav)
~ огня́, преде́льный maximum rate of fire
~ продвиже́ния rate of overall advance
~ стрельбы́ cyclic rate of fire, fire interval
температу́ра temperature
~, баллисти́ческая ballistic temperature
~ во́здуха, баллисти́ческая ballistic air temperature
~ воспламене́ния fire point (oil)
~ вспы́шки flash point
~ горе́ния combustion temperature
~ зажже́ния flash point
~ замерза́ния freezing point
~ заря́да charge temperature (gunn)
~ застыва́ния pour point (oil)
~ затвердева́ния solidification point
~ каплепаде́ния drop point (greases etc)
~ кипе́ния boiling point
~, окружа́ющая ambient temperature
~ окружа́ющей среды́ ambient temperature
~, отрица́тельная temperature below zero, sub-zero temperature
~ плавле́ния melting point
~ самовоспламене́ния self-ignition temperature, spontaneous ignition temperature
тенде́нция trend, tendency
~ автомоби́ля к изли́шнему повора́чиванию oversteer (MT)
~ автомоби́ля к недоста́точному повора́чиванию understeer (MT)
те́ндер 1. tender (rly)
2. tender (nav)
тенере́с lead trinitroresorcinate, lead syphnate (expl)
тенз(и)о́метр tensiometer, strain gauge
те́ни, гипсометри́ческие layer tints (carto)
тент canopy, awning
теодоли́т theodolite
~, геодези́ческий geodetic theodolite
~, опти́ческий optical scale theodolite
~, техни́ческий tacheometer (surv)
теоре́ма взаи́мности reciprocity theorem
тео́рия относи́тельности theory of relativity
~ рассе́ивания theory of dispersion
~ реакти́вного движе́ния rocket theory
~ стрельбы́ 1. theory of fire (SA)
2. dispersion and probability theory (gunn)
теплово́з diesel locomotive
~ с механи́ческой переда́чей diesel mechanical locomotive
~ с электри́ческой переда́чей diesel electric locomotive
теплово́й 1. heat, thermal
2. infra-red
3. calorific
теплоёмкость heat capacity
~, уде́льная specific heat
теплозащи́та thermal protection, heat protection
теплоизлуча́тель heat radiator
теплоизлуче́ние heat radiation, thermal radiation, radiation of heat
теплоизоля́тор heat screen
теплоизоля́ция heat insulation
теплонапряже́ние, объёмное heat-release rate
теплонепроница́емый heat-proof
теплоноси́тель heat transfer agent
теплообме́н heat exchange
теплообме́нник heat exchanger
теплообнаружи́тель thermograph (photo)
теплообразова́ние heat production
теплоотводя́щий heat-removing, cooling
теплоотда́ча 1. heat transfer
2. heat loss (med)
3. heat irradiation
теплопеленга́тор thermal direction finding set, thermal DF set
теплопеленга́ция infra-red direction finding, infra-red DF
теплопереда́ча heat transfer
теплопла́н thermal rocket-powered spacecraft
теплопоглоти́тель heat absorber
теплопоглоще́ние thermal absorption, heat absorption
теплопрово́дность heat diffusion, heat conductivity, thermal conductivity
теплопрово́дный heat-conducting, heat-conveying, heat-carrying
теплопрозра́чность diathermancy

теплопроизводи́тельность calorimetric value, calorific value
теплорассе́яние heat dispersion, heat dissipation, thermal dispersion, thermal dissipation
теплосодержа́ние heat content
теплосто́йкий heat-proof, heat-resisting
теплота́ воспламене́ния ignition heat
~ горе́ния calorimetric value, calorific value
~ излуче́ния radiant heat, radiation heat
~ сгора́ния heat of combustion, heat value
~, скры́тая latent heat
теплотво́рность calorimetric value, calorific value
~ то́плива fuel calorific value
теплотво́рный calorimetric, calorific
теплоулови́тель thermal direction finding set, thermal DF set
теплохо́д motor vessel, motor ship
теплочувстви́тельный heat-sensitive, sensitive to heat
теплу́шка 1. heated shelter 2. heated goods wagon (troop trains etc) (rly)
терапевти́ческий therapeutic
тёрка 1. grater, grinder 2. rasp 3. coated wire (friction igniter)
термисто́р thermistor (elec)
терми́т 1. thermite (chem) 2. termite (insect)
терми́ческий thermic, thermal, temperature-indicating
термоанемо́метр, про́волочный hot-wire anemometer
термоизвеща́тель 1. flame detector switch 2. fire warning device (ac)
термоизоляцио́нный heat-insulated, thermally insulated
термоизоля́ция heat insulation
термоио́нный thermionic
термо́метр thermometer
~, биметалли́ческий bimetallic thermometer
~, жи́дкостный liquid expansion thermometer
~, самопи́шущий recording thermometer
~ сопротивле́ния resistance thermometer
~, спиртово́й alcohol thermometer, spirit thermometer
термо́метр, сухо́й dry-bulb thermometer
термообрабо́тка heat treatment
термопа́ра thermo-couple (elec)
термопласти́чный thermoplastic
терморези́стор thermistor, thermal resistor
те́рмос 1. insulated food container 2. vacuum flask
термосифо́н thermosiphon
термостаби́льность thermal stability, heat stability
термоста́т thermostat
термосто́йкий heat-resistant
термосто́лбик thermopile
термосфе́ра thermosphere
термото́к thermoelectric current
термоэлектро́нный thermionic
термоэлеме́нт thermocouple (elec)
термоя́дерный thermonuclear
тёрочный friction
терпе́ть/потерпе́ть to suffer, to endure, to undergo; **терпе́ть/потерпе́ть ава́рию** to break down
терпу́г rasp
терра́са terrace
террито́рии, приполя́рные polar regions
террито́рия, закры́тая closed area, restricted area
~, отвоёванная recaptured territory
терча́сть territorial unit
теря́ть/потеря́ть to lose, to shed; **теря́ть/потеря́ть ориентиро́вку** to lose one's bearings; **теря́ть/потеря́ть управле́ние** to lose control, to get out of control
тёс thin planks, battens
теса́к infantry sword, broadsword
тесёмка, висо́чная head harness, longitudinal strap (of respirator)
~, заты́лочная rear band, head band (respirator)
~, ло́бная head harness cross strap (respirator)
тесёмки обмо́ток puttee tapes
тесло́ adze
тесни́на defile
тетра́дь по́вара, рабо́чая cook's work-book
тетранитромета́н tetranitromethane
тетранитропентаэритри́т pentaerythritol tetranitrate, PETN (expl)
тетра́эдр tetrahedron (A tk obstacle)
тетри́л tetryl, trinitro-phenyl-methyl-nitramine (expl)

тетри́л, порошкообра́зный powdered tetryl
тетрито́л HE consisting of 75% tetryl and 25% TNT
тетро́д tetrode (elec)
~, лучево́й beam tetrode
те́хник technician, mechanic, artificer
~, артиллери́йский armament artificer
~, оруди́йный gun fitter
~, оруже́йный armourer
~, ста́рший chief technician, chief artificer, senior technician
те́хника 1. technical equipment, equipment, material, materiel
~, автомоби́льная MT equipment
~, автотра́кторная MT vehicles and tractors
~ безопа́сности safety precautions, safety procedure
~ безопа́сности при хране́нии опа́сных гру́зов safeguarding of hazardous materials
~, боева́я fighting equipment, war equipment, war material, materiel
~, бронета́нковая armoured vehicles, AFV's (Sov term covers tks, SP's, APC's, ARV's and motorcycles used by armoured units)
~, вое́нная 1. war material 2. military weapons and technical equipment
~ вожде́ния driving technique (MT)
~, вычисли́тельная computing technology
~ доста́вленная по во́здуху, боева́я air-delivered war stores
~, зени́тная раке́тная anti-aircraft missiles, surface-to-air missiles
~, инфракра́сная infra-red equipment
~, ми́нно-подрывна́я 1. mine warfare technique 2. mine warfare equipment
~, однопо́лосная single side-band technique
~, передова́я advanced technology
~, перепра́вочная water crossing equipment, wet-gap crossing equipment
~, подви́жная 1. mobile plant, mobile equipment 2. vehicles
~ приспосо́бленная для перево́зки по во́здуху airportable equipment
те́хника произво́дства разруше́ний demolition technique
~, радиолокацио́нная radar equipment
~, та́нковая armoured fighting vehicles, AFVs
~, шта́тная organic equipment, authorized equipment
те́хник-испыта́тель test-point officer (GW)
те́хник-лабора́нт laboratory technician
те́хник-лейтена́нт technical lieutenant
~, мла́дший junior technical lieutenant (Sov)
~, ста́рший senior technical lieutenant (Sov)
те́хник-смотри́тель зда́ний clerk of works
те́хник-специали́ст по обезвре́живанию неразорва́вшихся бомб bomb disposal specialist
те́хникум technical school
техни́ческий technical, technological, mechanical
техноло́гия production process, technology, engineering
~ ремо́нтных рабо́т maintenance engineering
тече́ние 1. current, stream, course 2. tidal flow
течь flow, stream, run, escape, trickle, drip, leak, leakage
~ в гидросисте́ме hydraulic leak
~ ма́сла oil leak
ти́на mud, ooze
тип ду́льного то́рмоза, реакти́вный reactive type muzzle-brake (ord)
~ уда́рного механи́зма с боево́й пружи́ной, уда́рниковый spring-loaded striker firing mechanism
~ уда́рного механи́зма, уда́рниковый striker-type firing mechanism
типово́й standard, model
типогра́фия printing shop
тир indoor range
~, закры́тый indoor range
~, стрелко́вый smallbore range
тиратро́н thyratron (electron)
тире́ 1. dash (morse) 2. dash, hyphen
тиски́, верста́чные bench vice
~, паралле́льные parallel-jaw vice
~, ручны́е hand vice

тиф 1. typhus, typhus fever
2. typhoid fever, enteric fever
~, брюшно́й typhoid fever, enteric fever
~, возвра́тный relapsing fever, recurring fever, typhus recurrens
~, сыпно́й typhus fever
~, эпидеми́ческий сыпно́й epidemic typhus
тифо́н signal whistle, blast signal
тихохо́дный slow, slow-running
ткань, маскиро́вочная се́тчатая camouflage sheet (similar to Brit hessian sheet)
~, олифо́ванная oiled fabric, oil cloth
~, пролифо́ванная oilskin
~, проолифо́ванная oilskin
~, прорези́ненная rubberized fabric
тка́ньевая scrim
тле́ние smouldering, rotting, putrefaction
"Товсь!" "Ready!" (gunn)
тожде́ственность identity, sameness
тоже́ственность identity, sameness
ток current (elec)
~, ано́дный anode current
~ возбужде́ния excitation current
~, входно́й input current
~, вы́прямленный rectified current
~, индуктиро́ванный induced current
~ насыще́ния saturation current
~, обра́тный reverse current, back current
~, переме́нный alternating current, AC
~, постоя́нный direct current, DC
~, рабо́чий operating current
~, се́точный grid current (elec)
~, трёхфа́зный three-phase alternating current
~, электри́ческий electric current
то́ки, вихревы́е eddy currents
токовраща́тель pole changer, direct current ringer
токоотво́д terminal (elec)
токоприёмник current collector, pantograph
токопрохожде́ние circuit diagram
токосъёмник collector (elec)
токсеми́я toxaemia, blood poisoning
токси́л toxyl (expl)
токси́н toxin (med)
~ боту́линуса botulinum toxin
токси́чность toxicity
токси́чный toxic
тол trinitrotoluene, TNT
толка́тель 1. tappet (ICE)
2. pusher, push rod
3. hammer strut (pistol)
4. operating rod piston extension (SA)
~ возвра́тного механи́зма return spring rod
~ с автомати́ческой регулиро́вкой зазо́ра self-adjusting tappet
толка́ч 1. ram, rammer
2. pusher, tappet
3. pusher tug
~, гидравли́ческий 1. hydraulic ram
2. hydraulic jack (ship launching)
толма́ч translator, interpreter
толуо́л toluene
толчо́к 1. impetus, thrust
2. percussion
3. shock (elec)
толщина́ брони́ armour thickness
толщиноме́р feeler gauge
толь sanded roofing paper, asphalt roofing paper, roofing felt
~, кро́вельный tarred roofing paper
томпа́к copper zinc alloy, tombac
тон 1. tone
2. note (sigs)
тона́льный voice-frequency
тони́т tonite (guncotton)
тонконтро́ль tone control
тонкораспылённый finely pulverized, atomised
тонкосте́нный thin-walled
то́нна, америка́нская short ton (0.907 metric tonnes, 2,000 lbs)
~, англи́йская long ton (1.016 metric tonnes, 2,240 lbs)
~, больша́я long ton
~, дли́нная long ton (2,240 lbs)
~, коро́ткая short ton (2,000 lbs)
~, метри́ческая metric ton (2204.6 lbs)
то́пка furnace, fire-box
то́пливный fuel
то́пливо 1. fuel
2. propellant (rockets)
~, авиацио́нное aviation fuel
~, антидетонацио́нное anti-knock fuel
~, взрывоопа́сное hazardous propellant
~, высокоокта́новое high-octane fuel
~, гиперго́льное hypergolic propellant
~, двухкомпоне́нтное bipropellant
~, ди́зельное diesel oil, diesel fuel, DERV
~, жи́дкое 1. liquid fuel
2. liquid propellant (rockets)

то́пливо, иску́сственное synthetic fuel
~, коррозио́нно-акти́вное corrosive propellant
~, кре́кинговое cracked fuel
~, мото́рное heavy diesel fuel
~, недетони́рующее leaded fuel, petrol with anti-knock additive, anti-knock fuel
~, неокра́шенное undyed fuel
~, несамовоспламеня́ющееся non-hypergolic propellant, diergolic propellant, non-spontaneously ignitable propellant
~, низкока́чественное low grade fuel
~, низкосо́ртное low grade fuel
~, однокомпоне́нтное monopropellant
~, пожаробезопа́сное safety fuel
~, пылеви́дное powdered solid fuel, pulverized fuel
~, раке́тное rocket propellant
~, реакти́вное jet fuel (ac), rocket fuel
~, самовоспламеня́ющееся hypergolic propellant, spontaneously ignitable propellant
~, синтети́ческое synthetic fuel
~, сме́шанное blended fuel
~, спиртово́е alcohol fuel
~, твёрдое solid propellant, solid fuel
~, твёрдое на осно́ве каучу́ка rubber-base propellant
~, углеводоро́дное hydrocarbon fuel
~, унита́рное monopropellant
~, форса́жное reheat fuel, afterburner fuel (ac)
~, хими́ческое chemical propellant
~, я́дерное nuclear fuel, nuclear propellant
топливозамени́тель fuel substitute
топливозапра́вщик fuel supply vehicle, bowser
то́пливо-маслозапра́вщик fuel and oil bowser
топливоме́р fuel meter, fuel gauge
топливопода́ча fuel supply, fuel feed
топливопрово́д fuel pipe
топливоснабже́ние fuel supply
топливосодержа́щий fuel-carrying, fuel-bearing
топливохрани́лище fuel storage depot
топогеодези́ст topographic, geodetic surveyor
топо́граф surveyor, topographer
топогра́фия topography
топока́рта topographical map
топо́р axe
~, вое́нный hand axe
~, лесору́бочный felling axe
топори́ще axe handle, axe helve
топослу́жба topographical service
топь swamp, marsh
торе́ц 1. end face (shaft)
2. butt end
3. face, front
~ зу́ба шестерни́ face of gear tooth
торже́ственный ceremonial
то́рий thorium
торкре́т-бето́н gunite (eng)
торкрети́рование guniting (eng)
торможе́ние 1. braking, deceleration, retardation
2. damping
3. drag
~, авари́йное emergency braking
~ дви́гателем engine braking
~ нака́та control of run-out (ord)
~ плу́гом "snowplough" braking action (ski)
~ полуплу́гом "half-snowplough" braking action (ski)
~, рекуперати́вное regenerative braking
~, э́кстренное emergency braking
то́рмоз brake
~, автомати́ческий automatic brake
~ акти́вного ти́па, ду́льный positive-type muzzle brake
~ акти́вно-реакти́вного ти́па, ду́льный positive reactive-type muzzle brake
~, бараба́нный drum brake
~, башма́чный drag brake, shoe brake, chock, block brake, brake scotch (braking device)
~, ва́куумный vacuum brake
~, возду́шный air brake
~, вспомога́тельный auxiliary brake
~, гидромехани́ческий hydromechanical brake
~, го́рный mountain brake, parking brake
~, двухка́морный ду́льный double-baffle muzzle brake (gunn)
~, ди́сковый disc brake
~, ду́льный muzzle brake
~, клещево́й caliper brake
~, колёсный wheel brake
~, коло́дочный drum brake, block brake, shoe brake

то́рмоз, ле́нточный belt brake, band brake
~, механи́ческий mechanical brake
~, многоди́сковый multiple-disc brake
~ нака́та run-out speed controller (ord)
~, ножно́й foot brake
~, остано́вочный stopping brake
~ отка́та buffer, recoil absorber (ord)
~ отка́та, веретённый counter-acting piston type buffer (ord)
~ отка́та, гидравли́ческий hydraulic buffer
~ отка́та, золотнико́вый rotating-valve-type buffer
~ отка́та, кана́вочный tapered-groove-type buffer
~ отка́та, кла́панный retarding-valve-type buffer (ord)
~ отка́та, шпо́ночный tapered-rib-type buffer cylinder (ord)
~, педа́льный foot brake
~, пневмати́ческий pneumatic brake
~ прице́па trailer brake
~, реакти́вный inertia brake (power braking)
~, рекуперати́вный regenerative brake
~, ручно́й hand brake
~ с ва́куумным приво́дом vacuum brake
~ с коло́дками внутри́ тормозно́го бараба́на, колёсный internal expanding drum brake
~ с пневмати́ческим приво́дом air brake
~ с разде́льным гидравли́ческим приво́дом на пере́дние и за́дние колёса split front-rear hydraulic brake system
~ с тро́совым приво́дом cable brake
~ с усили́телем servo brake
~, стоя́ночный parking brake
~ стре́лки needle stop, needle plunger (compass)
~, центра́льный transmission brake (eg on prop shaft)
~ э́кстренного торможе́ния emergency brake
тормози́ть to brake
тормозно́й brake
тормозя́щий 1. braking
2. damping
торна́до tornado
тороида́льный toroidal
торо́с ice hummock (in pack ice)
торпе́да torpedo
~, управля́емая авиацио́нная guided glide-bomb, controlled glide-bomb
торпедоно́сец torpedo-bomber
торсио́н 1. torsion
2. torsion bar
торсио́нный torsional
торф peat
торфяни́к peat bog
торцево́й end, face
торцо́вый end, face
точи́ло grindstone
то́чка 1. point
2. dot (morse)
3. full stop (punctuation)
~, артиллери́йская опо́рная bearing picket (gunn)
~, бето́нная огнева́я concrete pill box
~, бронева́я огнева́я armoured fire position, armoured gun emplacement
~, брониро́ванная огнева́я armoured gun emplacement, armoured fire position
~, ве́рхняя мёртвая top dead centre, TDC (mech)
~ возникнове́ния детона́ции detonation point
~ встре́чи 1. point of impact (gunn)
2. first graze (SA)
~ вы́броски jump point (para)
~ вы́лета point of origin (ball)
~, вы́носная auxiliary aiming point
~ вы́стрела present position (AD gunn)
~, гла́вная principal point (photo, surv)
~, де́рево-земляна́я огнева́я earth-and-timber strongpoint, earth-and-timber pillbox
~, долговре́менная огнева́я permanent strong point (eg concrete pill box)
~, железобето́нная огнева́я concrete emplacement
~ замерза́ния freezing point
~, исхо́дная 1. present position of target (on observation) (AD)
2. initial position
3. initial station (surv)
~ к моме́нту по́лного выгора́ния то́плива all-burnt point (GW)
~ кипе́ния boiling point
~, коне́чная terminal point
~, контро́льная 1. aiming mark (zeroing)
2. reference point

то́чка, ко́нтурная 1. prominent landmark 2. spot height (surv, carto)
~ **крепле́ния** mounting point
~, **мёртвая** 1. dead spot (radio) 2. dead centre (mech)
~ **наблюде́ния** 1. observation point, observation station 2. instrument station (surv)
~ **наведе́ния, исхо́дная** start point for GCI
~ **наво́дки** aiming point (gunn)
~ **наво́дки, вспомога́тельная** reference point, datum aiming point (bombing)
~ **наво́дки, запасна́я** gun aiming point 2, GAP 2
~ **наво́дки, ночна́я** night firing mark
~ **наво́дки, оруди́йная** gun aiming point
~ **наво́дки, основна́я** gun aiming point 1, GAP 1
~ **нади́ра** plumb point (photo)
~, **настоя́щая** present position (gunn)
~ **насыще́ния** saturation point
~ **нача́ла вы́броски** drop starting point (para)
~ **нача́ла манёвра** initial point of manoeuvre
~ **нача́ла сниже́ния** jumping point (position where first parachutist of paradrop leaves aircraft)
~, **нача́льная сдви́нутая** false zero
~, **неподви́жная нулева́я** absolutely fixed zero
~, **неподви́жная огнева́я** static gun emplacement, static firing position
~, **ни́жная мёртвая** bottom dead centre (mech)
~, **нулева́я** null point, datum point
~ **нулевы́х искаже́ний** isocentre (photo)
~, **огнева́я** firing point, fire position
~, **опо́рная** control point (surv)
~ **опо́ры** 1. hold-fast, prop 2. fulcrum
~, **ориенти́рная** reference point (arty DF)
~ **отбо́ра мо́щности** power take-off point
~, **откры́тая огнева́я** open weapon emplacement
~ **отме́тки** reference point
~ **паде́ния** point of arrival, impact point (gunn)
~ **паде́ния, сре́дняя** mean point of impact, mpi
~ **паде́ния, табли́чная** point of graze (gunn)

то́чка перехва́та intercept point (msl)
~ **плавле́ния** melting point
~ **полови́нной мо́щности** half-power point (radar)
~ **попада́ний, сре́дняя** mean point of impact, mpi (gunn)
~ **попада́ния** point of impact (gunn)
~ **приземле́ния пе́рвого парашюти́ста** position for touch-down of first parachutist of paradrop
~ **приложе́ния нагру́зки** loading point
~ **прице́ливания** point of aim
~ **прице́ливания, вспомога́тельная** offset aiming point - point used as aiming point during radar bombing and parachute dropping
~, **пулемётная** machine-gun position, machine-gun nest
~, **пулемётная огнева́я** machine-gun position, machine-gun nest
~ **пу́ска** launching point (msl)
~ **разры́ва** point of burst
~ **разры́ва, сре́дняя** mean point of burst
~ **расходи́мости** point of divergence
~, **расчётная** reference point
~ **росы́** dewpoint
~ **сбра́сывания** bomb release point
~, **сре́дняя** mean point
~ **сры́ва пото́ка** burble point, stalling point (aeron)
~ **стоя́ния** station (surv)
~ **схо́да** point of convergence
~ **сходи́мости** point of convergence
~, **табли́чная** point of graze (gunn)
~, **теку́щая** current position (AD gunn)
~, **углова́я** nodal point
~, **узлова́я** 1. nodal point, node (elec) 2. central station (surv)
~, **упреждённая** future position (gunn)
то́чность accuracy, precision
~ **бо́я** accuracy (wpn)
~ **градуиро́вки** calibration accuracy
~, **достижи́мая** obtainable accuracy
~ **калибро́вки** calibration accuracy
~ **наведе́ния** guidance accuracy
~ **навига́ции** navigation accuracy
~ **определе́ния координа́т це́ли** accuracy of target location
~ **определе́ния положе́ния** positional accuracy
~ **ориента́ций** orientation accuracy
~, **полу́ченная** resultant accuracy
~ **прице́ливания** precision of aim

то́чность тариро́вки accuracy in calibration
~ управле́ния guidance accuracy
то́чный 1. exact, fine, accurate, precise
2. punctual
тра́верз beam (nav direction)
~, пра́вый starboard beam
тра́верс 1. traverse (fd wks)
2. beam (nav direction)
~, вися́чий bridge traverse (fd wks)
~, наве́сный bridge traverse (fd wks)
~, ты́льный rear parapet parados (fd wks)
тра́верса 1. cross-member, transverse member, tie beam
2. cross arm (teleg pole)
3. gantry crane
травле́ние etching (metall)
тра́вма contusion, contused wound, injury, trauma
~, психи́ческая 1. emotional shock
2. battle fatigue
траекто́рии, сопряжённые complementary trajectories (attaining the same range at different angles of departure)
траекто́рия trajectory, flight path
~, балладроми́ческая balladromic course (GW)
~, баллисти́ческая ballistic path
~, брахидроми́ческая brachydromic course (GW)
~, восходя́щая ascending trajectory
~ встре́чи с це́лью collision course (GW)
~, вы́численная 1. calculated trajectory (gunn)
2. calculated glide path (ac)
~, действи́тельная actual trajectory
~, за́данная desired trajectory, intended trajectory
~, клинодроми́ческая clinodromic course (GW)
~, крута́я steep trajectory
~, морти́рная high-angle trajectory
~, наве́сная high trajectory
~, накло́нная inclined trajectory
~, насти́льная flat trajectory
~, нисходя́щая descending trajectory
~, норма́льная standard trajectory
~, опо́рная reference trajectory
~, отло́гая shallow trajectory
~, параболи́ческая parabolic trajectory
~ перехва́та interception trajectory
~ полёта flight trajectory, flight path
траекто́рия, поло́гая flat trajectory
~, программи́рованная present flightpath (GW)
~, расчётная precalculated trajectory, predicted trajectory
~, рикошети́рующая ricochet trajectory
~ свобо́дного полёта free-flight path, ballistic trajectory
~, скоподроми́ческая scopodromic course (line-of-sight, GW)
~ сниже́ния descent trajectory
~, сопряжённая complementary trajectory
~, спира́льная spiral trajectory
~, сре́дняя trajectory to mean point of impact
~, табли́чная standard trajectory
~, томодроми́ческая tomodromic course, tomodromic flight path (GW)
трак track link, track shoe, track pin (tk)
тракт channel (elec)
~, волново́дный waveguide
~, охлажда́ющий coolant passage
тракта́т treaty, treatise
тра́ктор tractor
~, букси́рный towing tractor
~, гу́сеничный crawler tractor, tracked tractor
~, ди́зельный электри́ческий diesel electric tractor
~, оруди́йный gun tractor
~, пла́вающий amphibious tractor
тра́ктор-амфи́бия tracked amphibian, tracked landing vehicle, LVT
тра́ктор-бульдо́зер tractor dozer
трактори́ст tractor driver
тра́ктор-трубоукла́дчик pipe-laying tractor
тра́ктор-тяга́ч prime mover, tractor
трал 1. drag, trawl
2. mine-clearing tank attachment
3. sweep (min) (nav)
~, бойко́вый flail, mine clearing flail
~, катко́вый roller flail (mine clearing)
~, ми́нный mine clearer (device fitted to tanks)
~ молотко́вого ти́па, акусти́ческий acoustic hammer
~, неконта́ктный influence sweep (nav)
~, ножево́й blade mine-clearing device
~, ножево́й ми́нный mine plough
~, подрывно́й explosive sweep (nav)
~, придо́нный bottom sweep (nav)

трал, цепно́й flail
~, электромагни́тный magnetic sweep (nav)
тра́ление mine-sweeping, trawling
тра́льный 1. sweeper, sweeping
2. trawler
тра́льщик 1. mine-sweeper
2. trawler (rare)
~, ми́нный mine-sweeper (vessel)
~, та́нковый ми́нный mine-clearing tank
трамбова́ть to tamp, to ram
трамбо́вка 1. tamping
2. tamping machine, tamper
3. rammer
транзи́стор transistor
~, герма́ниевый germanium transistor
~, кре́мниевый silicon transistor
~, табле́точный tab transistor, wafer transistor
транзи́т transit, passage
транзи́том in transit
транзитро́н transitron
трансду́ктор transducer (electron)
трансли́ровать to transmit, to be broadcast, to relay
трансля́ция relay, retransmission, repeating, re-broadcast
~, радиолокацио́нная radar relay
~, регенерати́вная regenerative repeater (teleg)
~, реле́йная relaying, re-broadcasting
трансмиссио́нный transmission
трансми́ссия 1. transmission system (Brit)
2. drive system (US)
3. transmission compartment (tk)
4. transmission (of power)
~, гидравли́ческая hydraulic transmission
~, гидромехани́ческая hydromechanical transmission
~, гидрообъёмная hydrostatic transmission
транспо́рт figure carried forward (maths, book- keeping)
тра́нспорт 1. transport, conveyance
2. transportation
3. transport (nav) (ship)
~, автогужево́й road transport (MT and animal)
~, автодоро́жный road transport
тра́нспорт, автомоби́льный motor transport, mechanical transport, MT
~, арме́йский military train (rly)
~, артиллери́йский artillery transport, artillery train
~ боево́го пита́ния ammunition train
~, вездехо́дный cross-country transport
~, вну́тренний во́дный inland water transport
~, во́дный water transport
~, вое́нно-автомоби́льный motor transport, MT
~, возду́шный air transport
~, войсково́й troop transport, troopship
~, вью́чно-гужево́й animal transport
~, вью́чный animal transport, pack transport
~, городско́й city transport
~, гру́зо-деса́нтный cargo landing ship, landing ship logistic, LSL
~, гру́зо-пассажи́рский cargo troopship
~, гужево́й animal-drawn transport
~, гу́сеничный tracked transport
~, деса́нтный landing ship, amphibious warfare ship
~, деса́нтный войсково́й assault troop transport
~, деса́нтный грузово́й assault cargo ship
~, колёсный wheeled transport
~, междугоро́дный inter-city transport
~, морско́й 1. sea transport
2. transport (nav) (ship)
~, обще́ственный public transport
~, пехо́тно-деса́нтный infantry landing ship, LSI
~, поро́жний empty transport, empty vehicles
~, санита́рный ambulance, ambulance transport
~, скоростно́й high-speed transport
~, снегохо́дный oversnow transport
транспорта́бельность transportability
транспорта́бельный transportable
тра́нспорт-док, деса́нтный amphibious transport dock
транспортёр 1. carrier, personnel carrier, cargo carrier
2. conveyor
~, винтово́й worm conveyer, screw conveyer
~, грузово́й cargo carrier, load carrier
~, гу́сеничный tracked personnel carrier

транспортёр, гу́сеничный пла́вающий tracked amphibious transporter, tracked vehicle, tracked amphibian
~ **для перево́зки ли́чного соста́ва** personnel carrier
~, **кулачко́вый** shaker conveyor
~, **ле́нточный** belt conveyor
~ **о́бщего назначе́ния** general purpose carrier
~, **пла́вающий грузово́й** amphibious cargo carrier
~, **пла́вающий гу́сеничный** tracked amphibian (load carrying), amphibious tracked personnel carrier
~, **полугу́сеничный** half-track carrier
~, **та́нковый** tank carrier, tank transporter
транспортёр-устано́вщик transporter-erector (GW)
транспорти́р protractor
транспорти́рование transportation
транспорти́ровать to transport, to convey
транспортиро́вка transportation
~, **беста́рная** bulk transport
~ **поро́ды, назе́мная** surface disposal of spoil (mining)
~ **поро́ды, подзе́мная** underground removal of spoil (mining)
тра́нспортник transport worker
тра́нспортный transport
транспортоспосо́бный transportable
трансформа́тор transformer
~, **анте́нный** aerial transformer
~, **входно́й** input transformer
~, **высоково́льтный** high-voltage transformer
~ **высо́кого напряже́ния** high-voltage transformer
~, **высокочасто́тный** high-frequency transformer
~, **выходно́й** output transformer
~, **дифференциа́льный** differential transformer
~ **звуково́й частоты́** audio-frequency transformer
~, **и́мпульсный** pulse transformer (elec)
~, **лине́йный** line transformer
~, **ма́сляный** oil transformer (elec)
~, **нака́льный** heater transformer, filament supply transformer (electron)
~ **напряже́ния** voltage transformer
~, **повыша́ющий** step-up transformer
~, **понижа́ющий** step-down transformer
трансформа́тор, силово́й power transformer
~, **симметри́рующий** Balun, Balun transformer (electron)
~ **синхро́нной свя́зи** synchro-transformer
~, **упрощённый** simplified rectifier (photo)
трансформи́рование 1. transformation 2. rectification (photo)
трансформи́ровать to transform, to rectify (photo)
трансфу́зия transfusion
траншеекопа́тель trencher, trench digger
~, **колёсный** wheeled trench digger
~, **плу́жный** plough trench digger
~, **самохо́дный** mobile ditching machine, mobile trench digger
~, **цепно́й** ladder-type ditching machine
транше́йный trench
транше́я trench, ditch
~ **в сне́гу** snow trench
~, **дополни́тельная** alternative fire-trench
~, **исхо́дная** 1. departure trench 2. assembly trench
~ **непо́лного про́филя** hasty trench
~, **пе́рвая** front-line trench
~ **по́лного про́филя** fire trench, deep trench
~, **противопожа́рная** fire trench
~, **соедини́тельная** communication trench
~, **сплошна́я** continuous trench
~, **фильтрацио́нная** 1. filter trench 2. horizontal sand filter (water purification)
~, **штурмова́я** assault trench
трап 1. ladder 2. gangway
~, **въездно́й** loading and unloading ramp
~, **откидно́й** landing ramp, hinged ramp (landing craft)
~, **поса́дочный** mounting ladder
~, **штурмово́й** assault ramp
трапба́лка 1. davit 2. accommodation ladder (naut)
трапе́ция trapezium
тра́сса 1. path, route
2. trace (ammo)
3. trail (smoke)
4. location, projected route (surv)

тра́сса, авиацио́нная airline, air route
~, баллисти́ческая ballistic path
трассёр tracer compound
трасси́ровать to mark the route/track, to map the route/track, to trace the route/track
трассиро́вка tracing (fd eng)
трасси́рующий, оско́лочно-фуга́сно-зажига́тельно fragmentation high explosive incendiary tracer
трассшну́р tracing tape
тра́та, допуска́емая permissible losses, allowed loss
~, зако́нная permitted wastage, allowed loss
трафаре́т 1. stencil
2. template, pattern
трафаре́тка 1. title block (drawing)
2. stencil brush
трахеотоми́я tracheotomy
трахо́ма trachoma (eye disease)
тре́бование demand, requisition, requirement
~ на де́нежные сре́дства requisition for cash
~, отде́льное individual requirement
~, та́ктико-техни́ческое 1. military requirement
2. project requirement
~, техни́ческое technical requirement, technical specification
~, уста́вное requirement in accordance with regulations
~, че́ковое voucher, demand voucher
тре́бования, эксплуатацио́нные operating requirements, service requirements
тре́бовательность к подчинённым exacting attitude towards subordinates
треб-ремо́нта requiring repair, requires repair (heading on forms)
тре́бующий ремо́нта requiring repair, requires repair
трево́га 1. alarm
2. alert
~, а́томная warning of nuclear attack
~, боева́я 1. stand-to, battle alarm
2. alert
~, возду́шная air-raid warning
~, ло́жная false alarm
~, о́бщая general alert, general stand-to
~ по чрезвыча́йным происше́ствиям emergency alert (fire, flood, etc)
трево́га, пожа́рная fire alarm
~, та́нковая tank warning alarm
~, уче́бная practice alert
~, хими́ческая CW alarm, chemical alarm
~, ча́стная partial alert (involving only part of a unit or formation)
трево́жный alert, alarm
тре́йлер trailer
трена́ж training, training exercise
тренажёр trainer (eqpt), simulator
~, горя́чий live trainer, training vehicle
~ наво́дчиков, группово́й group gunners' simulator
~ Проничёва "Pronichev's Trainer" (parachute- jumping tower)
~, самолётный flight simulator
~, та́нковый tank trainer, tank simulator
~, та́нковый артиллери́йский tank gunnery trainer, tank gunnery simulator
~, та́нковый ба́шенный tank turret trainer, tank turret simulator
тренажёр-имита́тор training simulator
тре́ние friction
~ движе́ния dynamic friction
~, жи́дкостное fluid friction
~ каче́ния rolling friction
~, куло́новское Coulomb friction
~, пове́рхностное surface friction
~ поко́я static friction
~ скольже́ния sliding friction
трениро́ванность training, state of training
тренирова́ть/натренирова́ть to train
трениро́вка training, exercise, practice
~ по-та́нкому, пе́шая dismounted tank training (men on foot representing tanks)
~, противога́зовая respirator drill
~, совме́стная joint training, joint exercise
~, стрелко́вая shooting practice
~, штабна́я HQ exercise
трениро́вочный training
трено́га tripod, tripod stand
~, пулемётная machine-gun tripod
~, универса́льная universal MG tripod mounting, dual- purpose MG tripod mounting (AA and ground roles)
трено́жник, универса́льный tripod (gen)
тре́пел diatomite (expl)

треуго́льник 1. triangle
2. delta network (elec)
~, измери́тельный range triangle (range finding)
~ невя́зки triangle of errors (surv)
~, поворо́тный turning triangle (rly)
~ погре́шностей triangle of errors (surv)
~ попада́ний shot grouping
~, равнобе́дренный isosceles triangle
~, упреди́тельный azimuth prediction triangle (AD gunn)
треуго́льный triangular
трёхме́рный three-dimensional
трёхо́кись се́ры sulphur trioxide
трёхо́сный six-wheeled, three-axled
трёхпу́тный three-track
трёхто́нка three-ton truck
трёхходово́й 1. three-way
2. T-shaped
тре́щина crack
~, вну́тренняя internal crack
~, волосяна́я hair-line crack
~, несквозна́я surface crack
~, пове́рхностная surface crack
~, сквозна́я internal crack
трещо́тка 1. rattle (eg MG simulator rattle)
2. ratchet
триангуля́ция triangulation
три́бка driving pinion, elongated spur pinion
трибу́на 1. stand, platform, rostrum
2. saluting base
трибуна́л, вое́нный military tribunal
тригатро́н trigatron (electron)
три́ггер 1. trigger
2. trigger pulse
тридцатикилометро́вка map with scale 1cm = 30km
трикота́ж 1. stockinette (e.g. on respirator)
2. tricot, jersey
триметилалюми́ний trimethylaluminium
три́ммер tab, trimmer
тринитроксило́л trinitroxylene (expl)
тринитрорезорцина́т свинца́ lead styphynate, lead trinitroresorcinate (detonating expl)
тринитротолуо́л trinitrotoluene, TNT
тринитрофено́л trinitrophenol, lydite, picric acid, melinite
трио́д triode (elec)
~, ди́сковый disc-seal triode
~, кристалли́ческий crystal triode
трио́д, кристалли́ческий германиевый germanium crystal triode
~, полупроводнико́вый triode transistor
три́плекс triplex, triplex glass, safety glass
три́плексер triplexer (sigs)
три́тий tritium
трихино́з trichinosis (med)
трихлори́д бо́ра boron trichloride
трихлортривинила́рсин trichloro-trivinyl-arsine
трихлортриэтилами́н trichlortrietilamin
триэтилалюми́ний triethylaluminium
триэтилами́н triethylamine
триэтилбо́р triethylboron
тройни́к 1. T-shaped pipe joint, three-way pipe
2. T-connector, Y-connector
тропа́ 1. path, footpath, track
2. beam (radar)
~, вью́чная pack trail
~, ди́кая rough track
~, прото́птанная well-worn track
тро́пик Козеро́га tropic of Capricorn
~ Ра́ка tropic of Cancer
тропи́нка path, track
тропи́ческий tropical
тропопа́уза tropopause
тропосфе́ра troposphere
трос 1. rope
2. steel rope, steel wire rope, hawser, cable
~, боудено́вский Bowden cable (veh constr)
~, букси́рный tow rope
~, вытяжно́й rip-cord (para)
~, мани́льский manila rope
~, несу́щий 1. main cable (suspension br)
2. messenger (rope)
~ парашю́та parachute guyline
~, пенько́вый hemp rope
~, провиса́ющий sagging cable, slack
~ расчеко́вки quick-release strap
~, сиза́льский sisal rope
~, спаса́тельный lifeline
~, стально́й steel wire rope, SWR
~, тормозно́й brake cable
~ управле́ния, ги́бкий flexible control cable
~, швартóвный securing strap, tie-down
тросомёт, инжене́рный реакти́вный rocket-projected line-throwing equipment

тросомёт, реакти́вный rocket-projected line-throwing eqpt
тростни́к reed, cane (botanical)
~, са́харный sugar cane
троти́л trinitrotoluene, TNT
~, лито́й cast trotyl, cast TNT
~, порошкообра́зный powder TNT
трофе́й trophy, captured equipment
трофе́йный captured equipment
труба́ 1. pipe, tube
2. chimney, funnel, smoke stack, flue
3. trumpet
4. culvert (eng)
5. barrel (ord)
~, аэродинами́ческая wind-tunnel
~ венту́ри venturi tube
~, вертика́льная elevating screw tube (of motor)
~, визи́рная опти́ческая telescopic sight
~, вну́тренняя liner, lining tube
~, воздухопита́ющая 1. air supply pipe
2. underwater breathing pipe on AFV, snorkel
~, вса́сывающая suction pipe
~, выпускна́я exhaust pipe
~, выхлопна́я exhaust pipe
~, газопрово́дная gas pipe
~, газопрово́дная gas pipe
~, деревя́нная wooden culvert
~, доро́жная culvert
~, зри́тельная image amplification device (telescope, binoculars, etc)
~, ка́менная stone culvert (road)
~, канализацио́нная sewage pipe
~ карда́нного ва́ла universal-joint shaft tube, propeller shaft tube
~, кача́ющаяся swinging link (diesel ICE)
~ кожуха́, ве́рхняя barrel casing (of light machine-gun)
~ кожуха́, за́дняя rear radiator casing, rear jacket casing (MG)
~ кожуха́, ни́жняя piston casing (MG)
~ кожуха́, пере́дняя front radiator casing (MG)
~, ми́нная measle shaft (shaft for mined charge) (dml)
~ миномёта mortar barrel
~, наставна́я adapter, extension pipe
~, нефтепрово́дная oil pipeline
~, обво́дная by-pass pipe
~, обса́дная driving pipe (mining)
~, перегово́рная voice pipe
труба́, подзо́рная telescope
~, раздво́енная bifurcated tube
~, распо́рная spreader (pontoon)
~ реакти́вного противота́нкового гранатомёта anti-tank rocket launcher tube
~, ребри́стая finned tube
~, свобо́дная loose barrel (ord)
~ ствола́, вну́тренняя A-tube (ord)
~, стереоскопи́ческая battery commander's OP telescope, stereoscopic telescope, "donkey's ears"
~, стоя́чная stand pipe
~ теле́жки, попере́чная bogie-arm cross tube (tk)
труба́-лаз snorkel (veh)
труба́ч trumpeter
тру́бка 1. tube, pipe
2. fuse (ammo)
3. receiver (teleph)
~, бокова́я offset sleeve (gren)
~ венту́ри venturi tube
~, визи́рная 1. sighting collimator (of gun sight)
2. view-finder
~ возвра́тной боево́й пружи́ны, упо́рная return spring cap (SA)
~, воспламени́тельная 1. tube (vent)
2. flash tube (ammo)
~ воспламеня́ющего механи́зма striker mechanism sleeve
~, вытяжна́я friction tube (ammo)
~, га́зовая 1. gas pipe
2. gas cylinder (SA)
~, гальвани́ческая electric tube (ammo)
~, гильзоотво́дная cartridge-case ejection tube (MG)
~, головна́я nose fuse
~, головна́я уда́рная percussion nose fuse
~, гофриро́ванная corrugated tube (respirator)
~ двойно́го де́йствия double-action fuse
~ двойно́го де́йствия, механи́ческая mechanical double-action fuse
~, дистанцио́нная time fuse
~, до́нная base fuse
~, до́нная уда́рная base percussion fuse
~, дрена́жная vent pipe
~, зажига́тельная safety fuse detonator set
~, запа́льная vent tube, flash tube (ammo)

тру́бка, индика́торная detector tube
~, като́дная cathode ray tube, CRT
~, като́дно-лучева́я cathode ray tube, CRT
~, контро́льная radiator overflow.pipe
~ кругово́го обзо́ра PPI tube
~ кругово́го обзо́ра, электро́нно-лучева́я plan position indicator, PPI
~ магази́на magazine tube
~, механи́ческая дистанцио́нная mechanical time fuze
~, микротелефо́нная handset (telecomms) (obs)
~, направля́ющая primer guide tube
~, огнепрово́дная flash tube
~, окуля́рная eyepiece tube (optics)
~, пароотво́дная steam tube (MG)
~, перегово́рная interphone (tk)
~, перепускна́я by-pass pipe
~, порохова́я дистанцио́нная combustion time-fuse powder-burning time-fuse
~, предохрани́тельная safety collar (fuse)
~, разря́дная discharge tube
~, распо́рная distance piece
~ с темново́й за́писью dark-trace tube, skiatron
~, сливна́я overflow pipe
~, соедини́тельная connecting tube (respirator)
~ стабилиза́тора central tube (mor bomb tail)
~, та́нковая "tank pipe" (Sov nickname for a type of infantry anti-tank rocket launcher)
~, уда́рная 1. percussion fuse 2. percussion tube (ammo)
~ уда́рника striker case (ammo)
~ уда́рного механи́зма firing mechanism housing
~ фо́рмы Т, вытяжна́я friction "T" tube (ammo)
~ холо́дной пристре́лки boresight (tk gunn)
~, центра́льная 1. central tube (shell) 2. central piece (gren)
~ штыка́ bayonet barrel socket
~, штыкова́я bayonet barrel socket
~, электро́нно-лучева́я cathode ray tube, CRT
трубозахва́тчик calipers
трубола́з escape tube (tk)
трубообра́зный tubular
трубообре́з pipe-section
трубопрово́д pipeline, conduit
~, бензи́новый petrol (feed) pipe
~, вое́нный military pipeline
~, впускно́й intake manifold (ICE)
~, выпускно́й exhaust manifold (ICE)
~, дрена́жный drainage pipe(line), vent pipe(line)
~, магистра́льный bulk distribution pipeline
~ перека́чки то́плива fuel transfer pipeline
~, переливно́й overflow pipe
~ пода́чи то́плива fuel-feed pipe
~, подаю́щий delivery pipeline
~, подво́дный underwater pipeline
~, подзе́мный underground pipeline
~, полево́й field pipeline, military pipeline
~, полево́й складско́й field depot pipeline
~, самоте́чный gravity pipeline
~, сбро́сный discharge conduit
~, то́пливный fuel-feed pipeline
труборе́з pipe-cutter
тру́бочка small tube
~, дистанцио́нная safety delay fuze (gren)
~, предохрани́тельная safety sleeve (fuze, min)
тру́бчатый 1. tubular 2. made of tubes, consisting of tubes
тру́бы, зри́тельные sighting arms (periscope)
труднодосту́пный difficult of access
труднозатопля́емый virtually unsinkable
труднопроходи́мый negotiable with difficulty, almost impassable, impenetrable
трудоёмкий demanding much effort, laborious, labour-consuming, labour intensive
труп corpse
трус coward
трусы́ shorts, trunks, underpants
тру́щий rubbing, friction
трюм, та́нковый tank deck (landing ship)
тря́пка rag, duster
тряси́на quagmire, ooze
тря́ска shaking, vibration, flutter
ту́бус radar viewing hood

туго́й tight, taut, stiff
тугопла́вкий 1. refractory, infusible
2. hard (of glass)
тужу́рка jacket (uniform)
ту́ла-токарёв 7.62mm Tokarev pistol (Sov)
тулу́п long fur coat
тума́н 1. fog, mist
2. jamming, clutter (on radar screen)
~, иску́сственный tactical smoke
ту́мба pedestal
~, прича́льная mooring bollard
ту́мблер tumbler switch
~ аррети́рования гироско́па caging switch (gyro)
ту́мбочка small locker
ту́ндра tundra
тупи́к 1. cul-de-sac
2. stalemate, deadlock
3. spur (rly)
~, я́дерный nuclear stalemate
тур fascine, gabion, corbeil
турби́на turbine
~, возду́шная air turbine
~, га́зовая gas turbine
~, двухступе́нчатая double-stage turbine
~, осева́я axial-flow turbine
~, та́нковая га́зовая tank gas turbine
турбовинтово́й turboprop
турбокомпре́ссор turbocompressor
турбонасо́с turbine pump
турбоприво́д, подаю́щий delivery pipeline
турбореакти́вный turbojet
турбуле́нтность turbulence
туре́ль 1. ring mounting, ring mount (ord)
2. turret
~ лю́ка ба́шни turret-hatch ring
~, механи́ческая power driven turret (tk)
~, пу́шечная gun turret
~ с механи́ческим приво́дом power turret
турни́к horizontal bar (PT)
ту́ча storm cloud
ту́ша, мясна́я meat carcass
туше́ние extinguishing
тушь indian ink, drawing ink
тыл rear, rear area
~, арме́йский army rear services
~, войсково́й 1. immediate rear
2. communications zone units
3. administrative area
тыл, глубо́кий rear area, communications zone, deep rear
тылови́к "base wallah"
тылово́й 1. rear
2. administrative
тылы́ rear units, logistical units and establishments
ты́льный 1. rear
2. base
ты́сячная mil (Brit = 1/6,4ØØ X 36Ø degrees, Sov = 1/6,ØØØ X 36Ø degrees)
~, артиллери́йская mil
тэн pentryt, pentrite (expl)
тюрьма́ prison
~, вое́нная military prison
тя́га 1. pull, thrust, draught
2. stay, brace
3. traction
4. rod, connecting rod
5. control rod, control cable (tk)
~, а́нкерная tie rod
~, букси́рная tow bar
~, гу́сеничная tracked traction, crawling traction
~, двойна́я 1. double heading
2. four wheel drive
~ дро́ссельной засло́нки accelerator rod
~, жёсткая rigid tie-rod, rigid connecting-rod
~, ко́нная horse transport
~, осева́я axial thrust, end thrust
~ параллелогра́мма parallel motion link (gunsight)
~, парова́я steam traction
~, перево́дная regulating rod
~ подве́ски suspension link, suspension rod
~, попере́чная 1. track rod (Brit)(veh)
2. steering knuckle tie-rod (US)(veh)
~, попере́чная рулева́я track rod
~, постоя́нная constant thrust
~, приведённая corrected thrust
~ приво́да control linkage rod
~, продо́льная change speed rod
~, реакти́вная jet propulsion
~, рулева́я drag link (veh)
~ рычага́ приво́да то́рмоза brake operating lever rod
~ соба́чки то́рмоза brake-pawl rod
~, спускова́я 1. trigger bar (SA)
2. trigger lever, trip lever (SA)
~, стаби́льная constant thrust

тя́га ста́ртового реакти́вного дви́гателя boost thrust (msl)
~, стяжна́я brace rod
~, теорети́ческая theoretical thrust
~, теплово́зная diesel traction
~, уде́льная specific thrust
~ управле́ния фрикцио́ном clutch control rod
тяга́ч prime mover, tractor
~, авари́йный breakdown tractor
~, артиллери́йский artillery tractor, gun tractor, artillery prime mover
~, брониро́ванный armoured prime mover
~, вспомога́тельный auxiliary tractor (Sov classification denoting tracked tractors for unit domestic needs)
~, гу́сеничный tracked prime mover, tracked towing vehicle, tracked tractor
~, деж́урный duty tractor and crew
~ для буксиро́вки та́нков tank recovery vehicle, ARV
~ для морски́х деса́нтных опера́ций beach recovery vehicle
~ для эвакуа́ции та́нков с по́ля бо́я tank recovery vehicle, ARV
~, колёсный wheeled tractor
~ с полуприце́пом для перево́зки та́нков tank transporter
~, седе́льный tractor truck, tractor unit, goer (to tow semi-trailer)
~, та́нковый armoured recovery vehicle, ARV
~, тра́нспортный cargo tractor
тяговооружённость thrust loading, thrust-weight ratio
тя́говый draught, traction, towing
тяготе́ние gravity
тягу́честь 1. ductility, plasticity
2. tenacity, toughness
3. viscosity (of liquid)
тяж brace, tie rod
тяжелоброниро́ванный heavily armoured
тяжелове́сный heavy, heavyweight
тяжелора́неный 1. seriously wounded man
2. seriously wounded
тя́жесть 1. gravity
2. weight, heaviness, load
3. severity

У

уайт-спири́т white spirit, turpentine substitute
уба́вить (см убавля́ть)
убавля́ть/уба́вить to diminish, to lessen, to reduce; **убавля́ть/уба́вить ход** to reduce speed, to slow down
убе́жище shelter
~, а́томное atomic shelter, nuclear shelter
~, бето́нное concrete bunker
~, взрывоупо́рное blastproof shelter
~, де́рево-земляно́е timber-and-earth shelter
~, котлова́нное dug-out
~ лёгкого ти́па light shelter
~, назе́мное surface shelter
~, подбру́стверное trench shelter
~, подзе́мное underground shelter
~, противооско́лочное splinter-proof shelter
~, противорадиацио́нное fall-out shelter, radiation-proof shelter
~ сплошно́й ра́мной констру́кции log wall shelter
убира́ть/убра́ть to take away, to remove; **убира́ть/убра́ть газ** to throttle back (ac)
убира́ющийся 1. collapsible
2. retractable
3. disappearing
уби́тый в бою́ killed in action
убо́йность casualty effect, lethality
убо́йный lethal
убо́р, головно́й headgear, head-dress
убо́рка 1. cleaning, grooming
2. removal
3. collection (casualties)
убо́рная, похо́дная portable latrine, portable toilet
убо́рщик cleaner
убра́ть (см убира́ть)
у́быль 1. decrease, diminution
2. losses, casualties
3. wear and tear, depreciation
~, боева́я war wastages, battle casualties
уве́домить (см уведомля́ть)
уведомле́ние notification
уведомля́ть/уве́домить to inform, to notify; **уведомля́ть/уве́домить о получе́нии** to acknowledge (sigs)

увеличе́ние 1. magnification, magnifying power
2. increase, growth, enlargement
увели́чивать/увели́чить to increase, to enlarge (photo)
увеличи́тель magnifier, augmenter
"Увели́чить Диста́нцию!" "Open Out!", "Spread Out!"
"Увели́чить Ско́рость!" "Increase Speed!"
уве́чье serious injury, mutilation
увлече́ние pulling (elec)
уволи́ть (см увольня́ть)
увольне́ние 1. discharge, dismissal
2. leave
~ в запа́с transfer to the reserve
~ в отста́вку retirement
~ с вое́нной слу́жбы discharge, discharge from the service
увольня́ть/уво́лить to discharge, to dismiss, to cashier , to grant leave
увяза́ть (см увя́зывать)
увя́зка 1. organization, coordination, "tie-up"
2. lashing, roping, tying-up
увя́зывать/увяза́ть to co-ordinate, to lash, to tie up
уга́р 1. waste, loss (in combustion)
2. carbon monoxide fumes
углево́д carbohydrate
углеводоро́д hydrocarbon
углекислота́ carbonic acid
углеро́д carbon
~, радиоакти́вный radioactive carbon
углово́й angular
угло́м вперёд 1. arrowhead (tac)
2. one up (tk inf tac)
~ наза́д 1. inverted arrowhead (tac)
2. two up (tk inf tac)
угломе́р 1. protractor, angle gauge
2. field clinometer, gunner's quadrant
3. goniometer
4. line, deflection (arty fire orders)
~, ба́шенный gun position indicator (tk)
~ зени́тной артилле́рии azimuth instrument (AA)
~, исхо́дный initial deflection (gunn)
~, исчи́сленный corrected deflection
~ опа́сного направле́ния safety limit
~, опти́ческий optical protractor, optical bevel protractor
~, основно́й zero line (arty)
угломе́р панора́мы bearing scale of dial sight (arty)
угломе́р-квадра́нт 1. combined clinometer and azimuth scale
2. open sight and elevation quadrant
угломе́р-трансформа́тор position indicator
углуби́ть (см углубля́ть)
углубле́ние 1. draught (naut)
2. hole, cavity, recess
~ в колпачке́ hood cavity (min)
углубля́ть/углуби́ть to deepen
у́гол angle
~, азимута́льный azimuth angle
~ ата́ки 1. incidence, angle of incidence (GW)
2. angle of attack (tac)
~ безопа́сности safety angle (gunn)
~ боково́го отклоне́ния azimuth deviation, line error
~ боково́го упрежде́ния lateral lead (angle)
~ броса́ния angle of departure
~ броса́ния, оптима́льный optimum launching angle
~ в пла́не, вспомога́тельный end cutting-edge angle (machine tool)
~ в пла́не, гла́вный 90-degree-minus side cutting-edge angle (machine tool)
~, вектори́альный vector angle
~, ве́кторный vector angle
~ вертика́льного наведе́ния gun elevation
~ вертика́льного обстре́ла total elevation (difference between elevation angular limits)
~, вертика́льный angle of slope (in map reading)
~ ве́тра wind angle
~ ве́тра, курсово́й angle between heading and wind direction
~ виража́ banking angle
~ вну́треннего тре́ния angle of repose (material)
~ возвыше́ния quadrant elevation, angle of fire (gunn)
~ возвыше́ния, испра́вленный adjusted elevation
~ возвыше́ния, исчи́сленный corrected elevation
~ возвыше́ния, наиме́ньший safety angle (overhead fire)

у́гол возвыше́ния, пристре́лянный corrected elevation
~ **вперёд** 1. arrowhead (tac)
2. one up (tk inf tac)
~ **встре́чи** angle of impact
~, **встре́чный** angle of impact
~, **входя́щий** re-entrant angle (topo)
~ **въе́зда** angle of approach
~ **вы́лета** jump (gunn)
~ **вы́лета, горизонта́льный** throw-off (gunn)
~, **гла́вный за́дний** side relief angle (cutting tool)
~, **гониметри́ческий** goniometric angle
~ **горизонта́льного обстре́ла** total traverse (gunn)
~ **горизонта́льной наво́дки** angle of deflection
~ **дерива́ции** angle of drift (gunn)
~, **дирекцио́нный** grid bearing, grid azimuth
~ **дифферéнта** angle of pitch (GW)
~ **доворо́та** switch (gunn)
~, **за́данный и́стинный путево́й** predicted/planned true track (air navig)
~ **закру́чивания** angle of twist (eng)
~ **закры́тия** screening angle (radar)
~ **заостре́ния** tool angle
~ **засе́чки** angle of intersection
~ **захва́та** angular coverage (photo)
~ **зре́ния** angle of view
~, **изве́стный** known angle
~, **изме́ренный** measured angle
~, **исходя́щий** salient
~ **кабри́рования** angle of climb (aeron)
~ **кре́на** angle of roll, roll (GW)
~ **крутизны́** thread angle, helix angle
~ **крутизны́ наре́за** angle of rifling
~, **курсово́й** 1. angle of approach (gunn)
2. course angle (AA)
3. target angle (GW)
4. angle of movement (tac)
~ **ли́нии снаря́д-цель** angle of line of sight in space (GW)
~, **мёртвый** blind angle (AA gunn)
~ **ме́ста** angle of sight
~ **ме́ста разры́ва** angle of sight to burst
~ **ме́ста то́чки** angle of sight
~ **ме́ста це́ли** angle of sight
~ **ме́ста це́ли, отрица́тельный** angle of depression
~ **ме́ста це́ли, положи́тельный** positive angle of sight

у́гол ме́стности angle of sight
~, **наблюдённый** observed angle
~ **набо́ра высоты́** angle of climb
~ **наза́д** 1. reversed arrowhead (tac)
2. two up (tk inf tac)
~ **наибо́льшей да́льности** elevation for maximum range
~ **наиме́ньшего лобово́го сопротивле́ния** angle of minimum drag (aeron)
~ **накло́на** tilt angle, angle of slope
~ **накло́на гла́вной ре́жущей кро́мки** side rake angle (machine tool)
~ **накло́на каса́тельной** angle of arrival (trajectory)
~ **накло́на траекто́рии** flight path angle, angle of flight path vector in space (GW)
~ **накло́на шкво́рня вбок** king-pin inclination (veh)
~ **накло́на шкво́рня наза́д** castor angle (veh)
~ **направле́ния** directional angle
~ **наре́зки** angle of rifling
~, **нулево́й** zero angle
~ **обзо́ра** 1. look angle (radar, GW)
2. angle of visibility, arc of vision
~, **обра́тный дирекцио́нный** back bearing, reciprocal bearing
~ **обстре́ла** 1. arc of fire
2. traverse (gunn)
~ **обстре́ла, вертика́льный** 1. firing elevation (arty)
2. total elevation, difference between the elevation angular limits (gunn)
~ **ожива́ла** ogive angle
~ **опа́здывания** delay angle (ICE)
~ **опереже́ния** advance angle (ICE)
~ **опереже́ния зажига́ния** ignition advance angle (timing angle)
~ **ориента́ции** attitude angle (GW)
~ **ориенти́рования** reference angle (gunn)
~, **о́стрый** acute angle
~ **от норма́ли** obliquity, angle up to or away from normal (ball), angle of incidence (of a missile)
~ **отклоне́ния** angle of deflection, angular displacement
~ **отклоне́ния руля́** control surface incidence angle
~ **отклоне́ния часто́тной модуля́ции** phase excursion, phase deviation (radio)
~ **отко́са** angle of slope

у́гол отме́тки angle of observed error (tk gunn)
~ **отраже́ния** angle of reflection
~ **паде́ния** angle of descent (gunn)
~ **паде́ния, табли́чный** angle of arrival at the ballistic point of graze (gunn)
~ **паралла́кса** parallax angle
~, **параллакти́ческий** parallax angle
~, **пере́дний** front rake angle (machine tool)
~ **перено́са огня́** angle of shift, shift angle (gunn)
~ **пики́рования** angle of dive (aeron)
~ **плани́рования** glide angle, angle of glide
~ **поворо́та** angular displacement
~ **поворо́та ору́дия** angle of traverse (gunn)
~ **подсве́чивания** look angle (radar, GW)
~ **подхо́да** angle of approach
~ **подъёма** angle of climb
~ **по́ля зре́ния** angle of vision, field-of-vision angle
~ **пониже́ния** angle of depression
~ **попере́чного накло́на шкво́рня** king-pin inclination (veh)
~, **поса́дочный магни́тный путево́й** magnetic track on landing approach (air navig)
~ **преде́льной да́льности** angle of extreme range, elevation for maximum range (gunn)
~ **при верши́не** nose angle (machine tool)
~ **при обра́тной засе́чке** back bearing
~ **при основа́нии** base angle
~ **прице́ливания** 1. tangent elevation (angle of projection minus "jump") (gunn)
2. sighting angle (bombing)
3. elevation (SA fire)
~, **програ́ммный** programme angle (GW)
~ **продо́льного накло́на шкво́рня** castor angle (veh)
~ **пролёта** transit angle (electron)
~ **проника́ния, преде́льный** angle of penetration (from normal) (ball)
~ **протека́ния** conduction angle (electron)
~ **протека́ния ано́дного то́ка** angle of flow (electron)
~ **протя́жки, пере́дний** broach front rake angle

у́гол проходи́мости, за́дний angle of departure
~ **проходи́мости, пере́дний** angle of approach
~, **прямо́й** right angle
~ **пу́ска** launching angle (msl)
~, **рабо́чий** 1. conduction angle (electron)
2. angle of flow (electron)
~, **радиокурсово́й** relative bearing by DF
~ **радиоста́нции, курсово́й** true relative bearing of radio station from aircraft
~ **разва́ла** camber angle (wheels)
~ **разлёта** angle of fragmentation (shrapnel)
~ **раство́ра** 1. beam width (radar)
2. angle of lock (radar, GW)
~ **реза́ния** 9Ø-degree-minus top rake angle (machine tool)
~ **рефра́кции** refracting angle, angle of refraction (optics)
~ **рикоше́та** deflection angle (tk armour)
~ **ры́скания** angle of yaw (GW)
~ **с норма́лью** angle of incidence (ball)
~ **сближе́ния** angle of approach
~ **сближе́ния меридиа́нов** angle of convergence, convergent angle (topo)
~ **сбра́сывания** dropping angle
~ **ска́та** angle of incline
~ **склоне́ния** angle of depression
~ **скольже́ния** 1. angle of sideslip (GW)
2. angle of slide
~ **ско́са** 1. angle of skew
2. angle of downwash
~ **скру́чивания** twist angle, angle of torque
~ **смеще́ния** angle of parallax
~ **сниже́ния** angle of depression
~ **сно́са** drift angle
~ **спу́ска** angle of descent
~ **страхо́вки на безопа́сность** Standard Safety Factor (arty crest clearance)
~ **сходя́щегося ве́ера** concentration (line corrections required to cause guns to fire at same point)
~ **схожде́ния** toe-in (veh wheels)
~ **тангажа́** angle of pitch, attitude angle, attitude angle in space (GW)
~ **траекто́рии** trajectory angle
~, **тупо́й** obtuse angle
~ **угла́ встре́чи, дополни́тельный** angle of incidence (gunn)
~ **укло́на** angle of slope

у́гол укры́тия clearance angle
~ **упрежде́ния** lead (gunn)
~ **упрежде́ния по вертика́ли** lead angle (range)
~ **упрежде́ния по горизо́нту** lead.angle (line)
~ **упрежде́ния, расчётный** computed lead angle
~ **установки** angle of incidence (of wing, fan blades)
~ **це́ли, курсово́й** relative target bearing, target angle
у́голь 1. coal
2. charcoal
~, **активи́рованный** activated charcoal (respirator)
~, **бу́рый** brown coal
~, **древе́сный** charcoal
~, **ка́менный** hard coal, coal
уго́льник large central inverted V (AFV sight)
1. set square
2. try square
3. angle bracket, corner iron, angle bar
~, **большо́й** large central inverted V (AFV sight)
~, **ломо́чный** lever bar (pontoon)
~, **стри́нгерный** gunwale bar
угора́ние carbon monoxide poisoning
угрожа́ть/угрози́ть to threaten to menace, to endanger
угро́за threat, menace
~ **нанесе́ния стратеги́ческих я́дерных уда́ров** strategic nuclear threat
~ **нападе́ния бронета́нковых войск** armoured threat
угрози́ть (см угрожа́ть)
удале́ние 1. interval, distance
2. removal, disposal, withdrawal
~, **безопа́сное** safe range, safe distance
~, **максима́льное** maximum distance (topo)
~ **нага́ра** carbon removal, deposit removal
~ **омедле́ния** de-coppering (ord)
удали́ть (см удаля́ть)
удаля́ть/удали́ть to move off, to move away, to send away, to remove, to dismiss; **удаля́ть/удали́ть взрыва́тель** to defuse
уда́р 1. attack, strike, thrust (tac)
2. blow, hit
~ **авиа́ции** air strike
уда́р, а́томный atomic strike, nuclear strike
~, **бо́мбовый** bombing attack, air strike
~, **бомбоштурмово́й** bombing and strafing attack
~, **внеза́пный** 1. surprise strike (AF)
2. surprise assault
~, **вспомога́тельный** subsidiary attack
~, **встре́чный** encounter attack
~, **гла́вный** main thrust, main effort, main attack
~, **лобово́й** forward thrust, frontal attack
~, **ло́жный** feint, feint attack
~, **масси́рованный** thrust on a narrow front, attack on a narrow front, massed attack
~, **масси́рованный отве́тный** massive retaliation, massive counterstroke
~, **основно́й** main thrust
~, **отве́тный** retaliatory attack, counterstroke
~, **отве́тный я́дерный** retaliatory nuclear strike
~ **прикла́дом** butt stroke
~, **раке́тно-я́дерный** nuclear-missile strike
~, **раке́тный** missile strike
~ **с во́здуха** air strike
~, **сокруши́тельный** crushing blow, crippling blow
~, **со́лнечный** sunstroke
~, **тара́нный** ramming
~, **теплово́й** heat stroke
~, **упрежда́ющий** preventive attack
~, **упрежда́ющий я́дерный** pre-emptive nuclear strike
~, **штыково́й** 1. bayonet charge
2. bayonet thrust
~, **я́дерно-раке́тный** nuclear missile strike
~, **я́дерный** nuclear strike
~, **я́дерный, отве́тный** retaliatory nuclear strike
уда́рник 1. firing pin (SA)
2. striker (ord, min, gren)
3. shock worker
~ **взрыва́теля с ка́псюлем** set-back primer, detonator pellet
~, **двухбо́йковый** double-point striker
~, **дистанцио́нный** time needle (fuse)
~, **инерцио́нный** inertia pellet (ammo)
~, **пла́вающий** inertial firing pin, floating firing pin

уда́рник, уда́рный percussion pellet (fuse)
уда́рный 1. impact, percussion
2. strike (tac)
3. striker
уда́ры о сте́нки кана́ла ствола́ side slap (ammo)
удвое́ние частоты́ frequency doubling
удвои́тель частоты́ frequency doubler
уде́льный specific
удержа́ние holding, retention
удержа́ть (см уде́рживать)
уде́рживать/удержа́ть to retain, to hold, to hold by force (tac); **уде́рживать/удержа́ть плацда́рм** to hold a bridgehead; **уде́рживать/удержа́ть проти́вника от агре́ссии** to deter aggression
уде́ржник retaining pawl (ord)
удлине́ние 1. aspect ratio, aerodynamic fineness, fineness ratio, length-diameter ratio (msl), slenderness ratio (ball)
2. linear expansion, lengthening, extension
~, **относи́тельное** percentage elongation
удлини́тель artificial line
удлини́ть (см удлиня́ть)
удлиня́ть/удлини́ть to elongate, to lengthen
удобообтека́емость aerodynamic fineness, streamlining
удобообтека́емый streamlined
удобоуправля́емость manoeuvrability, controllability
удо́бство кла́дки workability (concrete)
~ **управле́ния** ease of control, ease of steering
удовлетвори́ть (см удовлетворя́ть)
удовлетворя́ть/удовлетвори́ть to satisfy, to content, to comply; **удовлетворя́ть/удовлетвори́ть зая́вку** to meet/to satisfy an indent; **удовлетворя́ть/удовлетвори́ть тре́бование** to meet a demand, to satisfy a demand
удостовере́ние 1. certificate
2. certification
~, **командиро́вочное** movement order
~ **ли́чности** identity card
~ **о го́дности к эксплуата́ции** certificate of fitness (mech)
~ **о сме́рти** death certificate
удостове́рить (см удостоверя́ть)
удостове́риться (см удостоверя́тся)
удостоверя́ться/удостове́риться to verify
удостоверя́ть/удостове́рить to certify, to attest; **удостоверя́ть/удостове́рить ли́чность** to prove one's identity
удуше́ние asphyxiation
у́зел 1. node (elec)
2. junction
3. knot (naut)
4. knot, loop
5. joint
6. assembly, complex, lay-out
~ **автомати́ческого опереже́ния** auto-advance assembly, automatic timing device (ICE ignition)
~, **аэродро́мный** airfield complex
~, **"ба́бий"** "granny" knot
~, **батальо́нный противота́нковый** battalion anti-tank layout
~, **вспомога́тельный** backup centre (sigs)
~ **доро́г** road junction
~, **доро́жный** road junction
~, **железнодоро́жный** railway junction
~ **коммуника́ции** communications centre, route centre
~, **компле́ктный** complete unit
~, **ма́лый** minor assembly, sub-assembly (mech)
~ **маши́ны** sub-assembly
~, **неусто́йчивый** unstable equilibrium node (cybernetics)
~ **оборо́ны** strong point, centre of resistance (prepared for all-round defence)
~, **противота́нковый** anti-tank strong point
~, **прямо́й** reef knot
~ **свя́зи** signal office, communications centre
~ **свя́зи деса́нта** assault signal office
~ **свя́зи, лине́йный** line communications centre
~ **свя́зи, опо́рный** hardened communications centre
~ **свя́зи, подви́жный** mobile signals centre
~ **сопротивле́ния** centre of resistance
~ **ствола́** barrel assembly
~, **укреплённый** fortified complex prepared for all-round defence

у́зел, шко́тный sheet-bend (knot)
узкоколе́йный narrow-guage
у́зкости narrows, narrow passage (naut)
у́зость defile
уйти́ (см уходи́ть)
укави́ст radio operator working on microwave
ука́з ukaze, edict, order
указа́ние directive, directions, instruction, order
~, оконча́тельное final order
~, пи́сьменное written instruction
~ по взаимоде́йствию co-ordination instructions
~, предвари́тельное warning order
указа́ния по светомаскиро́вке black-out regulations
ука́занный indicated, specified
указа́тель 1. sign, indicator, marker
2. lane marker (min)
3. index
~, азимута́льный traverse indicator (tk)
~ бараба́нчика прице́ла elevation micrometer
~ давле́ния ма́сла oil pressure indicator
~ да́льности range indicator
~ для обозначе́ния замини́рованного уча́стка minefield marking sign
~ доро́г road sign
~ заражённости ме́стности contamination marker, contamination warning sign
~ коли́чества то́плива fuel gauge
~ кре́на inclinometer
~, маршру́тный route marker
~, мига́ющий flashing (light) indicator
~ ми́ны mine marker, mine-marking sign
~ направле́ния а́томного взры́ва atomic burst direction indicator
~ направле́ния движе́ния direction indicator
~, опти́ческий optical indicator
~ отка́та recoil indicator (ord)
~ пе́ленга bearing indicator
~ поворо́та turn indicator
~ поворо́та и скольже́ния turn-and-bank indicator (ac)
~ прице́ла range reader
~ прохо́да lane marker (in minefield)
~, путево́й route sign
~ расстоя́ния distance indicator
указа́тель ско́рости 1. air speed indicator (ac)
2. speedometer
~ ско́рости ве́тра wind speed indicator
~ температу́ры охлажда́ющей жи́дкости coolant indicator (ICE)
~ угла́ возвыше́ния elevation indicator
~ угломе́ра main scale reader (gun dial-sight)
~ у́ровня depth gauge, level indicator
~ у́ровня воды́ water-level indicator
~ у́ровня то́плива fuel-level indicator, fuel gauge
~ ускоре́ний accelerometer
~ це́ли target indicator
~ числа́ оборо́тов revolution counter, tachometer, rev-counter
~, штыково́й dipstick
ука́зка 1. marker, sign, indicator, pointer
2. figure target
~, мише́нная target marker (device)
~, ручна́я aiming disc
~ с ди́ском для обуче́ния прице́ливанию sighting disc
ука́зка-самоотмеча́тель, универса́льная general-purpose aiming rest
ука́зчик marker
укла́дка 1. container, pack
2. stowage
3. packing (para)
~, боева́я battle stowage (tk)
~ боеприпа́сов ammunition stowage
~ в связь cross-stacking
~ в штабеля́ stacking
~ ка́беля cable laying
~ по-боево́му tactical stowage
~ пути́ track laying
укла́дчик парашю́тов parachute packer
укла́дывать/уложи́ть to pack, to stow
укло́н 1. taper
2. slope, incline, gradient
3. deviation
~, гидравли́ческий hydraulic gradient
~, обра́тный reverse slope
~, попере́чный cross slope
~, преде́льный maximum gradient
уклоне́ние 1. aberration, deviation
2. evasion
~, лине́йное боково́е lateral linear deviation (air navig)
уклони́ться (см уклоня́ться)
уклономе́р inclinometer
~, гироскопи́ческий gyroclinometer

уклоноуказа́тель grade marker, gradient board
уклоня́ться/уклони́ться to deviate, to avoid, to shun, to evade, : **уклоня́ться/уклони́ться от бо́я** to avoid contact with the enemy
уклю́чина rowlock
уко́л 1. prick, pricking
2. injection
3. thrust (bayonet)
укомплектова́ние outfitting, staffing
укомплекто́ванность 1. full strength, complete complement
2. state (numerical strength)
3. level of equipment
~, по́лная 1. full strength
2. full scale equipment
укомплектова́ть (см укомплекто́вывать)
укомплекто́вывать/укомплектова́ть to bring up to full strength, to equip fully **укомплекто́вывать/укомплектова́ть маши́ну** to equip a vehicle, kit a vehicle; **укомплекто́вывать/укомплектова́ть по шта́ту** to bring up to full strength, to bring up to establishment
укоси́на 1. cantilever
2. raker, batter post, raking strut (br)
укрепи́ть(ся) (см укрепля́ть(ся))
укрепле́ние 1. fortification
2. strengthening, reinforcing, reinforcement
~, берегово́е coastal fortification
~, бетони́рованное concrete reinforcement
~, долговре́менное permanent fortification, permanent field work
~, земляно́е earthwork
~ из ту́ров gabionage
~ ме́стности fortification of ground (for defence purposes)
~ ме́стности, заблаговре́менное deliberate fortification of ground
~ ме́стности, поспе́шное hasty fortification of ground
~, оборони́тельное defensive fortification
~, предмо́стное bridge-head
~ примо́рского плацда́рма consolidation of a beach-head
укреплённый fortified, reinforced
~ на шпо́нке keyed (eng)
укрепля́ть(ся)/укрепи́ть(ся) to fortify, to strengthen, to reinforce, : **укрепля́ться/укрепи́ться на ме́стности** to consolidate
укрепрайо́н fortified area
укрыва́ться/укры́ться to take shelter
укры́тие shelter, cover, concealment
~ внутри́ зда́ния indoor shelter
~, вре́менное emergency shelter
~ для ору́дия в око́пе gun pit
~ для расположе́ния та́нков tank harbour, tank leaguer, tank hide
~, есте́ственное natural cover
~ за гре́бнем defensive defilade
~ из сне́га snow shelter
~, индивидуа́льное individual shelter
~ котлова́нного ти́па cut-and-cover shelter
~, котлова́нное vehicle pit (field eng)
~,назе́мное surface shelter
~ от артиллери́йского огня́ shell-proof shelter
~ от радиоакти́вных оса́дков radiation fall-out shelter
~, подзе́мное underground shelter
~, поля́рное arctic shelter
~, противооско́лочное splinter-proof shelter
~, противорадиацио́нное radiation-proof shelter, fall-out shelter
~, самостоя́тельное коллекти́вное independent collective shelter
~, танконедосту́пное tank-proof cover, tank-proof shelter
укры́тие-нора́ snow-hole, snow-hole bivouac
укры́ться (см укрыва́ться)
уку́порка 1. packaging
2. carry box, carrying box
ула́вливатель 1. catcher, separator, interceptor
2. catcher cavity (electron)
3. detector
ула́вливать/улови́ть to capture
улови́ть (см ула́вливать)
уло́вка ruse, trick
уложи́ть (см укла́дывать)
ультима́тум ultimatum
ультразвуково́й supersonic, ultrasonic
"Уменьши́ть Диста́нцию!" "Close Up!"
"Уменьши́ть Ско́рость!" "Reduce Speed!"
умно́житель, втори́чно-электро́нный secondary (emission) electron multiplier
~ со скрещёнными поля́ми crossed-fields (electron-beam) multiplier

умно́житель, фотоэлектро́нный photomultiplier
~ **частоты́** frequency multiplier
~, **электро́нный** electron multiplier
умфо́рмер rotary converter (elec)
умыва́льная washroom, ablutions
у́мысел intent, intention
умягче́ние воды́ water softening
умягчи́тель softener, softening agent
универса́л engineer's transit (surv)
универса́льный general-purpose, universal
унифика́ция standardization
уничтожа́ть/уничто́жить to destroy, to annihilate; **уничтожа́ть/уничто́жить по́рознь** to destroy in detail
уничтоже́ние destruction, annihilation
~, **встре́чное** multi-path cancellation (radio), multi-path fading (electron)
~, **гаранти́рованное** assured destruction
~ **девиа́ции ко́мпаса** compass compensation
уничто́жить (см уничтожа́ть)
унтерофице́р non-commissioned officer, under-officer (used only of foreign armies)
упако́вка packing
~ **в мешки́** sacking
~ **в я́щики** boxing, crating
~, **водонепроница́емая** waterproof packing, moisture-proof packing
~, **карто́нная** carton
упако́вочный packing
упако́вщик packer
уплотне́ние 1. condensation, thickening 2. packing, sealing, seal 3. compression, contraction 4. multiplexing, channeling
~, **водонепроница́емое** watertight packing
~, **во́йлочное** felt seal
~, **вре́менное** time division multiplexing
~, **гермети́ческое** air seal
~, **лабири́нтовое** labyrinth formation (fluid seal)
~, **лабири́нтовое ма́сляное** labyrinth oil retainer
~ **ма́ха** mach diamond
~, **надувно́е** inflatable seal
~ **по вре́мени** time-division multiplexing
~ **по частоте́** frequency-division multiplexing
уплотне́ние подши́пника bearing packing
~, **са́льниковое** stuffing-box packing, oil seal, oil seal packing
~, **торцо́вое графи́товое** carbon face seal (eng)
~, **часто́тное** frequency division multiplexing
уплотни́тель packing gland, seal
уплотни́ть (см уплотня́ть)
уплотня́ть/уплотни́ть to pack, to compact, to multiplex
уполномо́ченный accredited agent, plenipotentiary, representative
уполномо́чивать/уполномо́чить to authorize, to empower
уполномо́чие authorization, delegation of authority
уполномо́чить (см уполномо́чивать)
упо́р 1. bearing surface (min) 2. rest, aiming rest 3. backstop, stop
~ **баланси́ра** rubber stop of road wheel arm
~, **боево́й** locking lug (hinged) (SA)
~ **бу́фера** buffer stop (arty)
~ **дро́сселя карбюра́тора** carburetter throttle stop
~ **затво́ра, боево́й** breech-lock, locking lug
~ **нажи́много рычага́** firing lever pawl (MG)
~, **патро́нный** cartridge stop
~, **плечево́й** shoulder rest (RL)
~ **подава́теля** feed pawl
~, **предохрани́тельный** safety stop
~ **то́рмоза отка́та** buffer stop (ord)
~, **шо́мпольный** cleaning rod fastening unit
~, **штыково́й** bayonet stud
упо́р-сошни́к, откидно́й retractable spade (ord)
упоря́дочение ordering, regulating, regularization
упра́вить (см управля́ть)
управле́ние 1. control, handling, steering 2. headquarters 3. directorate, department 4. administration 5. guidance (GW)
~, **автомати́ческое** automatic control
~, **автоно́мное** self-contained guidance, pre-set control (GW)

управле́ние, администрати́вно-хозя́йственное administration and supply directorate, administration
~, боево́е tactical control
~ боево́й подгото́вки, Гла́вное Chief Directorate of Military Training (Sov)
~ бо́ем tactical control
~ боепита́нием ammunition control
~ ве́ктором тя́ги thrust-vector control (GW)
~ во́енно-восстанови́тельных рабо́т, Гла́вное Chief Directorate of Military Reconstruction Work (rly) (Sov)
~, вое́нно-медици́нское military medical directorate
~, вое́нно-строи́тельное Military Construction Directorate
~ вое́нно-уче́бных заведе́ний Directorate of Higher Military Educational Establishments (Sov)
~ вое́нно-уче́бных заведе́ний, Гла́вное Chief Directorate of Higher Military Educational Establishments (Sov)
~ вое́нных сообще́ний, Гла́вное Chief Military Transport Directorate (Sov)
~ вое́нных трибуна́лов, Гла́вное Chief Directorate of Military Tribunals (Sov)
~ войска́ми control of troops, handling of troops
~ вооружённых сил, автомоби́льное Mechanical Transport Directorate of the Armed Forces (Sov)
~, гидравли́чески уси́ленное рулево́е hydraulic power-assisted steering
Управле́ние, Гла́вное Chief Directorate
управле́ние, Гла́вное Chief Directorate
~, Гла́вное автомоби́льное Chief Motor Transport Directorate (Sov)
~, Гла́вное артиллери́йское Chief Directorate of Missile Troops and Artillery (Sov)
~, Гла́вное бронета́нковое Chief Tank Directorate (Sov)
~, Гла́вное вое́нно- строи́тельное Chief Military Construction Directorate (Sov)
~, Гла́вное вое́нно-медици́нское Chief Military Medical Directorate (Sov)
~, Гла́вное вое́нно-фина́нсовое Chief Military Finance Directorate (Sov)
~, Гла́вное инжене́рное Chief Military Engineering Directorate (Sov)
~, Гла́вное интенда́нтское Chief Intendance Directorate (supply) (Sov)
управле́ние, Гла́вное полити́ческое Chief Political Directorate of the Army and Navy (Sov)
~, Гла́вное разве́дывательное Chief Intelligence Directorate (Sov)
~ движе́нием movement control
~, двойно́е dual control
~, диспе́тчерское centralised direction and control
~, дистанцио́нное remote control
~, изодро́мное proportional control (GW)
~, и́мпульсное bang-bang control (GW)
~, инерциа́льное inertial guidance (msl)
~ ка́дров personnel division
~, кварти́рно-эксплуатацио́нное Barrack Services Directorate, quartering directorate
~, кно́почное pushbutton control, pressbutton control
~ ко́рпуса corps headquarters
~, ле́вое left-hand drive
~, левосторо́ннее left-hand drive
~, маршру́тное route control
~ материа́льно-техни́ческого снабже́ния и зака́зов Directorate of technical equipment supply and orders
~ междунаро́дных возду́шных ли́ний, тра́нспортное international airlines traffic and schedules control
~ на расстоя́нии remote control
~, непосре́дственное direct control
~, ножно́е pedal control
~ огнём fire control
~ огнём, автомати́ческое automatic fire control
~ огнём, радиолокацио́нное radar fire control
~ огнём с передово́го наблюда́тельного пу́нкта control of fire from forward observation post
~ огнём, централизо́ванное centralized fire control
~, педа́льное pedal control
~ по кре́ну roll control
~ по провода́м wire guidance, wire control (GW)
~ по провода́м, кома́ндное wire command guidance (GW)
~ по ра́дио radio control
~ по радиолучу́ beam-rider guidance (GW)
~ по тангажу́ pitch control (GW)
~ полётом flight path control

управлéние, полуавтоматѝческое semi-automatic control
~, прáвое right-hand drive
~, правосторóннее right-hand drive
~, программѝрованное preset guidance (GW)
~, прогрáммное preset guidance, programme control (GW)
~, прямóе direct control
~ пýском launch control
~, развéдывательное Intelligence Directorate
~, располóженное слéва left-hand drive
~, располóженное спрáва right-hand drive
~, реактѝвное jet vane control (GW)
~, рулевóе steering (system)
~ ручнóго управлéния manually operated
~, ручнóе manual control, hand control, hand operation
~ с пóмощью радиолокáтора radar control
~ с пóмощью сервомеханѝзмов servomechanism control
~ стрельбóй fire control
~ тылом rear-services administration
~ угловы́м положéнием attitude control (GW)
~, централизóванное central control
~, центрáльное финáнсовое central finance directorate
~, электрогидравлѝческое electrohydraulic control
управля́емость handling characteristics, controllability
управля́емый 1. controllable, controlled
2. steerable
управля́ть/упрáвить to control, to handle, to operate, to run, to manage, to administer; **управля́ть/упрáвить машѝной** to drive; **управля́ть/упрáвить по проводáм** to control by wires
упражнéние exercise, practice
~, боевóе battle drill, exercise
~, групповóе training exercise (for officers)
~, учéбное training exercise
упрафóрм Directorate of Formation (obs)
упредѝть (см упреждáть)
упреждáть/упредѝть to forestall, to anticipate
упреждéние aim-off, lead
упреждéние, боковóе 1. lateral lead, lateral lead angle (AD)
2. lead (A tk gunnery)
~, вертикáльное vertical lead, vertical deflection
~, исхóдное initial lead
~, линéйное target travel, linear allowance (AA gunn)
~ разворóта, линéйное linear allowance for a turn
~, угловóе angular lead, lead angle
~ цéли target lead
упреждённый predicted, future (gunn)
упрочнéние toughening, hardening (tech)
упрощéнство oversimplification
упрýгий resilient, elastic, flexible
упрýгость elasticity, resilience
~ шѝны tyre flexibility
упря́жка, нáртовая sledge team (animals)
~, собáчья dog team
упущéние 1. omission
2. negligence, neglect
уравнéние equation
~, дифференциáльное differential equation
~ радиолокáции free-space radar equation
урáвнивание equalization, balancing
уравнѝтель 1. equalizer
2. balancing gear, equilibrator (ord)
3. balancer (elec)
4. synchronizer (mech)
~ тóрмоза brake compensator
~, тормознóй brake compensator
уравнѝтельный balance, balancing, equalizing, levelling
уравновéсить (см уравновéшивать)
уравновéшенность trim
уравновéшивание balancing
уравновéшивать/уравновéсить to balance
урагáн hurricane, wind force 12 (Beaufort scale)
~, песчáный sandstorm
~, пылевóй dust storm
урáн uranium
урáновый uranium
урегулѝрование regulation, adjustment
урéз (воды́) 1. cut-away slipway
2. bank (of river etc)
3. water's edge, waterline

у́ровень 1. level
2. spirit level, sight clinometer, levelling
3. angle of sight (arty)
~ **безотка́зности** degree of reliability
~, **бокова́й** side clinometer, sight clinometer (arty)
~ **грунтовы́х вод** ground water level, water table level
~, **за́данный** pre-set level
~ **квадра́нта** quadrant level
~, **контро́льный** field clinometer
~, **кру́глый** circular level, circular bubble
~ **мо́ря** sea level
~ **мо́ря, сре́дний** mean sea level, MSL
~, **наиме́ньший** crest angle, minimum elevation, minimum angle of sight (crest clearance, arty)
~, **нулево́й** datum level
~ **пово́зки** platform levelling bubble
~, **попере́чный** cross level (ord)
~, **преде́льно-допусти́мый** maximum permissible (radiation) level
~ **радиа́ции** radiation level
~ **сигна́ла** signal level
~, **спиртово́й** spirit level
~, **тру́бчатый** tubular level
~, **усло́вный** reference level
~, **шарово́й** circular level, circular bubble
~ **шу́ма** noise level
~, **энергети́ческий** energy level (nucl)
уровнеме́р level indicator, transmission level meter
уро́н losses, damage
уро́чище 1. natural boundary
2. isolated terrain feature
уса́дка shrinking, contraction, settling
~ **гру́нта** soil shrinkage
у́сик 1. cat's whisker (elec)
2. pressure prong (min)
у́сики, нажи́мные pressure prongs
усиле́ние 1. reinforcement, strengthening
2. increase
3. gain (electron)
4. amplification, intensification
~, **о́бщее** overall gain (electron)
~ **про́тив возде́йствия уда́рной волны́ (я́дерного) взры́ва** blast hardening (nucl)
уси́ленный reinforced, strengthened, increased, amplified, intensified
уси́ливать/уси́лить to reinforce, to strengthen, to increase, to amplify (elec), to intensify
уси́лие 1. stress, force
2. exertion, effort
~, **крутя́щее** torsional force
~ **на поворо́те, направля́ющее** cornering force
~, **окру́жное** circumferential load
~, **растя́гивающее** tensile force
~ **спу́ска** trigger pull, trigger pressure
~ **торможе́ния** braking effort, braking force, retarding force
~, **тормозно́е** braking effort, braking force, retarding force
~, **тя́говое** tractive effort
усили́тель 1. amplifier
2. booster
~, **анте́нный** antenna amplifier
~ **бегу́щей волны́** travelling-wave amplifier
~, **ва́куумный** vacuum servo-cylinder
~ **видеосигна́ла** video signal amplifier
~ **высо́кой частоты́** high-frequency amplifier, radio- frequency amplifier, RF amplifier
~, **двухта́ктный** push-pull amplifier
~ **детона́тора** booster (charge), booster explosive
~, **дифференци́рующий** differentiator (elec)
~, **звуково́й** audio-amplifier, sound amplifier
~ **звуково́й частоты́** audio-frequency amplifier, AF amplifier
~ **изображе́ния, электро́нно-опти́ческий** (electro-optical) image amplifier
~, **клистро́нный** klystron amplifier
~ **коммута́ции** switching amplifier
~, **кристалли́ческий** crystal amplifier
~, **ла́мповый** valve amplifier
~, **логарифми́ческий** logarithmic amplifier
~, **магни́тный** magnetic amplifier
~, **микромо́дульный** micromodule amplifier
~, **многокаска́дный** multistage amplifier
~ **мо́щности** power amplifier
~ **ни́зкой частоты́** low-frequency amplifier, audio-frequency amplifier, AF amplifier

усили́тель, оконе́чный 1. final amplifier 2. terminal repeater
~ опти́ческого диапазо́на, ква́нтовый laser (amplifier), optical maser
~, полосово́й band-pass amplifier
~, полупроводнико́вый transistor amplifier
~, порохово́й booster pellet (ammo)
~ постоя́нного то́ка direct-current amplifier, d.c. amplifier
~ промежу́точной частоты́ intermediate frequency amplifier, IF amplifier
~, резисти́вный resistance-coupled amplifier
~, резона́нсный resonance amplifier (electron)
~ с като́дной нагру́зкой bootstrap amplifier
~ с обра́тной свя́зью feedback amplifier
~ с переме́нным усиле́нием variable-gain amplifier
~ сигна́лов изображе́ния video amplifier
~, сумми́рующий summing amplifier (electron)
~ то́рмоза brake servo
~, узкополо́сный narrow-band amplifier
~, широкополо́сный wide-band amplifier
~, электро́нный electronic amplifier
усили́тель-прерыва́тель chopper-amplifier
уси́лить (см уси́ливать)
ускоре́ние acceleration
~, боково́е lateral acceleration (missile)
~, вертика́льное vertical acceleration
~ Кориоли́са Coriolis acceleration
~, кориоли́совое Coriolis acceleration (GW)
~, лине́йное linear acceleration
~, мгнове́нное instantaneous acceleration
~, нача́льное initial acceleration
~, норма́льное lateral acceleration
~, отрица́тельное deceleration
~, преде́льное acceleration limit
~, равноме́рное uniform acceleration
~, результи́рующее resulting acceleration
~ ры́скания yawing acceleration
~ си́лы тя́жести acceleration due to gravity
ускоре́ние тяготе́ния acceleration due to gravity
~, углово́е angular acceleration
~, центробе́жное centrifugal acceleration
~, центростреми́тельное centripetal acceleration
ускори́тель 1. accelerator 2. booster (GW)
~ за́пуска booster rocket
~ нака́та recuperator accelerator (gunn)
~, раке́тный booster rocket
~, ста́ртовый 1. booster (missile) 2. rocket-assisted take-off motor, RATO motor (ac)
~, хвостово́й tandem boost (msl)
ускори́тель-свя́зка wrap round boost, boost cluster (GW)
усло́вие 1. condition 2. term, stipulation 3. agreement, contract
~, предвари́тельное prior condition, pre-requisite
усло́вия, авари́йные emergency conditions
~, атмосфе́рные atmospheric conditions
~, баллисти́ческие ballistic conditions
~ бездоро́жья 1. roadless conditions 2. bad road conditions
~, благоприя́тные favourable conditions
~, боевы́е combat conditions
~ бо́я combat conditions
~ вое́нного вре́мени war-time conditions
~, войсковы́е service conditions
~, грунтовы́е soil conditions
~ капитуля́ции terms of surrender
~, климати́ческие climatic conditions
~ ме́стности, есте́ственные natural ground features, natural ground conditions
~, метеорологи́ческие meteorological conditions, weather conditions
~, неблагоприя́тные adverse conditions
~, норма́льные метеорологи́ческие standard meteorological conditions (gunn)
~, норма́льные топографи́ческие ideal ground conditions (gunn)
~, ночны́е night conditions
~, осо́бо тяжёлые very severe conditions

усло́вия переми́рия terms of truce, conditions of armistice
~, **полевы́е** field conditions
~, **по́чво-грунтовы́е** soil conditions, ground conditions
~ **прекраще́ния огня́** terms of cease-fire
~ **рабо́ты при максима́льной нагру́зке** peak load conditions
~, **расчётные** rated conditions
~, **сло́жные метеорологи́ческие** adverse weather conditions
~ **стрельбы́, баллисти́ческие** ballistic conditions for firing, non-standard conditions (excluding met) (gunn)
~, **табли́чные** range table conditions (arty)
~, **табли́чные баллисти́ческие** standard ballistic conditions (Sov term confined to MV and ammo factors)
~, **табли́чные метеорологи́ческие** standard meteorological conditions (gunn)
~, **табли́чные топографи́ческие** standard topographical conditions (level platform, angle of sight zero) (gunn)
~, **теку́щие** current conditions, present conditions
~, **техни́ческие** specification, standard specifications, technical specifications, technical provisions
~ **хране́ния** storage conditions
~, **эксплуатацио́нные** operating conditions, conditions of use, working conditions, service conditions (of a machine)
усло́вный 1. conditional, provisional
2. conventional
3. pre-arranged
усоверше́нствование improvement, modernization, perfecting
усоверше́нствовать (см совершенствовать
успе́х, такти́ческий tactical success
успокои́тель damper, shock absorber, stabilizer
уста́в manual, regulations
~, **боево́й** field manual
~ **вну́тренней слу́жбы** internal service regulations
~, **вое́нный** military regulations
~, **во́инский** army regulations, military regulations and manuals
уста́в, во́инский дисциплина́рный military disciplinary regulations
~ **гарнизо́нной слу́жбы** manual of garrison duties
~, **дисциплина́рный** disciplinary regulations
~ **карау́льной слу́жбы** manual of guard duties
~, **полево́й** field service regulations
~, **строево́й** drill regulations
уста́вный regulation, prescribed, normal
уста́лость мета́лла metal fatigue
устана́вливаемый 1. mountable
2. adjustable
устана́вливать/установи́ть to set up, to assemble, to erect, to mount, to adjust, to regulate, to lay down, to establish, to lay (mines); **устана́вливать/установи́ть а́зимут** to set the azimuth; **устана́вливать/установи́ть взрыва́тель** to set a fuse, to prime a fuse; **устана́вливать/установи́ть ми́ны** to lay mines; **устана́вливать/установи́ть на взрыв** to arm (ammo); **устана́вливать/установи́ть прице́л** to set sights
установи́ть (см устана́вливать)
устано́вка 1. setting up, assembly, erecting
2. installation, unit, plant
3. mounting (ord)
4. setting (instr)
~, **автодушева́я** mobile shower bath equipment
~, **автомоби́льная рентге́новская** vehicle mounted X-ray equipment
~, **артиллери́йская** gun and mounting
~, **артиллери́йская самохо́дная** 1. SU gun (Sov)
2. self-propelled gun, assault gun
~, **аэротранспорта́бельная пускова́я** air-portable launcher (msl)
~, **ба́нно-дезинфекцио́нно- пра́чечная** laundry and bath unit
~, **ба́шенная** turret mounting
~ **без направля́ющих, пускова́я** zero-length launcher
~, **бензоперека́чивающая** petrol dispenser
~, **бомбардиро́вочная** bombing gear
~, **бортова́я пускова́я** on-board launcher
~, **букси́руемая** towed load

установка, бучильная boiler installation (CW decontamination)
~, **вертлюжная** pivot mounting, bracket mounting (ord)
~ **взрывателя** fuse setting
~, **водоопреснительная** distilling plant, water purification plant
~, **водоочистительная** water purification plant
~, **водофильтровальная** filtration equipment, water filtering plant
~, **вращающаяся** rotatable mounting
~, **вспомогательная силовая** auxiliary power plant, auxiliary power unit
~, **выдвижная пусковая** retractable launcher
~, **грубая** coarse setting
~, **гусеничная** tracked mounting
~ **двигателя, поперечная** transverse engine mounting
~ **двигателя, продольная** longitudinal engine mounting
~, **дегазационная** CW decontamination plant
~, **дезинсекционная** delousing plant, disinfector
~, **дезинфекционная** disinfector, disinfestor, sterilizer
~, **дезинфекционно-душевая** mobile decontamination and shower unit (NBC)
~, **диатомитовая водоочистительная** diatomite water purification set, diatomite water purification plant
~, **дизель-генераторная** diesel-electric set
~ **дистанционного управления, пусковая** remote-controlled rocket launcher
~ **для запуска реактивных противотанковых снарядов** anti-tank rocket launcher
~ **для кондиционирования воздуха** air conditioner
~ **для обеззараживания воды** water sterilizing set
~, **душевая** shower, shower bath, shower unit
~, **дымопусковая** smoke discharger, smoke generator
~, **жёсткая** rigid mount
~ **зажигания** spark setting (ICE)
~ **запасного колеса** spare wheel mounting

установка, зенитная anti-aircraft gun and mounting
~, **зенитная самоходная** 1. self-propelled anti-aircraft mounting, self-propelled anti-aircraft gun
2. ZSU (Sov)
~, **испытательная** test rig
~, **казематная** casemate emplacement
~, **калибровочная** calibrator setting
~, **карбюраторная** carburettor setting
~, **колёсная** wheeled mounting
~, **колёсная пусковая** wheeled launcher
~, **комплексная** 1. multiple (gun) mounting
2. coaxial mounting
3. multiple mounting
~, **компрессорная** compressor
~, **купольная** bell mount
~ **курса и скорости** determination of course and speed
~, **лёгкая самоходно- артиллерийская**
1. light assault gun
2. light SU gun (Sov)
~, **лиже-санная** ski skids (arty attachment)
~, **лыжная** ski mounting (MG etc)
~, **лыжно-носилочная** ski stretcher
~, **лыжно-пулемётная** boat-shaped weapons sledge
~, **лыжно-санная** ski-sledge mounting, ski skid mounting
~, **малокалиберная** sub-calibre mounting
~, **метательная** launcher, projector (missiles etc)
~, **микрометрическая** micrometer mounting
~ **мин** mine laying
~ **мин внаброс** scattered laying (min)
~ **мин-ловушек** booby-trapping
~, **многозарядная пусковая** multiple rocket launcher
~, **многоствольная зенитная** AA multiple mount,AA multi-barrelled mounting
~, **многоствольная пулемётная** multi-barrel machine-gun
~, **многоствольная пусковая** multi-barrel rocket launcher
~, **многоствольная реактивная** multi-barrelled rocket launcher, multiple rocket launcher

установка, многоствольная реактивная пусковая multi-barrel rocket launcher
~, моторная power unit
~ на гусеничном ходу, пусковая tracked rocket launcher
~ на прицепе, пусковая trailer rocket launcher
~, наводимая пусковая traversing launcher
~, наземная пусковая ground launcher
~, неамортизированная rigid mount
~, неподвижная stationary mount, fixed mount
~, нулевая nil reading (gunn), zero setting (instr)
~ опережения зажигания ignition advance device
~ опережения зажигания, ручная manual ignition advance
~, орудийная artillery mounting, gun mounting
~, отопительная heating plant, heating stove
~ перископа periscope mounting
~, подвижная 1. mobile mounting 2. mobile unit, mobile equipment
~, подвижная гусеничная 1. tracked mobile launcher 2. tracked mobile mounting
~, подвижная дезинфекционная mobile disinfection unit, mobile decontamination unit (NBC)
~, подвижная душевая mobile shower unit
~, подвижная колёсная wheeled mobile launcher
~, подвижная компрессорная mobile air compressor unit
~, подвижная пусковая mobile launcher
~, подземная пусковая underground launching site, silo (GW)
~ пожарной сигнализации fire alarm installation
~, полевая душевая field bath unit, field shower unit
~, полуподвижная пусковая transportable launcher (GW)
~ поправки на снос ветром windage adjustment
~ прибора на нуль zeroing
~ прицела sight setting, setting for range
установка прицела, исходная initial sight setting
~ прицела, постоянная battle sight, battle mark
~, прицепная пусковая towed launcher
~ противотанковых управляемых ракет, пусковая anti-tank guided missile launcher
~ пулемёта, шаровая machine-gun ball mount
~, пульверизационная spray equipment
~, пусковая launcher (rocket), launching equipment
~, ракетная missile launcher (GW), rocket launcher, launcher, missile launching site
~, ракетная пусковая missile launcher
~ ракеты, двигательная rocket motor
~, рассчитанная calculated setting
~, расчленённая пусковая articulated launching platform (GW)
~, реактивная rocket launcher, field rocket launcher
~, реактивная пусковая rocket launcher
~ с круговым обстрелом, орудийная mounting with all-round traverse
~, самодвижущаяся auxiliary-propelled mounting
~, самоходная 1. self-propelled gun, self-propelled ordnance, assault gun 2. SU gun (Sov) 3. self-propelled mounting, carrier
~, самоходная артиллерийская 1. self-propelled gun, assault gun 2. SU gun (Sov)
~, самоходная зенитная 1. self-propelled AA gun 2. ZSU gun (Sov)
~, самоходно-артиллерийская 1. assault gun, self-propelled gun 2. SU gun (Sov)
~, санитарно-техническая technical medical vehicle (fitted with special medical equipment), laboratory and X-ray vehicle
~, санная sledge mounting (wpn etc), sledge runner mounting
~, санно-лыжная ski-sledge mounting
~, сатураторная carbonator, aerating plant
~, селекторная selector apparatus (small telephone exchange)
~, силовая power unit, power plant

устано́вка, спа́ренная 1. coaxial mounting (main armament and MG) 2. twin mounting
~, спа́ренная зени́тная twin anti-aircraft mounting
~, сре́дняя самохо́дно- артиллери́йская medium assault gun, medium SU gun (Sov)
~, ста́ртовая launcher
~, ста́ртовая пускова́я launcher
~, стациона́рная fixed installation
~, стациона́рная пускова́я permanent launching site, fixed launcher
~, стро́енная triple mounting
~, счетверённая quadruple mounting
~, счетверённая зени́тная quadruple AA mounting
~, счетверённая пускова́я quadruple launcher
~, съёмно-монта́жная lifting tackle
~ та́нка, силова́я tank power pack
~, та́нковая tank mounting
~, то́нкая fine adjustment
~, тру́бчатая пускова́я tubular launcher (GW)
~, ту́мбовая pedestal mounting (ord)
~, туре́льная ring mounting, ring mount (ord)
~, тяжёлая самохо́дно- артиллери́йская 1. heavy assault gun 2. heavy SU gun (Sov)
~ угла́ закру́тки торсио́на torsion bar spring tension
~ угломе́ра azimuth setting (gun)
~, умфо́рмерная mobile rotary converter (elec)
~, уче́бная оруди́йная drill, training, gun mounting
~, фи́льтро-вентиляцио́нная air filtration plant, ventilation filter
~, четырёхстфо́льная quadruple gun mounting
~, шарова́я 1. spherical mantlet (tk) 2. ball mount, spherical mounting
устано́вки, испра́вленные corrected settings, corrected gun data
устано́вленный 1. authorized, prescribed 2. set, corrected 3. installed, mounted
~ на це́пе trailer-mounted
устано́вщик 1. setter 2. fuse setter (gun number) 3. erector (GW)
устано́вщик взрыва́телей, механи́ческий mechanical fuse-setting machine
~ взрыва́теля fuse setter
~ взрыва́теля, автомати́ческий automatic fuse setter
~ взрыва́теля, ручно́й manual fuse setter
~ высоты́ height setter
~ да́льности range setter
~ прице́ла sight setter, range setter, elevation setter
~, самохо́дный erector-transporter vehicle (GW), self -propelled erector
устаре́вший obsolete
у́стный verbal
усто́й abutment (br)
~, берегово́й abutment (br)
~, мостово́й bridge abutment
усто́йчивость stability, stiffness, steadiness
~, аэродинами́ческая aerodynamic stability
~, баллисти́ческая ballistic stability
~ оборо́ны stability of defensive system
~ по ку́рсу yaw stability (GW)
~ по тангажу́ pitch stability (GW)
~, попере́чная lateral stability
~ при больши́х ско́ростях high-speed stability
~ при хране́нии storage stability
~, продо́льная longitudinal stability
~, противота́нковая anti-tank defensive solidity
~, путева́я yaw stiffness (GW)
усто́йчивый stable, steady
устра́ивать/устро́ить to arrange, to organise, to establish; **устра́ивать/устро́ить заса́ду** to ambush, to lay an ambush
устране́ние removal, elimination
~ заде́ржки immediate action, elimination of stoppage (SA)
~ препя́тствий obstacle clearing, removal of obstacles
устрани́ть (см устраня́ть)
устраня́ть/устрани́ть to remedy, to remove; **устраня́ть/устрани́ть неиспра́вности** to remedy faults
устраше́ние deterrent
~, реалисти́ческое credible deterrent
устремля́ться/устреми́ться to rush, hurry
устро́ить (см устра́ивать)

устро́йства, запасны́е предохрани́тельные backup safety devices
устро́йство 1. device, apparatus, unit 2. installation, establishment 3. working principles, mechanism
~, **автомати́ческое задающее** automatic positioning device
~, **анало́говое вычисли́тельное** analogue computer
~, **анте́нное** antenna system
~, **анте́нно-фи́дерное** antenna-feeder device
~ **анте́нны, перехо́дное** antenna matching device
~, **арифмети́ческое** arithmatic unit (computer)
~, **а́томное** nuclear device
~ **ба́шни, конта́ктное враща́ющееся** rotary turret collector ring
~ **безопа́сности** safety appliance
~, **блоки́рующее** blocking device
~, **букси́рное** towing gear
~, **бу́ферное** buffer assembly (ord, electron)
~, **быстроде́йствующее запомина́ющее** high-speed memory
~ **вво́да** input unit, input (electron)
~, **взрывно́е** explosive device
~, **видеоконтро́льное** video monitor, television monitor, picture monitor
~ **визуа́льного вы́вода** display unit
~, **внутрита́нковое перегово́рное** tank internal communication system, tank intercom
~, **воздухозабо́рное** air inlet, air intake
~, **воспламеня́ющее** initiating mechanism (fuse)
~, **враща́ющееся конта́ктное** rotary base junction (tk)
~, **вспомога́тельное дви́жительно-рулево́е** auxiliary steering motor device
~, **вхо́дное** RF section (radio receiver - antenna- to-valve coupling): input unit
~ **вы́вода** output unit, output (electron)
~ **выносно́го управле́ния** remote control equipment
~, **выносно́е** 1. remote control extension 2. remote control extension handset (sigs)
устро́йство, вычисли́тельное computer, computing machine, calculating machine
~ **вычисли́тельной маши́ны, запомина́ющее** computer memory
~, **га́зовое** gas (fired) device
~, **грозозащи́тное** lightning protector
~, **декоди́рующее** decoder
~, **демпфи́рующее** damping device
~, **детони́рующее** detonating device
~, **дифференци́рующее** differential device (servo mechanics)
~ **для за́пуска дви́гателя сжа́тым во́здухом** compressed air starter (ICE)
~ **для иниции́рования взры́ва** blast initiation device
~ **для наво́дки мо́ста** bridge launcher
~ **для обрабо́тки информа́ции** data handling unit
~ **для подавле́ния поме́х** noise suppressor (elec)
~ **для самоподры́ва** destructor, self-destruction equipment
~, **дози́рующее** metering device (NBC)
~ **загражде́ний** engineer obstacle building activities
~, **загру́зочное** loading device
~, **замедля́ющее** delay system
~, **запа́льное** primer, priming device
~, **запомина́ющее** memory, storage, store
~, **заря́дное** battery charger
~, **измери́тельное** metering device
~, **интегри́рующее** integrating device, integrator
~, **инфракра́сное следя́щее** infra-red tracker
~, **исполни́тельное** actuator (servo)
~, **калибро́ванное** calibrator (radar)
~, **коди́рующее** coder
~, **коммутацио́нное** switchboard installation
~, **коммути́рующее** switching equipment (electron)
~, **коно́идное множи́тельное** cam computing multiplier
~, **корректи́рующее** corrector
~, **крепёжное** fastening device
~, **меша́ющее** jamming device, radio jamming device
~, **модели́рующее** analogue computer
~, **монта́жно-погру́зочное** lifting tackle

устро́йство на конце́ ствола́ пу́шки, эжекцио́нное muzzle bore evacuator, fume extractor
~ на магни́тной ле́нте, запомина́ющее magnetic tape memory
~, нажи́мное pressure device
~, натяжно́е tensioner, tensioning device
~, нивели́рующее levelling equipment
~, облуча́ющее illuminator (radar)
~, обтюри́рующее obturating device, obturator (gunn)
~, оконе́чное terminal equipment, termination (elec)
~, отка́тное recoil system (gunn)
~, перегово́рно-вызывно́е tone generator (teleph)
~, передаю́щее transmitter unit
~, переключа́ющее switch gear, switching equipment
~, пересчётное scaler (nucl)
~, пневмати́ческое пусково́е compressed air starter
~, погру́зочно-разгру́зочное stores-handling equipment, loading equipment
~ подве́ски suspension arrangement
~, подвесно́е тра́нспортное overhead transporter
~, подъёмно-тра́нспортное portable lifting equipment (garage)
~, по́исковое detector-coil assembly, detector head (mine detector)
~, помехоподавля́ющее noise suppression device, anti-jamming device, interference suppression device
~, предохрани́тельное 1. safety device 2. fail-safe (GW)
~, приводно́е 1. homing device (GW) 2. driving gear, drive
~, приёмное receiver, receiving device, receiver unit
~, программи́рующее programmer (eqpt), programming unit
~, програ́ммное programming equipment, programme unit, programmer
~ про́тив запотева́ния demister
~, противооотка́тное recoil system (ord)
~, противопожа́рное fire extinguishing equipment
~, противопоме́ховое anti-jamming device
~, противорадиолокацио́нное anti-radar device

устро́йство, пусково́е launching equipment
~ рабо́ты ди́зелей под водо́й (РДП) snorkel
~, радиопроводно́е перехо́дное radio-to-line transfer unit
~, распо́рное spreading gear (eng)
~, распредели́тельное distributing device
~ резерви́рования stand-by equipment, reserve equipment
~, рулево́е steering gear
~, са́льниковое stuffing-box
~, самоликвиди́рующее self-destruction device
~ "свой-чужо́й" identification friend-or-foe equipment, IFF equipment
~ "свой-чужо́й", опознава́тельное IFF equipment
~ свя́зи communication device, communication system
~, сигна́льное warning device, tell-tale
~, следя́щее servomechanism
~, сма́зочное lubricating arrangement, lubricating device
~, согласу́ющее matching device
~, спусково́е trigger(ing) device
~, стабилизи́рующее stabilizer, stabilizing equipment, stabilizing gear
~, сто́порное locking device
~, сцепно́е coupling assembly
~, счётно-реша́ющее 1. computer 2. guidance unit
~, счётно-реша́ющее электро́нное electronic computer
~ та́нка, вне́шнее перегово́рное tank-infantry phone
~, та́нковое перегово́рное tank internal communication system, tank intercom
~, топливодози́рующее fuel-metering control
~, тормозно́е brake gear
~ ты́ла rear services organization, administrative planning
~, уплотни́тельное distance piece
~ управле́ния control unit
~, цифрово́е счётно-реша́ющее digital computer device (elec)
~, шарни́рное герметизи́рующее joint sealing device
~, эжекцио́нное bore evacuator, fume extractor (gunn)
~, экипиро́вочное handling equipment

устрóйство, электрóнное счётно-решáющее electronic computer
~ энергоснабжéния power installation
~, я́корное anchor gear, anchorage apparatus
устрóйство-анáлог, счётно-решáющее analogue computer
устýп 1. echelon
2. shelf, ledge, terrace
3. recess, step, shoulder
~, боевóй locking recess (SA)
~ влéво echelon left (tac) (form of "two up" with rear formation to left of forward formations), echelon to the left
~ впрáво echelon right (tac) (form of "two up" with rear formation to right of forward formations), echelon to the right
~ кáморы chamber mouth
~ куркá shoulder of cocking lever
~ лéйнера shoulder of liner (ord)
~, локтевóй berm (fd works)
~ орýдий range difference (gunn)
ýстье 1. estuary
2. mouth, opening
усýшка dryage, shrinkage due to drying, loss due to drying
"Утверждáю" "seen and approved" (on document)
утеплéние heating, insulation
утёс rock, cliff, rock face
утéчка 1. dispersion, ullage, dissipation
2. leakage, escape
3. leakance (elec)
~, внýтренняя internal leakage
~ воды́ water leak
~ вóздуха air leak
~, сéточная grid leak (elec)
утилизáция salvage, recovery of waste materials
ути́ль scrap, useful scrap, salvage
ýтка 1. cleat
2. canard (aeron)
утолщéние, центрýющее band, positioning band (ammo)
уточнéние clarification, detailing, amplification
уточни́ть (см уточня́ть)
уточня́ть/уточни́ть to clarify, to give in detail, to amplify, to elaborate
утрои́тель частоты́ frequency tripler
утыкáние jamming (ammo)
ухáб hole, rut (in a road)
ухáбистый bumpy, rutted (of roads)
ухищрéние, такти́ческое tactical ruse
~, техни́ческое technical device
ухóд 1. departure
2. drift (radio)
3. maintenance, care
4. deflection (bearing)
~, ежеднéвный last parade servicing
~ за рáнеными care of the wounded
~ и сбережéние maintenance
~, техни́ческий maintenance
уходи́ть/уйти́ to withdraw
уцелéвший 1. survivor
2. surviving
учáствовать to take part, to participate; **учáствовать в боя́х** to take part in operations
учáсток 1. section, part, portion
2. sector, area
3. parcel, plot, strip
4. region, district, locality
~ акти́вного полёта powered flight phase (GW)
~, акти́вный powered flight phase (GW)
~ бактериáльного заражéния bacteriologically contaminated area
~, батарéйный battery lane (arty)
~ ввóда gathering phase (GW)
~ вертикáльного подъёма vertical flight phase (GW)
~ вертикáльного полёта vertical flight phase (GW)
~ возвращéния в атмосфéру re-entry phase (GW)
~ возвращéния на зéмлю re-entry phase (GW)
~ выведéния launch phase (GW)
~ выведéния на траектóрию launch phase (GW)
~ выключéния burn-out phase, cut-out phase (GW)
~ вы́садки десáнта, промежýточный beach transit area (amph ops)
~ дожигáния re-ignition phase (GW)
~, дорóжно-комендáнтский section of railway under military control
~, заболóченный swampy area, marshy area
~ заграждéния obstacle belt
~, замини́рованный mined area
~ заражéния contaminated area
~, заражённый contaminated area
~, имитациóнный battle simulation area
~, испытáтельный test area

уча́сток ме́стности sector of the terrain
~ **ме́стности, заражённый** contaminated area
~ **ме́стности, танконедосту́пный** tank-proof locality
~ **ми́нного по́ля** mine belt
~ **наведе́ния** guided phase (GW)
~ **накло́нного полёта** tilt-over phase (GW)
~ **неподви́жного загради́тельного огня́** defensive fire task area
~ **оборо́ны** defended area (Sov concept is regimental defended area)
~ **оборо́ны, полково́й** regimental defended area
~ **огнево́го ва́ла, батаре́йный** battery lane (arty)
~**, опа́сный** danger zone
~ **пасси́вного полёта** free-flight phase (GW)
~**, пасси́вный** free-flight phase (GW)
~**, перекры́тый** covered-in section
~ **перепра́в** river crossing sector
~**, поражённый** affected area
~ **проры́ва** breakthrough sector
~**, протра́ленный** swept area (mine)
~ **пути́** leg of route, stage of route
~ **разго́на** boost phase, launch phase (GW)
~**, размини́рованный** swept area (mine)
~ **разруше́ний** demolished area
~ **самонаведе́ния** homing phase (GW)
~ **сбо́рки мо́ста** bridge assembly site
~ **свобо́дного неуправля́емого полёта** initial boost phase (GW)
~ **свобо́дного полёта** free-flight phase (GW)
~ **склоне́ния** tilt-over phase (GW)
~ **сосредото́ченного огня́** target area for concentrated fire
~ **спу́ска** descent phase (GW)
~ **стабилиза́ции** terminal phase (GW)
~**, ста́ртовый** vertical flight phase, launch phase (GW)
~ **траекто́рии** phase of trajectory
~ **траекто́рии, акти́вный** powered flight phase (GW)
~ **траекто́рии, пасси́вный** free-flight phase (GW)
~ **трубопрово́да** section of pipeline
~**, труднопреодоли́мый** difficult-to-negotiate sector
уча́сток управля́емого полёта guided flight phase (GW)
учёба 1. studying, learning, studies
2. drill
3. training
учéбник textbook
учéбно-трениро́вочный training
учéбный drill, training, practice
учéние 1. teaching, doctrine, training
2. drill, exercise
~**, боево́е** 1. action training, (nav)
2. battle exercise
~**, двухстепе́нное** two-level exercise
~**, двухстепе́нное кома́ндно-штабно́е** two-level command post exercise, HQ exercise involving superior and one or more subordinate HQs
~**, двухсторо́ннее** two-sided exercise with troops
~**, кома́ндно-штабно́е** headquarter exercise, command post exercise, CPX
~**, ночно́е** night exercise
~**, одино́чное боево́е** action drill
~**, одностепе́нное кома́ндно-штабно́е** HQ exercise (involving only one HQ), single-level command post exercise (CPX)
~**, односторо́ннее** one-sided exercise with troops
~**, оруди́йное** gun drill
~**, пожа́рное** fire practice, fire-fighting training
~**, смотрово́е** test exercise
~**, такти́ческое** tactical exercise
учéния с боево́й стрельбо́й, такти́ческие live firing exercise, field firing exercise
учёт 1. report
2. record, stock-taking
~ **боеприпа́сов** ammunition stock record
~**, вое́нный** registration for military service
~ **горю́чего** fuel accounting
~**, инвента́рный** stocktaking record
~**, коли́чественный** quantitative accounting
~**, материа́льный** stocktaking
~**, машинизи́рованный** machine accounting
~ **обстано́вки** appreciation (of the situation)
~ **осмо́тра** record of inspection
~ **расхо́да** expenditure record
учётный recording, registration

учи́лище, артиллери́йское gunnery school
~, вое́нное military school, military academy
~, вы́сшее общевойсково́е кома́ндное officers training school, military academy, officer cadet academy
учреди́ть (см учрежда́ть)
учрежда́ть/учреди́ть to found, to establish, to set up; **учрежда́ть/учреди́ть награ́ду** to institute an award
учрежде́ние establishment, institution
~, вое́нно-ветерина́рное military veterinary establishment
~, вое́нно-медици́нское service medical establishment
~, культу́рно-просвети́тельное cultural guidance centre, cultural and educational establishment
~, нау́чно-иссле́довательское scientific-research establishment
~, обеспе́чивающее supporting establishment
~ обслу́живания service installation
~, полево́е ба́нно-пра́чечное fied bath and laundry installation
~, ремо́нтное maintenance establishment, repair and maintenance establishment, repair and maintenance unit
~, тылово́е administrative or logistical installation
учрежде́ния ты́ла rear services, logistical services
~, тыловы́е rear services, logistical services
ушиб bruise, contusion
ушире́ние broadening, widening
~ ба́шни turret bulge
~ прохо́да gap widening
ушко́ 1. lug
2. eye (leaf spring)
~ крюка́ hook shackle
~, подвесно́е suspension lug
~ тормозно́й коло́дки brake-shoe return-spring lug
уще́лье gorge, ravine, canyon
уще́рб damage, injury, harm, loss
уязви́мость vulnerability
~ ты́ла logistical vulnerability
уязви́мый vulnerable
~ с во́здуха vulnerable to air attack
уясне́ние зада́чи clarification of mission, definition of mission

Ф

фа́за phase
~ обеспе́чения, берегова́я beach maintenance (amph ops) (logistics maintenance in the initial phase of a landing)
фази́рование phasing
фазитро́н phasitron
фазовраща́тель phase shifter, phase shifting network
фазовыра́вниватель phase equalizer
фазоинве́ртор phase inverter
фазорасщепи́тель phase splitter
фазорегуля́тор phase shifter
фа́кел, ма́гниевый magnesium flare
~ освети́тельного снаря́да star shell flare
~, освети́тельный star (of star shell), star shell flare
факси́миле facsimile
фа́ктор внеза́пности element of surprise
~, поража́ющий casualty-producing element, casualty producer (nucl)
факту́ра invoice
фал halyard, line
~, вытяжно́й static line (para)
фальшбо́рт bulwark
~, брезе́нтовый canvas bulwark, dodger (eng)
~ та́нка tank skirting plate
фане́ра 1. plywood
2. veneer
фантастро́н phantastron
фанфари́ст trumpeter
фа́ра headlamp
~, затемнённая blackout headlamp
~, инфракра́сная infra-red headlamp
~, подвижна́я spotlight
~, подкры́льная wing-tip flare
~, противотума́нная fog lamp
~, светомаскиро́вочная 1. blackout headlight
2. dimmed headlight
фара́да farad
фарва́тер fairway (naut)

фарва́тер, протра́ленный swept channel
фаренге́йт fahrenheit
фармаце́вт pharmacist
~, вое́нный military dispenser (med)
фармаци́я pharmacy
фа́ртук, грязево́й mud flap
~, защи́тный anti-gas apron
~, предохрани́тельный canopy sock (para)
фас face, side
фаса́д frontage, facade
фа́ска bevel edge
~ кла́пана valve face
фасо́нный shaped, fashioned, moulded
фа́у-2 V-2 (missile)
фаустпатро́н panzerfaust, faustpatrone (German bazooka-type weapon)
фаши́на fascine
~, двухко́мельная fascine with thick ends of brushwood to both edges
фе́динг fading (radio)
фельдма́ршал field marshal
фе́льдшер medical assistant
~ баталь́она, ста́рший Officer i/c battalion medical platoon
~, вое́нный medical assistant (trained junior medical officer, but not a fully-qualified doctor)
фельдъе́герь courier
фен foehn (wind)
фе́нол phenol
фе́рма 1. jib, girder
2. truss (br)
3. farm
~, ба́лочная beam girder
~, брусча́тая раско́сная flat-top pratt truss (br)
~ Га́у Howe truss
~, консо́льная cantilever girder
~ ла́нгера bowstring truss
~, откидна́я collapsible strut (heavy drop platform)
~, решётчатая lattice girder
~, ри́гельно-раско́сная double-lock frame-truss
~, сплошна́я доща́тая plank web girder (br)
фе́рменный lattice
фермово́й girder
ферри́т ferrite
ферри́товый ferrite
фи́бра fibre
фигу́ра, головна́я head and shoulders figure target
~, поясна́я waist figure target
фи́дер feeder (elec)
~, анте́нный antenna feeder
~, су́живающийся tapered feeder (electron)
физзаря́дка physical training
фи́зика твёрдого те́ла solid-state physics
физиотерапевти́ческий physiotherapeutic
физкульту́ра physical training
фикса́ж 1. fixer, fixing solution (photo)
2. fixing (photo)
фикса́тор 1. catch, lock
2. locator, index pin
3. fixer
~ затво́ра bolt stop
~ отсе́чек feed cylinder plunger (ord)
~ регуля́тора gas cylinder locking bar (SA)
~ ствола́ barrel catch
фикса́ция 1. fixing (photo)
2. fixation (surv)
3. clamping
фикси́рование clamping
фикси́ровать/зафикси́ровать to determine, to establish, to note down, to fix (photo); **фикси́ровать/зафикси́ровать ми́нное по́ле** to record a minefield
фикти́вный fictitious, imaginary
фила́та whaling board (min)
филёнка slat
фильм film
фильмоско́п projector
фильтр filter
~ во́здуха air filter
~, возду́шный air filter
~, вью́чный pack-horse-carried water purification filter
~ гру́бой очи́стки first-stage filter, coarse filter
~, заградительный band stop filter (electron)
~, загражда́ющий rejector (electron)
~, запира́ющий suppression filter
~, интерференцио́нный interference filter
~ кла́сса М M-derived filter
~, ма́сляный oil filter

фильтр, металли́ческий мембра́нный metal diaphragm filter
~ **ни́зких часто́т** low-pass filter
~, **опти́ческий** optical filter
~, **поло́сно-загражда́ющий** band-stop filter
~, **поло́сно-пропуска́ющий** band-pass filter
~, **полосово́й** band-pass filter
~ **пропуска́ния ве́рхних часто́т** high-pass filter
~, **пропуска́ющий** acceptor circuit
~, **пропуска́ющий высо́кие часто́ты** high-pass filter
~, **пропуска́ющий не́которую полосу́ часто́т** band-pass filter
~, **пропуска́ющий ни́зкие часто́ты** low-pass filter
~, **пропуска́ющий полосово́й** band-pass filter
~, **противоды́мный** smoke filter
~, **противоослепля́ющий** anti-dazzling screen
~, **пылеула́вливающий** dust filter
~ **развя́зки** decoupling filter
~, **развя́зывающий** decoupling filter (electron)
~, **резона́нсный** tuned filter
~, **сгла́живающий** smoothing filter circuit (elec)
~, **сетево́й** power supply filter
~, **се́тчатый** mesh filter, gauze filter
~, **узкополо́сный** narrow-band filter
~, **универса́льныйноси́мый** portable universal filter
~, **цепно́й** iterative network (electron)
~, **широкополо́сный** wide-band filter
фильтр-лову́шка wave trap (electron)
фильтрова́ние filtration, filtering
фильтрова́ть to filter
фильтр-отсто́йник fuel strainer
фильтр,П-обра́зный pi-section filter network
фильтр-поглоти́тель absorbent filter
фильтр-про́бка rejector
фи́нишер 1. finisher (road construction) 2. deck landing control officer, batsman (ac carriers)
фио́рд fiord
фитилёк small wick
фити́ль 1. wick 2. match, slow match
~, **воспламени́тельный** slow match
фити́ль, тле́ющий slow match
~, **тле́ющий пеньќовый** bitumenized hemp-covered slow match
фи́тинг fitting
~, **растру́бный** socketed fitting
~ **с фла́нцами** flanged fitting
флаг flag
~ **до полови́ны** flag at half mast
~, **развёрнутый** broken flag, unfurled flag
флагду́к bunting
флагшто́к flagstaff
флажо́к 1. small flag, pennant, guidon 2. signal flag, handflag 3. pin, catch
~ **кронште́йна** magazine catch (SA)
~ **подава́теля** ammunition follower release lever
~ **предохрани́теля** safety-catch thumb piece
фланг flank
~, **вне́шний** outer flank
~, **за́гнутый** retired flank
~, **заходя́щий** marching flank (drill), enveloping flank
~, **неподви́жный** pivot flank
~, **неприкры́тый** open flank, exposed flank
~, **откры́тый** exposed flank, open flank
~, **при́мкнутый** secured flank
~, **уязви́мый** vulnerable flank
фла́нги, при́мкнутые secured flanks
фланги́рование flanking, outflanking
фланго́вый flanking
фла́нец flange, collar
~, **впускно́й** inlet flange, intake flange
~, **враща́ющийся** rotating flange
~ **ги́льзы** cartridge case flange
~, **дро́ссельный** choke joint (waveguide)
~, **жёсткий** fixed flange
фланки́рование, огнево́е outflanking by fire
фланки́ровать to flank, to outflank
фланки́рующий flanking
фла́ттер flutter
флегматиза́тор desensitizer (expl)
флейц flat paint brush
флеп flap (tyre)
флот, вое́нно-морско́й navy
~, **возду́шный** air force
флоти́лия flotilla
флуктуа́ция jitter, scintillation, fluctuation

флуоресцéнция fluorescence
флю́гер 1. homing head
2. weathercock, weathervane, windvane
флюгеровáние винтá feathering of propeller
флю́ксметр fluxmeter
флюс 1. gumboil (med)
2. flux (welding)
~ в ви́де пáсты paste flux
~, жи́дкий liquid flux
~, порошкóвый powder flux
~, свáрочный welding flux
фля́га water bottle
фля́жка canteen, water bottle
фóкус 1. focus, focal point
2. trick
~, неустóйчивый unstable equilibrium focus (cybernetics)
фокуси́рование focussing
фокуси́ровать to focus
фокусирóвка focusing
фóльга foil, leaf (metal)
фон background
~, естéственный радиациóнный natural background radiation
~, инфракрáсный infra-red background, IR background
~, рассéянный diffuse background (nucl)
~, такти́ческий tactical background
~, тепловóй thermal background
~, шумовóй background noise
фонари́, боковы́е габари́тные clearance lights, side marker lights
~, ствóрные marker beacons (eg at entry and exit of a ford)
фонáрь 1. lantern, light
2. canopy (of transparent material), clerestory
~, боковóй стоя́ночный side parking lamp (MT)
~, габари́тный clearance light
~, габари́тный боковóй side-marker light
~, гакабóртный stern light, "tail light"
~, головнóй head-light
~ зáднего хóда reversing lamp (MT)
~, зáдний tail light (MT)
~, индикáторный convoy light (distance keeping indicator)
~, кармáнный pocket torch
~, сбрáсываемый jettisonable canopy (ac)
фонáрь, светомаскирóвочный габари́тный blackout clearance light, blackout marker lamp (for indicating width or height)
~, светсигнáльный signal lamp, Aldis lamp
~, сдвижнóй sliding canopy (ac)
~, сигнáльный signal lantern
~, стоя́ночный parking lamp
~, трёхцвéтный tricoloured lantern
~, хвостовóй marker lamp, tail lamp
~, щиткóвый panel light
~, электри́ческий electric torch, flash-light
фонд resources, stocks, fund
~, жили́щный housing resources
~, жилóй accommodation
~, казáрменно-жили́щный barracks accommodation
~, кварти́рно-жили́щный quartering and accommodation
~, обмéнный exchange pool
~, подмéнный exchange pool, replacement pool
~, премиáльный award fund
~, резéрвный reserve stocks
~, ремóнтный repair pool, spare parts pool
фоногрáмма phonogram
фонóграф phonograph, dictaphone
форкáмера precombustion chamber
фóрма 1. mould
2. shape
3. uniform
4. form (docs)
~, анкéтная questionnaire
~, аэродинами́ческая aerodynamic shape
~, воéнная military uniform
~ зернá grain form (ammo)
~, зи́мняя winter dress
~ и́мпульса pulse shape
~, караýльная guard mounting order (dress)
~, лéтняя summer dress
~ манёвра type of manoeuvre
~, обтекáемая streamlined shape
~ одéжды order of dress
~ одéжды, боевáя battle-dress
~ одéжды, выходнáя walking-out dress
~, парáдная full dress uniform
~, парáдно-выходнáя parade walking-out dress

фо́рма, повседне́вная everyday uniform, everyday dress
~, повседне́вно-полева́я everyday field dress
~, полева́я field service dress
~, похо́дная marching order, field dress
~, рабо́чая working dress, fatigue dress
~ релье́фа land form, topographic form, relief
фо́рменный uniform
формирова́ние forming, organizing
~ поездо́в marshalling (rly)
форми́ровать/сформи́ровать to form, to make up, to activate, to organise
формо́вка shaping
фо́рмула formula
~, колёсная wheeled configuration, wheel arrangement
формуля́р log book, equipment book
~ на боеприпа́сы ammunition data card
~ ору́дия gun book
формфа́ктор shape factor
форпо́ст outpost
форси́рование 1. boosting, forcing 2. assault crossing, forced crossing (usually water obstacle)
~ с хо́ду hasty crossing
~ снаря́да gripping of projectile by rifling, shot start (ball)
~ тя́ги thrust augmentation (msl)
форси́ровать to make an assault crossing, to force (usually water obstacles)
форсиро́вка overloading
форсу́нка 1. injector 2. burner
~, кла́панная outward-opening pintle-type nozzle (diesel)
форт fort
фортификацио́нный defence, fortification
фортифика́ция fortification
~, берегова́я coastal fortification
~, долговре́менная permanent fortification
~, полева́я field defences
фосге́н phosgene
фо́сфор phosphorous
~, бе́лый white phosphorous
фосфоресце́нция phosphorescence
фотоаппара́т camera
фотобума́га photographic paper
фотовспы́шка, электро́нная electronic flash
фотограмме́тр photogrammeter, phototheodolite
фотограмме́трия photogrammetry
фото́граф photographer
фотографи́рование photographing, photography
~, перспекти́вно-горизонта́льное high oblique photography
~ в инфракра́сных луча́х infra-red photography, IR photography
~, возду́шное air photography
~, возду́шное перспекти́вное oblique aerial photography
~, инфракра́сное infra-red photography
~, маршру́тное возду́шное strip aerial photography, line overlap (air photo)
~, непрямо́е инфракра́сное infra-red photography using converter and ordinary film
~, одино́чное возду́шное vertical pin-point (photo)
~, перспекти́вно- горизонта́льное high oblique photography
~, перспекти́вное oblique photography
~, пла́новое возду́шное vertical air photography
~, пла́ново-перспекти́вное low oblique photography
~, площадно́е возду́шное area aerial photography, mosaic aerial photography
~, прямо́е инфракра́сное photography using infra-red film
фотографи́ровать/сфотографи́ровать to photograph
фотографи́ческий photographic
фотогра́фия photography
~, инфракра́сная infra-red photography
~, тенева́я shadowgraphy (wind-tunnel obsn)
~, цветна́я colour photography
фотодио́д photodiode, solid-state photo-detector
~, герма́ниевый germanium photodiode
~, кре́мниевый silicon photodiode
~, лави́нный avalanche photodiode
фотодиссоциа́ция photodissociation (chem, nucl)
фотоза́пись photographic record
фотозатво́р shutter (camera)
~, самозаводя́щийся self-winding shutter (photo)
~, што́рный roller-blind shutter (camera)

фотокассе́та film badge (radiation dose indicator)
фотокато́д photocathode
фотокинопулемёт camera gun, cine-gun
фотоко́пия photostat, photographic print
~, конта́ктная contact print (photo)
фотолаборато́рия, полева́я инжене́рная field engineer photolaboratory
фотолюминесце́нция photoluminescence
фотоме́трия photometry
фотомонта́ж photomontage
фото́н photon
фотонейтро́н photoneutron
фотонноплá́н photon-powered spacecraft
фотопла́н photographic map, mosaic
фотопласти́нка photographic plate
фотоплёнка photographic film
фотопроводи́мость photoconduction, photoconductivity
фотопроводни́к photoconductor, photoresistor
фотопулемёт cine-gun, camera gun
фоторазве́дка photographic reconnaissance
фоторазложе́ние photodissociation (chem, nucl)
фоторекогносциро́вка photographic reconnaissance
фотосни́мок photograph
~, широкоуго́льный wide-angle photograph
фотосопротивле́ние 1. photoresistor, photocell
2. photoresistance
фотоста́т photostat
фотосхе́ма mosaic (photo)
~, маршру́тная strip mosaic (photo)
~, уточнённая semi-controlled mosaic (photo)
фотосъёмка, маршру́тная aerial strip mosaic (photo)
~, стереоскопи́ческая stereoscopic photography
фототелегра́мма facsimile, facsimile telegram
фототелеграфи́я facsimile telegraphy
фототеодоли́т phototheodolite
фототопогра́фия phototopography
фототравле́ние photo-etching
фототрансформа́тор rectifier, phototransformer
фотоумножи́тель photomultiplier (electron)
фотоэлеме́нт photoelectric cell
~, бромосере́бряный silver bromide photoelectric cell
~, ваку́умный vacuum photoelectric cell
~, ве́нтильный rectifier photoelectric cell
~, газонапо́лненный gas-filled photoelectric cell
~, герма́ниевый germanium photoelectric cell
~, кре́мниевый silicon photoelectric cell
~, полупроводнико́вый semi-conductor cell
~, пусто́тный vacuum photoelectric cell
~ резисти́вного де́йствия photoconductive cell
~ с вне́шним фотоэффе́ктом photo-emissive cell
~, селе́новый selenium photoelectric cell
~, эмиссио́нный photo-emissive cell
фотоэму́льсия photographic emulsion
фотоэффе́кт photoelectric effect
фра́за, ко́довая code phrase
фрактова́ние chartering, hiring
фракцио́нный fractional (chem)
фра́кция fraction (chem)
фрахт charter, hiring charges
фрега́т frigate
фрез cutter, milling cutter, mill
фре́за milling cutter, cutter, mill
~, доро́жная road harrow
фре́зер cutter, milling cutter
фре́он freon (refrigeration)
фрикцио́н 1. clutch
2. ball and disc, wheel and disc (AAFCE)
~, автомати́ческий automatic clutch
~, бортово́й steering clutch (tracked vehicle)
~, гла́вный engine clutch, master clutch
~, двойно́й double clutch
~ механи́зма поворо́та steering clutch
~, многоди́сковый (бортово́й) multiplate (steering) clutch
фронт 1. front
2. force, army group
~ батаре́й battery frontage
~, втори́чный secondary front (met)
~ и́мпульса leading edge of pulse
~ окруже́ния perimeter of envelopment
~ погру́зки loading bay
~, позицио́нный static front
~ проры́ва break-through frontage
~, развёрнутый extended front
~, сплошно́й continuous front line

фронт, сухопу́тный land front
~, тёплый warm front (met)
~ уда́рной волны́ shock-wave front
~, холо́дный cold front (met)
~, широ́кий extended front, broad front, wide front
фронта́льный frontal
фронтови́к front-line soldier
фронтово́й front-line
фронтогене́з frontogenesis (met)
фронтоли́з frontolysis (met)
фтор fluorine
фто́ристый fluorine, fluorous, fluoride
фуга́с fougasse, land mine
~, а́томный atomic landmine
~ заме́дленного де́йствия delayed action charge
~ натяжно́го де́йствия charge with pull-igniter (improvised mine etc)
~, полево́й field charge
~, противопехо́тный напа́лмовый anti-personnel napalm mine
~, управля́емый controlled mine, command-fire mine
~, хими́ческий chemical fougasse, chemical mine
~, я́дерный nuclear land mine
фуга́сный high explosive, HE
фуга́сы, до́нные bottom charges (underwater)
фульмина́т рту́ти fulminate of mercury (expl)
фунда́мент foundation, substructure, bed
фу́нкция function
~, баллисти́ческая ballistic function
~ корреля́ции correlation function
~, переда́точная transfer function
фура́ж forage, fodder
фура́жка peaked service cap
фура́н furan (chem)
фурго́н 1. van, enclosed truck
2. van body
фурниту́ра 1. accessories
2. dental appliances
футеро́вка, жаропро́чная combustion liner, combustion chamber liner (rocket engine)
футля́р case, casing, housing, container
футоме́р altimeter calibrated in feet (ac)
фу́торка threaded bushing, threaded sleeve
фуфа́йка 1. jersey, sweater
2. padded jacket
фюзеля́ж fuselage

X

ха́ки khaki
хала́т, маскиро́вочный denison smock, camouflage robe
хала́тно negligently
хала́тность neglect, negligence
хара́ктер 1. nature
2. type
~ бо́я nature of the battle
~ ме́стности nature of the terrain
характери́стика 1. performance, response
2. personal record, character
3. parameter, characteristic (electron)
~, амплиту́дная amplitude response
~, аэродинами́ческая aerodynamic performance
~, боева́я fighting qualities
~, вольт-а́мперная V-I characteristic (electron)
~, вре́менная transient response (servo system)
~ дви́гателя engine data
~ долгове́чности life performance
~ доро́г road data
~ запа́льных свече́й, теплова́я heat range of spark plugs
~ излуче́ния radiation pattern
~, и́мпульсная impulse characteristic
~, лине́йная linear response, linear characteristic
~ отраже́ний от по́чвы clutter pattern (radar)
~, перехо́дная transient response
~ плаву́чих ка́честв amphibious characteristic
~ приёмника, рабо́чая receiver operating characteristics
~ проходи́мости по пересе́ченной ме́стности cross-country performance
~, рабо́чая operating characteristics, performance
~, служе́бная confidential report, record of service (of civilian)
~, спектра́льная spectral characteristic, colour response

характери́стика, стати́ческая steady state response (control system)
~, фáзовая phase response
~ холостóго хóда no-load characteristic
~, частóтная frequency response
~, эксплуатациóнная field performance
характери́стики, баллисти́ческие ballistic characteristics
~, взры́вчатые explosive characteristics
~ дви́гателя engine characteristics
~, динами́ческие dynamic characteristics (electron)
~, нагрýзочные load characteristics
~ орýдия, оптимáльные gun's maximum performance
~, реáльные objective parameters
~, стати́ческие static characteristics (electron)
~, тáктико-техни́ческие combat characteristics
~ управлéния control characteristics
~, числовы́е numerical characteristics
характрóн charactron (CRT producing display in the form of letters or numbers)
хвóрост brushwood
хвост tail
~ затвóра feed arm actuating stud (MG)
~ колóнны tail of a column
~ спусковóго крючкá trigger tail
~ ствóльной корóбки butt socket
хвóстик, нарезнóй threaded tail-piece
хвостови́к shank, shaft, smallstem
хвостовóй tail
хвóя conifer branches
хемилюминесцéнция chemiluminescence
хемопáуза chemopause (atmosphere)
хемосфéра chemosphere
химвзвóд chemical platoon
химдетéктор chemical agent detector kit
химдозóр CW reconnaissance patrol
химзащи́та chemical defence
хи́мик-дегазáтор CW decontamination orderly
хи́мик-дезактивáтор orderly for radioactive decontamination
хи́мик-дозиметри́ст CW monitor (member of NBC monitoring branch)
хи́мики CW troops
хи́мик-развéдчик CW reconnaissance scout
химинстрýктор CW specialist
хими́ческий chemical, CW
~, биологи́ческий и радиологи́ческий nuclear, biological and chemical (NBC)
хи́мия chemistry
химснаря́д chemical shell
химснаряжéние CW equipment
химчáсть CW unit
хинизáция taking of quinine (med, before entering malarial zone)
хини́н quinine
хирýрг surgeon
хи́трость, воéнная stratagem (mil)
хладагéнт refrigerant, refrigerating medium, refrigerating agent
хладотрáнспорт refrigerated transport
хлебозавóд bakery plant
~, полевóй механизи́рованный mobile field bakery (truck mounted)
хлебопёк baker
хлебопекáрня bakery
хлебопечéние, полевóе field baking (bread etc)
хлеборéзка bread slicer
хлопóк bang, clap
хлóпок, нитрóванный guncotton, nitrated cotton
хлопóк, сверхзвуковóй sonic boom
хлор chlorine
хлорáт кáлия potassium chlorate
хлорацетофенóн chloracetophenone, CAP, tear gas (CW)
хлорвинилдихлóрарсин lewisite, chlorovinyl dichlorarsine (CW)
хлори́рование chlorination
хлóрокись oxychloride
хлороуглерóд carbon tetrachloride
хлорпикри́н aquinite, chloropicrin
хлорциáн cyanogen chloride (CW)
хлыст whip, horsewhip, riding whip
хля́стик half-belt (on greatcoats etc)
хóбот 1. trail, rear trail assembly (wpn) 2. trunk (of elephant)
~ лафéта carriage trail
~ лафéта, корóбчатый box trail (ord)

ход 1. motion, movement
2. entry, passage
3. run, running (mech)
4. move
5. thread
6. stroke (piston)
7. traverse (surv)
8. axle and wheel assembly, running gear
9. course (of events)
~, азимута́льный мёртвый azimuth backlash
~, азимута́льный мёртвый azimuth backlash
~, бесша́жный double-stroke advance (ski) (by sticks only - skis are kept together)
~, бесшу́мный silent running
~, боево́й wheels and axle assembly (ord)
~, буссо́льный compass traverse
~ в рулево́м управле́нии, свобо́дный steering backlash
~ вверх upstroke (ICE)
~, вися́чий flying traverse, open traverse (surv)
~ вниз downstroke (ICE)
~ вперекидку alternate step skiing
~ впу́ска intake stroke (ICE)
~ вса́сывания intake stroke (ICE)
~ вы́пуска exhaust stroke (ICE)
~ вы́хлопа exhaust stroke (ICE)
~, гу́сеничный tracked running gear
~, двухша́жный two-step walk (ski)
~, за́дний reverse gear, reverse movement
~, за́мкнутый closed traverse, circuit traverse (surv)
~ затво́ра, прямолине́йный straight blowback (SA)
~, колёсный wheeled mounting, wheeled running gear
~, ме́нзульный plane table traverse (surv)
~, мёртвый backlash
~ на лы́жках, одноша́жный one-step skiing
~, неза́мкнутый open traverse (surv)
~, неро́вный uneven running
~, обра́тный 1. reverse motion
2. flyback (CRT etc)
~ педа́ли pedal travel
ход пере́дний 1. forward movement
2. forward gear
3. front axle and wheels
~, полигона́льный traverse, traverse survey
~, полигонометри́ческий traverse (surv)
~, полугу́сеничный half track
~ по́ршня piston stroke
~ по́ршня, рабо́чий power stroke
~, преде́льный top speed
~, рабо́чий power stroke (engines)
~, равноме́рный balanced running
~, разо́мкнутый open traverse, continuous traverse
~, ру́сский walking step, one step (ski)
~, свобо́дный 1. free play, play
2. free wheeling
3. freewheel gear
~ сжа́тия compression stroke (ICE)
~, сжима́ющий compression stroke (ICE)
~, соедини́тельный connecting trench (fd wks)
~ сообще́ния communication trench
~ сообще́ния, соедини́тельный crawl trench
~, холосто́й idle running, idling
хода́тайство application, petition
ходи́ть/идти́ to go, to walk; **ходи́ть/идти́ за** to fetch, to go after; **ходи́ть за** to take care of **ходи́ть/идти́ на лы́жах** to ski; **ходи́ть/идти́ ю́зом** to skid, to slide
ходово́й running, moving, travelling
ходьба́ вы́падами alternate thrust (ski)
~ на лы́жах skiing
хожде́ние, подво́дное deep wading, snorkeling (tk)
хозя́йственник, ста́рший войсково́й senior Q officer
хозя́йственный administrative
хозя́йство 1. economy, management, housekeeping
2. internal economy
3. farm
4. equipment (farming or household)
5. establishment, outfit, unit
~, вое́нное 1. logistics, military economy
2. military formation or unit (sl)
~, войсково́е unit administration and services
~, ка́бельное cable harness
~, подсо́бное unit welfare farm
~, ро́тное company administration and services

холе́ра cholera
холм hill
~, песча́ный sand dune
холми́стый hilly
холоди́льник refrigerator
холосто́й 1. blank, idle, empty
2. unmarried
хому́т saddle, yoke, collar, clamp
~ крепле́ния ствола́ gun crutch
~ рессо́ры spring clamp
~ со́шки bipod sleeve
хому́тик holder, slide
~, закрепля́ющий positioning lug
~ прице́ла back-sight slide
~, прице́льный back-sight slide
хон hone
хонингова́ние honing
хо́рда chord, subtense
хране́ние 1. custody, safekeeping
2. stackage, storage, storeholding
3. stowage
4. care and maintenance
~, безопа́сное safe storage
~ боеприпа́сов ammunition storage
~, вре́менное temporary storage
~, дли́тельное extended storage, long-term storage
~, кратковре́менное short-term storage
~, складско́е warehousing
храни́лище storehouse, stores
~ боеприпа́сов magazine, ammunition store
~, огнеопа́сное store for inflammable items
~, откры́тое outdoor storage space, dump, stores compound
~, подзе́мное underground storage room, underground store room
~, полузакры́тое open-sided shed
храни́ть to keep, to store, to stow
храпови́к ratchet
~ ста́ртера starter dog
хребе́т 1. mountain range
2. ridge (topo)
~, го́рный mountain range, mountain ridge
хром chromium
~ ствола́ barrel chroming
хроми́рование chrome-plating
хромосфе́ра chromosphere (atmosphere of sun)
хромота́ lameness
хро́мпик bichromate
хрониза́тор timer
хронирова́ние timing
хроно́граф chronograph
~, счётный counter chronograph
хроно́метр chronometer, stop watch
хронометра́ж timing, timekeeping, stop-watch timing
хронометри́ровать to time
хронометри́ст time keeper, stop-watch operator
хру́пкость brittleness, friability, frangibility
хрящ 1. cartilage, gristle
2. gravel, shingle, coarse sand
ху́тор 1. farmstead
2. small village (in Ukraine)

Ц

ца́нга 1. tongs, tweezers, forceps
2. clamp
ца́пфа 1. journal, pivot, pin
2. trunnion
~, гла́вная main trunnion
~ лю́льки cradle trunnion (ord)
~, оруди́йная gun trunnion (ord)
~, поворо́тная king pin
цара́пина scratch, score
цвет побежа́лости tempering colour
цветно́й colour, coloured
цветоразличи́мость colour perception
цветочувстви́тельность colour sensitivity
цевьё 1. stock fore-end (rifle)
2. hand grip (LMG)
це́зий caesium
целеотыска́ние locating targets, target location
целераспределе́ние allocation of targets
целесообра́зность advisability, expedience, expediency, purposefulness
целесообра́зный desirable, expedient
целеуказа́ние target indication, target designation
~ от ориенти́ра target indication from a reference point
~ от репе́ра target indication from a reference point
~ по ка́рте target indication from a map

целеуказа́ние разры́вами снаря́дов target indication by reference from a salvo
~ трасси́рующими пу́лями target indication using tracer bullets
целеуказа́тель target indicator (helicopter)
целеустремлённый purposeful
це́лик 1. wind gauge
2. backsight
целина́ 1. virgin soil
2. open country
~, сне́жная virgin snow
це́лить(ся) to aim at, to point at, to direct at
целлуло́идный/целлюло́идный celluloid
целлюло́за cellulose
"Цель!" "Hit!" (gunn report)
цель 1. target
2. aim, objective, purpose
~, ближа́йшая immediate objective, immediate aim
~, брониро́ванная armoured target
~, быстроскрыва́ющаяся fleeting target (arty, tac)
~, быстрохо́дная high-speed target
~, быстроявля́ющаяся snap target (SA)
~, ви́димая visible target
"Цель Ви́жу!" "Seen!" (gunn report)
цель вне досяга́емости огня́ target out of range
~, внеза́пно появля́ющаяся surprise target
~, возду́шная air target
~, вре́менная snap target
~, второстепе́нная secondary target
~, вы́годная worthwhile target, profitable target, favourable target
~, группова́я multiple target, group target
~, дви́жуюаяся moving target
~, дополни́тельная alternate target
~, жива́я live target
~, закры́тая defiladed target
~, кабри́рующая climbing target
~, коне́чная final objective
~, ло́жная 1. false target (tac)
2. confusion reflector (EW, electron)
~, малова́жная low priority target, non-priority target
~, малоразме́рная small target, pin-point target
~, надво́дная surface target, sea target
цель,назе́мная ground target
~, наме́ченная intended target
~, неви́димая invisible target, indirect target
~, нежива́я material target
~, неконтра́стная target indistinguishable from the background
~, ненаблюда́емая unobserved target, indirect fire target
~, непла́новая opportunity target
~, неподви́жная stationary target
~, низколетя́щая low-flying target
~, обли́ческая crossing target
~, одино́чная single target, individual target
~, основна́я 1. main objective
2. primary target
~, откры́тая exposed target, direct target
~ пе́рвой о́череди primary target
~ первоочередно́й ва́жности high-priority target
~, пе́шая enemy (target) on foot
~, пла́новая preplanned target
~, площадна́я area target
~, пода́вленная neutralized target
~, подви́жная moving target
~, появля́ющаяся snap target, snap shooting target
~, приближа́ющаяся incoming target, approaching target
~, пристре́лянная registered target
~, простра́нственная extended target
~, радиолокацио́нная radar target
~, сопровожда́емая tracked target
~, теплоизлуча́ющая ло́жная heat-emitting decoy
~, тихохо́дная low-speed target
~, то́чечная pinpoint target
~, уходя́юая receding target, outgoing target
~, уче́бная aim of lesson
цель-квадра́т target co-ordinates
цельнометалли́ческий all-metal
цельносварно́й all-welded
цельнотя́нутый seamless, solid-drawn, weldless (pipes etc)
цеме́нт cement
~, быстросхва́тывающийся quick-setting cement, rapid-hardening cement
~, быстротверде́ющий quick-setting cement, rapid-hardening cement
~, высокоалюмини́стый high-alumina cement

цемента́ция 1. carburizing, case hardening
2. cementing
цементи́рованный 1. case-hardened, carburized
2. cemented
це́нзор censor
цензу́ра censorship
~, вое́нная military censorship
це́нность, боева́я military usefulness
центр акусти́ческой ба́зы sub-base centre (Srg)
~ величины́ centre of buoyancy
~ водоизмеще́ния centre of buoyancy
~ враще́ния 1. centre of rotation
2. fulcrum, fixed pivot, swivel
~ давле́ния centre of pressure
~ да́льней косми́ческой свя́зи long-range space communications centre
~ Земли́, географи́ческий Earth's geographic centre
~ Земли́, магни́тный Earth's magnetic centre
~ ине́рции centre of inertia
~, испыта́тельный testing centre
~, кома́ндный command centre, control centre
~, координацио́нно-вычисли́тельный co-ordinating and computing centre (space flights)
~ масс centre of mass
~ обрабо́тки информа́ции information processing centre
~ отраже́ния centre of reflectivity (radar)
~ ПВО, операти́вный air defence operations centre
~ по ликвида́ции после́дствий я́дерного нападе́ния damage control centre (nucl)
~ попада́ния mean point of impact, mpi
~, распредели́тельный distribution centre
~ рассе́ивания mean point of impact, mpi
~ рассе́ивания разры́вов mean point of burst, mpb
~ свя́зи communication centre
~ сопротивле́ния centre of pressure (ballistics)
~ тя́жести centre of gravity
~ управле́ния авиа́цией air control centre
центр управле́ния боевы́ми де́йствиями combat operations centre
~ управле́ния движе́нием на автомоби́льных доро́гах highway traffic control centre
~ управле́ния огнём fire control centre
~ управле́ния полётами плацда́рма десанти́рования airhead air traffic co-ordination centre
централи́т centralite (ammo)
центра́льный central
центри́рование centring
центрифу́га centrifuge
~, фильтрова́льная centrifugal filter
центрифугирова́ние centrifuging
центробе́жный centrifugal
центро́вка balance, trimming, centring, alignment
центропла́н wing centre section (ac)
центростреми́тельный centripetal
це́почка, искрогаси́тельная spark suppressor network
цепь 1. circuit, network (elec)
2. chain
3. extended line
4. chain, range (mountains)
~, ано́дная anode circuit
~, взрывна́я firing circuit, explosive chain
~ вну́тренней свя́зи intercommunication circuit
~, возду́шная open wire circuit
~ воспламене́ния ignition train
~, втори́чная secondary circuit
~, вту́лочно-роликова́я roller chain
~, входна́я input circuit
~, го́рная mountain range
~, гу́сеничная track (veh)
~ да́льнего обнару́живания long-range detection system
~, двухря́дная duplex chain
~, детонацио́нная detonating circuit, explosive circuit
~, за́мкнутая closed circuit
~, иску́сственная phantom circuit (comms)
~, ме́рная surveyor's chain
~ нака́ла filament circuit (valve)
~, нака́льная filament circuit
~, обесто́ченная dead circuit, de-energized circuit
~, огнева́я powder delay train (fuse)
~, паралле́льная parallel circuit

цепь, перви́чная primary circuit
~ пехо́ты extended line of riflemen
~, после́довательная series circuit
~, после́довательно-паралле́льная series-parallel circuit
~ противоскольже́ния anti-skid chain
~ прямо́й свя́зи intercom circuit
~ развя́зки decoupling circuit
~, разо́мкнутая open circuit
~, разо́мкнутая детонацио́нная unarmed explosive train (fuse)
~, се́точная grid circuit
~, силова́я power circuit
~, стрелко́вая extended line of riflemen
~, телефо́нная telephone circuit
~, тра́ковая track chain, overall chain (veh)
~, фанто́мная phantom circuit
~, электри́ческая electrical circuit
цета́н cetane
цех workshop, shop
~, лите́йный foundry
~, механи́ческий machine workshop
~, ремо́нтный maintenance workshop
~, сбо́рочный assembly shop
~, сталеплави́льный steel foundry
циа́н cyanide
~, хло́ристый cyanogen chloride (CW)
циано́метр cyanometer
цикл cycle
~, бина́рный binary cycle
~, водоро́дный hydrogen-helium cycle
~, двухта́ктный two-stroke cycle (ICE)
~ дезактива́ции radioaction decontamination cycle (NW)
~, за́мкнутый closed cycle
~, общевойсково́й combined arms course
~, рабо́чий duty cycle, cycle of operation
~, техни́ческий technical course
~, четырёхта́ктный four-stroke cycle (ICE)
циклогене́з cyclogenesis (met)
цикло́н 1. cyclone (met) 2. cyclone dust collector
циклотро́н cyclotron
циклофо́н cyclophon
цили́ндр cylinder
~ вене́льта control electrode (CRT)
~ взрыва́теля, оседа́ющий ferrule (fuse)
~, га́зовый gas cylinder
~, гидравли́ческий hydraulic cylinder
цили́ндр, гла́вный тормозно́й brake master cylinder
~, колёсный тормозно́й wheel brake cylinder
~ компре́ссора recoil cylinder (ord)
~ нака́тника recuperator cylinder (ord)
~, оседа́ющий ferrule (fuse), collapsible cylinder (mor)
~ рекупера́тора recuperator cylinder
~ с мо́крой ги́льзой wet liner
~, силово́й actuating cylinder, power cylinder, actuator
~, скрепля́ющий outer tube (barrel)
~ то́рмоза, гла́вный brake master cylinder
~ то́рмоза отка́та buffer cylinder (ord)
~, тормозно́й brake operating cylinder
~ уравнове́шивающего механи́зма balancing spring casing (ord)
цили́ндрик, переда́точный порохово́й perforated pellet powder train (fuse)
цилиндри́ческий cylindrical
цили́ндры, горизонта́льные располо́женные horizontally opposed cylinders
цинга́ scurvy
~, око́пная trench mouth (med)
цинк zinc
ци́нка moisture-proof metal liner
~, патро́нная airtight cartridge container
ци́нковый zinc
ци́ркуль compasses, dividers
~, разме́точный dividers
~, рыча́жный 1. beam caliper 2. beam compasses (for large circles)
ци́ркуль-измери́тель dividers
циркуля́ция воды́ water circulation
~ ма́сла oil circulation
~, принима́ющая forced circulation
~, принуди́тельная forced circulation
~, сифо́нная syphon circulation
~, терми́ческая thermal circulation
цисте́рна 1. tank (container) 2. cistern wagon
~, железнодоро́жная rail tank wagon, rail tank car, tanker
~ с питьево́й водо́й drinking water tank
цитаде́ль citadel
цифербла́т dial plate, dial, face
~, светя́щийся luminous dial
цифи́рь figures, numbers (obs)
ци́фра 1. number, figure, digit 2. cypher

ци́фра, двои́чная binary, binary number
цо́коль 1. base (electron valve)
2. lamp socket
цоцта́в, подвижно́й rolling stock (rly)
цуг волн wave train (nucl)
цы́пка chilblain

Ч

"Ч" час "H"-hour
чал mooring rope
ча́лка mooring rope
чан, бучи́льный boiling tank (used in NBC decontamination)
час "А" "A"-hour (of Atomic strike)
~, исхо́дный H-hour
~, коменда́нтский curfew
~ "Ч" "H"-hour
часово́й 1. sentry
2. clockwork
ча́сти, вози́мые запасны́е on-vehicle spare parts
~, вспомога́тельные accessories
~, деревя́нные furniture (SA)
~, доро́жные road building units (mil eng)
~, запасны́е spare parts
~ и подразделе́ния, авиацио́нные air force units and sub-units
~, инструме́нты и принадле́жности, запасны́е spare parts, tools and accessories
~, отка́тные recoiling parts
~, перекрыва́ющиеся overlap (air mosaics)
~, подвижны́е moving parts, recoiling parts (ord)
~, подде́рживающие supporting units
~, строи́тельные construction units (mil eng)
~, трудя́щиеся working parts
~, тру́щиеся friction parts (mech)
~, унифици́рованные запасны́е interchangeable parts, parts common to more than one equipment
части́ца particle
~ большо́й эне́ргии high-speed particle
~, бомбардиру́ющая incident particle (nucl)
части́ца, заряжённая charged particle (ion etc)
~, излучённая emitted particle
~, налета́ющая incident particle (nucl)
~, перви́чная precursor (eng, nucl)
части́цы particles
~, радиоакти́вные radioactive particles
части́чный partial, piecemeal
ча́стное quotient
частоко́л 1. palisade
2. railing interference (radar)
3. paling, fencing
частота́ frequency
~ бие́ний beat frequency
~, бокова́я side frequency
~, весьма́ высо́кая very high frequency, VHF (from 30-300 MHz)
~, высо́кая high frequency
~, гла́вная master frequency
~, грани́чная cut-off frequency
~, до́плеровская Doppler frequency
~, запасна́я reserve frequency
~, звукова́я audio frequency
~ ка́дров frame frequency
~, контро́льная pilot frequency (radio)
~, кра́йне высо́кая extremely high frequency, EHF (from 30-300 GHz)
~, крити́ческая critical frequency (radio)
~, максима́льная примени́мая maximum useable frequency, MUF
~, модули́рующая modulation frequency
~ модуля́ции modulation frequency
~, назна́ченная assigned frequency
~, наини́зшая приго́дная высо́кая lowest useful high frequency, LUHF
~, несу́щая carrier frequency
~, ни́зкая low frequency, LF (30-300 kHz)
~, нулева́я zero frequency
~, оптима́льная рабо́чая optimum working frequency, OWF, optimum traffic frequency, OTF
~, о́чень высо́кая very high frequency, VHF (from 30-300 MHz)
~, о́чень ни́зкая very low frequency, VLF (3-30 kHz)
~ переда́чи transmitting frequency
~ по отраже́нию, преде́льная cut-off frequency (ionospheric wave propagation)
~ по поглоще́нию, преде́льная absorbtion limiting frequency

частотá повторéния и́мпульсов pulse repetition rate, pulse repetition frequency, PRF
~ повторéния периоди́ческих и́мпульсов pulse repetition rate, pulse repetition frequency, PRF
~, поднесу́щая sub-carrier frequency
~ помéхи frequency modulated jamming
~, предéльная cut-off frequency
~, промежу́точная intermediate frequency, IF
~, рабóчая operating frequency (radio)
~, резонáнсная resonant frequency (electron)
~, сверхвысóкая super high frequency, SHF, microwave frequencies (from 3 to 30 GHz)
~ слéдования периоди́ческих и́мпульсов pulse repetition rate, pulse repetition frequency, PRF
~, сóбственная natural frequency
~ сóбственных колебáний natural frequency
~, срéдняя medium frequency (300 kHz to 3 MHz)
~ сры́ва колебáний quench frequency
~, топографи́ческая distance between grid lines (topo)
~, угловáя angular frequency
~, ультравысóкая ultra high frequency, UHF, microwave frequencies (from 300 to 3,000 MHz)
частотомéр frequency meter
часть 1. unit
2. part, section
3. matters, affairs
4. side
чáсть, авиациóнно-планёрная air force glider unit
часть, автобронетáнковая armoured unit
~, автомоби́льная MT unit
~, автонóмная self-contained unit
~, администрати́вно-хозя́йственная unit of Quartermaster General's branch, "Q" unit
~ артиллéрии, материáльная ordnance
~, áтомная боевáя atomic warhead
~, бактериологи́ческая боевáя bacteriological warhead
~ бáшни, выступáющая за борт turret overhang
часть, боевáя 1. military unit, fighting unit
2. warhead
~, боевáя кассéтная cluster warhead, sub-projectile warhead
~, бронебóйная боевáя armour-piercing warhead
~, броневáя armoured unit (Sov term includes armoured car unit)
~, бронетáнковая tank unit, armoured unit
~, брониро́ванная материáльная armoured materiel
~ вáла, шлицóванная splined portion of shaft
~, взаимозаменя́емая replacement part
~, воéнная military unit
~, войсковáя military unit
~, высокоэффекти́вная боевáя high lethality warhead
~, вью́чно-гужевáя трáнспортная animal transport unit
~, головнáя warhead, head (ammo)
~, голожнáя, моноблóчная single warhead
~, гóрная mountain unit
~, горнострелкóвая mountain rifle unit, mountain unit, mountain infantry
~, дегазациóнная CW decontamination unit
~, дежу́рная duty unit
~, довóльствуемая 1. administered unit
2. consumer unit, consuming unit, using unit, user unit, utilizing unit
~, ду́льная chase (muzzle end of barrel)
~, зáдняя отóгнутая backstrap (SA)
~, зажигáтельная боевáя incendiary warhead
~, запаснáя 1. spare part
2. reserve unit
~, запояскóвая base (shell)
~, запояснáя base (shell)
~, затвóрная breech piece
~, инженéрная engineer unit
~, истреби́тельно-противотáнковая tank destroyer unit
~, казённая breech, breech end (ord)
~, кáморная chamber (ord, SA)
~ канáла стволá, нарезнáя rifled portion of the bore (ord)
~, кассéтная боевáя sub-projectile warhead (GW), cluster warhead (GW)

часть, кассéтная головнáя warhead containing bomblets, cluster bomb warhead
~, квартúрно-эксплуатациóнная barrack services unit
~, конúческая запояснáя streamlined base (ammo)
~ кóрпуса, носовáя hull front section (AFV)
~, корпуснáя corps unit, unit of corps troops
~ кулачкá, цилиндрúческая cam dwell, back of cam
~, кумулятúвная боевáя hollow charge warhead
~ лафéта, хóботовая carriage trail
~, листевáя facepiece (respirator) (Sov term includes hose to container)
~, маскирóвочная camouflage unit (eng unit with cam duties)
~, материáльная materiel, equipment, technical equipment
~ материáльно-технúческого обеспéчения logistic support unit
~, медицúнская medical unit
~ медицúнского обеспéчения medical supplies section (of med bn, coy or pl)
~, многозарядная головнáя multiple warhead, multiple re-entry vehicle (MRV)
~, моноблóчная головнáя single warhead
~, мотомеханизúрованная motorized unit
~, моторизóванная motorized unit
~, напáлмовая боевáя napalm warhead
~, направляющая skirt (piston)
~, нарезнáя rifling
~, несýщая roadbearers (br)
~, носовáя hull nose (AFV)
~, обеспéчиваемая unit attached for administrative purposes, supported unit
~, обеспéчивающая supporting unit
~, обслýживающая service unit, servicing unit, supporting unit
~, обтекáемая запоясковáя streamlined base, boat-tail (shell)
~, обходящая flanking unit, enveloping unit
~, оживáльная ogive (nose portion of projectile), ogival section, curved taper
~, окулярная eyepiece (optics)

часть, оскóлочная боевáя fragmentation warhead
~, оскóлочно-фугáсная боевáя high-explosive fragmentation warhead
~, отбóрная "crack" unit
~, отделяющаяся головнáя jettisonable nose (missile)
~, откúдная folding part
~, отóгнутая backstrap (SA)
~, парашю́тная parachute unit
~, парашю́тно-десáнтная airborne assault unit
~, поддéрживающая supporting unit
~, подчинённая subordinate unit, unit under command
~, понтóнно-мостовáя pontoon bridge unit
~, посáдочная air-transported unit
~, похорóнная graves registration unit
~, прáчечная laundry unit
~, прúданная attached unit
~ прицéла, головнáя sight head
~, проéзжая carriageway, roadway
~, разбóрная detachable section (br)
~, ракéтная missile unit
~ ракéты, боевáя missile warhead
~, ремóнтная repair and maintenance unit
~ с ядерным зарядом мáлой мóщности, боевáя low yield nuclear warhead
~ самолёта, хвостовáя tailplane (ac)
~, сапёрная sapper unit
~, сбóрная composite unit
~, смéжная neighbouring unit, adjacent unit
~, смéшанная composite unit
~, снабжáемая user unit
~, сопловáя nozzle
~, составнáя component part, element, ingredient
~ станúны, хóботовая rear part of trail (ord)
~, строевáя 1. line unit 2. personnel and supply records section (of a unit)
~, строúтельная construction unit
~, счётная accounts branch, accounting branch
~ тактúческого назначéния, ядерная боевáя tactical nuclear warhead
~ тáнка, ходовáя tank running gear
~, тáнковая tank unit

часть, танкоистреби́тельная tank destroyer unit
~, термоя́дерная боева́я thermonuclear warhead
~, территориа́льная territorial unit
~, техни́ческая 1. technical unit
2. technical side, technical matters
~, топографи́ческая topographic unit
~, тра́нспортная transport unit
~, тылова́я administrative unit, rear unit, logistic unit
~, уда́рная monkey (weight of piling rig)
~, уче́бная training unit
~ учёта и отчётности accounts branch
~, фуга́сная боева́я blast warhead, high-explosive warhead
~, хвостова́я base, tail unit (mor bomb, missile)
~, хими́ческая боева́я chemical (agent) warhead
~, ходова́я 1. undercarriage
2. running gear
3. wheels and axles
4. tracks and suspension (tk)
~, цилиндри́ческая wall (shell)
~, цилиндри́ческая запоясна́я cylindrical base (ammo)
~, штрафна́я penal unit
~, я́дерная боева́я nuclear warhead
часы́, астрономи́ческие astronomical clock
~, а́томные atomic clock
~, ква́рцевые crystal clock
~, непи́ковые off-peak hours
~ пик rush hours, peak time
~ приёмные office hours, consulting hours
чафф chaff (radar)
чахо́тка consumption, pulmonary tuberculosis (med)
ча́шечка cup, recess
~ боево́й личи́нки bolthead recess
~ о́стова затво́ра bolthead recess
ча́шка га́зового регуля́тора gas-regulator cup (SA)
~, кони́ческая female cone
~, опо́рная base-plate socket, bearing socket (mor)
~ цепле́ния clutch cover plate
ча́ща thicket
"Ча́ще Шаг!" "Quicken Step!", "Step Faster!"
чека́ cotter, cotter pin, detent
чека́, боева́я 1. shear pin (fuse)
2. tension pin (dml)
3. firing pin
~ взрыва́теля, боева́я fuse shear pin
~ взрыва́теля, дополни́тельная igniter safety sleeve pin (fuse)
~ взрыва́теля, предохрани́тельная arming pin
~, натяжна́я release pin (push pull igniter, mine)
~, предохрани́тельная 1. safety pin, retaining pin, arming pin
2. shear wire (fuse)
~, предохрани́тельная устано́вочная safety pin (min)
~, про́волочная предохрани́тельная safety wire
~, скрепля́ющая securing pin
~, спускова́я release pin (mine)
~, срезна́я shear pin
~, сре́зывающаяся shear pin
~, центробе́жная centrifugal safety pin (fuse)
чекова́ть/зачекова́ть to secure, to tie down (a load)
челно́к 1. shuttle
2. small skiff
челно́чный shuttle
челове́ко-час man-hour
червя́к worm (gear)
~, глобоида́льный cone worm, globoid worm, globoidal worm, hourglass worm
~ горизонта́льного наведе́ния traversing gear worm (ord)
~ коро́бки прице́ла cross-levelling worm (gunsight)
~ поворо́тного механи́зма traversing worm
~, рулево́й steering worm (veh steering gear)
чередова́ние вспы́шек firing order, firing sequence (ICE)
черено́к handle, helve
чернозём chernozem soil, rich black soil
чёрный 1. black
2. unskilled, manual (labour etc)
3. rough, unfinished
черпа́к scoop, dredger bucket, bailer, pail
черпа́ть/черпну́ть to bail, to scoop, to ladle
черпа́чка dipper (lubrication)
черпачо́к, ма́сляный oil dipper (on ICE con rod)

черпачо́к шату́ка, ма́сляный connecting rod dipper
черпну́ть (см черпа́ть)
черта́ 1. line, boundary, limit
2. trace, characteristic, feature
~, берегова́я coastline
~, курсова́я lubber line (compass)
чертёж drawing
~ в натура́льную величину́ full-size drawing, full-scale drawing
~, габари́тный outline drawing
~, рабо́чий working drawing
~, схемати́ческий diagrammatic drawing, sketch
чертёжная drawing office
чертёжник draughtsman
черти́лка scriber, marking tool, scratch awl
черти́ть/начерти́ть to draw, to plot
чесо́тка scabies, mange
че́сти, тра́урные funeral honours
честь 1. honour
2. salute
~, во́инская 1. military salute
2. honour of a soldier
чёт even number
чета́ pair, couple
че́тверть кру́га quadrant
че́тверть-волнова́я quarter-wave
чёткость sharpness, precision, contrast, definition
~ изображе́ния image sharpness
четырёхло́пастный four-bladed
четырёхо́кись азо́та nitrogen tetroxide
четырёхопо́рный four-bearing (ICE)
четырёхо́сный four-axled
четырёхпо́люсник four-terminal network
~, акти́вный active quadripole, four-terminal active network (elec)
~, пасси́вный four terminal passive network (elec)
четырёхпу́тный four-lane
четырёхскоростно́й four-speed (gear)
чехо́л cover
~, ва́тно-ма́рлевый padded gauze covering (med) (used for retaining heat in wounded extremities of body)
~, ду́льный muzzle cover
~, защи́тный брезе́нтовый protective canvas cover
~ казённика breech cover
~ карда́на universal joint boot
чехо́л лопа́ты shovel case (on waistbelt)
~ на казённую часть ствола́ breech cover
~, наду́льный muzzle cover
~, оборони́тельный fragmentation jacket (ammo)
чехо́л-свёрток camouflage net case
чечеви́ца 1. lentil
2. lens
~ прожéктора searchlight lens, headlight lens
чин rank
чи́сленность numerical strength
~, боева́я effective strength, fighting strength
~ дово́льствуемых войск ration strength, dependency
~, спи́сочная listed strength (forces), strength report, strength state
~, шта́тная establishment strength
~, шта́тно-спи́сочная establishment and strength
число́ number, date
~, алгебраи́ческое algebraic number
~ бензи́на, окта́новое gasoline octane rating
~ веду́щего мо́ста, переда́точное axle ratio
~ вы́стрелов на ору́дие в мину́ту rounds per gun per minute
~ га́бера lethal figure (CW)
~, двои́чное bit, binary digit
~, двузна́чное two-digit number
~, десяти́чное decimal number
~ и́мпульсов pulse repetition rate, pulse recurrence frequency, PRF
~, иррациона́льное irrational number
~, кардина́льное cardinal number
~, коли́чественное cardinal number
~ коро́бки переда́ч, переда́точное transmission ratio
~, кра́тное multiple (maths)
~ "М" Mach number
~, ма́ссовое mass number (nucl)
~ ма́ха Mach number
~, нечётное odd number
~ оборо́тов в мину́ту revolutions per minute, r.p.m.
~, окта́новое octane number
~, переда́точное 1. gear ratio
2. gain, amplification factor (electron)
~ пери́одов колеба́ний mode number (electron)

число́ свечи́ зажига́ния, кали́льное spark plug heat value
~ **сме́си, окта́новое** blend octane value
~, **це́лое** integer, whole number
~, **цета́новое** cetane number, cetane rating (diesel oil)
~ **Циолко́вского** missile mass ratio
~, **чётное** even number
чи́стить/почи́стить to clean
чи́стка ору́жия weapon cleaning
чисто́ль brass cleaning fluid
чистота́ cleanness, finish
чита́льня reading room
член 1. member, limb, part 2. term (math) 3. member, membrum (anat)
~ **прогре́ссии** term of progression
~ **пропо́рции** term of proportion
~ **разложе́ния** expansion term
~ **экипа́жа** crew member
членовреди́тельство self-inflicted injury
чок choke (gunn)
чте́ние аэросни́мков air photo reading
~ **ка́рты** map reading
чувстви́тельность response, sensitivity
~ **взры́вчатого вещества́** sensitivity of an explosive
~ **отклоне́ния** deflection sensitivity (elec)
~, **светова́я** light sensitivity
~, **температу́рная** temperature sensitivity
чувстви́тельный sensitive
чугу́н cast iron, pig iron
~, **ста́листый** refined cast iron, gun iron
"чужо́й" hostile (IFF)
чулки́, защи́тные protective overboots, anti-gas overboots
чума́ plague
чу́чело mock-up figure, dummy
~ **для уда́ров** butt stroke dummy
~ **для уко́лов** thrusting dummy
чу́шка, чугу́нная "pig" of cast iron

Ш

ша́бер scraper (finishing tool)
шабло́н 1. template, stereotype, gauge 2. model copy
~, **путево́й** rail gauge
~, **фасо́нный** curve gauge
шабло́нный stereotyped, unoriginal, routine, trite
ша́брить (см шаброва́ть)
шаброва́ть/ша́брить to scrape
шабро́вка 1. scraping, scouring 2. bedding-in (eng)
шаг 1. step, pace, walk 2. pitch, spacing (mech)
~ **винта́** 1. pitch of propeller 2. pitch of screw thread
~ **ле́нты** distance between two rounds on ammunition belt (MG)
~ **ло́пастей** blade pitch
~ **мини́рования** distance between mines, mine spacing
~ **наре́за** twist (rifling)
~ **наре́зов в кали́брах** rifling twist in calibres (ord)
~, **о́бщий** collective pitch control (ac)
~, **похо́дный** marching easy, normal marching step
~, **строево́й** marching to attention
~ **угломе́ра** OT factor (azimuth switch adjustment to keep bursts on OT line when bracketing for range) (gunn)
~ **це́пи** chain pitch
шаги́стика footdrill
"Ша́гом Марш!" "Quick March!"
"Ша́гом Марш, На Ме́сте!" "Mark Time!"
"Ша́гом Марш, Строевы́м!" "Quick March!" (marching to attention)
шагоме́р pedometer
ша́йба washer, disk, collar
~, **замко́вая** lock washer
~, **замо́чная** lock washer
~, **коса́я** split washer
~, **кулачко́вая** cam ring
~ **на уда́рнике** percussion pellet washer (fuse)
~ **нака́тника** recuperator spring washer
~, **направля́ющая** guide washer
~, **опо́рная** thrust washer
~, **отги́бная** tab washer

ша́йба, пружи́нная зубча́тая spring lock washer
~, распо́рная spacing washer, distance washer
~ с зубца́ми, упру́гая spring lock washer
~, сто́порная, предохраня́ющая от самоотви́нчивания при тря́ске shake-proof washer
шала́нда 1. scow
2. hopper barge
~, вы́садочная landing barge
шала́ш improvised shelter, basha, hut
ша́нец field work, entrenchment
шанта́ж blackmail
~, я́дерный nuclear blackmail
ша́пка-уша́нка fur cap with ear flaps
шар, возду́шный air balloon
~, о́гненный fireball (nucl)
ша́р-зо́нд sounding balloon, sonde
ша́рик взрыва́теля retaining ball (fuse)
~, сто́порный retaining ball (fuse)
шарикоподши́пник ball bearing
ша́рик-предохрани́тель retaining ball (fuse)
шарни́р hinge, articulated joint
~, карда́нный universal joint
~, пло́ский pin joint
~, универса́льный gimbal, gimbal joint, universal joint
~, шарово́й ball joint, ball and socket joint
~, я́блочный ball coupling
шарни́рно-прикреплённый hinged, articulated
шарова́ры breeches (loose trousers tucked inside boots)
шаро́шка milling cutter, rasp
шарф scarf
шасси́ 1. chassis
2. landing gear, undercarriage
~ автомоби́ля motor vehicle chassis
~, гу́сеничное (full-)tracked chassis
~, колёсное wheeled chassis
~, колёсно-лы́жное wheel-ski undercarriage (ac)
~, многоколёсное multiple-wheel undercarriage (ac)
~, самохо́дное self-propelled chassis
~, убира́ющееся retractable undercarriage
шата́ться to work loose, to get loose, to reel, to stagger (imp only)
шату́н connecting rod (eng)
~, вну́тренний plain connecting rod (ICE)
~, гла́вный master connecting rod (ICE)
~, прицепно́й link rod, articulated rod
ша́хта 1. shaft
2. mine
~, пускова́я launch silo (GW)
~, ста́ртовая launching silo
ша́шка 1. slab, block, candle
2. sabre, sword
3. grain (solid propellant)
4. paving slab
~, бурова́я borehole charge
~ ВВ block of explosives
~ грана́ты auxiliary detonator (gren)
~, дымова́я 1. smoke candle, smoke pot
2. smoke generator (ammo)
~, запа́льная primer, priming charge
~, запа́льная пироксили́новая guncotton
~ имита́ции разры́вов артиллери́йских снаря́дов shell-burst simulator
~, многокана́льная multi-perforated grain (solid propellant)
~, пла́вленая cast demolition slab
~ пла́вленого троти́ла TNT slab
~ пла́вленого троти́ла TNT slab
~, подрывна́я demolition slab
~, по́лая hollow charge
~, порохова́я grain, propellant charge, propulsive charge
~ промежу́точного заря́да detonator pellet
~ со слезоточи́вым га́зом, хими́ческая tear gas generator
~, стержнева́я порохова́я rod grain (solid propellant)
~, то́ловая TNT demolition slab, TNT block
~, троти́ловая TNT block, TNT demolition block
~, ядовитодымова́я toxic smoke candle
ша́шки draughts, checkers
шва́бра swab, mop
швартóвка те́хнихи на параш́ютных платфо́рмах tying down of equipment on paradrop platforms
шве́ллер channel
шво́рень pintle, pivot, pin, bolt, draw bolt
ше́йка 1. journal, pin (eng)
2. neck, collar
~ ги́льзы cartridge neck
~, коренна́я main journal (mech)

ше́йка ло́жи small of the butt
~, **моты́левая** crankpin (engine)
~, **осева́я** axle journal
~ **о́си** axle journal
~ **пламегаси́теля** flash eliminator base
~ **полуо́си** axle journal
~ **прикла́да** small of the butt (SA)
~, **шату́нная** crankpin
шелла́к shellac
шепта́ло sear, sear noise (SA)
~, **промежу́точное** intermediate sear
шере́нга rank (drill)
шере́нги, разо́мкнутые open order (drill)
шерохова́тость roughness, lack of finish (eng)
шерсть wool
~, **стекля́нная** glass wool
шест pole, rod, staff
~, **оконе́чный** terminal pole (line comms)
~, **углово́й** angle pole (line comms)
шестерня́ pinion
~, **боева́я** spindled arc pinion (ord)
~, **ведо́мая** driven pinion
~, **веду́щая** driving pinion
~, **винтова́я** screw gear
~, **гипо́идная** hypoid gear
~ **дифференциа́ла, веду́щая** differential drive gear
~ **за́днего хо́да** reverse gear pinion
~, **кони́ческая** bevel wheel, bevel pinion, bevel gear
~, **косозу́бая** helical gear
~, **кулачко́вая** cam wheel, camshaft timing wheel, camshaft driving gear
~, **парази́тная** idle gear, intermediate gear
~ **пере́днего хо́да** forward gear
~, **планета́рная** planetary gear
~ **постоя́нного зацепле́ния** constant-mesh gear
~ **распределе́ния** camshaft drive gear
~, **распредели́тельная** timing gear (ICE)
~, **скользя́щая** sliding gear, "crash" gear (ICE)
~, **цепна́я** sprocket wheel
~, **цилиндри́ческая** spur pinion, elevating pinion
~, **червя́чная** worm wheel
~, **шевро́нная** herringbone gear
шестипо́люсник six-terminal network (elec)
ше́ст-тройни́к T-shaped pole
ши́ло awl, pricker
ши́льдик tally plate (on a piece of equipment)
ши́на 1. tyre
2. splint (med)
3. busbar, bus (elec)
~, **автомоби́льная** automobile tyre
~, **а́рочная** arched cross-country tyre, arched tyre, low profile tyre
~, **балло́нная** balloon tyre
~, **безве́нтильная** valveless tyre
~, **беска́мерная** tubeless tyre
~ **большо́го про́филя** cross-country tyre
~ **высо́кого давле́ния** high-pressure tyre
~ **ГК** foam rubber tyre
~, **гусмати́ческая** self-sealing tyre
~ **для вое́нных маши́н** combat tyre
~ **для го́ночного автомоби́ля** high speed tyre
~ **для движе́ния вне доро́г** cross-country tyre
~ **для движе́ния по песка́м** sand tyre
~ **для движе́ния по сне́гу** snow tyre
~ **для скоростно́го автомоби́ля** high speed tyre
~, **запа́сная** spare tyre
~, **запо́лненная пе́нистой рези́ной** foam-filled tyre
~, **лита́я** solid rubber tyre
~, **ло́пнувшая** burst tyre
~, **металлоко́рдная** steel-cord tyre
~, **нака́чанная** inflated tyre
~, **недока́чанная** underinflated tyre
~ **ни́зкого давле́ния** balloon tyre
~, **низкопро́фильная** low-profile tyre
~, **пневмати́ческая** pneumatic tyre
~ **повы́шенной проходи́мости** cross-country tyre
~, **полупневмати́ческая** cushion tyre
~, **проко́лотая** punctured tyre
~, **проколоусто́йчивая** puncture-proof tyre
~, **пулесто́йкая** bullet-proof tyre
~ **с изно́шенным прotéктором** bald tyre
~, **самозатя́гивающаяся** self-sealing tyre
~ **со стальны́ми шипа́ми в прotéкторе** steel-studded tyre
~, **сплошна́я** solid tyre
~, **спу́щенная** deflated tyre
~, **широкопро́фильная** large wide tyre
ши́на-балло́н balloon tyre
шине́ль greatcoat

ши́ны, двухска́тные twin tyres
шип journal, pivot pin, dowel pin, stud, tenon
~ **поворо́тной ца́пфы** stub axle
шира́с shell burst simulator
ширасомёт smoke bomb projector
"Ши́ре Шаг!" "Step Out!"
ширина́ width
~, **габари́тная** overall width, clearance width
~, **до́плеровская** Doppler width
~ **звена́ гу́сеницы** track width (AFV)
~ **и́мпульса** pulse width
~ **коле́и** 1. wheel gauge, track gauge
2. railway gauge
~ **луча́** beam width
~ **проте́ктора** tread width
~ **ре́льсовой коле́и** rail gauge
~ **строб-и́мпульса** gate width
~ **стро́я** frontage (drill)
~, **углова́я** angular width
~ **фро́нта** frontage
~ **фро́нта наступле́ния** attack frontage
ши́рма 1. screen, shield
2. cloak (figurative)
широкодиапазо́нный broad band
ширококоле́йный broad gauge
широкополо́сный broad band, wide-band
широта́ 1. width, breadth
2. latitude
~, **астрономи́ческая** astronomical latitude
~, **географи́ческая** geographical latitude, terrestrial latitude
широ́тно-модули́рованный width modulated
шифр 1. cipher
2. code name
3. classification number (libraries)
шифра́т ciphered text
шифра́тор coder, converter
шифрова́льщик cipher clerk
шифрова́ние enciphering
шифрова́ть to encipher (imp only)
шифро́вка 1. enciphering
2. cipher message
шифрогра́мма cipher telegram
шифро́рган cipher office
шифрпо́ст cipher post
ши́хта 1. schist (geol)
2. furnace charge (metall)
3. CW respirator absorbent
ши́хта, противога́зная respirator container filling
шкала́ 1. scale
2. dial
~ **боковы́х попра́вок** aim-off graticule
~ **Бо́форта** Beaufort scale
~, **высо́тная** height scale, height dial
~, **гипсометри́ческая** hypsometric scale, altitude scale (carto)
~, **градуи́рованная** graduated scale
~, **гру́бая** coarse scale
~ **гру́бого отсчёта** coarse scale
~, **дальноме́рная** range scale
~ **да́льности** range scale
~ **диста́нции** range scale
~, **дистанцио́нная** range scale
~, **кра́сная** upper register, high-angle range scale (arty)
~ **ку́рса** course bearing scale
~, **логарифми́ческая** logarithmic scale
~ **миномётного прице́ла** mortar sight scale
~, **прице́льная** 1. fixed ballistic graticule (tk gunn)
2. sight scale (AFV sight)
~ **рассе́ивания** length zone, zone of dispersion (ball)
~ **совмеще́ния** rate scale, rate dial (AAFCE)
~, **температу́рная** temperature scale
~, **то́чная** fine scale
~ **то́чного отсчёта** fine scale
~ **то́чных устано́вок угломе́ра** deflection drum micrometer scale
~, **ты́сячная** scale graduated in mils
~ **у́глов пики́рования и кабри́рования** angle-of-dive scale (AD gunn)
~ **устано́вок взрыва́теля** fuse-length scale, fuse setting scale
шкаф cupboard, rack
шкафу́т waist (aeron, naut)
шквал squall
~ **бе́глого огня́** burst of gun-fire
~, **огнево́й** intense concentration (using all available fire power), burst of rapid fire
шкво́рень, поворо́тный king pin
шкив pulley, sheave
~, **ведо́мый** driven pulley
~, **веду́щий** driving pulley
~ **вентиля́тора** fan pulley
~, **копро́вый** 1. pulley wheel (Brit)
2. hoisting sheave (US)

шкив, направля́ющий jockey pulley
~, приводно́й power take-off pulley
шкимушга́р spun yarn
шкиму́шка nettle stuff, small line (used for making gaskets etc)
шкиф pulley, sheave
шко́ла school
~, арме́йская реме́сленная army trades school, army apprentices school, army trade-training school
~, артиллери́йская gunnery school
шку́рка sandpaper, emery paper
~, нажда́чная emery paper
~, стекля́нная glass paper
шлагба́ум traffic control barrier
~, ми́нный mine sled barrier, mine necklace barrier
шлак slag, dross, clinker
шланг hose
~, возду́шный air hose
~, ги́бкий 1. flexible hose
2. stub (waveguide)
3. loop (elec)
~ для горю́чего fuel hose
~, дюри́товый reinforced rubber hose, high pressure hose
~, запра́вочный filling hose
~, подаю́щий delivery hose
~, пожа́рный fire hose
~, топливозапра́вочный fuelling hose
шлейф 1. multiple pair cable
2. stub (waveguide)
3. loop (elec)
~, согласу́ющий matching stub (waveguide, radio)
шлейф-анте́нна folded-dipole antenna
шлем helmet
~, вя́заный Balaclava helmet
~, защи́тный crash helmet, protective helmet
~, противоожо́говый anti-flash helmet
~, стально́й steel helmet
шлем-ма́ска 1. diving mask
2. helmet facepiece (CW)
шлемофо́н combined headset and laryngophone, interphone headset
шлифова́ть to polish up (imp only)
шлифо́вка grinding
шлиц spline
~ под отвёртку screwdriver slot
~, спира́льный helical spline
шлюз sluice, lock
шлюп sloop
шлю́пка, вы́садочная landing boat
~, рези́новая 1. rubber dinghy
2. rubber inflatable assault craft
~, спаса́тельная lifeboat
шля́па-пана́ма tropical hat
шля́пка 1. cap, head (of nail etc)
2. round cocking piece (rifle)
~ ги́льзы cartridge base, rimmed base
шнек worm conveyor, screw conveyor
шне́ллер hair-trigger
шно́ркел snorkel (both subs and tanks)
шнур flex, cord, lanyard
~ Би́кфорда Bickford safety fuse
~ Би́кфорда, асфалтиро́ванный asphalted Bickford fuse
~ Би́кфорда, бе́лый borehole Bickford fuse
~ Би́кфорда, гуттапе́рчевый waterproof Bickford fuse
~, быстрогоря́щий instantaneous fuse
~, быстрогоря́щий зажига́тельный instantaneous incendiary fuse
~, детони́рующий detonating cord, instantaneous detonating fuse
~, зажига́тельный burning fuse
~, координа́тный mine-spacing cord
~, медленногоря́щий огнепрово́дный slow-burning fuse, safety fuse
~, ми́нный mine cord
~, огнепрово́дный Bickford fuse, safety fuse
~, пентри́товый детони́рующий primacord (expl)
~, подрывно́й blasting fuse
~, спусково́й lanyard, firing lanyard
~, те́новый детони́рующий primacord
~, трассиро́вочный tracing tape (min)
~, трассиру́ющий tracing tape (min)
~, четырёхжи́льный four-conductor cable, four-conductor cord
шнуро́к грана́ты grenade connecting cord
шов joint, seam, weld
~, вне́шний outside weld
~, вну́тренний inside weld
~, во́гнутый concave weld
~, вы́пуклый convex weld
~, клеево́й glued seam
~, кругово́й circular seam
~, непреры́вный сварно́й continuous weld
~, полу́ченный при дугово́й сва́рке arc weld

шов, расчека́ненный caulked seam
~, сварно́й welded joint
~, сва́рочный weld, welded joint
~, стыково́й сва́рочный butt weld
~, то́чечный spot weld
шок shock
шо́ковая ward for shock cases (med)
шо́мпол ejector rod, extractor rod, cleaning rod
~, составно́й jointed cleaning rod (SA)
шор wet salt marsh
шо́рный saddler's
шо́рох rustle
шоссе́ surfaced highway
~, магистра́льное arterial road
~, щебёночное macadam road
шоссе́йный highway
шофёр driver, chauffeur
шофёр-крановщи́к crane driver-operator
шпа́га sword
шпага́т, натяжно́й trip wire
шпаклёвка putty, filler
шпа́ла sleeper (rly, br etc)
~, опо́рная support sleeper (br)
~, стыкова́я joint sleeper (timber baulk - part of ramp grillage - br)
шпалоразруши́тель tie-breaking machine, track-destroying machine (rly)
шпанго́ут 1. rib, frame (pontoon)
2. frame (nav)
3. former (aeron)
шпат, алма́зный corundum
шпенёк stud
~ штыка́ bayonet stud
шпи́лька stud-bolt, cotter-pin
~, а́нкерная retaining stud (ICE)
~, желе́зная steel chaining-arrow, iron marking pin (surv)
~ ко́мпаса compass pivot
~, ко́мпасная compass pivot
~ крепле́ния колеса́ wheel stud
~, легкосъёмная quick-release pin
~ спусково́го крючка́ trigger pin
~, сто́порная stop pin
~, устано́вочная fixing stud
шпи́ндель spindle
шпио́н spy
шпиона́ж espionage
~, вое́нный military espionage
шплинт pin, split pin, split cotter, cotter, cotter pin
~ звена́ гу́сеницы track pin (AFV)
~ па́льца тра́ка track pin (AFV)
шплинт, предохрани́тельный safety pin, safety cotter pin (fuze)
~, разводно́й split pin
шпо́нка spline, dowel, cotter, key, wedge
~, вытяжна́я spring key
~, забивна́я plain taper key
~, затяжна́я plain taper key
~, клинова́я plain taper key
~, направля́ющая sliding key, feather key
~, скользя́щая sliding key
шпо́ра 1. grouser (track)
2. claw (AFV)
~ гу́сеницы, зи́мняя ice grouser
~, дополни́тельная съёмная detachable grouser, detachable spud (tracked vehicle)
~, козлова́я trestle shoe (br)
~ тра́ка spud, grouser (tracked vehicle)
шприц 1. syringe, hypodermic needle
2. oil injector
~ для сма́зки grease gun
~, маслоразда́тельный oil pistol
~, ма́сляный oil gun
~ с атропи́ном atropine syrette (NBC)
~ с атропи́ном, тю́биковый atropine syrette (CW, med)
~, таво́тный grease gun
шприцева́ть to lubricate with a grease gun
шприц-пресс grease gun
шприц-тю́бик syrette, tube syringe
~ с атропи́ном atropine syrette (NBC)
шпур borehole (mining)
~, котлово́й camouflet (underground cavity formed by explosion)
~, отбо́йный ring hole, trimmer (mining)
шрапне́ль shrapnel shell, shrapnel
~, пулева́я shrapnel
~, разрывна́я high-explosive shrapnel
шрифт script, type-face
штаб 1. staff
2. headquarters, HQ
~, авиацио́нный 1. air force HQ
2. air force staff
~, артиллери́йский 1. artillery HQ
2. artillery staff
~ батальо́на battalion HQ
~ батаре́и battery HQ
~, вышестоя́щий superior HQ
~, генера́льный general staff
~ инжене́рных войск engineer headquarters

штаб, общевойсковóй combined arms HQ
~ **свóдного десáнтного соединéния** Joint Force Headquarters
~ **ты́ла** 1. administrative HQ, logistical HQ
2. logistical staff
~ **тыловóго обеспéчения** administrative support operation centre
штáбелер elevator-stacker
штабелеуклáдчик tiering machine, stacker (stores)
штабели́рование контéйнеров container stacking
штáбель stack, tier
~ **боеприпáсов** ammunition stack
штаби́ст staff officer
штабнóй staff, HQ
штади́в divisional HQ
штакéтник pale, paling
штакóр corps HQ
штамп 1. stamp, die
2. type, pattern
3. title block (tech drawing)
штамповáть to forge, to drop forge (imp only)
штампóвка drop-forging
штáнга rod, bar
~, **направля́ющая** guide rod
~, **поперéчная рулевáя** track rod (MT)
~, **продóльная рулевáя** steering drag link
~, **реакти́вная** torque rod
~, **скру́чивающая** torsion bar
~, **составнáя** 1. sectionalized bar
2. sectionalized search rod (mine detector)
штангензубомéр vernier gear caliper, gear tooth vernier guage
штангенинструмéнт vernier instrument
штангенци́ркуль vernier caliper
штáнги, направля́ющие guide rods, leads (pile driver)
штарм army HQ
штат establishment
~ **воéнного врéмени** war establishment
штати́в tripod, stand
~, **плечевóй** shoulder straps
штáтный organic, authorised, on the strength, on the establishment
штáтский civilian
штáты 1. establishment, manning table
2. equipment table
3. tables of organisation (US)
~ **воéнного врéмени** war establishment
~ **ми́рного врéмени** peace establishment
штáуфер compression grease cup
штéккер plug (elec)
~, **отрывнóй** quick release plug, quick disconnect plug, snatch plug
штéмпель stamp, imprint
штéпсель plug, socket (elec)
~, **банáновый** banana plug (elec)
шти́вка гру́зов stowage
штиль calm, wind force zero
штифт pin, peg
~, **выступáющий** drop pin
~ **запирáющего кли́на** breech-lock pin
~, **контровóй** locking pin
~, **ограничи́тельный** arresting pin
~, **останóвочный** detent pin
~, **предохрани́тельный** safety pin
~, **стóпорный** retaining pin
шти́хмас inside micrometer
штихпрóбер constant-volume fuel measuring device
шток rod, stem
~ **досылáтеля** rammer rod
~ **задви́жки** piston rod
~ **затвóрной рáмы** piston rod, piston stem (SA)
~ **клáпана** valve stem
~ **компрéссора** recoil piston rod
~ **накáтника** counterrecoil rod, recuperator piston rod
~, **направля́ющий** slide bar
~ **пóршня** piston rod
~ **пóршня накáтника** counterrecoil piston rod, recuperator piston rod
~ **противооткáтного устрóйства** recoil piston rod
~, **спусковóй** firing stem ignitor, striker stem (gren)
~ **тóрмоза откáта** recoil piston rod
штóльня drift, gallery (mining)
штóпор 1. corkscrew, screw picket
2. spin (aeron)
штопорéние spin (aeron)
штóпорить to spin (aeron) (imp only)
штóра blind, curtain
~, **маскирóвочная** apron, blind
~ **радиáтора** radiator blind
~, **светомаскирóвочная** black-out curtain

шторм strong gale, wind force nine
~, жёсткий storm, wind force eleven
~, крéпкий whole gale, wind force ten
~, сильный whole gale, Beaufort scale force ten
штрафнóй penal, punishment
штрих hachure
штриховáть to hachure (imp only)
штрихóвка hachuring, shading, hachure
строп strop, strap
~, причáльный mooring rope, ring line
штукатýрка plaster
штурвáл control column, steering column, steering wheel
~, рулевóй steering wheel
штурм assault, attack, storm
штýрман navigator, pilot
штýрман-навóдчик fighter control (GCI)
штурман-снайпер classification awarded to military navigators of exceptional skills
штурмовáть to assault, to storm (imp only)
штурмовик ground-attack aircraft
штурмóвка air strafing
штурмовóй assault, attack, shock
штýцер 1. union, adapter
2. nozzle
~, впускнóй inlet nozzle
~, выпускнóй discharge nozzle
~, заливнóй tank filler
~ клáпана valve connector
~ мáсляного манóметра oil pressure gauge connection
~, переливнóй overflow outlet
~ сапýна breather adapter
~, соединительный connecting sleeve, connecting pipe
~, спускнóй drainage fitting
~ топливопровóда fuel-pipe union
штык 1. bayonet
2. bend, hitch (knot)
~, гранёный fluted bayonet
~ игóльчатый needle bayonet
~, клинкóвый knife bayonet
~, несъёмный permanently fixed bayonet, non-detachable bayonet
~, примкнутый fixed bayonet
~, рéжущий knife bayonet
~, трёхгрáнный triangular-shaped rod bayonet
~, ширóкий клинкóвый trowel bayonet
штык-клинóк sword-bayonet
штык-тесáк sword bayonet, knife bayonet
штыр вертлю́га cradle pintle (ord)
штырек связи coupling probe (electron)
штырь 1. pin, cramp iron, nail, cup hook
2. spindle, drift hook
3. stub (electron)
~, боевóй saddle pivot (ord)
~, заземля́ющий earth pin, ground pin (elec)
~, козловóй trestle pin (br)
шум noise
~, акустический acoustic noise
~ в цéпи circuit noise (elec)
~ выхлопа exhaust noise
~, дробовóй shot noise (electron)
~, космический cosmic noise
~, лáмповый valve noise
~, линéйный line noise
~ мерцáния flicker noise
~, motóрный motorboating (electron, radio)
~ при качéнии шины tyre thump
~, сóбственный set noise (radio, radar)
шумоглушитель noise suppressor, silencer
шумопеленгáтор hydrophone listening gear, sound locator
шýмы, внятные crosstalk noise, intelligible crosstalk (electron)
~ от дробовóго эффéкта shot noise (electron)
~, перехóдные cross-talk noise
~, тепловые Johnson noise, thermal noise (elec)
шунт 1. shunt (elec)
2. bypass
шунтирование shunting, bridging (elec)
шунтировать to shunt (elec) (imp only)
шунтирующий shunt, shunting, bypass
шурýп wood screw
шурф 1. borehole (camouflet)
2. exploratory shaft (min)
шхéры sea cliffs, rocks, skerries

Щ

щéбень ballast, aggregate, crushed stone, chippings
щекá 1. cheek (anat)
2. web, crank arm (ICE)
3. side, sidepiece, flange
4. jaw (of vice or crusher)
~, вертикáльная side plate of saddle (ord)
~ основáния прицéла backsight ramp
~ рáмы пружи́ны spring frame endplate
щекóлда catch, latch, locking bar, pawl
щёлочь alkali
щелчóк 1. click (on gunsight etc)
2. flick, snap (of fingers)
щель 1. crack, slit, slot, chink
2. slit trench
~ коллимáтора collimating slit
~ (механика-)води́теля, смотровáя driver's slit (AFV)
~ навóдчика, смотровáя gunner's slit (AFV)
~, перекры́тая shell slit (covered slit trench)
~ свя́зи coupling aperture (radar)
~, смотровáя vision slit
~, ячéйковая slit trench
щель-убéжище slit trench
щель-укры́тие для огневóго срéдства weapon slit
щётка brush (general and electrical)
~ бáнника bore brush (gunn)
~, волосянáя bristle brush
~ генерáтора generator brush
~, графи́товая graphite brush
~, контáктная brush (elec)
~, прóволочная wire brush
~ распредели́теля, ýгольная distributor brush, carbon brush
~ стеклоочисти́теля windscreen wiper blade
~, ýгольная carbon brush (elec)
щеткодержáтель brushholder (elec)
щёчка stock plate, grip (pistol)
щипцы́ pliers
~, обжи́мные crimper
~, хирурги́ческие forceps
щит 1. shield
2. panel, board
~, амбразýрный mantlet plate (ord)
щит, áнкерный cable anchorage, log anchorage (br)
~, броневóй armour shield
~, въезднóй ramp (br etc)
~, генерáторный generator switchboard
~, глáвный распредели́тельный main switchboard
~, группово́й distribution board
~, грязевóй mudguard
~ дви́жителя, броневóй hull skirting
~ для холоднóй пристрéлки boresighting screen (gunn)
~, дощáтый plank wheeltrack
~ из камышá cane hurdle, reed hurdle
~ из плетня́ wattle hurdle
~ из хвóроста brushwood hurdle
~, колéйный plank wheeltrack, wheel track plank
~, контрóльный sights-testing target
~, лобовóй front deck panel (br)
~ навóдчика, прикривáющий gun layer's shield
~, насти́лочный decking (br)
~, ору́ди́йный gun shield
~ от дýльного плáмени blast screen
~, откиднóй flap plate (gun shield)
~, перенóсный mantelet
~ плáвких предохрани́телей fuse board
~, плетнёвый wattle hurdle
~, пролётный duckboard
~, распредели́тельный switchboard
~, решётчатый corduroy road mat
~, снéжный snow fence
~ телеуправлéния remote control panel
~, торцóвый end panel (eng)
~ управлéния control board
щи́тик shield
~, выдвижнóй sliding shield
щит-мишéнь screen target
щитóк panel
~, аккумуля́торный battery panel (tk)
~ бáшни, распредели́тельный turret junction box (tk)
~, воздушнонаправля́ющий air deflector
~ выводнóго окнá ejection opening cover (SA)
~, выдвижнóй movable shield
~, грязевóй mud guard, anti-splash guard
~ гýсеницы, защи́тный track guard
~, инструментáльный instrument panel
~, клéмный terminal panel
~, нажимнóй pressure plate

щито́к отража́теля гильз cartridge deflector (automatic weapon)
~, предохрани́тельный 1. safety cover 2. fuse board (elec)
~ приёмного окна́ magazine opening cover (SA)
~ смотрово́го лю́чка eyeshield
~, теново́й sun vizor
~ управле́ния генера́тором generator control panel
щит-ра́ма, подъёмная frame cover (eg for trench camouflage; raised for firing)
щуп feeler gauge, prodder
~ Влади́мирова "Vladimir's prodder" (deep sectionalized mine prodder, Sov)
~, маслоуказа́тельный oil dip-stick
~, ма́сляный oil dip stick
~, ми́нный mine probe (rod)
~, ручно́й hand probe
~, укоро́ченный short mine prodder

Э

эвакоподразделе́ние recovery sub-unit
эвакоприёмник evacuation reception station (med)
эвакуацио́нный 1. evacuation 2. recovery
эваку́ация 1. evacuation 2. recovery, backloading
~, медици́нская medical evacuation (of casualties)
~ по во́здуху air evacuation
~ по во́здуху, медици́нская aeromedical evacuation
~ ра́неных санита́рным тра́нспортом ambulance evacuation
~, санита́рная evacuation of wounded, casualty evacuation
эвакуи́ровать to evacuate, to backload
э́днатол ednathol
эже́ктор ejector (gunn)
эже́кция ejection (ord)
эква́тор, магни́тный magnetic equator
~, небе́сный celestial equator
эквивале́нт анте́нны dummy antenna
~ Рентге́на, биологи́ческий man roentgen equivalent, roentgen equivalent mammal, RME, MRE (NBC)
эквивале́нт Рентге́на, физи́ческий roentgen equivalent physical (NBC)
~, троти́ловый TNT equivalent yield (NBC)
экза́мен, выпускно́й passing-out examination, final exam
экзамена́тор teaching machine
экземпля́р, контро́льный file copy
экзосфе́ра exosphere (layer of the atmosphere)
экзотерми́ческий exothermic
экипа́ж crew
~ боево́й маши́ны combat vehicle crew
~, сколо́ченный efficient crew
~, сла́женный efficient crew
экипи́ровать to fit out, to equip
экипиро́вка equipment, equipping, outfitting
экли́метр clinometer
эконома́йзер economiser (carb)
эконо́мика сил economy of effort
эконо́мия то́плива fuel economy
экра́н 1. apron plate, skirting plate (tk) 2. screen 3. baffle, deflector, shield
~, проекцио́нный projection screen, viewing screen
~, просве́чивающий fluoroscopic screen
~, светя́щийся fluorescent screen
~, телевизио́нный television screen
~, флуоресци́рующий fluorescent screen
~, электромагни́тный electromagnetic screen
экраниза́ция screening, depiction on a screen
экрани́рование screening, shielding
экрани́рованный screened
экрани́ровать to screen, to shield
эксга́устер vacuum fan, exhaust fan
экскава́тор excavator
~, гре́йферный grab excavator
~, ковшо́вый bucket excavator
~, многочерпа́ковый endless-bucket excavator
~, одноковшо́вый single-bucket excavator
~, плову́чий dredger
~, ро́торный rotary excavator
~, ро́торный транше́йный rotary trencher
~, транше́йный trencher, trench digger, excavator

экскава́тор, универса́льный одноковшо́вый universal single-bucket excavator
~, цепно́й endless-bucket excavator
экскава́тор-кра́н excavator-mounted crane
экспеди́ровать to forward, to expedite
экспеди́тор dispatcher, dispatch clerk
экспеди́тор-пи́сарь dispatch clerk
экспеди́ция 1. expedition
2. traffic office, dispatch office (sigs)
~, вое́нная military expedition, dispatch of an expeditionary force, mounting of an expeditionary force
~, кара́тельная punitive expedition
эксперимента́льный experimental, trial
экспе́рт expert, examiner, investigator
эксперти́за 1. examination
2. commission of experts
экспли́ка́ция explanation, interpretation (of symbols), legend, key
эксплуатацио́нник operator, user
эксплуатацио́нный operational, maintenance
эксплуата́ция service, operation, running, exploitation, use
~, безавари́йная trouble-free service
~, бесперебо́йная trouble-free operation
~, повседне́вная day-to-day running
~, централизо́ванная pool operation (of MT etc)
эксплуати́ровать to operate (imp only)
экспози́метр, фотоэлектри́ческий photoelectric exposure meter
экспози́ция 1. exposition, show, exhibit
2. exposure (photo)
3. time of exposure (GW)
экспоно́метр exposure meter
~, фотоэлектри́ческий photoelectric exposure meter
экстраги́рование extraction, ejection
экстраги́ровать to extract
экстра́ктор extractor
~ разорва́вшихся гильз ruptured cartridge extractor
экстра́кция extraction
экстраполя́ция extrapolation
эксце́нтрик eccentric, cam
эксцентрисите́т eccentricity
~ реакти́вной си́лы thrust misalignment (rocket)
эксцентри́ческий out of centre, off-centre, eccentric
эксцентри́чность eccentricity
эласти́чность elasticity
эласти́чный elastic, resilient, flexible
элево́н elevon (aeron)
электриза́ция electrification
электризо́ванный electrified
электризу́емый electrified
эле́ктрик 1. electrician
2. power generator specialist
электрифика́ция 1. electrification
2. energizing
электрифици́рованный electrified
электрифици́ровать to electrify
электровзрыва́тель electric fuse
электровзрывпаке́т electrical battle-noise simulator
электрово́з electric locomotive
электровоспламене́ние electric firing
электровоспламени́тель electric detonator
электрогенера́тор dynamo, generator
электро́д electrode
~, кони́ческий conical electrode
~, поджига́ющий trigger electrode
~, пусково́й igniter electrode
~, разря́дный trigger electrode
~ свечи́ зажига́ния sparkplug electrode
электродви́гатель electric motor
электродетона́тор electric detonator
~, заме́дленный delayed action electric detonator
электродолбёжник electric chisel
электродре́ль electric drill
электрозагражде́ние electrified obstacle
электрозамыка́тель electric (type) fuse
электрозапа́л electric fuze
электроинжене́р electrical engineer
электрока́р electric dolly, electric trolley
электро́лиз electrolysis
электроли́т electrolyte
электролити́ческий electrolytic
электролюминесце́нция electroluminescence
электромагни́т electromagnet
электромете́лка electric wiper
электро́метр electrometer
электромеха́ник electrician
электромига́лка 1. flashing traffic indicator
2. electric gun-flash simulator
электромоби́ль electrically driven car
электро́н electron
~, быстролетя́щий high-speed electron

электро́н, втори́чный secondary electron
~, **ме́дленный** low velocity electron
~, **перви́чный** primary electron
электро́н-вольт electron-volt
электро́ника electronics
~, **интегра́льная** integrated electronics
~, **ква́нтовая** quantum electronics
~, **криоге́нная** cryogenic electronics, low-temperature electronics
~, **молекуля́рная** molecular electronics
~, **полупроводнико́вая** semi-conductor electronics
~, **тонкоплёночная** thin-film electronics
~, **транзи́сторная** transistor electronics
электро́нный electronic
электрообогрева́тель electric heater, electric demister
электрообору́дование electrical equipment
электроо́смос electro-osmosis
электропереда́ча electrical transmission, electric power transmission
электропиропатро́н electrically detonated cartridge
электропита́ние electricity supply
электропла́н electrically-powered space-craft
электропневмокла́пан solenoid-operated pneumatic valve
электропогру́зчик electric fork-lift truck
электроподогрева́тель electric heater
электропо́езд electric (multiple unit) train
электропрово́дка electric wiring
электропу́ск electric firing gear
электроразъём, автомати́ческий 1. electrical quick-disconnect connection 2. electrical snatch plug
электросва́рка electro-welding, electric welding
электросверло́ electric drill
электросвя́зь telecommunications
электроскребо́к electric wiper
электроснабже́ние electric power supply
электроспецобору́дование specialist electronic equipment
электроста́нция power station
~, **вью́чная** portable generator set
~, **подви́жная** mobile generator
электроста́ртер electric starter
электростати́ческий electrostatic, static
электроте́хник electrician
электрофона́рь electric lamp, electric torch
электрохозя́йство electrical complex
электроце́пь зажига́ния electrical firing circuit
электроэне́ргия electrical energy
элеме́нт 1. element 2. cell (elec)
~, **батаре́йный** battery cell
~ **внеза́пности** element of surprise
~, **гото́вый** prefabricated component
~, **деля́щийся** fissile element
~, **до́черный** daughter element (nucl)
~, **железобето́нный** pre-fabricated concrete framework, reinforced concrete section
~ **жёсткости** reinforcing member, strengthening member
~, **исхо́дный** parent element (nucl)
~ **констру́кции, несу́щий** load-carrying unit
~, **логи́ческий** logic element
~, **матери́нский** parent element (nucl)
~, **мо́крый** wet cell
~ **неизвлека́емости** anti-lift device (min)
~ **необезвре́живаемости** anti-disturbance device (min)
~, **несу́щий** supporting member
~ **па́мяти** memory element
~ **подве́ски** suspension member
~, **поража́ющий** contact element, fragment (shell)
~, **радиоакти́вный** radioactive element
~ **релье́фа** terrain feature (topo)
~, **сма́чивающий фильтру́ющий** wet-type filter element
~, **сухо́й** dry cell
~, **фильтру́ющий** filter element
~, **функциона́льный** functional element
~, **хими́ческий** chemical element
~ **ча́стной боево́й зада́чи** element of a particular operational task
элеме́нты, гото́вые prefabricated components
~, **загото́вленные** prefabricated components
~, **сбо́рные** prefabricated sections
элеро́н aileron
э́ллипс ellipse
~ **рассе́ивания** 1. beaten zone 2. zone of dispersion

э́ллипс рассе́ивания, по́лный 1. full ellipse of dispersion (Sov)
2. 1ØØ percent zone (gunn, ball)
эллипсо́ид ellipsoid
эллипти́ческий elliptical
эма́ль enamel
эмбле́ма insignia
эми́ссия emission
~, **втори́чная** secondary emission
~ **като́да** cathode emission
~, **первичная** primary emission
~, **спонта́нная** spontaneous emission
~, **термоэлектро́нная** thermionic emission
~, **фотоэлектро́нная** photoelectric emission
~, **электро́нная** electron mission
эми́тер emitter (transistor)
эмульги́рование emulsifying, emulsification
эмульги́ровать to emulsify
эмульса́ция emulsifying, emulsification
эму́льсия emulsion
~, **ма́сляная** oil emulsion
энерге́тик power engineer
энерге́тика power engineering
энергети́ческий energy (physics), power (eng)
эне́ргия energy
~, **а́томная** atomic energy
~, **внутрия́дерная** nuclear forces, nuclear binding energy
~, **ду́льная** muzzle energy
~ **излуче́ния** radiation energy
~ **излуче́ния, поглощённая** absorbed radiation energy
~, **инфракра́сная** infra-red energy
~, **кинети́ческая** kinetic energy
~ **лобово́го сопротивле́ния** frontal drag
~, **лучева́я** radiant energy
~, **лучи́стая** radiant energy
~ **нейтро́на** neutron energy
~ **отда́чи** recoil energy
~, **светова́я** thermal energy (nucl)
~ **свя́зи** binding energy (nucl)
~, **скоростна́я** kinetic energy
~, **со́лнечная** solar energy
~, **теплова́я** thermal energy
~ **уда́ра** impact energy
~, **хими́ческая** chemical energy
~, **я́дерная** nuclear energy
энерговооружённость power supply per producing unit
энергоснабже́ние power supply
энтальпи́я enthalpy (thermodynamics)
энтропи́я entropy
эпидемиоло́гия epidemiology
эпиде́мия epidemic
эпидиаско́п epidiascope
эпизоди́ческий spasmodic
эпизооти́ческий epizootic
эпице́нтр epicentre
~ **а́томного взры́ва** ground zero
~ **взры́ва** ground zero
~ **я́дерного взры́ва, действи́тельный** actual ground zero
эпю́р curve, diagram
эрг erg
эроди́ровать to erode
эро́зия erosion
~ **кана́ла ствола́** barrel erosion, barrel wear
эрсте́д oersted (unit of magnetic force)
эска́дра squadron (nav)
эскадри́лья squadron (ac)
~, **вертолётная** helicopter squadron
эскадри́лья-вы́лет squadron sortie
эскадро́н squadron
~, **кавалери́йский** cavalry squadron
эска́рп escarpment, one-way tank ditch
эскарпи́рование scarping
эско́рт escort
~, **почётный** military funeral escort
эсми́нец destroyer
эстака́да 1. trestle, platform
2. service ramp (veh inspection)
3. flyover, elevated roadway
эстафе́та 1. relay race
2. baton (used in relay)
эсти́ст operator of ST (signals telegraph) teleprinter set (Sov)
этало́н standard
этано́л ethanol, ethyl alcohol
эта́п 1. phase stage
2. halting place
~ **боевы́х де́йствий** operational stage
~ **нападе́ния** attack phase
~ **неуправля́емого полёта** ballistic flight phase (msl)
~, **подготови́тельный** preparatory phase
~ **разго́на** acceleration phase
~ **эвакуа́ции, пе́рвый** first stage evacuation
этапи́рование staging, phasing
этике́тка 1. cardboard disc, millboard disc
2. label, price tag

эти́л ethyl
этилами́н ethylamine
этилдихлорарси́н ethyldichlorarsine (NBC)
этиле́н ethylene
этиленглико́ль ethylene glycol
этилнитра́т ethyl nitrate
эти́ловый ethyl
этилцентрали́т ethyl centralite
эфе́с штыка́ bayonet handle
эфи́р ether
~, сло́жный ester
эффе́кт effect
~, анте́нный antenna effect
~, гидроскопи́ческий gyroscopic effect
~, го́рный high-terrain effect (airborne radar)
~, дальноде́йствующий long-range effect
~, динатро́нный secondary emission effect (electron)
~ дли́нной ли́нии long-line effect
~, дли́тельный long term effect
~ До́плера Doppler effect
~, дробово́й shot effect (elec)
~, ква́нтовый quantum effect
~, Комптоно́вский Compton effect, Compton recoil effect (NW)
~ Кори́олиса Coriolis effect
~, кратковре́менный short-term effect, immediate effect
~, кумуляти́вный shaped-charge effect, cumulative effect, hollow-charge effect
~, ла́зерный lazer effect
~ Ма́ха mach effect
~, микрофо́нный microphony (electron)
~ Ми́ллера Miller effect (electron)
~, мостово́й "bridge effect" (wear on underinflated tyre)
~, ночно́й night effect
~, пове́рхностный skin effect (elec)
~, полево́й field effect (elec)
~, поро́говый threshold effect
~, противоме́стный side reduction effect (radio)
~, пьезоопти́ческий piezooptic effect
~, пьезоэлектри́ческий piezo-electric effect
~ разры́ва burst effect
~, результи́рующий net effect
~, светово́й luminous effect
~, фото́нный photon effect
эффе́кт, фотоэлектри́ческий photoelectric effect
~, фотоя́дерный photonuclear effect
~, фуга́сный blast effect
эффекти́вность effectiveness
~ де́йствия lethality, effectiveness
~ излуче́ния radiation efficiency
~ огня́ fire effect
~, относи́тельная relative efficiency
~ сгора́ния combustion efficiency
~, токси́ческая toxic effect
~ торможе́ния braking efficiency
эффекти́вный effective
эхобо́кс echo box (radar)
эхоиндика́тор, доплеро́вский Doppler echo indicator
эхоло́т sounding device, echo-sounder
эхрезона́тор echo box
эшело́н 1. echelon
2. wave (of assault)
~, боево́й fighting echelon, Sov
~ возду́шного деса́нта, поса́дочный air-landed follow-up element of air-borne assault force
~, возду́шно-деса́нтный airborne wave
~, во́инский troop train, military transport, troop convoy
~, второ́й 1. follow-up echelon, rear echelon
2. second line (tpt)
~, железнодоро́жный train
~ морско́го деса́нта amphibious assault wave
~, пе́рвый assault echelon, A echelon
~, похо́дный serial, column serial (movement admin)
~, та́нковый tank echelon
~ штади́ва, второ́й Rear Div HQ
эшелони́рование 1. disposition in depth
2. stacking (ac)
эшелони́рованный in echelon, disposed in echelon
эшелони́ровать to echelon, to dispose in depth

Ю

ю́бка skirt
~ капо́та cowl flap
~ кла́пана valve petticoat
~ по́ршня piston skirt
~ цили́ндра cylinder skirt
юг South
югво́ст south-east
югза́п south-west
юго-восто́к south-east
юго-за́пад south-west
ю́жный south, southern
ю́ркий brisk, nimble
юстирова́ть to adjust finely
юстиро́вка 1. orientation (radar)
2. adjustment
юсти́ция, вое́нная military justice, legal service
юфть russian leather

Я

я́блоко мише́ни bull, bull's eye (tgt)
я́вка 1. appearance, presence, act of reporting
2. presentation (of docs)
3. rendezvous point
4. secret meeting place
5. secret meeting, rendezvous
явле́ние захва́тывания capture effect
~ ослабле́ния теплово́го излуче́ния thermal radiation attenuation effect
~, резона́нсное resonance phenomenon (electron)
явле́ния, прили́во-отли́вные tide conditions
явля́ться/яви́ться к to report to
яд poison
~, каталити́ческий catalyst poison, anticatalyst
я́дерное нападе́ние nuclear attack
~ ору́жие nuclear weapons
~ то́пливо nuclear fuel, nuclear propellant
я́дерно-раке́тный уда́р nuclear missile strike
я́дерные боеприпа́сы 1. nuclear warheads
2. nuclear weapons
3. nuclear shells
я́дерный nuclear
ядови́тый poisonous, toxic
ядохимика́т pesticide
ядро́ 1. nucleus
2. centre, core
3. cannon ball, shot
4. main body of troops
~, а́томное atomic nucleus
~, исхо́дное parent nucleus
~ повы́шенного давле́ния high pressure centre (met)
~, промежу́точное compound nucleus
~, составно́е compound nucleus
~, тяжёлое heavy nucleus
я́зва, коррозио́нная corrosion pit
~, сиби́рская anthrax (med)
язы́к 1. language, tongue
2. prisoner for interrogation
3. bolt (of lock)
я́корь 1. anchor
2. armature (elec)
~, гла́дкий smooth-core armature
~, зубча́тая slotted armature
~, плаву́чий drift anchor
я́ма pit, hole
~, возду́шная air pocket
"я́ма, во́лчья" concealed pit tank trap, "wolf's hole" (tank trap: pit, covered & camouflaged)
я́ма, противота́нковая anti-tank pit, pit tank trap
~, смотрова́я inspection pit
я́ма-лову́шка concealed pit, concealed-pit tank-trap
я́мка "костоло́м" concealed anti-personnel pit
яр steep bank
я́ркий bright, brilliant, luminous
я́ркость brightness, brilliance, luminance
~ фо́на background brightness
ярлы́к tag, label
~, авари́йный breakdown card
~ да́льности range indicator
~, порохово́й powder tag
~, стелла́жный bin card
ярмо́ yoke (on electromagnet)
я́рус layer (met, geological)
я́русный stepped, staged, layered
я́сень ash (timber and tree)

яче́истый cellular, honeycombed
яче́йка 1. nucleus
2. foxhole, small trench, weapon pit
3. unit, compartment
~, вы́несенная forward traverse (fd wks)
~, гранатомётная rocket launcher fire trench (A tk), anti-tank rocket-launcher fire trench
~ для веде́ния огня́ лёжа shell scrape
~ для стрельбы́ weapon pit, fire trench
~, запомина́ющая memory cell, storage cell
~ запомина́ющего устро́йства memory cell, storage cell
~, одино́чная foxhole, weapon slit
~, одино́чная стрелко́вая one-man fire trench, foxhole, one-man fire position
~ па́мяти memory cell, storage cell
~, па́рная two-man trench, two man fire trench, two-man fire position
~, при́мкнутая стрелко́вая integral firing position
~, пулемётная machine-gun pit
~, стрелко́вая weapon pit
~, три́ггерная flip-flop circuit (elec)
~ управле́ния command trench
~, фотоэлектри́ческая photo-electric cell, photocell
~, электроопти́ческая electrooptic cell
я́щик 1. box, case, container
2. drawer
~, аккумуля́торный battery box
~, возду́шный buoyancy tank (naut)
~ для инструме́нтов tool box
~ для панора́мы dial sight case
~, заря́дный caisson (arty)
~, инструмента́льный tool box
~, оруди́йный распредели́тельный gun junction box (AD, AAFCE)
~, оруди́йный соедини́тельный gun junction box (AD, AAFCE)
~, патро́нный ammunition box
~, пусково́й combined container and launcher (GW)
~ с боеприпа́сами ammunition box
~ с запасны́ми частя́ми spare parts case
~ с песко́м sand table (trg aid)
~, укла́дочный stowage box
~, укупо́рочный missile carry-box, carry box, carrying box
~, упако́вочный packing box, packing case
я́щик, центра́льный распредели́тельный main distribution box
~, шпа́льный ballast box
я́щик-заде́рживатель shell trap, bullet trap (in testing of projectiles)

Printed in the UK for HMSO
Dd736704 C40 6/83 (1238)

RONALD G. MOORE
CHIEF TARGETTING BRANCH
CM & TD USAADC
(AV) 923-2615/